# Innovative Earthquake Soil Dynamics

# Innovative Earthquake Soil Dynamics

Takaji Kokusho

CRC Press
Taylor & Francis Group
Boca Raton London New York Leiden

CRC Press is an imprint of the
Taylor & Francis Group, an **informa** business
A BALKEMA BOOK

Cover credentials:

1995 Kobe EQ | 1999 Chi-Chi EQ | 2008 Iwate-Niyagi Inland EQ
1995 Kobe EQ | 1999 Kocaeli EQ | 2011 Tohoku EQ
1995 Kobe EQ | 2004 Niigataken Chuetsu EQ | 2011 Tohoku EQ

2011 Tohoku EQ | 2016 Kumamoto EQ | 2016 Kumamoto EQ

*CRC Press/Balkema is an imprint of the Taylor & Francis Group, an informa business*



Typeset by MPS Limited, Chennai, India
Printed and bound in Great Britain by CPI Group (UK) Ltd, Croydon, CR0 4YY



*Library of Congress Cataloging-in-Publication Data*

Applied for

Published by: CRC Press/Balkema
Schipholweg 107C, 2316 XC Leiden, The Netherlands
e-mail: Pub.NL@taylorandfrancis.com
www.crcpress.com – www.taylorandfrancis.com

ISBN: 978-1-138-02902-6 (Hbk)
ISBN: 978-1-315-64505-6 (eBook)

# Dedications

*To my late parents*
*&*
*To my wife Noriko.*

# Table of contents

# Acknowledgments

Professor Kenji Ishihara is gratefully acknowledged for his long-time guidance and stimulation to date in soil dynamics since the present author was the graduate student in the University of Tokyo in 1960's. Also acknowledged are a number of PhD, Master and Undergraduate students who were engaged in various research topics in the author's laboratory in Chuo University in past 20 years and contributed to create huge database that this book largely depends upon. The electric power companies in Japan, the Japan Society for the Promotion of Science, and the Central Research Institute of Electric Power Industry, who supported the author's research for many decades are gratefully appreciated. National Research Institute for Earth Science and Disaster Resilience, Kansai Electric Power Company, Kobe Municipal City Office, and Tokyo Electric Power Company are acknowledged for valuable earthquake records including vertical array strong motion data used in this book.

# Acknowledgements

# Preface

This book deals with soil dynamics in earthquake engineering including almost all aspects of soil behavior from the bedrock up to the ground surface necessary for engineering design of structures, wherein generally accepted basic knowledge as well as advanced and innovative views are accommodated.

In recent years, a great number of earthquake observation data in surface soil deposits have been accumulated by newly deployed earthquake observation systems, particularly in Japan, demonstrating the significance of soil profiles and properties in site amplification and damage during strong earthquakes. Recent earthquakes have also presented a great number of case histories with new findings which may not be compatible with conventional knowledge, inspiring different views among investigators on the mechanisms of site amplification, liquefaction and slope failure.

Major topics discussed in this book of earthquake geotechnical engineering are (i) seismic site amplification, (ii) liquefaction and (iii) earthquake-induced slope failure. Associated with the above three topics, basic theories and knowledge on wave propagation/attenuation, soil properties, laboratory tests, numerical analyses, model tests are addressed in the earlier Chapters. Some of the advanced research findings are addressed, and associated recent laboratory data as well as field case history data are incorporated in this book.

Another important feature characterizing this book is an energy perspective to these topics in addition to conventional views based on the force-equilibrium perspective. It is because the present author strongly believes through his long-time experiences in reconnaissance of earthquake damage and model tests that the energy is a very relevant though simple index in determining seismic failures of structures, particularly soils and soil structures.

The book is intended to cover major recent research advances in this field during recent earthquakes such as the 1995 Kobe earthquake and the 2011 Tohoku earthquake in Japan. Many research results are originated from Japan, rich of earthquake records and case histories though isolated to international investigators and engineers because of the language barrier. It is written for international readers; graduate students, researchers and practicing engineers, interested in this field, to be able to understand as easily as possible.

# About the author

BS Degree from the University of Tokyo Japan in 1967.
MS Degree from the University of Tokyo Japan in 1969.
MS Degree from Duke University USA in 1975.
PhD. (Doctor of Engineering) from the University of Tokyo 1982.
"Dynamic soil properties and nonlinear seismic response of ground".
Central Research Institute of Electric Power Industry (CRIEPI) from 1969 to 1995.
Professor, Department of Civil & Environ. Eng., Chuo University, from 1996 to 2015.
Professor Emeritus, Chuo University, since 2015.

# Nomenclature

| | |
|---|---|
| $A_i$, $B_i$, $C_i$, $D_i$ | Wave amplitudes of $i$-th layer for Rayleigh wave. |
| $A_s$, $A_b$ | Spectrum amplitude of upward wave at ground surface and outcropping base, respectively. |
| $B_b$ | Spectrum amplitude of downward wave at downhole base. |
| $c$ | Surface wave velocity $(=\omega/k)$. |
| $c$ | Dashpot constant. |
| $C_c$ | Clay content. |
| $CC$ | Cement content. |
| $C_R$ | Compliance ratio for undrained soil testing system. |
| $C_u$ | Uniformity coefficient. |
| $d_i$ | Thickness of $i$-th soil layer. |
| $D$, $D_0$, $D_{\max}$ | Damping ratio, Small-strain damping ratio, and Maximum damping ratio, respectively. |
| $D_L$ | Damage level in fatigue theory. |
| $D_R$ | Relative density. |
| $e$, $e_{cr}$ | Void ratio, and Critical void ratio, respectively. |
| $E$, $E'$ | Deformation modulus of soil or Young's modulus, and Deformation modulus of soil in terms of effective stress, respectively. |
| $E$, $E_e$, $E_k$ | Wave energy, Strain wave energy, and Kinetic wave energy, respectively. |
| $E_u$, $E_d$, $E_w$ | Energies of upward, downward, and dissipated waves, respectively. |
| $E_{IP}$ | Incident earthquake wave energy at a base layer. |
| $E_{EQ}$, $E_{DP}$, $E_k$, $-\delta E_p$ | Earthquake energy, Dissipated energy, Kinetic energy, and Potential energy for slope failures, respectively. |
| $f$ | Frequency. |
| $f_0$ or $f_1$ | First peak frequency. |
| $F$ | Force. |
| $F_c$ | Fines content. |
| $F_L$ | Factor for liquefiability. |
| $F_s$ | Safety factor against slope failure. |
| $G$, $G_0$ | Shear modulus $(=\mu)$, Initial shear modulus, respectively. |
| $G_{eq}$ | Equivalent shear modulus. |
| $G^*$ | Complex shear modulus. |
| $G_c$ | Gravel content. |
| $G'$ | Frequency-independent constant for the Nonviscous damping. |

| | |
|---|---|
| $H$, $H_i$ | Surface layer thickness of 2-layer system, and Layer thickness of $i$-th layer, respectively. |
| $I_p$ | Plasticity index. |
| $k$, $k^*$ | Wave number, and Complex wave number, respectively. |
| $k$, $k_0$ | Spring constant, and Initial spring constant, respectively. |
| $k$ | Permeability coefficient. |
| $k$, $k_{cr}$ | Seismic coefficient, and Critical seismic coefficient, respectively, in horizontal direction. |
| $k$, $m$, $n$ | Exponents for confining pressure-dependency of S-wave velocity or shear modulus. |
| $K_\sigma$, $K_\alpha$ | Overburden correction factor, and Correction factor for initial shear stress, respectively. |
| $K$, $K_s$, $K_w$, $K_f$, $K_c$ | Bulk moduli of soil skeleton, soil solid particles, pure water, water with air, and pressurized water reflecting system-compliance effects, respectively. |
| $L_i$, $M_i$ | Wave amplitudes of $i$-th layer for Love wave. |
| $m$, $M$ | Mass. |
| $M$, $M_J$, $M_w$ | Earthquake magnitude in general, Earthquake magnitude used in the Japanese Meteorological Agency, and Earthquake moment magnitude, respectively. |
| [M], [C], [K] | Matrices of mass, damping and stiffness, respectively. |
| $N_c$, $N_L$ | Number of load cycles, and Number of cycles to initial liquefaction, respectively. |
| $N_{eq}$ | Equivalent number of cycles in earthquake motions. |
| $p$ | Mean stress in triaxial test $(=(\sigma_1+2\sigma_3)/3)$. |
| $p$ | Pressure. |
| $p_0$ | Unit pressure $(=98\ \text{kPa})$. |
| $p'_y$ | Consolidation yield stress. |
| $P_L$ | Parameter for depth-dependent cumulative effect of liquefaction. |
| $q$ | Deviatoric stress in triaxial test $(=\sigma_1-\sigma_3)$. |
| $Q$ | Amplification factor (=1/(2D)). |
| $r_d$, $r_n$ | Stress reduction coefficients in terms of soil depth and irregular seismic motion in liquefaction potential evaluations, respectively. |
| $r_N$ | Ratio of number of cyclic loading to initial liquefaction $(=N_c/N_L)$. |
| $r_u$ | Pore-pressure ratio $(=u/\sigma'_c)$. |
| $R$ | Hypocenter distance of earthquake. |
| $S_r$ | Saturation. |
| $t$ | Time. |
| $T$ | Wave period. |
| $u$ | Pore-pressure. |
| $u$, $v$, $w$ | Displacements in $x$, $y$, $z$-directions, respectively. |
| $\dot{u}$, $\dot{v}$, $\dot{w}$ | Particle velocities in $x$, $y$, $z$-directions, respectively. |
| $\ddot{u}$, $\ddot{v}$, $\ddot{w}$ | Accelerations in $x$, $y$, $z$-directions, respectively. |
| $\dot{u}_0$, $\dot{w}_0$ | Surface particle velocities in horizontal and vertical directions. |
| $\dot{u}_a$ | Particle velocity amplitude. |
| $V_p$, $V_r$ | Velocities of P and Rod-waves, respectively. |
| $V_s$, $V_s^*$ | S-wave velocity, and Complex S-wave velocity, respectively. |

| | |
|---|---|
| $\overline{V}_s$ | S-wave velocity averaged over the equivalent surface layer. |
| $V_{sb}$ | S-wave velocity of base layer. |
| $V_{s30}$ | S-wave velocity averaged over top 30 m from ground surface. |
| $w_L$, $w_p$ | Water contents in % for liquid limit and plastic limit, respectively. |
| $W$ | Maximum elastic strain energy. |
| $\Delta W$ | Dissipated energy in one-cycle of loading. |
| $2W_{-}$ | Strain energy in one cycle loading considering the energy recycling effect. |
| $x$, $y$, $z$ | Three-dimensional orthogonal axes. |
| $\alpha$, $\alpha^*$ | Impedance ratio, and Complex impedance ratio, respectively. |
| $\alpha$ | Reduction coefficient modifying irregular motion to equivalent harmonic motion. |
| $\alpha$, $\beta$ | Exponent constants. |
| $\alpha$, $\beta$ | Constants for Modified Hardin-Drnevich model. |
| $\alpha$, $\beta$ | Constants in Rayleigh damping for mass matrix $[M]$ and stiffness matrix $[K]$, respectively. |
| $\alpha_{RO}$, $\beta_{RO}$ | Constants for Ramberg-Osgood model. |
| $\beta$ | Wave attenuation coefficient for internal damping ($\approx \omega D/V_s$). |
| $\beta$, $\beta_0$ | Slope gradient, and Initial slope gradient, respectively. |
| $\gamma$ | Shear strain. |
| $\gamma_{DA}$ | Double amplitude shear strain. |
| $\gamma_{yz}$, $\gamma_{zx}$, $\gamma_{xy}$ | Shear strains on planes perpendicular to $x$, $y$, $z$-directions, respectively. |
| $\gamma_r$, $\gamma_y$ | Reference and yield strains, respectively. |
| $\gamma_a$ | Shear strain amplitude in cyclic loading. |
| $\gamma_{\text{eff}}$ | Effective strain in equivalent linear analysis. |
| $\delta$ | Phase delay angle. |
| $\delta_r$ | Residual horizontal displacement in slope failure. |
| $\delta_{rn}$ | Residual horizontal displacement in terms of centroid. |
| $\varepsilon$ | Axial strain. |
| $\varepsilon_v$ | Volumetric strain. |
| $\varepsilon_{DA}$ | Double amplitude axial strain. |
| $\varepsilon_x$, $\varepsilon_y$, $\varepsilon_z$ | Axial strains in $x$, $y$, $z$-directions, respectively. |
| $\theta$, $\theta_0$ | Slope angle and Initial slope angle, respectively. |
| $\theta_1$, $\theta_2$ | Incident and reflected/refracted wave angles, respectively. |
| $\theta_{cr}$ | Critical angle for wave refraction. |
| $\lambda$ | Wave length. |
| $\lambda$, $\mu = G$ | Lame's constants. |
| $\mu$ | Friction coefficient. |
| $\nu$, $\nu'$ | Poisson's ratio, and Poisson's ratio in terms of effective stress, respectively. |
| $\xi$ | Viscosity. |
| $\rho$, $\rho_t$, $\rho_{sat}$ | Soil density in general, Unsaturated soil density, and Saturated soil density, respectively. |
| $\sigma$, $\sigma'$ | Total normal stress, and Effective normal stress, respectively. |
| $\sigma_1$, $\sigma_3$ | Axial and lateral stresses in triaxial test, respectively. |
| $\sigma_v$, $\sigma_h$ | Vertical and horizontal stresses, respectively. |

| | |
|---|---|
| $\sigma_x$, $\sigma_y$, $\sigma_z$ | Normal stresses in $x$, $y$, $z$-directions, respectively. |
| $\sigma'_a$, $\sigma'_p$ | Effective stresses in the directions of wave propagation and wave vibration, respectively. |
| $\sigma_d$ | Dynamic axial shear stress amplitude. |
| $\tau$, $\tau_d$, $\tau_s$ | Shear stress, Dynamic shear stress, and Initial shear stress, respectively. |
| $\tau_{yz}$, $\tau_{zx}$, $\tau_{xy}$ | Shear stresses on planes perpendicular to $x$, $y$, $z$-directions, respectively. |
| $\phi$, $\phi^*$ | Friction angle for unsaturated and saturated slopes, respectively. |
| $\phi$, $\psi$ | Volumetric strain, and Rotation angle, respectively. |
| $\Phi$ | Wave amplitudes in wave scattering theory. |
| $\psi$ | State parameter on State diagram. |
| $\omega$ | Angular frequency. |
| $\omega_0$ | Resonant angular frequency. |
| $\omega_z$, $\omega_x$, $\omega_y$ | Rotation angles around $z$, $x$, $y$-axes, respectively. |

# Introduction of the book

## TOPICS COVERED IN THIS BOOK

Seismic ground motions and earthquake damage greatly reflect the local soil conditions. People intuitively understood from old times the significant effect of local soils on earthquake damage and selected better locations for their living. With the urbanization of human society worldwide, however, buildings and infrastructures are increasingly built in poor soil conditions, in waterfront, lowland and steep slopes. This has necessitated an interdisciplinary engineering field between geotechnical engineering and earthquake engineering and created "Earthquake Geotechnical Engineering" or "Earthquake Soil Dynamics".

The topics covered in this book are conceptually depicted in Fig. 1: (i) site amplification in surface soils above bedrocks including soil-structure interactions, (ii) soil failures near ground surface typically by liquefaction and (iii) slope failures under the gravitational effect.

The site amplification reflects soil profiles as well as soil properties. Though the incident wave at the bedrock level may not be so different within a certain horizontal area, the seismic motions at the ground surface may vary considerably depending on the

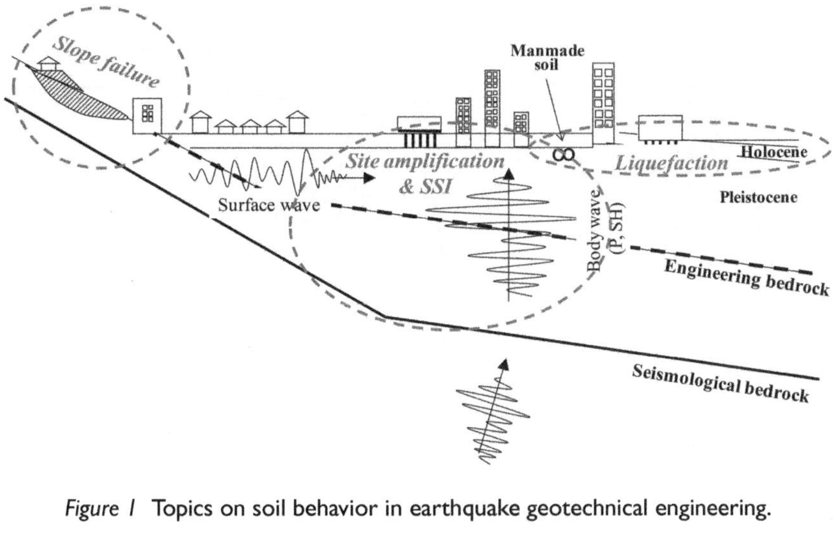


*Figure 1* Topics on soil behavior in earthquake geotechnical engineering.

local soil conditions. Superstructures resting on different surface soils tend to exhibit different seismic response accordingly. Structural damage may strongly reflect not only structural properties but also soil properties of the supporting ground including the effect of soil-structure-interaction (SSI).

With greater seismic shaking during destructive earthquakes, induced soil strain tends to be larger, getting surface soils approaching to failure. Typical soil failure caused by seismic loading is liquefaction, wherein non-cohesive soils saturated with water develop excess pore-water pressure due to cyclic loading, approaching to a zero-effective stress condition. This means considerable loss of the shear stiffness and strength, causing the settlement/tilting of superstructures on liquefied ground or the uplift of buried lifelines. Severe liquefaction may trigger lateral spreading or flow in loose deposits and manmade fills in gentle slopes and behind displaced retaining walls.

Another type of seismically-induced soil failure occurs in slopes. Not only liquefiable saturated sandy soils but also unsaturated clayey, gravelly or rocky materials are often involved in slope failures. The seismic cyclic loading contributes to reduce soil shear resistance and the inertial effect drives the soil along the slip surface, causing residual deformation with various scales. For large-scale slope failures, though triggered by the seismic inertia, the gravity plays a major role in driving large-volume soil mass in a long runout-distance, causing a great damage in downslope directions.

## INDUCED SOIL STRAIN AS A PERTINENT PARAMETER

Thus, the overview of soil behavior during strong earthquakes highlights a variety of earthquake geotechnical effects where the induced soil strain is a key parameter.

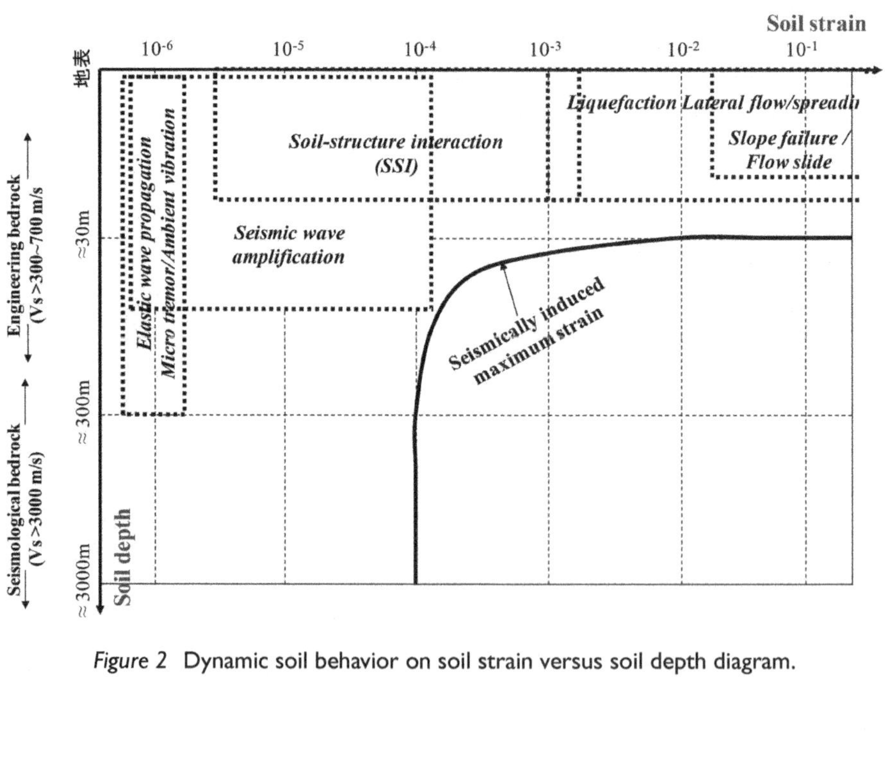


*Figure 2* Dynamic soil behavior on soil strain versus soil depth diagram.

Fig. 2 shows the soil behaviors illustrated on the soil strain versus soil depth diagram. The soil behaves as an elastic body at a small strain level of around $10^{-6}$, when elastic waves propagate in soils during small earthquakes. Ground vibrations in microtremors and environmental problems due to traffic and other vibration sources are also associated with the elastic wave propagation. With increasing strains to $10^{-4}$–$10^{-3}$, the soil behaves non-elastically in a shallower depth, changing site amplifications different from those during small earthquakes. Further increase in strain during strong earthquakes leads to the onset of soil liquefaction with $10^{-2}$ strain. This may give a considerable influence on the stability and integrity of superstructures and foundations. In slope failures, induced shear strains in shear zones may exceed well over 100%. The soil strains induced in the ground tend to be larger and larger in shallow depths in soft soils during destructive earthquakes. The upper bound of the induced strain may be drawn as in Fig. 2 schematically as the solid line against the depth; $10^{-4}$ or smaller in deeper ground because of stiffer soils and higher overburden stresses, while it drastically increases way over $10^{-2}$ with decreasing depth in soft soils in particular. Thus, induced soil strain is the pertinent parameter in dealing with seismic soil behavior from soft soil at a ground surface to a stiff soil of deep bedrock in earthquake geotechnical engineering.

## SOIL STRAINS IN VIEW OF PERFORMANCE BASED DESIGN

Performance-Based Design (PBD) is increasingly employed recently in structural design of buildings and infrastructural facilities in many countries. The performance of foundation ground and soil structures under earthquake loading has long been a major topic of discussion. Despite such long-lasting efforts, the PBD has not yet been established sufficiently in geotechnical engineering practice. Seismically induced ground deformation essential to PBD is not easy to evaluate mainly because, in contrast to superstructures, the ground is the 3-dimentional continuum with tremendous spatial variability and strong nonlinearity involving soil dilatancy. A rapid development in practical and reliable PBD is needed not only for foundation ground but also for superstructures resting on incompetent soils.

## BRIEF OVERVIEW OF THE BOOK

This book is composed of six Chapters; Chapter 1 deals with seismic wave propagation and attenuation in soil layers above the engineering bedrock. Basic theories on one/three-dimensional propagation of body waves and surface waves, are followed by viscoelastic theories needed to understand the attenuation and damping mechanism of seismic waves including the energy perspective. In Chapter 2, dynamic soil properties are focused in terms of small strain-properties, strain-dependent property variations and how to determine the properties in the field and in the laboratory. Chapter 3 addresses the modeling/formulation of soil properties; equivalent linear modeling and truly nonlinear stress–strain modeling together with dilatancy modeling in large strains for soil failures. How to conduct different type numerical analyses using different soil models and scaled model tests for different purposes are also discussed. In Chapter 4, seismic site amplifications are discussed in the light of the multi-reflection theory of the SH-wave utilizing vertical array records during many strong earthquakes recently occurred in Japan. SSI (Soil-Structure-Interaction) is also addressed briefly in the realm of the one-dimensional SH-wave propagation. The site amplification is then interpreted as the flow of wave energy, and the general trends in energy flow in shallow depths are discussed based on a number of strong motion records in conjunction with earthquake damage evaluations. Chapter 5 deals with several aspects of seismically-induced soil liquefaction; basic mechanisms, associated geotechnical/seismic conditions, liquefaction potential evaluations by the currently employed stress-based method and also by a newly proposed energy-based method, effects of incomplete saturation as well as initial shear stress, cyclic softening of cohesive soils, post-liquefaction residual deformations and structural effects, countermeasures and liquefaction-induced base-isolation effects. In Chapter 6, seismically-induced slope failures are discussed in terms of conventional stability analyses, dynamic analyses using Newmark-method, deformation analyses by degraded shear moduli. Then, a number of case history data during recent earthquakes in Japan are reviewed to look into what happened in the actual slopes. An energy-based evaluation based on a simple energy principle is applied to the case histories to understand the slope failure mechanism in the energy perspective.

# Chapter 1

# Elastic wave propagation in soil

When seismic wave propagates in deep or stiff ground, it behaves as an elastic wave with induced soil strain within an elastic range. Even in shallow depths, small amplitude waves by small earthquakes, machines or traffic vibrations propagate as the elastic wave. In this Chapter, basic theories on the elastic wave propagation for earthquake geotechnical engineering will be addressed in terms of one-dimensional and three-dimensional body waves as well as surface waves. In the latter part, basic theories on viscoelastic models are addressed to deal with soil damping incorporated in the attenuation of elastic waves.

## 1.1 INTRODUCTION

The seismic waves consist of body waves and surface waves as classified in Fig. 1.1.1. The body waves travel inside medium boundaries, while the surface waves transmit along the medium boundaries. The body waves are classified into P-wave (Primary wave) and S-wave (Secondary or Shear wave), while the surface waves are classified into Rayleigh wave and Love wave. During earthquakes, one first feels the P-wave of shorter period motions, then the S-wave of stronger motions carrying major earthquake energy, followed by the surface waves of longer period motions.

The body waves propagate into a seismological bedrock below a site as illustrated in Fig. 1.1.2 with an incident angle to the layer boundary. The wave front advances as a single plane perpendicular to the direction of wave propagation. In the

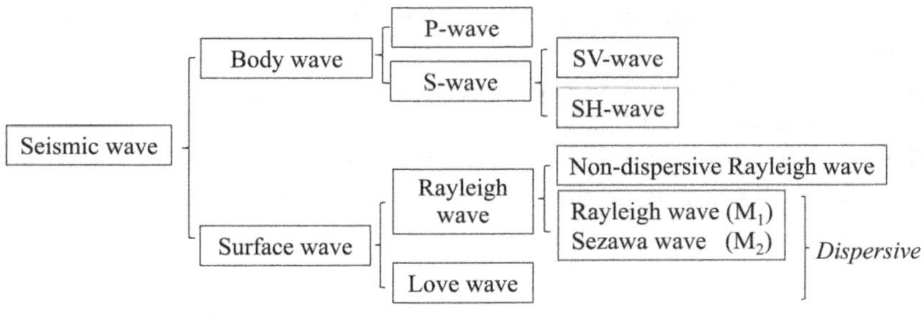


*Figure 1.1.1* Classification of seismic waves.

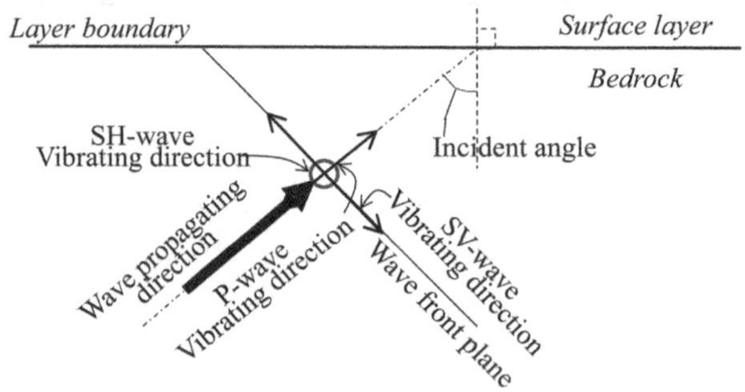


*Figure 1.1.2* Propagating direction, vibrating direction and wave front plane of body waves.

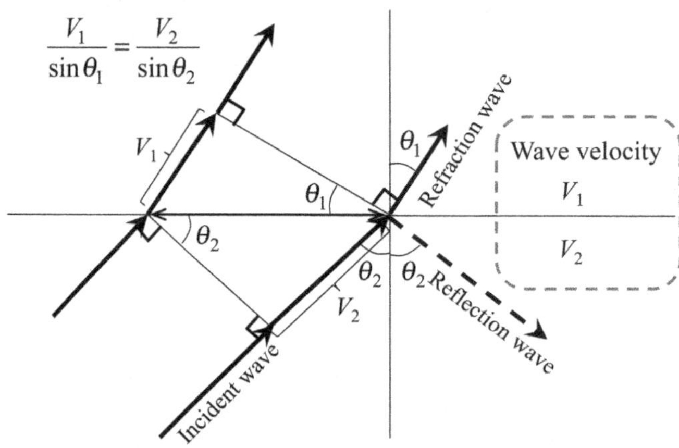


*Figure 1.1.3* Refraction and reflection of elastic waves at a horizontal boundary based on Snell's law.

P-wave, soil particles vibrate in the wave propagation direction, while, in S-wave, they vibrate normal to the propagating direction with shear distortion. The P-wave accompanies the change of soil volume, while the S-wave causes only shear distortion without volumetric strain. The S-wave is further classified into SV-wave (the vibration includes the vertical component) and SH-wave (the vibration only in the horizontal component).

When the plane body wave comes across a horizontal soil boundary with the incident angle $\theta_2$ as shown in Fig. 1.1.3, refraction and reflection waves are generated there. Then, the angles for refraction $\theta_1$ and reflection $\theta_2$ are correlated with the incident angle $\theta_2$ by the Snell's law using the wave propagation velocities $V_1$ and $V_2$ ($V_1 < V_2$ or $\theta_1 < \theta_2$) as:

$$\frac{\sin\theta_1}{\sin\theta_2} = \frac{V_1}{V_2} \tag{1.1.1}$$

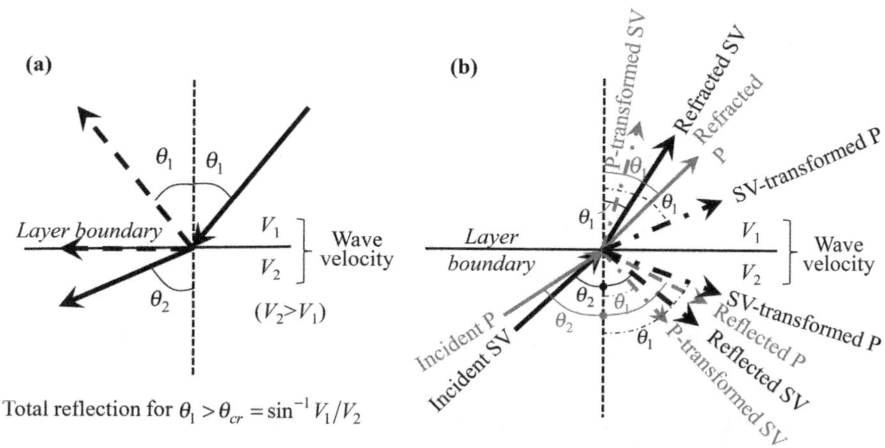


*Figure 1.1.4* Refraction and reflection at a horizontal boundary based on Snell's law: (a) Critical angle for total reflection, (b) Refraction, reflection and transformation of incident SV and P-waves.

This equation can be readily obtained from the fact that the angle of the wave front plane changes from $\theta_2$ to $\theta_1$ due to the difference in the wave velocity.

Fig. 1.1.4(a) illustrates the incident wave with the angle $\theta_1$ generating the refracted wave with the angle $\theta_2$ at the boundary between two layers with wave velocities $V_1$ and $V_2$, respectively ($V_1 < V_2$). The reflected angle $\theta_1$ is the same as the incident angle because $V_s$-values for the two waves are the same. It is easy to understand that the refracted angle corresponding to $\theta_2 = 90°$ is expressed as:

$$\sin\theta_1 = \frac{V_1}{V_2} \tag{1.1.2}$$

$\theta_1 \equiv \theta_{cr} = \sin^{-1} V_1/V_2$ is called as critical angle, beyond that ($\theta_1 > \theta_{cr}$) the total reflection without refraction occurs. At the critical angle, the refracted wave transmits along the layer boundary with the velocity $V_2$, which is incorporated conveniently in an in situ seismic wave exploration.

Among the body waves, the incident SH-wave generates only the SH-wave in refraction and reflection at the horizontal layer boundary because it vibrates only in the horizontal direction parallel with the boundary. In contrast, the SV-wave generates SV-wave and transformed P-wave as well, in both refraction and reflection, because the vibrating direction in the SV-wave has some vertical component. Similarly, the P-wave generates both P and transformed SV-wave except for the incident angle $\theta_2 = 0$. Fig. 1.1.4(b) summarizes in general how incident P or SV-wave generates refracted and reflected P and SV-waves. In any case, Eq. (1.1.1) can be used universally to correlate $\theta_1$ and $\theta_2$ if wave velocities $V_1$ and $V_2$ are used properly corresponding to the different wave types.

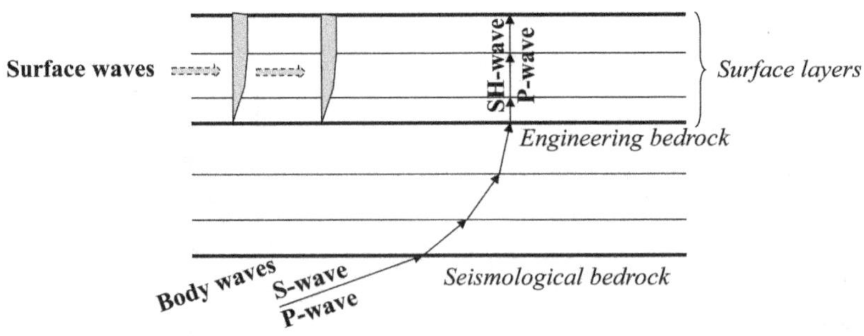


*Figure 1.2.1* One-dimensional body wave propagation in horizontal layers from bedrock to surface.

## 1.2 ONE-DIMENSIONAL WAVE PROPAGATION AND WAVE ENERGY

A soil profile of a site may be idealized as a one-dimensional soil column consisting of multiple horizontal layers as illustrated in Fig. 1.2.1. In order to determine site-dependent earthquake motions, seismological and engineering bedrocks are chosen at depths. The seismological bedrock is normally defined as an upper surface of the earth crust with S-wave velocity $V_s \approx 3$ km/s or larger, while the engineering bedrock is a shallower base layer with $V_s \approx 0.4$–$0.7$ km/s distinctively stiffer than overlying surface soft layers. As the body waves arriving at the seismological bedrock with a certain incident angle propagate upward, the incident angles are getting smaller at the upper layer boundaries due to refractions following the Snell's law, because the wave velocities of layers tend to be lower with decreasing depths as a global trend. Hence the propagation of body waves above the engineering bedrock can be simplified as one-dimensional vertical propagation. In this case, the S-wave is assumed to be composed of SH-wave without SV-component as a major seismic motion in engineering design. Thus, the one-dimensional vertical propagation of SH-wave and P-wave is widely accepted in simplifying dynamic loading in earthquake engineering as discussed in the following.

### 1.2.1 One-dimensional propagation of SH and P-waves

Let us consider wave propagations in a soil column as illustrated in Fig. 1.2.2 wherein $x$ = direction of vibration, $z$ = direction of wave propagation, and $u$, $w$ = displacements in $x$ and $z$-direction, respectively. For the SH-wave in Fig. 1.2.2(a), the force equilibrium in the horizontal direction for a thin slice with a thickness $dz$ and a horizontal area $A$ considering the inertial force in $x$-direction $X = -\rho A dz(\partial^2 u/\partial t^2)$ balancing with the shear stresses $\tau$ yields the following equation.

$$A\left\{\tau + \frac{\partial \tau}{\partial z}\frac{dz}{2} - \left(\tau - \frac{\partial \tau}{\partial z}\frac{dz}{2}\right)\right\} - \rho A dz \frac{\partial^2 u}{\partial t^2} = 0 \tag{1.2.1}$$

Then,

$$\rho \frac{\partial^2 u}{\partial t^2} = \frac{\partial \tau}{\partial z} \tag{1.2.2}$$

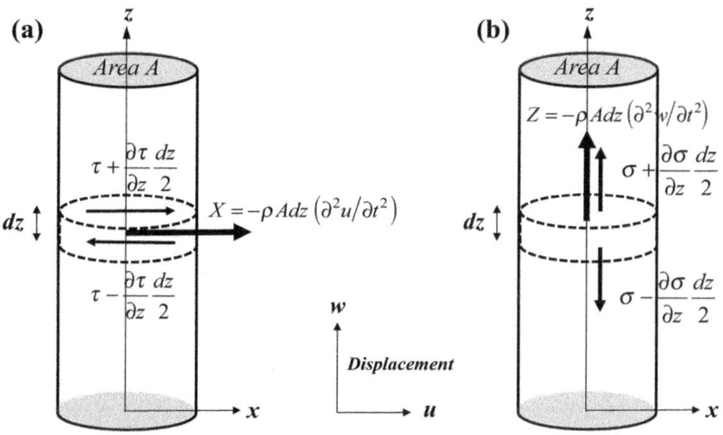


*Figure 1.2.2* Force equilibrium in one-dimensional body wave propagations in soil column: (a) SH-wave, (b) P-wave (Rod-wave).

Substituting $\tau = G(\partial u/\partial z)$ into Eq. (1.2.2) gives

$$\frac{\partial^2 u}{\partial t^2} = V_s^2 \frac{\partial^2 u}{\partial z^2} \tag{1.2.3}$$

where $V_s$ is the S-wave velocity expressed as

$$V_s = \sqrt{G/\rho} \tag{1.2.4}$$

The equation for one-dimensional P-wave propagation can be similarly derived as shown in Fig. 1.2.2(b). The force equilibrium for a thin slice with a thickness $dz$ and a horizontal area $A$ considering the inertial force in the $z$-direction $Z = -\rho A dz \times (\partial^2 w/\partial t^2)$ balancing with the stresses yields the following equation.

$$A\left\{\sigma + \frac{\partial \sigma}{\partial z}\frac{dz}{2} - \left(\sigma - \frac{\partial \sigma}{\partial z}\frac{dz}{2}\right)\right\} - \rho A dz \frac{\partial^2 w}{\partial t^2} = 0 \tag{1.2.5}$$

Then,

$$\rho \frac{\partial^2 w}{\partial t^2} = \frac{\partial \sigma}{\partial z} \tag{1.2.6}$$

Substituting $\sigma = E(\partial w/\partial z)$ into Eq. (1.2.6) gives

$$\frac{\partial^2 w}{\partial t^2} = V_r^2 \frac{\partial^2 w}{\partial z^2} \tag{1.2.7}$$

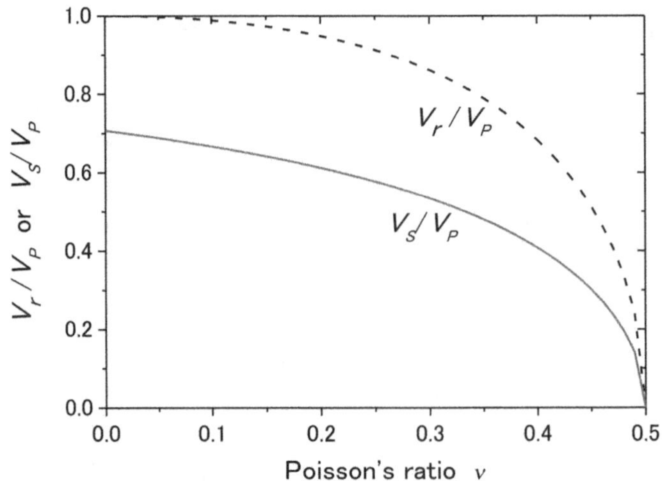


*Figure 1.2.3* Wave velocity ratios of Rod-wave to P-wave $V_r/V_p$ and S-wave to P-wave $V_s/V_p$ versus Poisson's ratio $\nu$.

where $V_r$ is the wave velocity expressed as:

$$V_r = \sqrt{E/\rho} \tag{1.2.8}$$

This velocity corresponds to P-wave propagating vertically in a rod (Rod-wave). Here, the lateral dimension of the rod is comparable with the wave length, wherein Poisson's ratio $\nu$ is not involved because the lateral displacement is allowed freely. With decreasing wave length relative to the rod lateral dimension, it changes to P-wave propagating in an infinitely large medium, wherein Poisson's ratio has a significant effect. The P-wave travelling in the $z$-direction in an infinitely large medium is formulated in Eq. (1.3.13) of Sec. 1.3 as:

$$\frac{\partial^2 w}{\partial t^2} = V_p^2 \frac{\partial^2 w}{\partial z^2} \tag{1.2.9}$$

with the P-wave velocity written as

$$V_p = \sqrt{(\lambda + 2G)/\rho} \tag{1.2.10}$$

wherein $\lambda = \nu E/[(1+\nu)(1-2\nu)]$ and $G \equiv \mu = E/[2(1+\nu)]$ are Lame's constants. The $V_p$-value in Eq. (1.2.10) is obviously different from $V_r$ in Eq. (1.2.8) and the velocity ratio is written as

$$\frac{V_r}{V_p} = \sqrt{E/(\lambda + 2G)} = \sqrt{(1+\nu)(1-2\nu)/(1-\nu)} \tag{1.2.11}$$

On the other hand, the ratio between $V_s$ in Eq. (1.2.4) and $V_p$ is written as

$$V_s/V_p = \sqrt{G/(\lambda + 2G)} = \sqrt{(1-2\nu)/\{2(1-\nu)\}} \tag{1.2.12}$$

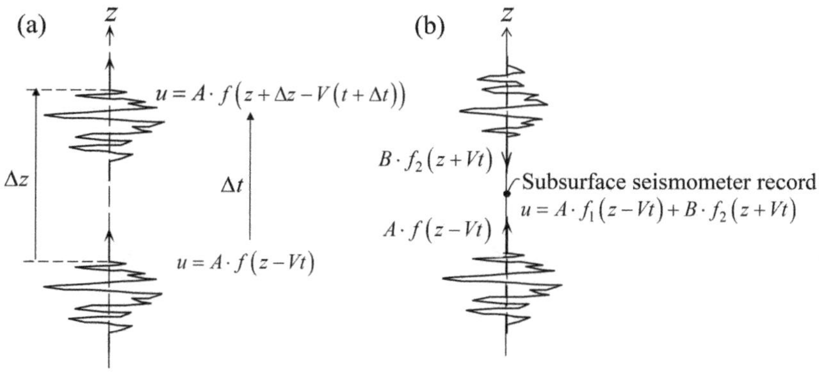


*Figure 1.2.4* Concept of one-dimensional SH-wave propagation: (a) Upward propagation, (b) Upward and downward propagation.

These velocity ratios are shown in Fig. 1.2.3 versus Poisson's ratio $\nu$. It is obvious that $V_r = V_p$ if Poisson's ratio $\nu$ is zero. The ratio $V_r/V_p$ monotonically decreases from 1.0 to 0 with increasing Poisson's ratio $\nu = 0$ to 0.5. The ratio $V_s/V_p$ also decreases monotonically from $1/\sqrt{2}$ to 0 with $\nu$ changing from 0 to 0.5. This indicates that $V_s$ is infinitely small in comparison with $V_p$ in an incompressible material of $\nu = 0.5$.

### 1.2.2 Basic formulation of wave propagation

Let us confirm that the wave equations such as Eq. (1.2.3) actually express the SH-wave propagation. The displacement $u$ in one-dimensional S-wave propagating to the positive direction of $z$-axis as illustrated in Fig. 1.2.4(a) can be expressed in the following form.

$$u = A \cdot f(z - V_s t) \tag{1.2.13}$$

Here, $f(\ )$ is an arbitrary function, and $A$ is the wave amplitude. It is easy to understand that Eq. (1.2.13) represents a propagating wave if $z - V_s t$ is constant with time. If a wave amplitude $u$ at $z = z_1$ and $t = t_1$ propagates to $z = z_1 + \Delta z$ at $t = t_1 + \Delta t$, then $z_1 - V_s t_1 = z_1 + \Delta z - V_s(t_1 + \Delta t) =$ constant should hold, from which a quite reasonable result defining the wave velocity; $V_s = \Delta z/\Delta t$, can be drawn, confirming that Eq. (1.2.13) actually expresses a wave propagation.

Furthermore, particle velocity $\dot{u} = \partial u/\partial t$ and acceleration $\ddot{u} = \partial^2 u/\partial t^2$ can be formulated, by using a new variable $\eta = z - V_s t$, as:

$$\dot{u} = \frac{\partial u}{\partial t} = \frac{\partial u}{\partial \eta}\frac{\partial \eta}{\partial t} = -V_s \frac{\partial u}{\partial \eta} \tag{1.2.14}$$

$$\ddot{u} = \frac{\partial^2 u}{\partial t^2} = \frac{\partial}{\partial t}\frac{\partial u}{\partial t} = \frac{\partial}{\partial \eta}\frac{\partial \eta}{\partial t}\left(-V_s \frac{\partial u}{\partial \eta}\right) = V_s^2 \frac{\partial^2 u}{\partial \eta^2} \tag{1.2.15}$$

On the other hand, the shear strain $\gamma = \partial u/\partial z$ and $\partial^2 u/\partial z^2$ is formulated as:

$$\gamma = \frac{\partial u}{\partial z} = \frac{\partial u}{\partial \eta}\frac{\partial \eta}{\partial z} = \frac{\partial u}{\partial \eta} \tag{1.2.16}$$

$$\frac{\partial^2 u}{\partial z^2} = \frac{\partial}{\partial z}\frac{\partial u}{\partial z} = \frac{\partial}{\partial \eta}\frac{\partial \eta}{\partial z}\left(\frac{\partial u}{\partial \eta}\right) = \frac{\partial^2 u}{\partial \eta^2} \tag{1.2.17}$$

From Eqs. (1.2.15) and (1.2.17), the following equation identical to Eq. (1.2.3) can be obtained,

$$\frac{\partial^2 u}{\partial t^2} = V_s^2 \frac{\partial^2 u}{\partial z^2} \tag{1.2.18}$$

confirming that Eq. (1.2.13) can be a general solution of the wave equation Eq. (1.2.18).

Moreover, Eqs. (1.2.14) and (1.2.16) yield the following important formula.

$$\gamma = -\frac{\dot{u}}{V_s} \tag{1.2.19}$$

This indicates that the shear strain $\gamma$ induced by wave propagation in one direction can be evaluated in such a simple equation from the particle velocity $\dot{u}$ and the wave velocity $V_s$.

Eq. (1.2.13) can be generalized as:

$$u = A \cdot f_1(z - Vt) + B \cdot f_2(z + Vt) \tag{1.2.20}$$

to express the superposition of upward ($+z$ direction) and downward ($-z$ direction) waves with the amplitudes $A$ and $B$ in the first and second terms, respectively. Thus, it is generally considered that the SH wave recorded at subsurface is the superposition of upward and downward waves as shown in Fig. 1.2.4(b) based on the assumption of one-dimensional wave propagation. By using a harmonic function in place of the arbitrary functions $f_1(\ )$ and $f_2(\ )$, a general solution of the wave equation can be written as:

$$u = Ae^{i(kz-\omega t)} + Be^{i(kz+\omega t)} \tag{1.2.21}$$

Here, $\omega = 2\pi f$ is the angular frequency ($f$ = frequency), and $k = \omega/V_s$ is the wave number with a dimension of the inverse of length.

### 1.2.3 Basic formulation of wave energy

Let us consider wave energy carried by the upward SH-wave with the wave velocity $V_s$ passing through a horizontal plane A-A′ of a unit area as illustrated in Fig. 1.2.5. Kinetic energy in a soil element of a unit horizontal area times a small thickness

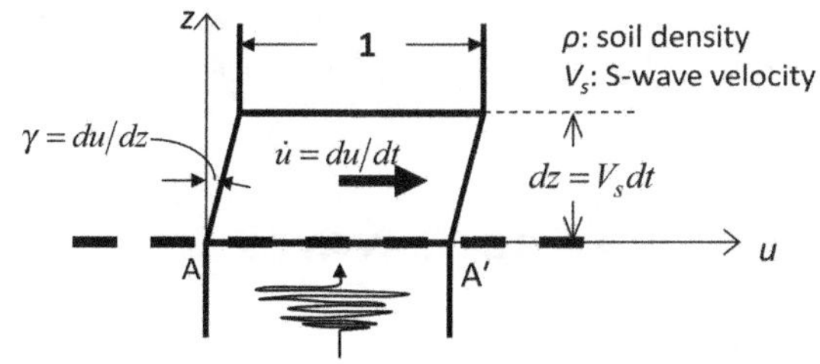


*Figure 1.2.5* Schematic illustration on wave energy in upward SH-wave propagation.

$dz = V_s\Delta t$ (a travel distance in a short time increment $\Delta t$) having particle velocity $\dot{u}$ can be expressed as:

$$\Delta E_k = \frac{1}{2}\rho V_s \Delta t(\dot{u})^2 \tag{1.2.22}$$

Strain energy simultaneously induced by the wave propagation in the same thin soil element is expressed by shear stress $\tau = G\gamma$ and shear strain $\gamma$, and using $\gamma = -\dot{u}/V_s$ in Eq. (1.2.19) as:

$$\Delta E_e = \int_0^{\gamma} (V_s\Delta t)\tau d\gamma = (V_s\Delta t)G\int_0^{\gamma} \gamma d\gamma = \frac{1}{2}\rho V_s^3 \Delta t\gamma^2 = \frac{1}{2}\rho V_s \Delta t(\dot{u})^2 \tag{1.2.23}$$

Hence, $\Delta E_k = \Delta E_e$ and this equality always holds for the one-directional propagating wave at the same wave section. Thus, the wave energy passing through the unit area in the time increment $\Delta t$ is their sum.

$$\Delta E = \Delta E_k + \Delta E_e = \rho V_s \Delta t(\dot{u})^2 \tag{1.2.24}$$

Accumulated energy in a time interval $t = t_1 \sim t_2$ can be expressed as the sum of the kinetic wave energy and strain wave energy, $E_k$ and $E_e$, of the equal amount (Timoshenko and Goodier 1951, Bath 1956, Sarma 1971) as:

$$E = E_k + E_e = \rho V_s \int_{t_1}^{t_2} (\dot{u})^2 dt \tag{1.2.25}$$

Note that the unit of $E$ is *Energy* divided by *Area* and kJ/m$^2$ will be used in this book. Time derivative of the energy is called as energy flux or energy flow rate and written as

$$\frac{dE}{dt} = \frac{dE_e}{dt} + \frac{dE_k}{dt} = \rho V_s(\dot{u})^2 \tag{1.2.26}$$

Thus, the wave energy is defined for the unilaterally propagating wave. In order to calculate the energy flow from earthquake records at or below the ground surface assuming the one-dimensional vertical propagation of P or SH-wave, it is necessary to separate the recorded wave motion into upward and downward waves. Energy flows for actual seismic records will be calculated in Sec. 4.6.

## 1.3 THREE-DIMENSIONAL BODY WAVES

In order to consider a wave equation in a three-dimensional elastic medium, the force-equilibrium in a small rectangular element with its edge lengths $dx$, $dy$, $dz$ and density $\rho$ is considered in the orthogonal $x$, $y$, $z$ space as shown in Fig. 1.3.1. The normal stresses $\sigma_x$, $\sigma_y$, $\sigma_z$ and tangential stresses $\tau_{xy}=\tau_{yx}$, $\tau_{yz}=\tau_{zy}$, $\tau_{zx}=\tau_{xz}$ are working at individual faces of the element, and the body forces X, Y, Z are working at the center of the element. The equilibrium in the $x$-direction for example yields:

$$\left\{\left(\sigma_x+\frac{\partial\sigma_x}{\partial x}dx\right)-\sigma_x\right\}dydz+\left\{\left(\tau_{yx}+\frac{\partial\tau_{yx}}{\partial y}dy\right)-\tau_{yx}\right\}dzdx$$
$$+\left\{\left(\tau_{zx}+\frac{\partial\tau_{zx}}{\partial z}dz\right)-\tau_{zx}\right\}dxdy+Xdxdydz=0$$

By simplifying it, the first formula in (1.3.1) is obtained, and the second and third formulas are also given in the same way.

$$\frac{\partial\sigma_x}{\partial x}+\frac{\partial\tau_{yx}}{\partial y}+\frac{\partial\tau_{zx}}{\partial z}+X=0,\quad \frac{\partial\tau_{xy}}{\partial x}+\frac{\partial\sigma_y}{\partial y}+\frac{\partial\tau_{zy}}{\partial z}+Y=0,$$
$$\frac{\partial\tau_{xz}}{\partial x}+\frac{\partial\tau_{yz}}{\partial y}+\frac{\partial\sigma_z}{\partial z}+Z=0 \tag{1.3.1}$$

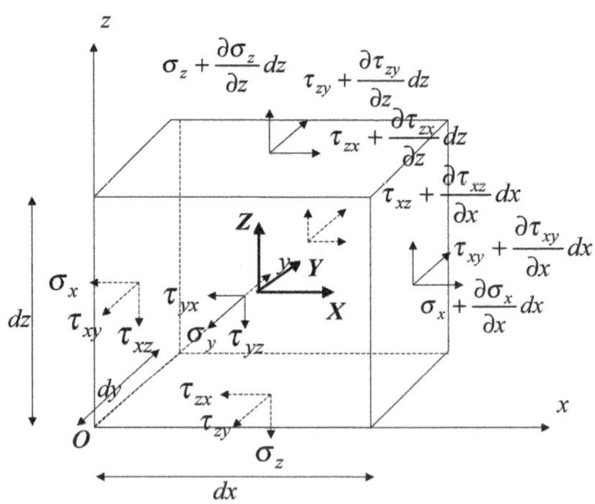


*Figure 1.3.1* Force equilibrium in three-dimensional soil element.

If the body forces $X$, $Y$, $Z$ associated with the wave propagation are inertial forces, they are expressed by using displacements $u$, $v$, $w$, in the $x$, $y$, $z$-axis, respectively, as:

$$X = -\rho\frac{\partial^2 u}{\partial t^2},\quad Y = -\rho\frac{\partial^2 v}{\partial t^2},\quad Z = -\rho\frac{\partial^2 w}{\partial t^2} \tag{1.3.2}$$

Thus, the three-dimensional equilibrium equation is obtained as follows.

$$\begin{aligned}&\frac{\partial\sigma_x}{\partial x}+\frac{\partial\tau_{xy}}{\partial y}+\frac{\partial\tau_{xz}}{\partial z}-\rho\frac{\partial^2 u}{\partial t^2}=0,\quad \frac{\partial\tau_{xy}}{\partial x}+\frac{\partial\sigma_y}{\partial y}+\frac{\partial\tau_{yz}}{\partial z}-\rho\frac{\partial^2 v}{\partial t^2}=0,\\ &\frac{\partial\tau_{xz}}{\partial x}+\frac{\partial\tau_{yz}}{\partial y}+\frac{\partial\sigma_z}{\partial z}-\rho\frac{\partial^2 w}{\partial t^2}=0\end{aligned} \tag{1.3.3}$$

Next, the stress components in Eq. (1.3.3) are expressed by the displacements $u$, $v$, $w$, using the following.

$$\begin{aligned}&\varepsilon_x=\frac{\partial u}{\partial x},\quad \varepsilon_y=\frac{\partial v}{\partial y},\quad \varepsilon_z=\frac{\partial w}{\partial z},\\ &\gamma_{xy}=\frac{\partial v}{\partial x}+\frac{\partial u}{\partial y},\quad \gamma_{yz}=\frac{\partial w}{\partial y}+\frac{\partial v}{\partial z},\quad \gamma_{zx}=\frac{\partial u}{\partial z}+\frac{\partial w}{\partial x}\end{aligned} \tag{1.3.4}$$

Here, $\varepsilon_x$, $\varepsilon_y$, $\varepsilon_z$ are the axial strains in the $x$, $y$, $z$-directions and $\gamma_{xy}$, $\gamma_{yz}$, $\gamma_{zx}$ are the shear strains on the planes normal to $z$, $x$, $y$-axes. If the material is of isotropic elasticity, the strain components are correlated with the stress components by the Hook's law as:

$$\begin{aligned}&\varepsilon_x=\frac{\sigma_x-\nu(\sigma_y+\sigma_z)}{E},\quad \varepsilon_y=\frac{\sigma_y-\nu(\sigma_z+\sigma_x)}{E},\quad \varepsilon_z=\frac{\sigma_z-\nu(\sigma_x+\sigma_y)}{E}\\ &\gamma_{xy}=\frac{\tau_{xy}}{G},\quad \gamma_{yz}=\frac{\tau_{yz}}{G},\quad \gamma_{zx}=\frac{\tau_{zx}}{G}\end{aligned} \tag{1.3.5}$$

Here, $E$ = Young's modulus, $\nu$ = Poisson's ratio, and $G$ = shear modulus. If the stresses are expressed by the strains:

$$\begin{aligned}&\sigma_x=\lambda e+2G\varepsilon_x=K\varepsilon_v+2G\{\varepsilon_x-(\varepsilon_v/3)\}\\ &\sigma_y=\lambda e+2G\varepsilon_y=K\varepsilon_v+2G\{\varepsilon_y-(\varepsilon_v/3)\}\\ &\sigma_z=\lambda e+2G\varepsilon_z=K\varepsilon_v+2G\{\varepsilon_z-(\varepsilon_v/3)\}\\ &\tau_{xy}=G\gamma_{xy},\quad \tau_{yz}=G\gamma_{yz},\quad \tau_{zx}=G\gamma_{zx}\end{aligned} \tag{1.3.6}$$

Here, $\varepsilon_v=\varepsilon_x+\varepsilon_y+\varepsilon_z$ is the volumetric strain, $\lambda$ and $\mu$ (=$G$) are Lame's constants, and $K$ is bulk modulus, written as:

$$\lambda=\frac{\nu E}{(1+\nu)(1-2\nu)},\quad \mu=G=\frac{E}{2(1+\nu)},\quad K=\frac{E}{3(1-2\nu)} \tag{1.3.7}$$

Substituting Eq. (1.3.6) to (1.3.3) gives the equilibrium equation in terms of displacements, $u$, $v$, $w$. In the $x$-direction for example:

$$-\lambda\frac{\partial\varepsilon_v}{\partial x}-2G\frac{\partial^2 u}{\partial x^2}-G\frac{\partial}{\partial y}\left(\frac{\partial u}{\partial y}+\frac{\partial v}{\partial x}\right)-G\frac{\partial}{\partial z}\left(\frac{\partial w}{\partial x}+\frac{\partial u}{\partial z}\right)+\rho\frac{\partial^2 u}{\partial t^2}=0$$

Hence, the following equation can be obtained in the $x$, $y$, $z$-directions.

$$(\lambda+G)\frac{\partial\varepsilon_v}{\partial x}+G\nabla^2 u-\rho\frac{\partial^2 u}{\partial t^2}=0,\quad (\lambda+G)\frac{\partial\varepsilon_v}{\partial y}+G\nabla^2 v-\rho\frac{\partial^2 v}{\partial t^2}=0,$$
$$(\lambda+G)\frac{\partial\varepsilon_v}{\partial z}+G\nabla^2 w-\rho\frac{\partial^2 w}{\partial t^2}=0 \tag{1.3.8}$$

where $\nabla^2=\frac{\partial^2}{\partial x^2}+\frac{\partial^2}{\partial y^2}+\frac{\partial^2}{\partial z^2}$ is the Laplacian operator.

Here, let us consider two types of waves. The first is a wave without volumetric strain, i.e. $\varepsilon_v=\varepsilon_x+\varepsilon_y+\varepsilon_z=0$, wherein Eq. (1.3.8) reduces to:

$$G\nabla^2 u-\rho\frac{\partial^2 u}{\partial t^2}=0,\quad G\nabla^2 v-\rho\frac{\partial^2 v}{\partial t^2}=0,\quad G\nabla^2 w-\rho\frac{\partial^2 w}{\partial t^2}=0 \tag{1.3.9}$$

It is obvious that this represents three-dimensional S-wave with the wave velocity $V_s=\sqrt{G/\rho}$. The second type is a wave where the displacements have no rotational component. Rotations of small rectangles normal to the $z$, $x$, $y$-axes $\omega_z, \omega_x, \omega_y$ are defined as the average rotations (anti-clockwise) of two sides for individual rectangles.

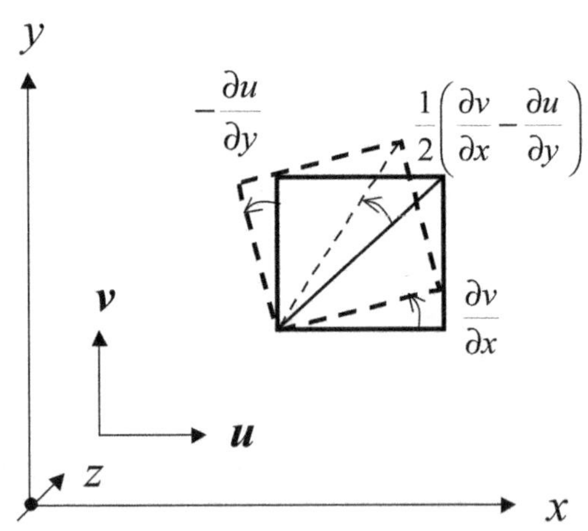


*Figure 1.3.2* Rotation of rectangular element around z-axis.

For example, as illustrated in Fig. 1.3.2, the rotation angle around $z$-axis $\omega_z$ is obtained as the average of $\partial v/\partial x$ and $-\partial u/\partial y$, and hence they are written as:

$$\omega_z = \frac{1}{2}\left(\frac{\partial v}{\partial x} - \frac{\partial u}{\partial y}\right), \quad \omega_x = \frac{1}{2}\left(\frac{\partial w}{\partial y} - \frac{\partial v}{\partial z}\right), \quad \omega_y = \frac{1}{2}\left(\frac{\partial u}{\partial z} - \frac{\partial w}{\partial x}\right) \tag{1.3.10}$$

From the condition that all the rotations are zero:

$$\frac{\partial v}{\partial x} = \frac{\partial u}{\partial y}, \quad \frac{\partial w}{\partial y} = \frac{\partial v}{\partial z}, \quad \frac{\partial u}{\partial z} = \frac{\partial w}{\partial x} \tag{1.3.11}$$

Utilizing the above relationships, $\partial \varepsilon_v/\partial x$ for example can be expressed as:

$$\begin{aligned}\frac{\partial \varepsilon_v}{\partial x} &= \frac{\partial}{\partial x}\left(\frac{\partial u}{\partial x} + \frac{\partial v}{\partial y} + \frac{\partial w}{\partial z}\right) = \frac{\partial^2 u}{\partial x^2} + \frac{\partial}{\partial x}\frac{\partial v}{\partial y} + \frac{\partial}{\partial x}\frac{\partial w}{\partial z} \\ &= \frac{\partial^2 u}{\partial x^2} + \frac{\partial}{\partial y}\frac{\partial v}{\partial x} + \frac{\partial}{\partial z}\frac{\partial w}{\partial x} = \frac{\partial^2 u}{\partial x^2} + \frac{\partial}{\partial y}\frac{\partial u}{\partial y} + \frac{\partial}{\partial z}\frac{\partial u}{\partial z} = \nabla^2 u\end{aligned}$$

Hence, the same operations are possible for all the terms to yield:

$$\frac{\partial \varepsilon_v}{\partial x} = \nabla^2 u, \quad \frac{\partial \varepsilon_v}{\partial y} = \nabla^2 v, \quad \frac{\partial \varepsilon_v}{\partial z} = \nabla^2 w \tag{1.3.12}$$

Substituting this into Eq. (1.3.8) finally gives the equation.

$$\begin{aligned}&(\lambda + 2G)\nabla^2 u - \rho\frac{\partial^2 u}{\partial t^2} = 0, \quad (\lambda + 2G)\nabla^2 v - \rho\frac{\partial^2 v}{\partial t^2} = 0, \\ &(\lambda + 2G)\nabla^2 w - \rho\frac{\partial^2 w}{\partial t^2} = 0\end{aligned} \tag{1.3.13}$$

This is the equation of three-dimensional wave without rotational displacements, representing P-wave with the wave velocity $V_p = \sqrt{(\lambda + 2\mu)/\rho}$. P-wave is sometimes considered as the wave consisting of volumetric strain only and no shear strain. However, P-wave is actually defined as the wave without rotation and can include shear strain (Timoshenko and Goodiers 1951). For instance, soils undergo shear strain in one-dimensional propagation of plane P-wave, because the axial strain in the direction of wave propagation is obviously different from zero axial strain in the perpendicular directions except for Poisson's ratio $\nu = 0.5$.

## 1.4 SURFACE WAVES

If body waves come across a free boundary or a boundary in contact with a different medium, surface waves are generated and propagate with the wave energy concentrating near the boundary surface. The surface waves near the ground surface are divided into Rayleigh wave and Love wave as explained in Fig. 1.1.1. The surface waves are rich in long period motions compared to the body waves and tend to be dominant in

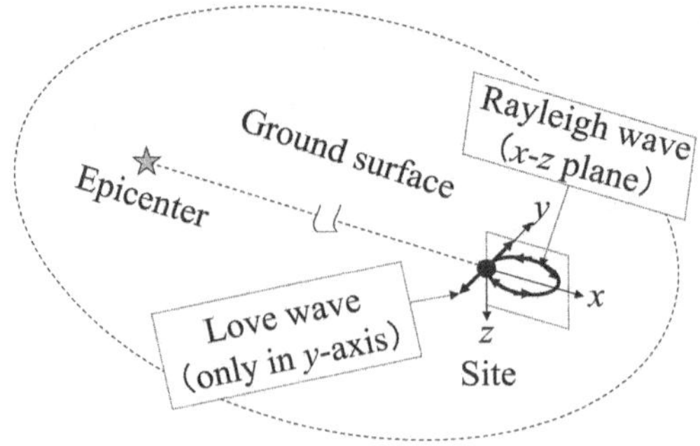


*Figure 1.4.1* Vibration directions by Rayleigh wave and Love wave.

the later part of seismic shaking. Unlike the body waves coming up from below, the surface waves transmit horizontally to a site.

Fig. 1.4.1 schematically illustrates the directions of vibration of the surface waves. Rayleigh wave is generated by combining SV-wave and P-wave. It vibrates in a two-dimensional vertical plane including the horizontal $x$-axis originating from the epicenter and the vertical $z$-axis, and no vibration occurs in the $y$-axis. In contrast, Love wave, which is actually the SH-wave travelling laterally in the $x$-direction, vibrates horizontally only in the $y$-axis with no component in the $x$ and $z$-axes. Rayleigh wave is further classified into non-dispersive Rayleigh wave propagating in a half-space uniform medium and dispersive Rayleigh wave propagating in multi-layered media. Love wave can propagate not in a uniform medium but in layered media only and is dispersive.

Surface waves propagating in layered media are characterized by the dispersion of waves. Namely, waves with longer periods travel faster than those of shorter periods, because longer period waves tend to be more influenced by deeper and stiffer soil properties. Hence, surface waves composed of various period motions tend to disperse as they propagate because longer period motions travel faster than those of shorter period. In order to define the surface wave velocity, phase velocity is normally used, and the variation of the phase velocity depending on the period, frequency or wave length is called a dispersion curve. In the following, Rayleigh wave and Love wave propagating in layered media are dealt with theoretically in matrix forms based on a well-referred previous paper by Haskell (1953).

## 1.4.1 Rayleigh wave

### *1.4.1.1 General formulation*

Fig. 1.4.2 shows a horizontally-layered soil model in the two-dimensional $x$-$z$ plane where Rayleigh wave propagates horizontally along the $x$-axis. The $z$-axis is defined downward from the origin O at the ground surface. The model consists of $n$-layers with individual density $\rho_i$, thickness $d_i$ and Lame's constants $\lambda_i, \mu_i = G_i$ for $i = 1$–$n$. Based on Eqs. (1.3.13) and (1.3.9) for P and S-wave, the wave equations for volumetric

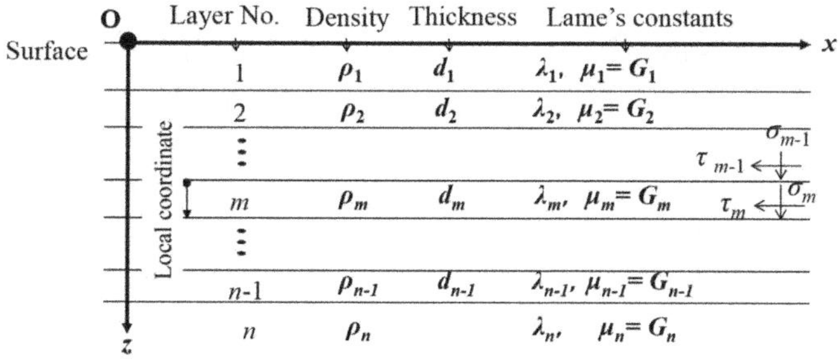


*Figure 1.4.2* Horizontally layered model near ground surface for Rayleigh wave propagating in x-axis.

strain $\phi$ and rotation angle $\psi$ can be expressed in the following form using the Laplace operator $\nabla^2 = \partial^2/\partial x^2 + \partial^2/\partial z^2$.

$$\left\{\nabla^2 + \left(\frac{\omega}{V_p}\right)^2\right\}\phi = 0 \tag{1.4.1}$$

$$\left\{\nabla^2 + \left(\frac{\omega}{V_s}\right)^2\right\}\psi = 0 \tag{1.4.2}$$

Here, $u$, $w$ = displacements in $x$ and $z$ direction, $\omega$ = angular frequency, and the volumetric strain $\phi$ and the rotation angle $\psi$ are defined as follows.

$$\phi = \frac{\partial u}{\partial x} + \frac{\partial w}{\partial z}, \quad \psi = \frac{1}{2}\left(\frac{\partial u}{\partial z} - \frac{\partial w}{\partial x}\right) \tag{1.4.3}$$

Then, let us assume the following formulas for general solutions of Eqs. (1.4.1) and (1.4.2).

$$\phi = Ae^{i(\omega t - kx - \alpha kz)} + Be^{i(\omega t - kx + \alpha kz)} = (Ae^{-i\alpha kz} + Be^{i\alpha kz})e^{i(\omega t - kx)} \tag{1.4.4}$$

$$\psi = Ce^{i(\omega t - kx - \beta kz)} + De^{i(\omega t - kx + \beta kz)} = (Ce^{-i\beta kz} + De^{i\beta kz})e^{i(\omega t - kx)} \tag{1.4.5}$$

Here, $k = \omega/V_s$ is the wave number, $c = \omega/k$ is the surface wave velocity in $x$-direction, and $\alpha, \beta, A, B, C, D$ are the constants to determine in order to characterize the surface wave properly. Substituting Eqs. (1.4.4) and (1.4.5) into Eqs. (1.4.1) and (1.4.2) gives

$$\alpha^2 = \left(\frac{c}{V_p}\right)^2 - 1 \tag{1.4.6}$$

$$\beta^2 = \left(\frac{c}{V_s}\right)^2 - 1 \tag{1.4.7}$$

The square roots of $\alpha$ and $\beta$ are chosen as follows, respectively (Haskell 1953).

$$\alpha=+\left[\left(\frac{c}{V_p}\right)^2-1\right]^{0.5} : c>V_p \quad \alpha=-\left[\left(\frac{c}{V_p}\right)^2-1\right]^{0.5} : c<V_p \tag{1.4.8}$$

$$\beta=+\left[\left(\frac{c}{V_s}\right)^2-1\right]^{0.5} : c>V_s \quad \beta=-\left[\left(\frac{c}{V_s}\right)^2-1\right]^{0.5} : c<V_s \tag{1.4.9}$$

If $c>V_p$ and the positive real value is chosen for $\alpha$ as in the first case of Eq. (1.4.8), Eq. (1.4.4) expresses P-wave with the amplitude $A$ and $B$ propagating upward and downward, respectively, as it transmits into the $x$-direction. If $c<V_p$ and the negative imaginary value is chosen for $\alpha$ as in the second case of Eq. (1.4.8), Eq. (1.4.4) expresses surface wave transmitting in the $x$-direction with the amplitude $A$, because the wave amplitude tends to decrease with increasing $z$. The constant $B$ in Eq. (1.4.4) has to be zero in this case as will be addressed later because the amplitude tends to increase with increasing depth. Similar observations for $\beta$ in Eqs. (1.4.9) and (1.4.5) indicate that surface wave is possible in the second case of Eq. (1.4.9) only if $c<V_s$ and the negative imaginary value is chosen for $\beta$, and the constant $D$ in Eq. (1.4.5) has to be zero.

The general solutions of Eqs. (1.4.1) and (1.4.2) can be written in terms of displacements $u$, $w$:

$$u=-\frac{V_p^2}{\omega^2}\frac{\partial\phi}{\partial x}-2\frac{V_s^2}{\omega^2}\frac{\partial\psi}{\partial z}, \quad w=-\frac{V_p^2}{\omega^2}\frac{\partial\phi}{\partial z}+2\frac{V_s^2}{\omega^2}\frac{\partial\psi}{\partial x} \tag{1.4.10}$$

This is possible because by first implementing the operation $\nabla^2=\partial^2/\partial x^2+\partial^2/\partial z^2$ for each term in Eq. (1.4.10), and then by replacing $\nabla^2\phi$ and $\nabla^2\psi$ using Eqs. (1.4.1) and (1.4.2), the following formulas are obtained.

$$\begin{aligned}\nabla^2 u&=-\frac{V_p^2}{\omega^2}\frac{\partial\nabla^2\phi}{\partial x}-2\frac{V_s^2}{\omega^2}\frac{\partial\nabla^2\psi}{\partial z}=\frac{\partial\phi}{\partial x}+2\frac{\partial\psi}{\partial z},\\ \nabla^2 w&=-\frac{V_p^2}{\omega^2}\frac{\partial\nabla^2\phi}{\partial z}+2\frac{V_s^2}{\omega^2}\frac{\partial\nabla^2\psi}{\partial x}=\frac{\partial\phi}{\partial x}-2\frac{\partial\psi}{\partial z}\end{aligned} \tag{1.4.11}$$

Substituting $\phi$ and $\psi$ defined in Eq. (1.4.3) into Eq. (1.4.11) immediately shows that these equations actually hold.

Hence, Eqs. (1.4.4) and (1.4.5) are substituted into Eq. (1.4.10) to obtain equations for the displacements, $u$, $w$.

$$\begin{aligned}u&=\left\{ik\frac{V_p^2}{\omega^2}(Ae^{-i\alpha kz}+Be^{i\alpha kz})+2i\beta k\frac{V_s^2}{\omega^2}(Ce^{-i\beta kz}-De^{i\beta kz})\right\}e^{i(\omega t-kx)}\\ w&=\left\{i\alpha k\frac{V_p^2}{\omega^2}(Ae^{-i\alpha kz}-Be^{i\alpha kz})-2ik\frac{V_s^2}{\omega^2}(Ce^{-i\beta kz}+De^{i\beta kz})\right\}e^{i(\omega t-kx)}\end{aligned} \tag{1.4.12}$$

In place of $u$, $w$, particle velocities $\dot{u}$, $\dot{w}$ normalized by the phase velocity of surface wave $c=\omega/k$ are written (omitting the common term $e^{i(\omega t-kx)}$) hereafter.

$$\begin{aligned}
\dot{u}/c &= -\frac{V_p^2}{c^2}(Ae^{-i\alpha kz}+Be^{i\alpha kz}) - 2\beta\frac{V_s^2}{c^2}(Ce^{-i\beta kz}-De^{i\beta kz}) \\
&= -(V_p/c)^2(\cos\alpha kz)(A+B) + i(V_p/c)^2(\sin\alpha kz)(A-B) \\
&\quad -2\beta(V_s/c)^2(\cos\beta kz)(C-D) + 2i\beta(V_s/c)^2(\sin\beta kz)(C+D) \\
\dot{w}/c &= -\alpha\frac{V_p^2}{c^2}(Ae^{-i\alpha kz}-Be^{i\alpha kz}) + 2\frac{V_s^2}{c^2}(Ce^{-i\beta kz}+De^{i\beta kz}) \\
&= i\alpha(V_p/c)^2(\sin\alpha kz)(A+B) - \alpha(V_p/c)^2(\cos\alpha kz)(A-B) \\
&\quad -2i(V_s/c)^2(\sin\beta kz)(C-D) + 2(V_s/c)^2(\cos\beta kz)(C+D)
\end{aligned} \tag{1.4.13}$$

As for the stresses, normal and shear stresses $\sigma_z$, $\tau_{zx}$ are expressed as:

$$\begin{aligned}
\sigma_z &= \lambda\phi + 2G\frac{\partial w}{\partial z} = (\lambda+2G)\phi - 2G\frac{\partial u}{\partial x} = \rho\left(V_p^2\phi - 2V_s^2\frac{\partial u}{\partial x}\right), \\
\tau_{zx} &= G\left(\frac{\partial u}{\partial z}+\frac{\partial w}{\partial x}\right) = \rho V_s^2\left(\frac{\partial u}{\partial z}+\frac{\partial w}{\partial x}\right)
\end{aligned} \tag{1.4.14}$$

By substituting Eqs. (1.4.4), (1.4.12) into Eq. (1.4.14), the following equations are obtained (omitting the common term $e^{i(\omega t-kx)}$).

$$\begin{aligned}
\sigma_z &= \rho V_p^2(Ae^{-i\alpha kz}+Be^{i\alpha kz}) \\
&\quad -2\rho V_s^2\left\{\frac{V_p^2}{c^2}(Ae^{-i\alpha kz}+Be^{i\alpha kz}) + 2\beta\frac{V_s^2}{c^2}(Ce^{-i\beta kz}-De^{i\beta kz})\right\} \\
&= -\rho V_p^2(2(V_s/c)^2-1)(\cos\alpha kz)(A+B) + i\rho V_p^2(2(V_s/c)^2-1)(\sin\alpha kz)(A-B) \\
&\quad -4\beta\rho c^2(V_s/c)^4(\cos\beta kz)(C-D) + 4i\beta\rho c^2(V_s/c)^4(\sin\beta kz)(C+D) \\
\tau_{zx} &= 2\rho V_s^2\left[\alpha\frac{V_p^2}{c^2}(Ae^{-i\alpha kz}-Be^{i\alpha kz}) + \frac{V_s^2}{c^2}(\beta^2-1)(Ce^{-i\beta kz}+De^{i\beta kz})\right] \\
&= -2i\rho\alpha V_p^2(V_s/c)^2(\sin\alpha kz)(A+B) + 2\rho\alpha V_p^2(V_s/c)^2(\cos\alpha kz)(A-B) \\
&\quad -2i\rho c^2(\beta^2-1)(V_s/c)^4(\sin\beta kz)(C-D) \\
&\quad +2\rho c^2(\beta^2-1)(V_s/c)^4(\cos\beta kz)(C+D)
\end{aligned} \tag{1.4.15}$$

Eqs. (1.4.13), (1.4.15) can be expressed in a matrix form as:

$$\{\dot{u}/c \quad \dot{w}/c \quad \sigma_z \quad \tau_{zx}\}^T = [E]\{A+B \quad A-B \quad C-D \quad C+D\}^T \tag{1.4.16}$$

where the matrix $[E]$ is as follows.

$$\begin{bmatrix} -(V_p/c)^2\cos\alpha kz & i(V_p/c)^2\sin\alpha kz & -2\beta(V_s/c)^2\cos\beta kz & 2i\beta(V_s/c)^2\sin\beta kz \\ i\alpha(V_p/c)^2\sin\alpha kz & -\alpha(V_p/c)^2\cos\alpha kz & -2i(V_s/c)^2\sin\beta kz & 2(V_s/c)^2\cos\beta kz \\ -\rho V_p^2(2(V_s/c)^2-1)\cos\alpha kz & i\rho V_p^2(2(V_s/c)^2-1)\sin\alpha kz & -4\beta\rho V_s^2(V_s/c)^2\cos\beta kz & 4i\beta\rho V_s^2(V_s/c)^2\sin\beta kz \\ -2i\alpha\rho V_s^2(V_p/c)^2\sin\alpha kz & 2\alpha\rho V_s^2(V_p/c)^2\cos\alpha kz & -2i(\beta^2-1)\rho V_s^2(V_s/c)^2\sin\beta kz & 2(\beta^2-1)\rho V_s^2(V_s/c)^2\cos\beta kz \end{bmatrix} \tag{1.4.17}$$

Let us apply this equation to the top and bottom boundaries of the $m$-th layer of thickness $d_m$ in Fig. 1.4.2. Hereafter, $\sigma_z$, $\tau_{zx}$ are abbreviated as $\sigma$, $\tau$ and the subscript $m$ is used for layer $m$, and a local coordinate starting from the top of each layer is employed. At the top of the $m$-th layer (at the bottom of the $(m-1)$-th layer), Eq. (1.4.16) is written as:

$$\begin{aligned} &\{\dot{u}_{m-1}/c \quad \dot{w}_{m-1}/c \quad \sigma_{m-1} \quad \tau_{m-1}\}^T \\ &\quad = [E_m]\{A_m + B_m \quad A_m - B_m \quad C_m - D_m \quad C_m + D_m t\}^T \end{aligned} \tag{1.4.18}$$

Here, $[E_m]$ is obtained as below by setting local coordinates starting from $z=0$ in Eq. (1.4.17).

$$[E_m] = \begin{bmatrix} -(V_p/c)^2 & 0 & -2\beta(V_s/c)^2 & 0 \\ 0 & -\alpha(V_p/c)^2 & 0 & 2(V_s/c)^2 \\ -\rho V_p^2(2(V_s/c)^2-1) & 0 & -4\beta\rho V_s^2(V_s/c)^2 & 0 \\ 0 & 2\alpha\rho V_p^2(V_s/c)^2 & 0 & 2(\beta^2-1)\rho V_s^2(V_s/c)^2 \end{bmatrix} \tag{1.4.19}$$

At the bottom of the $m$-th layer, Eq. (1.4.16) is written as:

$$\{\dot{u}_m/c \quad \dot{w}_m/c \quad \sigma_m \quad \tau_m\}^T = [F_m]\{A_m + B_m \quad A_m - B_m \quad C_m - D_m \quad C_m + D_m\}^T \tag{1.4.20}$$

Matrix $[F_m]$ is obtained by substituting $z = d_m$ in Eq. (1.4.17). Eliminating the vector $\{A_m + B_m \quad A_m - B_m \quad C_m - D_m \quad C_m + D_m\}^T$ from Eqs. (1.4.18) and (1.4.20) yields

$$\{\dot{u}_m/c \quad \dot{w}_m/c \quad \sigma_m \quad \tau_m\}^T = [a_m]\{\dot{u}_{m-1}/c \quad \dot{w}_{m-1}/c \quad \sigma_{m-1} \quad \tau_{m-1}\}^T \tag{1.4.21}$$

$$[a_m] = [F_m][E_m]^{-1} \tag{1.4.22}$$

This recursive formula gives a relationship between the values $\dot{u}/c$, $\dot{w}/c$, $\sigma$, $\tau$, at the top and bottom of the $m$-th layer. Then, a relationship between 1st and $(n-1)$-th layer can be written in the following form.

$$\{\dot{u}_{n-1}/c \quad \dot{w}_{n-1}/c \quad \sigma_{n-1} \quad \tau_{n-1}\}^T = [a_{n-1}] \cdots [a_1]\{\dot{u}_0/c \quad \dot{w}_0/c \quad \sigma_0 \quad \tau_0\}^T \tag{1.4.23}$$

Here, $\dot{u}_0/c$, $\dot{w}_0/c$, $\sigma_0$, $\tau_0$ are particle velocities and stresses at the ground surface.

If Eqs. (1.4.18) and (1.4.23) are applied in the $n$-th layer, the constants $A_n + B_n$, $A_n - B_n$, $C_n - D_n$, $C_n + D_n$ at the bottom of the $n$-th layer can be formulated as

$$\begin{aligned}
&\{A_n + B_n \quad A_n - B_n \quad C_n - D_n \quad C_n + D_n\}^T \\
&= [E_n]^{-1}\{\dot{u}_{n-1}/c \quad \dot{w}_{n-1}/c \quad \sigma_{n-1} \quad \tau_{n-1}\}^T \\
&= [E_n]^{-1}[a_{n-1}] \cdots [a_1]\{\dot{u}_0/c \quad \dot{w}_0/c \quad \sigma_0 \quad \tau_0\}^T = [J]\{\dot{u}_0/c \quad \dot{w}_0/c \quad \sigma_0 \quad \tau_0\}^T
\end{aligned} \tag{1.4.24}$$

where the 4 by 4 matrix $[J]$ stands for:

$$[J] = [E_n]^{-1}[a_{n-1}] \cdots [a_1] \tag{1.4.25}$$

In the inverse matrix $[E_m]^{-1}$ for an arbitrary $m$-th layer, the individual elements are written as follows.

$$[E_m]^{-1} = \begin{bmatrix} -2(V_s/V_p)^2 & 0 & \dfrac{1}{\rho V_p^2} & 0 \\ 0 & \dfrac{2(V_s/c)^2 - 1}{\alpha(V_p/c)^2} & 0 & \dfrac{1}{\alpha\rho V_p^2} \\ \dfrac{2(V_s/c)^2 - 1}{2\beta(V_s/c)^2} & 0 & -\dfrac{1}{2\beta\rho V_s^2} & 0 \\ 0 & 1 & 0 & \dfrac{1}{2\rho V_s^2} \end{bmatrix} \tag{1.4.26}$$

Here, correlations, $\beta^2 + 1 = (c/V_s)^2$, $(\beta^2 - 1)/(\beta^2 + 1) = -[2(V_s/c)^2 - 1]$ derived from Eq. (1.4.9) are used. Then, 16 elements of the 4 by 4 matrix $[a_m] = [F_m][E_m]^{-1}$ in Eq. (1.4.22) are written as follows.

$$\left.\begin{aligned}
(a_m)_{11} &= 2(V_s/c)^2\cos\alpha kd - [2(V_s/c)^2-1]\cos\beta kd\\
(a_m)_{12} &= i\frac{2(V_s/c)^2-1}{\alpha}\sin\alpha kd + 2i\beta(V_s/c)^2\sin\beta kd\\
(a_m)_{13} &= -\frac{1}{\rho c^2}\cos\alpha kd + \frac{1}{\rho c^2}\cos\beta kd\\
(a_m)_{14} &= i\frac{1}{\alpha\rho c^2}\sin\alpha kd + i\frac{\beta}{\rho c^2}\sin\beta kd\\
(a_m)_{21} &= -2i\alpha(V_s/c)^2\sin\alpha kd - i\frac{2(V_s/c)^2-1}{\beta}\sin\beta kd\\
(a_m)_{22} &= -[2(V_s/c)^2-1]\cos\alpha kd + 2(V_s/c)^2\cos\beta kd\\
(a_m)_{23} &= i\frac{\alpha}{\rho c^2}\sin\alpha kd + i\frac{1}{\beta\rho c^2}\sin\beta kd\\
(a_m)_{24} &= -\frac{1}{\rho c^2}\cos\alpha kd + \frac{1}{\rho c^2}\cos\beta kd = (a_m)_{13}\\
(a_m)_{31} &= 2[2(V_s/c)^2-1]\rho V_s^2\cos\alpha kd - 2[2(V_s/c)^2-1]\rho V_s^2\cos\beta kd\\
(a_m)_{32} &= i\frac{[2(V_s/c)^2-1]^2}{\alpha}\rho c^2\sin\alpha kd + 4i\beta\rho V_s^2(V_s/c)^2\sin\beta kd\\
(a_m)_{33} &= -[2(V_s/c)^2-1]\cos\alpha kd + 2(V_s/c)^2\cos\beta kd = (a_m)_{22}\\
(a_m)_{34} &= i\frac{2(V_s/c)^2-1}{\alpha}\sin\alpha kd + 2i\beta(V_s/c)^2\sin\beta kd = (a_m)_{12}\\
(a_m)_{41} &= 4i\alpha(V_s/c)^2\rho V_s^2\sin\alpha kd + i\frac{[2(V_s/c)^2-1]^2}{\beta}\rho c^2\sin\beta kd\\
(a_m)_{42} &= 2[2(V_s/c)^2-1]\rho V_s^2\cos\alpha kd - 2[2(V_s/c)^2-1]\rho V_s^2\cos\beta kd = (a_m)_{31}\\
(a_m)_{43} &= -2i\alpha(V_s/c)^2\sin\alpha kd - i\frac{2(V_s/c)^2-1}{\beta}\sin\beta kd = (a_m)_{21}\\
(a_m)_{44} &= 2(V_s/c)^2\cos\alpha kd - [2(V_s/c)^2-1]\cos\beta kd = (a_m)_{11}
\end{aligned}\right\} \tag{1.4.27}$$

Note that the properties involved in Eq. (1.4.27) belong to the $m$-th layer. In Eq. (1.4.24), the stresses at the free ground surface ($z=0$) are zero; $\sigma_0=\tau_0=0$, hence the right half elements of the $4\times 4$ matrix $[J]$ are not involved in the calculation. Furthermore, the constants $B$ and $D$ in Eqs. (1.4.4) and (1.4.5) have to be zero; $B=D=0$, because otherwise surface wave amplitudes tend to increase with increasing depth ($z$). Hence, Eq. (1.4.24) can be written as:

$$\{A_n \quad A_n \quad C_n \quad C_n\}^T = [J]\{\dot{u}_0/c \quad \dot{w}_0/c \quad 0 \quad 0\}^T \tag{1.4.28}$$

Thus, using the elements of the matrix $[J]$

$$\begin{aligned}
A_n &= J_{11}\dot{u}_0/c + J_{22}\dot{w}_0/c = J_{21}\dot{u}_0/c + J_{22}\dot{w}_0/c\\
C_n &= J_{31}\dot{u}_0/c + J_{32}\dot{w}_0/c = J_{41}\dot{u}_0/c + J_{42}\dot{w}_0/c
\end{aligned} \tag{1.4.29}$$

From this, the following formula can be drawn.

$$\frac{J_{22}-J_{12}}{J_{11}-J_{21}}=\frac{J_{42}-J_{32}}{J_{31}-J_{41}}=\frac{\dot{u}_0}{\dot{w}_0} \tag{1.4.30}$$

Because $J_{11}$ to $J_{42}$ are functions of $c, k, \alpha, \beta, \rho, V_p, V_s$, it is possible to obtain a direct relationship between $c$ and $k$ (phase velocity dispersion relationship) by numerically solving the above equation consisting of 8 terms $J_{11}$ to $J_{42}$, by using the values$\alpha, \beta, \rho, V_p, V_s$ for individual layers in the layered ground. It should be noted here that the thickness of the bottom layer $d_n$ (= infinitely large) is not included in the elements in $[J]$, as $[E_n]^{-1}$ in Eq. (1.4.25) does not include $d_n$. Then using these values, the ratio $\dot{u}_0/\dot{w}_0$ between horizontal and vertical particle velocities of Rayleigh wave at the ground surface can be calculated for various $c$ or $k$.

If a 4 by 4 matrix $[Q]=[a_{n-1}]\cdots[a_1]$ with 16 elements $q_{11}, q_{12}, \ldots, q_{43}, q_{44}$ is used in Eq. (1.4.25):

$$[J]=[E_n]^{-1}[a_{n-1}]\cdots[a_1]=[E_n]^{-1}[Q] \tag{1.4.31}$$

Then, $J_{11}-J_{21}$, $J_{22}-J_{12}$, $J_{31}-J_{41}$, $J_{42}-J_{32}$ in Eq. (1.4.30) can be expressed in the following forms.

$$\left.\begin{aligned}
J_{11}-J_{21}&=-2(V_s/V_p)^2q_{11}-\frac{2(V_s/c)^2-1}{\alpha(V_p/c)^2}q_{21}+\frac{1}{\rho V_p^2}q_{31}-\frac{1}{\alpha\rho V_p^2}q_{41}\\
J_{22}-J_{12}&=2(V_s/V_p)^2q_{12}+\frac{2(V_s/c)^2-1}{\alpha(V_p/c)^2}q_{22}-\frac{1}{\rho V_p^2}q_{32}+\frac{1}{\alpha\rho V_p^2}q_{42}\\
J_{31}-J_{41}&=\frac{2(V_s/c)^2-1}{2\beta(V_s/c)^2}q_{11}-q_{21}-\frac{1}{2\beta\rho V_s^2}q_{31}-\frac{1}{2\rho V_s^2}q_{41}\\
J_{42}-J_{32}&=-\frac{2(V_s/c)^2-1}{2\beta(V_s/c)^2}q_{12}+q_{22}+\frac{1}{2\beta\rho V_s^2}q_{32}+\frac{1}{2\rho V_s^2}q_{42}
\end{aligned}\right\} \tag{1.4.32}$$

These 4 formulas are further arranged into the following forms and represented each by $-L$, $K$, $-N$ and $M$.

$$\left.\begin{aligned}
\alpha\left(\frac{V_p}{c}\right)^2(J_{11}-J_{21})&=-2\alpha(V_s/c)^2q_{11}-[2(V_s/c)^2-1]q_{21}+\frac{\alpha}{\rho c^2}q_{31}-\frac{1}{\rho c^2}q_{41}\equiv -L\\
\alpha\left(\frac{V_p}{c}\right)^2(J_{22}-J_{12})&=2\alpha(V_s/c)^2q_{12}+[2(V_s/c)^2-1]q_{22}-\frac{\alpha}{\rho c^2}q_{32}+\frac{1}{\rho c^2}q_{42}\equiv K\\
2\beta\left(\frac{V_s}{c}\right)^2(J_{31}-J_{41})&=[2(V_s/c)^2-1]q_{11}-2(V_s/c)^2\beta q_{21}-\frac{1}{\rho c^2}q_{31}-\frac{\beta}{\rho c^2}q_{41}\equiv -N\\
2\beta\left(\frac{V_s}{c}\right)^2(J_{42}-J_{32})&=-[2(V_s/c)^2-1]q_{12}+2(V_s/c)^2\beta q_{22}+\frac{1}{\rho c^2}q_{32}+\frac{\beta}{\rho c^2}q_{42}\equiv M
\end{aligned}\right\} \tag{1.4.33}$$

Then from Eq. (1.4.30), next two equations can be written.

$$-\frac{\dot{u}_0}{\dot{w}_0}=\frac{K}{L}=\frac{M}{N} \tag{1.4.34}$$

$$ML-KN=0 \tag{1.4.35}$$

Let us consider now about the phase difference between $\dot{u}_0$ and $\dot{w}_0$ in Eq. (1.4.34). As already mentioned, surface wave is possible to exist only if $c < V_s$, hence negative imaginary $\alpha$ and $\beta$ should be chosen. In this case, it is obvious that $\cos\alpha kd = (e^{i\alpha kd}+e^{-i\alpha kd})/2$ is real and $\sin\alpha kd=(e^{i\alpha kd}-e^{-i\alpha kd})/2i$ is imaginary. In the similar manner $\cos\beta kd$ is real and $\sin\beta kd$ is imaginary. If we go back to Eq. (1.4.27) and examine each element $(a_m)_{ij}$, it is easily found that if $i+j$ is even, then $(a_m)_{ij}$ is real $(R)$ and if $i+j$ is odd, then $(a_m)_{ij}$ is imaginary $(I)$. Thus the elements of 4 by 4 matrix $[a_m]$ has the following formation.

$$[a_m]=\begin{bmatrix} R & I & R & I \\ I & R & I & R \\ R & I & R & I \\ I & R & I & R \end{bmatrix} \tag{1.4.36}$$

It is clear that this $R$-$I$ formation in the matrix is sustained in the recursive operation $[a_m][a_{m-1}]$, hence the matrix $[Q]$ in Eq. (1.4.31) has the same formation as Eq. (1.4.36). Then, it is concluded from Eq. (1.4.33) that $L\to I$, $K\to R$, $M\to R$, and $N\to I$. This leads $\dot{u}_0/\dot{w}_0\to I$ in Eq. (1.4.34), indicating that horizontal and vertical velocity has a phase difference $\pm 90°$. This means that Rayleigh wave has an elliptical locus with horizontal and vertical axes in the $x$-$z$ plane. It is because if $\dot{u}_0=\pm iU_0e^{i\omega t}$ and $\dot{w}_0= W_0e^{i\omega t}$ so that $\dot{u}_0/\dot{w}_0\to I$, then:

$$\dot{u}_0=\pm iU_0e^{i\omega t}=U_0e^{\pm i\pi/2}e^{i\omega t}=U_0e^{i\omega(t\pm\pi/2)} \tag{1.4.37}$$

and a phase lag angle $\pm 90°$ in $\dot{u}_0$ occurs compared to $\dot{w}_0$. This means that a particle motion in Rayleigh wave is elliptical in retrograde or reverse with respect to the propagation direction. If internal damping in wave-transmitting media is considered, it is necessary to deal $c$ and $k$ as complex values. Then, the ratio $\dot{u}_0/\dot{w}_0$ is not pure imaginary, meaning that a phase difference other than $\pm 90°$ occurs and the axes of the ellipse is inclined from the vertical axis (Haskell 1953).

#### 1.4.1.2 Uniform semi-infinite layer

If the wave length $\lambda=c/f$ of Rayleigh wave becomes large enough, the wave number $k=\omega/c=2\pi/(c/f)$ becomes very small and $kd\to 0$ can be assumed in Eq. (1.4.27). Then for the elements in $[a_m]$, $(a_m)_{ij}=0$ for $i\neq j$, and $(a_m)_{ij}=1$ for $i=j$, indicating that $[a_m]$ is a unit matrix $[I]$. Then, Eq. (1.4.35) reduces to the following formula.

$$[2(V_s/c)^2-1]^2+4\alpha\beta(V_s/c)^4=0 \tag{1.4.38}$$

The same equation for Rayleigh wave in a uniform semi-infinite layer was obtained by Timoshenko and Goodier (1951) through a different approach. Using $\eta=(c/V_s)^2$,

Eq. (1.4.38) can be written as a cubic equation of an unknown $\eta$, where $R = (V_s/V_p)^2 = (1-2\nu)/[2(1-\nu)]$ is a constant depending on Poisson's ratio $\nu$.

$$\eta^3 - 8\eta^2 + 8(3-2R)\eta + 16(R-1) = 0 \tag{1.4.39}$$

For $\nu = 0.25$ for example, out of the three solutions for $\eta$ in Eq. (1.4.39), one with $\alpha$ and $\beta$ being both imaginary based on Eqs. (1.4.8), (1.4.9) is chosen as the wave velocity ratio.

$$\frac{c}{V_s} = \sqrt{\eta} = \sqrt{2 - \frac{2}{\sqrt{3}}} = 0.9194 \tag{1.4.40}$$

Namely, the Rayleigh wave velocity $c$ of a semi-infinite uniform layer with $\nu = 0.25$ is constant (non-dispersive) and about 92% of the S-wave velocity.

In order to examine the amplitude ratio between the horizontal and vertical directions, let us rewrite Eqs. (1.4.13), (1.4.15) wherein $B = D = 0$ as already explained.

$$\begin{aligned}\dot{u}/c &= -A\frac{V_p^2}{c^2}e^{-i\alpha kz} - 2C\beta\frac{V_s^2}{c^2}e^{-i\beta kz}\\ \dot{w}/c &= -A\alpha\frac{V_p^2}{c^2}e^{-i\alpha kz} + 2C\frac{V_s^2}{c^2}e^{-i\beta kz}\end{aligned} \tag{1.4.41}$$

$$\begin{aligned}\sigma_z &= A\rho V_p^2 e^{-i\alpha kz} - 2\rho V_s^2\left(A\frac{V_p^2}{c^2}e^{-i\alpha kz} + 2C\beta\frac{V_s^2}{c^2}e^{-i\beta kz}\right)\\ \tau_{zx} &= 2\rho V_s^2\left[A\alpha\frac{V_p^2}{c^2}e^{-i\alpha kz} + C\frac{V_s^2}{c^2}(\beta^2-1)e^{-i\beta kz}\right]\end{aligned} \tag{1.4.42}$$

From the condition of zero-stress at the ground surface:

$$\begin{aligned}\sigma_z\Big|_{z=0} &= A\rho V_p^2 - 2\rho V_s^2\left\{A\frac{V_p^2}{c^2} + 2C\beta\frac{V_s^2}{c^2}\right\} = 0\\ \tau_{zx}\Big|_{z=0} &= 2\rho V_s^2\left[A\alpha\frac{V_p^2}{c^2} + C\frac{V_s^2}{c^2}(\beta^2-1)\right] = 0\end{aligned} \tag{1.4.43}$$

The ratio $C/A$ is given and the associated equations are drawn as follows.

$$\frac{C}{A} = \frac{V_p^2}{V_s^2}\frac{\alpha}{(1-\beta^2)} = \frac{V_p^2}{V_s^2}\frac{(c^2-2V_s^2)}{4\beta V_s^2} \tag{1.4.44}$$

$$\frac{(c^2-2V_s^2)}{4\beta V_s^2} = \frac{\alpha}{(1-\beta^2)} \tag{1.4.45}$$

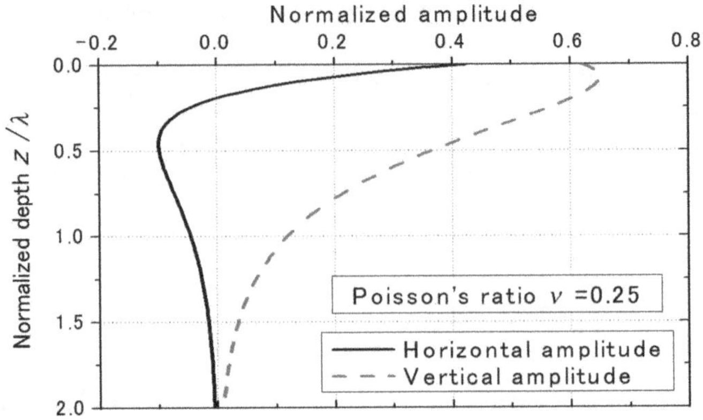


*Figure 1.4.3* Normalized horizontal and vertical amplitudes versus normalized depth for Rayleigh wave in uniform semi-infinite medium.

Using $\beta^2 = (c/V_s)^2 - 1$ in Eq. (1.4.7), Eq. (1.4.45) becomes Eqs. (1.4.46).

$$(c^2 - 2V_s^2)(2 - (c/V_s)^2) = 4\alpha\beta V_s^2 \tag{1.4.46}$$

From Eq. (1.4.46), the same equation as Eq. (1.4.38) can be drawn. Substituting Eq. (1.4.44) into Eq. (1.4.41) gives the next equation wherein $A$ is a constant.

$$\begin{aligned} \frac{\dot{u}/V_s}{AV_p^2/V_s^2} &= -\frac{1}{c/V_s}\left(e^{-2i\alpha\pi(z/\lambda)} + \frac{2\alpha\beta}{1-\beta^2}e^{-2i\beta\pi(z/\lambda)}\right) \\ \frac{\dot{w}/V_s}{AV_p^2/V_s^2} &= -\frac{\alpha}{c/V_s}\left(e^{-2i\alpha\pi(z/\lambda)} - \frac{2}{1-\beta^2}e^{-2i\beta\pi(z/\lambda)}\right) \end{aligned} \tag{1.4.47}$$

Note that $\alpha\beta$ is a negative real value because both $\alpha$ and $\beta$ are negative imaginary. Fig. 1.4.3 shows normalized amplitudes numerically calculated on the right sides of Eq. (1.4.47) using Poisson's ratio $\nu = 0.25$ in the horizontal axis versus the depth normalized by surface wave length $z/\lambda$ in the vertical axis. It is seen for the Rayleigh wave that the vertical motion is larger than the horizontal motion for all depth and the effect of the surface wave becomes almost negligible at the depth twice the wave length ($z = 2\lambda$).

### *1.4.1.3 Two-layer system*

Let us consider a two-layer system shown in Fig. 1.4.4 (a) consisting of a soft surface layer with thickness $H$, soil density $\rho$, Poisson's ratio $\nu$, P-wave velocity $V_p$, and S-wave velocity $V_s$, and a underlying stiffer base layer with infinite thickness, the corresponding constants $\rho_b$, $\nu_b$, $V_{pb}$, $V_{sb}$. Here, let $\alpha_b$, $\beta_b$ correspond to $\alpha$, $\beta$. Eq. (1.4.31) in this case becomes

$$[J] = [E_n]^{-1}[a_1] \tag{1.4.48}$$

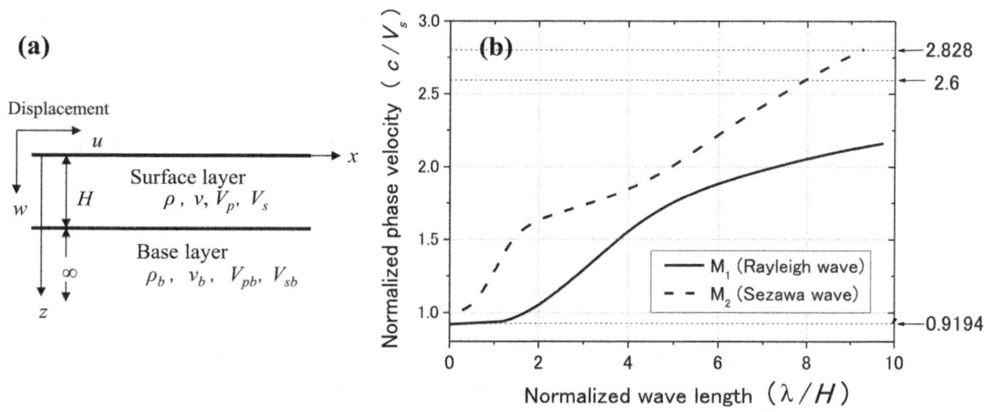


*Figure 1.4.4* Two-layer system (a), and Dispersion curves (b), for dispersive Rayleigh waves.

Using the elements $J_{11}$ to $J_{42}$ in the matrix $[J]$ in Eq. (1.4.48), Eq. (1.4.35) or the next equation is solved.

$$(J_{11}-J_{21})(J_{42}-J_{32})-(J_{22}-J_{12})(J_{31}-J_{41})=0 \tag{1.4.49}$$

wherein each term is written as:

$$\begin{aligned}
J_{11}-J_{21} &= -2\left(\frac{V_{sb}}{V_{pb}}\right)^2(a_1)_{11}-\frac{2(V_{sb}/c)^2-1}{\alpha_b(V_{pb}/c)^2}(a_1)_{21}+\frac{1}{\rho_b V_p b^2}(a_1)_{31}-\frac{1}{\alpha_b\rho_b V_p b^2}(a_1)_{41}\\
J_{22}-J_{12} &= 2\left(\frac{V_{sb}}{V_{pb}}\right)^2(a_1)_{12}+\frac{2(V_{sb}/c)^2-1}{\alpha_b(V_{pb}/c)^2}(a_1)_{22}-\frac{1}{\rho_b V_p b^2}(a_1)_{32}+\frac{1}{\alpha_b\rho_b V_p b^2}(a_1)_{42}\\
J_{31}-J_{41} &= \frac{[2(V_{sb}/c)^2-1]}{2\beta_b(V_{sb}/c)^2}(a_1)_{11}-(a_1)_{21}-\frac{1}{2\beta_b\rho_b V_s b^2}(a_1)_{31}-\frac{1}{2\rho_b V_{sb}^2}(a_1)_{41}\\
J_{42}-J_{32} &= -\frac{2(V_{sb}/c)^2-1}{2\beta_b(V_{sb}/c)^2}(a_1)_{12}+(a_1)_{22}+\frac{1}{2\beta_b\rho_b V_{sb}^2}(a_1)_{32}+\frac{1}{2\rho_b V_{sb}^2}(a_1)_{42}
\end{aligned} \tag{1.4.50}$$

Here, $(a_1)_{ij}$ is the element of the matrix $[a_1]$ for the surface layer given by Eq. (1.4.27). Fig. 1.4.4 (b) shows the dispersion curves for Rayleigh wave calculated by Eqs. (1.4.49), (1.4.50) for $V_s/V_{sb}=1/\sqrt{8}$, $\rho/\rho_b=1$, and $\nu=\nu_b=0.25$. The wave length $\lambda$ is normalized by the layer thickness $H$ in the horizontal axis, and the phase velocity $c$ is normalized by $V_s$ in the surface layer in the vertical axis. Two different dispersion curves are given in this way corresponding to two different types of Rayleigh waves, named as dispersive Rayleigh wave ($M_1$) and Sezawa-wave ($M_2$). For the solid curve for the $M_1$-wave, the velocity ratio takes $c/V_s=0.9194$ for infinitely short wave length $\lambda/H=0$ corresponding to Rayleigh wave in an uniform semi-infinite layer having the same $V_s$ as in the surface layer as already shown in Eq. (1.4.40). For infinitely long wave length $\lambda/H=\infty$, the solid curve approaches to

$c/V_s = c/(V_{sb}/\sqrt{8}) = \sqrt{8} \times 0.9194 = 2.60$ or $c/V_{sb} = 0.9194$, corresponding to Rayleigh wave in a uniform semi-infinite layer with the $V_{sb}$ as in the base layer. In the dashed curve for $M_2$-wave, wherein the velocity ratio with respect to the S-wave velocity in the surface layer $c/V_s$ changes from 1 to 2.828 (while the velocity ratio with respect to $V_{sb}$ in the base layer $c/V_{sb}$ changes from 0.354 to 1.0) for $\lambda/H$ changing from 0 to a cut-off value corresponding to $c/V_{sb} = 1.0$ (Sezawa 1927). In addition to the dispersion curves for the first mode shown in Fig. 1.4.4 (b), dispersion curves for the higher mode of $M_1$ and $M_2$-wave can also be obtained (Kanai 1951).

### 1.4.2 Love wave

Fig. 1.4.5 shows a horizontally-layered soil model in a three-dimensional $x$-$y$-$z$ space where Love wave vibrates in $y$-direction and propagates in $x$-direction horizontally. In Love wave, displacements $u = w = 0$ and $v$ in $y$-direction can be written as:

$$v = Le^{i(\omega t - kx - \beta kz)} + Me^{i(\omega t - kx + \beta kz)} = e^{i(\omega t - kx)}(Le^{-i\beta kz} + Me^{i\beta kz}) \tag{1.4.51}$$

Here, $\beta$ is the square root of $\beta^2 = (c/V_s)^2 - 1$ shown in Eq. (1.4.9) wherein $c = \omega/k$ is the phase velocity of Love wave, and $L$ and $M$ are the wave amplitudes. In the same manner as in Rayleigh wave, if a positive real value is chosen for $\beta$ as in the first case of Eq. (1.4.9), Eq. (1.4.51) becomes a S-wave equation with the amplitude $L$ and $M$ propagating upward and downward, respectively, as it transmits into the $x$-direction. If $c < V_s$ and the negative imaginary value is chosen for $\beta$, Eq. (1.4.51) expresses surface wave transmitting into $x$-direction with the surface amplitude $L$ because the wave amplitude tends to decrease with increasing $z$. The constant $M$ in Eq. (1.4.51) has to be zero in this case because the amplitude tends to increase with increasing depth.

By similar procedures as in Eqs. (1.4.13), (1.4.15) for Rayleigh wave, the next equation is obtained for the upper boundary of $m$-th layer in the local coordinate $z = 0$ (omitting the common term $e^{i(\omega t - kx)}$).

$$\begin{aligned} (\dot{v}/c)_{m-1} &= ik(L_m + M_m) \\ \tau_{m-1} &= -ikG_m\beta_m(L_m - M_m) \end{aligned} \tag{1.4.52}$$

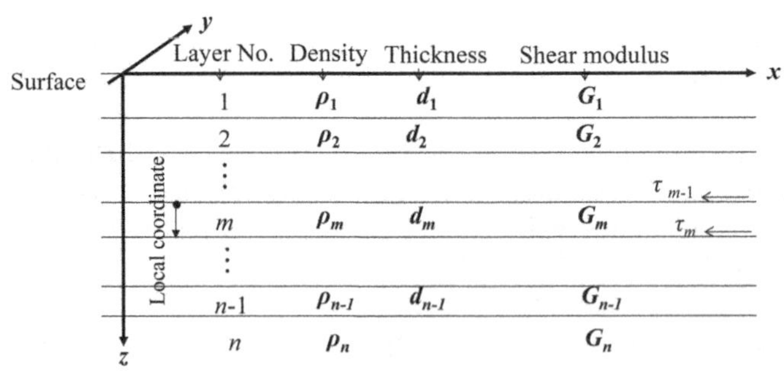


*Figure 1.4.5* Horizontally layered model for Love wave propagating in $x$-direction.

At the lower boundary of the $m$-th layer in the local coordinate $z = d_m$ (omitting the common term $e^{i(\omega t-kx)}$),

$$\begin{aligned}(\dot{v}/c)_m &= ik(L_m e^{-i\beta_m k d_m} + M_m e^{i\beta_m k d_m}) = ik(L_m + M_m)\cos\beta_m k d_m \\ &\quad + k(L_m - M_m)\sin\beta_m k d_m \\ \tau_m &= ikG_m\beta_m(-L_m e^{-i\beta k d_m} + M_m e^{i\beta k d_m}) \\ &= -ikG_m\beta_m(L_m - M_m)\cos\beta_m k d_m - kG_m\beta_m(L_m + M_m)\sin\beta_m k d_m\end{aligned} \quad (1.4.53)$$

Eliminating $(L_m + M_m)$ and $(L_m - M_m)$ from Eqs. (1.4.52) and (1.4.53) yields:

$$\begin{aligned}(\dot{v}/c)_m &= (\dot{v}/c)_{m-1}\cos\beta_m k d_m + i(\tau_{m-1}/G_m\beta_m)\sin\beta_m k d_m \\ \tau_m &= \tau_{m-1}\cos\beta_m k d_m + i((\dot{v}/c)_{m-1}G_m\beta_m)\sin\beta_m k d_m\end{aligned} \quad (1.4.54)$$

which can be expressed in the following recursive form:

$$\{(\dot{v}/c)_m, \tau_m\}^T = [a_m]\{(\dot{v}/c)_{m-1}, \tau_{m-1}\}^T \quad (1.4.55)$$

Here, $[a_m]$ represents the following.

$$[a_m] = \cos\beta_m k d_m \begin{bmatrix} 1 & iG_m^{-1}\beta_m^{-1}\tan(\beta_m k d_m) \\ iG_m\beta_m\tan(\beta_m k d_m) & 1 \end{bmatrix} \quad (1.4.56)$$

If a $2 \times 2$ matrix using the recursive formula in Eq. (1.4.55) from $m = 1$ to $(n-1)$ is introduced as:

$$[Q] = \begin{bmatrix} q_{11} & q_{12} \\ q_{21} & q_{22} \end{bmatrix} = [a_{n-1}][a_{n-2}]\cdots[a_2][a_1] \quad (1.4.57)$$

Then, a relationship between 1st and $(n-1)$-th layer is expressed as:

$$\begin{aligned}(\dot{v}/c)_{n-1} &= q_{11}(\dot{v}/c)_0 + q_{12}\tau_0 \\ \tau_{n-1} &= q_{21}(\dot{v}/c)_0 + q_{22}\tau_0\end{aligned} \quad (1.4.58)$$

By using Eq. (1.4.52) for $m = n$, the next equation is obtained.

$$\begin{aligned}L_n + M_n &= (ik)^{-1}(\dot{v}/c)_{n-1} = (ik)^{-1}q_{11}(\dot{v}/c)_0 + (ik)^{-1}q_{12}\tau_0 \\ L_n - M_n &= -(ikG_n\beta_n)^{-1}\tau_{n-1} = -(ikG_n\beta_n)^{-1}q_{21}(\dot{v}/c)_0 - (ikG_n\beta_n)^{-1}q_{22}\tau_0\end{aligned} \quad (1.4.59)$$

In the above equation, shear stress at the ground surface $\tau_0 = 0$, and $M_n$ should be zero because otherwise the wave amplitude tends to increase with increasing depth as already mentioned. Hence the next simple formula is obtained from Eq. (1.4.59).

$$q_{11} = -(G_n\beta_n)^{-1}q_{21} \quad (1.4.60)$$

For the two-layer system shown in Fig. 1.4.6 (a), wherein $H$ = surface layer thickness, $\rho$ = soil density, $V_s$, $V_{sb}$ = S-wave velocity in the surface and base layer, respectively,

Eq. (1.4.60) becomes $q_{11} = -(G_2\beta_2)^{-1}q_{21}$, and the substitution of $q_{11} = \cos\beta_1 kd_1$, $q_{21} = iG_1\beta_1 \sin(\beta_1 kd_1)$ gives the dispersion curve as follows.

$$\tan(\beta_1 kd_1) = -\frac{iG_2\beta_2}{G_1\beta_1} \tag{1.4.61}$$

This equation is rewritten by using $\beta^2 = (c/V_s)^2 - 1$, $k = \omega/c = 2\pi f/c$, and also considering the condition $V_s < c < V_{sb}$ as:

$$\tan\left[2\pi\frac{H}{\lambda}((c/V_s)^2 - 1)^{1/2}\right] = -i\frac{G_2}{G_1}\frac{((c/V_{sb})^2 - 1)^{1/2}}{((c/V_s)^2 - 1)^{1/2}} = \left(\frac{V_{sb}}{V_s}\right)\frac{((V_{sb}/V_s)^2 - (c/V_s)^2)^{1/2}}{((c/V_s)^2 - 1)^{1/2}} \tag{1.4.62}$$

Fig. 1.4.6(b) shows a numerically calculated results of Eq. (1.4.62) for three $V_s$-ratios between surface and base $V_s/V_{sb} = 1/2$, 1/4, 1/8, wherein the phase velocity ratio $c/V_s$ in the vertical axis is plotted versus normalized wave length $\lambda/H$. Wave dispersion in Love-wave is clearly seen in that the longer the wave length relative to the surface layer thickness, the faster the phase velocity is. In the extreme, for $\lambda/H \to 0$, the Love-wave velocity $c$ becomes equal to $V_s$ of the surface layer, while for $\lambda/H \to \infty$, $c$ approaches to $V_{sb}$ of the base layer.

On the other hand, group velocity is sometimes used as the velocity corresponding to a representative wave period. Because the group velocity of surface wave can be defined as $d\omega/dk$ (Ewing et al. 1957), it is calculated from the dispersion curves such as in Figs. 1.4.4(b) and 1.4.6(b) for Rayleigh and Love waves using $\omega = ck$ as:

$$\frac{d\omega}{dk} = \frac{d(ck)}{dk} = c + k\frac{dc}{dk} \tag{1.4.63}$$

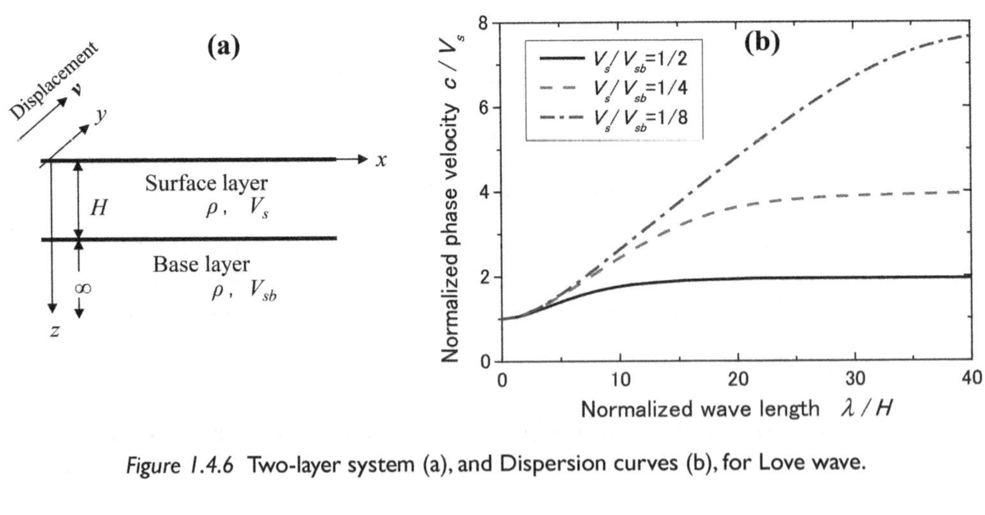


*Figure 1.4.6* Two-layer system (a), and Dispersion curves (b), for Love wave.

## 1.5 VISCOELASTIC MODEL AND SOIL DAMPING FOR WAVE PROPAGATION

So far, soil materials have been assumed to be perfectly elastic with no internal dissipation of energy in wave propagation. In reality, soil dissipates wave energy due to internal damping even for very small seismic waves. The energy dissipation in soil materials can be idealized by viscoelasticity models wherein elastic springs are coupled with viscous dashpots (Ishihara 1996). Here, the basic theory on viscoelastic models is dealt with for cyclic loading.

### 1.5.1 General stress-strain relationship of viscoelastic material

In viscoelastic materials loaded cyclically, induced strain has a phase lag from applied stress. The cyclic stress $\tau$ with the angular frequency $\omega = 2\pi f$ ($f$ = frequency) and the amplitude $\tau_a$, and the corresponding cyclic strain $\gamma$ with the amplitude $\gamma_a$, and the phase delay angle $\delta$ may be expressed as follows (Ishihara 1996).

$$\tau = \tau_a \sin \omega t, \quad \gamma = \gamma_a \sin(\omega t - \delta) \tag{1.5.1}$$

Eliminating $\omega t$ from the above yields:

$$\left(\frac{\tau}{\tau_a}\right)^2 - 2\left(\frac{\tau}{\tau_a}\right)\left(\frac{\gamma}{\gamma_a}\right)\cos\delta + \left(\frac{\gamma}{\gamma_a}\right)^2 = \sin^2\delta \tag{1.5.2}$$

According to Eq. (1.5.2), a hysteresis loop is drawn on a $\tau$–$\gamma$ plane as in Fig. 1.5.1. The area enclosed by the elliptical loop $\Delta W$ represents the energy dissipated during one cycle of loading and expressed as:

$$\Delta W = \int \tau d\gamma = \omega \tau_a \gamma_a \int_0^{2\pi/\omega} \sin \omega t \cos(\omega t - \delta) dt = \tau_a \gamma_a \pi \sin \delta \tag{1.5.3}$$

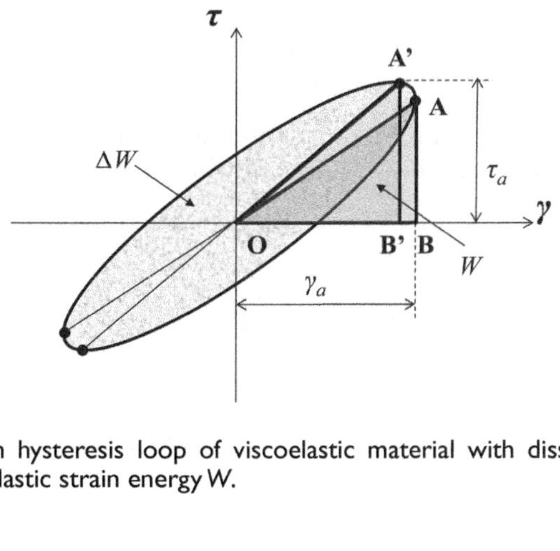


*Figure 1.5.1* Stress-strain hysteresis loop of viscoelastic material with dissipated energy $\Delta W$ and maximum elastic strain energy $W$.

On the other hand, the elastic strain energy stored during cyclic loading takes a maximum value $W$ at the peak of stress or strain, corresponding to the area of the triangle OAB or OA′B′ in the figure.

$$W = \frac{\tau_a \gamma_a \cos\delta}{2} \tag{1.5.4}$$

A ratio of the dissipated energy to the maximum elastic strain energy per cycle characterizes a potential of energy dissipation in the material and can be written as:

$$\frac{\Delta W}{W} = 2\pi \tan\delta \tag{1.5.5}$$

The angle of phase delay $\delta$ is correlated with the energy ratio $\Delta W/W$ by the following formula.

$$\tan\delta = \frac{\Delta W}{2\pi W} \tag{1.5.6}$$

As will be explained in Sec.1.5.2.4, the damping ratio $D$ defined in the one-degree-of-freedom vibration system can be correlated as $D = (\tan\delta)/2$, and hence $D$ and $\Delta W/W$ are correlated as:

$$D = \frac{\tan\delta}{2} = \frac{\Delta W}{4\pi W} \quad \text{or} \quad \frac{\Delta W}{W} = 4\pi D \tag{1.5.7}$$

## 1.5.2 Viscoelastic models

Among several viscoelastic models combining springs and dashpots, Kelvin-model (or Voigt-model) and Maxwell model are representative in engineering mechanics. In addition, Nonviscous Kelvin model is also used in soil dynamic problems in particular.

### *1.5.2.1 Kelvin model*

This model consists of a linear spring (representing the shear modulus) and a dashpot in parallel as illustrated in Fig. 1.5.2(a). The applied stress $\tau$ is the sum of the spring stress $\tau_1$ and the dashpot stress $\tau_2$, and expressed as:

$$\tau = \tau_1 + \tau_2 = G\gamma + \xi\frac{d\gamma}{dt} = \left(G + \xi\frac{d}{dt}\right)\gamma \tag{1.5.8}$$

Here, $\gamma$ = strain of the model, $G$ = shear modulus of the spring and $\xi$ = viscosity of the dashpot. If the applied cyclic stress $\tau$ and the induced strain $\gamma$ in this model are written as:

$$\tau = \tau_0 e^{i\omega t}, \quad \gamma = \gamma_0 e^{i\omega t} \tag{1.5.9}$$

and substituted into Eq. (1.5.8), the next is obtained.

$$\tau_0 = (G + i\omega\xi)\gamma_0 \tag{1.5.10}$$

Substituting this into the first equation of Eq. (1.5.9) gives

$$\tau = (G + i\omega\xi)\gamma = G^*\gamma \tag{1.5.11}$$

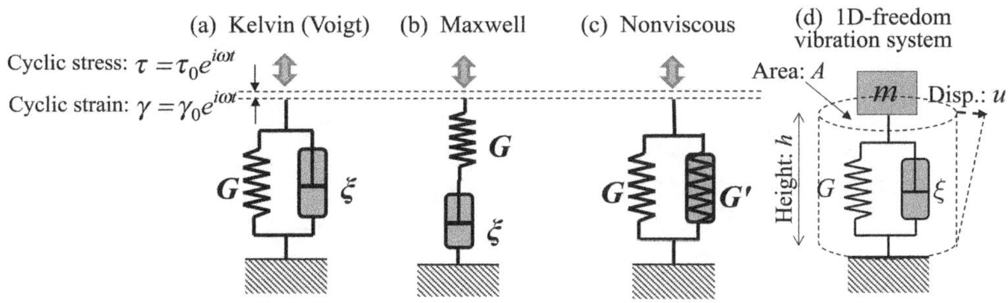


*Figure 1.5.2* Three viscoelastic models employed in dynamic problems (a)~(c), and 1D-freedom vibration system with viscoelastic spring (d).

Here, $G^*$ is named as complex modulus and expressed as:

$$G^* = G + i\omega\xi = |G^*|e^{i\delta} \tag{1.5.12}$$

Absolute normalized modulus $|G^*/G|$ and $\tan\delta$ are written as:

$$|G^*/G| = \sqrt{1 + (\omega\xi/G)^2} \tag{1.5.13}$$

$$\tan\delta = \omega\xi/G \tag{1.5.14}$$

Thus, Eq. (1.5.11) is written eventually as

$$\tau = |G^*|\gamma_0 e^{i(\omega t + \delta)} \tag{1.5.15}$$

This indicates that the maximum stress amplitude is $|G^*|\gamma_0$ and $\gamma$ is delayed by the phase angle $\delta$ relative to $\tau$.

#### *1.5.2.2 Maxwell model*

In this model, the spring is connected in series with the dashpot as in Fig. 1.5.2(b). Then, the induced strain $\gamma$ is the sum of $\gamma_1$ in the spring and $\gamma_2$ in the dashpot and can be written as:

$$\gamma = \gamma_1 + \gamma_2, \quad \tau = G\gamma_1, \quad \tau = \xi\frac{d\gamma_2}{dt} \tag{1.5.16}$$

Combining these relationships yields the next equation.

$$\frac{1}{\xi}\tau + \frac{1}{G}\frac{d\tau}{dt} = \frac{d\gamma}{dt} \quad \text{or} \quad \tau = \left[\frac{d}{dt}\Big/\left(\frac{1}{\xi} + \frac{1}{G}\frac{d}{dt}\right)\right]\gamma \tag{1.5.17}$$

Substituting Eq. (1.5.9) into Eq. (1.5.17) yields:

$$\left(\frac{1}{\xi} + \frac{i\omega}{G}\right)\tau_0 = i\omega\gamma_0 \tag{1.5.18}$$

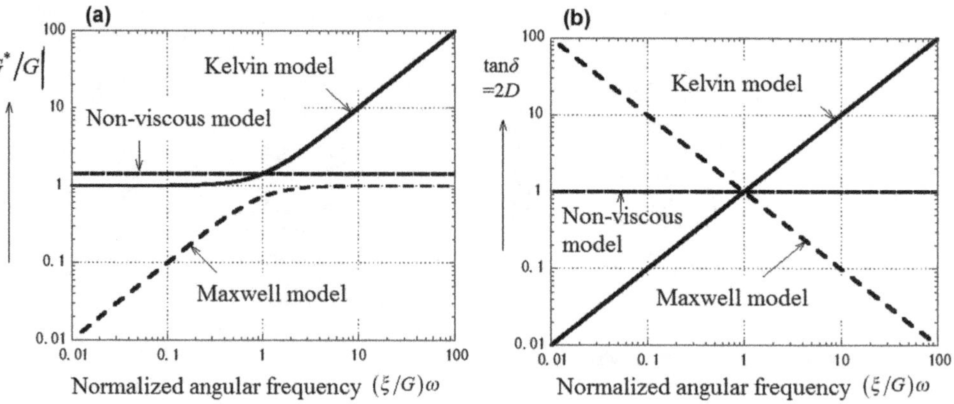


*Figure 1.5.3* Frequency-dependent property variations of three models: (a) $|G^*/G|$ versus $(\xi/G)\omega$, (b) $\tan\delta$ versus $(\xi/G)\omega$.

$$\tau_0 = \frac{i\omega}{\dfrac{1}{\xi} + \dfrac{i\omega}{G}} \gamma_0 = G^* \gamma_0 \tag{1.5.19}$$

$$G^* = \frac{i\omega}{\dfrac{1}{\xi} + \dfrac{i\omega}{G}} = |G^*| \, e^{i\delta} \tag{1.5.20}$$

wherein $G^*$ is the complex shear modulus, and $|G^*/G|$ and $\tan\delta$ are written as follows.

$$|G^*/G| = \frac{\omega\xi/G}{\sqrt{1 + (\omega\xi/G)^2}} \tag{1.5.21}$$

$$\tan\delta = \frac{G}{\omega\xi} \tag{1.5.22}$$

Then, the stress–strain relationship is expressed by the same formula as in Eq. (1.5.15).

In Fig. 1.5.3, the values $|G^*/G|$ and $\tan\delta$ are plotted versus normalized angular frequency $(\xi/G)\omega$ on log-log charts for the Kelvin and Maxwell models based on Eqs. (1.5.13), (1.5.14), (1.5.21) and (1.5.22) as already shown by Ishihara (1996). As for the shear modulus, $|G^*|$ in the Kelvin model significantly increases with increasing $\omega$ for $(\xi/G)\omega \geq 1.0$, while it changes in the completely opposite way in the Maxwell model. As for $\tan\delta$ which is equal to $2D$ (twice the damping ratio) as shown in Eq. (1.5.7), it increases in the Kelvin model and decreases in the Maxwell model linearly with increasing $\omega$ on the log-log diagram. Thus, the shear modulus and damping ratio derived from the two models are very much dependent on frequency $\omega$ because of the viscous dashpots incorporated in the models.

#### 1.5.2.3 Nonviscous Kelvin model

As will be addressed in Sec. 2.3.3, laboratory soil tests indicate that dynamic soil properties, the damping ratio in particular, are essentially independent of frequency and almost unchanged for a wide range of frequency. In order to get rid of the frequency-dependency in viscoelastic models shown in Fig. 1.5.3, Nonviscous Kelvin model is often used in soil dynamics. In this model, the viscous dashpot in the Kelvin model is replaced by a nonviscous dashpot as shown in Fig. 1.5.2 (c), wherein $i\omega\xi$ in Eq. (1.5.10) is replaced by $iG'$, leading to a modified stress-strain relationship without $\omega$ as follows.

$$\tau = (G + iG')\gamma = G^*\gamma \tag{1.5.23}$$

$$G^* = G + iG' = |G^*|e^{i\delta} \tag{1.5.24}$$

with $|G^*/G|$ and $\tan\delta$ written as follows.

$$|G^*/G| = \sqrt{1 + (G'/G)^2} \tag{1.5.25}$$

$$\tan\delta = G'/G \tag{1.5.26}$$

These relationships are also shown in Fig. 1.5.3, where $|G^*|$ and $\tan\delta = 2D$ are independent of $\omega$.

#### 1.5.2.4 Comparison with 1D-of-freedom vibration system

The viscous Kelvin model is often coupled with a mass $m$ as a one-degree-of-freedom lumped mass-spring system in vibration problems. Free-vibration of the mass supported by the Kelvin spring as shown in Fig. 1.5.2 (d) can be formulated in terms of the displacement $u$ as:

$$m\frac{d^2u}{dt^2} + c\frac{du}{dt} + ku = 0 \tag{1.5.27}$$

Here, $m$ = mass, $c = \xi a/h$ is a dashpot constant and $k = Ga/h$ is a spring constant, where $a$ and $h$ are the cross-sectional area and height of the Kelvin spring, respectively. If a solution $u = Ae^{i\omega t}$ is assumed and substituted into Eq. (1.5.27), then:

$$-m\omega^2 + ic\omega + k = 0 \tag{1.5.28}$$

Thus, $\omega$ is obtained as:

$$\omega = i\frac{c}{2m} \pm \frac{\sqrt{4mk - c^2}}{2m} \tag{1.5.29}$$

Substituting this into $u = Ae^{i\omega t}$ and using the resonant frequency $\omega_0 = \sqrt{k/m}$, the final solution for a decayed free vibration is written as

$$u = e^{-D\omega_0 t}(A_1 e^{i\sqrt{1-D^2}\omega_0 t} + A_2 e^{-i\sqrt{1-D^2}\omega_0 t}) \tag{1.5.30}$$

wherein damping ratio $D$ ($D \leq 1.0$ to vibrate) defined below is used.

$$D = c/2\sqrt{mk} = \omega_0\, c/2k = \omega_0\, \xi/2G \tag{1.5.31}$$

If $D$ in Eq. (1.5.31) is compared with $(\tan\delta)/2 = \omega\xi/2G$ using Eq. (1.5.14), it is clear that the right sides of the two equations are identical for $\omega = \omega_0$. This indicates that the damping ratio $D$ defined in the one-degree-of-freedom vibration system is correlated with the phase delay angle $\delta$ in the hysteretic stress-strain relationship of the Kelvin model as written below and also in Eq. (1.5.7).

$$D = \omega\xi/2G = (\tan\delta)/2 \tag{1.5.32}$$

Consequently, the damping ratios for the above three viscoelastic models can be expressed using the correlation $D = (\tan\delta)/2$ derived above.

$$\text{Kelvin model: } D = (\tan\delta)/2 = \omega\xi/2G \tag{1.5.33}$$

$$\text{Maxwell model: } D = (\tan\delta)/2 = G/2\omega\xi \tag{1.5.34}$$

$$\text{Non-viscous Kelvin model: } D = (\tan\delta)/2 = G/2G' \tag{1.5.35}$$

## 1.6 WAVE ATTENUATION BY INTERNAL DAMPING

The attenuation of seismic waves during their propagation from a wave source is attributed to two mechanisms; geometric damping and internal damping. The geometric damping also named as radiation damping occurs because the wave energy density per unit area is getting lower as the area of wave front increases, although the total energy is unchanged. The internal damping also named as material damping occurs due to energy dissipation as the wave propagates in the viscous or frictional medium. Here, the wave attenuation due to the internal damping is dealt with in terms of wave amplitude and wave energy based on the viscoelastic models explained in the previous Section. The geometric damping is also discussed as well to be integrated in the wave attenuation as a whole.

### 1.6.1 Viscoelastic models and wave attenuation

Here, the attenuation of plane waves transmitting unilaterally in viscoelastic media is considered for the three viscoelastic models addressed in Sec. 1.5.

#### *1.6.1.1 Attenuation for Kelvin model*

If the stress-strain relationship of Kelvin model Eq. (1.5.8) is substituted into Eq. (1.2.2), the wave equation is obtained.

$$\rho\frac{\partial^2 u}{\partial t^2} = \left(G + \xi\frac{\partial}{\partial t}\right)\frac{\partial^2 u}{\partial z^2} \tag{1.6.1}$$

If the displacement $u$ is expressed as:

$$u = U(z)e^{i\omega t} \tag{1.6.2}$$

and substituting into Eq. (1.6.1), the amplitude $U$ as a function of $z$ is obtained as:

$$\frac{\partial^2 U}{\partial z^2} + \frac{\rho\omega^2}{G + i\omega\xi}U = 0 \tag{1.6.3}$$

and the general solution is written as:

$$U = Ae^{ik^*z} + Be^{-ik^*z} \tag{1.6.4}$$

where complex wave number $k^*$ is defined, using the complex shear modulus $G^* = G + i\omega\xi$ in Eq. (1.5.12), as:

$$k^* = \left(\frac{\rho\omega^2}{G^*}\right)^{1/2} = \left(\frac{\rho\omega^2}{G + i\omega\xi}\right)^{1/2} \tag{1.6.5}$$

Substituting Eq. (1.6.4) into Eq. (1.6.2) yields the general solution for Kelvin model.

$$u = Ae^{i(k^*z + \omega t)} + Be^{-i(k^*z - \omega t)} \tag{1.6.6}$$

The second term in the right side of the above equation is chosen here as the wave unilaterally propagating toward the positive direction of $z$-axis, and transformed as:

$$u = Be^{i(\omega t - k^*z)} = Be^{i\omega[t - (z/V_s^*)]} \tag{1.6.7}$$

Here, $V_s^*$ is introduced as a complex S-wave velocity as:

$$V_s^* = \sqrt{(G + i\omega\xi)/\rho} = (G^2 + \omega^2\xi^2)^{1/4}e^{i\delta/2}/\rho^{1/2} = V_s(\cos\delta)^{-1/2}e^{i\delta/2} \tag{1.6.8}$$

wherein $\tan\delta = \omega\xi/G$ as already shown in Eq. (1.5.14). Eq. (1.6.7) is further transformed as:

$$u = Be^{-\beta z}e^{i\omega(t - z/V_s')} \approx Be^{-\beta z}e^{i\omega(t - z/V_s)} \tag{1.6.9}$$

$$V_s' = \frac{V_s}{(\cos\delta)^{1/2}(\cos(\delta/2))} \tag{1.6.10}$$

$$\beta = (\omega/V_s)(\cos\delta)^{1/2}\sin(\delta/2) = (\omega/V_s')\tan(\delta/2) \tag{1.6.11}$$

Eq. (1.6.9) indicates that $u$ consists of two parts; $e^{i\omega(t - z/V_s')}$ representing harmonic vibration and $e^{-\beta z}$ representing wave attenuation due to increasing $z$ because $\beta$ is a positive value from Eq. (1.6.11). The constant $\beta$ is named here as the wave attenuation coefficient by internal damping. The $V_s'$ value in Eq. (1.6.10) is the S-wave velocity in Kelvin model. For smaller $\delta$, $\cos\delta \approx \cos(\delta/2) \approx 1$ and $V_s' \approx V_s$ in Eqs. (1.6.9) and

(1.6.10). In Eq. (1.6.11), $\beta$ may be approximated for smaller $\delta$ by using $\tan(\delta/2) \approx (\tan\delta)/2$ and also Eq. (1.5.33) as:

$$\beta \approx (\omega \tan\delta)/(2V_s) = \omega D/V_s = \omega^2 \xi/(2\rho V_s^3) \tag{1.6.12}$$

Thus, in the Kelvin-type viscoelastic medium, the wave attenuation coefficient $\beta$ is in proportion to $\omega^2$, the square of angular frequency.

#### *1.6.1.2 Attenuation for Maxwell model*

If, the stress-strain relationship of Maxwell model, Eq. (1.5.17) or the next equation

$$\tau = \left[\frac{\partial}{\partial t} \Big/ \left(\frac{1}{\xi} + \frac{1}{G}\frac{\partial}{\partial t}\right)\right] \frac{\partial u}{\partial z} \tag{1.6.13}$$

is substituted into Eq. (1.2.2), the wave equation is obtained.

$$\rho \frac{\partial^2 u}{\partial t^2} = \left[\frac{\partial}{\partial t} \Big/ \left(\frac{1}{\xi} + \frac{1}{G}\frac{\partial}{\partial t}\right)\right] \frac{\partial^2 u}{\partial z^2} \tag{1.6.14}$$

The solution of this equation has the same form as in Eq. (1.6.7) for a unilaterally propagating wave toward the positive direction of $z$-axis, wherein $k^*$ and $V_s^*$ are defined as:

$$k^* = \left(\frac{\rho\omega^2}{G^*}\right)^{1/2} = \left(\frac{\rho\omega^2}{i\omega G\xi/(G + i\omega\xi)}\right)^{1/2} \tag{1.6.15}$$

$$V_s^* = \sqrt{G^*/\rho} = V_s(\cos\delta)^{1/2} e^{i\delta/2} \tag{1.6.16}$$

and $\tan\delta = G/\omega\xi$ as in Eq. (1.5.22). Corresponding to Eqs. (1.6.10) and (1.6.11), the following formulas are obtained for Maxwell model.

$$V_s' = V_s(\cos\delta)^{1/2}/\cos(\delta/2) \tag{1.6.17}$$

$$\beta = \sin(\delta/2)(\cos\delta)^{-1/2}(\omega/V_s) = (\omega/V_s')\tan(\delta/2) \tag{1.6.18}$$

For small $\delta$, $\beta$ may be approximated in the similar manner by using Eq. (1.5.34) as:

$$\beta \approx (\omega \tan\delta)/(2V_s) = \omega D/V_s = G/2\xi V_s \tag{1.6.19}$$

Thus, in the Maxwell-type medium, the wave attenuation coefficient $\beta$ is independent of angular frequency $\omega$.

#### *1.6.1.3 Attenuation for Nonviscous Kelvin model*

The complex wave number and the complex S-wave velocity for the Nonviscous Kelvin model are written respectively using $\tan\delta = G/G'$ in Eq. (1.5.26), as:

$$k^* = \left(\frac{\rho\omega^2}{G^*}\right)^{1/2} = \left(\frac{\rho\omega^2}{G + iG'}\right)^{1/2} \tag{1.6.20}$$

$$V_s^* = \sqrt{(G + iG')/\rho} = V_s(\cos\delta)^{-1/2}e^{i\delta/2} \tag{1.6.21}$$

$V_s'$ and $\beta$ in Eq. (1.6.9) are expressed as:

$$V_s' = \frac{V_s}{(\cos\delta)^{1/2}(\cos(\delta/2))} \tag{1.6.22}$$

$$\beta = (\omega/V_s)(\cos\delta)^{1/2}\sin(\delta/2) = (\omega/V_s')\tan(\delta/2) \tag{1.6.23}$$

For small $\delta$, $\beta$ may be approximated by using Eq. (1.5.35) as:

$$\beta \approx (\omega\tan\delta)/(2V_s) = \omega D/V_s = \omega G'/(2\rho V_s^3) \tag{1.6.24}$$

Thus, in the Nonviscous Kelvin medium, the wave attenuation coefficient $\beta$ is in proportion to angular frequency $\omega$.

To summarize the above, it has been shown that plane wave attenuates with distance $z$ in proportion to $e^{-\beta z}$ as:

$$u = Be^{-\beta z}e^{i\omega(t-z/V_s)} \tag{1.6.25}$$

The coefficient of wave attenuation by internal damping $\beta$ is frequency-dependent or independent; $\beta \propto \omega^2$ in Kelvin model, $\beta \propto \omega$ in Non-viscous Kelvin model, and $\beta \propto \omega^0$ in Maxwell model.

### 1.6.2 Energy dissipation in wave propagation

Let us consider the wave energy in the SH-wave propagating upward in the vertical $z$-direction as illustrated in Fig. 1.6.1 (a). If the wave displacement is expressed using a sine function in place of Eq. (1.6.25) as:

$$u = Be^{-\beta z}\sin\omega(t - z/V_s) \tag{1.6.26}$$

the particle velocity $\dot{u} = du/dt$ becomes

$$\dot{u} = \omega Be^{-\beta z}\cos\omega(t - z/V_s) \tag{1.6.27}$$

According to Eq. (1.2.25), the energy of SH-wave in one wave length $\lambda = V_s/f = 2\pi V_s/\omega$ passing through a unit horizontal area at $z$ during the time from $t = 0$ to one period $T = 1/f = 2\pi/\omega$ is calculated as:

$$E = \rho V_s \int_0^T (\dot{u})^2 dt = \rho V_s\omega^2 B^2 e^{-2\beta z}\int_0^{2\pi/\omega}\cos^2\omega(t - z/V_s)dt = \pi\rho V_s\omega B^2 e^{-2\beta z} \tag{1.6.28}$$

From Eq. (1.2.19), the velocity amplitude $\dot{u}_a$ and the strain amplitude $\gamma_a$ of the harmonic wave at $z$ are correlated as:

$$\dot{u}_a = [\omega Be^{-\beta z}\cos\omega(t - z/V_s)]_{\max} = \omega Be^{-\beta z} = -V_s\gamma_a \tag{1.6.29}$$

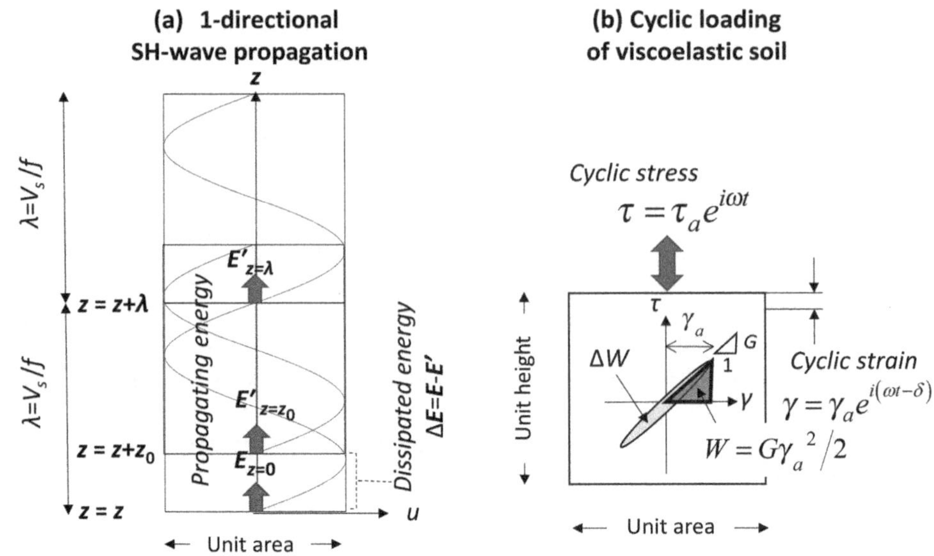


*Figure 1.6.1* Comparison of wave energies $E$ and $E'$ in one-directional wave propagation (a), and Associated stress-strain curve in cyclic loading test (b), in viscoelastic medium (Kokusho 2016).

The energy $E$ in Eq. (1.6.28) and the energy density per unit volume $E/\lambda$ can be expressed using $W = G\gamma_a^2/2$ defined by the triangular area illustrated in Fig. 1.6.1(b) and also using Eq. (1.6.29) as follows.

$$E = \pi\rho V_s\omega B^2 e^{-2\beta z} = [\rho(\dot{u}_a)^2/2]\lambda = W\lambda \tag{1.6.30}$$

$$E/\lambda = \rho(\dot{u}_a)^2/2 = G\gamma_a^2/2 = W \tag{1.6.31}$$

Because $E$ is the energy per unit area, it is correlated with $W$ the energy per unit volume or the energy density using the wave length $\lambda$ as in Eqs. (1.6.30) and (1.6.31). The energy density for the harmonic wave is expressed as $\rho(\dot{u}_a)^2/2$ or $W = G\gamma_a^2/2$, wherein the wave energy is shared evenly between the kinetic and strain energies as explained in Sec. 1.2.3.

The energy, $E'$, at $z = z + z_0$ shown in Fig. 1.6.1(a) can be calculated using Eq. (1.6.28) and written in a similar manner as:

$$E' = \pi\rho V_s\omega B^2 e^{-2\beta(z+z_0)} = \pi\rho V_s\omega^2 B^2 e^{-2\beta z} e^{-2\beta z_0}/\omega = [\rho(\dot{u}_a)^2/2]e^{-2\beta z_0}\lambda = We^{-2\beta z_0}\lambda \tag{1.6.32}$$

Then, the difference of wave energy in one wave length between $E$ and $E'$ is:

$$\Delta E = E - E' = (1 - e^{-2\beta z_0})[\rho(\dot{u}_a)^2/2]\lambda = (1 - e^{-2\beta z_0})W\lambda \tag{1.6.33}$$

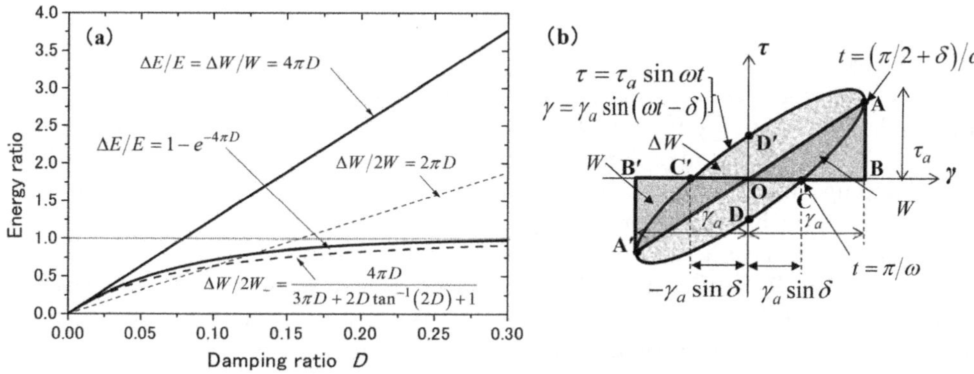

*Figure 1.6.2* ΔE/E–D curve in wave propagation compared with ΔW/2W–D curve by cyclic loading (a) and Schematic stress-strain hysteresis loop of ideal viscoelastic material (b) (Kokusho 2016).

Hence, the dissipated energy per unit volume is written using $\beta = \omega D/V_s$ in Eq. (1.6.24) as:

$$\Delta E/\lambda = (1 - e^{-2\beta z_0})W = (1 - e^{-2(\omega D/V_s)z_0})W \tag{1.6.34}$$

The ratio of the dissipated energy to the original wave energy is expressed using Eqs. (1.6.31) and (1.6.34) as:

$$\Delta E/E = 1 - e^{-2\beta z_0} = 1 - e^{-2(\omega D/V_s)z_0} = 1 - e^{-4\pi D(z_0/\lambda)} \tag{1.6.35}$$

Hence, the dissipated energy ratio in one wave-length is written by substituting $z_0 = \lambda$ as:

$$\Delta E/E = 1 - e^{-4\pi D} \tag{1.6.36}$$

If $\beta$ or $D$ is small enough, $e^{-2\beta\Delta z} \approx 1 - 2\beta\Delta z$ using the Taylor series, and also $\beta = \omega D/V_s = 2\pi D/\lambda$, Eq. (1.6.35) becomes:

$$\Delta E/E = 4\pi D \times z_0/\lambda \tag{1.6.37}$$

Hence for one wave-length propagation:

$$\Delta E/E = 4\pi D \tag{1.6.38}$$

The dissipated energy ratios$\Delta E/E$ formulated in Eqs. (1.6.36) and (1.6.38) for one wave length are plotted versus the damping ratio $D$ with thick solid lines in Fig. 1.6.2(a). It is obviously seen that the two equations coincide at $D = 0$ and tend to diverge with increasing $D$, because $\Delta E/E$ in Eq. (1.6.36) approaches to unity, an upper limit for increasing $D$-values (Kokusho 2016).

### 1.6.3 Energy dissipation in wave propagation compared with cyclic loading

Eq. (1.6.38) has the same form as the dissipated energy ratio $\Delta W/W = 4\pi D$ in Eq. (1.5.7) during cyclic loading in the viscoelastic material. This indicates that the ratio of dissipated energy for one wave length as illustrated in Fig. 1.6.1(a) is determined by the same function of damping ratio $D$ in cyclic loading shown in Fig. 1.6.1(b), if the damping ratio $D$ is small. Note that $W$ is the maximum elastic strain energy in unit volume per a half cycle of loading, while $\Delta W$ is the dissipated energy density in one cycle. The reason $\Delta E/E$ is expressed by the same function as $\Delta W/W$ is that the strain energy $W$ in the first half cycle can be mostly recovered to be recycled in the second half cycle because the dissipated energy $\Delta W$ is sufficiently small. This is what happens in the wave propagation, too, wherein the wave energy $E = W\lambda$ passing through a unit area in one wave length $\lambda$ is dissipated by $\Delta E = \Delta W\lambda$.

As the dissipated energy increases with increasing damping ratio, it has to be compensated by the wave energy $E$ in wave propagations or by the strain energy density $W$ in cyclic loading tests. In the stress-strain curve of the viscoelastic material illustrated in Fig. 1.6.2(b), the strain energy provided in one-cyclic loading is *Area*(ABCDA'B'C'D'A), while the energy $\Delta W = Area(\text{ACDA'C'D'A})$ is dissipated in the specimen during the same cycle. Out of the one-cycle strain energy, the energy corresponding to *Area*(ABC) in the first 1/2 cycle can be recovered and recycled in the second 1/2 cycle for *Area*(A'B'C'). In Sec. 1.5.1, the dissipated energy $\Delta W$ is given in Eq. (1.5.3) as $\Delta W = \tau_a \gamma_a \pi \sin\delta$. By revisiting the same viscoelastic theory wherein shear stress $\tau = \tau_a \sin\omega t$ is loaded to induce strain $\gamma = \gamma_a \sin(\omega t - \delta)$ with a phase-delay angle $\delta$, the *Area*(ABC) is calculated by referring to Fig. 1.6.2(b) as:

$$Area(ABC) = \omega\tau_a\gamma_a \int_{(\pi/2+\delta)/\omega}^{\pi/\omega} \sin\omega t \cos(\omega t - \delta)dt = \Delta W[1 - 2(\pi/2 - \delta)D]/4\pi D \tag{1.6.39}$$

The energy denoted here as $2W_-$ supplied in one cycle loading considering the energy recycling effect is thus obtained from $\Delta W$ and *Area*(ABC) or $Area(ABCDA'B'C'D')$ alternatively as:

$$\begin{aligned} 2W_- &= \Delta W + Area(ABC) = [\Delta W + Area(ABCDA'B'C'D')]/2 \\ &= (\Delta W/4\pi D) \times \{3\pi D + 2D\tan^{-1}(2D) + 1\} \end{aligned} \tag{1.6.40}$$

Then, the ratio of the dissipated energy $\Delta W$ to the supplied energy $2W_-$ is written as:

$$\Delta W/2W_- = 4\pi D/\{3\pi D + 2D\tan^{-1}(2D) + 1\} \tag{1.6.41}$$

In Fig. 1.6.2(a), the energy ratio $\Delta W/2W_-$ in Eq. (1.6.41) versus damping ratio $D$ is superposed with the dashed curve and compared with $\Delta E/E = 1 - e^{-4\pi D}$ in Eq. (1.6.36). The two curves are very similar to each other, both have almost the same initial tangent and tend to approach to the asymptote $\Delta W/2W_- = \Delta E/E = 1.0$ with increasing $D$. This indicates that the energy dissipation mechanism during wave propagation is very similar to and almost reproducible in cyclic loading. However, there is a small gap of maximum 10%, which cannot be explained by less than a few percent error caused by the approximation in Eq. (1.6.24); $\beta \approx \omega D/V_s$. Instead, it may be attributed to the difference in loading; namely, simultaneous cyclic loading on the whole soil specimen versus phase-delayed loading on the transmission medium during wave propagation with wave attenuation.

In cyclic loading, the maximum elastic strain energy $W = Area(\text{OAB})$ is normally employed to compare with the dissipated energy $\Delta W$ as in Eq. (1.5.47). If the wave energy ratio $\Delta E/E = 1 - e^{-4\pi D}$ is compared with energy ratios using $\Delta W$ and $W$ in Fig. 1.6.2(a), $\Delta E/E$ is more closely approximated by $\Delta W/2W = 2\pi D$ than $\Delta W/W = 4\pi D$ for $D$-value of 5% to 15% as indicated by the dotted straight line in the diagram. Nevertheless, it should be noted that the wave attenuation mechanism for larger $D$-value or larger internal damping during strong earthquakes can best be represented by the cyclic loading mechanism expressed by Eq. (1.6.41) (Kokusho 2016).

## 1.7 WAVE ATTENUATION INCLUDING GEOMETRIC DAMPING

Waves propagating in three-dimensional media may be categorized as (a) plane wave, (b) cylindrical wave, and (c) spherical wave as illustrated in Fig. 1.7.1. In the plane wave (a), the wave front translates with the same plane area with the same energy density as in Fig. 1.7.1(a), and thus the wave does not attenuate by geometric damping but by internal damping only. In the cylindrical wave from a point source (b) such as surface waves propagating axis-symmetrically, the wave front expands as a cylinder. The gross energy on the wave front should be unchanged, if no internal damping is considered. Let the wave displacement at $r = r_0$ be $u = U_0 e^{i\omega t}$, and that at $r$ be $u = Ue^{i\omega t}$, and also remembering that the wave energy is proportional to the square of wave amplitude as in Eq. (1.2.25), then:

$$2\pi r_0 U_0^2 = 2\pi r U^2 \tag{1.7.1}$$

$$U/U_0 = (r/r_0)^{-1/2} \tag{1.7.2}$$

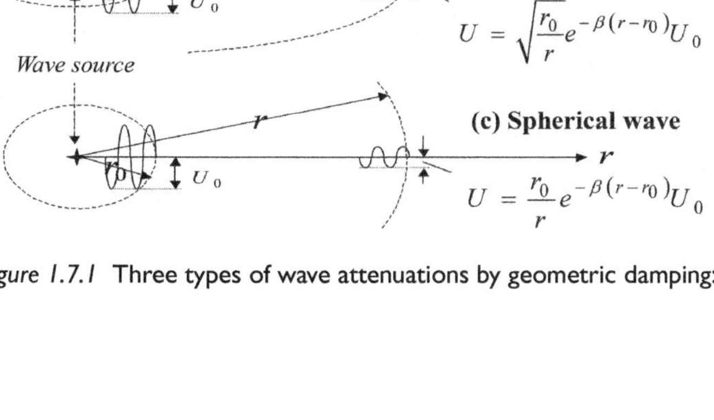


*Figure 1.7.1* Three types of wave attenuations by geometric damping:

Integrating this with the effect of internal damping yields the following equation considering the two damping mechanisms.

$$u = B(r/r_0)^{-1/2} e^{-\beta r} e^{i\omega\{t-(r/V_s)\}} \tag{1.7.3}$$

In the spherical wave (c) such as body waves from a point source propagating spherically, the wave front expands as a sphere. Let the wave displacement at $r = r_0$ be $u = U_0 e^{i\omega t}$, and that at $r$ be $u = Ue^{i\omega t}$, then:

$$4\pi r_0^2 U_0^2 = 4\pi r^2 U^2 \tag{1.7.4}$$

$$U/U_0 = (r/r_0)^{-1} \tag{1.7.5}$$

Hence, the equation considering the two damping mechanisms is written as

$$u = B(r/r_0)^{-1} e^{-\beta r} e^{i\omega\{t-(r/V_s)\}} \tag{1.7.6}$$

## 1.8 SUMMARY

1. Seismic waves consist of body waves and surface waves. The body waves consist of S-wave without volumetric strain and P-wave without rotation.
2. The surface waves consist of Rayleigh wave and Love wave, and propagate horizontally along ground surface. Surface waves propagating in layered media are characterized by wave dispersion, wherein waves with longer wave length tend to travel faster.
3. The amplitude $u$ of body waves attenuates with the travel distance $R$ as $u \propto R^{-1}$, while that of the surface waves does as $u \propto R^{-1/2}$, indicating that the latter tends to be more dominant in longer distance than the former.
4. Body waves coming up from deeper ground tend to propagate almost vertically in shallow soil layers due to refractions at multiple layer boundaries (the Snell's law). The vertically propagating SH-wave are normally chosen as a major seismic wave to be considered in engineering design because it carries major seismic energy.
5. The magnitude of induced shear strain is essential to soil behavior during earthquakes. Shear strain $\gamma$ induced in soils of S-wave velocity $V_s$ by one-directionally propagating SH-wave with the particle velocity $\dot{u}$ is obtained as $\gamma = -\dot{u}/V_s$, indicating that soil strain tends to be larger in softer soils even for the same particle velocity.
6. Wave energy $E$ for the one-directionally propagating SH-wave with the particle velocity $\dot{u}$ passing through a unit horizontal area with the impedance $\rho V_s$ in a time interval $t_1 \sim t_2$ is expressed as $E = E_k + E_e = \rho V_s \int_{t_1}^{t_2} (\dot{u})^2 dt$ and evenly shared by kinetic energy $E_k$ and strain energy $E_e$ at exactly the same wave section.
7. Among three viscoelastic models to idealize internal damping of soils for wave attenuation, the dependency of damping ratio $D$ on the angular frequency $\omega$ is different. The $D$-value can be expressed as; $D \propto \omega$ for Kelvin model, $D \propto 1/\omega$ for Maxwell model, and $D \propto \omega^0$ for Nonviscous Kelvin model.

8 When the plane wave amplitude $u$ attenuates with the distance $z$ as $u \propto e^{-\beta z}$, the coefficient of wave attenuation by internal damping $\beta$ is accordingly written as: $\beta \propto \omega^2$ for Kelvin model, $\beta \propto \omega^0$ for Maxwell model, and $\beta \propto \omega$ for Nonviscous Kelvin model.

9 Dissipated energy ratio $\Delta E/E$ in wave propagation with one wave-length travel distance can be formulated as $\Delta E/E = 1 - e^{-4\pi D}$, wherein $\Delta E =$ dissipated wave energy and $E =$ gross wave energy. For a very small $D$-value, it is approximated as $\Delta E/E = 4\pi D$.

10 $\Delta E/E = 4\pi D$ has the same form as $\Delta W/W = 4\pi D$ in a stress-strain hysteresis loop in cyclic loading where $\Delta W =$ dissipated energy in one-cycle loading and $W =$ maximum elastic strain energy in the half cycle. This is because $W$ in the first half cycle can be mostly recovered to be recycled in the second half cycle if the dissipated energy $\Delta W$ is sufficiently small, as actually occurring in the small strain wave propagations. The trend of $\Delta E/E = 1 - e^{-4\pi D}$ converging to unity with increasing $D$ can mostly be reproduced by $\Delta W/2W_{-}$ in an ideal viscoelastic material wherein one-cycle strain energy $2W_{-}$ is evaluated considering the effect of recycling the elastic energy in cyclic loading.

# Chapter 2

# Soil properties during earthquakes

In this chapter, variations on soil properties from small strain to medium and large strain exhibited during earthquakes are first outlined in Sec. 2.1, followed by methodologies how to measure the properties for different strain ranges in situ and in the laboratory in Sec. 2.2. Typical soil properties in small strain level are exemplified in Sec. 2.3, while those in medium to large strain levels approximated by equivalent linearization are shown in Sec. 2.4 together with some comments in view of their applicability in actual problems. Strongly nonlinear soil properties corresponding to soil failures are discussed from the viewpoint of modelling in Sec. 3.1 and applicability in liquefaction problems in Chapter 5.

## 2.1 CHARACTERIZATION OF DYNAMIC SOIL PROPERTIES

### 2.1.1 Small-strain properties

Soil behaves as an elastic body in small strain levels of $10^{-6}$ in problems such as site amplification during small earthquakes, ground vibrations in microtremors and environmental problems due to traffic/machine vibrations. The viscoelastic models studied in Secs. 1.5 and 1.6 can directly apply to these problems. The small strain properties are also important because they provide a very basic condition for large-strain soil properties.

A set of basic soil properties necessary for earthquake engineering problems are; soil density, $\rho$ ($\rho_t$ for unsaturated soil and $\rho_{sat}$ for saturated soil), S-wave velocity $V_s$, P-wave velocity $V_p$, Poisson's ratio $\nu$ and soil damping ratio $D$. Among them, $V_s$, $V_p$ and $\nu$ are readily obtained from in situ wave logging tests. The S-wave velocity $V_s$ is of utmost importance for earthquake geotechnical problems, because $V_s$ governs seismic amplification in horizontal motions, and the shear modulus $G$ calculated from $V_s$ as $G = \rho V_s^2$ determines the shear deformation. The P-wave velocity is sometimes used in evaluating the dynamic ground response in vertical direction, Poisson's ratio and the degree of saturation of in situ soils.

The soil damping ratio cannot be measured in the field as readily as the wave velocities. However, it may be evaluated under some favorable conditions where geometric damping can be postulated during in situ wave measurements. Back-calculations of vertical array earthquake records using the multi-reflection theory of the SH wave

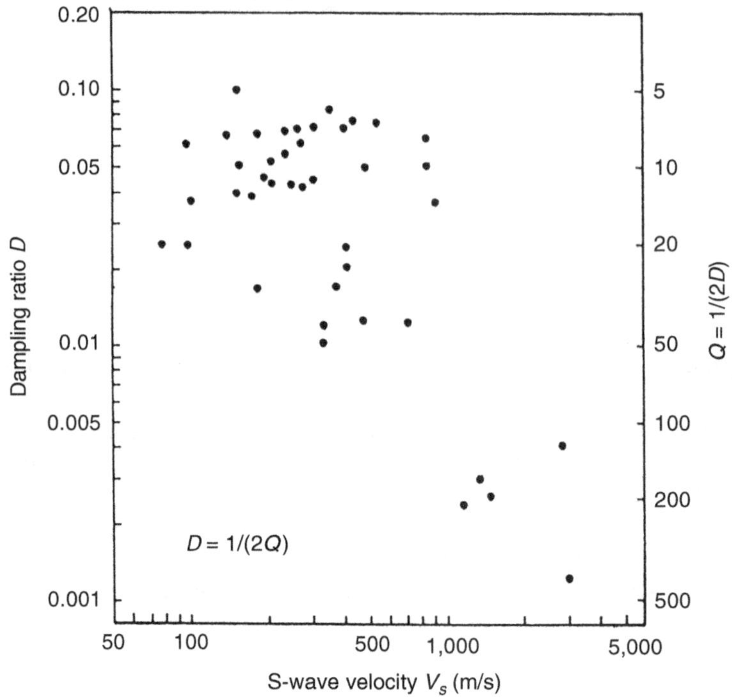


*Figure 2.1.1* Damping ratio $D$ versus S-wave velocity $V_s$ obtained from in situ wave measurements (modified from Kudo 1976).

may also allow to optimize not only the S-wave velocity but also the soil damping ratio.

As a typical example, in situ damping ratios $D$ or $Q = 1/(2D)$ measured in wave logging tests or optimized from earthquake records (all for small strain levels) are plotted in Fig. 2.1.1 versus S-wave velocities $V_s$ in corresponding layers on a full logarithmic diagram (Kudo 1976). Despite tremendous data scatters, a weak correlation may be recognized so that $D$ tends to decrease with increasing $V_s$ for the wide velocity range $V_s = 100$–$3000$ m/s. Note that quite a few plots of small-strain damping ratios larger than 5% are obtained for $V_s < 1000$ m/s. In Fig. 2.1.2, damping ratios $D$ (in %) back-calculated in previous researches from vertical array earthquake records are depicted versus ground depth $z$ on a log-log scale (Kokusho and Mantani 2002). The soil strain $\gamma$ associated with the earthquakes is $10^{-4} \leq \gamma \leq 10^{-3}$ shallower than 100 m while $\gamma \leq 10^{-4}$ deeper than that. There is a clear decreasing trend in the optimized damping ratio $D$ with increasing depth; $D < 1\%$ deeper than several hundred meters, while $D$ tends to increase up to several percent at depths shallower than 100 m. Note that, at a very shallow depth less than a few meter from the ground surface, in situ $D$-value as large as around 10% is optimized.

As for the soil density $\rho$, it is normally judged from past experience according to the soil type and the depth relative to water table. Saturated density $\rho_{sat}$ is taken 1.8–2.0 t/m$^3$ for typical sandy soils and 1.4–1.6 t/m$^3$ for typical cohesive soils. For other

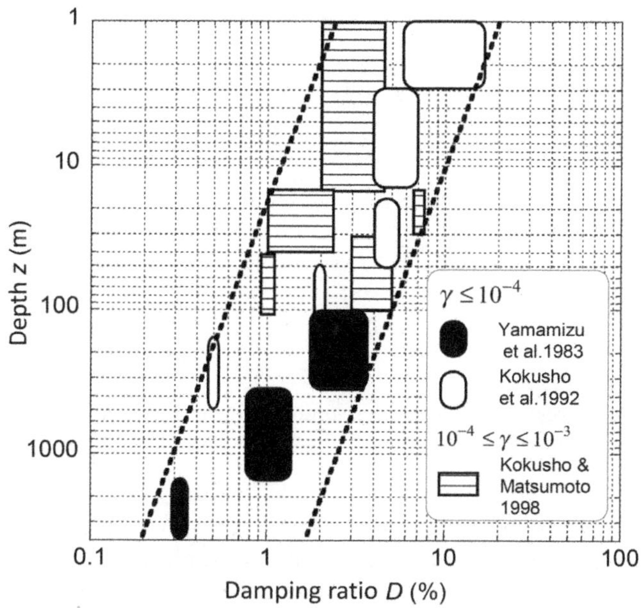


*Figure 2.1.2* Damping ratio D versus ground depth z optimized from vertical array earthquake records at different depths (modified from Kokusho and Mantani 2002).

materials, $\rho_{sat} = 2.0\,\mathrm{t/m^3}$ or larger for gravelly soils and $\rho_{sat} = 2.3\,\mathrm{t/m^3}$ or larger for rocks. The soil density is different due to a degree of saturation $S_r$, and the unsaturated soil density $\rho_t$ is determined assuming $S_r$, though the difference between $\rho_{sat}$ and $\rho_t$ is around $0.1\,\mathrm{t/m^3}$ in most case.

Poisson's ratio $\nu$ is needed in computing 2 or 3-dimensional dynamic response of soil structures. From Eq. (1.2.12), the $\nu$-value for a small strain level corresponding to the elastic wave propagation can be correlated with the wave velocity ratio $V_s/V_p$ as

$$\nu = \frac{1/2 - (V_s/V_p)^2}{1 - (V_s/V_p)^2} \tag{2.1.1}$$

If the soil is below water table and saturated, Poisson's ratio thus obtained corresponds to the undrained condition and approximately $\nu = 0.5$ as an incompressible material.

### 2.1.2 Strain-dependent nonlinearity in soil properties

Soil behaves as an inelastic nonlinear material except for a small strain level. With increasing soil strains during earthquakes, soils change from linear to nonlinear materials. During strong earthquakes, the soil nonlinearity strongly affects the seismic response of soft soil ground and structures on it, and hence the occurrence of structural damage and soil failures such as liquefaction and slope instability. Fig. 2.1.3 depicts

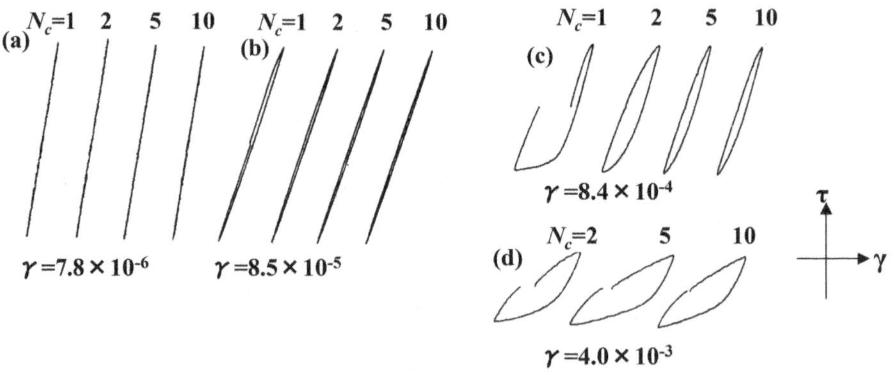


*Figure 2.1.3* Changes in stress-strain curves measured for Toyoura sand in improved cyclic triaxial test for wide strain ranges (Kokusho 1980).

stress-strain ($\tau \sim \gamma$) curves measured by a triaxial test of clean sand wherein 10 cycles of harmonic waves are given with various stress amplitudes inducing a wide range of shear strains $\gamma$ (Kokusho 1980). Obviously, the soil behaves linear-elastically for $\gamma$ smaller than $10^{-5}$ strain in (a), and almost linearly with some hysteresis area for $\gamma$ of $10^{-5}$ order strain in (b). If the strain exceeds $10^{-4}$ in (c) and (d), the stress-strain curves become evidently nonlinear with spindle-shapes, and the secant shear moduli connecting the edges of the hysteresis loops tend to decrease with increasing strains. At the same time, the loops tend to be wider, increasing the areas or energy dissipations in individual cycles. The number of loadings $N_c$ makes a clear difference in the stress-strain curve as the strain level increases. Such changes in soil properties tend to make the difference in ground response during earthquakes of different intensities.

As mentioned in Eq. (1.2.19), shear strain induced in soils due to a vertically propagating SH wave is expressed as $\gamma = -\dot{u}/V_S$. This indicates that the absolute value of seismically-induced strain becomes larger with increasing particle velocity $\dot{u}$ and decreasing S-wave velocity $V_S$. In other words, the soil strain tends to be larger in softer soils and for larger particle velocity. For small strains induced during small tremors, the soil vibrates linearly with linear shear modulus and minimal internal damping. With increasing shaking intensity, the dynamic response in soft soils tends to change to have lower resonant frequency and higher soil damping due to increasing soil strains. If the seismic intensity becomes still stronger, the soil tends to behave more nonlinearly and approach to failure.

Fig. 2.1.4 shows the dynamic soil response of a saturated clean sand to irregular seismic loading in an undrained torsional simple shear test. The seismic stress given to the specimen in (a) is made from an acceleration time history recorded during the 2011 Tohoku earthquake in Japan. The stress–strain response in (b) indicates how the soil behaves in small strains at the start and in very large strains corresponding to liquefaction failures at the end. A nearly linear relationship, at the start of loading, changes to a spindle-shape with a hysteretic area. It further changes to a strongly nonlinear banana-shape curve called "cyclic mobility", reflecting 100% excess pore-pressure buildup followed by a drastic recovery in shear stiffness

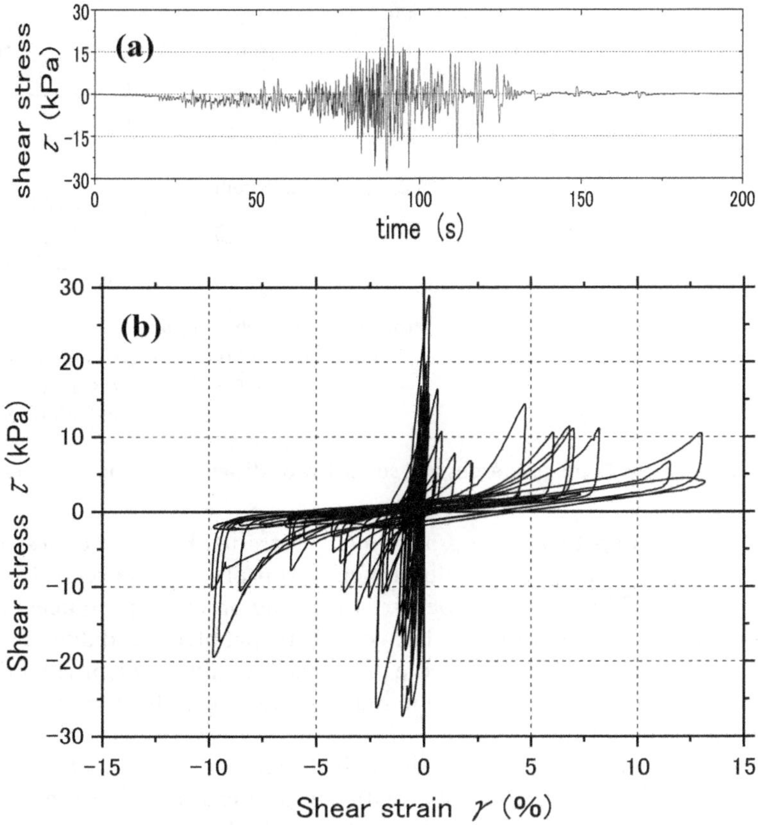


*Figure 2.1.4* Irregular seismic loading (a), and Corresponding stress-strain curves (b), in hollow cylindrical undrained torsional shear test for the 2011 Tohoku earthquake motion in Japan (Kaneko 2015).

due to pore-pressure decrease in shifting from the negative to positive dilatancy in individual cyclic loadings.

Thus, the following two basic mechanisms govern large-strain properties of soil during strong earthquakes. First is the strain-dependent soil property change as typically shown in Fig. 2.1.3, reflecting a change from elasticity to plasticity with increasing strain. Secondly, the pore-pressure variations under undrained cyclic loading have a significant effect on large-strain soil properties as vividly seen in Fig. 2.1.4. Normally, the soil dilatancy effect becomes evident for a strain level $\gamma > 10^{-4}$–$10^{-3}$ and changes the pore pressure in undrained conditions. For loose soils, the negative dilatancy generates positive pore pressure, which tends to decrease effective confining stress and promote soil nonlinearity further in the undrained condition.

In seismic loadings, nonlinear soil properties should also be considered in the context of loading rate and the number of load cycles (Ishihara 1996). It is normally accepted that a stress-strain relationship is not so sensitive to the loading rate associated with earthquake problems (the loading frequency is typically 0.1~10 Hz) for

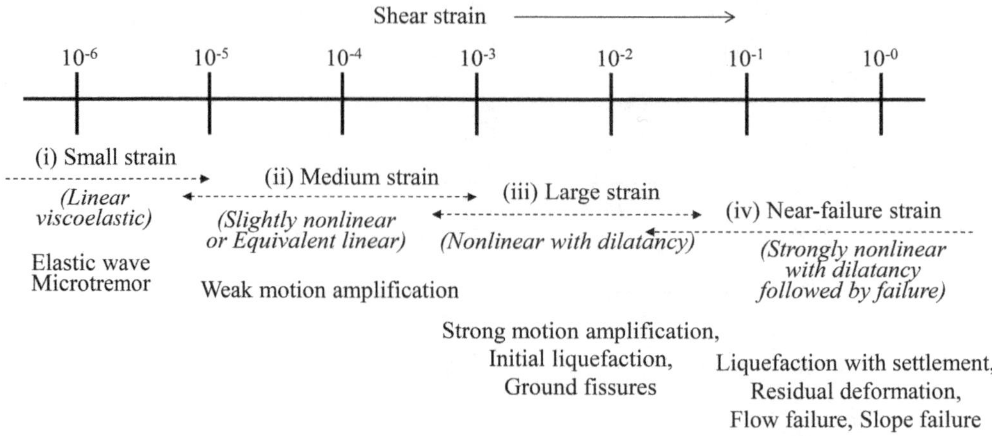


*Figure 2.1.5* Variations in soil properties and soil behavior depending on induced shear strains.

non-cohesive soils in particular for small to large strains. Even for cohesive soils, the loading rate makes an insignificant difference in soil properties except for very large strains near failures. There seems to be a trend among practicing engineers to consider that soil properties are quite different between static problems and dynamic problems for earthquake engineering. However, the properties are almost identical if the induced strain is identical despite the difference in loading rate except for the creep effects near failures in static problems.

In Fig. 2.1.5, the induced shear strain as the fundamental parameter is taken in the horizontal axis for $10^{-6}\sim 10^{-0}$, together with soil properties, where a set of associated soil behavior is embedded. Here, the strain axis is divided into ranges; (i) small strain (linear viscoelastic) for elastic wave propagation and microtremor, (ii) medium strain (slightly nonlinear or equivalent linear) for amplification of weak earthquake motions, (iii) large strain (nonlinear with dilatancy) for strong motion amplifications, initial liquefaction and ground fissures, and (iv) near failure strain (strongly nonlinear with dilatancy followed by failure) for liquefaction with settlement, residual deformation, flow failure and slope failures.

### 2.1.3 Equivalent linearization

Nonlinear stress-strain relationship not so strongly nonlinear may be approximated by equivalent linearization. A hysteretic stress-strain curve for shear strain around $\gamma = 10^{-4}$ or larger as exemplified in Fig. 2.1.3 is schematically shown as a loop in Fig. 2.1.6. The loop is assumed to close in one cycle though it may not be precisely so in actual soil response. The loop also tends to slightly change with the number of cycles, and hence that of 5th to 10th may be chosen to represent equivalent linear properties, considering that the equivalent number of cycles of harmonic motions representing seismic motions is assumed as 10 to 20 in normal engineering practice.

In a given loop, the secant modulus $G$ is defined from a gradient of line AA′ and the hysteretic damping ratio is calculated by Eq. (1.5.7), $D = \Delta W/(4\pi W)$, where $\Delta W$

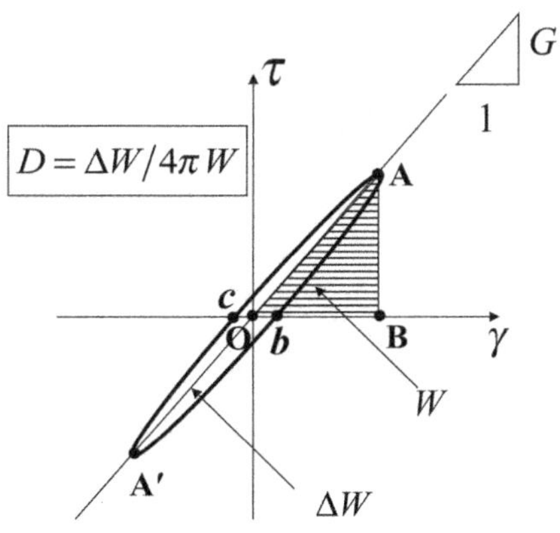


*Figure 2.1.6* Stress-strain hysteresis loop and definition of equivalent linear properties; Secant modulus G and damping ratio D.

stands for the dissipated energy during one cycle loading AbA′c, and $W$ is the maximum elastic strain energy corresponding to the area of triangle OAB, respectively. Equivalent linear properties $G$ and $D$ thus determined are normally plotted versus single amplitude shear strains $\gamma$ in a horizontal log axis as shown in Fig. 2.1.7. The secant modulus in a small strain range for $\gamma = 10^{-6}$ is called initial shear modulus or small-strain shear modulus, denoted as $G_0$, and correlated with the S-wave velocity $V_s$ as $G_0 = \rho V_s^2$. As for the damping ratio $D$, the initial $D$-value for small strains associated with elastic wave propagation takes non-zero small values and tends to increase with increasing strains.

The shear modulus $G$ is often normalized by the initial shear modulus $G_0$ as $G/G_0$ and plotted versus log $\gamma$ as shown in Fig. 2.1.7(b) to show shear modulus degradation with increasing shear strain. Strain values corresponding to $G/G_0 = 0.5$ are called reference strains $\gamma_r$ and used to characterize the degradation rates among different soils. In numerical analyses using the equivalent linear properties, the degraded shear moduli are determined from $G/G_0 \sim \gamma$ curves using initial moduli $G_0 = \rho V_s^2$ where $V_s$ is evaluated by in situ S-wave logging tests. However, it sometimes occurs that $G_0$ measured in laboratory tests disagrees with that evaluated from in situ $V_s$ due to many different reasons including in situ soil heterogeneity, differences in stress conditions from in situ, and mechanical disturbances during soil sampling. Thus, the problem is if it is acceptable to incorporate $G/G_0 \sim \gamma$ curves obtained by laboratory tests to determine in situ $G$-values by using $G_0 = \rho V_s^2$ from wave logging test results at specific sites.

In this respect, Fig. 2.1.8 shows shear modulus degradation versus shear strain plots obtained from seismic response observed at a ground surface of loose sand underlain by a stiff gravel layer (Tokimatsu and Midorikawa 1981). The ground was idealized as a simple two-layer system to determine shear moduli reproducing the observed predominant frequencies and corresponding induced strain. Most of the plots are in between the pairs of degradation curves measured in independent laboratory tests for sands, indicating that satisfactory matching in the modulus degradation may be seen between in situ and laboratory.

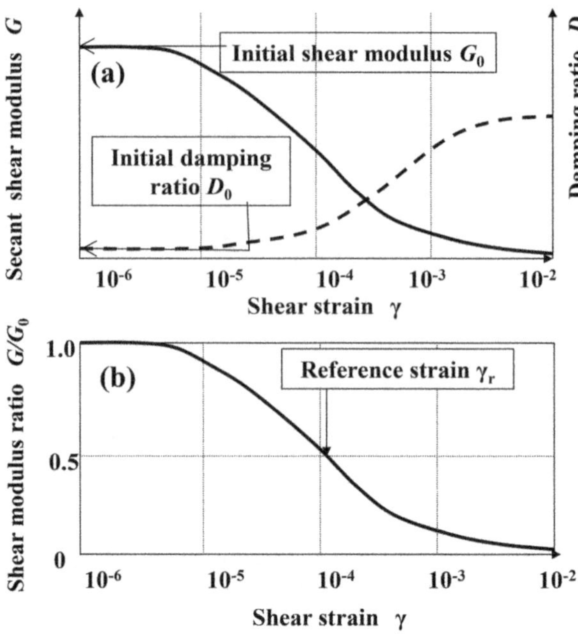


*Figure 2.1.7* Strain-dependent variations of secant shear modulus $G$ and damping ratio $D$ (a), and Shear modulus ratio $G/G_0$ (b), versus shear strain in semi-log diagrams.

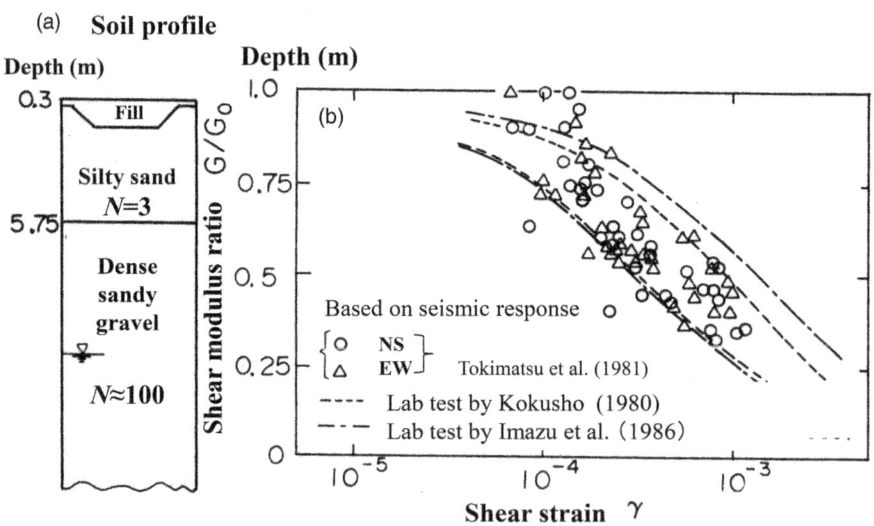


*Figure 2.1.8* Soil profile of 2-layer system (a), and Back-calculated modulus degradation $G/G_0$ versus $\gamma$ plots compared with laboratory tests (b) (Tokimatsu and Midorikawa 1981).

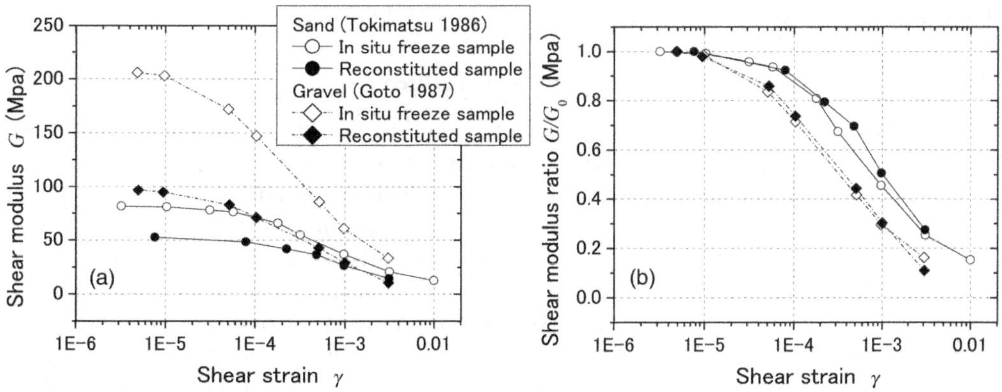


*Figure 2.1.9* Secant shear modulus $G$ (a), and Shear modulus ratio $G/G_0$ (b), versus shear strain $\gamma$ for sand and gravel of intact and reconstituted samples (Kokusho 1987).

Furthermore, Fig. 2.1.9 shows the curves of shear moduli in (a) and modulus degradations in (b) versus shear strains, respectively, for sand and gravel sampled intact by in situ freezing technique to compare with the same soils completely remolded and reconstituted to have the identical densities in the laboratory. Despite the big difference in absolute values of $G$ including the initial modulus $G_0$ between the intact (open symbols) and reconstituted specimens (close symbols) as shown in (a) due to the different soil fabric, the modulus ratios $G/G_0$ in (b) show almost identical degrading trends. Because the intact soils are considered to retain in situ soil properties, this figure indicates that, despite the pronounced reductions in absolute $G$-values in the reconstituted samples, the $G/G_0 \sim \gamma$ curves derived from them may be acceptable to determine in situ $G$-values by using $G_0 = \rho V_s^2$ from in situ wave logging tests. The above test results may justify a normal engineering practice in that soil-specific modulus degradation curves based on laboratory tests are incorporated in equivalent linear numerical analyses combined with initial $G_0$-values determined from in situ S-wave velocity measurements.

### 2.1.4 Strong nonlinearity toward failure

With increasing soil strain up to $10^{-3} \sim 10^{-2}$, soils become more plastic, where the loading histories in previous cycles make differences in the subsequent soil response. The stress-strain curves become far from stationary, making the equivalent linear approximation almost inadequate.

In this strain level, soil dilatancy becomes dominant, causing ground settlement in unsaturated loose soils. In saturated loose soils, pore-pressure builds up in the undrained condition, because seismic loading is normally too fast for the pore-pressure to dissipate. The soil dilatancy, negative in loose soils, reduces the effective stress and shear stiffness of soil in the undrained condition, leading to still larger strains with increasing cycles of seismic loading. This eventually triggers soil liquefaction in non-cohesive soils causing uneven settlements and tilting of structures, and slope failures where not only seismic cyclic stresses but also sustained stresses of overlying structures have substantial roles in generating large strains and ultimate failures.

#### 2.1.4.1 Basic mechanism of seismic soil failure

Fig. 2.1.10 illustrates schematically how soils tend to fail during strong earthquakes. In earthquake engineering, a major seismic impact on soils is normally represented by the SH-wave. The simplest stress condition to be considered for soil failure is that a soil element in level ground under the $K_0$-consolidation shown in (a) is loaded by cyclic shear stresses of the SH-wave and deforms with certain induced strain amplitudes. In this case, the soil strain for failure specified in design may be determined according to the seismic performance of pile foundations, buried structures and lifelines embedded in the level ground. Such soil behaviors in the level ground with stress conditions and induced strain amplitudes are simulated in laboratory element tests, typically in liquefaction tests in the undrained conditions wherein pore-pressure buildup and strain-development are measured. Fig. 2.1.11(a) exemplifies an outcome of an undrained cyclic shear test, where a sand specimen of relative density $D_r \fallingdotseq 50\%$ isotropically consolidated with $\sigma_c' = 98$ kPa is loaded in a torsional simple shear test. As a constant-amplitude cyclic shear stress representing the SH-wave is applied to the specimen, the pore-pressure builds up equal to the initial confining stress 98 kPa, when the cyclic strain amplitude starts to grow rapidly. At the moment when the strain reaches to a prescribed value, the soil is judged to fail. That strain value is usually taken as $\gamma_{DA} = 7.5\%$ in shear strain or $\varepsilon_{DA} = 5\%$ in axial strain both in the double amplitude in the normal liquefaction evaluation practice, because soils tend to develop almost 100% excess pore-pressure at that strain as exemplified in Fig. 2.1.11(a).

Another type of seismic soil failure occurs due to not only cyclic loads but also dead loads from structures resting on soils or from sloping ground as illustrated in Fig. 2.1.10(b). In this case, the sustained initial shear stresses working on soil elements

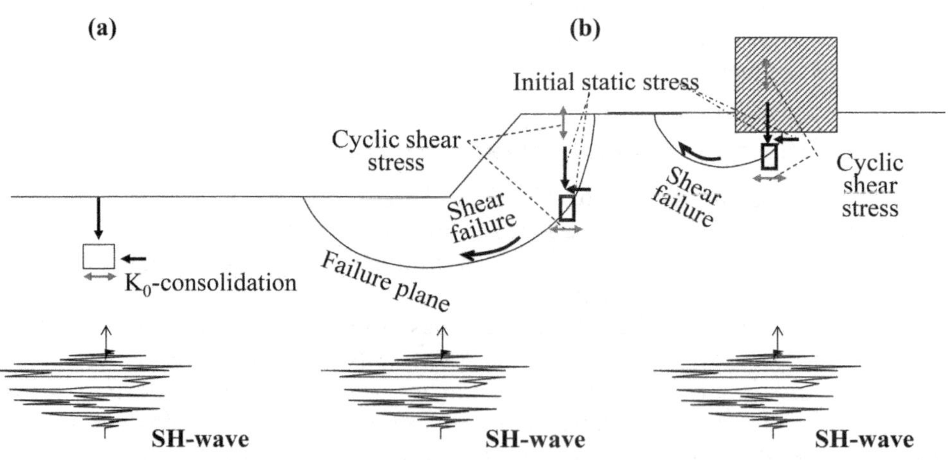


*Figure 2.1.10* Different failure modes of soils during earthquakes: (a) Level free ground in $K_0$-consolidation, (b) Sloping or near-structure ground with working initial shear stresses.

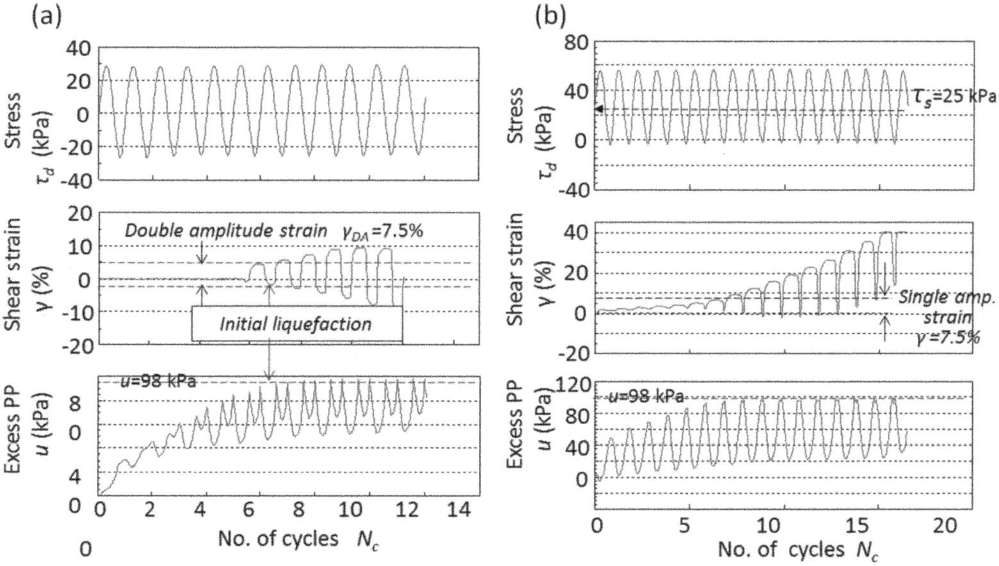


*Figure 2.1.11* Undrained torsional cyclic shear test results of isotropically consolidated clean sand specimen $\sigma_c' = 98$ kPa, $D_r \fallingdotseq 50\%$: (a) Without initial shear stress, (b) With initial shear stress (Ito 2011).

together with the cyclic stresses have great impacts on seismically induced soil strains. Fig. 2.1.11(b) shows an undrained cyclic shear test result similar to (a) but first loaded with the static shear stress of $\tau_s = 25$ kPa in the drained condition and then cyclically loaded with the dynamic stress $\tau_d = 27$ kPa to reproduce a soil element under the initial shear stress. As the excess pore-pressure approaches to 98 kPa with increasing number of cycles, the soil develops not only cyclic strain but also unilateral strain in this case reflecting the initial shear stress. While the cyclic loading tends to induce a gradual strain accumulation into the direction of the static stress, it does not accompany a brittle shear failure along a clear slip plane but a spatially continuous shear deformation.

The soil failure reproduced in the test specimens shown above may be incorporated in design wherein seismically induced soil strains are compared with threshold strains considering the performance of designed structures during earthquakes (such as ultimate safety, reparability, serviceability). Fig. 2.1.12 shows a typical relationship between the cyclic stress amplitude $\tau_d$ versus the number of cycles of sinusoidal loading $N_c$ to attain a certain threshold strain, $\gamma_{DA} = 7.5\%$ for example, in the two cases without and with initial shear stresses. In both cases, $\tau_d$ tends to monotonically decrease with increasing $N_c$, and these relationships are approximated by the lines shown in the diagram which may be interpreted by the fatigue theory as addressed later in Sec. 3.1.6. An equivalent number of loading cycles $N_c$ corresponding to design earthquake loading is chosen ($N_c = 20$ for example) and the corresponding cyclic stress amplitude $\tau_d$ is determined as a dynamic strength for the threshold strain.

Unlike the gradual or ductile failure under the effect of initial effective stress shown in Fig. 2.1.11(b), another type of failure is exemplified in Fig. 2.1.13. Here, a loose

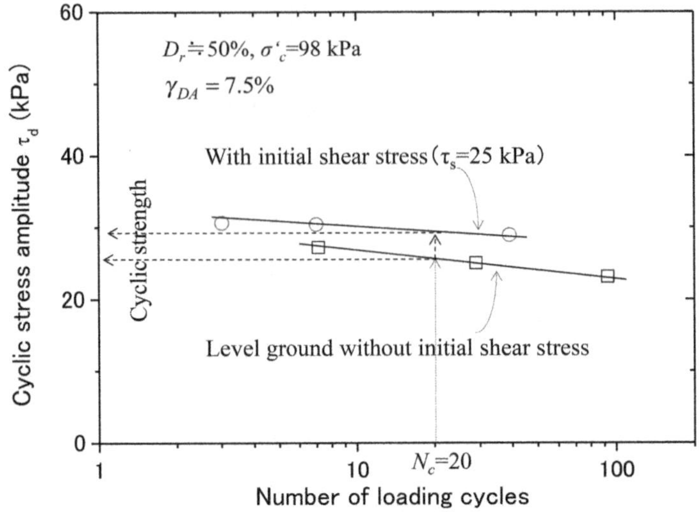


*Figure 2.1.12* Typical relationships between cyclic stress amplitudes versus number of loading cycles of sinusoidal wave and how to define cyclic strength due to seismic loading.

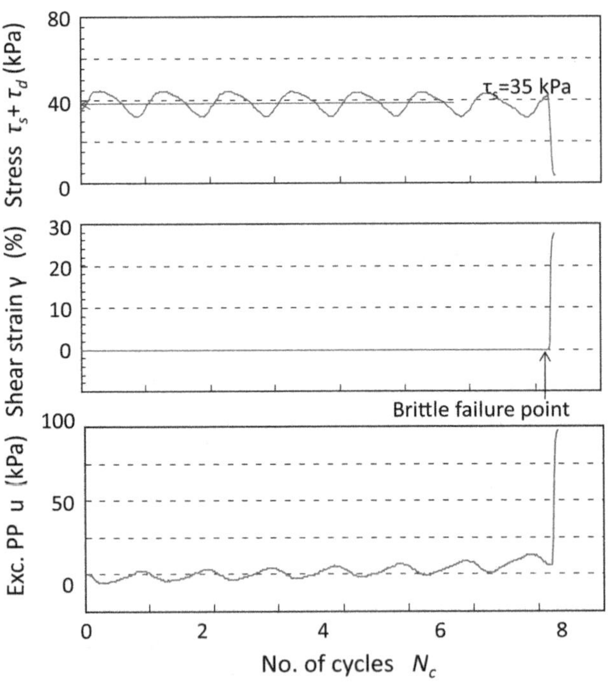


*Figure 2.1.13* Undrained torsional cyclic shear test results of isotropically consolidated clean sands $\sigma'_c = 98$ kPa, $D_r \fallingdotseq 30\%$, $F_c = 30\%$ with initial shear stress $\tau_d = 35$ kPa (Ito 2011).

sand of $D_r \fallingdotseq 30\%$ with non-cohesive fines (fines content $F_c = 30\%$) isotropically consolidated with $\sigma'_c = 98$ kPa is first loaded monotonically with the initial shear stress $\tau_s = 35$ kPa in the drained condition, and then cyclically sheared in the undrained condition with a dynamic stress amplitude $\tau_d = 17$ kPa. At $N_c = 7$ when the excess pore-pressure is rising only slightly, a drastic flow failure occurs with unilateral shear strain exceeding $\gamma = 25\%$ (a limitation by the test apparatus) in the direction of the initial shear stress. Such a sudden or brittle failure may occur in soil types of significant volume contractility as will be discussed in Sec. 5.8. In such a failure mode, the ultimate safety of structures should be focused in design without evaluating residual deformations.

Thus, soil failures during earthquakes may be classified into the three types below.

i) Cyclic soil strain grows with cyclic loading in a level ground and reaches a double amplitude threshold strain corresponding to structural performance.
ii) Unilateral strain developing gradually with cyclic loading due to the initial shear stress is superposed on cyclic strain to reach a threshold value corresponding to the structural performance.
iii) Unilateral strain increase occurs suddenly at some stage of cyclic loading in the direction of initial shear, leading to a brittle soil failure with very large strain.

Among the three, the last failure type has to be paid special attention in design, because it may result in catastrophic consequences of structures resting on the failed soil. As will be described later in Sec. 5.8, the volume contractility of soils has a lot to do in discriminating between ii) and iii) above.

#### 2.1.4.2 Effects of loading rate and loading cycle

A seismic load is quite different from a static load in terms of loading rate and loading cycles (Ishihara 1996). It is essential to know how these effects will affect soil strength in a large strain range near failures. Fig. 2.1.14(a) shows triaxial test results of a clay with the plasticity index $I_p = 35$, isotropically consolidated, and cyclically loaded in the undrained condition with frequencies $f = 0.01$, 0.1 and 1 Hz. The cyclic stress amplitude $\tau_d$ to attain double amplitude axial strain $\varepsilon_{DA} = 10\%$ in cycles $N_c = 10$ normalized by the corresponding stress amplitude for $f = 0.1$ Hz is taken in the vertical axis against the frequency $f$ in the horizontal axis. It clearly indicates that the change in frequency (loading rate) of 100 times increase the cyclic shear strength for $\varepsilon_{DA} = 10\%$ by only 10–20% (Kanatani et al. 1989).

In Fig. 2.1.14(b), triaxial test results on the same clay explained above are exemplified to show the effect of loading cycles under the effect of initial stress. The specimen with various OCR (overconsolidation ratio) are anisotropically consolidated with vertical and lateral stresses $\sigma_1$ and $\sigma_3$, respectively, introducing initial shear stress $\tau_s = (\sigma_1 - \sigma_3)/2$, and then cyclically loaded with the shear stress amplitude $\tau_d$ in the undrained condition. In the vertical axis of the chart, the failure stress defined by $\tau_d + \tau_s$ corresponding to the axial strain $\varepsilon_{DA} = 10\%$ in the number of cycles $N_c = 10$ is plotted versus the initial stress $\tau_s$ in the horizontal axis, both of which are normalized by the static failure stress of the same clay $\tau_{sf}$. Considering that the diagonal line in the diagram represents the condition $\tau_d = 0$, the cyclic stress $\tau_d$ for the soil to fail tends to decrease with increasing $\tau_s$. Though the effect of the initial shear stress

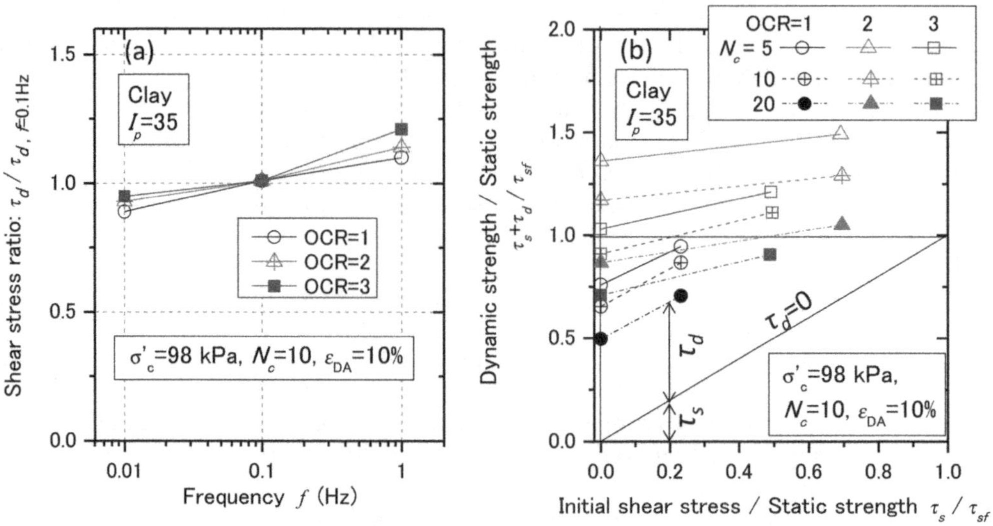


*Figure 2.1.14* Triaxial test results of clay showing effects of (a) loading rate and (b) number of cycles, on cyclic shear strength $\tau_d$ ($\varepsilon_{DA}$ =10%, $N_c$ = 10) (modified from Kanatani et al. 1989).

is another important issue in liquefaction to be discussed in Sec. 5.8, the number of cycles obviously dominates the failure stress so that $\tau_d + \tau_s$ tends to decrease systematically with increasing $N_c$ for all the test conditions. It may readily be inferred that the pore-pressure buildup has a major role in bringing about this effect even in clayey soils. The values in the vertical axis lower than unity, $\tau_d + \tau_s < \tau_{sf}$, indicates that cyclic loading tends to reduce the strength lower than the static value for soil specimens with lower OCR and smaller initial shear stress even for a cohesive soil. As for sandy soils, the effect of loading cycle is obviously seen in the ductile failure of clean sands in Fig. 2.1.12.

Thus, the effects of loading rate and loading cycle in seismic loading on soil strength may be summarized as follows. The increase in loading rate will increase the cyclic strength corresponding to prescribed large strain only to a minor extent even for clay soils. In contrast, the effect of loading cycle is dominant in reducing the strength not only in sands but also in cohesive soils, wherein the initial shear stress has a lot to do with how the strength reduction occurs.

## 2.2 HOW TO MEASURE SOIL PROPERTIES

### 2.2.1 In situ wave measurement for small strain

In situ wave velocities are the most popular and significant variables to evaluate small-strain soil properties. There are several different options to measure them as listed in Table 2.2.1. Some of them in the table need boreholes while the others do not. Those with the bold letters seem to be more often used than the others.

*Table 2.2.1* Various in situ wave measurements with/without boreholes.

| *With borehole* | *Without borehole* |
|---|---|
| **Down-hole method** | **Refraction method** |
| Up-hole method | Reflection method |
| **Cross-hole method** | **Surface-wave method** |
| **Suspension-method** | |

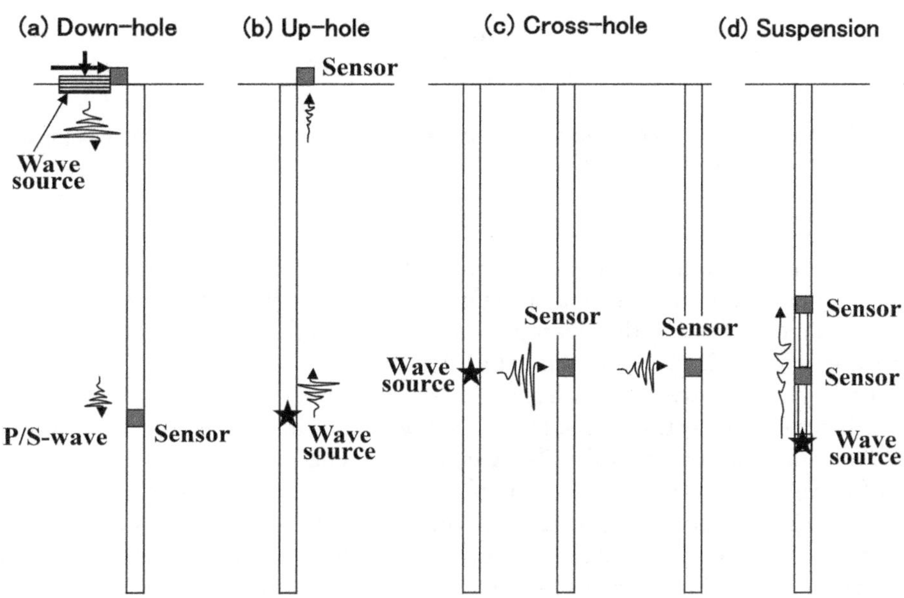


*Figure 2.2.1* In situ wave velocity measurements with boreholes: (a) down-hole, (b) up-hole, (c) cross-hole and (d) suspension.

### 2.2.1.1 *Measurements using boreholes*

As illustrated in Fig. 2.2.1, the methods of in situ wave measurement using boreholes include; (a) down-hole, (b) up-hole, (c) cross-hole and (d) suspension. The brief descriptions are as follows.

*(a) Down-hole method*

This is the most conventional and popular method using a single bore-hole as (a). An artificial wave with frequency of several tens to a hundred Hz is generated by hitting a half-buried wooden plank in most case (both P and S-wave can be generated depending on how the plank is hit). The elastic waves travelling from the wave source at ground surface to depth along the hole are detected by a down-hole wave sensor suspended at variable depths to have a travel-time curve as schematically illustrated in

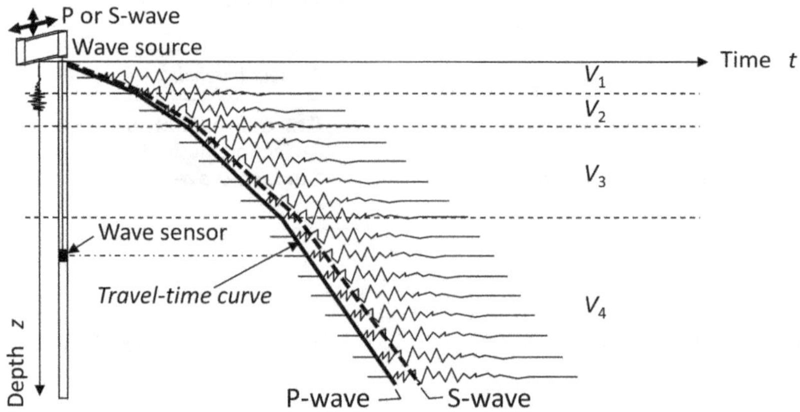


*Figure 2.2.2* Schematic travel-time curve from wave source at ground surface to depth along a borehole by a down-hole wave sensor suspended at variable depths.

Fig. 2.2.2. Sometimes, the wave sensor is integrated in the electronic cone penetrometer in CPT tests to detect waves propagating in a similar path from the wave source at the ground surface, obtaining the stratigraphic and mechanical information at the same time (Robertson et al. 1986, Nishida et al. 1999).

Care is needed however if underground shafts or piles are beside the bore hole, or rigid casing pipes protecting the hole are used, wherein the elastic waves may transmit faster than surrounding natural soils. A travel-time curve is drawn by connecting the plots of wave arrival time versus distance from the wave source. The gradient of this curve (the depth increment $\Delta z$ divided by the time increment $\Delta t$) gives the velocity of the P or S-wave. The wave travel curve is normally approximated by multilinear lines with various gradients or wave velocities, and the soil is idealized by a set of layers with different velocities separated by layer boundaries.

The small-strain shear modulus $G_0$ can be determined in each layer as:

$$G_0 = \rho V_s^2 \tag{2.2.1}$$

Combining $V_p$ and $V_s$, Poisson's ratio for a low strain level can be obtained as:

$$\nu = \left[1 - 2\left(\frac{V_s}{V_p}\right)^2\right] \Big/ \left\{2\left[1 - \left(\frac{V_s}{V_p}\right)^2\right]\right\} \tag{2.2.2}$$

Note that the $\nu$-value is highly dependent on the degree of saturation except for stiff rocks; typically $\nu \approx 0.5$ in saturated soils versus $\nu \approx 0.3$ in unsaturated soils. This is because $V_p$ is very sensitive to degree of saturation, and tends to drastically decrease from around 1500 m/s in fully saturated soils by a slight decrease in saturation as addressed in Sec. 5.7.3. Ground water tables can often be detected by clear changes of P-wave velocity or Poisson's ratio.

*(b) Uphole method*

Though not as popular as the downhole method, it is used together with dynamic penetration tests such as Standard Penetration Test (SPT, addressed in Sec. 5.5.1), in which waves generated from the penetrating tip at different depths are detected by a wave sensor at the ground surface to have the travel time curve economically as in Fig. 2.2.1(b).

*(c) Cross-hole method*

Multiple bore-holes, tens of meters apart in between, as illustrated in Fig. 2.2.1(c) are utilized to measure velocities of waves propagating horizontally from a wave source to receivers at the same depth in the neighboring holes. By changing the depths in those holes step by step simultaneously, a continuous wave speed variation can be obtained. Two receivers at neighboring holes in addition to the wave source hole are recommended for higher reliability, though a simpler measurement with one receiver and one source in two neighboring holes is also possible. If a thin soft layer is in contact with stiff layers, a care is needed so that the wave speed is not influenced by the adjacent stiffer layers.

*(d) Suspension method*

In this method increasingly employed recently, a sounding equipment integrating a set of wave receivers and a transmitter as illustrated in Figs. 2.2.1(d) and 2.2.3(a) is suspended by wire in a bore-hole to measure elastic waves starting from the transmitter propagating to the receivers through adjacent soils. The transmitter and receivers are separated by a given distance (4–5 m) with wave dampers in between and the velocity of P or S-wave vertically travelling that distance is measured. The borehole has to be filled with water in this method because the waves have to transmit between the sensors

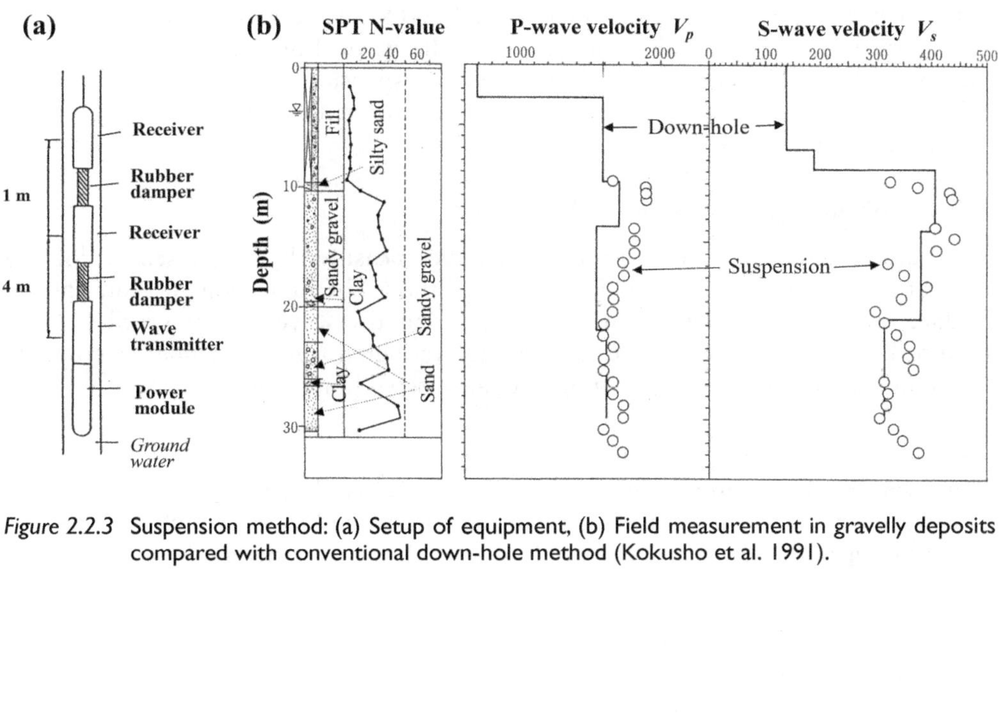


*Figure 2.2.3* Suspension method: (a) Setup of equipment, (b) Field measurement in gravelly deposits compared with conventional down-hole method (Kokusho et al. 1991).

and surrounding soil. Fig. 2.2.3(b) exemplifies typical measurements of $V_p$ and $V_s$ in gravelly deposits by this method. It demonstrates that the suspension method can measure the continuous changes of wave velocities in a short interval more in detail than the downhole method. It is noted however that the frequency of waves employed in this method is about 1 kHz or higher, much greater than that in the downhole method or of actual earthquake waves, in order to make high-precision wave-velocity measurement in the short travel distance. This may mislead the velocity to slightly higher values than that for actual earthquake waves in some soils. The frequency of waves in a conventional downhole method is normally several tens of Hz. If the S-wave velocity in gravelly soil is 300 m/s, the wave length $\lambda$ is around 5 m, long enough to average over the variations in wave velocity due to gravel particles. In the suspension method with the frequency around 1 kHz, however, $\lambda$ is around 0.3 m being comparable with gravel size, indicating that the measured velocity can possibly be faster than the velocity for low-frequency earthquake waves, because the waves of small $\lambda$ tend to transmit faster through stiffer portions.

Thus, the various methods are available for measuring in situ P and S-wave velocities utilizing bore holes. Among the S-wave velocity measurements, the down-hole, up-hole and suspension methods utilize the SH-wave, which oscillates horizontally and propagates vertically. In contrast, the cross-hole method uses the SV-wave, vibrating vertically and propagating horizontally. It is known that $V_s$ is dependent on the orthogonal 3-dimensional stress system as formulated in the next equation (Roesler 1979).

$$V_s \propto \left(\frac{\sigma'_a}{p_0}\right)^{\alpha} \left(\frac{\sigma'_p}{p_0}\right)^{\beta} \tag{2.2.3}$$

Here, $\sigma'_a$, $\sigma'_p$ are the effective stresses in the direction of wave propagation and wave vibration, respectively, and $\alpha$, $\beta$ are exponent constants. Because in situ soil is anisotropically consolidated with different vertical and horizontal stresses; $\sigma'_a \neq \sigma'_p$, which may result in different $V_s$-values for SH and SV-waves if $\alpha \neq \beta$. Besides, in situ soil fabric can be intrinsically anisotropic in the vertical and horizontal directions (Tanaka 2001). Therefore, the down-hole or suspension method using the vertically propagating SH-wave may be recommended more if the similarity with seismic SH-waves is considered.

#### 2.2.1.2 *Measurements without boreholes*

There are geophysical methods using elastic waves measured by a set of sensors located only at the ground surface; (a) Refraction method, (b) Reflection method and (c) Surface wave method. Not only wave velocities but also wave velocity profiles can be obtained economically only by surface sensors, though it is recommended to compare the results with wave-velocity logging data at selected boreholes.

##### *(a) Refraction method*

Wave sensors are set up in a line on ground surface as illustrated in Fig. 2.2.4(a). P or S-wave starting from a wave source at the end of the line A or B propagates in a surface layer of the wave velocity $V_1$ ($V_{p1}$ or $V_{p1}$). If it is underlain by a stiffer layer

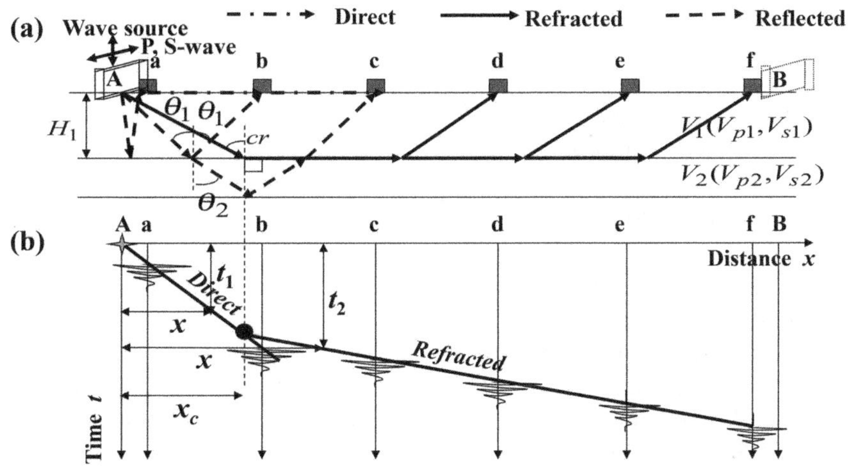


*Figure 2.2.4* Refraction method: (a) Setup of measuring system, (b) Schematic travel time curve in two-layer system.

with higher velocity $V_2$ ($V_{p2}$ or $V_{p2}$), the wave can be refracted at the interface with the critical angle $\theta_{cr} = \sin^{-1}(V_1/V_2)$ as already studied in Sec. 1.1. For waves with the incident angles larger than $\theta_{cr}$, no refraction but reflection occurs. The wave refracted at the critical angle propagates horizontally along the boundary of the lower layer as a head wave with the velocity $V_2$ and emits the refracted wave into the surface layer with the radiation angle $\theta_{cr}$. Because the wave travels faster in the lower layer as the head wave than in the surface layer, it arrives earlier at wave sensors located farther than a threshold distance. Thus, a travel-time curve obtained by multiple sensors at a ground surface allows to yield a two-dimensional multi-layered wave-velocity profile. It is clear that the refraction method is valid only if the lower wave velocity is greater than the upper velocity, indicating that a sandwiched softer layer cannot be detected because no head-wave migration can occur. The edge-to-edge (A-B) span of the ground surface wave sensors has to be much longer than the depth of the layers to be investigated in this method (more than five times).

In Fig. 2.2.4(b), a travel time curve is schematically shown for a simple two-layer system. Let $t_1$ and $t_2$ denote the travel times of direct and refracted waves, respectively, for a horizontal distance $x$ from the wave source. Then, the velocities $V_1$ and $V_2$ can be read off from the travel time chart using the next equations.

$$t_1 = \frac{x}{V_1}, \quad t_2 = \frac{2H_1/\cos\theta_{cr}}{V_1} + \frac{x - 2H_1 \tan\theta_{cr}}{V_2} \tag{2.2.4}$$

In this example, the direct wave can arrive earlier from the source A if the sensor is nearer than around Point b, while the refracted wave travel faster beyond that point. The distance to the corner of the travel-time chart $x_c$ can be formulated as:

$$\frac{x_c}{V_1} = \frac{2H_1/\cos\theta_{cr}}{V_1} + \frac{x_c - 2H_1 \tan\theta_{cr}}{V_2} \tag{2.2.5}$$

The thickness of the surface layer $H_1$ can be obtained from $V_1$, $V_2$ and $x_c$ as:

$$H_1 = \frac{x_c}{2}\sqrt{\frac{V_2 - V_1}{V_2 + V_1}} \tag{2.2.6}$$

For more realistic soil profiles consisting of more than three layers, essentially the same algorithm can be followed. The measurement by the same sensor array is carried out twice with the wave starting from two opposite ends to detect the dip angles of layer interfaces.

*(b) Reflection method*

Waves starting from a source at a ground surface and reflecting at a layer boundary to arrive at surface sensors are measured as illustrated in Fig. 2.2.5(a) to analyze soil profiles and associated wave velocities. This method started historically as an exploratory technology of natural resources in deeper ground and expanded its applications to shallower soils later. Compared to the refraction method where only the first wave arrival is concerned, a higher expertise is needed because the reflected waves are normally later to arrive than the direct or refracted waves and more difficult to discern.

The arrival time $t$ and the distance $x$ from the wave source to a particular wave sensor can be correlated as

$$t^2 = \frac{x^2 + 4H_1^2}{V_1^2} \tag{2.2.7}$$

In Fig. 2.2.5(b), a $t$ versus $x$ diagram is shown for the refracted, direct and reflected waves, wherein the correlation in Eq. (2.2.7) is shown with a chain-dotted curve.

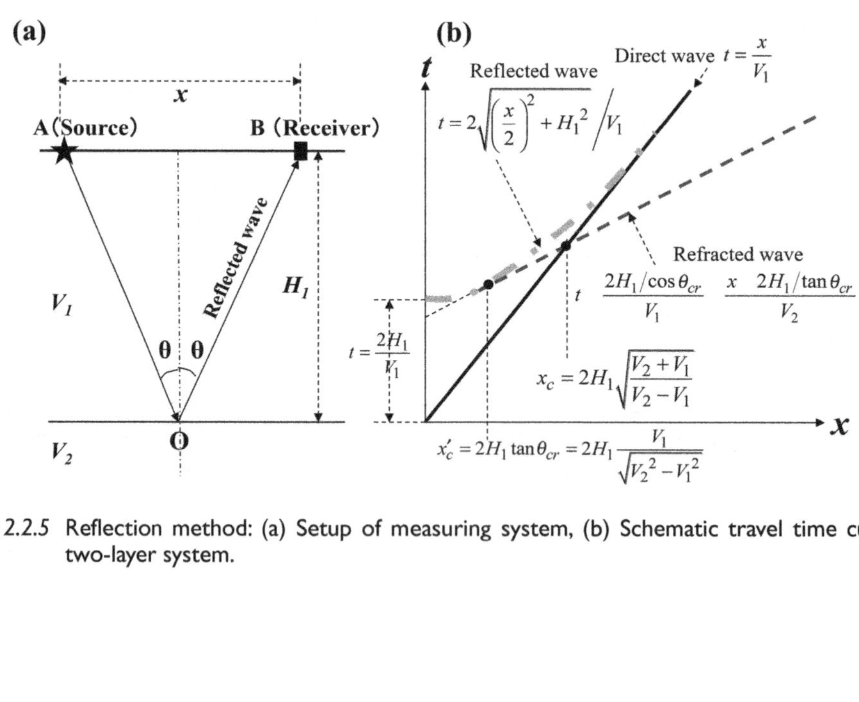


*Figure 2.2.5* Reflection method: (a) Setup of measuring system, (b) Schematic travel time curve in two-layer system.

It indicates that the arrival time of the reflected wave is always longer, though it approaches to that of the direct wave $t = x/V_1$ with increasing distance. Using $\sin\theta_{cr} = V_1/V_2$ based on Eq. (1.1.2), the critical distance $x'_c$ for the generation of refracted wave arriving earlier than the reflected wave is formulated as

$$x'_c = 2H_1 \tan\theta_{cr} = 2H_1 \frac{V_1}{\sqrt{V_2^2 - V_1^2}} \tag{2.2.8}$$

If reflected waves of small incident angles are used for a greater depth with a shorter surface sensor line, the refracted wave is hard to be observed, leading the reflection-wave analysis easier. It is also clear in Fig. 2.2.5(b) that refracted wave travels faster than direct wave beyond the critical distance $x_c = 2H_1\sqrt{(V_2 + V_1)/(V_2 - V_1)}$ as already indicated in Eq. (2.2.6). Eq. (2.2.7) implies that a linear correlation between $x^2$ versus $t^2$ plotted in the horizontal and vertical axes, respectively gives $1/V_1^2$ as the gradient, and the intercept $t^2 = 4H_1^2/V_1^2$ at $x^2 = 0$ determines the layer thickness $H_1$. This analysis works for reflected waves at deeper layer boundaries whether or not softer layers are sandwiched in between. The reflection method needs considerable computer analyses to filter noises, discern reflected waves at various interfaces, and properly interpret complicated wave records consisting of different wave types.

### *(c) Surface wave method*

Rayleigh waves of various wave length are measured by a set of wave sensors aligned at a ground surface. As already studied in Sec. 1.4, the phase velocity of dispersive surface wave is dependent on the wave velocity profile of a site, because waves with longer periods tend to travel faster due to higher wave velocities at greater depths in layered deposits. In this method, a dispersion curve of the phase velocity corresponding to varying wave length is used to back-calculate the most probable soil profile together with corresponding wave velocities. This method originates from a geophysical investigation on the earth crust, and became popular for geotechnical investigations for shallower soils, by the name Spectral Analysis of Surface Wave (SASW) or Multi-channel Analysis of Surface wave (MASW), in USA (e.g. Stokoe and Nazarian 1984). Artificial wave sources such as a mechanical vibrator or a falling weight are used (an active method) to generate Rayleigh waves of relatively short wave length to investigate shallower depths, whereas natural microtremors of longer wave lengths are sometimes used (a passive method) to investigate deeper depths. Fig. 2.2.6 exemplifies a test setup composed of a vertical vibrator and a pair of two-dimensional wave sensors (the mutual distance $D$) with the center point A distant from the vibrator by $L$. The selection of $L$ and $D$ relative to the wave length $\lambda$ is recommended (Tokimatsu et al. 1991) as:

$$\frac{\lambda}{4} \leq L, \quad \frac{\lambda}{16} \leq D < \lambda \tag{2.2.9}$$

Measurement of phase velocity $c$ between the two sensors with stepwise-varying frequency $f$ allows to develop a Rayleigh wave dispersion curve between the phase velocity $c$ versus the wave length $\lambda = c/f$. From that, the dimensions of soil profile together with the corresponding wave velocities in individual layers are optimized by back-calculation schemes based on the theories studied in Sec. 1.4 (Haskell 1953,

| Layer number | Thick-ness | Density | P-wave velocity | S-wave velocity |
|---|---|---|---|---|
| 1 | $H_1$ | $\rho_1$ | $V_{P1}$ | $V_{S1}$ |
| 2 | $H_2$ | $\rho_2$ | $V_{P2}$ | $V_{S2}$ |
| 3 | $H_3$ | $\rho_3$ | $V_{P3}$ | $V_{S3}$ |
| ⋮ | ⋮ | ⋮ | ⋮ | ⋮ |
| N-1 | $H_{N-1}$ | $\rho_{N-1}$ | $V_{PN-1}$ | $V_{SN-1}$ |
| N | ∞ | $\rho_N$ | $V_{PN}$ | $V_{SN}$ |

(a)

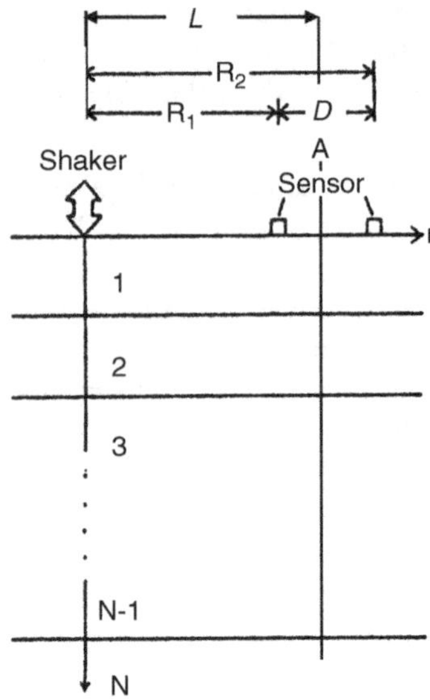


(b)

*Figure 2.2.6* Multi-layered ground (a), and Measuring system of active surface wave method (b) (Tokimatsu et al. 1991).

Harkrider 1964). The depth investigated by artificial wave generations is around 20 m maximum, while that by using natural microtremors would be over 100 m (Tokimatsu 1995). In the latter, many wave sensors are arrayed in circle so that surface waves migrating from arbitrary directions can be analyzed.

In Fig. 2.2.7(a), the phase velocities $c$ and the amplitude ratios $u/w$ (horizontal/vertical) depending on the wave length $\lambda$ observed in sand deposits in Niigata city Japan by the active method are compared with the optimized solutions. A good correspondence of $c$ and $u/w$ can be recognized between the observation and back-calculation. In Fig. 2.2.7(b), the optimized $V_s$-profile is compared with those obtained by the downhole method indicating a satisfactory agreement between them. In the back-calculation, the layer thickness, the velocities $V_s$, $V_p$ and the soil density $\rho$ of individual layers are variables to optimize, though the latter two may be roughly determined without significant influence on the optimization results (Tokimatsu et al. 1991).

## 2.2.2 Laboratory tests for small-strain properties

The most fundamental dynamic soil properties for small strain, $V_s$ and $V_p$, are normally evaluated by in situ wave velocity measurements. However, laboratory tests are sometimes conducted for the velocities or corresponding small-strain moduli to examine the effects of pertinent parameters or to compare with the in situ values to know the intactness of sampled soils. The laboratory tests for small-strain properties

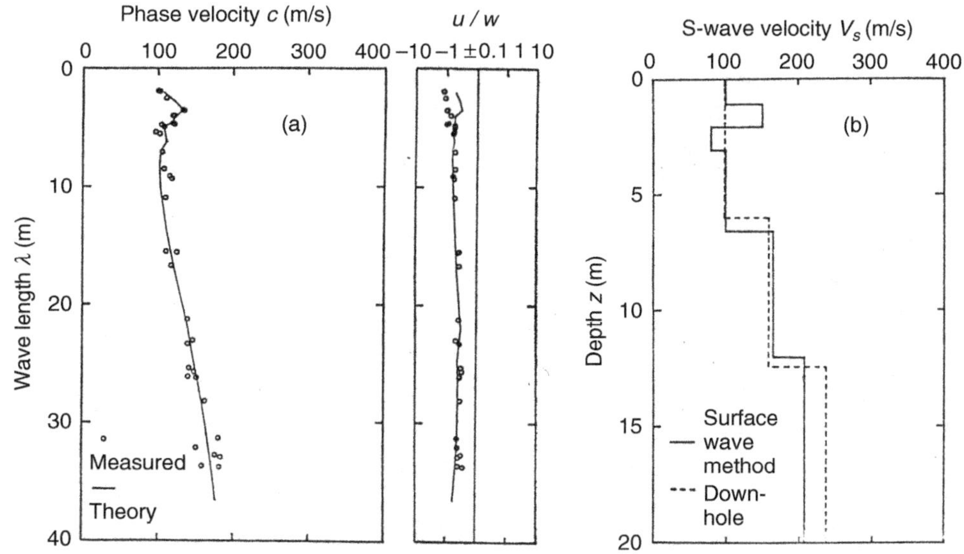


*Figure 2.2.7* Surface wave method in sand deposit: (a) Phase velocity $c$ and amplitude ratio $u/w$ versus wave length $\lambda$, (b) Back-calculated $V_s$ versus depth $z$ compared with downhole method.

are classified into two groups; wave transmission tests and small-strain cyclic loading tests.

### 2.2.2.1 *Wave transmission tests*

#### *(a) Resonant column test*

This method was first used on dry sand under the atmospheric pressure by a Japanese researcher (Iida 1938). It was followed by American investigators (Richart et al. 1970), wherein a column-shaped sand specimen in a pressurized chamber was vibrated by electro-magnetic torque. In this test, rotational vibration is given to a soil specimen to excite it in resonance by sweeping the frequency. Fig. 2.2.8(a) illustrates a typical test equipment using a hollow cylindrical soil specimen, wherein the vibratory torque is given from above to the mass on the top of the specimen with the fixed bottom end. There are other options in the boundary condition, "fixed top and free bottom" or "free top and free bottom". The wave velocity is determined from the resonant frequency by using the equation of S-wave propagation in the soil column with the various boundary conditions (Richart et al. 1970, Ishihara 1996). The damping ratio $D$ of the soil specimen is obtained from free decay vibration schematically shown in Fig. 2.2.8(b) just after the driving torque is cut during the resonance, wherein $u_k$ and $u_{k+1}$ are amplitudes of arbitrary two neighboring peaks of the vibration, as

$$D = \frac{1}{2\pi} \ln \frac{u_k}{u_{k+1}} \tag{2.2.10}$$

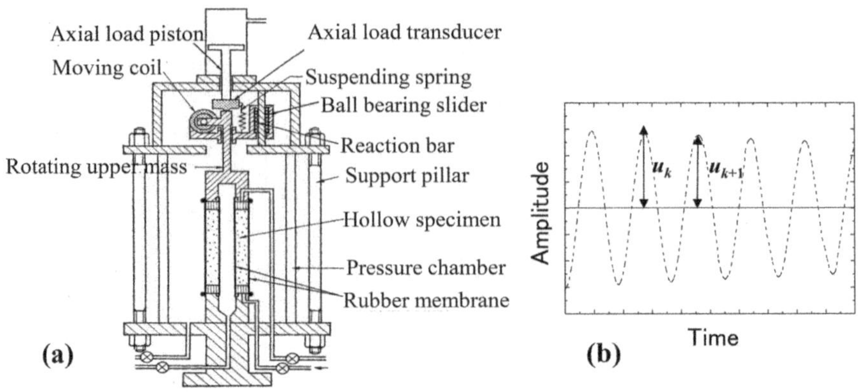


*Figure 2.2.8* Resonant column test: (a) test device, (b) Decay vibration for damping measurement.

In the resonant column test, the S-wave velocity can be directly determined in steady-state wave propagation for low-strain waves. On the other hand, tens of thousands of wave cycles given to a specimen may have measurable effects on the soil properties with increasing strain amplitudes. In addition, calibration tests are indispensable using a dummy specimen of known properties to know the inertial effect of the end mass (e.g. Tatsuoka and Silver 1980), and further to know if the radiational or mechanical damping in the test equipment itself is not large enough to influence the soil damping evaluated in the above equation (Kokusho 1982).

#### *(b) Pulse wave test (Bender element test)*

Pulse wave signals emitted from one end of a specimen are received at the other end to measure the wave velocity of a soil specimen in a testing device such as a triaxial apparatus. Piezoelectric elements are incorporated as the transmitter as well as the receiver of the wave. Since 1980's, bender elements (BE) are increasingly employed as one of the piezoelectric elements (Shirley and Hampton 1977). The bender element is a thin, two-layer piezo-electric plate (about 10 mm wide and 1 mm thick penetrating into a soil specimen by several millimeters) that can be built in most soil test devices. There are two types of bender elements due to different electric circuits; series and parallel. For the same applied voltage, the parallel-type provides twice the BE displacement of the series-type connection, hence, the parallel-type BE as a transmitter versus the series-type BE as a receiver is recommended (Lee and Santamarina 2005).

There are several technical details for reliable measurement and they are mostly standardized internationally as comprehensively summarized by Yamashita et al. 2009. Fig. 2.2.9(a) shows a typical BE test setup built in a triaxial test apparatus with the transmitter and the receiver on both ends of the specimen under confining pressure. As the emitting waves, harmonic waves of several kHz or step waves are used. In Fig. 2.2.9(b), step-shaped waves are transmitted in opposite directions and recorded in the receivers to confirm that they are completely reverse as a S-wave signal. The travel time $\Delta t$ is read off at the first wave arrival to determine S-wave velocity as

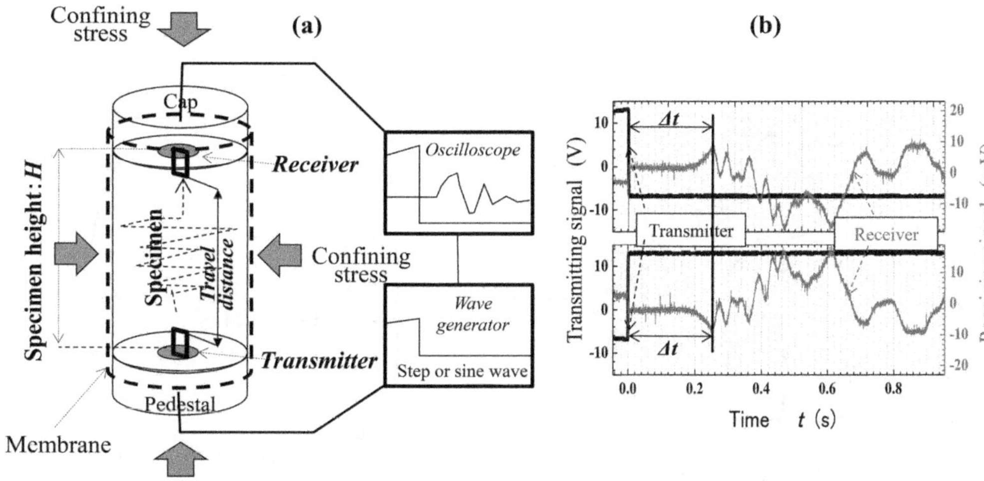


Figure 2.2.9 Bender element test: (a) Test setup in triaxial specimen, (b) SH-wave records by transmitter and receiver.

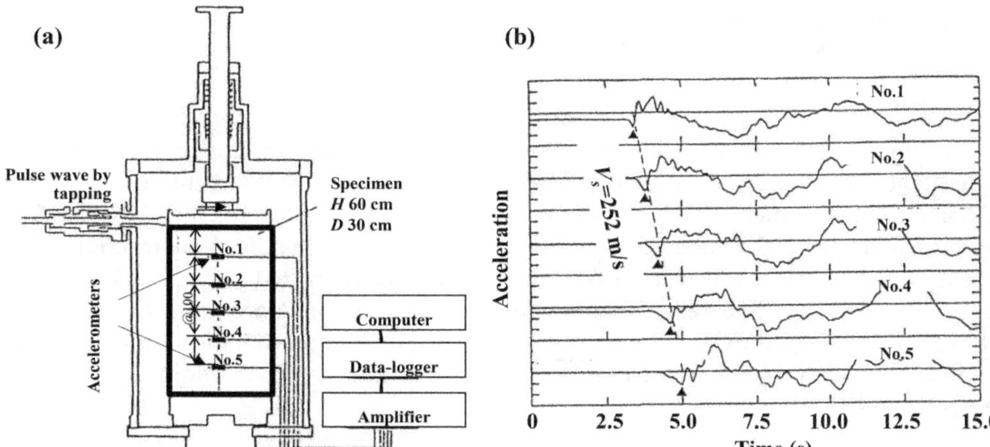


Figure 2.2.10 Pulse wave test using wave sensors: (a) Test setup in large-scale triaxial specimen, (b) SH-wave records by accelerometers attached on specimen side (Nishio et al. 1987).

$V_s = \Delta H/\Delta t$. As the travel distance $\Delta H$, the tip to tip length between the sensors is chosen normally (Yamashita et al. 2009) though some calibration tests for the effective travel distance are recommended if possible.

### (c) Pulse wave test using wave sensors at specimen sides

Pulse wave tests are sometimes carried out as a more conventional and robust scheme during triaxial tests to have correlations between medium to large strain soil properties and $V_s$ or initial shear modulus $G_0$. Fig. 2.2.10(a) shows a large-scale triaxial

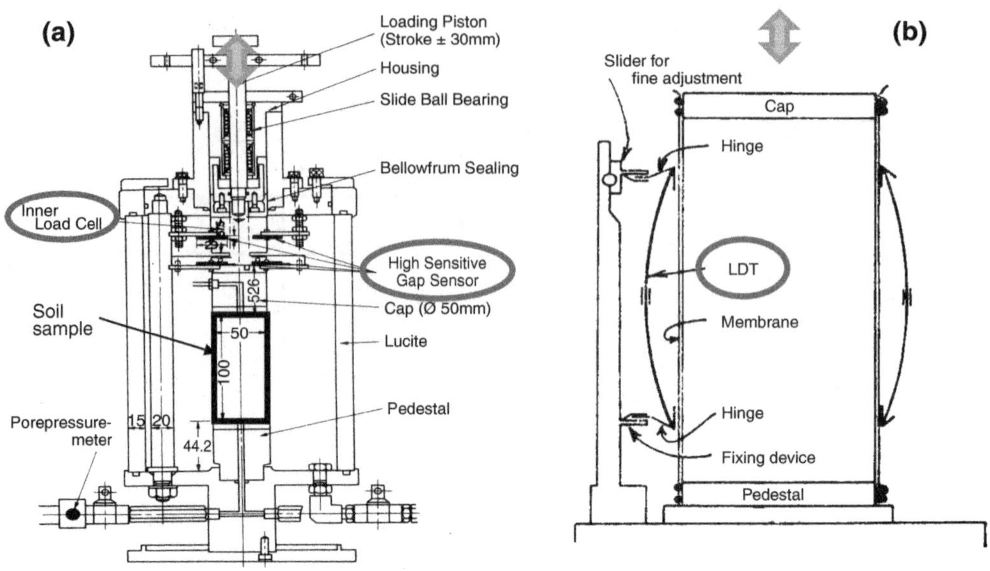


*Figure 2.2.11* Small-strain cyclic loading triaxial tests: (a) Using non-contact type gap sensors (Kokusho 1980), (b) Using LDTs (Goto et al. 1991).

apparatus with the specimen size 60 cm in height and 30 cm in diameter, where a set of small and high-sensitivity accelerometers are glued on both sides of the soil specimen (indirectly via the membrane) to measure SH-wave (e.g. Nishio et al. 1987). Lateral or torsional SH-wave is generated by faintly hitting a loading piston of the triaxial apparatus and recorded as it travels through the specimen as depicted in Fig. 2.2.10(b). It is recommended that two reversely-polarized S-waves are generated and the recorded waves are confirmed to be completely reverse with little contamination by other types of waves. Also recommended is to read off travel times corresponding to the first arrival not to the wave peak.

#### 2.2.2.2 *Small-strain cyclic loading tests*

*(a) Cyclic loading triaxial test with high-sensitivity gap-sensor*

Triaxial tests were developed historically to measure strength or deformation properties of large to medium strain levels. Major technical difficulty to apply this to measure small-strain properties was to precisely measure small-strain soil properties in the pressure chamber. Fig. 2.2.11(a) shows an advanced-type triaxial apparatus started by Kokusho (1980), where a pair of high-sensitivity axial-deformation transducers (gap-sensors or proximeters) were installed on both sides of the loading axis in the pressure chamber. The deformation transducers were non-contact type in order to avoid the effect of mechanical frictions contaminating cyclic stress-strain hysteresis in small-strain levels. An axial load transducer was also introduced inside the pressure chamber to avoid the effect of friction along the loading piston. A set of stress-strain

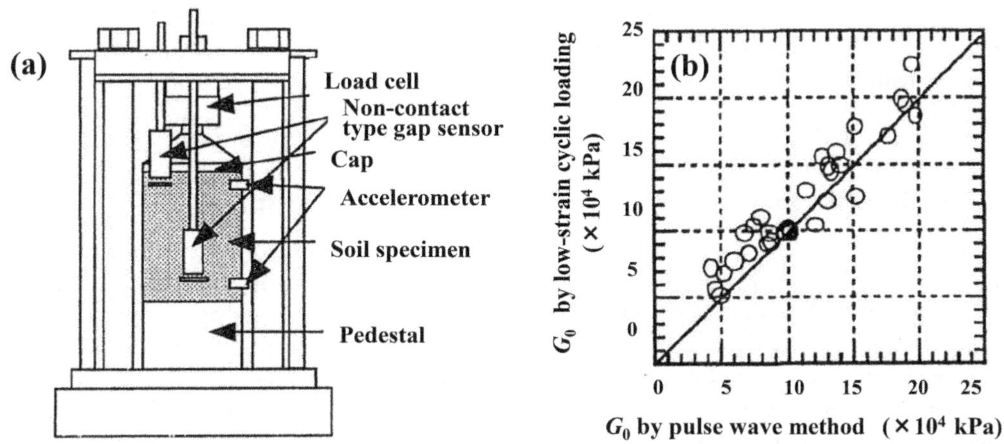


*Figure 2.2.12* Small-strain cyclic triaxial tests avoiding end effects (Kokusho et al. 1999): (a) Test setup with gap sensors at specimen sides, (b) Small-strain moduli $G_0$ compared between two methods.

hysteresis obtained by this advanced triaxial test is shown in Fig. 2.1.3. It is clear that this test can readily capture the elastic behavior of sand with linear modulus and almost no hysteretic damping in $10^{-6}$ strain level. Also obvious is that properties of not only $10^{-6}$ strain but also $10^{-4}$ or higher can be measured in the same test specimen.

In this test method, the axial soil strain is calculated from the axial displacement of the upper loading cap relative to the lower pedestal. In order to avoid the effect of poor contact at the ends of soil specimen (end effects) for coarse granular or stiff specimens in particular, the axial displacements can be measured directly on the specimen sides by non-contact type sensors. Fig. 2.2.12(a) exemplifies this type of cyclic triaxial tests, where the targets of gap sensors are glued at two different levels on the specimen side to measure the relative displacement in between. The measured small-strain moduli $G_0$ of coarse decomposed granite are compared fairly well with those obtained by the pulse wave method using tiny acceleration sensors also glued on the specimen sides as indicated in Fig. 2.2.12(b) (Kokusho et al. 1999).

### *(b) Cyclic loading triaxial test with LDT*

Another type of displacement sensor used in a cyclic loading triaxial test to measure small-strain properties directly on soil specimens avoiding unfavorable effects of poor contact between soil specimen and loading plate is a Local Deformation Transducer (LDT) (Goto et al. 1991). As shown in Fig. 2.2.11(b), a pair of the LDT glued on two opposite sides of a specimen are a sort of clip gauges made from thin flexible metal strip with strain gauges glued at its center on both sides. They are pinched between two hinges attached at the top and bottom of the specimen sides to allow free flexural deformations. The measured relative displacement between the two hinges is obtained for a wide range of strain from $10^{-6}$ to $10^{-2}$ by electronically linearizing the strain gauge readings because they are nonlinearly correlated with the axial strain. Note that this gauge is not a non-contact type, which may possibly reflect unfavorable mechanical frictions of the gauge itself particularly in cyclic loading tests, though such deficiency in measuring the hysteretic damping ratio is not recognized in the literature (Goto et al. 1991).

### 2.2.3 Laboratory tests for medium to large strain

It is difficult to evaluate dynamic soil properties for medium to large strain directly in situ. Those properties are normally measured in the laboratory by means of cyclic loading tests, wherein soil samples under the same static stresses as in the field are loaded with cyclic stresses representing earthquake effects to monitor corresponding variations in strain and pore-pressure. The quality of test samples is immensely important in laboratory tests in small to medium strains to have reliable soil properties, because soil properties in those strain levels are very much sensitive to soil fabric that is easily affected by mechanical disturbance. Hence, a great care is needed to preserve not only soil densities but also microscopic soil fabric in test specimens during in situ soil sampling and preparatory works in the laboratory.

In cyclic loading tests, the dynamic stress is normally considered to represent SH-wave among others, because it is the major seismic effect on soil elements in shallow depths. In contrast, the effect of P-wave is normally ignored in these tests for saturated soils because the isotropic stress change by P-wave does not change effective stresses in fully saturated in situ soils theoretically. Thus, P-wave is considered to have no effect on deformation and failure in saturated soft soils where the principle of effective stress holds.

As for the loading rate of seismic stresses, the period of cyclic motion is normally chosen as 1 to 10 seconds slightly lower than the dominant period of seismic motions to make soil tests technically easier. During earthquakes, saturated soils in situ are considered essentially undrained even for permeable sandy or gravelly soils because of the short loading duration and the high loading rate. Hence the soil specimen is covered by a rubber membrane and the drainage of pore-water is prohibited during cyclic loading to simulate the undrained condition. Before cyclic loading, the specimen is consolidated with cell pressure (total stress) together with back-pressure (pore-pressure) to reproduce in situ stress conditions. The back-pressure is sometimes intentionally raised higher than the actual pore-pressure with the effective stress being kept constant in order to secure higher saturation or higher pore-pressure coefficient $B$-value in undrained cyclic loading tests of saturated soils. Different types of cyclic loading tests are used in order to reproduce in situ stress conditions in soil elements during earthquake loading.

#### 2.2.3.1 *Simple shear test*

In situ stress condition is normally idealized as follows. A soil element in a level ground is initially consolidated with effective vertical and horizontal stresses, $\sigma'_v$ and $\sigma'_h = K_0\sigma'_v$, respectively as illustrated in Fig. 2.2.13. Here, $K_0$ is the coefficient of earth pressure at rest. Then it is loaded with cyclic shear stress $\tau_d$ on the horizontal plane when SH-wave is coming up. The simple shear test device is supposed to directly reproduce

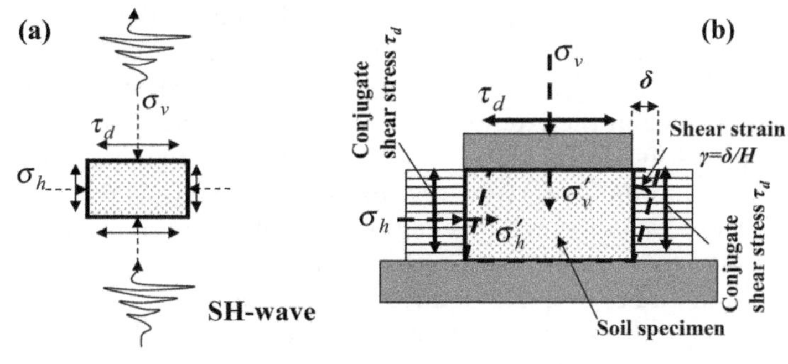


*Figure 2.2.13* In situ stress condition in level ground (a) and Simple shear test device (SGI-type) (b).

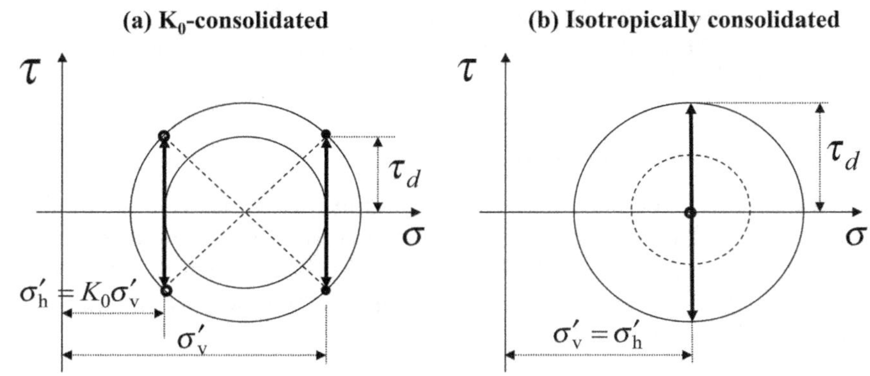


*Figure 2.2.14* Mohr's stress circle in simple-shear specimen: (a) $K_0$-consolidated specimen, (b) Isotropically consolidated specimen.

this stress condition in the test specimen. In the test, the ratio $\tau_d/\sigma_v'$ is called a cyclic stress ratio (*CSR*) and used as the normalized value of cyclic stress.

As illustrated in Fig. 2.2.13(b), a low-height column-shape soil specimen confined in a special case is consolidated by vertical and horizontal effective stresses $\sigma_v'$ and $\sigma_h' = K_0\sigma_v'$. In order to secure lateral confinement for $K_0$-consolidation and free shear deformation at the same time, the specimen case is made of wire-meshed rubber (the Norwegian Institute type) or teflon-coating donut-shape thin laminar plates in which soil specimen covered by rubber membrane is stored (the Swedish Geotechnical Institute type). Cyclic shear stress $\pm\tau_d$ is given at the upper or lower side of the specimen in either undrained or drained condition, and averaged shear strain is obtained as $\gamma = \delta/H$ dividing horizontal relative displacement $\delta$ by the specimen height $H$. As illustrated in the Mohr's stress circles in Fig. 2.2.14(a), the principal stress plane rotates continuously with the shear stress $\tau_d$ in a $K_0$-consolidated soil specimen. In contrast, if the soil is isotropically consolidated with $K_0 = 1.0$, the Mohr's circle turns to be Fig. 2.2.14(b), where no rotation of the principal stress but discontinuous alternation of the major and minor principal stresses occurs.

There exists a serious mechanical drawback in the simple shear device such as the NGI or SGI type, however, in that the stress distribution in the specimen is difficult to be uniform in its edge portions in particular. It is because a pair of conjugate shear stresses shown in Fig. 2.2.13(b) is difficult to work on the specimen sides due to the lack of shear resistance in the lateral confinement of the meshed rubber membranes or the laminar plates. It is difficult to solve the conflicting requirements between free shear deformation and uniform stress distribution while allowing vertical settlement of the soil specimen. Nevertheless, this device is simple in mechanism and easy to test intact soils sampled from in situ, and often used to investigate large strain problems such as liquefaction and earthquake-induced soil settlements.

There is a modified test method named as a constant volume simple shear test wherein the vertical movement of the upper loading plate is constrained, which can simulate undrained shear behavior of saturated soils by using dry soil specimens. In this test (named as an equal-volume simple shear test), the vertical stress tends to change because of the constraint of the loading plate in the same manner as if it is the effective stress in saturated soils with no volume change in the undrained condition (e.g. Pickering et al. 1973).

### 2.2.3.2 *Torsional simple shear test*

The test specimen has a cylindrical shape with a thin wall as shown in Fig. 2.2.15, and loaded vertically with $\sigma_v'$ by a loading cap, and horizontally with $\sigma_o'$ from outside and $\sigma_i'$ from inside of the specimen by hydraulic pressure. Then the cyclic shear stress $\pm\tau_d$ is applied on the top or bottom face of the sample, while the other face is fixed. The shear strain is calculated from

$$\gamma = \frac{r\theta}{H} \tag{2.2.11}$$

where $r$ = radius from the specimen center, $\theta$ = angle of specimen distortion and $H$ = specimen height. The shear strain is not uniform, and the value at the middle of the wall thickness at $r = (a+b)/2$ is normally used. Because of the circular shape of the specimen, the radial and tangential stresses $\sigma_r$, $\sigma_\theta$ are variable with the radius $r$ and written as:

$$\begin{aligned} \sigma_r &= -\frac{a^2b^2(\sigma_o - \sigma_i)}{b^2 - a^2}\frac{1}{r^2} + \frac{(b^2\sigma_o - a^2\sigma_i)}{b^2 - a^2} \\ \sigma_\theta &= \frac{a^2b^2(\sigma_o - \sigma_i)}{b^2 - a^2}\frac{1}{r^2} + \frac{(b^2\sigma_o - a^2\sigma_i)}{b^2 - a^2} \end{aligned} \tag{2.2.12}$$

In practice, they are averaged at the center of the specimen thickness; $r = (a+b)/2$, where $a$ and $b$ are inner and outer thicknesses, respectively. It is possible to have these three stresses, $\sigma_v$, $\sigma_r$, $\sigma_\theta$ chosen independently in this test by choosing inner cell pressure $\sigma_i$ different from outer cell pressure $\sigma_o$. Also note that the conjugate shear stresses can transit to neighboring soils smoothly as in situ because of the circular shape of the specimen. Thus, this test device, though not simple in the mechanism, can eliminate the stress disturbance in the simple shear device.

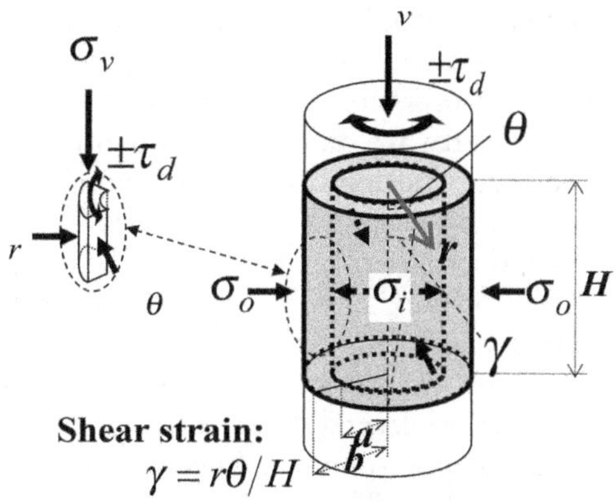


*Figure 2.2.15* Hollow-cylindrical torsional simple shear specimen.

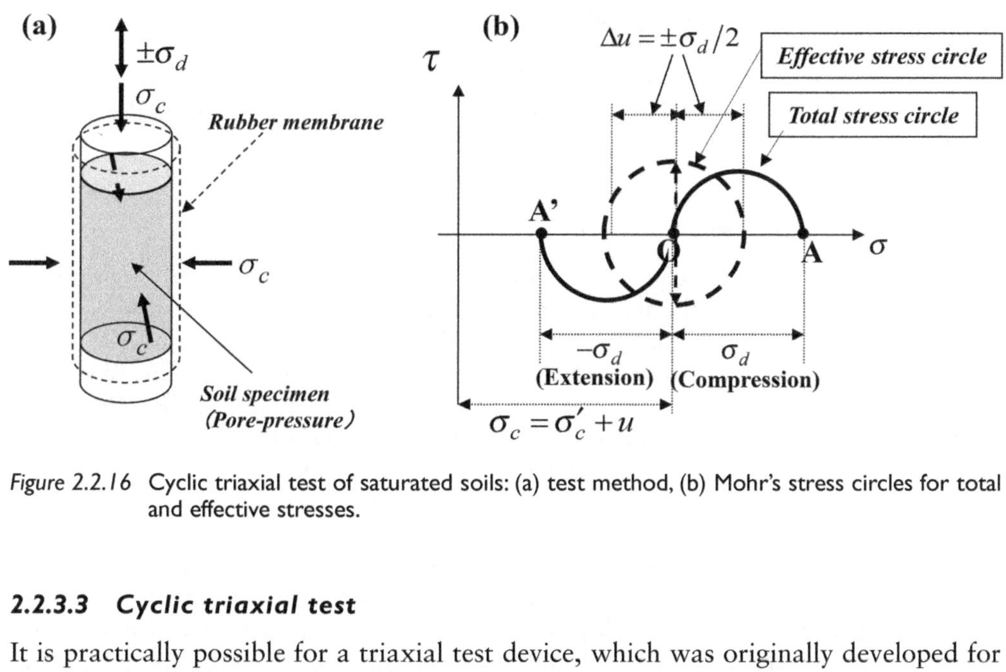


*Figure 2.2.16* Cyclic triaxial test of saturated soils: (a) test method, (b) Mohr's stress circles for total and effective stresses.

### 2.2.3.3 Cyclic triaxial test

It is practically possible for a triaxial test device, which was originally developed for static loading tests, to make cyclic loading tests simulating seismic stress conditions, though the stress condition appears to be different from that of the in situ $K_0$-consolidated soil shown in Fig. 2.2.13(a). Fig. 2.2.16(a) shows a typical cyclic triaxial test device where the lateral stress is kept constant to be $\sigma_c'$, and the vertical stress is fluctuating in compression and extension loading as $\sigma_c' \pm \sigma_d$. The stress ratio is defined as $\tau_d/\sigma_v' = (\sigma_d/2)/\sigma_c' = \sigma_d/2\sigma_c'$ because the maximum shear stress is $\tau_d = \sigma_d/2$ in this test and named as cyclic stress ratio (*CSR*), again.

As the Mohr's stress circle in Fig. 2.2.16(b) indicates, the specimen first isotropically consolidated at the stress point O ($\sigma = \sigma_c' + u_0$: $u_0$ = initial pore pressure) is loaded by increasing the axial stress by $+\sigma_d$ to have the stress circle OA. Because the axial stress is loaded in the undrained condition, excess pore-pressure develops in accordance with the mean stress increment, which is $\sigma_d/2$ in the two-dimensional stress condition assumed in the Mohr's circle, because the soil is fully saturated and the pore-pressure coefficient $B=1.0$. If this pore-pressure increment by $\Delta u=\sigma_d/2$ is considered, the effective stress circle translates leftward on the stress diagram to the dashed circle centered at Point O. The same occurs both in the compression and extension loadings by $\pm\sigma_d$, and the pressure changes $\Delta u=\pm\sigma_d/2$, respectively. Hence, the Mohr's circle in terms of the effective stress reduces to a set of concentric dashed circles with its center always at O. This stress condition is exactly the same as that shown in Fig. 2.2.14(b) for the isotropically consolidated specimen in the simple shear test, though the maximum shear stress plane is different by 45° in the two tests. If the three-dimensional stress condition in the actual world is concerned different from the virtual two-dimensional stress condition above, the cyclic pore-pressure fluctuates by $\Delta u=\pm\sigma_d/3$ corresponding to the change of axial stress $\pm\sigma_d$. This pore-pressure change has to be subtracted from recorded $\Delta u$ to obtain pressure change exclusively due to soil dilatancy.

If the cyclic stress ratio becomes larger than 0.5; $\sigma_d/2\sigma'_c > 0.5$, $\sigma_d$ becomes negative in the extension side of loading, hence the cyclic triaxial test becomes impossible to conduct if initial pore-pressure is $u_0=0$ ($\sigma_c=\sigma'_c$) as easily understood in Fig. 2.2.16(b). Hence, in triaxial tests for stiff soils with higher *CSR*, it is necessary to choose the initial pore-pressure (named as backpressure) $u_0$ much higher than in situ to make the test possible.

#### 2.2.3.4 *Membrane penetration effect in undrained tests*

In laboratory undrained cyclic loading tests, soil specimens are covered by thin rubber membranes to make the undrained condition. In liquefaction tests, the excess pore-pressure tends to increase during cyclic shearing, which may relax the membrane initially penetrated into voids between sand particles and retard the process of pore-pressure buildup. This is named as "Membrane Penetration" (MP) effect, one of the system-compliance problems where flexible boundaries in the undrained system, such as diaphragms of pressure-gages and rubber membranes, tend to make a significant effect on the pore-pressure buildup because the water itself is of very low compressibility.

As conceptually illustrated on the MP-effect in Fig. 2.2.17(a), nothing changes on both sides of a virtual membrane in in situ soil before and after pressure-buildup because the pressure changes equally on both sides, while in the soil specimen sealed by the membrane to separate from constant water pressure outside in (b), the membrane tend to loosen by rising pore-pressure in the specimen, creating extra volume in the specimen which never occurs in situ in (a). This effect tends to be more significant as the side face of the specimen becomes rougher. As a countermeasure to reduce the MP-effect, the rough specimen face may be smeared with fine sands.

A modification of liquefaction test results for the MP-effect may be possible based on compliance ratio $C_R$ defined as follows (Lade and Hernandez 1977, Martin et al. 1978).

$$C_R=\frac{K}{K_c} \tag{2.2.13}$$

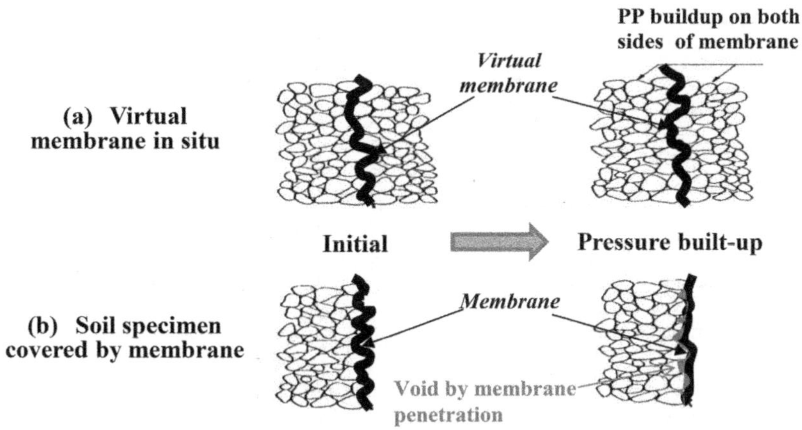


*Figure 2.2.17* Schematic illustration of Membrane Penetration (MP)-effect: (a) Virtual membrane in situ, (b) Soil specimen covered by membrane in laboratory.

Here, $K$ = bulk modulus of soil skeleton in terms of the effective stress and $K_c$ = bulk modulus of pressurized water reflecting the MP and other system-compliance effects. If the undrained test system is perfectly incompressible; $K_c = \infty$ and $C_R = 0$, pore-pressure increment $\Delta u_0$ in the undrained condition due to contractive volumetric strain $\Delta\varepsilon_{vd}$ by negative dilatancy can be expressed (see Sec. 3.1.5.3) as:

$$\Delta u_0 = K\Delta\varepsilon_{vd} \tag{2.2.14}$$

If $C_R \neq 0$, the pressure increment is written as:

$$\Delta u = \frac{\Delta u_0}{1 + C_R} \tag{2.2.15}$$

indicating that the pressure increment $\Delta u$ is smaller than the case $C_R = 0$. The effect of $C_R$ on the pore-pressure buildup of saturated sands was actually quantified by specially designed cyclic triaxial tests by Tokimatsu and Nakamura (1986). It indicated that the MP-effect can be evaluated in terms of the number of loading cycles $N_c$ to attain 100% pressure buildup more easily than directly in terms of cyclic stress ratios *CSR*. Fig. 2.2.18(a) shows $C_N = N_c/N_0$; the ratio of $N_c$ for a given test result to $N_0$ for an ideal test result of $C_R = 0$, plotted versus $C_R$ on the semi-log diagram which may be approximated by the straight line for sands with different grain sizes and relative densities. As an example, uncorrected original data in Fig. 2.2.18(b) can be modified using the ratio $C_N$ read off from $C_R$ in Fig. 2.2.18(a) to have the corrected *CSR* versus $N_c$ curve (Tokimatsu 1990).

Hence, it is necessary to know the compliance ratio $C_R$ for particular test results. Two methods are addressed here to determine the $C_R$-value experimentally (Tokimatsu 1990, Tanaka et al. 1991). Suppose a triaxial test specimen of an isotropic elastic soil

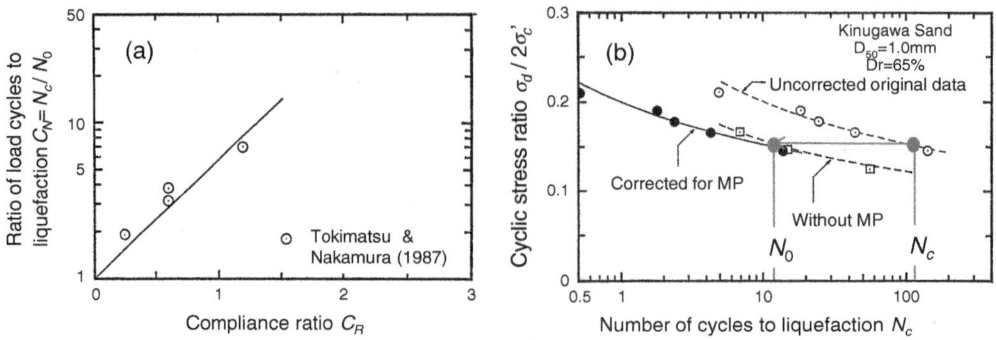


*Figure 2.2.18* How to correct test data considering MP-effect: (a) $C_R$ versus $C_N$ relationship (Tokimatsu and Nakamura 1987), (b) $N_c$-*CSR* curves, uncorrected and corrected for MP-effect (Tokimatsu 1990).

skeleton, with Young's modulus $E'$ and Poisson's ratio $\nu'$ in terms of effective stress, is loaded by axial and lateral stress increments $\Delta\sigma_1$ and $\Delta\sigma_3$ respectively, and develops axial and lateral strain increments $\Delta\varepsilon_1$ and $\Delta\varepsilon_3$ and pore-pressure increment $\Delta u$. Then, the volumetric strain increment (positive for contraction) of the soil skeleton is written as:

$$\begin{aligned}\Delta\varepsilon_v = \Delta\varepsilon_1 + 2\Delta\varepsilon_3 &= \frac{1}{E'}[\Delta\sigma_1 - \Delta u - 2\nu'(\Delta\sigma_3 - \Delta u)] \\ &+ \frac{2}{E'}[\Delta\sigma_3 - \Delta u - \nu'(\Delta\sigma_3 - \Delta u) - \nu'(\Delta\sigma_1 - \Delta u)]\end{aligned} \tag{2.2.16}$$

If $\Delta\sigma_1$ increases under constant $\sigma_3$, the above equation reduces as follows by using the bulk modulus of the soil skeleton $K = E'/3(1-2\nu')$.

$$\Delta\varepsilon_v = \frac{1}{E'}[\Delta\sigma_1 - \Delta u + 2\nu'\Delta u] - \frac{2}{E'}[\nu'\Delta\sigma_1 + \Delta u - 2\nu'\Delta u] = \frac{\Delta\sigma_1 - 3\Delta u}{3K} \tag{2.2.17}$$

On the other hand, the positive pressure increment $\Delta u$ generates volumetric strain decrement of pressurized water $-\Delta u/K_c$ due to system compliance including the MP-effect. In the undrained test, the volumetric strain in Eq. (2.2.17) should be compensated by that created by the system compliance as;

$$\Delta\varepsilon_v - \frac{\Delta u}{K_c} = \frac{\Delta\sigma_1 - 3\Delta u}{3K} - \frac{\Delta u}{K_c} = 0 \tag{2.2.18}$$

which yields the same formula as Eq. (2.2.15).

$$\Delta u = \frac{\Delta\sigma_1/3}{1 + K/K_c} = \frac{\Delta\sigma_1/3}{1 + C_R} = \frac{\Delta u_0}{1 + C_R} \tag{2.2.19}$$

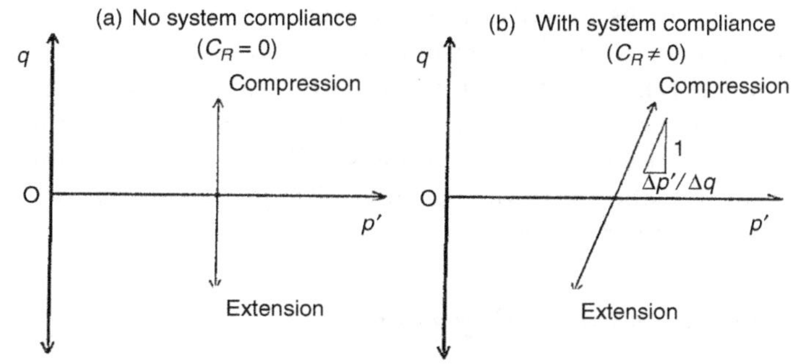


*Figure 2.2.19* Determination of $C_R$-value using gradient of effective stress path (Tokimatsu 1990): (a) With no system compliance $C_R = 0$, (b) With system compliance $C_R \neq 0$.

Here, $\Delta u_0 = \Delta\sigma_1/3$ because $\Delta\sigma_1/3$ is the isotropic stress increment all carried by pore-pressure in undrained condition. From Eq. (2.2.19), the next equation can be obtained.

$$C_R = \frac{1}{3}\frac{\Delta\sigma_1}{\Delta u} - 1 \tag{2.2.20}$$

This formula can be used to determine the $C_R$-value in undrained triaxial tests by measuring the pore-pressure increment $\Delta u$ when the axial stress is increased by $\Delta\sigma_1$ with $\Delta\sigma_3$ unchanged (Tanaka et al. 1991). The increment $\Delta\sigma_1$ had better be small enough to minimize the effect of soil dilatancy.

Another simple method to determine $C_R$ uses effective stress paths on the $p'$ versus $q$ plane in undrained cyclic loading tests, wherein $C_R$, by using Eq. (2.2.20), is correlated with the increments of stresses $\Delta p' = \Delta\sigma_1/3 - \Delta u$ and $\Delta q = \Delta\sigma_1$ as follows.

$$C_R = \frac{3\Delta p'/\Delta q}{1 - 3\Delta p'/\Delta q} \tag{2.2.21}$$

The ratio $\Delta p'/\Delta q$ in the above equation is obtained from an effective stress path as illustrated in Fig. 2.2.19 corresponding to the gradient of the stress path crossing the $p'$-axis ($q = 0$) to minimize the effect of dilatancy. If the path is normal to the $p'$-axis and $\Delta p'/\Delta q = 0$ as in (a), then $C_R = 0$ with no MP-effect, while if $\Delta p'/\Delta q$ has some non-zero value as in (b), $C_R$ can be determined from Eq. (2.2.21). The test result exemplified in Fig. 2.2.18(b) is corrected from original plots using the ratio $C_R$ obtained in this method to have a modified *CSR* versus $N_c$ curve, which was found coincidental with the test result of the $C_R = 0$ condition as indicated in the same figure (Tokimatsu 1990).

## 2.3 TYPICAL SMALL-STRAIN PROPERTIES

Small-strain soil properties were investigated by researchers using various soil materials reconstituted in the laboratory to know the effect of soil type, density, grain size, confining stress and other parameters. Soils in situ may differ in many ways from soil specimens reconstituted in the laboratory, and in situ test values such as P, S-wave velocities have to be focused in design more than lab-test based empirical formulas. However, it is important to understand the basic mechanisms how the properties change depending on various in situ parameters based on systematic laboratory tests in simplified conditions. In the following, small-strain shear moduli and damping ratios measured in the laboratory for various soil types are revisited to discuss their dependency on those influencing factors.

## 2.3.1 $V_s$ and $G_0$ for sand and gravel

### 2.3.1.1 *Effects of void ratio and confining stress*

The initial shear modulus or small-strain modulus of clean sands $G_0$ is normally formulated (Hardin and Richart 1963) as:

$$\frac{G_0}{p_0} = A \cdot f(e) \left(\frac{\sigma_c'}{p_0}\right)^m \tag{2.3.1}$$

Here $f(e)$ = a function of void ratio $e$, $\sigma_c'$ = effective confining stress, $m$ = an exponent, $A$ = a constant and $p_0$ = unit pressure (98 kPa). This formula was developed by a set of experimental results by resonant column tests. Two dry sands were tested; roundish Ottawa sand and angular quartz sand, and it was found that $V_s$ (m/s) is essentially a linear function of $e$ and proportional to an exponent of the effective confining stress $\sigma_c'$. The initial shear modulus $G_0$ was calculated from $G_0 = \rho_d V_s^2$ wherein $\rho_d$ was determined from void ratio $e$ and soil particle density $\rho_s$ = 2.71 t/m$^3$. The empirical formulas on $V_s$ and $G_0$ for the two sands was given respectively as:

Roundish Ottawa sand:

$$V_s = (346 - 159e)\left(\frac{\sigma_c'}{p_0}\right)^{0.25}, \quad \frac{G_0}{p_0} = 700\frac{(2.17 - e)^2}{1+e}\left(\frac{\sigma_c'}{p_0}\right)^{0.5} \tag{2.3.2}$$

Angular quartz sand:

$$V_s = (324 - 109e)\left(\frac{\sigma_c'}{p_0}\right)^{0.25}, \quad \frac{G_0}{p_0} = 330\frac{(2.97 - e)^2}{1+e}\left(\frac{\sigma_c'}{p_0}\right)^{0.5} \tag{2.3.3}$$

The function of void ratio employed in Eq. (2.3.2) $f(e) = (2.17 - e)^2/(1+e)$ has actually been used in practice for a variety of sands not only this particular sand. In Japan for example, quite a few similar tests were conducted on Toyoura sand (e.g. Tatsuoka et al. 1979) and similar formula to Eq. (2.3.2) with the same function of $e$ was proposed (e.g. Kokusho 1980) as:

$$\frac{G_0}{p_0} = 840\frac{(2.17 - e)^2}{1+e}\left(\frac{\sigma_c'}{p_0}\right)^{0.5} \tag{2.3.4}$$

As for the effect of confining stress $\sigma_c'$, it is widely accepted experimentally that $V_s$ and $G_0$ are almost proportional to the power of 1/4 and 1/2 of $\sigma_c'$, respectively, as already indicated in Eqs. (2.3.2) to (2.3.4). Most of such experiments were however conducted under isotropic stress condition. In order to apply these equations to $K_0$-consolidated in situ soils, the confining stress is equalized with effective mean stress; namely; $\sigma'_c = \sigma'_m = (\sigma'_1 + 2\sigma'_3)/3$ in normal practice.

The effect of stresses on the S-wave velocity $V_s$ was basically investigated by Roesler (1979) in terms of the three-dimensional stress; $\sigma'_a$ in wave propagation direction, $\sigma'_p$ in vibration direction and $\sigma'_s$ perpendicular to the above two directions. It revealed that $V_s$ is almost equally dependent on the first two directions and independent of the last direction, so that it is expressed in the following power function where the power constants concerning the three directions are $\alpha \approx 0.14$ and $\beta \approx 0.11$.

$$V_s \propto \left(\frac{\sigma'_a}{p_0}\right)^{\alpha} \left(\frac{\sigma'_p}{p_0}\right)^{\beta} \left(\frac{\sigma'_s}{p_0}\right)^{0} \tag{2.3.5}$$

Similar experimental research (Yu and Richart 1984) showed $\alpha \approx \beta \approx 0.125$ and proposed the next formula in that $V_s$ is dependent only on stresses in the directions of wave propagation and vibration.

$$V_s \propto \left(\frac{\sigma'_a}{p_0}\right)^{0.125} \left(\frac{\sigma'_p}{p_0}\right)^{0.125} \tag{2.3.6}$$

On the other hand, if $V_s$ is assumed to be dependent on the mean stress or confining stress $\sigma'_c$, it should be written as:

$$V_s \propto \left(\frac{\sigma'_c}{p_0}\right)^{0.25} = \left(\frac{\sigma'_a + \sigma'_p + \sigma'_s}{3p_0}\right)^{0.25} \tag{2.3.7}$$

If $\sigma'_p = \sigma'_s$ and $K_0 = \sigma'_p/\sigma'_a = 0.5$ is postulated, then the difference in $V_s$ between Eqs. (2.3.6) and (2.3.7) is only 3%. Hence in engineering practice the dependency of $V_s$ and $G_0$ on the stresses may be approximated respectively as:

$$V_s \propto \left(\frac{\sigma'_c}{p_0}\right)^{0.25}, \quad G_0 \propto \left(\frac{\sigma'_c}{p_0}\right)^{0.5} \tag{2.3.8}$$

In Fig. 2.3.1, not only the initial shear moduli $G_0$ but also the shear moduli $G$ (both normalized by $f(e) = (2.17 - e)^2/(1+e)$) for various strain levels measured on saturated Toyoura sand are plotted versus effective confining stress $\sigma'_c$ on the full logarithmic diagrams. In the diagram (a) by Iwasaki et al. (1978a), the moduli measured by cyclic torsional shear tests and resonant column tests are almost coincidental, while those in (b) by Kokusho (1980) measured by cyclic triaxial tests are also coincidental with the data in (a) except for very small strain levels. It is observed from both diagrams that the exponent $m$ in Eq. (2.3.1) for initial shear moduli $G_0$, corresponding to the shear stress amplitude $\gamma = 10^{-6}$, is not exactly 0.5 but slightly smaller to be 0.40–0.47. Also obvious is that $m$ defined for degraded moduli for various shear strain levels $\gamma > 10^{-5}$ tends to clearly increase with increasing $\gamma$.

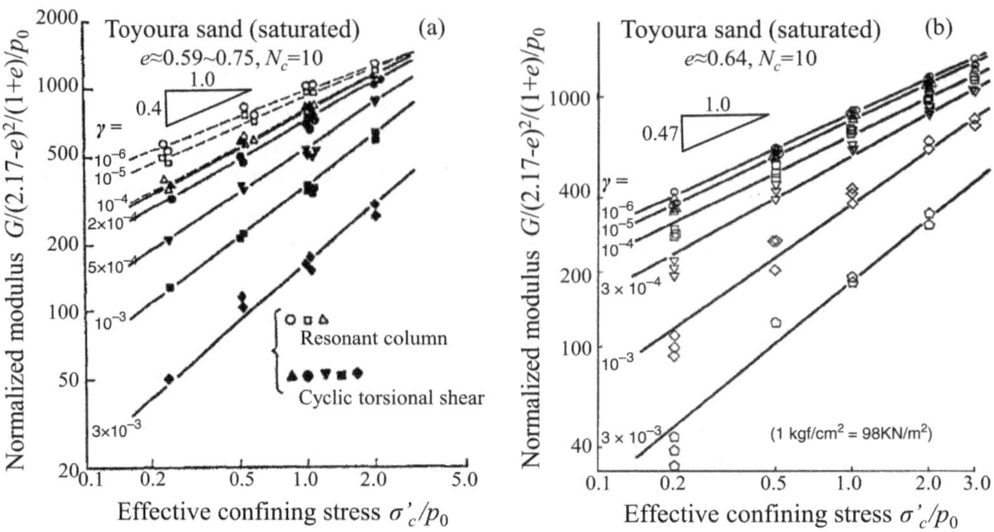


*Figure 2.3.1* Initial shear moduli $G_0$ and $G$ versus effective confining stress $\sigma_c'$ for saturated sand: (a) By cyclic torsional shear test (Iwasaki et al. 1978a), (b) By cyclic triaxial test (Kokusho 1980).

### 2.3.1.2 *Effect of particle grading*

In empirical formulas on the initial shear modulus $G_0$, the effect of void ratio $e$ is normally evaluated by the function $f(e) = (2.17 - e)^2/(1 + e)$ as in Eqs. (2.3.2) to (2.3.4). This function was originally derived from series of tests on Ottawa sand (Hardin and Richrt 1963) with the maximum and minimum void ratios $e_{\max} = 0.66$–$0.89$ and $e_{\min} = 0.32$–$0.54$, respectively, though it is frequently used for sandy soils with different particle grading, particle shapes and different values of $e_{\max}$ and $e_{\min}$. In order to know the effect of particle grading on the $G_0$ versus $e$ relationship, a series of pressure chamber tests were undertaken as illustrated in Fig. 2.3.2(a) to measure S-wave velocities of various soils with various particle gradations artificially compacted to have different void ratios (Kokusho and Yoshida 1997, Kokusho 2007). Five types of fluvial sandy gravelly soils composed of fresh hard grains with different uniformity coefficients $C_u$ depicted in Fig. 2.3.2(b) were tested under parametrically changing overburden stresses given by an overlying pressurized rubber bag.

The S-wave velocity measured for the five tested soils are normalized by the vertical and horizontal stresses, $\sigma_v' \equiv \sigma_p'$, $\sigma_h' \equiv \sigma_a'$, as $V_{s0} = V_s/\{(\sigma_v'/p_0)(\sigma_h'/p_0)\}^m$ considering the correlation in Eq. (2.3.6), and the normalized velocity $V_{s0}$ is plotted versus void ratio $e$ in Fig. 2.3.3, where $p_0 = 98$ kPa. The power $m$ here was determined by the regression analysis to be approximated as $m = 0.125$ for all the materials (Kokusho and Yoshida 1997). Two series of tests using two chambers with different specifications, HC and LC shown in the figure, were conducted, the results of which were not significantly different. It is clearly seen that $V_{s0}$ is almost linearly related to void ratio but the relationship is widely different from one soil to another. Similar relationships derived by Hardin and Richart (1963) for Ottawa sand and quartz sand (Eqs. (2.3.2) and

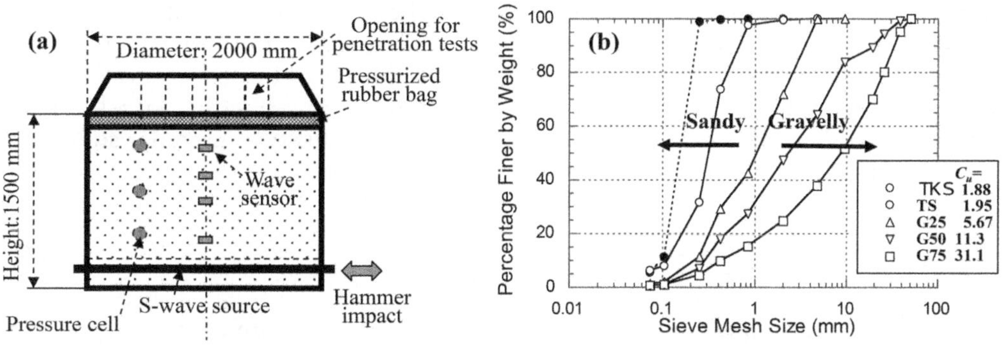


*Figure 2.3.2* Pressure chamber test measuring S-wave velocity on soils with various particle gradations: (a) Test setup, (b) Five tested sandy and gravelly soils with different $C_u$ (Kokusho and Yoshida 1997).

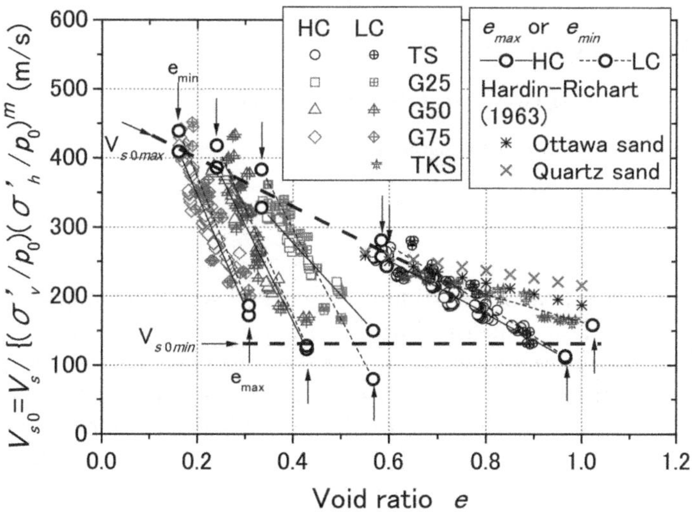


*Figure 2.3.3* Normalized S-wave velocities $V_{s0}$ versus void ratios e for five granular soils with different particle gradations (Kokusho 2007).

(2.3.3)) are also superposed in the figure. Their relationships are located near the sands tested here, but quite different from well graded gravelly soils not only in void ratio but also in the range of S-wave velocity. Thus, it is obvious that the S-wave velocities of granular soils are not determined by a unique function of void ratio but highly variable depending on their particle gradations.

In order to formulate the S-wave velocity of granular soils considering particle gradations, normalized S-wave velocities corresponding to $e_{max}$ and $e_{min}$, denoted as $Vs_{0\,min}$ and $Vs_{0\,max}$, respectively, may be introduced (Kokusho 2007). These values are determined at the cross points of the diagonal straight lines approximating the data points of the five soils with the vertical lines of $e = e_{max}$ or $e_{min}$ (indicated by the arrows) and plotted with the large open circles in Fig. 2.3.3 for the HC and LC tests. It is observed in the figure that $Vs_{0\,min}$, despite some scatters tends to be nearly constant, while $Vs_{0\,max}$ obviously increases with increasing uniformity coefficient $C_u$ from poorly-graded sands to well-graded gravels as indicated by dashed straight lines in the figure. By approximating $Vs_{0\,min} = 136$ m/s and $V_{s0\,max} = 440C_u/(C_u + 1.4)$ as a function of the uniformity coefficient $C_u$, the S-wave velocity of granular soils with various particle gradations may be formulated by the equation.

$$V_s = \left[136 + \left\{\frac{440C_u}{C_u + 1.4 - 136}\right\} D_r\right] \left\{\left(\frac{\sigma'_v}{p_0}\right)\left(\frac{\sigma'_h}{p_0}\right)\right\}^{0.125} \tag{2.3.9}$$

It was shown that Eq. (2.3.9) can evaluate the $V_s$-values measured in the tests with a factor of 1.2 to 1/1.2 despite the wide variety of particle gradations (Kokusho 2007). Accordingly, initial shear moduli of widely varying particle gradations with given uniformity coefficient $C_u$ can be evaluated from Eq. (2.3.9) as $G_0 = \rho V_s^2$, if soil densities $\rho$ are known.

## 2.3.2 $G_0$ for cohesive soil

### 2.3.2.1 *Effects of void ratio and confining stress*

Initial shear modulus of cohesive soils is expressed as:

$$\frac{G_0}{p_0} = A \cdot f(e) \left(\frac{\sigma'_c}{p_0}\right)^m (OCR)^k \tag{2.3.10}$$

similar to sandy soils but with the additional term of over consolidation ratio ($OCR$). Not only the consolidation history but also the consolidation duration has another significant effect on the modulus. These effects of cohesive soils are very variable depending on electrochemical properties of fine clay particles.

For normally-consolidated low to medium plasticity kaolin clay of the void ratio $e = 0.5\sim1.4$ and the plasticity index $I_p = 0\sim52$, it was shown that the modulus can be approximated by the same correlation as proposed for the angular quartz sand in Eq. (2.3.3) (Hardin and Black 1968).

$$\frac{G_0}{p_0} = 330\frac{(2.97 - e)^2}{1 + e}\left(\frac{\sigma'_c}{p_0}\right)^{0.5} \tag{2.3.11}$$

On the other hand for high plasticity bentonite clay of $e > 1.5$, another correlation was obtained by Marcuson and Wahls (1972) as:

$$\frac{G_0}{p_0} = 44.5\frac{(4.4 - e)^2}{1 + e}\left(\frac{\sigma'_c}{p_0}\right)^{0.5} \tag{2.3.12}$$

For intact samples of soft normally-consolidated alluvial clays in Japan with the void ratios up to $e = 3.8$ with various plasticities, initial shear moduli for shear strain of $\gamma = 10^{-5}\sim2 \times 10^{-4}$ (where the strain-dependent modulus degradations were marginal) were measured by the advanced small-strain triaxial tests under different confining

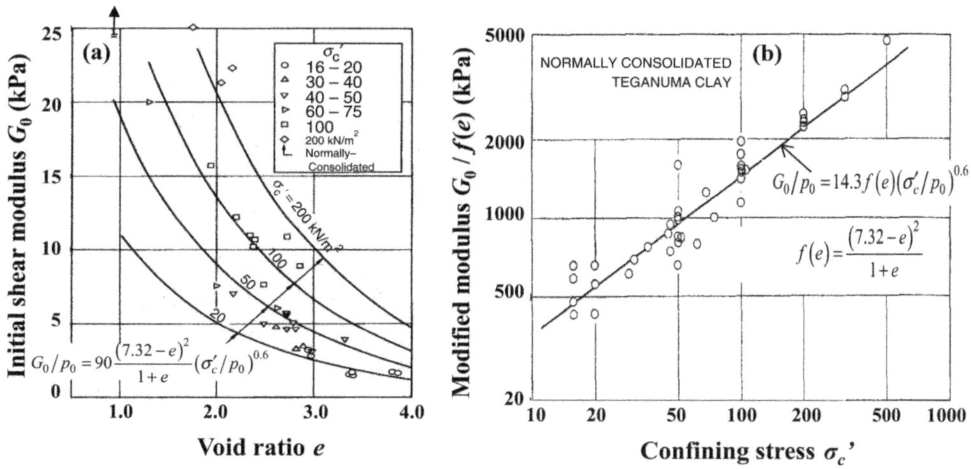


*Figure 2.3.4* Initial shear modulus $G_0$ for intact normally consolidated alluvial clays: (a) $G_0$ versus e (Kokusho 1982), (b) $G_0/f$(e) versus $\sigma_c'$ (Kokusho et al. 1982).

stresses and plotted versus void ratios in Fig. 2.3.4(a) (Kokusho et al. 1982). The soil was sampled intact by thin-wall tube samplers and isotropically consolidated by in situ effective overburden stress. The plots were approximated by the following formula.

$$\frac{G_0}{p_0} = 90\frac{(7.32-e)^2}{1+e}\left(\frac{\sigma_c'}{p_0}\right)^{0.6} \quad (2.3.13)$$

The same initial moduli are divided by the function of void ratio $f(e) = (7.32 - e)^2/(1+e)$ and plotted versus the confining stresses $\sigma_c'$ in Fig. 2.3.4(b). Note that the gradient of the plots is obviously larger than 0.5 and approximated as 0.6, probably due to the compressibility of normally consolidated clays which is not fully reflected by the change of void ratio in the function $f(e)$. The moduli of various clays versus the void ratios by Eqs. (2.3.11)~(2.3.13) are superposed for the confining stresses $\sigma_c' = 20$, 50, 100 kPa in Fig. 2.3.5. It is understood that the initial moduli for clayey soils and the corresponding void ratios are widely varied so that multiple empirical formulas have been proposed depending on types of clays, and hence care is needed to choose one appropriate for particular clay depending on the range of void ratio.

### 2.3.2.2 *Long-term consolidation effect*

Fig. 2.3.6(a) exemplifies test results on the time-dependent variation of the initial shear modulus $G_0$ of the same alluvial clays mentioned above plotted versus the logarithmic time of isotropic consolidation by $\sigma_c' = 50$ or 100 kPa in triaxial tests (Kokusho et al. 1982). The completion of the primary consolidation can be clearly identified at the kink of volume change of the specimen also plotted on the same diagram near 100 minutes, when $G_0$ exhibits a clear kink, too. The dashed line indicates the variation of

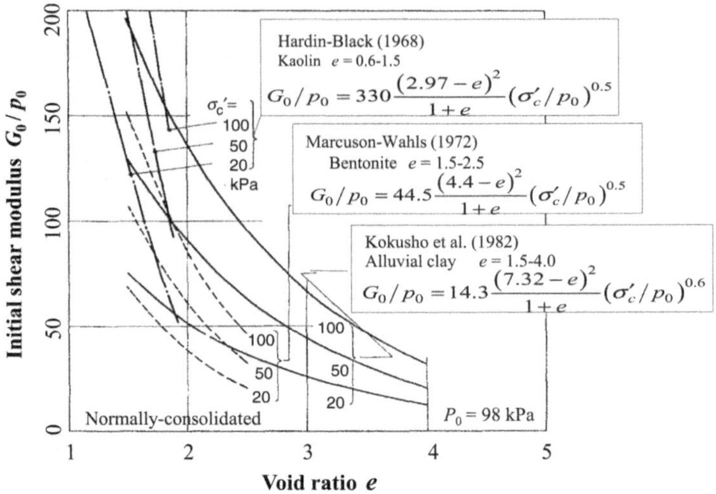


*Figure 2.3.5* Empirical relationships on Initial shear modulus $G_0$ versus void ratio e for various types of clayey soils (Kokusho et al. 1982).

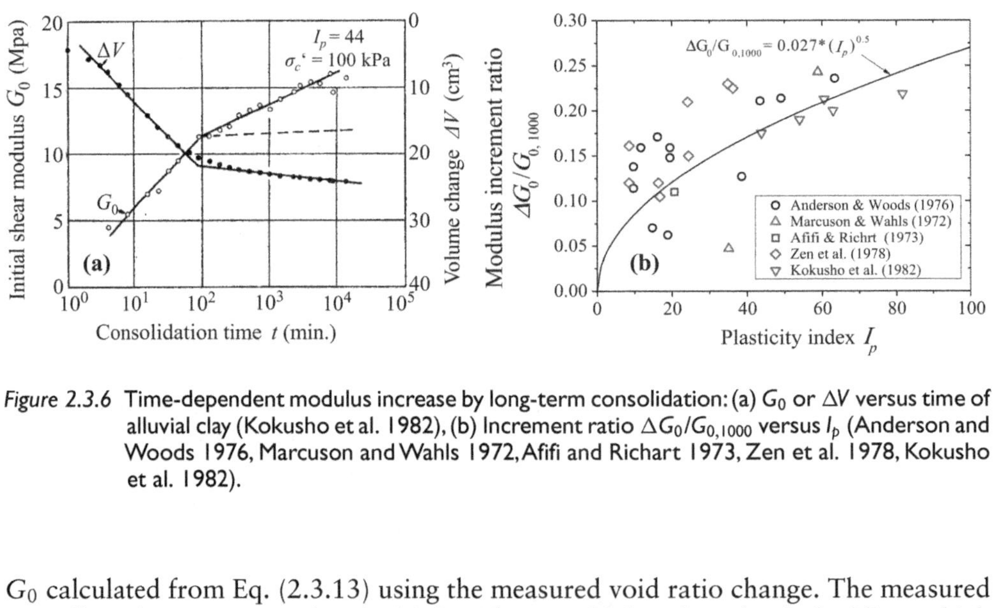


*Figure 2.3.6* Time-dependent modulus increase by long-term consolidation: (a) $G_0$ or $\Delta V$ versus time of alluvial clay (Kokusho et al. 1982), (b) Increment ratio $\Delta G_0/G_{0,1000}$ versus $I_p$ (Anderson and Woods 1976, Marcuson and Wahls 1972, Afifi and Richart 1973, Zen et al. 1978, Kokusho et al. 1982).

$G_0$ calculated from Eq. (2.3.13) using the measured void ratio change. The measured $G_0$ still tends to increase after the kink with a rate higher than the dashed line, which seems to reflect a long-term consolidation effect including the secondary consolidation.

Though this long-term consolidation effect may have some dependency on the mean grain size of soils (Anderson and Woods 1976, Afifi and Richart 1973), it seems to be related more closely to the electrochemical properties of clays represented by the plasticity index $I_p = w_L - w_p$ ($w_L$, $w_p$: water contents in % for liquid limit and plastic limit, respectively). In Fig. 2.3.6(b), values $\Delta G_0$, defined as the modulus increment in ten-times of the time increment (one log-span) are divided by the modulus at $t = 1000$ min. $G_{0,1000}$ and plotted in the vertical axis versus $I_p$ in the horizontal axis for various soils used by different researchers (Anderson and Woods 1976, Marcuson and Wahls 1972, Afifi and Richart 1973, Zen et al. 1978, Kokusho et al. 1982). Despite the large data scatters, a positive correlation may be recognized between the modulus increment ratios $\Delta G_0/G_{0,1000}$ and $I_p$, which may be practically approximated by a simple formula shown with the solid curve in the diagram as:

$$\frac{\Delta G_0}{G_{0,1000}} = 0.027\sqrt{I_P} \tag{2.3.14}$$

### 2.3.2.3 *Effect of overconsolidation*

Fig. 2.3.7(a) shows log-log plots of the initial shear moduli versus the effective confining stresses $\sigma_c'$ for natural intact clays of relatively low void ratios with $I_p = 0$–$52$ sampled in situ (Hardin and Black 1969). In the vertical axis $G_0$ is modified by the function of void ratio $e$, $f(e) = (2.97 - e)^2/(1 + e)$ already introduced in Eq. (2.3.11). The close and open plots are of relatively high and low $I_p$-values, respectively, and those connected by the dashed curves are in overconsolidated conditions. It is observed in the diagram that the slopes of the dashed curves for the soils with higher $I_p$ tend to be gentler due to the overconsolidation effect. Based on the test results in Fig. 2.3.7(a), Hardin and Black (1969) modified the shear moduli in Eq. (2.3.10) for OC clays using $p_y' =$ consolidation yield stress as:

$$\frac{G_0}{p_0} = A \cdot f(e) \left(\frac{\sigma_c'}{p_0}\right)^{0.5} \left(\frac{p_y'}{\sigma_c'}\right)^k \tag{2.3.15}$$

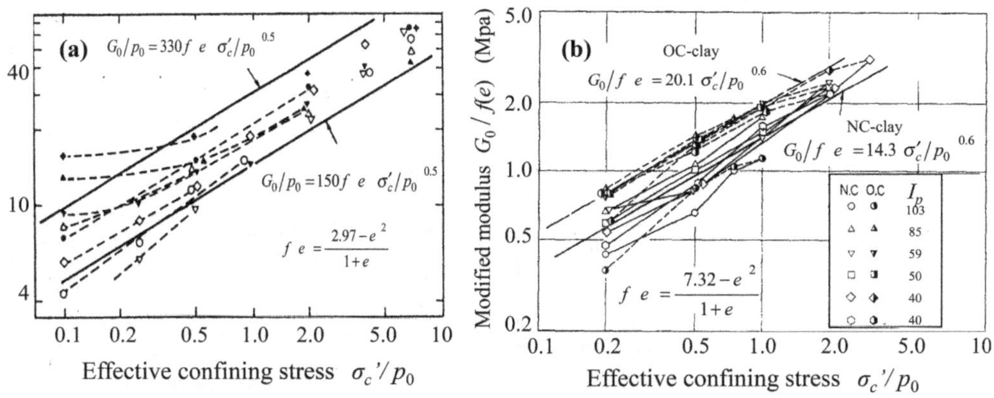


*Figure 2.3.7* Modified shear modulus $G_0/f(e)$ versus confining stress $\sigma_c'$ of NC and OC clays: (a) Natural intact clays of $I_p = 0\sim52$ (Hardin and Black 1969) with permission from ASCE, (b) soft alluvial clays (Kokusho et al. 1982).

and proposed to use $k=0$ and $G_0/p_0 = A \cdot f(e)(\sigma'_c/p_0)^{0.5}$ for $I_p < 40$, while $k=0.5$ and $G_0/p_0 = A \cdot f(e)(p'_y/p_0)^{0.5}$ for $I_p > 40$.

Fig. 2.3.7(b) shows similar test results on the soft alluvial clays in Japan mentioned before with higher void ratios how the initial moduli are affected by overconsolidation. The moduli $G_0$ in the vertical axis is modified by the function of void ratio $e$, $f(e) = (7.32 - e)^2/(1+e)$ as already introduced in Eq. (2.3.13). Here, the specimens were consolidated with stepwise stresses first from $\sigma' = 20$ kPa up to 200–300 kPa, then stepping back to 20 kPa, and the $G_0$-values were measured at the individual steps. The moduli in the down-steps (the OC-conditions) indicated by the dashed curves are evidently larger than those in the up-steps (the NC-condition) by the solid curves, and the two types of curves are essentially parallel. Thus, the trend is considerably different from that shown in Fig. 2.3.7(a), probably because the soils here were overconsolidated artificially without long geological aging effects.

### 2.3.3 Frequency-dependency of damping ratio in the laboratory

Before dealing with damping ratio of soils in view of its strain-dependency in Sec. 2.4, the frequency-dependency of damping ratio investigated in laboratory tests are addressed here. Fig. 2.3.8(a) shows the damping ratios of a clean sand for various frequencies for double amplitude shear strain $\gamma_{DA} = 0.2\text{–}0.3 \times 10^{-4}$ measured by resonant column tests or low-frequency cyclic loading tests plotted versus the frequency of dynamic loading (Hardin 1965). The vertical axis is not the absolute value of damping ratio but is divided by the average of all damping ratios measured here. Despite huge difference in the frequency of smaller than 1 Hz to several hundred Hz, the difference in the measured damping values are within ±20%. In Fig. 2.3.8(b), damping ratios of a silty soil measured in small to medium strain ranges by two test methods with different frequencies (Hardin and Drnevich 1972a) are plotted versus shear strain amplitudes.

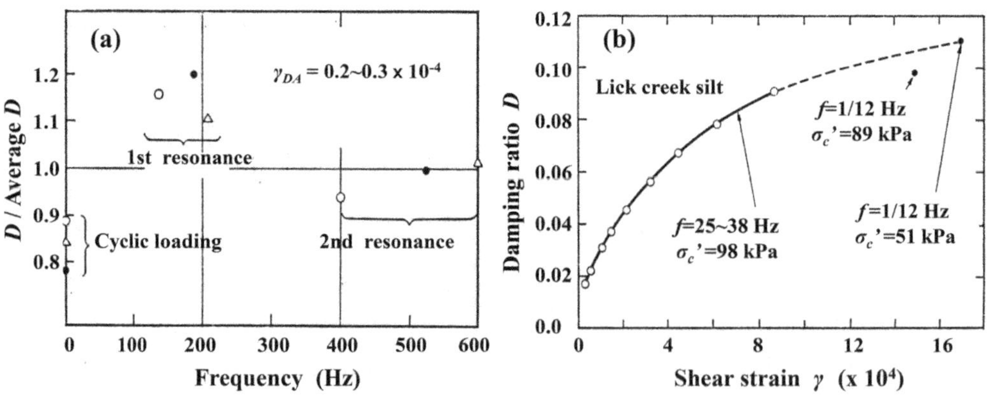


*Figure 2.3.8* Effect of frequency on soil damping in laboratory tests: (a) Variation of damping ratio of sand depending on frequency (Hardin 1965), (b) Damping ratio versus shear strain measured for silty soil in various frequencies (Hardin and Drnevich 1972a), with permission from ASCE.

It shows that the $D$-values measured in the low frequency cyclic loading test are essentially consistent with the $D$ versus $\gamma$ plots obtained in frequencies more than 100 times higher in the resonant column test.

Thus, as far as laboratory soil tests are concerned, it is generally accepted that soil damping is essentially independent of frequency or loading rate. Based on this experimental finding, a non-viscous damping model (such as Nonviscous Kelvin model addressed in Sec. 1.5) is generally employed in modelling dynamic soil properties among viscoelastic models. This is probably because soil materials are essentially frictional rather than viscous, justifying the use of non-viscous damping in large-strain problems during destructive strong earthquakes. On the other hand, actual seismic soil response records suggest that in addition to the frequency-independent damping a frequency-dependent damping mechanism is also involved as will be discussed in Sec. 4.3.3.

## 2.4 STRAIN-DEPENDENT EQUIVALENT LINEAR PROPERTIES

As shear strain increases from small strains of $10^{-6}\sim10^{-5}$ to medium strains of $10^{-5}\sim10^{-3}$, shear modulus tends to degrade and damping ratio tends to increase. The transition of soil properties in these strain ranges is equivalently linearized as already explained in Sec. 2.1.3. Namely, from hysteretic stress-strain curves in Fig. 2.1.6, secant shear modulus $G$ is defined from the gradient of a straight line connecting the two ends of the hysteretic loop and normalized by the initial shear modulus for small strain $G_0$ as $G/G_0$. Damping ratio is defined as $D=\Delta W/4\pi W$ where $\Delta W=$ dissipated energy per one cycle of loading and $W=$ maximum elastic strain energy. In the following, the variations of shear modulus ratio $G/G_0$ and damping ratio $D$ depending on the shear strain amplitude (single amplitude) obtained from cyclic loading triaxial tests on various types of soils are exemplified to overview their soil-specific characteristics.

### 2.4.1 Modulus degradation

#### *2.4.1.1 Sand and gravel*

In Fig. 2.4.1 the shear moduli $G$ of Toyoura clean sand measured under stepwise varying effective confining stresses $\sigma_c'$ are plotted versus shear strain single amplitudes $\gamma$ in the semi-logarithmic diagram (Kokusho 1980). As the initial shear moduli $G_0$ corresponding to the strain level of $\gamma=10^{-6}$ tend to increase in proportion approximately to the square root of $\sigma_c'$, $G$ for a given strain tends to increase with increasing $\sigma_c'$, too. Fig. 2.4.2(a) shows the modulus degradation curves, that is the normalized modulus $G/G_0$ versus $\gamma$ relationships. The curves evidently change their relative positions corresponding to the confining stresses $\sigma_c'$; shifting leftward and showing stronger degradations with decreasing $\sigma_c'$. This trend is commonly observed for non-cohesive granular soils such as sands and gravels.

The strain for $G/G_0=0.5$ corresponds to the reference strain $\gamma_r$ in the hyperbolic stress-strain model as will be addressed in Sec. 3.1.1, and $\gamma_r$ is defined by:

$$\gamma_r=\frac{\tau_f}{G_0} \tag{2.4.1}$$

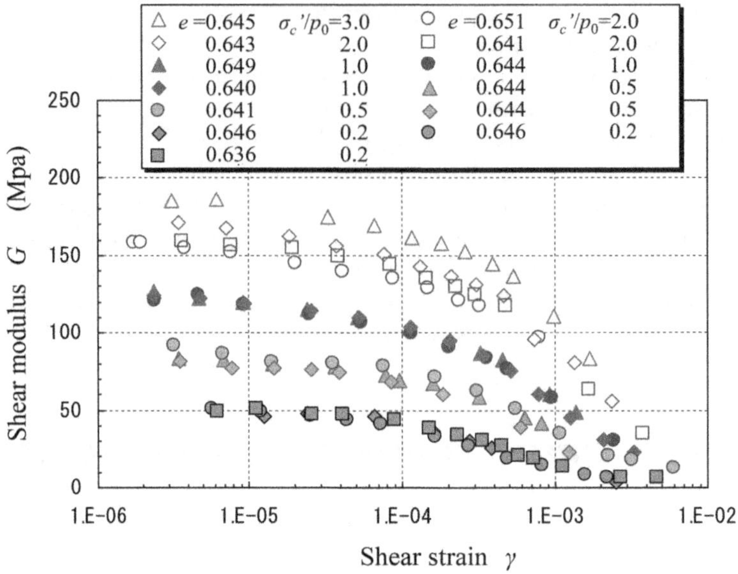


*Figure 2.4.1* Shear modulus $G$ versus shear strain single amplitude $\gamma$ of Toyoura sand (Kokusho 1980).

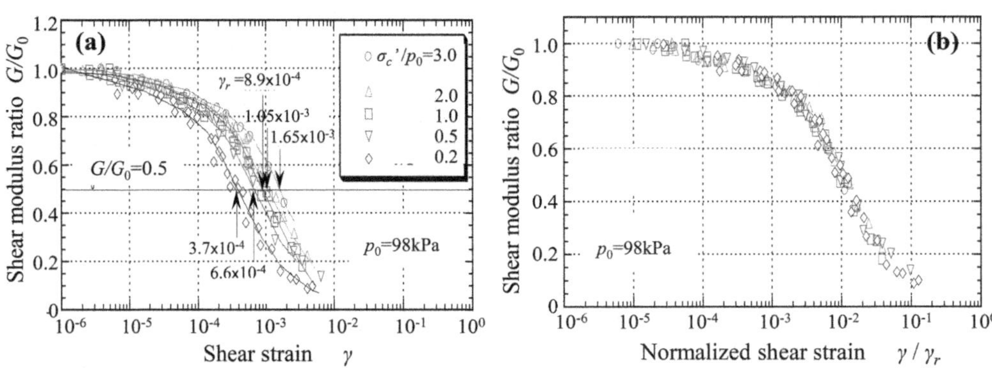


*Figure 2.4.2* Normalized expressions of shear modulus degradations versus shear strain $\gamma$ (Kokusho 1980): (a) $G/G_0 \sim \gamma$, (b) $G/G_0 \sim \gamma/\gamma_r$.

where $\tau_f$ = shear strength of the hyperbolic model. In Fig. 2.4.2(b), the horizontal axis of Fig. 2.4.2(a) is normalized as $\gamma/\gamma_r$, where $\gamma_r$-values are read off at $G/G_0 = 0.5$ from the individual curves. Obviously, all the curves are almost unified for the wide range of confining stress $\sigma_0'/p_0 = 0.2{\sim}3.0$ so that the modulus ratio can be approximated in a simple formula as:

$$\frac{G}{G_0} = f\left(\frac{\gamma}{\gamma_r}\right) \tag{2.4.2}$$

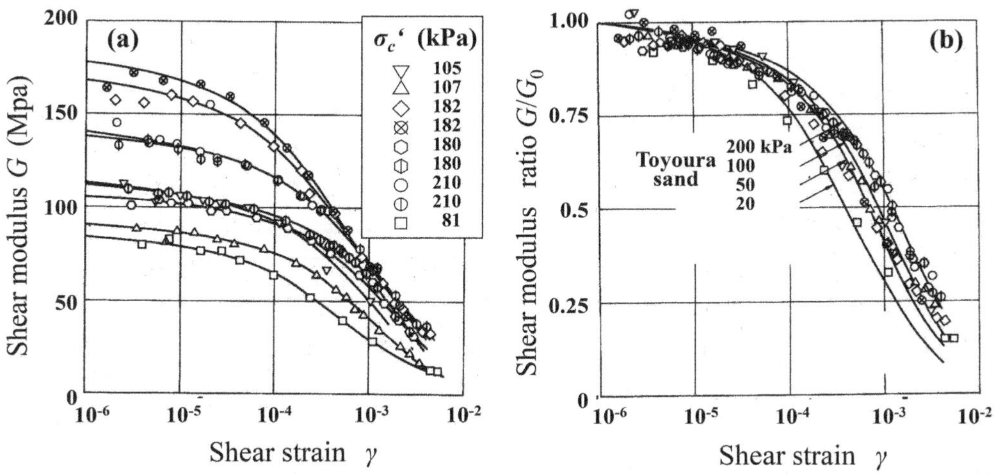


*Figure 2.4.3* Shear modulus degradations of intact Pleistocene stiff sands (Kokusho 1982): (a) $G{\sim}\gamma$, (b) $G/G_0{\sim}\gamma$ compared with Toyoura sand.

This indicates that the shape of the $G/G_0$ versus $\gamma$ curve defined by a function $f(\ )$ is almost unchanged, whereas its relative position tends to shift with confining stress (Kokusho 1980).

In Fig. 2.4.3(a), similar results of cyclic loading tests on intact sands sampled by tube sapling from stiff Pleistocene dense sand deposits (SPT $N$-values over 50) are depicted. The specimens were consolidated with the isotropic confining stresses of $\sigma_c' = 81{\sim}210$ kPa calculated from $\sigma_c' = (1+2K_0)\sigma_v'/3$ using in situ vertical stresses $\sigma_v'$ for individual samples and $K_0 = 0.67$. Shear modulus degradations can be clearly measured for the wide strain range of $10^{-6}{\sim}10^{-2}$. The shear moduli in the same dataset are normalized by $G_0$ at $\gamma = 1\times 10^{-6}$ and replotted on the $G/G_0{\sim}\gamma$ diagram in Fig. 2.4.3(b). Again, the trend of the $G/G_0{\sim}\gamma$ curves shifting rightward and showing less degradations with increasing $\sigma_c'$ can evidently be observed and is mostly compatible with that of the reconstituted Toyoura sand as indicated with a set of solid curves.

Fig. 2.4.4 shows modulus degradation curves for intact fluvial gravelly soils sampled in situ by in situ freezing technique. Because of large gravel sizes, large size specimens of 30 cm in diameter and 60 cm in height were tested in a large scale cyclic loading triaxial apparatus (Kokusho and Tanaka 1994). The $G/G_0{\sim}\gamma$ curves under four stepwise confining stresses $\sigma_c'$ tend to shift rightward with increasing $\sigma_c'$ in a similar manner as the sandy soils.

Fig. 2.4.5 depicts the reference strains $\gamma_r$ read off from $G/G_0{\sim}\gamma$ curves of sands and gravels plotted in the vertical axis versus the effective confining stresses $\sigma_c'$ in the horizontal axis in the log-log scale. It is obviously seen that for all the granular soils including here the decomposed granite and Pleistocene gravels, intact or reconstituted, the reference strain can be approximated in the following form.

$$\gamma_r \propto (\sigma_c')^{0.5} \tag{2.4.3}$$

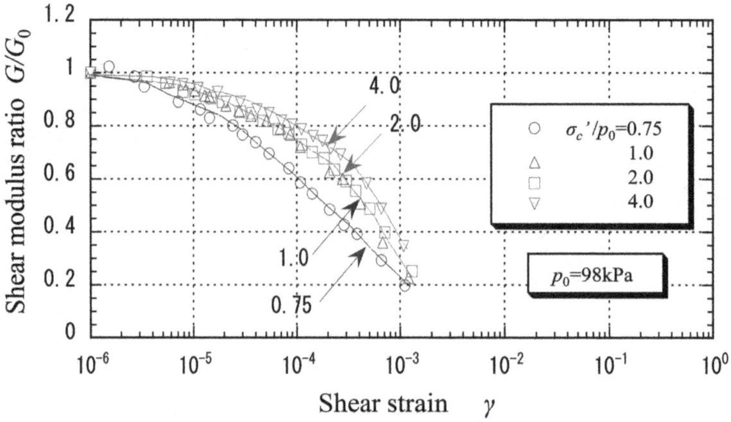


*Figure 2.4.4* Shear modulus degradation $G/G_0$ versus shear strain $\gamma$ for intact gravel samples (Kokusho and Tanaka 1994).

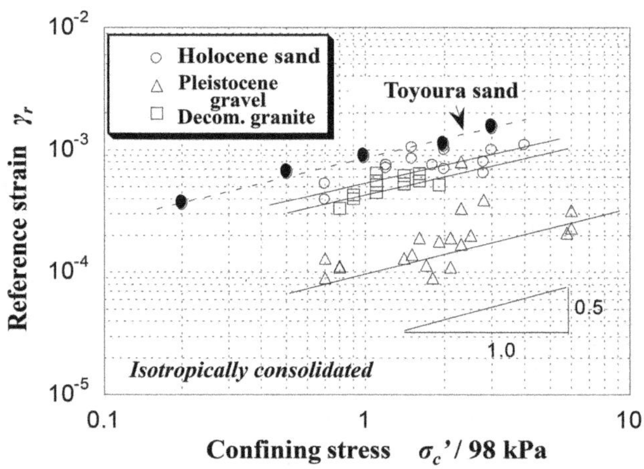


*Figure 2.4.5* Reference strains $\gamma_r$ versus effective confining stresses $\sigma_c'$ of sandy and gravelly soils.

This trend seems to be compatible with the hyperbolic model as mentioned in Sec. 3.1.1.

### 2.4.1.2 *Cohesive soil*

Fig. 2.4.6 exemplifies the typical modulus degradation curves of soft intact clays sampled by a thin wall tube sampler and tested in the improved cyclic triaxial apparatus (Kokusho et al. 1982). Although the confining stresses are widely varied stepwise as $\sigma_c'/p_0 = 0.45$–$5.0$, the $G/G_0$–$\gamma$ plots for the clays of similar plasticity indices $I_p = 38$–$56$ are almost overlapping, showing almost no effect of confining stress. Thus, a remarkable difference exists in $\sigma_c'$-dependency of $G/G_0$–$\gamma$ curves between cohesive

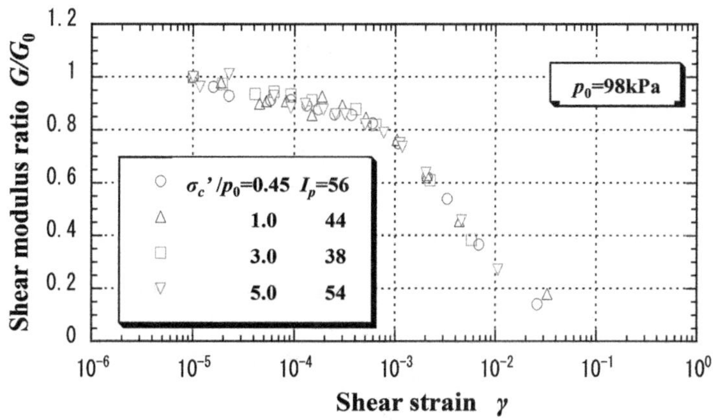


*Figure 2.4.6* Shear modulus degradation $G/G_0$ versus shear strain $\gamma$ for soft alluvial clays under different confining stresses (Kokusho et al. 1982).

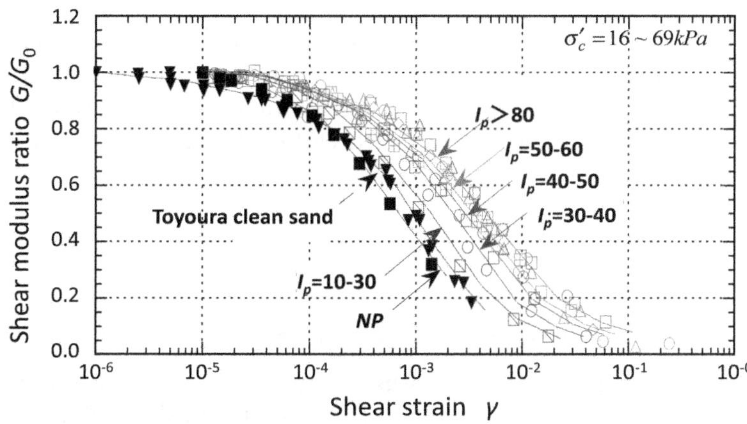


*Figure 2.4.7* Shear modulus degradation $G/G_0$ versus shear strain $\gamma$ for soft alluvial clays of different plasticity indexes (Kokusho et al. 1982).

and non-cohesive soils. Also note that the overconsolidation and long-term consolidation histories which tend to give significant influence on the initial shear modulus does not seem to have measurable effect on $G/G_0$–$\gamma$ curves (Kokusho et al. 1982).

In Fig. 2.4.7, modulus degradation curves of cohesive soils with widely varying plasticity index $I_p = 0$–over 80 obtained from the same alluvial clays are shown. The confining stresses are $\sigma_c' = 16$–$69$ kPa corresponding to in situ overburden stresses, which give only a marginal effect on the degradations as mentioned above. The test results however show a significant effect of the plasticity index $I_p$ on the degradation curves. The curves tend to shift rightward with increasing $I_p$, particularly between $I_p = 0$ (NP) to around 40, and the NP curve almost matches with that of Toyoura clean sand as indicated in the diagram. Thus, the plasticity index is the key parameter

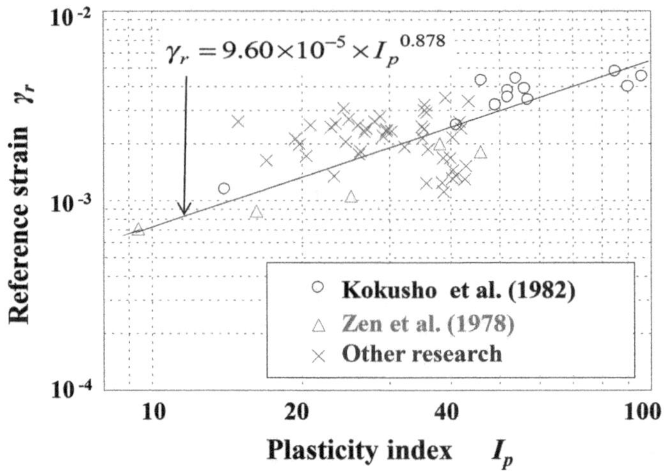


*Figure 2.4.8* Reference strains $\gamma_r$ versus plasticity index $I_p$ of clays.

in cohesive soils to decide the modulus degradation in place of the confining stress in non-cohesive soils. In Fig. 2.4.8, the reference strains $\gamma_r$ read off from the $G/G_0$–$\gamma$ curves of cohesive soils including those from different sites are plotted versus the corresponding $I_p$-values on the log-log chart. Despite the large data scatters, $\gamma_r$ tends to increase with increasing $I_p$ and may be approximated as:

$$\gamma_r = 9.60 \times 10^{-5} \times I_p^{0.878} \tag{2.4.4}$$

### 2.4.1.3 *Overview of cohesive/non-cohesive soil*

Fig. 2.4.9 shows an overview of modulus degradation curves obtained from laboratory cyclic loading tests to compare different types of soils (Kokusho 1982). Here, the confining stresses for non-cohesive soils are 98 kPa and the plasticity indices for cohesive soils are chosen $I_p > 40$. It is observed that for the same induced strain the modulus degradation tends to be most manifested in gravels, then sands and least in clays. The reference strain $\gamma_r$ corresponding to $G/G_0 = 0.5$ for clays is about 10 times larger than that for gravels.

There seems to be a certain correlation between the reference strains and soil densities valid for all types of soils universally, because Fig. 2.4.9 indicates that cohesive soils of lower densities tend to have larger $\gamma_r$ than higher density gravels, and also clays with higher $I_p$ (this normally means high void ratio and low density) tend to have larger $\gamma_r$ in Fig. 2.4.8. In Fig. 2.4.10, the reference strains $\gamma_r$ read off from a number of degradation curves for various types of soils are plotted versus either dry soil densities on the semi-log chart in (a) or void ratio on the log-log chart in (b). In the vertical axes in these charts, $\gamma_r^*$ is taken in place of $\gamma_r$ as:

$$\gamma_r^* = \frac{\gamma_r}{(\sigma_c'/p_0)^{0.5}} \tag{2.4.5}$$

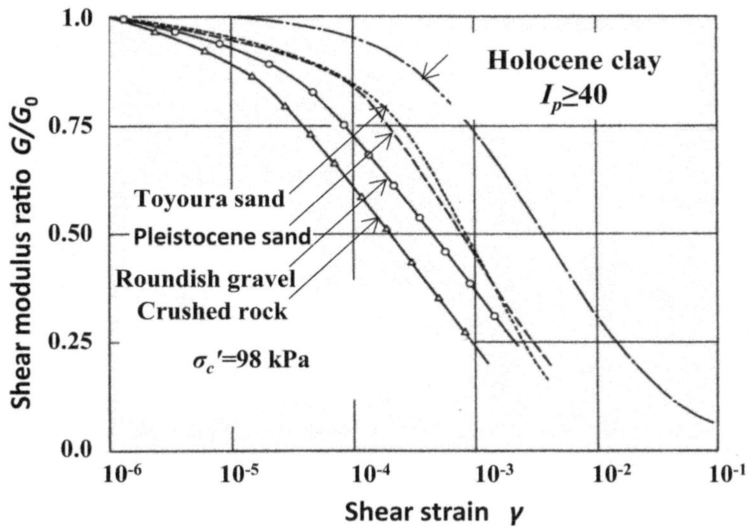


*Figure 2.4.9* Modulus degradation curves to compare different types of soils (Kokusho 1982).

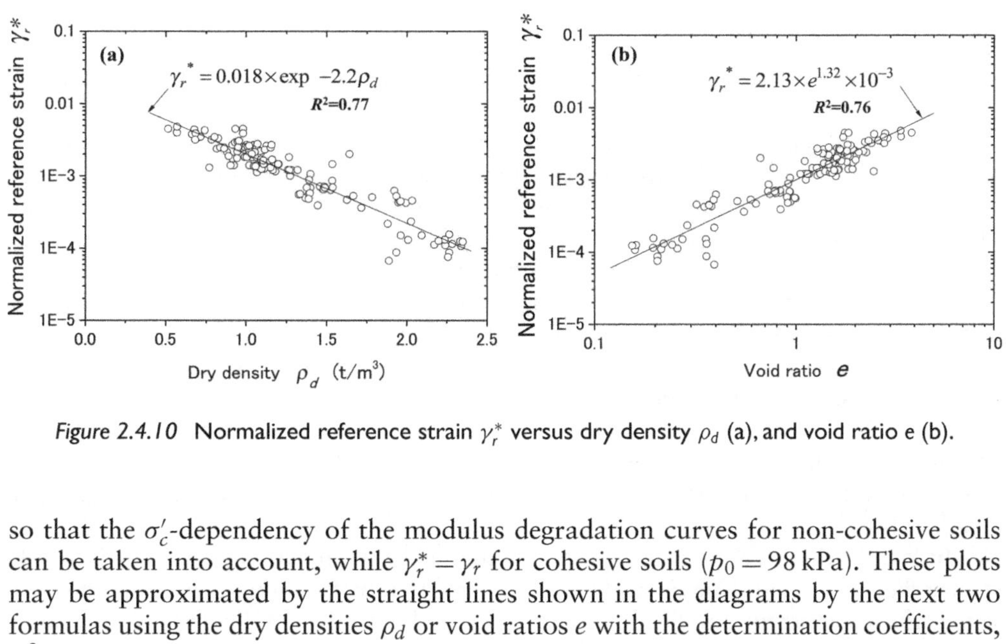


*Figure 2.4.10* Normalized reference strain $\gamma_r^*$ versus dry density $\rho_d$ (a), and void ratio e (b).

so that the $\sigma_c'$-dependency of the modulus degradation curves for non-cohesive soils can be taken into account, while $\gamma_r^* = \gamma_r$ for cohesive soils ($p_0 = 98$ kPa). These plots may be approximated by the straight lines shown in the diagrams by the next two formulas using the dry densities $\rho_d$ or void ratios $e$ with the determination coefficients, $R^2 = 0.76{\sim}0.77$.

$$\gamma_r^* = 0.018 \times \exp(-2.2\rho_d), \quad \gamma_r^* = e^{1.32} \times 10^{-3} \tag{2.4.6}$$

These empirical relationships may be conveniently used in estimating the reference strains or modulus degradations for various soil types, regardless the soils are cohesive or non-cohesive.

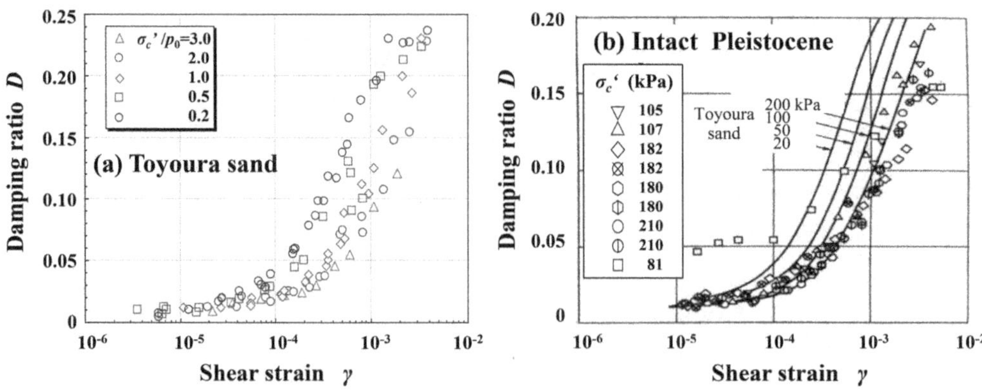


*Figure 2.4.11* Damping ratio $D$ versus shear strain $\gamma$ for sandy soils under different confining stresses: (a) Reconstituted Toyoura sand (Kokusho 1980), (b) Intact stiff Pleistocene sands (Kokusho 1982).

## 2.4.2 Damping ratio

### *2.4.2.1 Sand and gravel*

In Fig. 2.4.11(a), hysteretic damping ratios $D$ measured on Toyoura clean sand measured in low-frequency cyclic triaxial tests under confining pressures of $\sigma_c'/p_0 = 0.2{\sim}3$ stepwise varied ($p_0 = 98$ kPa), are plotted versus single strain amplitude $\gamma$ on the semi-log chart. The $D$-value evidently increases with increasing $\gamma$, and the $D{\sim}\gamma$ plots tend to shift rightward with increasing $\sigma_c'$ in a similar manner as the $G/G_0{\sim}\gamma$ plots in Fig. 2.4.2. Also noted is that as the strain approaches zero, the damping ratios are not inclined to converge to zero but to non-zero small values; $D \approx 1\%$ in the case of Toyoura sand used here. The small-strain $D$-value is important in evaluating the seismic response of soil deposits during small earthquakes or even during strong earthquakes in great soil depths. A potential reason for the non-zero internal damping at an infinitely small strain level may be attributed to pore-fluids, gas or liquid, in soil materials. This is suggested by an observation that moonquakes, which occur in the vacuum environment presumably without any fluid, are known to keep vibrating for a few hours much longer than earthquakes probably because of very low damping in small strain.

In Fig. 2.4.11(b), the damping ratios $D$ measured on intact soils sampled in block from a Pleistocene sandy layer and tested under different confining stresses are plotted versus the shear strain amplitude $\gamma$, and compared with the solid curves representing the test results for Toyoura sand. Though the intact soils are not so systematically plotted as the reconstituted Toyoura sand, the $D{\sim}\gamma$ plots are essentially compatible and still reflecting the effect of $\sigma_c'$ in a similar manner as the reconstituted sand, except for one of the test specimens of extraordinarily high damping ratios in small strains presumably reflecting the heterogeneity of in situ soils.

Fig. 2.4.12 shows the similar plots of intact gravelly soils sampled by in situ freezing technique and tested under four steps of confining stresses. The $D{\sim}\gamma$ plots are

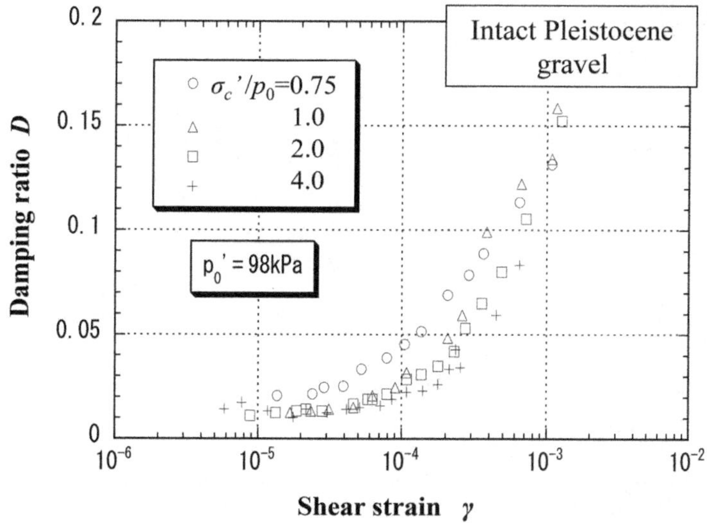


*Figure 2.4.12* Damping ratio $D$ versus shear strain $\gamma$ for intact Pleistocene gravels (Kokusho and Tanaka 1994).

essentially compatible with the sandy soils and the effect of confining stresses is also visible. Also note that the small-strain $D$-values of intact gravels tend to converge to nearly 1% as Toyoura sand.

#### 2.4.2.2 Cohesive soil

Fig. 2.4.13(a) shows the $D\sim\gamma$ plots of soft Alluvial intact clays of $I_p = 38\sim56$ (Kokusho et al. 1982). Despite the widely varying confining stresses $\sigma_c' = 75\sim400$ kPa in which the test was conducted, very little effect of $\sigma_c'$ is actually observed in accordance with the $G/G_0\sim\gamma$ plots already mentioned. In Fig. 2.4.13(b), the $D\sim\gamma$ plots are depicted for the clays sampled intact from the same alluvial deposits with variable plasticity indices $I_p = 0$ (NP) to over 90. Despite some data dispersions presumably due to heterogeneous in situ soils, the rightward shifting trend of the plots with increasing $I_p$ can be recognized, being compatible with the shift of $G/G_0\sim\gamma$ curves in Fig. 2.4.7. It can also be observed that the small-strain damping ratios of clayey soils tend to converge to around $D \approx 3\sim5\%$ slightly larger than those of sands or gravels. In this regard, Kokusho et al. (1982) also showed that the small-strain $D$-values tend to decrease linearly with time in log-scale in long-term consolidation tests, indicating that in situ damping ratios may be smaller than the laboratory values actually.

### 2.4.3 Strain-dependent property variations compared with in situ

It is of a great concern that if strain-dependent variations of shear moduli and damping ratios thus measured and accumulated as a database in the form of $G/G_0$–$\gamma$ curves and $D$–$\gamma$ curves are consistent with actual dynamic behavior of in situ soils during

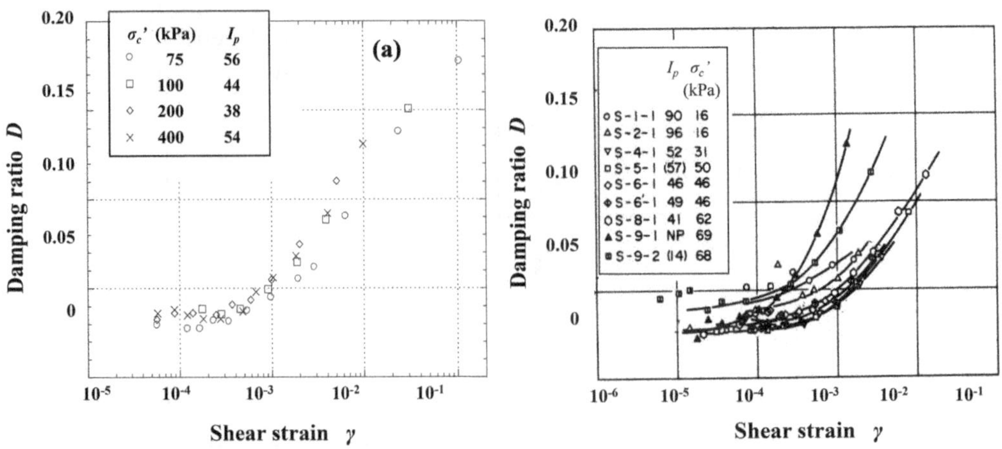


*Figure 2.4.13* Damping ratio $D$ versus shear strain $\gamma$ for intact clayey soils (Kokusho 1982): (a) $D$–$\gamma$ under different confining stresses $\sigma_c'$, (b) $D$–$\gamma$ for clayey soils with various plasticity index $I_p$.

earthquakes. Dynamic soil properties in medium to large strain ranges are not easy to directly measure in situ, but can be estimated from the dynamic response of surface soils if strong motion seismic records are available. A typical example is already shown in Fig. 2.1.8 where earthquake records obtained at a surface of loose sand layer underlain by a stiff base were utilized to evaluate the variations of the predominant frequency of the surface layer with increasing induced shear strain. More sophisticated back-calculations are also possible to evaluate strain-dependent equivalent linear soil properties if a set of vertical array records during strong earthquakes are available together with basic site characterization data. In the following, a systematic back-calculation study utilizing vertical array records at four sites during the 1995 Kobe earthquake in Japan is addressed to see how the back-calculated shear modulus ratios and damping ratios are compared with laboratory data (Kokusho et al. 2005a).

Based on the multi-refection theory of SH-wave as mentioned in Sec. 3.2.3, S-wave velocities $V_s$ or shear moduli $G = \rho V_s^2$ and damping ratios of individual layers at a vertical array site are optimized so that the residuals between observed and calculated spectrum ratios between the ground surface and the stiff base layer are minimized based on the soil models in vertical array sites wherein different types of soils involved are categorized. On the other hand, dynamic response analyses are implemented for each earthquake by incorporating the back-calculated soil properties to obtain induced maximum strains $\gamma_{max}$ and effective strains $\gamma_{eff} = 0.65\gamma_{max}$ in individual layers in accordance with the approximation coefficient 0.65 as will be discussed in Sec. 3.2.4. The strain-dependent variations of soil properties thus back-calculated are compared with laboratory test curves for different soil types in the following.

### 2.4.3.1 Modulus degradations

Fig. 2.4.14 shows the shear modulus ratios $G/G_0$ back-calculated at four vertical array sites for the main shock and multiple aftershock records of the 1995 Kobe earthquake

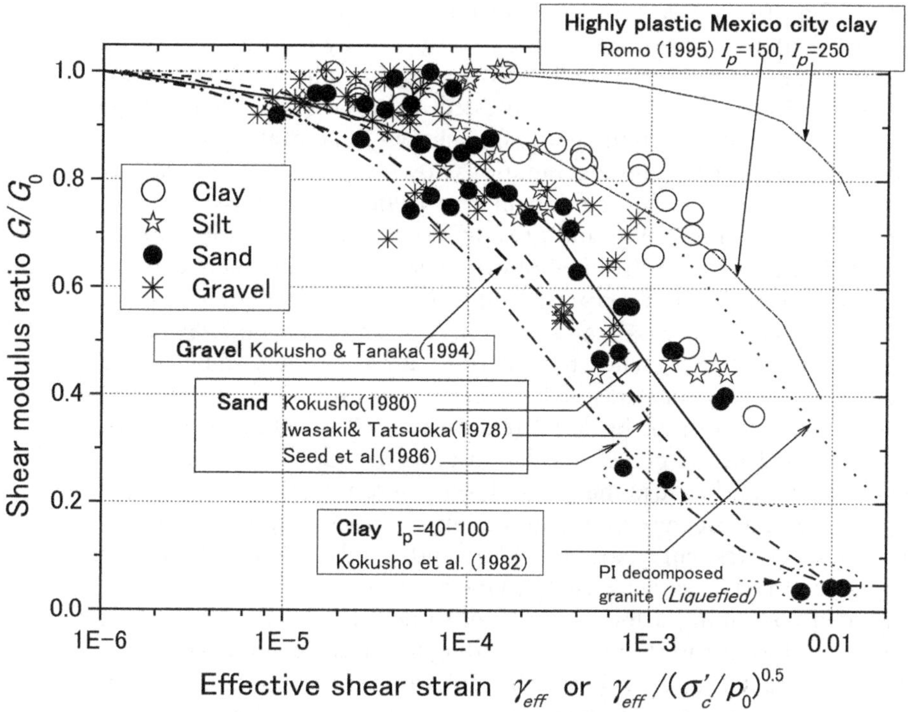


*Figure 2.4.14* Back-calculated modulus ratio $G/G_0$ versus effective shear strain $\gamma_{\text{eff}}$ for various soils compared with those by laboratory tests of corresponding soil types (Kokusho et al. 2005a), with permission from ASCE.

plotted versus the effective shear strains $\gamma_{\text{eff}}$ (Kokusho et al. 2005a). The plots are categorized into four different soil types with different symbols; clays, silts, sands and gravelly soils. For sands and gravelly soils, the effective shear strains $\gamma_{\text{eff}}$ in individual layers are normalized as $\gamma_{\text{eff}}/(\sigma_c'/p_0)^{0.5}$ to consider the effect of in situ confining stresses. For clays insensitive to $\sigma_c'$, in situ soils of interest near Kobe areas have the plasticity index $I_p$ over 40 where the modulus degradation curves will not shift so much compared to those for $I_p$ under 40 as indicated in Fig. 2.4.7.

A brief overview of the back-calculated $G/G_0{\sim}\gamma$ plots in Fig. 2.4.14 reveals that the plots tend to be positioned in the similar manner as in the laboratory tests in accordance with the soil types. Namely clayey soils are most rightward, then silts are slightly left. Plots for sandy and gravelly soils are located still left averagely, though they are considerably scattered. On the same chart, the curves from laboratory tests are superposed for clayey, sandy and gravelly soils, among which the confining stresses for sandy and gravelly soils are chosen as $\sigma_c' = p_0 = 98$ kPa.

With regard to clayey soils, the back-calculated plots show a fair coincidence with the dashed curve of the laboratory tests for $I_p = 40{\sim}100$. Besides the clayey soils in Japan above, a pair of chain-dotted curves are drawn for highly plastic Mexico-city clay of $I_P = 150{\sim}250$ which incurred severe vibration damage during the 1985

Mexican earthquake (Romo 1995). These laboratory data for Mexican clay investigated by laboratory cyclic loading tests were reported to have a good coincidence with the back-calculated properties based on vertical array records during the earthquake (Taobada et al. 1999). Thus, the back-calculated results also demonstrate that modulus degradation tend to be less manifested up to large strains for clayey soils with higher plasticity indexes up to $I_P = 250$ as shown in laboratory tests.

As for sands, the modulus degradations down to $G/G_0 = 0.04$ are back-calculated. Most of the low $G/G_0$-values are from Port Island where decomposed granite (well-graded sand containing fines and gravels) used as the back-fill material extensively liquefied. The sand curves from laboratory tests are not uniquely determined because there still seem to be some gaps among representative research results indicated by three different curves (Iwasaki and Tatsuoka 1978a, Kokusho 1980, Seed et al. 1986). The curve by Seed et al. tends to underestimate the $G/G_0$-values for $G/G_0 > 0.50$, while it seems consistent with the back-calculation for $G/G_0 < 0.20$, while the other curves match well with the back-calculation for $G/G_0 > 0.50$ in particular.

With regard to the gravelly soils, all plots are $G/G_0 > 0.50$, showing relatively minor degradation because the gravel layers were mostly from Pleistocene and stiff. The plots are very much scattered and many of them are positioned on the right side of the laboratory test curve, and mixed up with sands and silts. This seems to indicate that, unlike gravel specimens used in laboratory tests, natural gravelly soils are very well-graded containing a lot of sands and fines. If the soil particles finer than the gravels exceed some threshold corresponding to the critical fines content, around 20% as will be discussed in Sec. 5.4.1, the soil structure tends to change from "coarse-grain supporting" to "matrix supporting" because the overflowing finer soils form the matrix of the gravel particles.

Thus, the $G/G_0 \sim \gamma$ curves back-calculated in situ during strong earthquakes are found to be essentially compatible with those in laboratory tests so that the relative positions of the individual soil types are basically the same. Among the soils, clays show a good coincidence, while gravelly soils tend to exhibit a variety of degradation than in laboratory tests, probably because actual gravelly soils contain a lot of finer particles unlike gravels in laboratory tests.

#### 2.4.3.2 *Damping ratios*

In Fig. 2.4.15, the damping ratios $D$ optimized in the same back-calculations as the shear moduli are plotted versus the effective shear strain $\gamma_{\mathrm{eff}}$ for the four types of soils. Note that the damping ratio is assumed here as non-viscous and frequency-independent. In order to compare with them, the curves of laboratory test results for different soil types are superposed.

As a whole, the back-calculated $D$-values for all soil types tend to increase with increasing strain $\gamma_{\mathrm{eff}}$ mainly for $\gamma_{\mathrm{eff}} > 10^{-4}$–$10^{-3}$, demonstrating a qualitative compatibility of strain-dependent variation with the laboratory tests. Quantitatively however, the back-calculated $D$-values are widely dispersed in a small strain range $\gamma_{\mathrm{eff}} < 10^{-5}$–$10^{-4}$ even for the same soil type. Among them, a majority of plots are concentrating around $D \approx 5\%$, evidently larger than the laboratory test results irrespective of the soil types, though some minor data are around $D \approx 1\%$. The similar trends having larger $D$-values in situ than in the laboratory are reported in other literatures, too (e.g. in

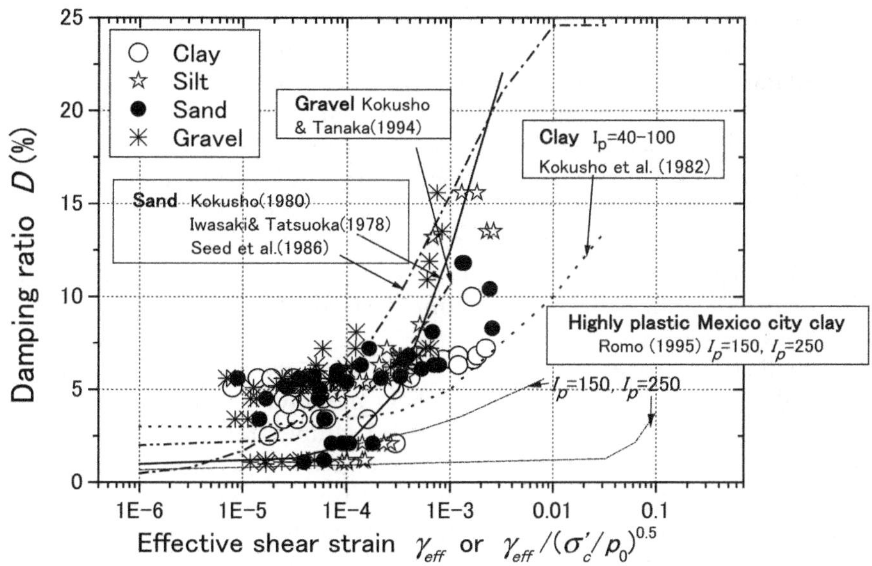


*Figure 2.4.15* Back-calculated damping ratios D versus effective shear strain $\gamma_{eff}$ for various soils compared with those by laboratory tests of corresponding soil types (Kokusho et al. 2005a), with permission from ASCE.

Fig. 2.1.1 by Kudo 1976). One of the possible explanations for this discrepancy may be frequency-dependency of the damping ratio observed in earthquake response records to be discussed in Sec. 4.3.3, not considered here in back-calculating the *D*-values.

As for the soil types, clays with $I_p > 40$ exhibit smaller damping ratios than other soils even for larger strains. This seems to be compatible with the laboratory test curves qualitatively at least. For the pair of curves of highly plastic Mexico city clay of $I_p = 150$ and 250 drawn in the same chart, this trend is further extended to the lower *D*-value (Romo 1995). This may explain why the ultra soft clay ground in Mexico city amplified the seismic motion so much during the 1985 earthquake (Tabaoda et al. 1999).

Thus, the in situ damping ratios back-calculated from strong earthquake records seem to be qualitatively compatible with the laboratory test results. However, the *D*-values applicable in situ for a small strain range in particular may possibly be larger (back-calculated as around 5%) than laboratory test results. More study is needed to have a complete picture of in situ damping mechanism reflecting the frequency-dependent damping.

## 2.5 SUMMARY

1 Soil properties during earthquakes are highly dependent on seismically induced strain. Large-strain properties during destructive earthquakes are considerably different from those for small-strain properties during weak earthquakes, though the small-strain properties are essential to determine the large-strain properties.

2 Soil properties during earthquakes are also characterized by higher loading rate and the number of loading cycles, though the effect of loading rate is not so much significant as to differentiate dynamic and static failures.
3 Soil dilatancy tends to be dominant with increasing strain toward failures. It causes pore-pressure changes in undrained saturated soils, giving tremendous effects on soil properties, induced strains and the resistance to failures, wherein the number of loading cycles during earthquakes plays an important role.
4 Failures of soils during earthquakes may be defined by threshold strains corresponding to performances of particular structures induced by earthquake cyclic loading. However, failures may occur suddenly in brittle modes in some circumstances under the effect of initial shear stress, which requires special considerations if necessary.
5 The small-strain properties obtained by in situ P, S-wave tests are most fundamental for design. Soil properties, secant shear modulus $G$, in small to large strains are obtained in laboratory soil tests on soil samples taken from in situ, and the continuous variations of the modulus ratio $G/G_0$ with induced strains may be used in practice by equating $G_0$ in the laboratory test with $G_0 = \rho V_s$ determined from in situ $V_s$. The small-strain damping ratio $D$ tends to converge to a non-zero small value with decreasing induced strain, suggesting a possible involvement of viscous damping by fluid.
6 Damping ratios measured in laboratory tests are almost frequency-independent, as idealized by Nonviscous Kelvin damping. For small-strain damping, laboratory test results and in situ values may not be compatible well in some cases.
7 As a convenient parameter of the modulus degradation, the reference strain $\gamma_r$ for $G/G_0 = 0.5$ in equivalent linear $G/G_0 \sim \gamma$ curves tends to correlate with effective confining stress $\sigma_c'$ as $\gamma_r \propto (\sigma_c')^{0.5}$ for non-cohesive soils, indicating stronger modulus degradations for lower confining stresses. In cohesive soils, $\gamma_r$ is dependent not on $\sigma_c'$ but on plasticity index $I_p$.
8 Equivalent linear $D \sim \gamma$ curves tend to shift along the $\gamma$-axis in the similar manner as $G/G_0 \sim \gamma$ curves with varying $\sigma_c'$ in non-cohesive soils and with varying $I_p$ in cohesive soils, respectively. The reference strains $\gamma_r$ may be approximately evaluated from dry densities or void ratios regardless of soil types.
9 Laboratory test-based $G/G_0 \sim \gamma$ curves are found essentially compatible with in situ properties back-calculated from vertical array earthquake records for different soil types individually. Among the soil types, clay shows better compatibility in particular, while gravels tend to give largely scattered in situ properties different from lab tests presumably because in situ gravels are normally well-graded and contain a lot of fines and sands.
10 Laboratory test-based $D \sim \gamma$ curves are similar to in situ properties back-calculated from earthquake records for different soil types. However, the small-strain $D$-values tend to be larger in the back-calculation than lab tests, presumably due to different reasons including a frequency-dependent mechanism in in situ damping.

Chapter 3

# Soil modeling for analyses and scaled model tests

In designing structures against earthquakes, the resistance of foundation soils and the induced deformations during strong earthquakes have to be evaluated. In a simplified design, a quasi-static horizontal force representing the earthquake inertia is applied to foundation soils or soil structures and a safety factor against failure is evaluated. However, dynamic response analyses are increasingly carried out in many cases with the advent of the performance-based design (PBD), wherein soil properties have to be modeled properly in numerical analyses in order to quantify the displacements in soils and foundations.

As already seen in Chapter 2, stress-strain relationships are strongly nonlinear and the soil dilatancy becomes dominant as soils are approaching failures. It is of utmost importance to model the nonlinear soil properties to be incorporated in the numerical methods so that the soil behaviors for large deformations during strong earthquakes can be properly reproduced. In some cases, in important projects in particular, scaled model tests along with numerical analyses are also conducted to make sure that the key mechanisms involved are properly taken into account. In the model tests, what kind of soil to use and how to evaluate the soil properties are essential to implement and interpret test results properly.

In this chapter, the modelling of stress-strain relationships of soils; hysteretic stress-strain curves and their equivalent linearization, the modeling of dilatancy, and the definition of cyclic strength in seismic loading are discussed. Then numerical methods using these models are outlined in terms of their aims, distinctions and numerical schemes. Equivalent linear analyses and stepwise nonlinear analyses are compared with the dynamic response of sand layers in shaking table tests to characterize the analytical results. Furthermore, shaking table model tests and associated similitudes are addressed together with soil properties in scaled models under very low confining pressures.

## 3.1 MODELLING OF SOIL PROPERTIES

### 3.1.1 Nonlinear stress-strain curves

Nonlinear stress-strain curves of soils under cyclic loading are idealized by simple functions. Among them, bilinear, hyperbolic and Ramberg-Osgood models are representative in many engineering problems. Their skeleton curves for monotonic loading

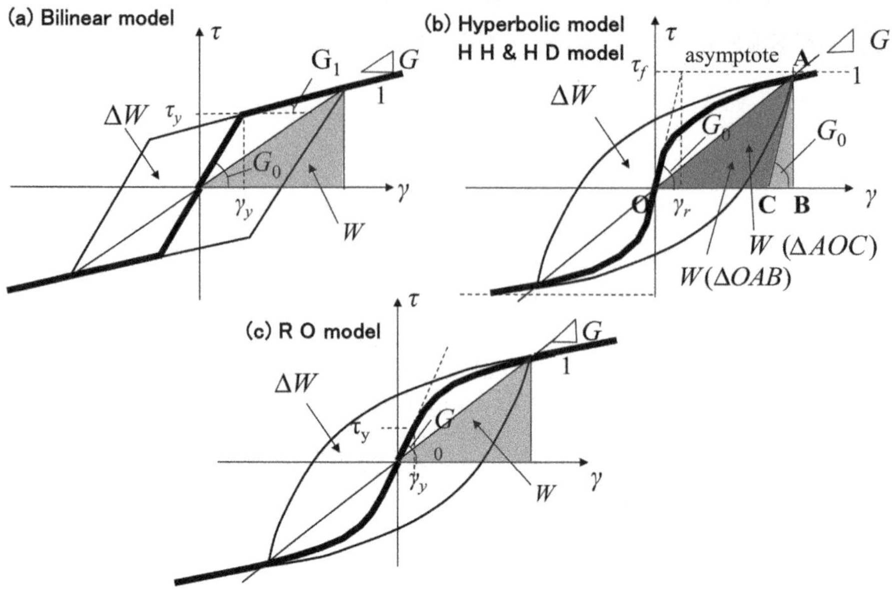


*Figure 3.1.1* Three representative nonlinear stress-strain models.

are symmetric with respect to the origin as illustrated with thick lines in Fig. 3.1.1, and those on the positive side are formulated in the following.

(a) Bilinear model: It simplifies the stress-strain curve with a straight line of the initial shear modulus $G_0$ up to a yield strain $\gamma_y$ and then another line of the second modulus $G_1$ after that. The skeleton curve is formulated by the three constants, $\gamma_y$, $G_0$ and $G_1$, as follows.

$$\left.\begin{aligned} \tau &= G_0\gamma & ;\quad \gamma \leq \gamma_y \\ \tau &= G_0\gamma_y + G_1(\gamma - \gamma_y); & \gamma > \gamma_y \end{aligned}\right\} \tag{3.1.1}$$

(b) Hyperbolic model: It is often used in modeling monotonic stress-strain relationships of soils and formulated as follows.

$$\tau = \frac{G_0\gamma}{1 + \gamma/\gamma_r} \tag{3.1.2}$$

It consists of only two parameters, the initial shear modulus $G_0$ and the reference strain $\gamma_r$, expressed as:

$$\gamma_r = \frac{\tau_f}{G_0} \tag{3.1.3}$$

where $\tau_f$ = shear strength. The reference strain $\gamma_r$ is correlated with the confining stress $\sigma'_c$ as:

$$\gamma_r \propto \sigma_c'^{0.5} \tag{3.1.4}$$

because $\tau \propto \sigma'_c$ for uncemented non-cohesive soils, and $G_0 \propto \sigma_c'^{0.5}$. With the strain level increasing to infinity $\gamma \to \infty$, the stress $\tau$ approaches to $\tau_f = G_0\gamma_r$ in Eq. (3.1.2), indicating that the shear stress always stays lower than the asymptote $\tau_f$.

(c) Ramberg-Osgood model (R-O model): The bilinear model with the initial modulus $\tau = G_0\gamma$ is modified by adding a nonlinear stress term as:

$$G_0\gamma = \tau\left\{1 + \alpha_{RO}\left(\frac{\tau}{G_0\gamma_y}\right)^{\beta_{RO}}\right\} \tag{3.1.5}$$

including four parameters, $G_0$, $\gamma_y$ and constants $\alpha_{RO}$, $\beta_{RO}$ ($\geq 0$). The yield stress $\tau_y$ is defined as:

$$\tau_y = G_0\gamma_y \tag{3.1.6}$$

For $\beta_{RO} = 0$, Eq. (3.1.5) becomes a linear model, while for $\beta_{RO} = \infty$ it reduces to a linear-perfect plastic model, because $G_0\gamma = \tau$ for $\tau < \tau_f$ and $G_0\gamma = \infty$ for $\tau > \tau_f$. With the four parameters, the R-O model has better curve fitting capability than the hyperbolic model, though the stress is difficult to formulate as an explicit function of strain.

### 3.1.2 Masing rule for cyclic loading

In order to construct the stress–strain relationship under cyclic loading, the Masing rule (Newmark and Rosenblueth 1971) is incorporated to extend the skeleton curve to the hysteretic curve corresponding to cyclic stress changes. If the skeleton curve on the $\tau$–$\gamma$ plane in Fig. 3.1.2 is expressed by an arbitrary function $f(\ )$ as:

$$\tau = \pm f(\pm\gamma) \tag{3.1.7}$$

where $\pm$ indicates the curves in positive and negative directions, the descending curve turning from the point P($\gamma_a$, $\tau_a$) is formulated by the Masing rule as:

$$\frac{\tau - \tau_a}{2} = -f\left(-\frac{\gamma - \gamma_a}{2}\right) \tag{3.1.8}$$

If this descending curve turns up at S($\gamma_b$, $\tau_b$), then

$$\frac{\tau_b - \tau_a}{2} = -f\left(-\frac{\gamma_b - \gamma_a}{2}\right) \tag{3.1.9}$$

and if the ascending curve is given as

$$\frac{\tau - \tau_b}{2} = f\left(\frac{\gamma - \gamma_b}{2}\right) \tag{3.1.10}$$

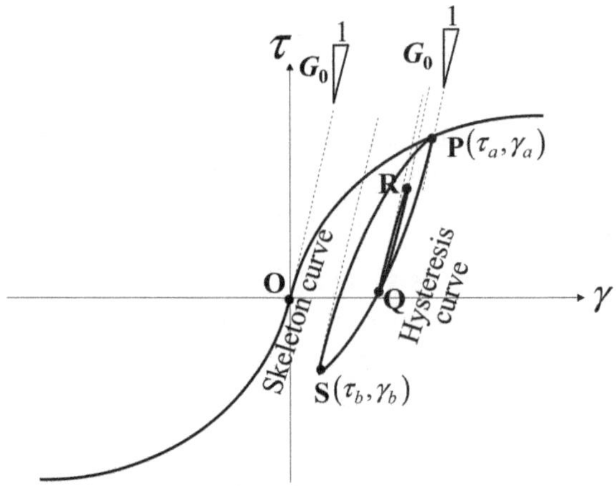


*Figure 3.1.2* Schematic stress-strain skeleton curve and hysteretic curve following Masing rule.

then, obviously the curve returns to the original point P($\gamma_a$, $\tau_a$) to make the loop P-S-P. The same holds in the local loop Q-R-Q. Thus, the succession of upward and downward curves in this manner eventually leads to the original point however complex the stress history may be, and the skeleton curve or hysteresis curve is followed thereafter.

The initial tangent modulus at O is expressed as:

$$\left.\frac{d\tau}{d\gamma}\right|_{\gamma=0} = f'(\gamma=0) = G_0 \tag{3.1.11}$$

and the modulus at a turning point with the strain $\gamma_R$ is

$$\left.\frac{d\tau}{d\gamma}\right|_{\gamma=\gamma_R} = 2\left.\frac{df((\gamma-\gamma_R)/2)}{d\gamma}\right|_{\gamma=\gamma_R} = 2\left.\frac{df(\gamma)}{d\gamma}\times\frac{1}{2}\right|_{\gamma=0} = G_0 \tag{3.1.12}$$

indicating that the modulus is equal to the initial modulus at all the turning points.

For the hyperbolic model, the ascending and descending curves from the turning points ($\gamma_a$, $\tau_a$) and ($\gamma_b$, $\tau_b$) are formulated respectively as,

$$\tau-\tau_a = \frac{G_0(\gamma-\gamma_a)}{1-(\gamma-\gamma_a)/2\gamma_r}, \quad \tau-\tau_b = \frac{G_0(\gamma-\gamma_b)}{1+(\gamma-\gamma_b)/2\gamma_r} \tag{3.1.13}$$

For the R-O model in the same manner, they are formulated as,

$$\gamma-\gamma_a = \frac{\tau-\tau_a}{G_0}\left\{1+\alpha_{RO}\left(-\frac{\tau-\tau_a}{2G_0\gamma_y}\right)^{\beta_{RO}}\right\}, \quad \gamma-\gamma_b = \frac{\tau-\tau_b}{G_0}\left\{1+\alpha_{RO}\left(\frac{\tau-\tau_b}{2G_0\gamma_y}\right)^{\beta_{RO}}\right\} \tag{3.1.14}$$

It is pointed out that, in actual soil tests, the hysteresis curve does not return to exactly the same point but tends to flow to a larger strain and the tangent shear modulus at the turning points tends to decrease as the soil approaches to failure. In order to take this effect into consideration, some modifications may be possible to the Masing rule. Pyke (1979) proposed to replace 2 in the denominators of Eq. (3.1.8) by a constant $c < 2$, where $c$ is defines as:

$$c = \left| \pm 1 - \frac{\tau_a}{\tau_f} \right| \tag{3.1.15}$$

so that those equations are written as:

$$\frac{\tau - \tau_a}{c} = \mp f\left( \mp \frac{\gamma - \gamma_a}{c} \right) \tag{3.1.16}$$

### 3.1.3 Hysteretic models for cyclic loading

The Masing rule can construct stationary hysteresis loops corresponding to the various skeleton curves. From the loops, equivalent linear properties such as secant shear moduli and hysteretic damping ratios are determined for various strain levels.

#### *3.1.3.1 Bilinear model*

The stationary loop based on the bilinear skeleton curve forms a parallelogram as illustrated in Fig. 3.1.1(a). The secant modulus $G$ is correlated with the shear strain $\gamma$ and the initial shear modulus $G_0$ as,

$$\begin{aligned} \frac{G}{G_0} &= 1 && : \quad \gamma \leq \gamma_y \\ \frac{G}{G_0} &= \frac{\gamma_y}{\gamma} + \left( \frac{G_1}{G_0} \right) \left( 1 - \frac{\gamma_y}{\gamma} \right) && : \quad \gamma > \gamma_y \end{aligned} \tag{3.1.17}$$

The equivalent damping ratio $D$ is calculated using the ratio of the dissipated energy to the strain energy $\Delta W/W$ as

$$\begin{aligned} D &= 0 && : \quad \gamma \leq \gamma_y \\ D &= \frac{\Delta W}{4\pi W} = \frac{2}{\pi} \frac{(1 - G_1/G_0)(\gamma/\gamma_y - 1)}{1 + (G_1/G_0)(\gamma/\gamma_y - 1)(\gamma/\gamma_y)} && : \quad \gamma > \gamma_y \end{aligned} \tag{3.1.18}$$

The variations of $G/G_0$ and $D$ are shown versus the normalized strain in Figs. 3.1.3(a), (b) with solid lines. With increasing strain $\gamma \to \infty$, the modulus becomes $G \to G_1$ while the damping ratio converges to zero, having a peak value $D = (2\pi)(\sqrt{G_0/G_1} - 1)/(\sqrt{G_0/G_1} + 1)$ when $\gamma/\gamma_y = \sqrt{G_0/G_1} + 1$. If $G_1 = 0$ (perfect plastic), then $D \to 2/\pi$ for $\gamma \to \infty$.

### 3.1.3.2 *Hysteretic hyperbolic (HH) model and Hardin-Drnevich (HD) model*

A stationary loop based on the hyperbolic skeleton curve is shown in Fig. 3.1.1(b). The secant modulus is formulated as

$$\frac{G}{G_0} = \frac{1}{1 + (\gamma/\gamma_r)} \tag{3.1.19}$$

The equation indicates that at the reference strain $\gamma = \gamma_r$ the modulus ratio $G/G_0$ becomes 0.5 and with infinitely increasing strain it approaches to zero. From the hysteresis curves in Eq. (3.1.13), the equivalent damping ratio for the hysteretic hyperbolic model (HH model) is obtained (Kokusho 1982) as:

$$D = \frac{2}{\pi}\left[2\left(\frac{\gamma_r}{\gamma} + 1\right)\left\{1 - \frac{\gamma_r}{\gamma}\ln\left(1 + \frac{\gamma}{\gamma_r}\right)\right\} - 1\right] \tag{3.1.20}$$

The secant modulus ratio and damping ratio are illustrated with dashed curves in Figs. 3.1.3(a) and (b), respectively, versus normalized strain $\gamma/\gamma_r$. Note that the $D$-value tends to increase monotonically with the strain ratio $\gamma/\gamma_r$ and approach to $2/\pi$ for infinitely large strain. It is obvious that this trend $D \to 2/\pi$ for $\gamma \to \infty$ occurs in cases where the shear strength approaches to a constant, such as in the elastic-perfect plastic bilinear model $G_1/G_0 = 0$.

Because $D = 2/\pi$ is too large to be comparable with actual soil data obtained in laboratory tests, the Masing rule used to derive Eq. (3.1.20) is modified in such a way that, as shown in Fig. 3.1.1(b), a ratio of the triangular area AOC ($W'$) to the area of the hysteresis loop ($\Delta W$) is always the same. Considering that the line AC has a

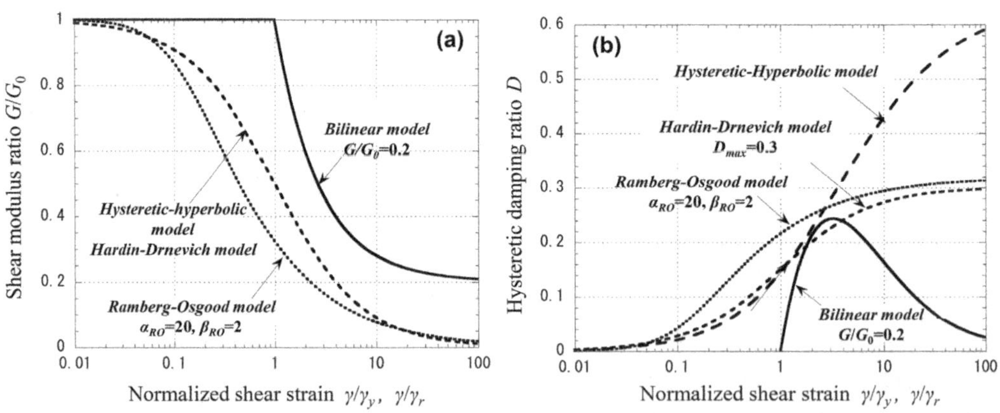


*Figure 3.1.3* Shear modulus ratio (a) and Damping ratio (b), versus normalized shear strain for various stress-strain models.

gradient $G_0$ for the turning point, $W'$ can be written as $W' = \tau^2(1/G - 1/G_0)/2$ and $W = \tau^2/2G$ for a stress amplitude $\tau$, it is obvious that the next equation holds.

$$D = \frac{\Delta W}{4\pi W} = \frac{\Delta W/W'}{4\pi W/W'} = \frac{\Delta W/W'}{4\pi}\left(1 - \frac{G}{G_0}\right) \tag{3.1.21}$$

Using the notation $D_{max} = \Delta W/4\pi W'$, the following formula is derived.

$$\frac{D}{D_{max}} = 1 - \frac{G}{G_0} \tag{3.1.22}$$

A model combining the secant modulus in Eq. (3.1.19) and the modified damping ratio Eq. (3.1.22) was proposed by Hardin and Drnevich (1972b) and named as the Hardin-Drnevich model (HD model), in which $D_{max}$ the maximum damping ratio for $G \to 0$ can be given so that $D$ can be adjusted to be more compatible with experimental soil data.

#### 3.1.3.3 *Ramberg-Osgood (RO) model*

A hysteresis loop shown in Fig. 3.1.1(c) is obtained by applying the Masing rule to the skeleton curve in Eq. (3.1.5). The modulus ratio $G/G_0$ is expressed in terms of the stress ratio $\tau/\tau_y$ as,

$$\frac{G}{G_0} = \frac{1}{1 + \alpha_{RO}(\tau/\tau_y)^{\beta_{RO}}} \tag{3.1.23}$$

or in terms of the strain ratio

$$\left(\frac{G_0}{G} - 1\right)\left(\frac{G_0}{G}\right)^{\beta_{RO}} = \alpha_{RO}\left(\frac{\gamma}{\gamma_y}\right)^{\beta_{RO}} \tag{3.1.24}$$

and the damping ratio is expressed as:

$$D = \frac{2}{\pi}\frac{\beta_{RO}}{\beta_{RO} + 2}\left(1 - \frac{G}{G_0}\right) \tag{3.1.25}$$

As shown with the dotted curves in Fig. 3.1.3, $G_0/G \to 0$ and $D \to (2/\pi)[\beta_{RO}/(\beta_{RO} + 2)]$, and their strain-dependent variations can be adjusted by the constants $\alpha_{RO}$ and $\beta_{RO}$ for better fitting with soil test data.

### 3.1.4 Comparison of laboratory test data with equivalent linear model

As a typical equivalent linear model, the HD model or its modified form is compared with laboratory test data of various types of soils (Kokusho 1982). Figs. 3.1.4 shows shear modulus ratios versus normalized shear strains for three types of non-cohesive soils (a)–(c) under three steps of confining stresses and one clayey soil (d). The shear strain in the horizontal axis of the diagram is normalized with the reference strain $\gamma_r$, where $\gamma_r \propto (\sigma_c')^{0.5}$ is assumed for non-cohesive soils in (a)–(c), and $\gamma_r = \text{constant}$

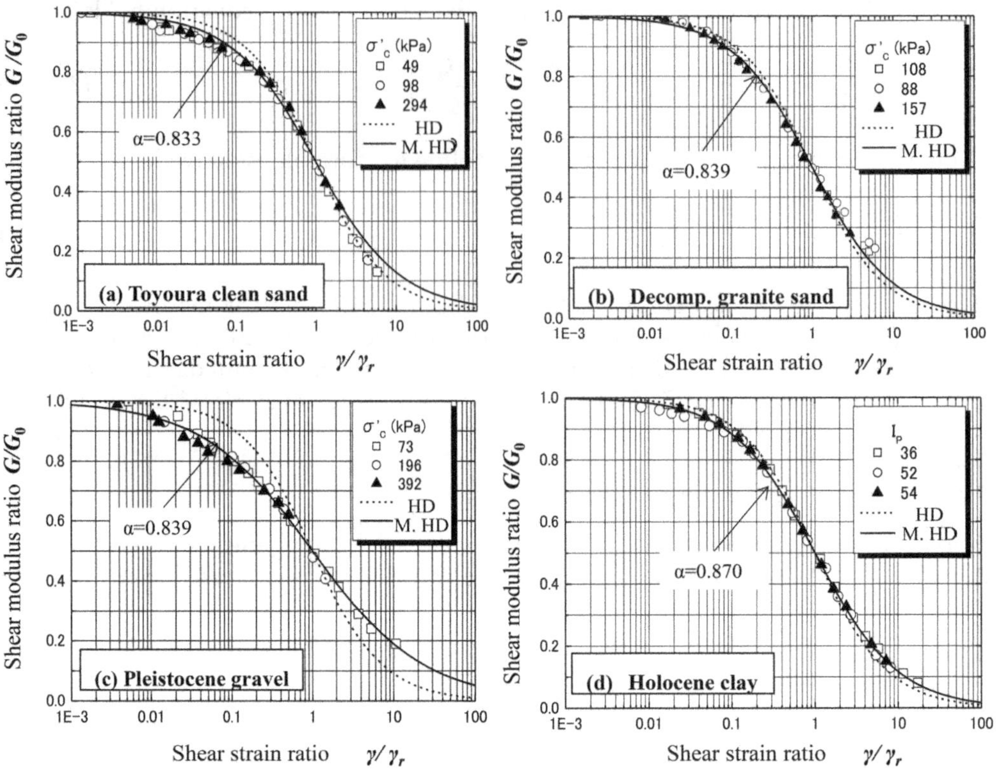


*Figure 3.1.4* Shear modulus ratio $G/G_0$ versus shear strain ratio $\gamma/\gamma_r$ under different confining stress or plasticity index: (a) Toyoura sand, (b) Decomp. granite sand, (c) Pleistocec gravel, (d) Holocene clay.

for cohesive soil in (d), respectively. The dashed curve in each diagram represents the modulus degradation by the HD model, showing that the plots can be approximated mostly by Eq. (3.1.19) for all soil types (a)–(d), though some gaps are recognizable for $\gamma/\gamma_r$ much smaller or larger than unity. For better agreement between the empirical curves and the test data, a slight modification of Eq. (3.1.19) by introducing an exponent $\alpha$ seems to be effective (Kokusho 1982) to fill the gap as:

$$\frac{G}{G_0} = \frac{1}{1 + (\gamma/\gamma_r)^{\alpha}} \qquad (3.1.26)$$

For the data in Fig. 3.1.4, the modified HD model with $\alpha = 0.833$–$0.870$ shown with the solid curves tends to give better fitting for gravels in (c) in particular.

Figs. 3.1.5(a)–(d) depict relationships between damping ratio versus shear modulus ratio for soils corresponding to those in Figs. 3.1.4(a)–(d). The plots tend to converge to a unique curve, though some scatters are visible for clays (d), and may be approximated by the dashed curve in Eq. (3.1.22) of the HD model in each diagram despite visible gaps

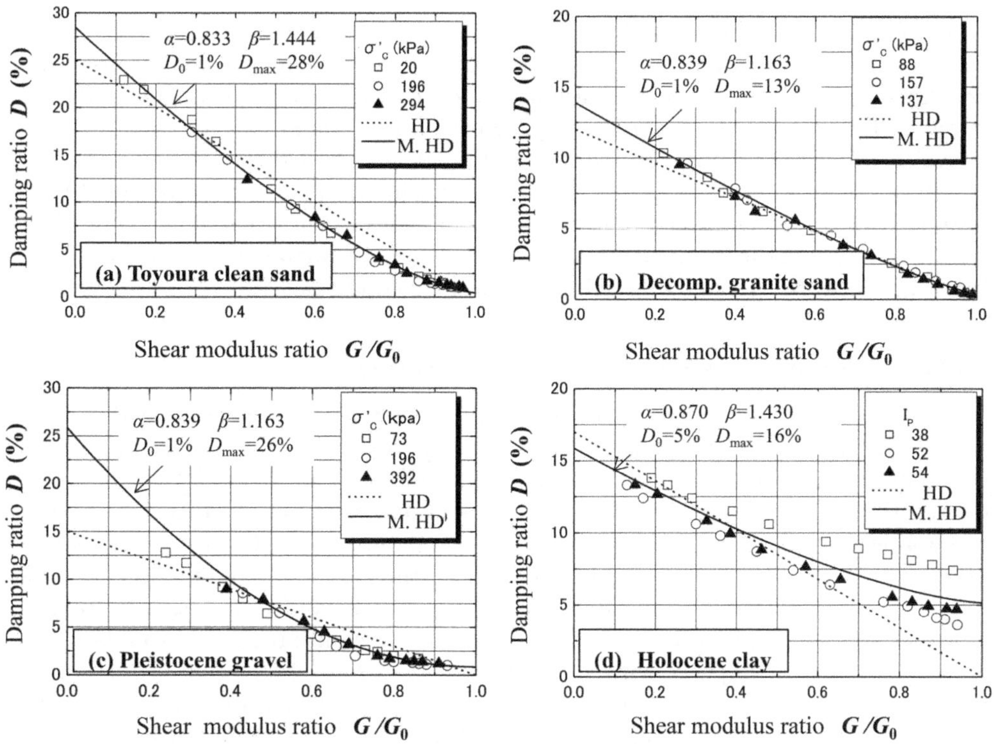


*Figure 3.1.5* Damping ratio $D$ versus shear strain ratio $\gamma/\gamma_r$ under different confining stress or plasticity index: (a) Toyoura sand, (b) Decomposed granite sand, (c) Pleistocene gravel, (d) Holocene clay.

for some soils. The major causes of the gap come from that $D/D_{\max}$ is a linear function of $G/G_0$ and converges to zero for $G/G_0 \rightarrow 0$ in the HD model as in Eq. (3.1.22), while the test data obviously show nonlinearity and converge to non-zero values. The same problem exists for the RO model as easily understandable from Eq. (3.1.25). In order to fill the gap, another modification of the HD model may be possible by introducing an exponent $\beta$ and an initial damping ratio $D_0$ for infinitely small strain to Eq. (3.1.22) (Kokusho 1982) as:

$$\frac{D - D_0}{D_{\max} - D_0} = \left(1 - \frac{G}{G_0}\right)^{\beta} \tag{3.1.27}$$

For the data in Fig. 3.1.5, the modified HD model with $\alpha = 0.833$–$0.870$ and $\beta = 1.163$–$1.444$ shown with the solid curves tends to give better fitting with the test data for gravels (c) and clays (d) in particular.

### 3.1.5 Modeling of soil dilatancy

Non-cohesive soils such as sands and gravels exhibit soil dilatancy for induced strain larger than $10^{-4}$~$10^{-3}$, and develop either ground settlement in a drained condition or excessive pore-pressure in an undrained condition leading to liquefaction if effective stress is all lost. For cohesive soils, too, the pressure buildup tends to occur, resulting in cyclic softening and post-seismic long lasting ground settlement.

In order to understand the soil dilatancy mechanically and how to model it for seismic behavior of soils, fundamental characteristics of the volume change of soils during monotonic shearing are first addressed. Then, the dilatancy effect during cyclic loading is discussed for the drained and undrained conditions.

#### *3.1.5.1 Dilatancy in drained monotonic shearing*

Typical results of monotonic shear test by the Swedish-type simple shear device are addressed here to see the effect of soil dilatancy in the drained condition. The test was implemented for four kinds of sandy or gravelly soils S1 to S4 with the grain size distributions in Fig. 3.1.6(a) under the effective vertical stress of 98 kPa. Fig. 3.1.6(b) exemplifies changes of void ratio versus shear strain up to $\gamma \approx 40\%$ for poorly graded sand S1 with variable initial void ratios between the maximum and minimum values $e_{\max}$ and $e_{\min}$. Obviously, the sands with larger initial void ratios (smaller relative densities $D_r$) tend to contract and those with smaller void ratios tend to dilate with increasing strain if compared with the final void ratios at $\gamma \approx 40\%$. Thus, a critical void ratio $e_{cr} \approx 0.91$ may be identified where the initial and final void ratios are unchanged as indicated with the dashed horizontal line in the middle. The $e_{cr}$-value tends to decrease with the change of soils from poorly-graded S1 with a smaller uniformity coefficient $U_c$ to well-graded S4 of larger $U_c$, while the relative densities are almost the same,

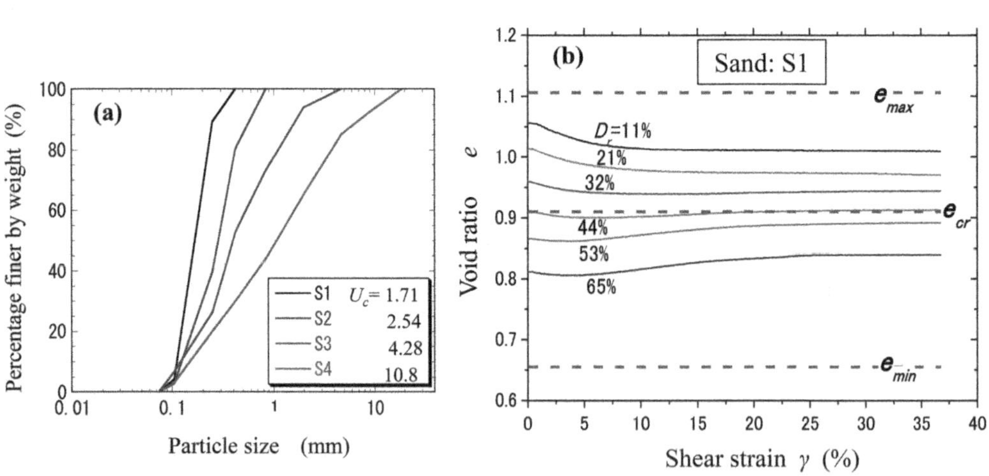


*Figure 3.1.6* Grain size curves of 4 granular soils S1~S4 (a), and Volume changes versus shear strain by drained simple shear tests for S1-soil with different initial void ratios (b) (Iwamoto et al. 2003).

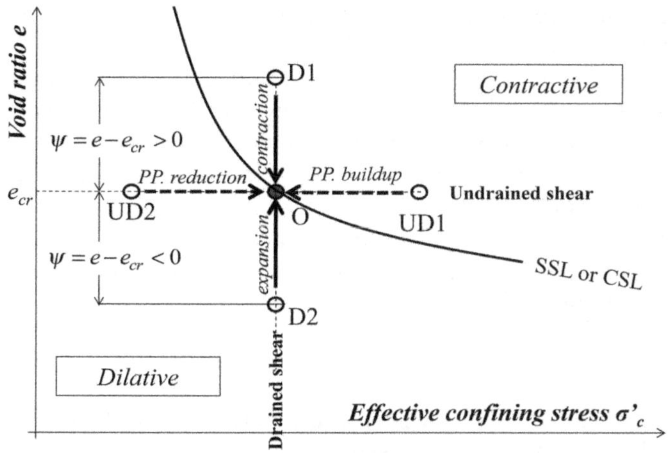


*Figure 3.1.7* Dilatancy behavior in drained and undrained conditions on e–$\sigma'_c$ state diagram.

$D_r \approx 40\%$ (Iwamoto et al. 2003). There exists a particular correlation between the critical void ratio $e_{cr}$ and the effective confining stress $\sigma'_c$ which is named as Steady State Line (SSL) or Critical State Line (CSL).

The corresponding dilatancy behavior of sand was discussed in detail referring to SSL under undrained monotonic and cyclic loading (Casagrande 1971, Castro 1975, Poulos et al. 1985). The SSL-curve is shown on the $e$–$\sigma'_c$ plane named as the state diagram in Fig. 3.1.7 to understand the dilatancy behaviors in both drained and undrained conditions collectively. The right-upside and left-downside of the line correspond to contractive and dilative zones due to shearing, respectively. In the drained shearing, the void ratio $e$ moves vertically with $\sigma'_c = $ constant, while in the undrained shearing, the confining stress $\sigma'_c$ moves horizontally with $e = $ constant.

In the drained shearing under the constant $\sigma'_c$ for example, an initial point D1 for a loose sand supposedly moves down to O on the SSL-curve eventually, whereas D2 for a dense sand goes up to the same point O in the steady state. This conceptual trend is consistent with the experimental results shown in Fig. 3.1.6(b), in that looser sands tend to contract and denser sands tend to dilate, though the test results do not converge to a unique point for the steady state O but to different points corresponding to individual initial void ratios. In this respect, more microscopic researches conducted by Desrues et al. (1996) and Finno and Rechenmacher (2003) suggested that the formation of shear bands make the sand behavior complicated. According to them, sand if dense tends to develop a thin shear band with the thickness equivalent to 10 to 20 sand particles, wherein the void ratio measured by a X-ray analysis approaches a constant $e_{cr}$ independent of initial void ratios. Namely, the local void ratio in the shear band where the shear strain concentrates during shearing in denser sands seems to take a unique value $e_{cr}$ locally corresponding to Point O, though the average void ratio takes the ultimate value quite different from $e_{cr}$, depending on the initial void ratio as indicated in Fig. 3.1.6(b).

In the undrained shearing of sand with a given void ratio, an initial point UD1 under high confining stress $\sigma'_c$ on the contractive side of SSL tends to move left horizontally as the pore-pressure builds up ultimately to the point O on SSL, while UD2 under a low $\sigma'_c$ value tends to move right to the same point O eventually, generating the negative pore-pressure. In the case starting from UD1 where soil is particularly contractive, a catastrophic flow failure by spontaneous liquefaction (e.g. Seed 1987) due to static initial shear stress may be triggered in the approach to the point O.

The $e$–$\sigma'_c$ diagram in Fig. 3.1.7 is helpful in systematically understanding the seismic liquefaction mechanism in a broader spectrum as will be explained in Sec. 5.8. The difference in the void ratio at present $e$ and the corresponding critical void ratio $e_{cr}$ under the same $\sigma'_c$-value is defined as

$$\psi = e - e_{cr} \tag{3.1.28}$$

and named as the state parameter by Been and Jefferies (1985). Contractive and dilative soils correspond to $\psi > 0$ and $\psi < 0$, respectively.

### *3.1.5.2 Dilatancy in drained cyclic shearing*

Fig. 3.1.8(a) shows a typical volumetric strain versus shear strain curve for the S1-sand of $D_r = 45\%$, cyclically sheared under the drained condition and $\sigma'_c = 98$ kPa in the stress-controlled test with the cyclic stress ratio $CSR = 0.3$. The shear strain initially about 5% tends to decrease with the number of cycles and so does the void ratio converging to a certain value. Fig. 3.1.8(b) summarizes several such test results for the same sand in the drained cyclic shear with $CSR = 0.3$ starting with different initial void ratios. It is remarkable that, unlike the monotonic tests shown in Fig. 3.1.6(b), the soil volume tends to decrease regardless of the initial void ratios. A close look at the test results further reveals that the sand with the initial void ratio larger than the critical void ratio $e_{cr} = 0.91$, $e > e_{cr}$, tends to exhibit much larger volume reduction

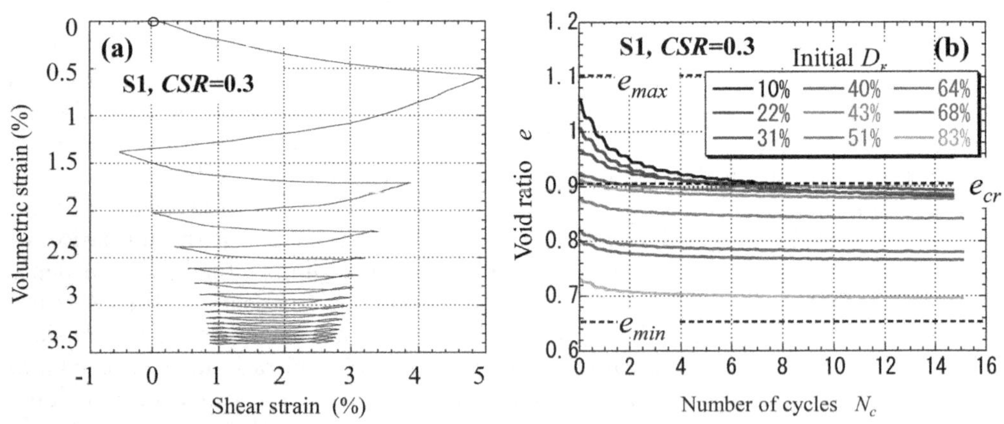


*Figure 3.1.8* Volumetric strain versus shear strain (a), and Void ratio versus number of cycles (b), for S1-soil with different initial relative density by drained cyclic shearing (Iwamoto et al. 2003).

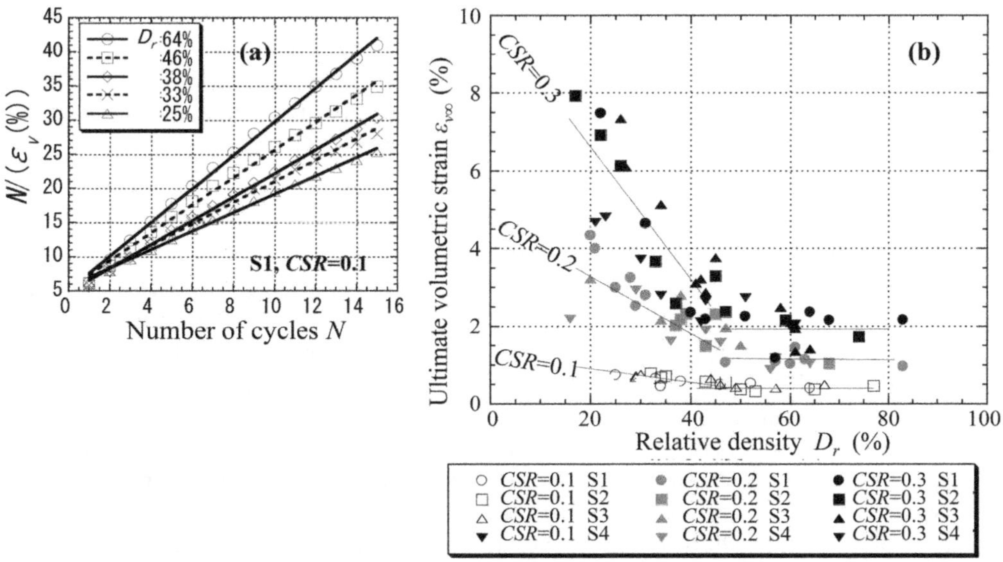


*Figure 3.1.9* Values $N_c/\varepsilon_v$ versus $N_c$ for S1-soil (a), and Ultimate volumetric strain $\varepsilon_{v\infty}$ versus relative densities $D_r$ for 4 soils, S1∼S4, by $CSR = 0.1, 0.2$ and $0.3$ (b) (Iwamoto et al. 2003).

than those $e < e_{cr}$. The same trend can be observed not only for the sand but also for all the soils S1 to S4 studied by Iwamoto et al. (2003) as demonstrated in Fig. 3.1.9.

The volumetric strains $\varepsilon_v$ in such drained cyclic loading tests may be approximated by a hyperbolic function of the number of cycles $N_c$ and constants $a$ and $b$ as follows (Yagi 1978).

$$\varepsilon_v = -\frac{\Delta e}{1+e} = \frac{N_c}{a+bN_c} \tag{3.1.29}$$

The values $N_c/\varepsilon_v$ plotted versus $N_c$ in Fig. 3.1.9(a) for the S1-sand with different initial $D_r$-values sheared by $CSR = 0.1$ show linear relationships, from which the constants $a$ and $b$ in Eq. (3.1.29) can be decided. For $N_c \to \infty$, $\varepsilon_v \to 1/b$, hence the ultimate volumetric strain for infinite number of cycles $\varepsilon_{v\infty}$ can be obtained as $\varepsilon_{v\infty} = 1/b$. The $\varepsilon_{v\infty}$-values thus obtained are plotted in Fig. 3.1.9(b) versus the relative densities $D_r$ for the 4 types of soil materials, S1 to S4, cyclically sheared by $CSR = 0.1$, 0.2 and 0.3. It shows unique decreasing trends of the ultimate volumetric strains $\varepsilon_{v\infty}$ with increasing relative densities despite the difference in the soil materials, although they are different for the different $CSR$-values. They may be approximated by bilinear relationships with kinks at $D_r \fallingdotseq 45\%$, having almost constant $\varepsilon_{v\infty}$-values for $D_r$ larger than that. This $D_r$-value seems to correspond to the critical void ratio of each soil material, $e_{cr} = 0.91$ for the S1-sand for example, according to the research results by Iwamoto et al. (2003). This indicates that the ultimate volumetric strain $\varepsilon_{v\infty}$ induced by the drained cyclic loading is always positive no matter how dense the soil is, though $\varepsilon_{v\infty}$ tends to be evidently larger with decreasing $D_r$, if $D_r$ is smaller than the threshold corresponding to the critical void ratio. Based on the finding, it may be generally interpreted for drained cyclic loading that the contractive volumetric strain occurs all over the $e$–$\sigma_c'$ state diagram in Fig. 3.1.7, and tends to be larger above SSL with increasing state parameter $\psi = e - e_{cr}$. For the monotonic drained loading, in contrast, the contraction occurs above SSL only, while the dilative strain occurs elsewhere.

### 3.1.5.3 Dilatancy in undrained cyclic shearing

Fig. 3.1.10(a) shows excess pore-pressure responses during undrained cyclic torsional shear tests on clean sand specimens. The pore pressure buildup ratio, $u$ normalized by $\sigma_c'$, in the vertical axis tends to rise to 100% as shown in the diagram versus the number of cycles $N_c$ normalized by $N_L$ (=$N_c$ for 100% pore-pressure buildup) in the horizontal axis. For low relative densities, the pressure ratio tends to rise gradually in the first part and then drastically increase in the final loading stage, whereas it tends to rise relatively faster from the first if the cyclic stress amplitude is larger. The pressure fluctuations cycle by cycle caused by the soil dilatancy are manifested in denser sands compared to looser sands, wherein their peak values correspond to the stepwise pressure buildup. The shape of the curves also reflects the differences in $CSR$ and $N_L$ of the cyclic loading tests.

Based on the measured pressure buildup curves exemplified in (a), Fig. 3.1.10(b) illustrates the pressure buildup curves (corresponding to the stepwise peak values in (a)) idealized by the following function

$$r_N = \frac{N_c}{N_L} = \left[\frac{1}{2}(1 - \cos \pi r_u)\right]^{\alpha} = \left(\sin \frac{\pi}{2} r_u\right)^{2\alpha} \tag{3.1.30}$$

or its reverse function proposed by Seed et al. (1976).

$$r_u = \frac{u}{\sigma_c'} = \frac{1}{2} + \frac{1}{\pi} \sin^{-1}(2 r_N^{1/\alpha} - 1) = \frac{2}{\pi} \sin^{-1} r_N^{1/(2\alpha)} \tag{3.1.31}$$

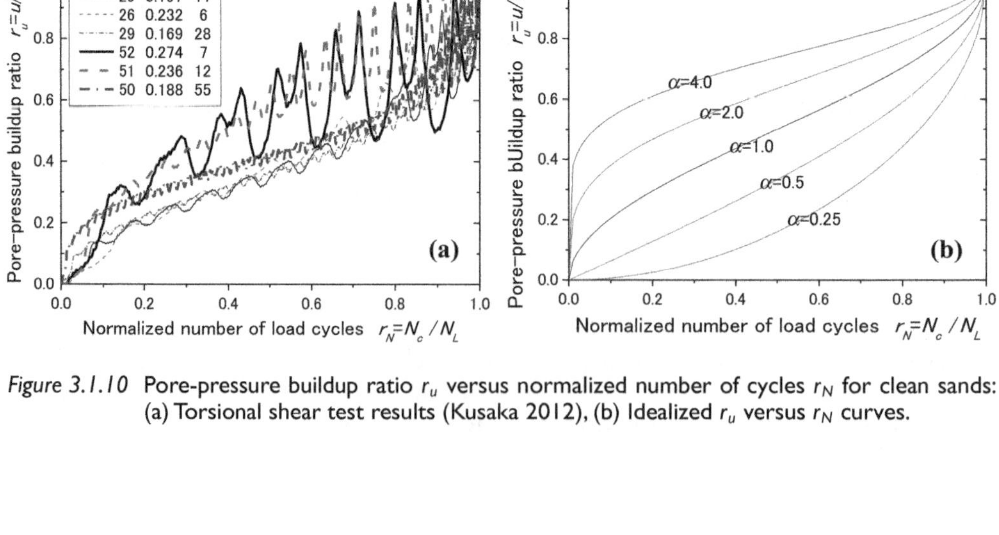


*Figure 3.1.10* Pore-pressure buildup ratio $r_u$ versus normalized number of cycles $r_N$ for clean sands: (a) Torsional shear test results (Kusaka 2012), (b) Idealized $r_u$ versus $r_N$ curves.

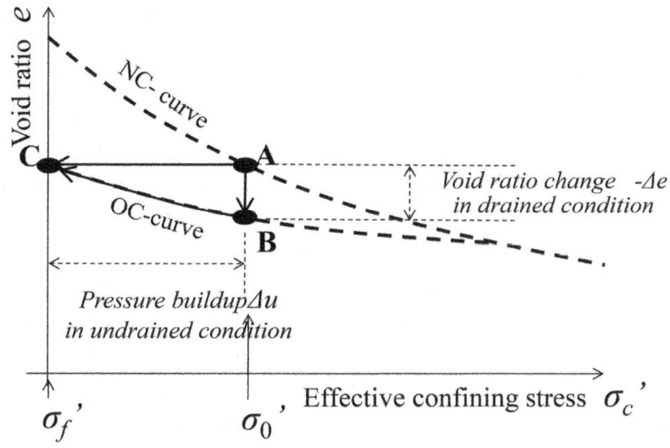


*Figure 3.1.11* Conceptual diagram showing drained void ratio change and undrained pressure buildup on void ratio versus effective confining stress diagram.

Here, $r_N = N_c/N_L$, $r_u = u/\sigma_c'$, and $\alpha$ = an exponent determining the shape of the curves.

The pore-pressure change in the undrained condition shown here and the volume change in the drained condition shown in Fig. 3.1.9 both reflect the same dilatancy characteristics of soils, hence the two changes should be closely correlated to each other. Fig. 3.1.11 illustrates a diagram showing a conceptual relationship between the two dilatancy behaviors in drained and undrained cyclic loading. If the void ratio reduces in drained cyclic shearing from Point A to B by $-\Delta e$, it increases the pore-pressure by $\Delta u = \sigma_f' - \sigma_0'$ by undrained cyclic shearing from A to C, where C represents the point with the same void ratio as A and on the swelling curve through Point B as pointed out by Martin et al. (1975). This indicates that the volumetric strain $\varepsilon_v = -\Delta e/(1+e)$ can be correlated with the pore-pressure change $\Delta u$ as

$$\varepsilon_v = -\frac{\Delta e}{1+e} = -\frac{\Delta u}{K} + \frac{n\Delta u}{K_w} - \frac{n\Delta u}{K_s} \approx -\frac{\Delta u}{K} \tag{3.1.32}$$

where $K$, $K_w$, $K_s$ = bulk moduli of soil skeleton (for swelling), water, and soil particle, respectively, and $n$ = porosity. Because of big differences among the moduli as $K \ll K_w, K_s$, the volumetric strain can be simplified as $\varepsilon_v = -\Delta u/K$ in the last term of the above equation. This means that $\varepsilon_v$ in the drained loading can be compensated by the pore-pressure increase, $\Delta u$, or the effective stress decrease in the undrained loading.

If loose sand is liquefied, however, not only the residual effective stress $\sigma_f'$ is zero but also the soil particles are not in contact with each other at C in Fig. 3.1.11 and suspended in water in the extreme. In such circumstances, Eq. (3.1.32) may not be applicable at Point C, because the soil behavior is too nonlinear to be idealized by the swelling curve of soil skeleton. Instead, the soil behavior at Point C seems to be simulated better by a sedimentation process with sand particles falling down freely as demonstrated in tube tests for liquefied sand layer and associated numerical analyses (Kokusho 2003, Tsurumi et al. 2003). Eq. (3.1.32) connecting the dilatancy behaviors in the drained and undrained loadings becomes valid, once contacts between the particles are recovered.

As already mentioned, the drained cyclic loading always brings about contractive volume change irrespective of its position in the state diagram in contrast to the drained monotonic loading. This fact together with Eq. (3.1.32) indicates that, in the undrained condition, cyclic loading always generates positive excess pressure unlike monotonic loading no matter where the initial state of sand is located relative to SSL in Fig. 3.1.7. Hence, even a very dense sand under low confining stress can approach to the state of 100% pore-pressure buildup called as the initial liquefaction if the cyclic shear stress amplitude and the number of cycles are large enough. Thus, the cyclic loading during earthquakes tends to manifest the dilatancy behavior quite different from the monotonic loading in both drained and undrained conditions.

### 3.1.6 Dynamic strength in cyclic loading based on fatigue theory

In many engineering problems, soil failures during earthquakes are expected to occur in the undrained conditions. As already discussed in Sec. 2.1.4, the dynamic soil strength in such circumstances is defined by a single amplitude of stress which induces strain magnitude corresponding to various structural performance levels in a given cyclic loading. The cyclic loading effect is represented by a sinusoidal motion equivalent to a given earthquake motion working on a soil element, which generates the threshold strain in a given number of cycles, such as $N_c = 10, 15, 20$, corresponding to particular seismic motions. The dynamic soil strength is normally defined by the cyclic resistance ratio ($CRR$) as the cyclic stress amplitude divided by the initial effective confining stress.

In liquefaction problems, the following strain magnitude is normally taken as a default value; $\varepsilon_{DA} = 5\%$ in the double amplitude axial strain or $\gamma_{DA} = (1+\nu)\varepsilon_{DA} = 7.5\%$ in the double amplitude shear strain considering Poisson's ratio $\nu = 0.5$ in undrained condition. It is because around those strain-values the excess pore-pressure tends to buildup 100% (pore-pressure ratio $r_u = 1.0$). Threshold strains other than the above values may be used according to various levels of the performance of structures such as serviceability, reparability and ultimate failure.

#### *3.1.6.1 Regular and irregular cyclic loading*

In converting an irregular seismic motion to the equivalent sinusoidal motion with a constant amplitude and a given number of cycles, the concept of fatigue is sometimes employed to interpret cyclic loading soil test results. In this concept, the cyclic stress ratio $CSR$ of a sinusoidal motion to attain a given induced strain is correlated with the number of cycles $N_c$ so that the $CSR$ quite reasonably decreases linearly with increasing $N_c$ on a log-log diagram. Fig. 3.1.12(a) exemplifies the $CSR$ versus $N_c$ plots obtained from cyclic torsional simple shear tests by constant shear stress amplitudes on clean sands of loose to medium densities to attain threshold double amplitude

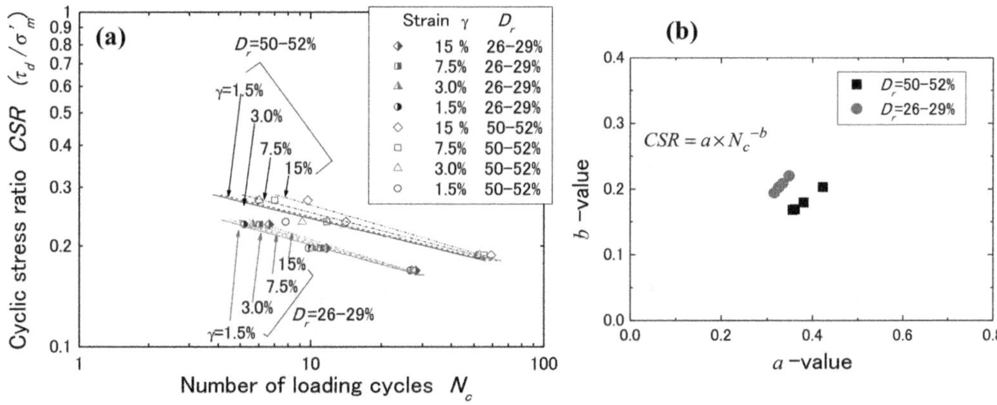


*Figure 3.1.12* Cyclic stress ratio *CSR* versus number of cycles $N_c$ relationship on log-log scale from torsional simple shear tests on clean sand (a), and Variations of constants *a*, *b* in $CSR = a \times N_c^{-b}$ (b).

strains $\gamma_{DA} = 1.5$, 3, 7.5, 15%. The $CSR$–$N_c$ relationships for different $\gamma_{DA}$ can be approximated by a set of parallel lines as:

$$CSR = a \times N_c^{-b} \tag{3.1.33}$$

where, $a =$ a constant, and $b =$ an exponent determining the gradient of the *CSR* versus $N_c$ line on the log-log chart. The *a*, *b*-values obtained in the tests are plotted in the horizontal and vertical axes in Fig. 3.1.12(b), showing that for the loose to medium dense clean sands $a \fallingdotseq 0.3$–$0.4$ and $b \approx 0.2$ not so sensitive to the relative densities and the threshold strains.

Based on the *CSR* versus $N_c$ relationship, it becomes possible in the light of the fatigue theory (e.g. Annaki and Lee 1977) to convert an irregular stress history with the maximum stress $S_{max}$ schematically shown in Fig. 3.1.13(a) into the sinusoidal motion with a certain amplitude $S_e$ and the number of cycles $N_{eq}$ shown in (b), which incurs the soil damage equivalent to the irregular loading. In the theory, the irregular stress history is first decomposed into a set of single cycles, that are grouped into discrete stress amplitudes $S_i$ ($i = 1$–$n$) changing stepwise as shown in (a). The soil damage is assumed to be dependent only on the stress amplitudes $S_i$ and the corresponding numbers of cycles $N_i$ for individual amplitudes and not on their sequence of appearance. Fig. 3.1.13(c) is a schematic chart similar to Fig. 3.1.12(a) showing a linear relationship named as a *S-N* line for a damage level $D_L$. Here, *S* is the stress amplitude, *N* is the number of cycles to reach the damage level $D_L$ wherein $D_L = 1.0$ means failure. Similar to Eq. (3.1.33), the *S-N* line for $D_L = 1.0$ is expressed as:

$$S = a \times N^{-b} \tag{3.1.34}$$

It is known for many engineering materials including soils that the *S-N* line is represented by a straight line on the log-log diagram descending rightward. For a constant

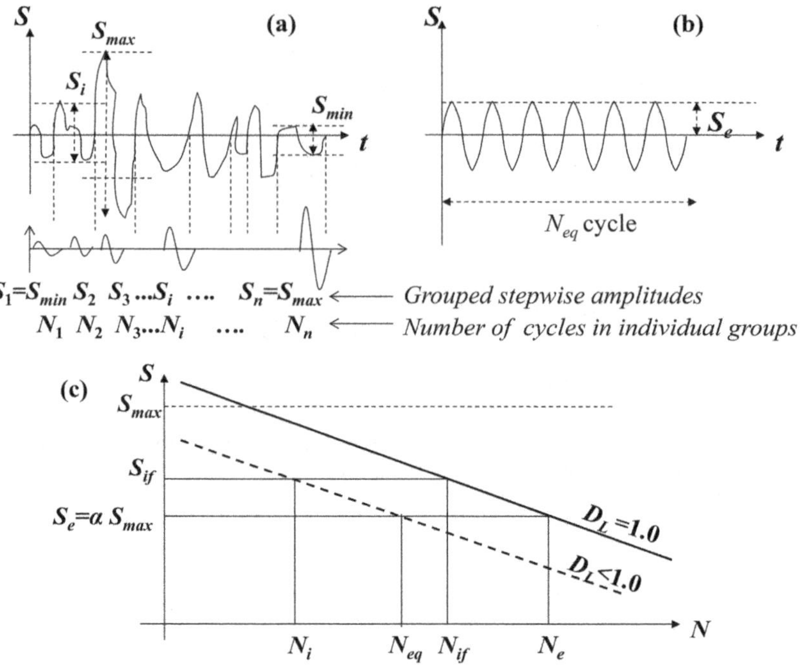


*Figure 3.1.13* Fatigue theory on how to convert irregular stress history (a) into regular cyclic stress of a given amplitude and number of cycles (b), and Damage levels on S–N plane (c).

stress amplitude $S = S_{if}$, the number of cycles to the failure $D_L = 1.0$ is $N_{if}$ as shown in Fig. 3.1.13(c), and the damage by a single cycle is $1/N_{if}$ based on the previously mentioned postulate. Hence, the damage by the number of waves $N_i$ of that amplitude group is $N_i/N_{if}$, and the total damage is obtained by the sum of all the amplitude groups $i = 1$–$n$ as:

$$D_L = \sum_i \left(\frac{N_i}{N_{if}}\right) \tag{3.1.35}$$

and the failure occurs for $D_L \geq 1.0$.

Next, let us consider how to convert the irregular stress wave with the maximum stress $S_{\max}$ in Fig. 3.1.13(a) into the regular wave with the amplitude $S_e$ and the number of cycles $N_{eq}$ shown in (b). If $S_e$ is prescribed using a constant $\alpha$ ($\alpha < 1.0$) as

$$S_e = \alpha S_{\max} \tag{3.1.36}$$

then, $N_e$ corresponding to $D_L = 1.0$ is determined as indicated in (c). Let the damage level by the irregular wave be $D_L < 1.0$, then its *S-N* line (the dashed line) can be drawn in parallel with the line of $D_L = 1.0$. In the liquefaction problem as indicated in Fig. 3.1.12(a), the *S-N* lines corresponding to different induced strains (different damage levels) are actually recognized to be almost in parallel to each other.

If the damage level of the irregular loading $D_L$ in Eq. (3.1.35) is to be identical to that of sinusoidal loading with the amplitude $S_e$ and the number of cycles $N_{eq}$, then Fig. 3.1.13(c) shows that

$$D_L = \frac{N_{eq}}{N_e} = \sum_i \left(\frac{N_i}{N_{if}}\right) \tag{3.1.37}$$

Hence, the equivalent number of cycles $N_{eq}$ can be obtained as

$$N_{eq} = N_e \sum_i \left(\frac{N_i}{N_{if}}\right) \tag{3.1.38}$$

In Eq. (3.1.36), the stress amplitude $S_e$ is dependent on $\alpha$, and so is $N_{eq}$. This seems to indicate that the choice of $\alpha$ is critical and may differentiate the result. However, whatever $\alpha$ may be chosen, it is clear that the ratio $N_{eq}/N_e$ is always the same as $D_L = \sum_i (N_i/N_{if})$, indicating that the same damage level as the irregular motion can be obtained no matter how the $\alpha$-value is chosen. In other words, if smaller $S_e$ is chosen, $N_{eq}$ becomes larger to secure the same value $D_L = N_{eq}/N_e$.

If two sinusoidal motions with different amplitudes, $S_e$ and $S_{if}$, are compared, the number of cycles to $D_L = 1.0$ are $N_e$ and $N_{if}$, respectively, as indicated in Fig. 3.1.13(c), and the ratio $r_e = N_e/N_{if}$ can be written from Eq. (3.1.34) using the gradient of the *S-N* line $b$ as

$$r_e = \frac{N_e}{N_{if}} = \left(\frac{S_e}{S_{if}}\right)^{-1/b} \tag{3.1.39}$$

Considering $r_e = N_e/N_{if} = (1/N_{if})/(1/N_e)$ and $1/N_{if}$ and $1/N_e$ means damages by single cycles of amplitude $S_i$ and $S_e$, respectively, it is obvious that $r_e$ in the above equation also represents a coefficient to replace a single cycle of $S_{if}$ by $r_e$-cycles of $S_e$. By using $r_e$ for each amplitude group $i$ in the irregular motion in Fig. 3.1.13(a), the total number of cycles of a sinusoidal motion with the amplitude $S_e$, having the same $D_L$ as the irregular motion can be calculated.

The above fatigue theory may be utilized to liquefaction problems, because a damage level $D_L$ can be uniquely correlated with a certain pore-pressure buildup ratio or induced strain. In Sec. 5.6, it will be shown that the damage level is closely correlated with cumulative dissipated energy in sand during liquefaction.

#### 3.1.6.2 Two-directional loading

Seismic loading by the SH-wave in a horizontal plane is intrinsically two directional and can be divided into two orthogonal components with a certain phase lag, because the two directional loading without a phase lag results in one-directional loading. In laboratory simple shear tests, soil specimens are normally loaded in one direction and the effect of the two-directional loading has to be properly accounted for. It is experimentally known in the two-directional shear tests that a loading history in one

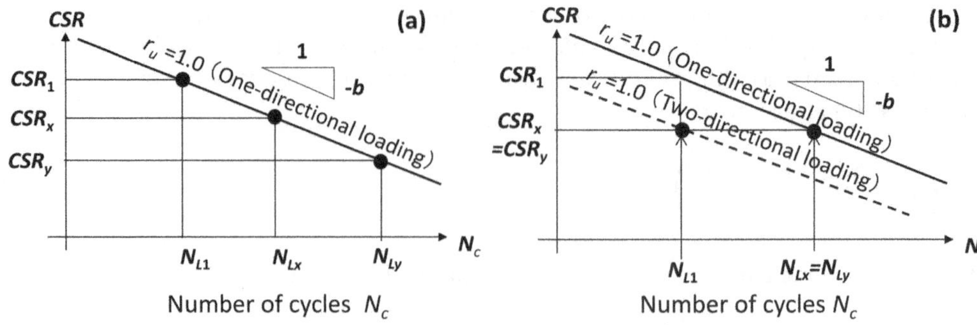


*Figure 3.1.14* Fatigue theory for two-directional loading calculated from one-directional loading test: (a) *CSR–$N_c$* line of one-directional loading, (b) Comparison of *CSR–$N_c$* lines of one/two-directional loading.

direction gives only minimal effect on the other direction normal to that (e.g. Seed et al. 1978). Based on the independency of two orthogonal loading histories, the effect of two-directional loading may be quantitatively calculated from one-dimensional test results with the help of the fatigue theory (Tokimatsu and Yoshimi 1982).

Fig. 3.1.14(a) schematically illustrates the $CSR$-$N_c$ line for the onset of liquefaction (the excess pore-pressure ratio $r_u = 1.0$) corresponding to the $S$-$N$ line in the fatigue theory. Let $N_{L1}$ be the number of cycles for $r_u = 1.0$ in the one-directional cyclic shear test of stress ratio $CSR_1$. If the same soil is loaded in the two orthogonal directions $x$, $y$ with the stress ratios, $CSR_x$ and $CSR_y$, and liquefied in $N_{L1}$ cycles, then the next equation holds because of the independency of the two orthogonal loading histories.

$$r_u = r_{ux} + r_{uy} = 1 \tag{3.1.40}$$

Here, $r_{ux}$ and $r_{uy}$ are the pressure buildup ratios by the $N_{L1}$ cycle loading in the two directions, individually. The substitution of Eq. (3.1.31) into Eq. (3.1.40) gives

$$\frac{1}{\pi}\sin^{-1}(2r_{Nx}^{1/\alpha} - 1) + \frac{1}{\pi}\sin^{-1}(2r_{Ny}^{1/\alpha} - 1) = 0 \tag{3.1.41}$$

where $r_{Nx}$, $r_{Ny}$, the ratios of the number of cycles to that for 100% pore-pressure buildup ($r_u = 1.0$) in $x$, $y$-directions by the stress ratios $CSR_x$ and $CSR_y$, respectively, are written as:

$$r_{Nx} = \frac{N_{L_1}}{N_{Lx}}, \quad r_{Ny} = \frac{N_{L_1}}{N_{Ly}} \tag{3.1.42}$$

If the expressions $\sin A = 2r_{Nx}^{1/\alpha} - 1$ and $\sin B = 2r_{Ny}^{1/\alpha} - 1$ are used, $A + B = 0$ from Eq. (3.1.41), and the next equation is obtained.

$$r_{Nx}^{1/\alpha} + r_{Ny}^{1/\alpha} = \frac{\sin A + \sin B}{2} + 1 = \sin\left(\frac{A+B}{2}\right)\cos\left(\frac{A-B}{2}\right) + 1 = 1 \tag{3.1.43}$$

Substituting $r_{Nx}$ and $r_{Ny}$ in Eq. (3.1.42) into Eq. (3.1.43) yields the following.

$$\frac{N_{Lx}}{N_{L_1}} = \left[1 + \left(\frac{N_{Lx}}{N_{Ly}}\right)^{1/\alpha}\right]^{\alpha} \tag{3.1.44}$$

In Fig. 3.1.14(b), the solid and dashed lines schematically represent liquefaction onset lines ($r_u = 1.0$) in the one and two directional loadings, respectively. For the equal stress ratios in the two directions, $CSR_x = CSR_y$, Eq. (3.1.44) gives $N_{Lx}/N_{L_1} = 2^{\alpha}$. Hence, in comparison with the number of cycles for liquefaction onset $N_{L1}$ by the two-directional loading with the same stress ratio $CSR_x = CSR_y$, the number of cycles for liquefaction onset $N_{L_x} = N_{Ly}$ in the one-directional loading with the same stress ratio $CSR_x = CSR_y$ is $2^{\alpha}$ (or 2 if $\alpha = 1$) times larger.

Then, the $CSR_1$ to liquefy by one-direction loading with the number of cycles $N_{L_1}$ and $CSR_x$ and $CSR_y$ combined for two-directional loading liquefied with the same $N_{L_1}$ is compared. It is obviously seen from Fig. 3.1.14(a) that the following equations hold in this case.

$$\frac{N_{Lx}}{N_{Ly}} = \left(\frac{CSR_x}{CSR_y}\right)^{-1/b}, \quad \frac{N_{Lx}}{N_{L_1}} = \left(\frac{CSR_x}{CSR_1}\right)^{-1/b} \tag{3.1.45}$$

If these are substituted to Eq. (3.1.44), a ratio of $CSR$-values $CSR_x/CSR_1$ for liquefaction onset between two-directional and one-directional loadings indicated in Fig. 3.1.14(b) is expressed as

$$\frac{CSR_x}{CSR_1} = \left[1 + \left(\frac{CSR_y}{CSR_x}\right)^{1/\alpha b}\right]^{-\alpha b} \tag{3.1.46}$$

In case the $CSR$-values are the same in two directions $CSR_x = CSR_y$, the above equation becomes

$$\frac{CSR_x}{CSR_1} = 2^{-\alpha b} \tag{3.1.47}$$

Thus, $CSR$ in the two directional loading can be calculated using the constant $\alpha$ (determining the pressure buildup curve in Eq. (3.1.30)) and $b$ (the gradient of the $CSR$–$N_c$ line in Eq. (3.1.33)) based on the fatigue theory (Tokimatsu and Yoshimi 1982).

## 3.2 DYNAMIC SOIL ANALYSES

Historically speaking, seismic designs of structures were carried out by pseudo-static analyses using seismic coefficients for a long time. Seismic stabilities of earth-structures, slopes and foundation ground are still calculated basically by the pseudo-static methods using slip surface analyses. With the development of computer technologies, dynamic response analyses have been increasingly employed in important projects in particular. More recently, Performance-Based Design (PBD) has been providing a further impetus for the dynamic response analyses. PBD is increasingly employed recently in the structural design of buildings and infrastructures in many countries. Seismically induced ground deformation is critical to the PBD in terms of serviceability, reparability and ultimate safety of structures.

### 3.2.1 Distinctions of dynamic analyses on soils

Dynamic response analyses of soils and earth-structures as distinct from those of superstructures may be summarized as follows.

a) Soils are strongly nonlinear materials including dilatant properties with drastically changing properties depending on the induced strain levels. Accordingly, the type of analytical tools has to be properly chosen so that soil behaviors are best reproduced to meet the goals of the analyses.
b) Foundation soils or earth-structures are composed of bulky heterogeneous materials and difficult to efficiently analyze the whole three-dimensional body. Hence it is essential how to convert them to simplified models easier to analyze. A layered soil ground is simplified as one-dimensional column, and slopes, embankments, retaining structures and tunnels are idealized as two-dimensional plane strain models in normal engineering practice.
c) Despite the spatial variability of in situ soils, soil investigation data to make reliable analytical models are normally too few. Detailed numerical models become meaningless unless they are supported by high density reliable soil data.
d) The damping mechanism involved in the soil response analyses is classified into the radiational damping and the internal damping. The former is also called as the geometrical damping which occurs due to the radiation of wave energy to the outside of the analytical model. The latter, called as the material (soil) damping, occurs due to internal energy dissipation mostly by the soil friction. The internal damping is very much dependent on the induced soil strain during strong earthquakes, while the radiation damping is affected by the impedance ratio between the model and the surrounding soils outside the boundary.
e) In this, respect, it is essential to eliminate as far as possible the effect of outer boundaries artificially introduced in the analytical model on the dynamic response and residual deformations. As for the dynamic response, the boundaries have to transmit or absorb radiated waves so that the waves artificially reflected at the boundaries are minimized.

### 3.2.2 Goals of dynamic soil analyses

Goals of dynamic soil analyses may be summarized as follows.

(a) Earthquake response of foundation ground; acceleration, velocity and displacement given to superstructures, seismic site amplification, and soil-structure interaction.
(b) Earthquake-induced settlement and residual deformation of foundation ground.

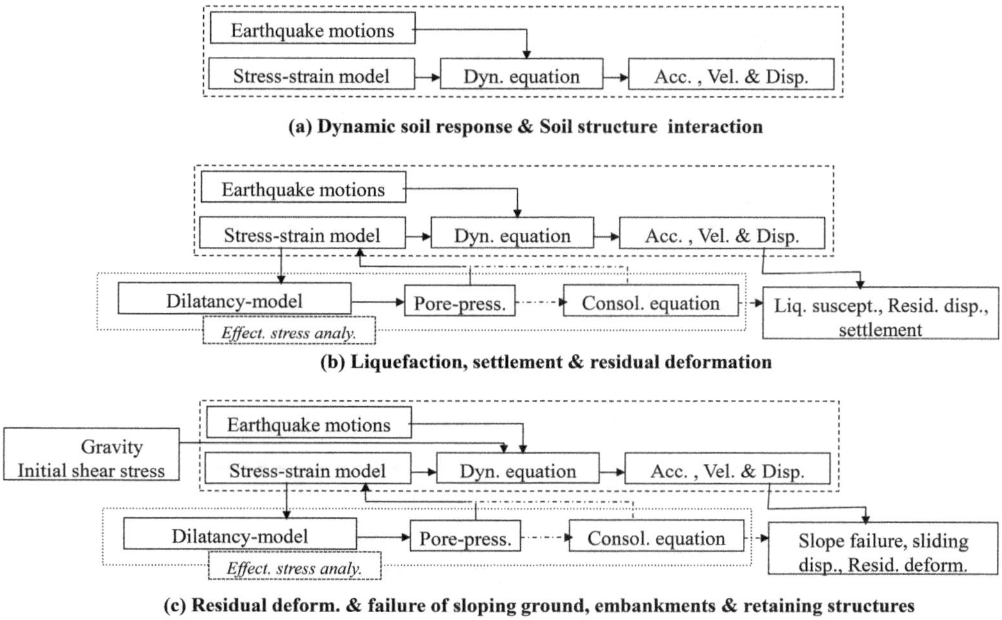


*Figure 3.2.1* Flow chart for dynamic soil analyses (a), (b) and (c).

(c) Earthquake-induced instability and residual deformation of foundations and slope failures under the effect of initial shear stress.

The flow charts for the analyses aiming (a) to (c) are shown in Fig. 3.2.1. The analysis (a) is the common core part to calculate the dynamic soil response or soil-structure interaction using linear/nonlinear constitutive stress-strain relationships during given earthquake motions. In the linear and equivalent linear analyses, the dynamic response is calculated normally in the frequency domain for computation efficiency. In the nonlinear analyses, the tangent moduli of stress-strain curves are followed step by step using the Masing rule, and incorporated in the stiffness matrices in calculating the dynamic equation in the time domain. This analysis is called "total stress analysis" wherein no variation of effective stresses due to dilatancy is considered. In the analysis (b), dilatancy models are added to the core part (a) to calculate the pore-pressure buildup and the effective stress changes and reflect them in modifying the tangent moduli step by step in the time-domain dynamic response analyses. This type is called "effective stress analysis". Sometimes, pore pressure variation is calculated in parallel by coupling consolidation or seepage equations. Liquefaction susceptibility, post-liquefaction settlements and residual deformations are evaluated in the effective stress analysis. The total stress analysis may also be applied to approximately evaluate the dynamic response of liquefied ground by modifying stress-strain curves reflecting the pore-pressure buildup. In the analysis (c), the effect of the gravity or initial shear stress due to the self-weight is taken into account along with

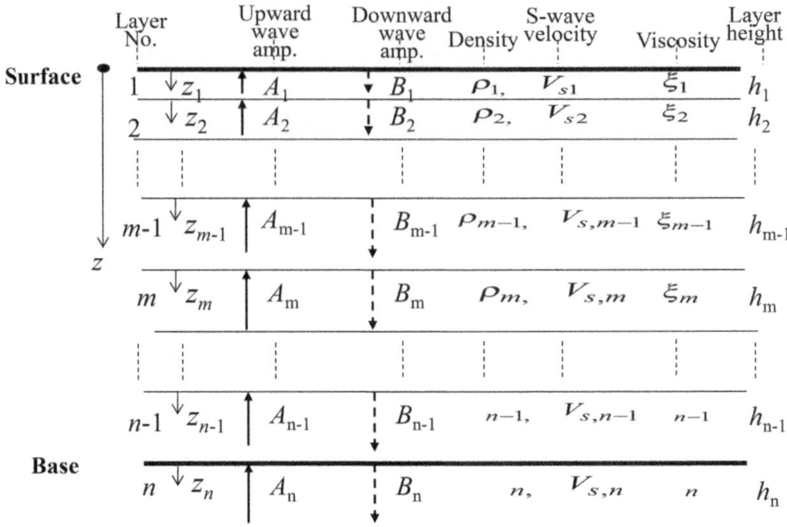


*Figure 3.2.2* One-dimensional model of horizontal layers for vertically travelling SH-wave.

the seismic motions to evaluate slope stabilities, sliding displacements and residual deformations.

### 3.2.3 Outline of dynamic response analyses

In order to calculate the common core part of the flow chart Fig. 3.2.1(a), a soil ground or soil structure to analyze is idealized with a continuum or discrete elements with various properties. Horizontally layered soils can be idealized by one-dimensional soil columns, easy to analyze by wave equations using linear and equivalent-linear soil properties. Foundation soils, slopes, embankments and earth-structures are idealized with two to three dimensional models consisting of discrete elements to be analyzed by the finite element or finite different method. The dynamic equation of the discrete element models is efficiently solved in the frequency domain if the soil properties are assumed linear or equivalent linear. It is also efficiently solved in the time-domain by the mode-superposition analysis. If the soil properties are treated to be truly nonlinear, it has to be solved by the step by step nonlinear analysis in the time domain.

#### *3.2.3.1 One-dimensional wave propagation analysis in continuum model*

In this analysis, one dimensional dynamic soil response is calculated based on the multi-reflection theory of one-dimensional SH-wave (Shnabel et al. 1972). Fig. 3.2.2 shows a soil model in this analysis consisting of multiple horizontal layers with prescribed thicknesses $h$, densities $\rho$, S-wave velocities $V_s$, shear modulus $G=\rho V_s^2$ and viscosity $\xi$. The model is underlain by a stiff base-layer named as an engineering bedrock where input earthquake motions are defined as outcropping motions.

*(i) General formulation*

As already shown in Eq. (1.6.1), the wave equation of upward propagating S-wave in the Kelvin-model is expressed as

$$\rho \frac{\partial^2 u}{\partial t^2} = \left( G + \xi \frac{\partial}{\partial t} \right) \frac{\partial^2 u}{\partial z^2} \tag{3.2.1}$$

where $u$ is the horizontal displacement and the $z$-axis is heading downward. The general solution for harmonic wave of the angular frequency $\omega$ is obtained as:

$$u = Ae^{i(k^* z + \omega t)} + Be^{-i(k^* z - \omega t)} \tag{3.2.2}$$

where $A$ and $B$ are the amplitudes of the upward and downward waves, respectively, and $k^*$ is the complex wave number, and its square is written as:

$$k^{*2} = \frac{\rho \omega^2}{G + i\omega\xi} = \frac{\rho \omega^2}{G^*} \tag{3.2.3}$$

Here, $G^*$ is the complex shear modulus for the Kelvin model defined in Eq. (1.5.12). In the case of the Nonviscous Kelvin model, Eq. (3.2.3) is replaced by the next equation by using $G^*$ in Eq. (1.5.24).

$$k^{*2} = \frac{\rho \omega^2}{G + iG'} = \frac{\rho \omega^2}{G^*} \tag{3.2.4}$$

Shear stress is written as:

$$\tau = G^* \frac{\partial u}{\partial z} = ik^* G^* [Ae^{i(k^* z + \omega t)} - Be^{-i(k^* z - \omega t)}] \tag{3.2.5}$$

If local coordinates $z_1 \sim z_n$ are employed in the individual layers $1 \sim n$ as shown in Fig. 3.2.2, the displacement $u$ and stress $\tau$ at the upper boundary ($z_m = 0$) and lower boundary ($z_m = h_m$) of the $m$-th layer are

$$\left.\begin{aligned}
u_{m,z_m=0} &= (A_m + B_m) e^{i\omega t} \\
u_{m,z_m=h_m} &= (A_m e^{ik^*_m h_m} + B_m e^{-ik^*_m h_m}) e^{i\omega t} \\
\tau_{m,z_m=0} &= ik^*_m G^*_m (A_m - B_m) e^{i\omega t} \\
\tau_{m,z_m=h_m} &= ik^*_m G^*_m (A_m e^{ik^*_m h_m} - B_m e^{-ik^*_m h_m}) e^{i\omega t}
\end{aligned}\right\} \tag{3.2.6}$$

At the ground surface ($m = 1$), $\tau_{1,z_1=0} = ik^* G^*_1 (A_1 - B_1) e^{i\omega t} = 0$ from the third formula in Eq. (3.2.6), resulting in:

$$A_1 = B_1 \tag{3.2.7}$$

so that the amplitudes of the upward and downward waves at the ground surface are the same. Furthermore, from the continuity in the displacement and stress between the lower boundary of the $(m-1)$-th layer and the upper boundary of the $m$-th layer, the following recursive matrix formula is derived for $A_m$ and $B_m$.

$$\begin{Bmatrix} A_m \\ B_m \end{Bmatrix} = \begin{bmatrix} \frac{k_m^* G_m^* + k_{m-1}^* G_{m-1}^*}{2k_m^* G_m^*} e^{ik_{m-1}^* h_{m-1}} & \frac{k_m^* G_m^* - k_{m-1}^* G_{m-1}^*}{2k_m^* G_m^*} e^{-ik_{m-1}^* h_{m-1}} \\ \frac{k_m^* G_m^* - k_{m-1}^* G_{m-1}^*}{2k_m^* G_m^*} e^{ik_{m-1}^* h_{m-1}} & \frac{k_m^* G_m^* + k_{m-1}^* G_{m-1}^*}{2k_m^* G_m^*} e^{-ik_{m-1}^* h_{m-1}} \end{bmatrix} \begin{Bmatrix} A_{m-1} \\ B_{m-1} \end{Bmatrix} \quad (3.2.8)$$

If the matrix in Eq. (3.2.8) is expressed by $[T_m]$ and this equation is recursively used together with Eq. (3.2.7), the constants $A_m$ and $B_m$ can be formulated as:

$$\begin{Bmatrix} A_m \\ B_m \end{Bmatrix} = [T_m][T_{m-1}]\cdots[T_2]\begin{Bmatrix} A_1 \\ A_1 \end{Bmatrix} \quad (3.2.9)$$

and further expressed by using functions of $\omega$, $a_m(\omega)$ and $b_m(\omega)$, as:

$$\left.\begin{aligned} A_m &= a_m(\omega)A_1 \\ B_m &= b_m(\omega)A_1 \end{aligned}\right\} \quad (3.2.10)$$

Using the corresponding functions for arbitrarily chosen $n$-th layer, $a_n(\omega)$ and $b_n(\omega)$, the transfer function of the composite waves (superposing upward and downward waves) between the $m$-th and $n$-th layer $T_{m,n}$ is expressed as:

$$T_{m,n} = \frac{u_m}{u_n} = \frac{a_m(\omega) + b_m(\omega)}{a_n(\omega) + b_n(\omega)} \quad (3.2.11)$$

Thus, the amplitude at any layer boundary can be calculated from Eq. (3.2.11), if it is given at any other layer boundary for stationary vibration of angular frequency $\omega$ (Schnabel et al. 1972). It is obvious that such a formulation becomes possible because $A_1 = B_1$ holds in Eqs. (3.2.7), (3.2.9) and (3.2.10).

A recursive expression different from Eq. (3.2.8) is also possible by eliminating $A_m$ and $B_m$ in Eq. (3.2.6) and using the continuity of displacement $u_{m,z_m=h_m} = u_{m+1,z_{m+1}=0}$ and shear stress $\tau_{m,z_m=h_m} = \tau_{m+1,z_{m+1}=0}$ at the boundary.

$$\begin{Bmatrix} u_{m+1,z_{m+1}=0} \\ \tau_{m+1,z_{m+1}=0} \end{Bmatrix} = \begin{bmatrix} \cos k_m^* h_m & \frac{\sin k_m^* h_m}{k_m^* G_m^*} \\ -k_m^* G_m^* \sin k_m^* h_m & \cos k_m^* h_m \end{bmatrix} \begin{Bmatrix} u_{m,z_m=0} \\ \tau_{m,z_m=0} \end{Bmatrix} \quad (3.2.12)$$

If the matrix of this equation is written as $[S_m]$, displacement and stress at the upper boundary of arbitrarily chosen $n$-th layer are correlated with those of $m$-th layer:

$$\begin{Bmatrix} u_{n,z_n=0} \\ \tau_{n,z_n=0} \end{Bmatrix} = [S_{n-1}][S_n]\cdots[S_m]\begin{Bmatrix} u_{m,z_m=0} \\ \tau_{m,z_m=0} \end{Bmatrix} = [S_{n-1,m}]\begin{Bmatrix} u_{m,z_m=0} \\ \tau_{m,z_m=0} \end{Bmatrix} \quad (3.2.13)$$

This indicates that displacement and stress at any boundary can be obtained from those at any arbitrary boundary by using $[S_{n-1,m}] = [S_{n-1}][S_n]\cdots[S_m]$.

All above equations for the stationary response for a harmonic motion of frequency $\omega$ can be applicable to irregular seismic motions by superposing the harmonic motions of Fourier series. Namely, a digitized acceleration data at the $m$-th layer boundary, $\ddot{u}_m$, discretized into N data points (even number) with the time increment $\Delta t$, is expressed by the finite Fourier series of harmonic waves as:

$$\ddot{u}_m = \sum_{s=0}^{N/2} (a_s e^{i\omega_s t} + b_s e^{-i\omega_s t}) \tag{3.2.14}$$

where frequency $\omega_s$ of the $s$-th term is written as:

$$\omega_s = \frac{2\pi s}{N\Delta t} \quad \left(s = 0 - \frac{N}{2}\right) \tag{3.2.15}$$

The complex Fourier constants $a_s$ and $b_s$ in Eq. (3.2.14) are expressed as:

$$\left.\begin{aligned} a_s &= \frac{1}{N}\sum_{j=0}^{N-1} \ddot{u}_m(j\Delta t)\, e^{-i\omega_s (j\Delta t)} \\ b_s &= \frac{1}{N}\sum_{j=0}^{N-1} \ddot{u}_m(j\Delta t)\, e^{i\omega_s (j\Delta t)} \end{aligned}\right\} \tag{3.2.16}$$

In the light of Eq. (3.2.11), irregular acceleration response at the upper boundary of an arbitrary $n$-th layer can be calculated in the next equation if a seismic irregular motion $\ddot{u}_m$ is given at the upper boundary of an arbitrary $m$-th layer.

$$\ddot{u}_n(t) = \sum_{s=0}^{N/2} T_{m,n}(\omega_s)(a_s e^{i\omega_s t} + b_s e^{-i\omega_s t}) \tag{3.2.17}$$

*(ii) Input wave at engineering bedrock*

The composite wave at the upper boundary of the $n$-th layer superposing upward and downward waves is obtained from Eqs. (3.2.8) and (3.2.10) as:

$$\begin{aligned} A_n + B_n &= e^{ik^*_{n-1}h_{n-1}} A_{n-1} + e^{-ik^*_{n-1}h_{n-1}} B_{n-1} \\ &= [a_{n-1}(\omega)e^{ik^*_{n-1}h_{n-1}} + b_{n-1}(\omega)e^{-ik^*_{n-1}h_{n-1}}]A_1 \end{aligned} \tag{3.2.18}$$

The amplitude ratio between the observed motions at the ground surface and at the top of the $n$-th layer is written as:

$$\frac{2A_1}{A_n + B_n} = \frac{2}{a_{n-1}(\omega)e^{ik^*_{n-1}h_{n-1}} + b_{n-1}(\omega)e^{-ik^*_{n-1}h_{n-1}}} \tag{3.2.19}$$

It is clear that Eq. (3.2.19) does not depend on properties of the $n$-th layer, because $a_{n-1}(\omega)$ and $b_{n-1}(\omega)$ consist of the properties of up to the $(n-1)$-th layers and the properties of the $n$-th layer ($k^*_n$ and $G^*_n$) are cancelled in calculating $A_n + B_n$ as obviously seen in Eqs. (3.2.8) to (3.2.10). Consequently, the amplitude ratio of the observed motions between the ground surface and a certain depth is independent of properties of layers deeper than that depth.

On the other hand, the amplitude ratio of the upward waves ($A_1/A_n$) or the outcropping waves ($2A_1/2A_n$) between the surface and the top of the $n$-th layer is written as follows which is dependent on the properties ($k_n^*$ and $G_n^*$) of the $n$-th layer.

$$\frac{A_1}{A_n} = \frac{2A_1}{2A_n} = \frac{1}{a_n(\omega)} \tag{3.2.20}$$

Both Eqs. (3.2.19) and (3.2.20) represent site amplifications during earthquakes defined between the ground surface and a base layer (engineering bedrock) due to the vertical propagation of the SH-wave, if the $n$-th layer is assumed as the base layer as shown in Fig. 3.2.2. Eq. (3.2.19) corresponds to the amplification of observed motions in the vertical arrays, wherein the properties of the base layer have nothing to do with the amplification. In contrast, $2A_1/2A_n$ in Eq. (3.2.20) corresponds to the amplification of outcropping motions between the ground surface and the outcropping $n$-th layer (assuming the overburden soils were all removed), and depends on the properties of the base layer ($k_n^*$ and $G_n^*$), though being independent of layer thickness $h_n$ as can be seen from Eq. (3.2.8). These observations tell that, in a site amplification analysis using a soil model precisely reproducing the site condition, one can forget soil properties in the base layer ($n$-th layer) if the earthquake motions observed at the base ($A_n + B_n$) are given in the analysis. However, if the outcropping motions at that depth ($2A_n$) are to be given to the same model, the properties of the base layer should be prescribed.

*(iii) Decomposition of subsurface motion to upward and downward waves*

If the earthquake record is given at the ground surface $u_1 = U_1 e^{i\omega t}$, it is easy to decompose it into upward and downward components as:

$$A_1 = B_1 = \frac{U_1}{2} \tag{3.2.21}$$

Although $A_m$ and $B_m$ in an arbitrary $m$-th layer can be theoretically determined from the surface record, its reliability largely depends on the applicability of the multi-reflection theory of vertically propagating SH-waves to site-specific actual ground responses, and the applicability tends to be poorer with increasing depth. If there are multiple subsurface down-hole records at different depths available, the calculations of $A_m$ and $B_m$ in deeper ground can be more reliable.

The displacement at the top of the $m$-th layer is given from the first formula in Eq. (3.2.6) as:

$$u_m = U_m e^{i\omega t} = (A_m + B_m) e^{i\omega t} \tag{3.2.22}$$

where $A_m$ and $B_m$ are complex amplitudes of upward and downward waves in the $m$-th layer. If, in addition to $u_m$ in Eq. (3.2.21), another records $u_n$ at the top of the $n$-th layer is available as:

$$u_n = U_n e^{i\omega t} = (A_n + B_n) e^{i\omega t} \tag{3.2.23}$$

then, in a similar manner as Eq. (3.2.9), the amplitudes between the two depths are correlated as:

$$\begin{Bmatrix} A_n \\ B_n \end{Bmatrix} = [T_n][T_{n-1}]\cdots[T_{m+1}]\begin{Bmatrix} A_m \\ B_m \end{Bmatrix} = [T_{n,m+1}]\begin{Bmatrix} A_m \\ B_m \end{Bmatrix} \tag{3.2.24}$$

If the two by two matrix above is expressed as:

$$[T_{n,m+1}] = \begin{bmatrix} T_{11} & T_{12} \\ T_{21} & T_{22} \end{bmatrix} \tag{3.2.25}$$

Then, Eqs. (3.2.22), (3.2.23), (3.2.24) give the amplitudes $A_m$ and $B_m$ as follows.

$$\begin{Bmatrix} U_n \\ U_m \end{Bmatrix} = \begin{bmatrix} T_{11} + T_{21} & T_{12} + T_{22} \\ 1 & 1 \end{bmatrix}\begin{Bmatrix} A_m \\ B_m \end{Bmatrix} = [P_{n,m+1}]\begin{Bmatrix} A_m \\ B_m \end{Bmatrix} \tag{3.2.26}$$

$$\begin{Bmatrix} A_m \\ B_m \end{Bmatrix} = [P_{n,m+1}]^{-1}\begin{Bmatrix} U_n \\ U_m \end{Bmatrix} \tag{3.2.27}$$

Accordingly, the corresponding amplitudes for the $n$-th layer are given by substituting Eq. (3.2.27) to Eq. (3.2.24) (Kokusho and Motoyama 2002) as:

$$\begin{Bmatrix} A_n \\ B_n \end{Bmatrix} = [T_{n,m+1}][P_{n,m+1}]^{-1}\begin{Bmatrix} U_n \\ U_m \end{Bmatrix} \tag{3.2.28}$$

#### 3.2.3.2 *Complex response analysis of discretized model*

With the rapid development in computer technology, two/three-dimensional earthquake response analyses of complicated models are often implemented in geotechnical engineering, too. By discretizing foundation soils and structures into the finite element or finite difference models consisting of a numerous number of nodes and elements, the following multi-dimensional dynamic matrix equation can eventually be derived.

$$[M]\ddot{U} + [C]\dot{U} + [K]U = -\{m\}\ddot{z} \tag{3.2.29}$$

Here, $\ddot{U}$, $\dot{U}$, $U$ are the vectors for accelerations, velocities and displacements of the nodes, and $[M]$, $[C]$, $[K]$ are the matrices of mass, damping and stiffness, respectively. The right side term $-\{m\}\ddot{z}$ represents the earthquake loads wherein $\ddot{z}$ = input acceleration and $\{m\}$ = mass vector consisting of the diagonal components of the mass matrix $[M]$.

There exist a variety of methods in numerically solving Eq. (3.2.29). A typical method is the complex response analysis, wherein the stationary response in the frequency domain for harmonic motions is computed with variable input frequencies, and the dynamic response to irregular earthquake motions in the time domain are efficiently calculated using the Fourier and inverse-Fourier transforms (e.g. Lysmer et al. 1975). The input acceleration time history $\ddot{z}$ discretized into $N$ data points (even numbers) with the time increment $\Delta t$ is first expressed by the finite Fourier series of harmonic waves with the terms $s = 0$ to $N/2$ as:

$$\ddot{z} = \mathrm{Re}\sum_{s=0}^{N/2} \ddot{z}_s e^{i\omega_s t} \tag{3.2.30}$$

Accordingly, the response displacement vector $U$ is also expanded in a Fourier series as:

$$U = \mathrm{Re}\sum_{s=0}^{N/2} U_s e^{i\omega_s t} \tag{3.2.31}$$

Substituting Eqs. (3.2.30) and (3.2.31) into Eq. (3.2.29) yields the next equation for each term of Fourier transform $s = 0$–$N/2$.

$$(-\omega_s^2[M] + i\omega_s[C] + [K])U_s = -\{m\}\ddot{z}_s \tag{3.2.32}$$

The matrices $[M]$ and $[C]$ are real number matrices, while the stiffness matrix $[K]$ can be a complex number matrix to represent not only the soil stiffness but also the internal soil damping. As already mentioned in Sec. 1.5, the complex shear moduli $G^*$ in various viscoelastic models can be incorporated in $[K]$, though $G^*$ for the Nonviscous Kelvin model in Eq. (1.5.23) is normally employed in soil materials. Thus, Eq. (3.2.32) with the complex coefficient in the parenthesis on the left side is directly solved. The stiffness and damping properties can be prescribed individually on the element by element basis. The damping matrix $[C]$ is not actually necessary for material damping in this method because the complex matrix $[K]$ represents not only the stiffness but also the damping characteristics.

Eq. (3.2.32) is solved first by assuming $\ddot{z}_s = 1.0$ to have the solutions of the equation for the number of terms of the Fourier series $s = 0$–$N/2$ as:

$$U_{s1} = (\omega_s^2[M] - i\omega_s[C] - [K])^{-1}\{m\} \tag{3.2.33}$$

The computation is implemented actually not for all the terms $s$ but with some skips and interpolations for the computational efficiency (Lysmer et al. 1975). The time history of earthquake response is obtained by using Eq. (3.2.33) as:

$$U = \mathrm{Re}\sum_{s=0}^{N/2} U_s e^{i\omega_s t} = \mathrm{Re}\sum_{s=0}^{N/2} U_{s1}\ddot{z}_s e^{i\omega_s t} \tag{3.2.34}$$

Because Eqs. (3.2.33) and (3.2.34) are calculated by the computer algorithm named Fast Fourier Transform efficiently, the number of data points in the time axis $N$ should be the powers of 2 such as 2048, 4096, 8192, etc. including trailing zeros following actual records to secure the sufficient length for a subsequent quiet zone.

The complex response analysis in the frequency domain is convenient in dealing with boundary conditions. If a special treatment called as a transmitting boundary is built in by adding complex components at the vertical side nodes of a two-dimensional soil model, the surface wave can travel through the lateral boundary without reflection in a layered soil model, and thus reproducing the wave radiation to neighboring soils (Lysmer and Kuhlemeyer 1969, Lysmer and Waas 1972).

The complex response analysis utilizes the superposition of linear solutions, and cannot go beyond the linear properties. In order to apply this analysis to nonlinear problems in an approximate way, the equivalent-linear analysis is implemented.

#### 3.2.3.3 Mode-superposition analysis of discretized model

A mode superposition or modal analysis is another method based on the superposition of linear solutions, where nonlinear problems can be dealt with only approximately by the equivalent linear analysis. Here, the displacement vector $U$ of a $N$-degree of freedom system can be expressed as a linear summation of individual mode vectors $\{X_i\}$, which are orthogonal to each other, with the amplitude $q_i$ ($i=1$–$N$) as shown in Eq. (3.2.35).

$$U=\sum_{i=1}^{N}\{X_i\}q_i=[X]\{Q\} \tag{3.2.35}$$

$[X]$ is a $N$ by $N$ matrix with individual $\{X_i\}$ comprising $i$-th column, and $\{Q\}$ is a vertical vector composed of $q_i$. Substituting Eq. (3.2.35) into Eq. (3.2.29) and implementing some matrix calculations based on the orthogonality of $\{X_i\}$ yield the next $N$ independent equations.

$$\ddot{q}_i+2D_i\omega_i\dot{q}_i+\omega_i^2q_i=\frac{p_i}{m_i}=-\eta_i\ddot{z}\quad(i=1-N) \tag{3.2.36}$$

Here, $m_i=X_i^T[M]X_i$, $D_i=c_i/2\sqrt{m_ik_i}$, $\omega_i^2=k_i/m_i$, $p_i=-X_i^T\{m\}\ddot{z}=-\eta_im_i\ddot{z}$, $c_i=X_i^T[C]X_i$, $k_i=X_i^T[K]X_i$, and $\eta_i=X_i^T\{m\}/m_i$ is called a mode-participation factor, representing the degree of participation of individual modes in the global vibration. Eq. (3.2.36) represents one-degree of freedom vibration systems of $N$-numbers shaken by the acceleration $\eta_i\ddot{z}$. Calculating $q_i(t)$ for $i=1$–$N$ individually in the time domain and superposing them in Eq. (3.2.35) yields the ultimate solution $U$. In calculating Eq. (3.2.36) in actual problems, it is sufficient to calculate only the lower order modes with larger mode-participation factors, ignoring the higher order modes to have a solution with a sufficient accuracy.

As for the damping matrix [C], so-called Rayleigh damping in the next equation consisting of two components; one proportional to the mass matrix $[M]$ and the other proportional to the stiffness matrix $[K]$, with the proportionality constants $\alpha$ and $\beta$ with units $s^{-1}$ and s, respectively, is sometimes used.

$$[C]=\alpha[M]+\beta[K] \tag{3.2.37}$$

Substituting this into Eq. (3.2.29) and conducting the matrix calculations based on the orthogonality of $\{X_i\}$ using Eq. (3.2.35) yields the damping ratio $D_i$ for the $i$-th mode.

$$D_i=\frac{1}{2\omega_i}\alpha+\frac{\omega_i}{2}\beta \tag{3.2.38}$$

This indicates that, in the Rayleigh damping, the modal damping ratio of $i$-th mode $D_i$ is not constant but variable with the resonant frequency of $i$-th order $\omega_i$. It is obvious that $D_i$ takes a minimum value at $\partial D_i/\partial\omega_i=0$ ($\omega_i=\sqrt{\alpha/\beta}$), and tends to increase with increasing/decreasing $\omega_i$. Thus, care is needed how to choose the constants $\alpha$ and $\beta$ properly in implementing the modal analyses if the soil damping is postulated to be essentially frequency-independent.

In the modal analysis, the damping ratios can be prescribed not to the individual elements but to the individual vibration modes such as the modal damping ratios $D_i$ in Eq. (3.2.36). The Rayleigh damping is one of the methods how to determine the modal damping associated with the mass and stiffness matrices. This is quite different from the complex response analysis where the material damping ratios can be prescribed to individual elements of a model. However, it is possible in the modal analyses to determine the modal damping ratios which are equivalent to the material damping so that the same energy is dissipated in a single cycle of vibration in that particular vibration mode. Namely, the shear strain energy $W_{ij}$ and the dissipated energy $\Delta W_{ij}$ in the $i$-th vibration mode in one cycle in an element $j$ can be written by using shear strain $\gamma_{ij}$, shear modulus $G_{ij}$, material damping ratio $D_{ij}$ referring to Eq. (1.5.7) as:

$$W_{ij}=\frac{G_{ij}\gamma_{ij}^2}{2},\quad \Delta W_{ij}=4\pi W_{ij}D_{ij} \tag{3.2.39}$$

Using these energies per unit volume and each element volume $V_j$, the corresponding energies in the $i$-th vibration mode in all the elements are written respectively as follows.

$$\sum_j V_jW_{ij}=\sum_j\frac{V_jG_{ij}\gamma_{ij}^2}{2},\quad \sum_j V_j\Delta W_{ij}=\sum_j 4\pi V_jW_{ij}D_{ij} \tag{3.2.40}$$

In order to equalize the total dissipated energy in one cycle in the $i$-th vibration mode, the modal damping ratio in that particular vibration mode $D_{M_i}$ is expressed by using Eq. (1.5.7) again as:

$$D_{M_i}=\frac{\sum_j V_j\Delta W_{ij}}{4\pi\sum_j V_jW_{ij}}=\frac{\sum_j[D_{ij}V_j(G_{ij}\gamma_{ij}^2/2)]}{\sum_j[V_j(G_{ij}\gamma_{ij}^2/2)]} \tag{3.2.41}$$

This indicates that the modal damping $D_{M_i}$ is obtained as the weighting average of the material damping $D_{ij}$ in the $i$-th vibration mode in the individual elements with the weights $V_j(G_{ij}\gamma_{ij}^2/2)$, the strain energy in the individual elements. This allows to choose appropriate modal damping ratios for different modes considering the frequency-independent soil damping, without using the Rayleigh damping.

#### 3.2.3.4 *Time-domain stepwise nonlinear analysis of discretized model*

Eq. (3.2.29) is directly integrated in the time domain incorporating nonlinear soil properties which change in a small time increment. It is suitable to strongly-nonlinear problems such as liquefaction, though the computational efficiency is not as good as the methods mentioned above. For the stepwise calculation, the dynamic equation is expressed in the incremental form with the time increment $\Delta t$.

$$[M]\,\Delta\ddot{U} + [C]\,\Delta\dot{U} + [K_t]\,\Delta U = -\{m\}\Delta\ddot{z} \qquad (3.2.42)$$

The stiffness matrix $[K_t]$ changes step by step incorporating the tangent moduli determined in the stress-strain models in each step. Starting from the initial values of acceleration $\ddot{U}$, velocity $\dot{U}$ and displacement $U$ at $t=0$, the dynamic response in each step is calculated by adding the computed increments to corresponding values in the previous step as:

$$\left.\begin{array}{l} \ddot{U}(t) = \ddot{U}(t-\Delta t) + \Delta\ddot{U} \\ \dot{U}(t) = \dot{U}(t-\Delta t) + \Delta\dot{U} \\ U(t) = U(t-\Delta t) + \Delta U \end{array}\right\} \qquad (3.2.43)$$

As for the stress–strain curves to determine tangent moduli, several models such as the HH and RO models are used together with the Masing rule as mentioned in Sec. 3.1. At each time step, ascending (loading) or descending (unloading) is judged on the stress-strain curve to determine the tangent modulus using the Masing rule for the stepwise calculation. In this nonlinear analysis, the residual displacement can be calculated unlike other linear or equivalent linear analyses. The bulk modulus or Poisson's ratio necessary in two or three-dimensional analyses is assumed constant in a normal total stress analysis, although they may also behave nonlinearly.

In the effective stress analyses, the tangent moduli of submerged soils below water table are further modified step by step reflecting excess pore-pressure variations evaluated in another evaluation flow based on the soil dilatancy. Pore-water migration due to consolidation or seepage flow may also be considered in the effective stress analyses. Several constitutive relations including elastoplastic models are used to reproduce the soil dilatancy, the theoretical details of which are out of the scope here.

In the step-by-step nonlinear analysis, damping characteristics are intrinsically included in chasing hysteretic stress-strain curves in the stepwise calculations, and hence the damping matrix [C] for the hysteretic damping is not necessary. Nevertheless, the Rayleigh damping [C] in Eq. (3.2.37) is sometimes employed in the nonlinear analysis. Out of the force by the Rayleigh damping $[C]\Delta\dot{U} = \alpha[M]\Delta\dot{U} + \beta[K]\Delta\dot{U}$, $\alpha[M]\Delta\dot{U}$ is analogous to the inertial force with larger damping effect in lower frequency, while $\beta[K]\Delta\dot{U}$ is analogous to the spring force with larger damping effect in higher frequency as inferred from Eq. (3.2.38). Considering these trends, the Rayleigh damping is used to stabilize the dynamic response of strongly nonlinear system when the hysteretic damping is temporarily small because of small induced strains.

In order to properly reproduce the wave radiations through the model boundaries in the time domain analyses, a viscous boundary can be used. This boundary is equipped with a line of viscous dashpots as illustrated in Fig. 3.2.3, which is supposed to absorb the wave energy without reflection. The dashpot constant is equal to the wave impedance, $\rho V_s$ for the SH-wave or $\rho V_p$ for the P-wave of the soil at the boundary.

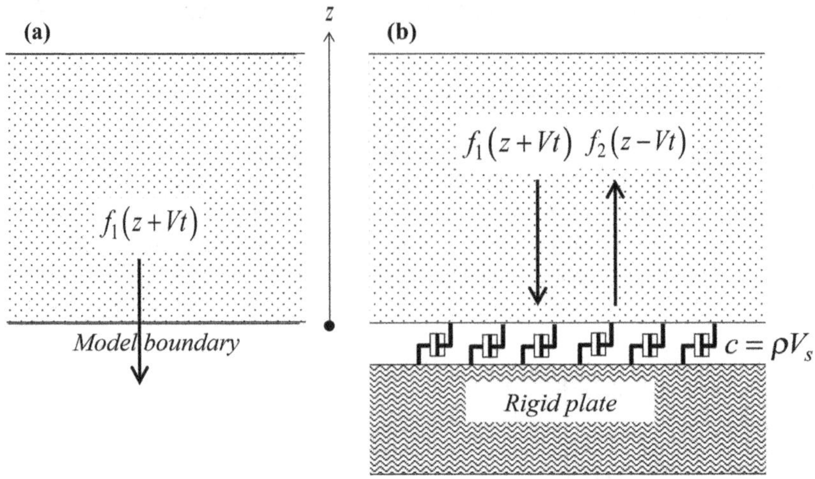


*Figure 3.2.3* Viscous boundary with dashpots without reflecting waves.

If the vertical $z$-axis is taken upward, the one-dimensional SH-wave propagation with the downward $f_1$ and upward reflected $f_2$ waves is expressed as:

$$u = f_1(z + V_s t) + f_2(z - V_s t) \tag{3.2.44}$$

If there is no reflected wave, it becomes $u = f_1(z + V_s t)$ and the shear stress at the boundary is written as:

$$\tau\Big|_{z=0} = G\frac{\partial}{\partial z}f_1(z + V_s t)\Big|_{z=0} = \frac{G}{V_s}\frac{\partial}{\partial t}f_1(z + V_s t)\Big|_{z=0} = \rho V_s \frac{\partial f_1}{\partial t}\Big|_{z=0} = \rho V_s \dot{u}\frac{\partial}{\partial z}\Big|_{z=0} \tag{3.2.45}$$

Here, $\rho V_s \dot{u}|_{z=0}$ is the product of the wave impedance $\rho V_s$ and the particle velocity at the boundary $\dot{u}|_{z=0}$, which is equivalent to the stress exerted in a dashpot with the viscosity $c = \rho V_s$ loaded with the particle velocity $\dot{u}$. This indicates that the viscous boundary connected with the rigid plate via the dashpots can eliminate the reflected waves. Likewise, for the P-wave the dashpot viscosity is $c = \rho V_p$. These are simple cases where the incident waves are normal to the boundaries. For oblique waves, it is difficult to completely eliminate the reflected waves by the viscous boundary.

It is not difficult to implement this kind of stepwise nonlinear analysis today on two or three dimensional models with a huge number of degree-of-freedom, though the analytical results are sometimes very diverted among different analytical schemes and sensitive to optional parameters. Nevertheless, their reliability cannot be verified by comparing with rigorous solutions unlike linear analyses. Consequently, case history studies with well-documented soil and earthquake data or well-organized model test studies are indispensable to demonstrate their applicability and reliability in various conditions.

### 3.2.4 Equivalent linear analysis

If soil properties are not so strongly nonlinear, equivalent linear analyses are often employed in engineering practice. In the analysis, equivalent linear properties compatible with induced strain levels are determined element by element by iterations, and the linear analysis using the properties ultimately converged yield the final solution. Because of the linearity, the equivalent linearization can be incorporated in the complex response analysis or the modal analysis to efficiently solve the systems of large degrees of freedom.

#### 3.2.4.1 *Analytical procedure*

The equivalent linear properties; the secant shear modulus ratio $G/G_0$ and the hysteretic damping ratio $D$, versus the effective shear strain $\gamma_{\mathrm{eff}}$ schematically illustrated in Fig. 3.2.4 are followed iteratively in this analysis (Seed and Idriss 1971). The calculation steps are as follows.

i) Estimate the effective strains $\gamma_{\mathrm{eff}}$ induced by a particular earthquake motion and determine initial tentative values of $G$ and $D$ using the $G{\sim}\gamma_{\mathrm{eff}}$ and $D{\sim}\gamma_{\mathrm{eff}}$ curves.
ii) Compute the dynamic response of the model by the earthquake motion using the predetermined $G$ and $D$. This is a linear analysis with the soil properties unchanged during the one iteration cycle of the analysis.
iii) From the calculated strain time-histories, the effective strains $\gamma_{\mathrm{eff}}$ are determined in individual elements. From the $\gamma_{\mathrm{eff}}$-values, the newly-iterated $G$ and $D$ are determined using the $G{\sim}\gamma_{\mathrm{eff}}$ curves and $D{\sim}\gamma_{\mathrm{eff}}$ curves as indicated in Fig. 3.2.4.
iv) Compare the newly-iterated $G$ and $D$ with the previously iterated values, and if their differences are all within certain allowable limits (e.g. ±5%), then these values are judged as finally converged properties corresponding to the converged effective strains $\gamma_{\mathrm{eff}}$. Otherwise, return to ii) above and the iteration continues.

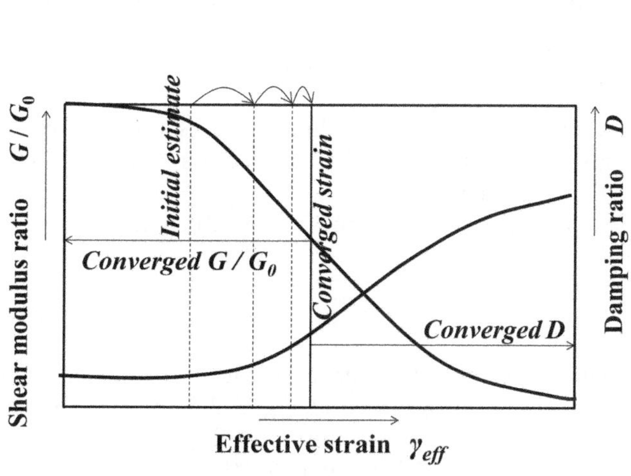


*Figure 3.2.4* Iterative procedure in equivalent linear analyses using $G/G_0$–$\gamma$ curves and $D$–$\gamma$ curves in semi-log diagram.

Thus, in the equivalent linear analysis, the effective strain $\gamma_{eff}$ is a key parameter to obtain the final results, though it is not so clear how to determine $\gamma_{eff}$ logically from the calculated strain time-histories in iii) above. The following simple formula is employed normally in practice (e.g. Schnabel et al. 1972), where $\gamma_{max}$ = maximum induced shear strain during a particular earthquake motion, and $\alpha$ = reduction coefficient.

$$\gamma_{eff} = a\gamma_{max} \tag{3.2.46}$$

The coefficient $\alpha$, providing optimum results reflecting nonlinear soil properties, is certainly dependent on the earthquake motions, the degree of soil nonlinearity, and what kind of design values (acceleration, strain, etc.) are needed. It is accepted that in many cases $\alpha$ is between 0.5 and 0.7, and $\alpha = 0.65$ is chosen as a default value (Seed and Idriss 1971).

Anyway, this analysis is actually a linear analysis which cannot follow the time-dependent variations of soil properties during seismic loading. Hence the discrepancies from actual soil behavior tends to widen with the increasing strain levels. The earthquake-induced strain calculated by the equivalent linear analysis tends to be significantly smaller than that by the stepwise nonlinear analysis as the soil nonlinearity gets stronger, while the difference in the exerted stresses may not differ so widely between the two analyses. Furthermore, the non-zero residual strains at the end of nonlinear soil response cannot be evaluated by the equivalent linear analysis.

#### 3.2.4.2 Modification of equivalent linear analysis

As one of the significant problems of the equivalent linear analysis, high-frequency accelerations in the computed response tend to be underestimated as the soil nonlinearity becomes stronger. This is because the induced strains by accelerations of higher frequency tend to be smaller than those of lower frequency involved in a given earthquake motion. Because a single damping ratio $D$ has to be chosen in each soil element in the one-iteration cycle of the analysis, larger damping ratio corresponding to a larger $\gamma_{max}$-value induced by the lower frequency acceleration motions is employed actually. It is sometimes too large for the higher acceleration motions, leading to the underestimation of acceleration amplifications in the higher frequency compared to the lower frequency.

In this regard, a modified equivalent linear analysis was proposed where the soil properties can be given differently considering the different induced strains in the different frequency ranges for the same earthquake motion (Sugito 1995). In this method, the coefficient $\alpha$ in Eq. (3.2.46) is not constant but variable with the frequency $\omega = 2\pi f$ as:

$$\gamma_{eff}(\omega) = C\gamma_{max}\frac{F_\gamma(\omega)}{F_{\gamma_{max}}} \tag{3.2.47}$$

Here, $C$ = a constant, $F_\gamma(\omega)$ = Fourier spectrum of induced strain and $F_{\gamma_{max}}$ = maximum value of $F_\gamma(\omega)$ as exemplified in Fig. 3.2.5. Eq. (3.2.47) enables the shear modulus and damping to be not only strain-dependent but also frequency-dependent, while it returns to original Eq. (3.2.46) if $F_\gamma(\omega)/F_{\gamma_{max}} = 1.0$ and $C = \alpha$. In the figure for example, the peak value $F_{\gamma_{max}}$ appears at around $f = 2$ Hz, and $F_\gamma(\omega)$ at higher frequencies is much smaller than $F_{\gamma_{max}}$, indicating that $\gamma_{eff}(\omega)$ becomes smaller in the

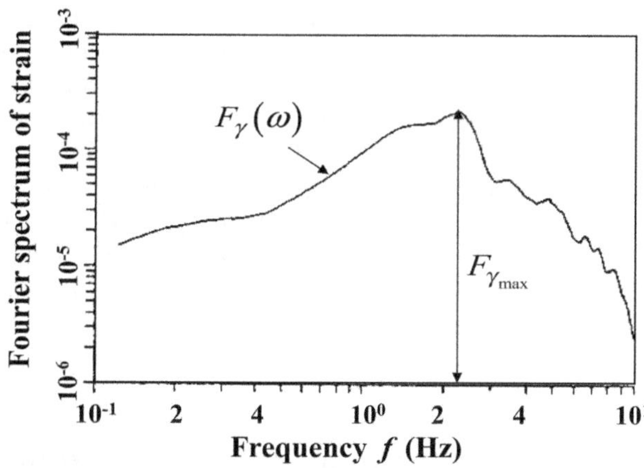


*Figure 3.2.5* Typical Fourier spectrum of strain for modified equivalent linear analysis (Sugito 1995).

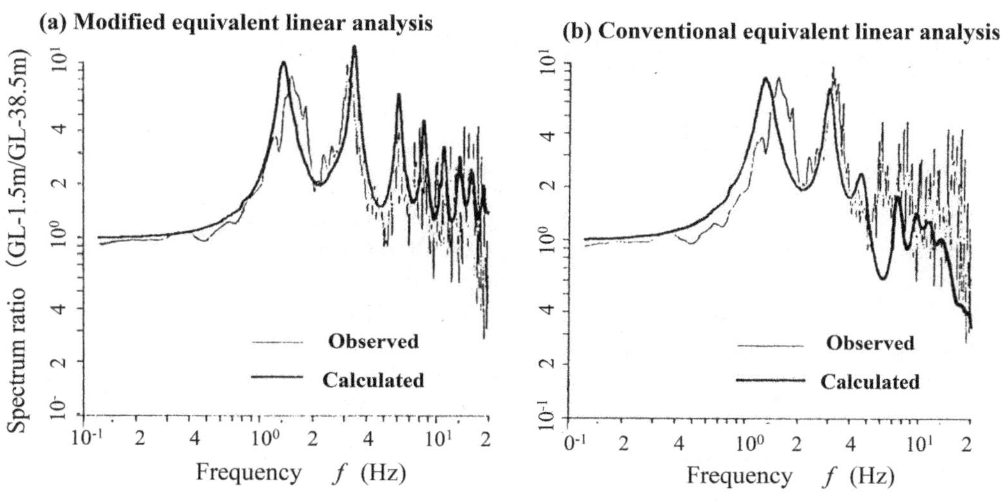


*Figure 3.2.6* Calculated acceleration response spectrum ratio compared with observation: (a) Modified equivalent linear analysis, (b) Conventional equivalent linear analysis (Sugito 1995).

higher frequencies in Eq. (3.2.47) and so does the corresponding damping ratio $D$. In this manner, the effective strains can be given differently in multiple frequency ranges using the Fourier strain spectrum such as Fig. 3.2.5 and the associated soil properties are obtained by using $G\sim\gamma_{eff}$ and $D\sim\gamma_{eff}$ curves of Fig. 3.2.4 in the individual frequency ranges.

Fig. 3.2.6 shows the ratios of acceleration response spectra between the ground surface and the depth of 38.5 m at a site calculated by the one-dimensional equivalent-linear response analysis, wherein the computed spectrum ratio with the thick curve by the modified analysis in (a) fits well with the observed spectrum of vertical array records in the thin curve. In contrast, the spectrum ratio by the conventional analysis in (b) obviously underestimates the observation in higher frequencies because the equivalent damping ratio is given too high. Thus, it is possible to improve the equivalent linear analysis to a certain extent by introducing the frequency-dependent properties.

The lower damping ratio in higher frequencies introduced in the modified equivalent linear analysis apparently looks similar to the frequency-dependency of damping by wave scattering to be addressed in Sec. 4.3.3. However, the frequency-dependency of damping discussed here is originated from the strain dependency of soil properties. Despite the similarity in the frequency-dependency, their mechanisms are completely different.

### 3.2.5 Equivalent linear and nonlinear analyses compared with model test

After addressing various numerical analyses, it is of interest how well or poorly they reproduce actual soil response. In the following, the dynamic soil response by 1G shaking table tests is simulated by two representative analyses; the equivalent linear analysis and the stepwise nonlinear analysis to understand their characteristics (Kokusho et al. 1979, Kokusho 1982).

#### 3.2.5.1 *Shaking table test and 1D soil model*

A dry or saturated uniform sand layer was made in a laminar shear box in Fig. 3.2.7 with depth 1005 mm and inner horizontal area 1200 mm and 800 mm. It was probably a pioneering work of shaking table model test using the laminar shear box to realize free shear mode vibration of a horizontal soil layer (Kokusho et al. 1979). The sand layer was shaken horizontally by irregular seismic motions with various intensities to observe the dynamic response and its variation due to soil nonlinearity. The one-dimensional soil layer was idealized either by a 4-lumped mass model for the nonlinear step-by-step analysis or by a 4-layer continuum model for the equivalent linear analysis

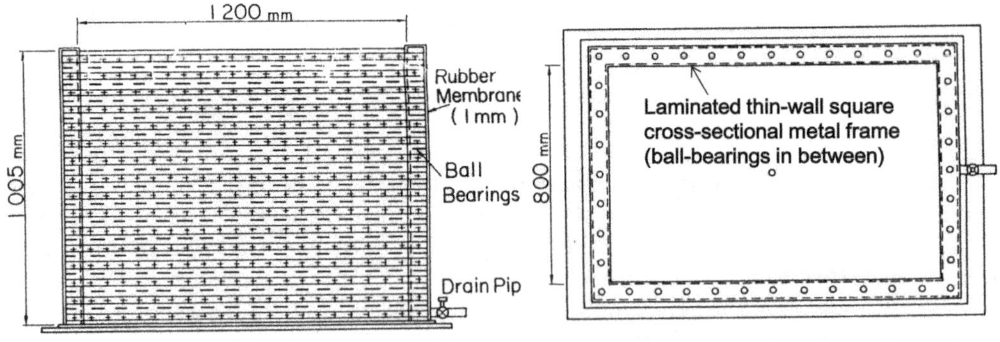


*Figure 3.2.7* Laminar shear box used in model shaking table test (Kokusho et al. 1979).

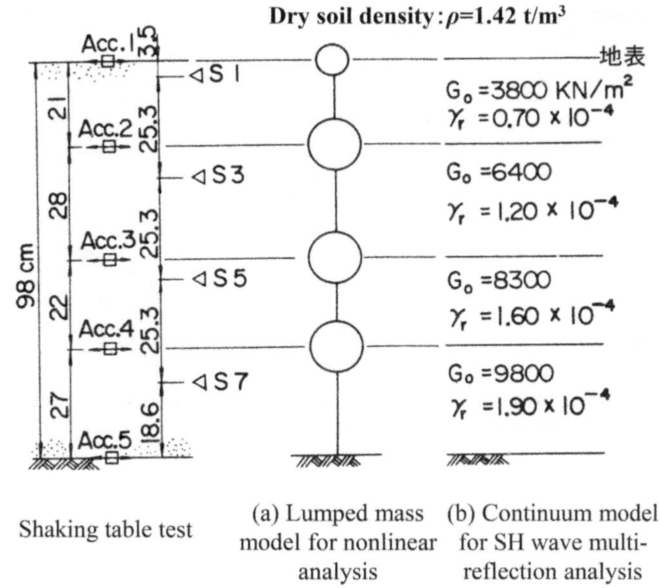


*Figure 3.2.8* Two types of analytical models for shaking table tests: (a) Lumped mass-spring model for nonlinear analysis, (b) continuum model for equivalent linear analysis (Kokusho 1982).

as illustrated in Fig. 3.2.8. The soil properties used in the analyses were measured for the sand under very low confining stresses. Namely, the initial shear moduli of individual layers were determined so that the first resonant frequency of the analytical model is equal to the observed frequency in the sand model for small strains. The strain-dependent variations of $G/G_0$ and $D$ were measured in torsional simple shear tests under the confining stresses from 100 kPa to very low 2 kPa as shown in Fig. 3.2.9. This figure demonstrates that the variations of $G/G_0$ and $D$ are almost uniquely correlated with the normalized shear strain $\gamma/\gamma_r$ or $\gamma/\gamma_y$ as mentioned in Sec. 2.4.1 even for very low confining stresses down to $\sigma_c' = 2$ kPa. The test results are compared fairly well with the equivalent linear HH model formulated in Eqs. (3.1.19), (3.1.20) and the RO model in Eqs. (3.1.23), (3.1.25). However, damping ratio by the HH model tends to considerably overestimate the experiments for $\gamma/\gamma_r > 1.0$ as already pointed out in Sec. 3.1.3. For the RO model, $\alpha_{RO} = 3.46$, $\beta_{RO} = 1.8$ were chosen among the three curves drawn in Fig. 3.2.9.

In the equivalent linear analysis, the HH and RO models were used to compare with the shaking table test results, though $D = 0.02$ was added in the two damping models to have better matching with the measured small-strain damping. The stress reduction coefficient in Eq. (3.2.46) was chosen as $\alpha = 0.65$.

In the step-by-step nonlinear analysis, the two hysteretic models were used to compare to each other. The stress-strain curves are constructed from the skeleton curves using Eqs. (3.1.7)~(3.1.10) in the Masing rule. The tangent moduli $G_t$ used in

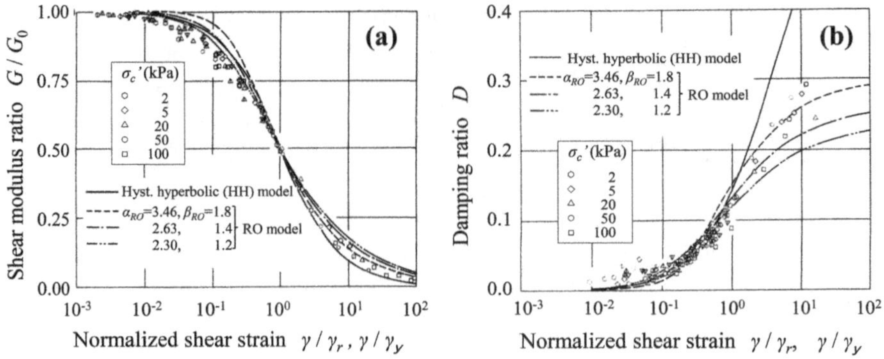

*Figure 3.2.9* Soil properties of clean sand from ultra-low to normal confining stresses $\sigma_c' = 2\sim100$ kPa (Kokusho 1982): (a) $G/G_0 \sim \gamma$, (b) $D \sim \gamma$.

the analysis in the HH model are formulated as:

$$\frac{G_t}{G_0} = \frac{1}{(1+\gamma/\gamma_r)^2} = \left(1 - \frac{|\tau|}{\tau_f}\right)^2 \quad \text{(skeleton curve)} \tag{3.2.48}$$

$$\frac{G_t}{G_0} = \frac{1}{[1 \mp (\gamma - \gamma_a)/\gamma_r]^2} = \left(1 \pm \frac{\tau - \tau_a}{2\tau_f}\right)^2 \quad \text{(hysteretic curves)} \tag{3.2.49}$$

and those in the RO model are formulated as:

$$\frac{G_t}{G_0} = \frac{1}{[1 + \alpha_{RO}(1+\beta_{RO})|\tau/\tau_y|^{\beta_{RO}}]^2} \quad \text{(skeleton curve)} \tag{3.2.50}$$

$$\frac{G_t}{G_0} = \frac{1}{[1 + \alpha_{RO}(1+\beta_{RO})|\mp(\tau - \tau_a)/\tau_y|^{\beta_{RO}}]^2} \quad \text{(hysteretic curves)} \tag{3.2.51}$$

where the upper and lower signs in $\mp$ or $\pm$ correspond to the descending and ascending curves, respectively, and $\gamma_a$ and $\tau_a$ are the strain and stress values at the turning point, respectively. The stiffness matrix is constructed with these tangent moduli at each time step, and the dynamic equation Eq. (3.2.42) incorporating these is solved. Here, the damping matrix such as the Rayleigh damping is not used. It is the total stress analysis considering the effect of pore-pressure buildup only implicitly even for saturated sand layers.

### 3.2.5.2 *Comparison of analyses and model test*

Figs. 3.2.10(a), (b) show acceleration responses at the surface of the dry sand model calculated by the stepwise nonlinear and equivalent linear analyses, respectively, with input accelerations given at the bottom of the layer, and compared with the test results. Here, the maximum acceleration at the surface is around 0.2 g with the maximum induced strain in the layer $\gamma = 5 \times 10^{-4}$, corresponding to the modulus degradation $G/G_0 = 0.4$–$0.5$. It is observed that the stepwise nonlinear analysis can reproduce

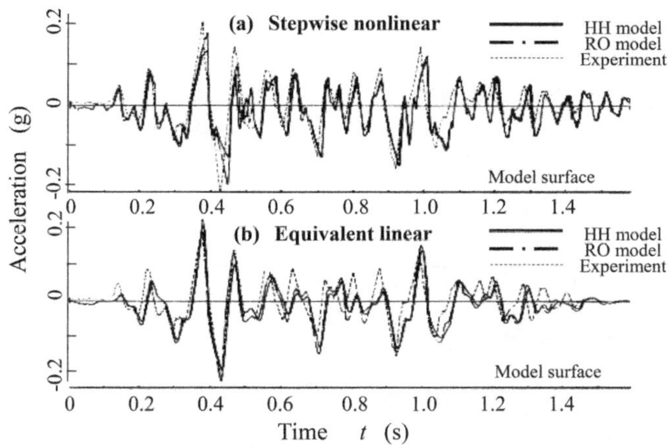


*Figure 3.2.10* Calculated acceleration responses at model surface compared with experiment results (Kokusho 1982): (a) Stepwise nonlinear analysis, (b) Equivalent linear analysis.

smaller amplitudes involved in the same seismic wave as good as larger amplitudes. In contrast, the equivalent linear analysis can reproduce larger amplitudes much better than smaller amplitudes. This is because the properties are varied stepwise in the time domain in the nonlinear analysis while they stay at constant values optimized to have better matching for larger amplitudes in the equivalent linear analysis. A closer look at the same figure also reveals that the calculated response contains high-frequency small-amplitude vibrations in the former, while it is very smooth with not much high frequency vibrations in the latter. It is because of the excessive damping of high-frequency motions in the equivalent linear analysis as already mentioned in Sec. 3.2.4.2 and also because no damping works literally in small amplitude vibrations in the nonlinear analysis, necessitating a mechanism such as Rayleigh damping. Also note that the difference in constitutive models has larger impact on the calculated response in the former than the latter. For the nonlinear HH model in particular, the response acceleration tends to be suppressed to a lower value because the HH model has an asymptotic maximum shear strength and an extraordinarily large damping ratio in large strains.

Figs. 3.2.11 depicts acceleration response spectra (damping ratio = 5%) at the model surface for small and large input accelerations for the nonlinear analysis in (a) and the equivalent linear analysis in (b), respectively. For the small input acceleration (0.045 g) with weak soil nonlinearity in the top diagram, the stepwise nonlinear analysis overestimates the model test in the higher frequency range probably due to too small damping ratio there, while the equivalent linear analysis gives a fair coincidence with the test for all the frequency range. For the large input acceleration (0.414 g) in the bottom diagram, the HH model predicts the smaller response than the RO model both for the nonlinear and equivalent linear analyses, reflecting significantly larger damping ratio in the HH model for large strains. Also can be seen from the figure that there is a single spectrum peak in the large-input equivalent linear analysis because it has a distinct resonant period determined from the converged moduli. In contrast, a single

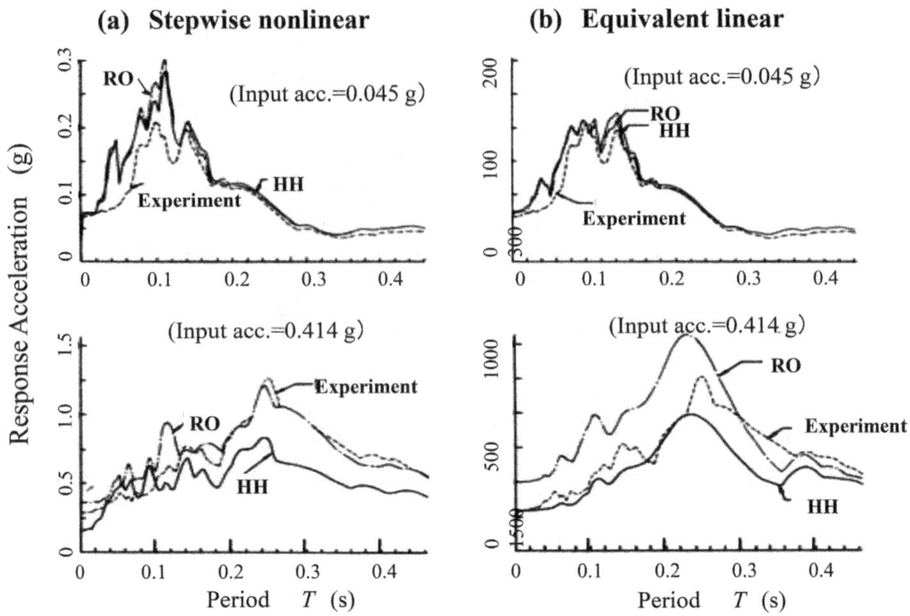


*Figure 3.2.11* Calculated acceleration response spectra (damping ratio = 5%) at model surface compared with test results for two input accelerations of 0.045 g (top) and 0.414 g (bottom) (Kokusho 1982): (a) Stepwise nonlinear analysis, (b) Equivalent linear analysis.

dominant peak is difficult to appear in the strong-input nonlinear analysis because soil properties are always changing drastically without any fixed resonant period. In the nonlinear analysis, the RO model, whose parameters $\alpha_{RO}$ and $\beta_{RO}$ in Eqs. (3.2.50) and (3.2.51) are adjustable to have better matching with the actual soil properties, can simulate the measured response spectrum much better than the HH model, though it tends to exaggerate ups and downs of the measured spectrum in the short period range.

Fig. 3.2.12 shows the shear strain time-histories at the 2/3 depth of the uniform sand layer for the strong input acceleration (0.414 g) calculated in the nonlinear and equivalent linear analyses using the two soil models. The strain amplitudes in the nonlinear analysis are larger than those in the equivalent linear analysis for the two models. Also note that the residual strains obtained in the nonlinear analyses due to the irregularity of input motions using the HH and RO models are very different, indicating that the proper choices of soil models and associated parameters are essential to have meaningful residual deformations critical to the Performance-Based Design.

Fig. 3.2.13 illustrates stress-strain hysteretic curves at the same depth obtained by the nonlinear analyses using the two soil models. Though the difference in the hysteresis looks small in the earlier stage, it becomes greater later on reflecting the asymptotic maximum stress in the HH model beyond that the shear stress cannot go.

In the Performance-Based Design, various numerical analyses (total and effective stress analyses) incorporating strong nonlinear properties of soils are already

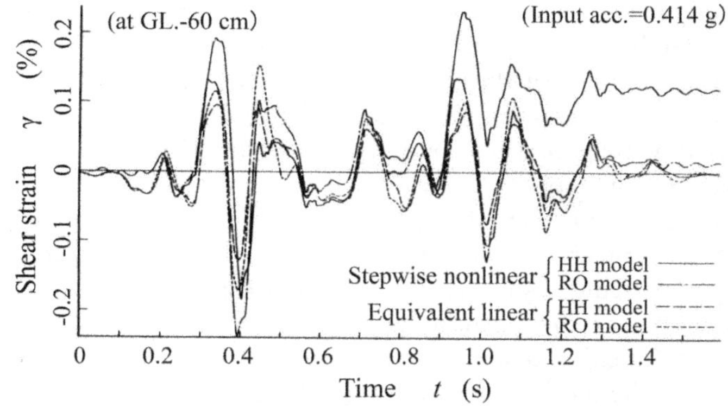


*Figure 3.2.12* Calculated strain time-histories for different soil models (Kokusho 1982).

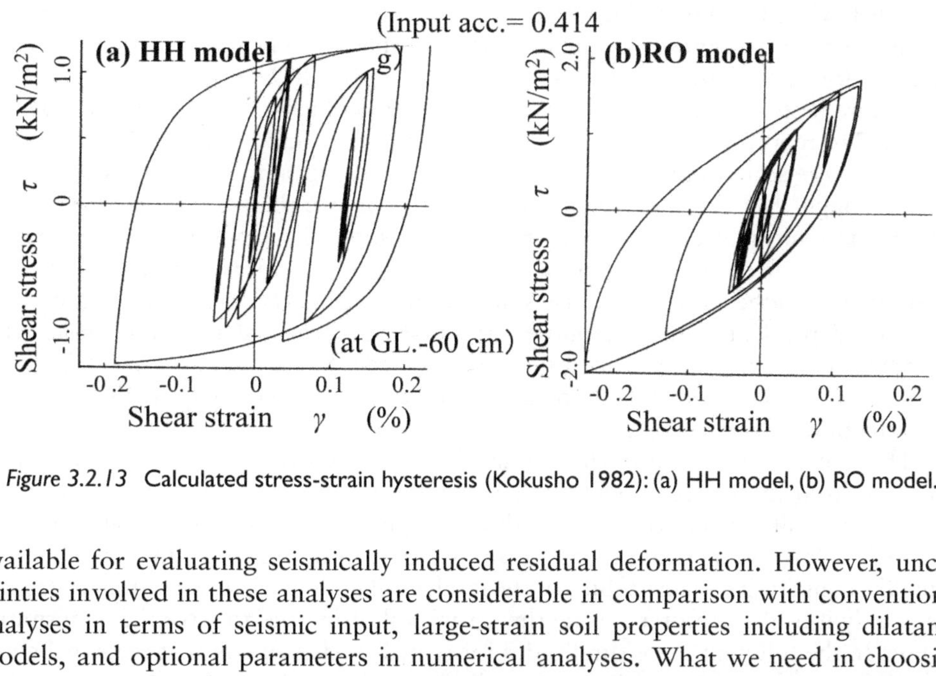


*Figure 3.2.13* Calculated stress-strain hysteresis (Kokusho 1982): (a) HH model, (b) RO model.

available for evaluating seismically induced residual deformation. However, uncertainties involved in these analyses are considerable in comparison with conventional analyses in terms of seismic input, large-strain soil properties including dilatancy models, and optional parameters in numerical analyses. What we need in choosing appropriate values for input parameters and judging the reliability of analytical outcomes is a sort of benchmark case histories with well-documented geotechnical and seismic conditions in situ and also benchmark model test results.

## 3.3 SCALED MODEL TESTS AND SOIL MODELS

### 3.3.1 Needs for model tests

Dynamic soil behaviors during earthquakes are sometimes investigated by means of model tests in addition to numerical analyses. Unlike soil element tests on uniform soil materials, the model tests aim to simulate soil behaviors in prototypes by incorporating geometrically similar models with soils and boundary conditions as realistic as possible. In the numerical analysis, the constitutive relationship of soil is specified in individual elements, and the response of the global system is solved with initial and boundary conditions for a given seismic motion. In contrast, the model test is undertaken to simulate the dynamic response of a total physical system without specifying particular soil constitutive relationship.

Prototype field tests or case-history records would serve as ultimate model tests, if associated detailed geotechnical and seismic data are available. Recently great efforts are increasingly producing good and high-density in situ records in many parts of the world and disseminating them among researchers. However, the case histories with high-density geotechnical and seismic data coupled are very limited in quantity and quality. In contrast, the model tests have a great advantage in that the necessary data can be repeatedly acquired in prescribed conditions. Hence, the aim of model tests is not to reproduce in situ soil behaviors in general but to focus on some specified aspects of soil behaviors to be investigated.

As the model tests in soil dynamics, shaking table tests in 1 g were carried out from old times for earth dams (e.g. Clough and Pirtz 1956). Model tests for horizontal soil layers were first conducted in a laminar shear box (Kokusho et al. 1979) as shown in Fig. 3.2.7 to reproduce the ground vibration in the shear mode by the SH-wave propagation. Large scale model tests of several meters in depth are sometimes conducted today, though they are still considerably smaller in size than the prototype. It may well be expected that a model with geometrical similarity made from the same soil materials tends to show similar behavior to the prototype. However, the quantitative similarity cannot be assured because soils are frctional materials, and their properties are highly dependent on the overburden stress. More recently, shaking table tests in the centrifugal environments are increasingly conducted wherein the overburden is raised, so that small models can be comparable with prototypes not only qualitatively but also quantitatively in the light of similitude.

There may be two approaches to draw the quantitative results on the prototype behavior from the scaled model tests: i) Establish a similitude law to interpret the model test results so that they can be directly applicable to prototype soil behaviors, and ii) Develop a numerical procedure and validate its applicability in a scaled model using the constitutive relationships of the model soil under very-low confining stresses, then apply it to the prototype using the actual soil properties to have the quantitative results. In the following, the similitudes for scaled models of soil materials in 1 g and centrifugal accelerations arc discussed, followed by the soil properties under very low confining stresses in 1 g scaled models.

## 3.3.2 Similitude for scaled model tests

### *3.3.2.1 How to derive similitude*

In soil-related problems, nonlinear stress-strain relationships and dilatancy in the soil materials make it difficult to rigorously describe the constitutive laws included in the governing equations. In such cases, the following three methods may be employed to develop the similitude.

A. Non-dimensional parameters are generated arbitrarily or in the light of "$\pi$-theorem" (Buckingham 1914), combining pertinent physical variables associated with a particular problem. The similitudes are derived by equating these parameters in the model and the prototype.
B. Forces involved in a particular problem are picked up, and the similitudes are derived by equating the ratios of the individual forces between the model and prototype (Kagawa 1978).
C. Governing equations associated with a particular problem are chosen, and the variables included in the equations are assumed to be all proportional between the model and prototype. The similitudes are derived by determining the constants of proportionality so that the same governing equations hold both in the model and in the prototype (Joseph et al. 1988, Iai 1989).

### 3.3.2.2 *Derivation of similitude by forces*

Here, a similitude for a shaking table model test conceptually illustrated in Fig. 3.3.1 is derived. Let $L$ and $t$ be length and time, and the subscripts $m$ and $p$ represent the model and prototype, respectively, then their scaling ratios are written here as:

$$\frac{L_m}{L_p} = \frac{1}{\lambda}, \quad \frac{t_m}{t_p} = \frac{1}{\tau} \tag{3.3.1}$$

In a similar manner, the scaling ratios for $\rho$ = soil density and $g$ = gravitational acceleration are:

$$\frac{\rho_m}{\rho_p} = \frac{1}{\eta}, \quad \frac{g_m}{g_p} = \frac{1}{\gamma} \tag{3.3.2}$$

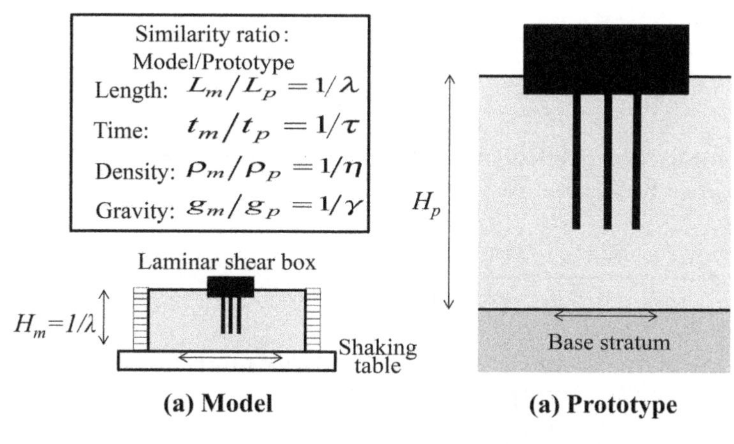


Figure 3.3.1 Schematic illustrations of (a) shaking table model test versus (b) prototype.

Following the method B above by Kagawa (1978), the six forces are focused as follows: i) self-weight, ii) inertia, iii) force for deformation, iv) force for internal damping, v) force for cohesion and vi) force for friction.

i) Force for self-weight can be expressed as $\rho \times g \times L^3$, hence the ratio $r_d$ between the model and prototype is written as:

$$r_d = \frac{\rho_m g_m L_m^3}{\rho_p g_p L_p^3} = \frac{1}{\eta \gamma \lambda^3} \tag{3.3.3}$$

ii) Force for inertia can be expressed, by using displacement $u = \varepsilon L$, as $\rho \times L^3 \times \varepsilon \times L \times t^{-2}$, hence the ratio $r_i$ can be written as:

$$r_i = \frac{\rho_m \varepsilon_m L_m^4 t_m^{-2}}{\rho_p \varepsilon_p L_p^4 t_p^{-2}} = \frac{\tau^2}{\eta \beta \lambda^4} \tag{3.3.4}$$

iii) Force for deformation can be expressed using deformation modulus $E$ as $\varepsilon \times E \times L^2$, hence the ratio $r_e$ can be written as:

$$r_e = \frac{\varepsilon_m E_m L_m^2}{\varepsilon_p E_p L_p^2} = \frac{1}{\beta \lambda^2} \frac{E_m}{E_p} \tag{3.3.5}$$

iv) Force for internal damping $D$ can be expressed as $D \times \varepsilon \times E \times L^2$, hence the ratio $r_D$ can be written as:

$$r_D = \frac{D_m \varepsilon_m E_m L_m^2}{D_p \varepsilon_p E_p L_p^2} = \frac{1}{\beta \lambda^2} \frac{D_m E_m}{D_p E_p} \tag{3.3.6}$$

v) Force for cohesion $c$ can be expressed as $c \times L^2$, hence the ratio $r_c$ can be written as:

$$r_c = \frac{c_m L_m^2}{c_p L_p^2} = \frac{1}{\lambda^2} \frac{c_m}{c_p} \tag{3.3.7}$$

vi) Force for friction coefficient $\tan\phi$ can be expressed as $\sigma \times L^2 \times \tan\phi$, hence the ratio $r_f$ can be written as:

$$r_f = \frac{\sigma_m L_m^2 \tan\phi_m}{\sigma_p L_p^2 \tan\phi_p} = \frac{1}{\alpha \lambda^2} \frac{\tan\phi_m}{\tan\phi_p} \tag{3.3.8}$$

By equating the above six force ratios, the similarity ratios of basic physical variables between the model and prototype can be determined. First, the equality of self-weight and inertia $r_d = r_i$ leads to a similarity ratio in time as:

$$t_m/t_p = 1/\tau = \sqrt{\gamma/\beta\lambda} \tag{3.3.9}$$

Similarly, $r_d = r_e$, $r_d = r_c$ lead to the similarity ratios for the deformation coefficient and cohesion, respectively, as:

$$\frac{E_m}{E_p} = \frac{\beta}{\eta\gamma\lambda}, \quad \frac{c_m}{c_p} = \frac{1}{\eta\gamma\lambda} \tag{3.3.10}$$

Likewise, $r_e = r_D$, $r_d = r_f$ lead, respectively, to:

$$\frac{D_m}{D_p} = 1.0, \quad \frac{\tan\phi_m}{\tan\phi_p} = \frac{\alpha}{\eta\gamma\lambda} \tag{3.3.11}$$

Next, the similarity ratios for $\sigma =$ stress and $\varepsilon =$ strain are expressed respectively as:

$$\frac{\sigma_m}{\sigma_p} = \frac{1}{\alpha} \tag{3.3.12}$$

$$\frac{\varepsilon_m}{\varepsilon_p} = \frac{1}{\beta} \tag{3.3.13}$$

The similarity ratio of stress can also be written as:

$$\frac{\sigma_m}{\sigma_p} = \frac{1}{\alpha} = \frac{\rho_m}{\rho_p}\frac{g_m}{g_p}\frac{L_m}{L_p} = \frac{1}{\eta\gamma\lambda} \tag{3.3.14}$$

That of seismic acceleration $a$ can be obtained, by using displacement $u = \varepsilon L$, as:

$$\frac{a_m}{a_p} = \frac{\varepsilon_m L_m t_m^{-2}}{\varepsilon_p L_p t_p^{-2}} = \frac{\tau^2}{\beta\lambda} \tag{3.3.15}$$

Nonlinearity in the stress-strain relationships is an important issue to deal with in developing the similitude for soil materials (Rocha 1957). Let us consider stress-strain relationships shown with the two solid curves in Fig. 3.3.2 for a model and a prototype, which is expressed with an identical function $f(\ )$ as:

$$\sigma_m = f(\varepsilon_m), \quad \sigma_p = f(\varepsilon_p) \tag{3.3.16}$$

In this regard, different types of similitude can be derived for a model of the length ratio $L_m/L_p = 1/\lambda$ in the following by choosing different ratios for the basic physical variables, $t_m/t_p = 1/\tau$, $\rho_m/\rho_p = 1/\eta$, $g_m/g_p = 1/\gamma$, $\sigma_m/\sigma_p = 1/\alpha$ and $\varepsilon_m/\varepsilon_p = 1/\beta$.

*(a) Similitude in 1 g without scaling strain ($\beta = 1.0$)*

In the shaking table model tests in 1 g, $g_m/g_p = 1/\gamma = 1.0$. The geometrical similarity is sometimes essential between the model and prototype not only in the initial condition but also during the failure, hence $\varepsilon_m/\varepsilon_p = 1/\beta = 1.0$ (Clough and Pirtz 1956). Then, Eq. (3.3.9) gives $1/\tau = 1/\sqrt{\lambda}$. If the same soil material is used in the model and prototype, $\rho_m/\rho_p = 1/\eta = 1.0$, then Eq. (3.3.14) gives $1/\alpha = 1/\lambda$, which is identical to that of length $L_m/L_p = 1/\lambda$. Then, Eq. (3.3.10) gives $E_m/E_p = c_m/c_p = 1/\lambda$ for the ratios of deformation modulus and cohesion to be satisfied in soil materials used in

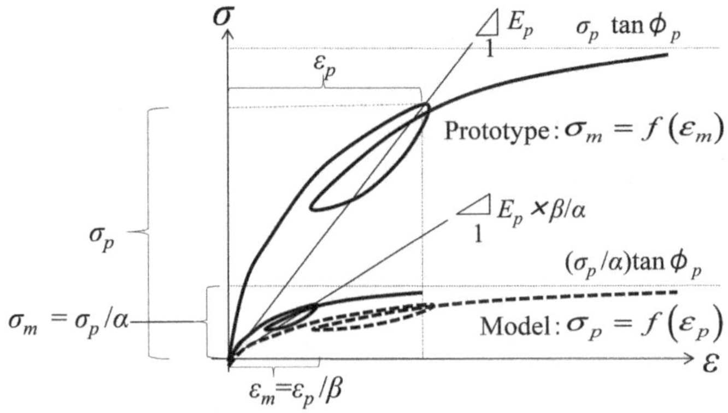


*Figure 3.3.2* Schematic stress-strain curves of soil material for scaled model and prototype.

the model. This indicates that the stress-strain curve for the model should be like a dashed curve in Fig. 3.3.2 with the stress scaled by $1/\alpha = 1/\lambda$ and the strain scaled by $\varepsilon_m/\varepsilon_p = 1/\beta = 1.0$, which may be very difficult to realize. Thus, the 1 g model test using the same soil as the prototype with the time scale $1/\tau = 1/\sqrt{\lambda}$ cannot satisfy the similitude in terms of stress and deformation modulus. On the other hand, Eq. (3.3.11) gives $\tan\phi_m/\tan\phi_p = 1.0$, which may be realized in the same soil if friction angle does not significantly change with confining stress. Hence, this type of model test may be usable in geometrically similar model tests using soils without cohesion to reproduce prototype behaviors not in the prefailure stage but in the ultimate failure mode (Clough and Pirtz 1956).

### *(b) Similitude in 1 g with scaling strain ($\beta > 1.0$)*

Unlike (a) above, the similarity ratio $\varepsilon_m/\varepsilon_p = 1/\beta$ wherein $\beta > 1.0$ is introduced in shaking table model tests in 1 g, $g_m/g_p = 1/\gamma = 1.0$. Also, the hyperbolic stress-strain relationship mentioned in Sec. 3.1.1 is assumed and expressed as:

$$\sigma = E_0 \frac{1}{1 + \varepsilon/\varepsilon_r} \varepsilon = E_0 \cdot f\left(\frac{\varepsilon}{\varepsilon_r}\right) \cdot \varepsilon \tag{3.3.17}$$

wherein $E_0$ = initial deformation modulus, $\varepsilon_r$ = reference strain corresponding to secant modulus $E = \sigma/\varepsilon$ being 50% of $E_0$, and $f(\varepsilon/\varepsilon_r) = 1/(1 + \varepsilon/\varepsilon_r)$. From Eq. (3.3.14), the similarity ratio of stress can be written as:

$$\frac{\sigma_m}{\sigma_p} = \frac{E_{0,m} \cdot f_m(\varepsilon_m/\varepsilon_{r,m})}{E_{0,p} \cdot f_p(\varepsilon_p/\varepsilon_{r,p})} \frac{1}{\beta} = \frac{1}{\eta\gamma\lambda} = \frac{1}{\lambda} \tag{3.3.18}$$

In the above equation, the following relationship holds as already seen in Sec. 2.4 in normal confining stress and also in very low confining stress as will be shown in Sec. 3.3.3.

$$\frac{f_m(\varepsilon_m/\varepsilon_{r,m})}{f_p(\varepsilon_p/\varepsilon_{r,p})} = 1.0 \tag{3.3.19}$$

Then, $E_{0,m}/E_{0,p} = \beta/\lambda$ is obtained from Eq. (3.3.18). As for the initial deformation modulus $E_0$, it is roughly proportional to the square root of the overburden stress or soil thickness, hence $E_{0,m}/E_{0,p} = \beta/\lambda = 1/\sqrt{\lambda}$ and the next equation holds.

$$\frac{1}{\beta} = \frac{1}{\sqrt{\lambda}} \tag{3.3.20}$$

Thus, a similitude considering the confining stress-dependency of the deformation modulus of soil is derived by taking the similarity ratio of strain $1/\beta$ as in Eq. (3.3.20). In this case, stress-strain curves in the model and prototype are conceptually drawn as the two solid curves in Fig. 3.3.2. The similarity ratio of time is accordingly obtained from Eq. (3.3.9) as $1/\tau = 1/\sqrt{\beta\lambda} = 1/\lambda^{3/4}$. That of acceleration is given by Eq. (3.3.15) as $a_m/a_p = \tau^2/\beta\lambda = 1.0$, indicating no scaling in seismic acceleration between model and prototype in this similitude. Substituting these into Eqs. (3.3.3)–(3.3.8), $r_d = r_i = r_f = r_c = r_e = r_D = 1/\lambda^3$ is confirmed provided that $\phi_m = \phi_p$, $c_m/c_p = 1/\lambda$, $D_m/D_p = 1.0$ hold. These conditions may be satisfied at least for clean sands in ultra-low confining stress as will be shown in Sec. 3.3.3. Then, the similarity of deformation holds to a certain degree in a 1G model test using the same soil material in this similitude if the time is scaled as $1/\tau = 1/\lambda^{3/4}$. However, it cannot apply to ultimate failures where the similarity of strain should be $1/\beta = 1.0$ because the geometric similarity of the failure mode is not satisfied here.

*(c) Similitude in centrifugal acceleration*

Soil properties are strongly dependent on confining stress, and not only shear modulus but also dilatant behavior varies considerably under very low confining stresses in model tests. The similitude (b) explained above can take account the effect of confining stress on the shear modulus but not on the dilatant behavior which becomes dominant as soils approach failures.

A centrifuge model test is a solution to overcome this difficulty. As schematically illustrated in Fig. 3.3.3, an arm with a swing at the end where a small-size shaking table is mounted. The shaking table test is conducted while the arm is rotating, under centrifugal acceleration of around 50 g typically applied to a test model on the table about 90 degrees inclined from the horizontal plane. Thus, the soil model under the body force around 50 times larger than 1 g can reproduce the confining stresses of the prototype in the scaled model. Though other types of body force may be usable for this purpose; such as seepage force (Zelikson 1969, Kato and Kokusho 2012), the centrifugal force is used as a more convenient measure.

In the centrifuge model tests, where the same soil material is used as in the prototype ($1/\eta = \rho_m/\rho_p = 1.0$) and the scaling ratios in stress and strain are unity ($\alpha = \beta = 1$), it is expected that the model can reproduce the prototype in terms of stress without

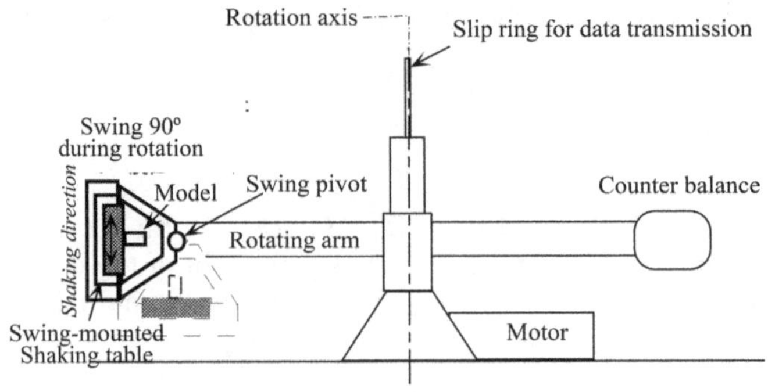


*Figure 3.3.3* Schematic view of centrifuge shaking table test facility.

knowing the soil properties. For $\alpha=1$ in Eq. (3.3.14), $1/\alpha=1/(\eta\gamma\lambda)=1.0$, hence $1/\gamma=g_m/g_p=\lambda$ which is realized by applying the centrifugal acceleration $\lambda g$. For $\beta=1$, the similarity ratio of time is given from Eq. (3.3.9) as $1/\tau=\sqrt{\gamma/\beta\lambda}=1/\lambda$, and that of deformation modulus is $E_m/E_p=\beta/(\eta\gamma\lambda)=1.0$ from Eq. (3.3.10). Similarly, Eqs. (3.3.11)~(3.3.13) give the similarity ratios of all unity for damping ratio, cohesion and friction coefficient. Furthermore, that of seismic acceleration $a$ is given from Eq. (3.3.15) as $a_m/a_p=\tau^2/\beta\lambda=\lambda$, which is identical to that of the gravitational acceleration $g_m/g_p=\lambda$. Thus in the centrifuge tests, a model under the centrifugal acceleration of $\lambda\times g$ is shaken by a wave motion with the seismic acceleration $\lambda$ times larger in amplitude and $1/\lambda$ times shorter in time length.

### 3.3.2.3 *Similitude for other variables*

*(i) Permeability*

The Darcy's law written as follows is a key governing equation in model tests where water is interacted with soil particles.

$$v=-\frac{k}{\rho_w g}\frac{\partial p}{\partial z} \tag{3.3.21}$$

Here, $v$ = water seepage velocity, $k$ = permeability coefficient, $\rho_w$ = water density, $p$ = water pressure, $z$ = flow distance. Use of the method C listed in Sec. 3.3.2.1 for constructing the similitude yields the following:

$$\frac{v_m}{v_p}=\frac{k_m/\rho_{w,m}g_m}{k_p/\rho_{w,p}g_p}\frac{\partial p_m/\partial z_m}{\partial p_p/\partial z_p}=\frac{k_m}{k_p}\frac{\eta\gamma\lambda}{\alpha} \tag{3.3.22}$$

Because the ratio in seepage velocity is also written as $v_m/v_p=(L_m/t_m)/(L_p/t_p)=\tau/\lambda$, the ratio of permeability coefficient is obtained as:

$$\frac{k_m}{k_p}=\frac{\alpha\tau}{\eta\gamma\lambda^2} \tag{3.3.23}$$

Another important phenomenon associated with model soil tests involving water, such as post-liquefaction pressure dissipation and associated ground settlement, is governed by the consolidation equation where the time factor $T$ defined as follows is a non-dimensional key parameter controlling the consolidation time.

$$T = \frac{C_v}{H^2} t = \frac{k}{m_v \rho_w g} \frac{t}{H^2} \tag{3.3.24}$$

Here, $C_v$ = coefficient of consolidation, $H$ = water drainage distance, $m_v$ = compressibility coefficient. The ratio of $T$ can be expressed in the next equation considering $m_{v,m}/m_{v,p} = \alpha$.

$$\frac{T_m}{T_p} = \frac{C_{vm}}{C_{vp}} \frac{\lambda^2}{\tau} = \frac{k_m/(m_{v,m}\rho_{w,m} g_m)}{k_p/(m_{v,p}\rho_{w,p} g_p)} \frac{\lambda^2}{\tau} = \frac{k_m}{k_p} \frac{\eta\gamma}{\alpha} \frac{\lambda^2}{\tau} \tag{3.3.25}$$

Following the method A in Sec. 3.3.2.1, $T_m/T_p = 1.0$ gives the same result as Eq. (3.3.23), indicating that the same similitude holds for both seepage and consolidation.

Hence, the similarity ratio of the permeability coefficient $k$ is given corresponding to the three kinds of similitude explained above as: (a) in the similitude in 1 g without scaling strain ($\beta = 1.0$), $\alpha = \lambda, \eta = 1, \gamma = 1, \tau = \lambda^{0.5}$ yield $k_m/k_p = 1/\lambda^{0.5}$, (b) in the similitude in 1 g with scaling strain ($\beta > 1.0$) $\alpha = \lambda, \eta = 1, \gamma = 1, \tau = \lambda^{0.75}$ yield $k_m/k_p = 1/\lambda^{0.25}$, and (c) in the similitude in centrifugal acceleration, $\alpha = \beta = \eta = 1, \gamma = 1/\lambda, \tau = \lambda$ yield $k_m/k_p = 1$.

By the way, the permeability coefficient $k$ was originally introduced in an equation for stationary tube flow as:

$$k = K \frac{\rho_w g}{\mu} \tag{3.3.26}$$

Here, $\mu$ = viscosity of water and $K$ = physical permeability of soil. In the model using the same soil and water as in the prototype, $\mu_m/\mu_p = 1$ and $K_m/K_p = 1$ (Joseph et al. 1988). Hence

$$\frac{k_m}{k_p} = \frac{K_m}{K_p} \frac{\rho_{wm} g_m/\mu_m}{\rho_{wp} g_p/\mu_p} = \frac{1}{\eta\gamma} \tag{3.3.27}$$

Combining Eqs. (3.3.27) and (3.3.23) gives

$$\frac{1}{\tau} = \frac{\alpha}{\lambda^2} \tag{3.3.28}$$

which is obviously different from the similarity ratio of time Eq. (3.3.9).

Hence, the similarity ratios of time and permeability coefficient for consolidation and seepage problems are given using Eqs. (3.3.28) and (3.3.27) differently from those for dynamic response problems, corresponding to the three kinds of similitude explained before as:

(a) In 1 g without scaling strain ($\beta = 1.0$), $\alpha = \lambda, \eta = 1, \gamma = 1$ yield $1/\tau = 1/\lambda$, $k_m/k_p = 1$.
(b) In 1 g with scaling strain ($\beta > 1.0$) $\alpha = \lambda, \eta = 1, \gamma = 1$ yield $1/\tau = 1/\lambda, k_m/k_p = 1$.
(c) In the centrifugal acceleration, $\alpha = \eta = 1.0, \gamma = 1/\lambda$ yield $1/\tau = 1/\lambda^2, k_m/k_p = \lambda$.

In the case of liquefaction tests under the centrifugal acceleration in particular, the time scale for seepage and consolidation $1/\tau = 1/\lambda^2$ has to be equalized to that for dynamic response $1/\tau = 1/\lambda$. In order to do that, the permeability coefficient $k_m$ is lowered by $1/\lambda$-times from $k_m/k_p = \lambda$ to $k_m/k_p = 1$ in centrifuge tests by adding some viscous fluid into water or by using essentially the same soil materials as prototype but with smaller particles.

*(ii) Axial and flexural rigidity of piles*

The force equilibrium of vertical piles in horizontal direction ($x$-axis) considering the flexural rigidity is expressed by the following equation.

$$EI\frac{\partial^4 u}{\partial x^4} + \rho_b hB\ddot{u} + B\sigma_x = 0 \tag{3.3.29}$$

Here $u$ = horizontal displacement, $EI$ = flexural rigidity, $\rho_b, B, h$ = density, thickness, width of piles with a rectangular cross-section, and $\sigma_x$ = horizontal earth pressure on the pile. Following the method C explained in Sec. 3.3.2.1 so that the above governing equation holds both in the model and the prototype, and also noting that the displacement is strain times length dimensionally, the next relationship can be derived.

$$\frac{(EI)_m}{(EI)_p}\frac{\lambda^4}{\beta\lambda} = \frac{\tau^2}{\eta\lambda^2\beta\lambda} = \frac{1}{\alpha\lambda} \tag{3.3.30}$$

The equality between the two terms on the right is obvious from Eqs. (3.3.9) and (3.3.14), and the next equation is derived from Eq. (3.3.30).

$$\frac{(EI)_m}{(EI)_p} = \frac{\beta}{\alpha\lambda^4} \tag{3.3.31}$$

As for the axial rigidity of piles with the cross-sectional area $A$, the ratio is obtained similarly as:

$$\frac{(EA)_m}{(EA)_p} = \frac{\beta}{\alpha\lambda^2} \tag{3.3.32}$$

Hence, in 1 g with scaling strain ($\beta > 1.0$), $\alpha = \lambda, \beta = \lambda^{0.5}, \eta = 1, \gamma = 1, \tau = \lambda^{0.75}$ yield $(EI)_m/(EI)_p = 1/\lambda^{4.5}$ and $(EA)_m/(EA)_p = 1/\lambda^{2.5}$. In centrifugal acceleration, $\alpha = \beta = 1$, $\eta = 1, \tau = \lambda, \gamma = 1/\lambda$ yield $(EI)_m/(EI)_p = 1/\lambda^4$ and $(EA)_m/(EA)_p = 1/\lambda^2$. These similitudes can be satisfied in model tests where the model piles with the same material as the prototype are scaled by $1/\lambda$.

*Table 3.3.1* Three types of similitude (a)~(c) for dynamic soil model tests.

| | *Similitude for dynamic soil model tests (model/prototype)* $\ll\gg$*: Similitude for seepage/consolidation* | | | |
|---|---|---|---|---|
| | *General* | *(a) Test in 1 g (no scaling in strain)* | *(b) Test in 1 g (scaling in strain)* | *(c) Centrifuge test* |
| Length: $L$ | $1/\lambda$ | $1/\lambda$ | $1/\lambda$ | $1/\lambda$ |
| Time: $t$ | $1/\tau=[\gamma/(\beta\lambda)]^{0.5}$ $\ll 1/\tau=\alpha/\lambda^2\gg$ | $1/(\lambda)^{0.5}$ $\ll 1/\lambda \gg$ | $1/(\lambda)^{0.75}$ $\ll 1/\lambda \gg$ | $1/\lambda$ $\ll 1/(\lambda)^2 \gg$ |
| Soil density: $\rho$ | $1/\eta$ | 1 | 1 | 1 |
| Gravitational acceleration: $g$ | $1/\gamma$ | 1 | 1 | $\lambda$ |
| Stress/pressure: $\sigma$ | $1/\alpha=1/(\eta\gamma\lambda)$ | $1/\lambda$ | $1/\lambda$ | 1 |
| Strain: $\varepsilon$ | $1/\beta$ | 1 | $1/(\lambda)^{0.5}$ | 1 |
| Displacement: $u$ | $1/(\beta\lambda)$ | $1/\lambda$ | $1/(\lambda)^{1.5}$ | $1/\lambda$ |
| Deformation modulus: $E$ | $\beta/(\eta\gamma\lambda)$ | $1/\lambda$ | $1/(\lambda)^{0.5}$ | 1 |
| Damping ratio: $D$ | 1 | 1 | 1 | 1 |
| Cohesion: $c$ | $1/(\eta\gamma\lambda)$ | $1/\lambda$ | $1/\lambda$ | 1 |
| Friction coefficient: $\tan\phi$ | $\alpha/(\eta\gamma\lambda)$ | 1 | 1 | 1 |
| Seismic acceleration: $a$ | $\tau^2/(\beta\lambda)$ | 1 | 1 | $\lambda$ |
| Permeability coefficient: $k$ | $\alpha\tau/(\beta\lambda^2)$ $\ll 1/(\eta\gamma)\gg$ | $1/(\lambda)^{0.5}$ $\ll 1 \gg$ | $1/(\lambda)^{0.25}$ $\ll 1 \gg$ | $1$ $\ll \lambda \gg$ |
| Flexural rigidity of pile: $EI$ | $\beta/(\alpha\lambda^4)$ | $1/(\lambda)^5$ | $1/(\lambda)^{4.5}$ | $1/(\lambda)^4$ |
| Axial rigidity of pile: $EA$ | $\beta/(\alpha\lambda^2)$ | $1/(\lambda)^3$ | $1/(\lambda)^{2.5}$ | $1/(\lambda)^2$ |

The three types of similitude discussed here are summarized in Table 3.3.1. Among them, the second similitude under 1 g depends on the stress versus strain relationships of soil materials in the model and prototype. In the similitude for centrifuge tests, no scaling rule in stress-strain relationships is needed. Thus, model tests may provide quantitative results applicable to prototype behavior with the help of similitudes. However, an unfavorable problem to be noted with scaled model tests is that not only the model size but also the soil particles are also scaled by $1/\lambda$. If the same soil is used in a scaled model, it corresponds to the similar soil of $\lambda$-times larger. The same occurs for pickups, sensors and signal wires used in the scaled model. The problem becomes particularly significant in centrifuge tests where $\lambda$ is as high as around 50. The viscous fluids sometimes employed in liquefaction problems may also change the soil properties being different from those of prototype soils. Hence it should be borne in mind that model tests cannot perfectly reproduce prototype behaviors, and it may be sound to conduct numerical analyses on the scaled models to compare with the test results and then evaluate the actual performance by resorting to the analytical methods calibrated by the model behaviors.

### 3.3.3 Soil properties for model test under ultra-low confining stress

Model shaking table tests in 1 g are sometimes conducted together with associated numerical analyses, where soil properties under ultra-low confining pressure are indispensable. Such soil properties were investigated experimentally in shaking table tests or torsional shear tests (Kokusho 1982, Okumura et al. 1985, Kong et al. 1986, Tatsuoka

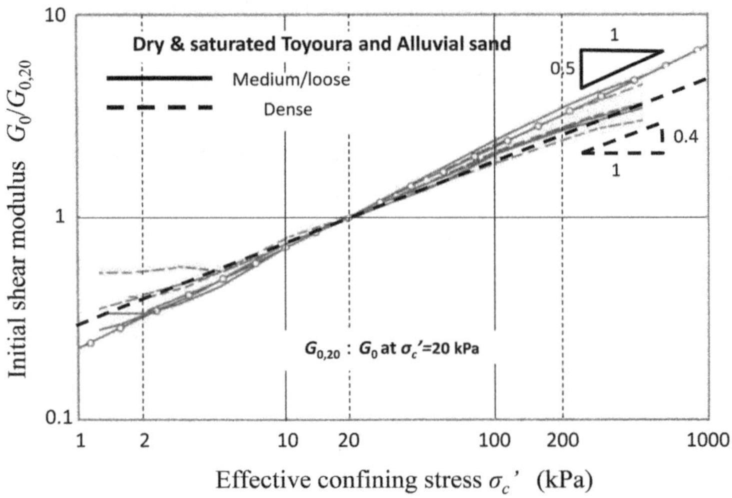


*Figure 3.3.4* Variations of initial shear modulus $G_0$ of dry clean sand for widely varied effective confining stress $\sigma_c'$ (Kokusho 1982).

et al. 1986b). The lowest confining stress attained in these studies was 1–2 kPa which corresponds to 10–20 cm depth in a model soil layer. In the following, the ultra-low pressure properties of clean sands are addressed as a typical soil material easy to handle and often used in model tests.

Fig. 3.3.4 shows the initial shear modulus $G_0$ versus the isotropic confining stress $\sigma_c'$ in the log-log diagram, measured by resonant column tests on air-dried Toyoura and alluvial clean sands. The $G_0$-value in the vertical axis is normalized by that for $\sigma_c' = 20$ kPa, $G_{0,20}$. The relationships, though slightly different for different sand densities, can obviously be approximated by the straight line of the following formula for $\sigma_c'$ widely spanning from about 1 kPa to 500 kPa.

$$\frac{G_0}{G_{0,20}} = \left(\frac{\sigma_c'}{20\,\text{kPa}}\right)^n \tag{3.3.33}$$

The exponent $n$, being slightly dependent on the sand density, takes $n = 0.4$–$0.5$ down to $\sigma_c' = 1$–$2$ kPa in most test results. This indicates that the confining stress-dependency of shear modulus identical to the normal stress level may be applicable to the low stress level corresponding to 10–20 cm depth from a model surface at least for clean sands. Similar test results were obtained by other researchers (Kong et al. 1986) that even for moist clean sands of water content 1.8%, the linear relationship holds between $G_0$ and $\sigma_c'$ on the log-log diagram for $\sigma_c' = 2$–$30$ kPa.

Fig. 3.2.9 indicates the variations of shear modulus ratio $G/G_0$ and damping ratio $D$ versus the normalized shear strain $\gamma/\gamma_r$ obtained from a series of resonant column and cyclic loading tests on saturated alluvial clean sand. Here, $\gamma_r$ stands for the reference strain corresponding to the shear strain $\gamma$ for $G/G_0 = 0.5$. The data points plotted with different symbols are concentrated along the unique curve both for $G/G_0$ and $D$

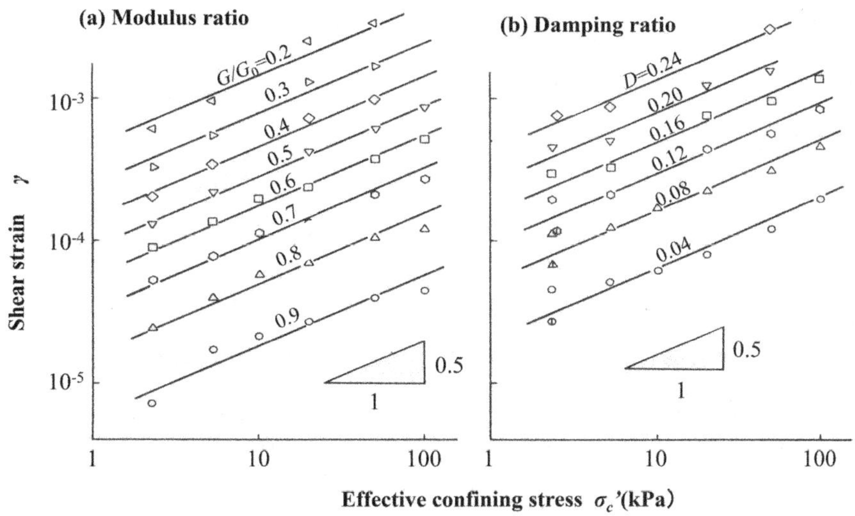


*Figure 3.3.5* Shear strain for shear modulus ratio $G/G_0 = 0.2\sim0.9$ (a), and Damping ratio $D = 0.04\sim0.24$ (b), versus effective confining stress $\sigma_c'$ for saturated clean sand (Kokusho 1982).

even for widely varied isotropic effective confining stresses $\sigma_c' = 2\sim100$ kPa. In order to examine the confining stress-dependency in the same test data set, shear strains corresponding to $G/G_0 = 0.2\sim0.9$ or $D = 0.04\sim0.24$ are plotted versus $\sigma_c'$ on a log-log diagram in Figs. 3.3.5(a) and (b), respectively. This indicates that not only the reference strain corresponding to $G/G_0 = 0.5$ but also strains corresponding other values of $G/G_0$ and $D$ are also roughly proportional to $(\sigma_c')^{0.5}$. Namely, it is obvious that the $G/G_0$ and $D$ versus log $\gamma$ curves keep almost the same forms in the normal-stress level and in the ultra-low stress level as well and tend to translate leftward on the diagrams with decreasing $\sigma_c'$ as assumed in Eq. (3.3.19) for the similitude in 1 g with scaled strain.

Fig. 3.3.6(a) shows similar test results of $G/G_0$ versus log $\gamma$ for moist clean sand measured by cyclic torsional shear tests under very low confining pressures (Kong et al. 1986). All the data for $\sigma_c' = 5\sim80$ kPa are plotted along the solid $G/G_0$ versus $\gamma/\gamma_r$ curve obtained for the confining stress $\sigma_c' = 98$ kPa. Damping ratios $D$ plotted versus $G/G_0$ in Fig. 3.3.6(b) are also concentrating along the solid average curve drawn in the diagram as formulated by the HD model in Eq. (3.1.22). Thus, it is demonstrated that strain-dependent variations of $G/G_0$ and $D$ can be expressed in the same formula in very low confining stresses at least for clean sands.

Different from the cyclic loading test results with pre-failure shear strain levels $\gamma < 10^{-2}$, Fig. 3.3.7(a) shows monotonic loading test results of a clean sand in much larger strain levels (axial strain $\varepsilon_1$ up to 10%) under widely varied confining stresses of $\sigma_c' = 5\sim400$ kPa (Tatsuoka et al. 1986b). The test was conducted using a plane strain triaxial test apparatus wherein the axial compression stress $\sigma_1'$ was given in the drained condition on a rectangular column sand specimen with the strain confined at one pair of lateral faces and the stress $\sigma_3'$ kept constant at the other pair of faces, wherein the axial and lateral strains $\varepsilon_1$ and $\varepsilon_3$ are measured. The group of curves at the upper part of

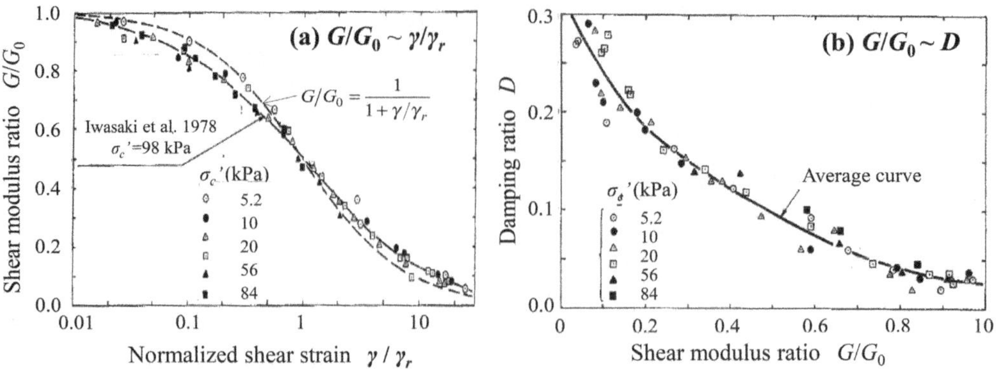

*Figure 3.3.6* Shear modulus ratio $G/G_0$ versus normalized shear strain $\gamma/\gamma_r$ (a), and Damping ratio $D$ versus shear modulus ratio $G/G_0$ (b), for moist clean sand (Ko et al. 1986).

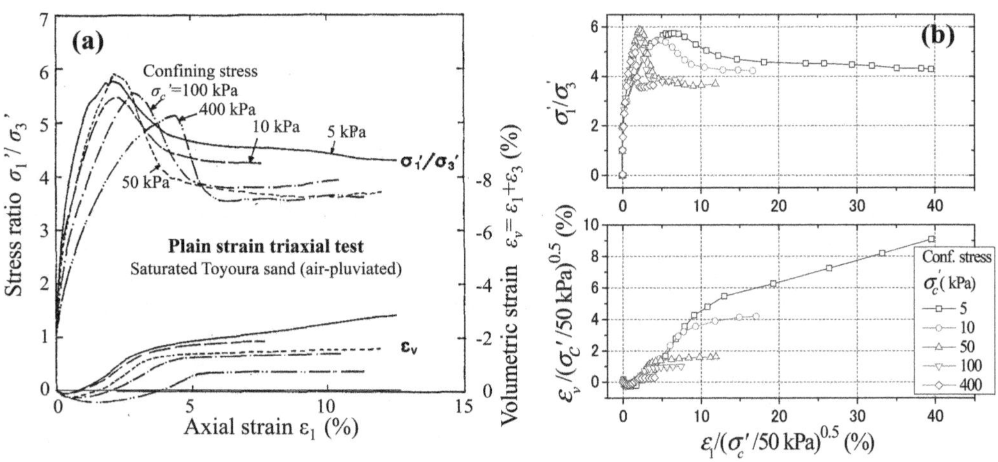

*Figure 3.3.7* Monotonic loading plane-strain tests: (a) Stress ratio or volumetric strain versus axial strain, (b) Stress ratio or normalized volumetric strain versus normalized axial strain (modified from Tatsuoka et al. 1986b).

the diagram in Fig. 3.3.7(a) represent $\sigma_1'/\sigma_3'$ versus $\varepsilon_1$ relationships, which correspond to stress–strain relationships in normal triaxial compression tests with constant lateral stress $\sigma_3'$. It is noted that the peak values of $\sigma_1'/\sigma_3'$ (or $\phi_d = \sin^{-1}[(\sigma_1' - \sigma_3')/(\sigma_1' + \sigma_3')]$ the drained friction angle calculated from $\sigma_1'/\sigma_3'$) are not much different despite widely varied confining stress $\sigma_c' = 5{\sim}400$ kPa. Also noted is that the initial gradients of the curves are steeper for lower confining stresses with peaks appearing at smaller strains. As shown in the volumetric strain $\varepsilon_v$ $(=\varepsilon_1 + \varepsilon_3)$ versus axial strain $\varepsilon_1$ relationships with the lower curves in Fig. 3.3.7(a), the sand is obviously more dilative and exhibits larger volumetric strain in lower confining stresses.

As already seen in the cyclic loading test results, the modulus ratio $G/G_0$ can be formulated as a unique function of the normalized strain $\gamma/\gamma_r$ where $\gamma_r \propto (\sigma'_c/p_0)^{0.5}$. In the similar manner, the same set of curves are redrawn in Fig. 3.3.7(b) wherein the axial and volumetric strains are normalized respectively as $\varepsilon_1 \rightarrow \varepsilon_1/(\sigma'_c/50)^{0.5}$ and $\varepsilon_v \rightarrow \varepsilon_v/(\sigma'_c/50)^{0.5}$ in terms of the confining stress $\sigma'_c$ relative to $\sigma'_c = 50$ kPa. It is interesting to observe in the upper diagram that the normalization tends to make the curves of $\sigma'_1/\sigma'_3$ closer and overlapped for small strains, though the curves for $\sigma'_c = 5\sim10$ kPa tend to separate from those for $\sigma'_c = 50\sim400$ kPa at a certain strain before the peaks of $\sigma'_1/\sigma'_3$. Correspondingly, the normalized curves of $\varepsilon_v$ for different $\sigma'_c$ in the lower diagram are almost overlapping up to a certain strain before the peaks of $\sigma'_1/\sigma'_3$. However, the curves tend to separate thereafter and individually develop larger volumetric expansion particularly in $\sigma'_c = 5$ and 10 kPa. The separation points are read off as $\varepsilon_1/(\sigma'_c/50)^{0.5} \approx$ 1% in Fig. 3.3.7(b), indicating that the stress-strain or dilative behavior of clean sands corresponding to strain $\varepsilon_1$ larger than about 1% in prototype soils of $\sigma'_c \approx 50$ kPa or larger cannot be reproduced by model soils confined by $\sigma'_c$ much lower than 50 kPa. If the strain is much less than around 1%, the soil response to cyclic loading may be reproduced in model tests for confining stress down to $\sigma'_c \approx 2$ kPa by using the similitude with scaled strain.

## 3.4 SUMMARY

1 Nonlinear stress-strain curves of soils under cyclic loading are idealized by simple functions to be used in numerical analyses such as hyperbolic and Ramberg-Osgood models. In order to construct a stress~strain relationship under cyclic loading, the Masing rule is incorporated to extend the skeleton curve to the hysteretic curve corresponding to cyclic stress changes.

2 The Masing rule can construct stationary hysteresis loops to given stress amplitudes corresponding to different soil models, from which equivalent linear properties; secant shear moduli and hysteretic damping ratios are defined for various strain levels. Among the equivalent linear properties, care is needed that the damping ratio for the hyperbolic model tends to approach to $2/\pi$ with increasing strain, much larger than actually measured in soils. In order to avoid this problem, Hardin-Drenevich model is often used in the equivalent linear analyses wherein maximum damping ratio can be specified.

3 Dilatancy behavior in large soil strains is divided into dilative and contractive sides by the Steady State Line (SSL) on the state diagram ($e\sim\sigma'_c$ plane) under the monotonic loading in both drained and undrained conditions. In contrast, the cyclic loading during earthquakes makes soils always contractive on both sides of SSL, though the contractility tends to be definitely smaller on the dilative side.

4 In the drained cyclic loading, larger volume contraction occurs in looser soils with void ratios larger than the critical void ratio $e_{cr}$ than in denser soils. In the undrained cyclic loading, pore-pressure buildup occurs more rapidly for void ratios larger than $e_{cr}$, though pressure tends to buildup up gradually to 100% even for very dense soils. The volume contraction in the drained cyclic loading and the pressure buildup in the undrained cyclic loading may be theoretically correlated

using the bulk modulus for swelling of soil skeleton, except a fully-liquefied state of loose soils wherein particle contacts are completely lost.

5 In many engineering problems, soil failures in the undrained conditions tend to occur gradually with the increasing number of cyclic loading. In such cases, the soil strength are defined by the stress amplitude of a harmonic motion (normally expressed by *CRR*: the cyclic stress amplitude divided by the initial effective confining stress) with an equivalent number of cycles, which induces soil strain corresponding to various structural performance levels. However, special care is needed to another flow-type brittle soil failures, which may potentially occur during cyclic as well as monotonic loading on the contractive side of SSL due to strain-softening in post-peak stress~strain relationships.

6 In order to convert an irregular earthquake motion to a harmonic motion incurring the equivalent damage, combinations of the amplitude and number of cycles can be chosen based on the fatigue theory. The same theory can also be employed to convert the effect of two-dimensional shearing in situ into one-dimensional shearing in laboratory tests.

7 Numerical analyses considering the strain-dependent soil nonlinearity are classified into equivalent linear and stepwise nonlinear analyses. The former is actually the linear analysis using strain-compatible nonlinear properties, that can be conveniently incorporated in numerical schemes such as complex response methods and mode-superposition methods wherein the solutions are efficiently obtained by the superposition of linear solutions. However, the calculated strains tend be evaluated smaller, and strongly nonlinear problems are beyond the scope of this analysis. Also, the amplification of high-frequency acceleration motions tends to be evaluated smaller because the equivalent damping ratio is chosen compatible with low-frequency major earthquake motions. This deficiency may be improved by determining individual strain-dependent damping ratios at several different frequency ranges according to frequency-dependent strain spectra.

8 In the stepwise nonlinear analysis, tangent moduli of soils in the stiffness matrices of models are varied in each time increment to be able to pursue strongly nonlinear stress-strain curves. Pore-pressure changes evaluated by dilatancy models and also considering seepage and consolidation effects can be coupled to modify the tangent moduli at each step to conduct effective stress analyses. The dynamic response tends to be instable due to high frequency noises generated by near-zero soil damping near the turning points of stress-strain curves and the Rayleigh damping has to be incorporated in order to stabilize the analyses.

9 The hysteretic hyperbolic (HH) model used in the stepwise nonlinear total stress analyses, wherein asymptotic shear strength is defined, tends to give inappropriately low amplifications during strong earthquake shakings due to very high damping ratio compared with measured values in soil tests in large strain levels. The Ramberg-Osgood (RO) model, though the maximum shear stress cannot be defined, has a larger freedom of parameters to have better fitting with the measured soil damping.

10 In the stepwise nonlinear analyses, though it can numerically determine residual soil deformations for the PBD, uncertainties involved are considerable in comparison with conventional analyses, and well-documented benchmark case histories

and model test results are essential in judging the reliability of the analytical outcomes.

11 In model tests in 1g with the scale ratio of length $1/\lambda$, the similarity of deformation holds to a certain degree using the same soil material if the induced strain is scaled as $1/\lambda^{1/2}$ and the time is scaled as $1/\lambda^{3/4}$. Soil properties under very low overburden can satisfy this similitude in the case of clean sands. However, it cannot apply to the ultimate failure where the scale ratio of strain should satisfy 1.0 because of the geometrical similarity in the failure mode.

12 In the similitude for centrifuge model tests, no scaling rule in stress-strain relationships is required. The model tests may provide quantitative results applicable to the prototype behavior with the help of similitudes. However, not only the model size but also the soil particle size are scaled by $1/\lambda$ quite unfavorably. Viscous fluids sometimes employed in liquefaction problems may also change the soil properties. Thus, the model tests cannot perfectly reproduce prototype behavior, and numerical analyses may help the test results be understood better in evaluating the actual performance in the prototype.

Chapter 4

# Seismic site amplification and wave energy

In the first stage of seismic design of structures, a seismic ground motion at a site has to be evaluated. It consists of body waves (P-wave and S-wave) and surface waves (Rayleigh wave and Love wave) as already mentioned. Among them, the SH-wave in a horizontally-layered soil deposit is considered to have the most significant effect on geotechnical engineering problems, although other wave types have to be considered in more heterogeneous or topographically complex site conditions. Site amplifications in horizontally-layered soil systems are discussed in this chapter using the multi-reflection theory of the SH-wave with a special emphasis on the effects of soil profiles and nonlinear soil properties during strong earthquakes.

In order to measure the site amplification, two earthquake observation systems may be available; a horizontal array system at different surface geologies, and a vertical array system consisting of the ground surface and downhole depths of the different geologies in the same place. Besides the earthquake observations, microtremor measurements are often implemented as a convenient and cost-efficient way to roughly evaluate site-specific amplification characteristics, predominant frequencies in particular, in small strain vibrations.

For strong motion earthquake observations, a considerable numbers of vertical arrays have been deployed particularly in Japan since the 1995 Kobe earthquake, and numerous data have been accumulated since then. The strong motion records obtained by the vertical array systems are examined here in comparison with the idealized one-dimensional SH-wave multi-reflection theory for two-layer and multi-layer systems. A special emphasis is placed on the effect of soil profiles and soil properties during strong earthquakes.

The same one-dimensional SH-wave multi-reflection theory for a two-layer system is applied to review the basic mechanism of soil-structure interaction and the associated radiation damping of a shear-vibration structure resting on a semi-infinite uniform soil.

Furthermore, the site amplification mechanism by the SH-wave is interpreted in terms of energy flow in the vertical direction. The energy flow is first calculated in a simplified two-layer system to understand how the energy demand to be used in design is related with the pertinent parameters in site amplification. Then, in situ wave energy flows are summarized from a number of vertical array records during recent strong earthquakes, and the general trends in incurred damage in soils and superstructures are discussed from the viewpoint of energy-based design.

## 4.1 SOIL CONDITION AND SITE AMPLIFICATION

From old times, it was intuitively understood that structural damage during strong earthquakes tend to reflect local soil conditions. Fig. 4.1.1 depicts (TCEGE 1999) distributions of seismic intensities in Tokyo during the 1923 Kanto earthquake as one of the typical examples (originally from Imamura 1925). Considerable differences in the earthquake damage and associated seismic intensity, larger in the Holocene lowland in the east area versus smaller in the Pleistocene terrace in the west area, are believed to reflect the differences in surface geologies. Thus, soil conditions and soil properties may make big differences in seismic site amplification.

The evaluation of seismic amplifications due to different site conditions is named as micro-zonation. It constitutes an important task in developing seismic hazard maps for local governments. In determining the design motions for superstructures, the site amplifications should be evaluated on soil layers above an engineering bedrock at which the design motion is generally prescribed, while it may sometimes be directly given at the ground surface depending on the surface geologies. The engineering

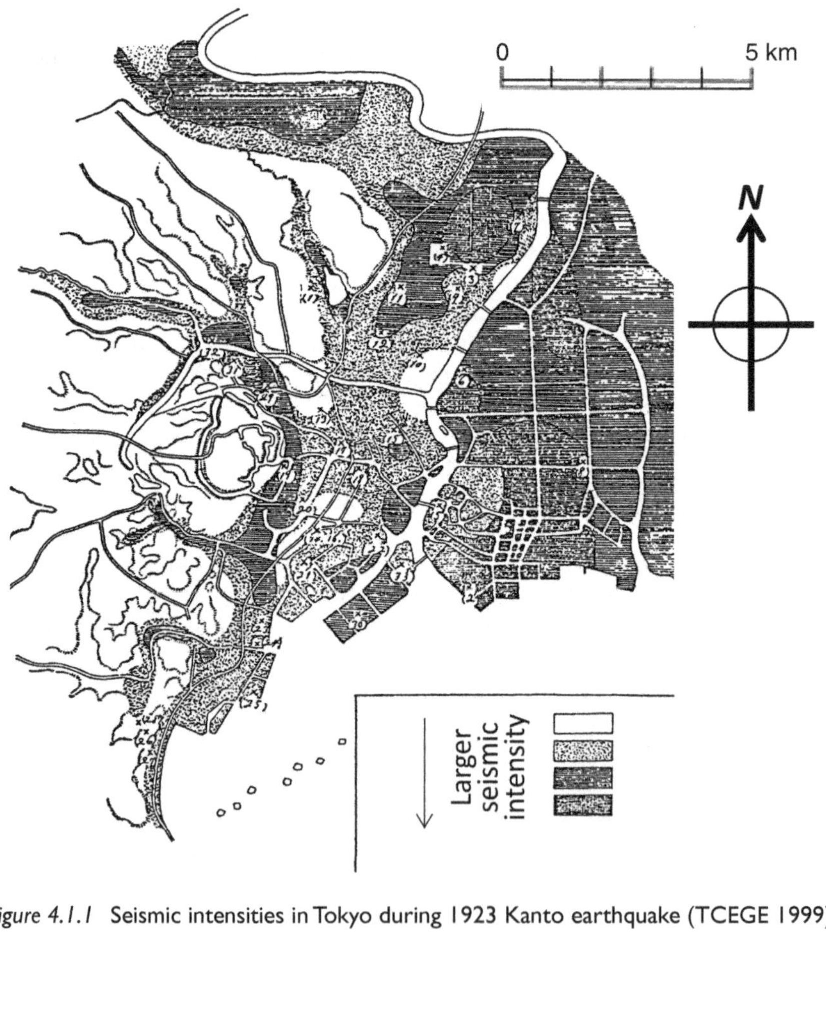


*Figure 4.1.1* Seismic intensities in Tokyo during 1923 Kanto earthquake (TCEGE 1999).

bedrock is defined for structural design at stiff bearing strata with S-wave velocities of around $V_s = 400$–$700$ m/s. The incident earthquake motions there are evaluated from the fault mechanisms and attenuation characteristics along the wave paths between faults and particular sites.

Earthquake motions measured in multiple sites at the ground surface of different geologies including outcropping engineering bedrock are utilized as a set of the horizontal array records in order to evaluate site amplifications (e.g. BSSC 2003). Additionally, the vertical array earthquake observation systems are increasingly deployed recently particularly in Japan, wherein site-specific amplifications can be investigated at the same sites between the ground surface and the base layer for site conditions having various soil profiles. As a pioneering work of this kind, Kanai et al. (1959, 1966) made simultaneous earthquake observations at the ground surface and a subsurface level below in a tunnel, demonstrating that seismic ground motions can be idealized essentially by the vertical propagation of SH-wave.

Fig. 4.1.2(a) illustrates a simplified model of a soft soil layer on an underlying base layer (engineering bedrock) at a site A and the same base layer outcropping at a distant location B. Though it has a two/three-dimensional soil structure globally, it can be locally assumed to be as a horizontally layered system consisting of the surface soft soil and the stiff base layer at the site A, because the variation of soil profile is normally moderate in the horizontal direction. If the upward seismic wave amplitude $A_2$ propagating in the base layer is assumed to be identical everywhere as in Fig. 4.1.2(a), the amplitude $A_2$ in the base is amplified to be $A_s$ in the surface soil. Correspondingly, the amplitudes of surface ground motions measured by the seismometers A and B deployed on the soft soil layer and the outcropping base layer are $2A_s$ and $2A_2$, respectively, as will be mentioned later. Hence, the seismic amplification between the two locations is expressed as $2A_s/2A_2$ or $A_s/A_2$ and named as the horizontal array amplification or the incident wave amplification. On the other hand, if the observed amplitude at A on the surface soil $2A_s$, is compared with that measured by the downhole seismometer B in the vertical array as illustrated in Fig. 4.1.2(b), the amplification becomes $2A_s/(A_2+B_2)$, because not only the upward but also downward wave is superposed in the downhole records at B. This value $2A_s/(A_2+B_2)$ is called the vertical array amplification or the composite wave amplification.

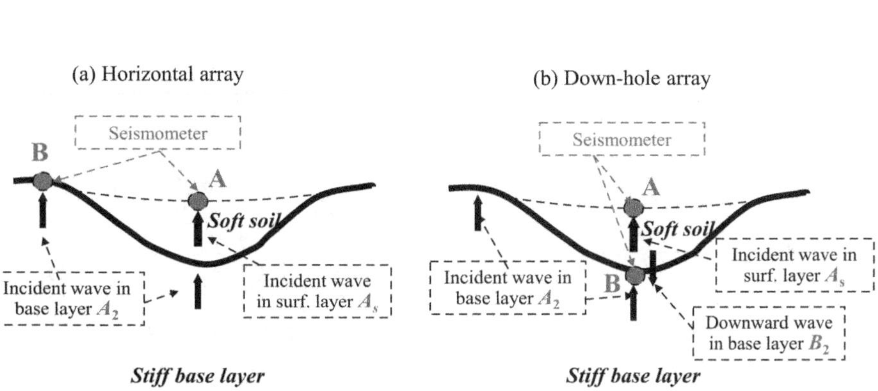


*Figure 4.1.2* Two types of earthquake observation array systems for site amplification between ground surface and base layer (Kokusho and Sato 2008).

In the micro-zonation study, what we need is $2A_s/2A_2$ rather than $2A_s/(A_2+B_2)$, because relative surface amplifications among different sites with various soil conditions are to be used for earthquake damage evaluations. It is therefore necessary to clearly differentiate the two types of seismic arrays in discussing the seismic site amplification by the vertical arrays increasingly available recently.

## 4.2 AMPLIFICATION IN TWO-LAYER SYSTEM

### 4.2.1 Two-layer system without internal damping

Near-surface soft soils consisting of multiple layers with different properties underlain by the engineering bedrock are often simplified by a two-layer system of a soft surface layer of thickness $H$ above a stiff base layer of infinite depth. This two-layer model can reproduce the essence in the dynamic response of actual soil layers underlain by a stiff bearing strata, the engineering bedrock or base layer, despite the simplification.

Let us start from horizontal displacement $u_1$ of the vertically-propagating SH wave in the uniform ground in Fig. 4.2.1(a) of S-wave velocity $V_{s1}$ expressed as:

$$u_1 = A_s e^{i(\omega t - k_1 z)} + B_s e^{i(\omega t + k_1 z)} \tag{4.2.1}$$

Here, $z$-axis is taken upward from the ground surface and $A_s$, $B_s$ = amplitudes of upward and downward waves, $\omega$ = angular frequency, and $k_1 = \omega / V_{s1}$ is the wave number. Because the shear stress at the ground surface is always zero,

$$\tau\Big|_{z=0} = G_1 \frac{\partial u_1}{\partial z}\Big|_{z=0} = G_1 k_1 e^{i\omega t}\left(-A_s e^{-ik_1 z} + B_s e^{ik_1 z}\right)\Big|_{z=0} = iG_1 k_1 e^{i\omega t}(-A_s + B_s) = 0 \tag{4.2.2}$$

Hence, $A_s = B_s$ and Eq. (4.2.1) becomes

$$u_1 = 2A_s e^{i\omega t} \tag{4.2.3}$$

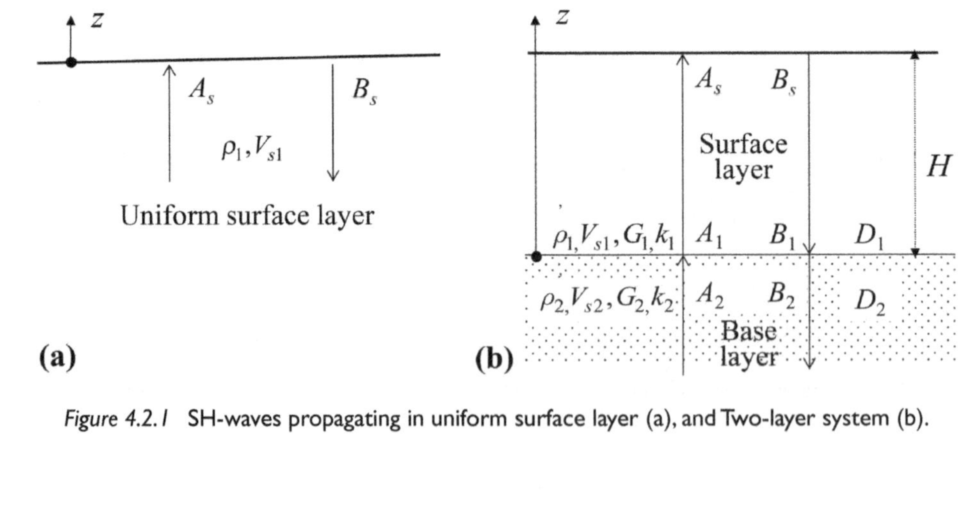


*Figure 4.2.1* SH-waves propagating in uniform surface layer (a), and Two-layer system (b).

This indicates that the measured amplitude at the ground surface is twice as large as that of the upward or downward wave.

Next, the two-layer system in Fig. 4.2.1(b) is considered, where $\rho_1$, $\rho_2$ = soil densities of the surface and base layers, respectively, $V_{s1}$, $V_{s2}$ = corresponding S-wave velocities, $G_1$, $G_2$ = corresponding shear moduli and the $z$-axis is defined upward originated from the layer boundary. The horizontal displacements in the surface and base layers, $u_1$ and $u_2$, are expressed respectively as:

$$u_1 = A_1 e^{i(\omega t - k_1 z)} + B_1 e^{i(\omega t + k_1 z)} \tag{4.2.4}$$

$$u_2 = A_2 e^{i(\omega t - k_2 z)} + B_2 e^{i(\omega t + k_2 z)} \tag{4.2.5}$$

Here, $A_1$, $B_1$ are the amplitudes for the upward and downward waves in the surface layer, respectively, $A_2$, $B_2$ are those in the base layer, and $k_1 = \omega/V_{s1}$, $k_2 = \omega/V_{s2}$. At the layer boundary $z = 0$, the displacements and stresses in the two layers are the same so that

$$u_1|_{z=0} = u_2|_{z=0} \tag{4.2.6}$$

$$G_1 \left.\frac{\partial u_1}{\partial z}\right|_{z=0} = G_2 \left.\frac{\partial u_2}{\partial z}\right|_{z=0} \tag{4.2.7}$$

Since the shear stress is zero at the ground surface $z = H$,

$$G_1 \left.\frac{\partial u_1}{\partial z}\right|_{z=H} = -iG_1 k_1 \left(A_1 e^{-ik_1 H} - B_1 e^{ik_1 H}\right) e^{i\omega t} = 0 \tag{4.2.8}$$

From Eq. (4.2.6) and Eq. (4.2.7), the following two equations are obtained, respectively.

$$A_1 + B_1 = A_2 + B_2 \tag{4.2.9}$$

$$A_1 - B_1 = \frac{k_2 G_2}{k_1 G_1}(A_2 - B_2) \tag{4.2.10}$$

From Eq. (4.2.8), the following equation is obtained.

$$\frac{B_1}{A_1} = e^{-2ik_1 H} \tag{4.2.11}$$

The ratios between the amplitudes can be obtained from Eqs. (4.2.9), (4.2.10), (4.2.11) as:

$$\frac{A_1}{A_2} = \frac{2}{(1 + e^{-2ik_1 H}) + (k_1 G_1)/(k_2 G_2)(1 - e^{-2ik_1 H})} = \frac{2}{(1+\alpha) + (1-\alpha)e^{-2ik_1 H}} \tag{4.2.12}$$

$$\frac{A_1}{B_2} = \frac{2}{(1-\alpha) + (1+\alpha)e^{-2ik_1 H}} \tag{4.2.13}$$

Here, $\alpha$ is the impedance ratio between the two layers.

$$\alpha = \frac{k_1 G_1}{k_2 G_2} = \frac{\rho_1 V_{s1}}{\rho_2 V_{s2}} \tag{4.2.14}$$

The absolute value of Eq. (4.2.12) represents the amplitude ratio between the upward waves in the surface and base layers, and expressed as follows.

$$\left|\frac{A_1}{A_2}\right| = \frac{2}{|(1+\alpha)+(1-\alpha)e^{-2ik_1H}|} = \frac{2}{\{((1+\alpha)+(1-\alpha)\cos 2k_1H)^2 + ((1-\alpha)\sin 2k_1H)^2\}^{1/2}} \tag{4.2.15}$$

Fig. 4.2.2 shows the calculation of Eq. (4.2.15) for three values of the impedance ratio $\alpha$, where the amplitude ratio in the vertical axis represents the amplification of harmonic motions at the ground surface relative to the virtual base surface because $|A_1/A_2| = |2A_1/2A_2|$ if the base layer were outcropped without the overlying surface layer. It takes the maximum value at $2k_1H = (2n-1)\pi$ and the corresponding wave number and resonant frequency are written respectively as:

$$k_1 = \frac{\omega}{V_{s1}} = \frac{2\pi f}{V_{s1}} = \frac{(2n-1)\pi}{2H} \tag{4.2.16}$$

$$f = \frac{\omega}{2\pi} = (2n-1)\frac{V_{s1}}{4H} \tag{4.2.17}$$

with the integer numbers $n = 1$–$\infty$. The frequency in the horizontal axis of Fig. 4.2.2 is normalized by the first resonant frequency $f_1 = V_s/(4H)$. Substituting Eq. (4.2.17) to Eq. (4.2.15) gives the peak amplification as:

$$\left|\frac{A_1}{A_2}\right| = \frac{1}{\alpha} \tag{4.2.18}$$

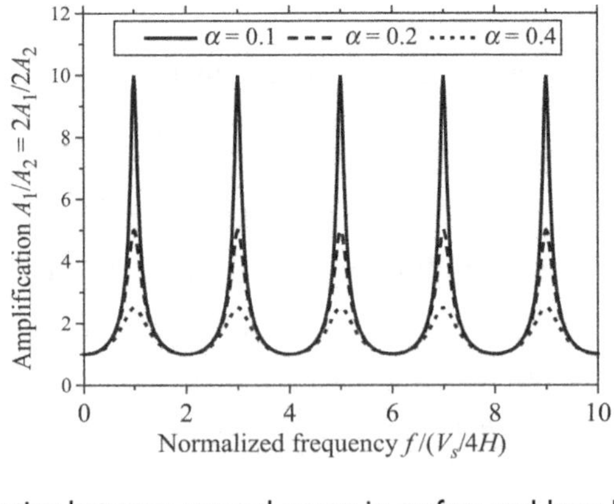


*Figure 4.2.2* Amplitude ratios between upward waves in surface and base layers in two-layer model versus normalized frequency for three impedance ratios $\alpha$.

which is identical for all the resonance $n=1$–$\infty$, taking $|A_1/A_2|=10, 5, 2.5$ for $\alpha=0.1$, 0.2, 0.4, respectively.

The displacement in the surface layer can be formulated by using Eqs. (4.2.4) and (4.2.11) as:

$$\begin{aligned} u_1 &= A_1 e^{i(\omega t - k_1 z)} + B_1 e^{i(\omega t + k_1 z)} = A_1 e^{i(\omega t - k_1 z)} + A_1 e^{-2ik_1 H} e^{i(\omega t + k_1 z)} \\ &= A_1 e^{i\omega t} e^{-ik_1 H}[e^{-ik_1(z-H)} + e^{ik_1(z-H)}] = 2A_1 e^{i\omega t} e^{-ik_1 H} \cos k_1(z-H) \end{aligned} \quad (4.2.19)$$

Substituting Eq. (4.2.17) yields the displacement in the surface layer in the resonant vibrations for $n=1$–$\infty$.

$$u_1 = 2A_1 e^{i\omega t} e^{-ik_1 H} \cos k_1(z-H)\Big|_{k_1=\frac{2n-1}{2H}\pi} = \pm 2iA_1 e^{i\omega t} \cos(2n-1)\frac{z-H}{H}\frac{\pi}{2} \quad (4.2.20)$$

The shear vibration modes thus calculated from Eq. (4.2.20) are illustrated for $n=1$, 2, 3 in Fig. 4.2.3. The first mode $n=1$ corresponds to a 1/4 wave-length of the cosine wave, while $n=2$ or 3 does to 3/4 or 5/4 wave-length with their nodes and antinodes at the bottom and surface of the layer.

Among the infinite numbers $n$, the first resonance for $n=1$ is the most important actually, where the frequency is $f=V_s/4H$ and the period is $T=1/f=4H/V_s$. This period corresponds to the time for the wave travelling 4 times the layer thickness $H$, which is equal to the one-wave length $\lambda$, or the layer thickness $H$ corresponds to a quarter wave length $\lambda/4$. Hence the following is called "the quarter wave-length formula" and conveniently used in engineering practice.

$$f=\frac{V_s}{4H} \quad \text{or} \quad T=\frac{4H}{V_s} \quad (4.2.21)$$

Actual site conditions may not be as simple as the two-layer model but consisting of multiple surface layers of variable properties above the stiff base layer as illustrated in Fig. 4.2.4. The quarter wave-length formula is still applied to such conditions

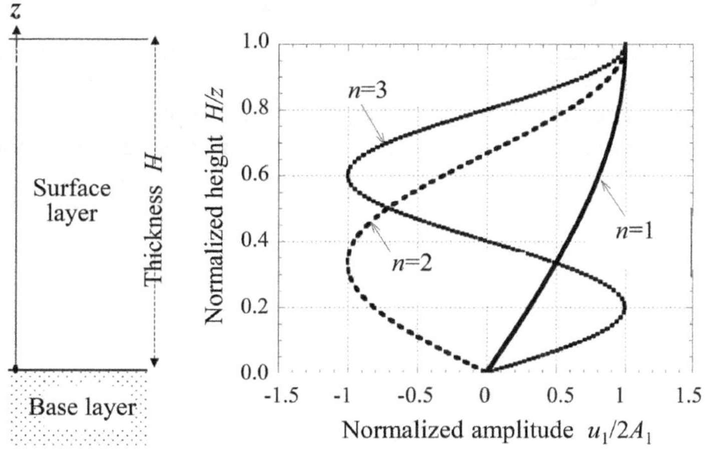


*Figure 4.2.3* Shear vibration modes of surface layer in first three resonances, $n=1, 2, 3$.

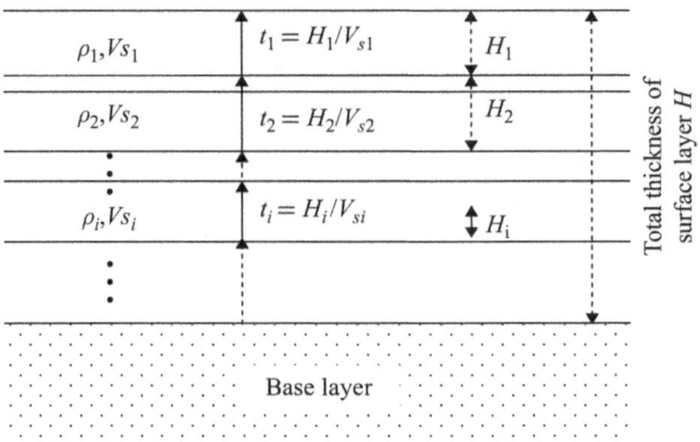


*Figure 4.2.4* Approximation of actual soil condition by idealized two-layer system.

by modifying it as follows (JRA 2002), though it is not the exact solution but only approximation.

$$f = \frac{1}{T} = \frac{1}{(4\sum_i H_i/V_{si})} \tag{4.2.22}$$

Here $H_i/V_{si}$ is the travelling time for the SH-wave in the $i$-th layer of the thickness $H_i$ and wave velocity $V_{si}$. Eq. (4.2.22) can also be expressed as $f = \overline{V}_s/4H$ using the total thickness $H$ and the average wave velocity in the surface layer:

$$\overline{V}_s = \frac{H}{\sum_i H_i/V_{si}} \tag{4.2.23}$$

### 4.2.2 Two-layer system with internal damping

In order to consider the internal damping in the amplification, the shear modulus $G$ is replaced by the complex shear modulus $G^*$, and the S-wave velocity $V_s^*$, wave number $k^*$, impedance ratio $\alpha^*$ are all redefined as complex numbers incorporating $G^*$ as already formulated in Sec. 1.5.2. If the non-viscous damping is assumed, the complex modulus in the surface layer is written as:

$$G_1^* = G_1 + iG_1' \tag{4.2.24}$$

and accordingly,

$$V_{s1}^* = \sqrt{G_1^*/\rho_1} = \sqrt{(G_1 + iG_1')/\rho_1} \tag{4.2.25}$$

$$k_1^* = \frac{\omega}{V_{s1}^*} = \omega\left(\frac{\rho}{G_1 + iG_1'}\right)^{1/2} = \frac{\omega}{V_{s1}}\left(\frac{1}{1 + 2iD_1}\right)^{1/2} \tag{4.2.26}$$

$$\alpha^* = \frac{k_1^* G_1^*}{k_2^* G_2^*} = \frac{\rho_1 V_{s1}^*}{\rho_2 V_{s2}^*} \tag{4.2.27}$$

Eqs. (4.2.12), (4.2.13) are rewritten correspondingly in the case with the internal damping as:

$$\frac{A_1}{A_2} = \frac{2}{(1+\alpha^*) + (1-\alpha^*)e^{-2ik_1^* H}} \tag{4.2.28}$$

$$\frac{A_1}{B_2} = \frac{2}{(1-\alpha^*) + (1+\alpha^*)e^{-2ik_1^* H}} \tag{4.2.29}$$

The amplitude ratios in the above equations are defined between the bottom of the surface layer and the top of the base layer. If the amplitude $A_1$ in the upward wave $u_1$ at the bottom ($z=0$) decays to $A_s$ at the top ($z=H$), then, the following holds

$$u_1 = A_1 e^{i(\omega t - k_1^* z)} = A_s e^{i(\omega t - k_1^*(z-H))} = A_s e^{ik_1^* H} \cdot e^{i(\omega t - k_1^* z)} \tag{4.2.30}$$

Then, $A_1$ and $A_s$ are correlated reflecting the internal damping as:

$$A_1 = A_s e^{ik_1^* H} \tag{4.2.31}$$

Substituting this into Eq. (4.2.28) gives the amplification for the horizontal array in the two-layer model as illustrated in Fig. 4.1.2(a).

$$\frac{A_s}{A_2} = \frac{2A_s}{2A_2} = \frac{2}{(1+\alpha^*)e^{ik_1^* H} + (1-\alpha^*)e^{-ik_1^* H}} \tag{4.2.32}$$

In the vertical array of the two-layer model shown in Fig. 4.1.2(b) where the seismometers are at the ground surface and downhole at the top of the base layer, the amplification of motions is calculated as the ratio of $2A_s$ versus $A_2 + B_2$. Combining Eq. (4.2.32) with the following equation obtained by the substitution of Eq. (4.2.31) to Eq. (4.2.29),

$$\frac{A_s}{B_2} = \frac{2}{(1-\alpha^*)e^{ik_1^* H} + (1+\alpha^*)e^{-ik_1^* H}} \tag{4.2.33}$$

the next formula can be obtained for the amplification of the vertical array.

$$\frac{2A_s}{A_2 + B_2} = \frac{2}{A_2/A_s + B_2/A_s} = \frac{2}{e^{ik_1^* H} + e^{-ik_1^* H}} \tag{4.2.34}$$

#### 4.2.2.1 Amplification in horizontal array versus vertical array

The amplifications for the horizontal array $2A_s/2A_2$ and vertical array $2A_s/(A_2 + B_2)$ formulated in Eqs. (4.2.32) and (4.2.34) are calculated for the impedance ratios $|\alpha^*| \approx \alpha = 0.1$, 0.3, 0.5 and their absolute values are shown versus normalized input frequency of harmonic motion $f/f_1$ where $f_1 = V_s/4H$ is the first resonance frequency. Fig. 4.2.5(a) is for $D_1 = 0\%$ and (b) is for $D_1 = 5\%$ in the surface layer, while $D_2 = 0\%$

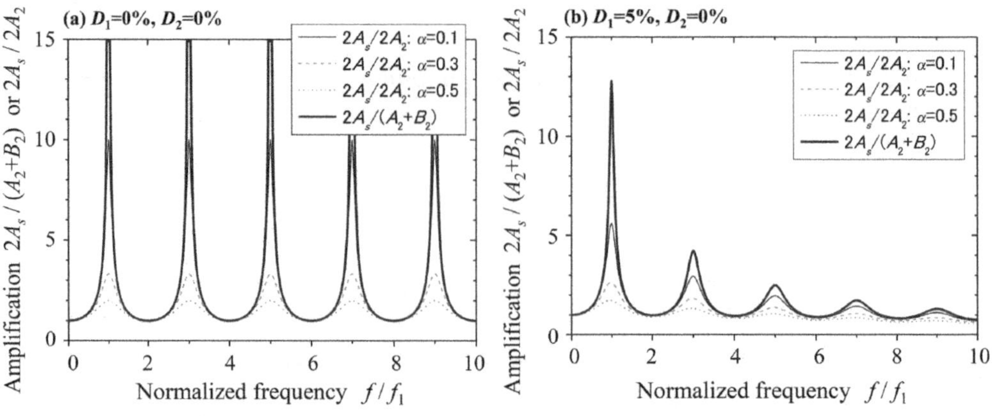


*Figure 4.2.5* Amplifications for vertical array $2A_s/(A_2+B_2)$ and horizontal array $2A_s/2A_2$ versus normalized frequency $f/f_1$ in two-layer model with $D_1=0\%$ (a) and $D_1=5\%$ (b).

in the base layer in both cases though $D_2$ does not significantly affect the results. The resonance appears at the same frequencies as the case without soil damping, and the associated peak amplifications for the horizontal array are obtained by substituting Eq. (4.2.17) into Eq. (4.2.32) as:

$$\left|\frac{2A_s}{2A_2}\right| = \frac{1}{\sinh(\pi\delta_1(2n-1)/4) + \alpha\cosh(\pi\delta_1(2n-1)/4)} \quad n=1-\infty \tag{4.2.35}$$

where $\delta_1 = \tan^{-1}2D_1$. For the vertical array, Eq. (4.2.34) substituted by Eq. (4.2.17) yields:

$$\left|\frac{2A_s}{A_2+B_2}\right| = \frac{1}{\sinh(\pi\delta_1(2n-1)/4)} \quad n=1-\infty \tag{4.2.36}$$

If Eqs. (4.2.35) and (4.2.36) are compared, the former involves $\alpha$, while the latter does not, implying that $|2A_s/2A_2|$ depends on impedance ratio, while $|2A_s/(A_2+B_2)|$ does not and always takes a larger value than the former. Also note that Eq. (4.2.33) or Eq. (4.2.35) becomes Eq. (4.2.34) or Eq. (4.2.36) if $\alpha=\rho_1 V_{s1}/\rho_2 V_{s2}=0$ or $\rho_2 V_{s2}\to\infty$, indicating that the amplifications of the two types of arrays become identical when the base layer is infinitely stiff. These observations may be confirmed in Figs. 4.2.5, because the curves of $|2A_s/2A_2|$ tend to approach to those of $|2A_s/(A_2+B_2)|$ with decreasing $\alpha$-value.

Fig. 4.2.6 shows the amplifications in the horizontal array $|2A_s/2A_2|$ in (a) and the vertical array $|2A_s/(A_2+B_2)|$ in (b) for parametrically varying damping ratios in the surface layer $D_1=2.5$, 5, 10% versus the normalized frequency $f/f_1$ where $f_1$ = first resonant frequency. Obviously, the amplification in (a) is strongly affected not only by the internal damping ratio $D_1$ but also by the radiation damping determined by the impedance ratio chosen here as $\alpha=\rho_1 V_{s1}/\rho_2 V_{s2}=0.3$, while in (b) it corresponds

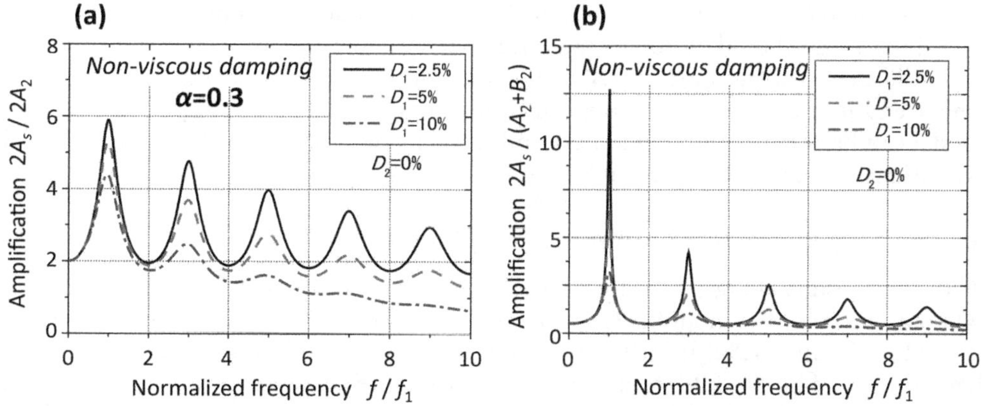


*Figure 4.2.6* Amplifications for horizontal array $2A_s/2A_2$ (a), and Vertical array $2A_s/(A_2+B_2)$ (b), versus normalized frequency $f/f_1$ in two-layer model with parametrically changing $D_1 = 2.5$, 5, 10% and $\alpha = 0.3$.

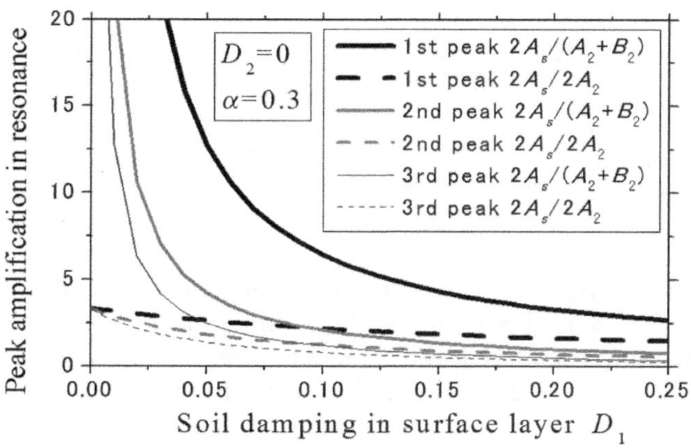


*Figure 4.2.7* Peak amplification values versus damping ratio $D_1$ for vertical array $2A_s/(A_2+B_2)$ and horizontal array $2A_s/2A_2$.

to $\alpha = 0$ in (a) as already mentioned and hence $|2A_s/(A_2+B_2)|$ is much larger than $|2A_s/2A_2|$.

In Figs. 4.2.7, the peak values in the two types of amplifications $|2A_s/2A_2|$ and $|2A_s/(A_2+B_2)|$ shown in Fig. 4.2.6 are plotted versus the damping ratio $D_1$. The peak values for $|2A_s/(A_2+B_2)|$ by the solid curves which are much larger than those for $|2A_s/2A_2|$ by the dashed curves for smaller $D_1$ tend to rapidly decrease with increasing $D_1$ and approach to $|2A_s/2A_2|$ not only in the first but also in the higher order peaks. On the other hand, the decreasing trends of $|2A_s/2A_2|$ due to increasing $D_1$ are more gradual than $|2A_s/(A_2+B_2)|$ due to the dominant effect of the impedance ratio $\alpha$. This indicates that the difference in the two types of peak amplifications tends to be narrower with increasing soil damping in the surface layer. In other words, the amplification in the vertical array $|2A_s/(A_2+B_2)|$ tends to be much larger in small earthquakes than in large earthquakes due to the difference in damping ratios, while the amplification in the horizontal array $|2A_s/2A_2|$ is not so sensitive to earthquake intensities. This trend in the two types of amplifications obtained from the simplified two-layer model can be actually observed in ground motions recorded in more complicated soil conditions as will be discussed in Sec. 4.4.3.

#### 4.2.2.2 Amplification by different damping models

The above results in Figs. 4.2.6 and 4.2.7 are based on the Nonviscous Kelvin damping wherein the complex modulus $G^*$ is defined as Eq. (4.2.24) or Eq. (1.5.24). If $G^*$ in Eqs. (1.5.12) and (1.5.20) for the Kelvin and Maxwell models, respectively, are used here in place of Eq. (1.5.24) together with the associated complex variables $V_{s1}^*$, $k_1^*$ and $\alpha^*$ in Eqs. (4.2.25)–(4.2.27), the corresponding amplifications of the same two-layer model can be readily derived. Figs. 4.2.8(a), (b) compare the three damping models for the horizontal and vertical array amplifications, respectively, with $D_1=5\%$, $D_2=0\%$ and $\alpha=0.1$. It is noted here based on Eqs. (1.5.33)–(1.5.35) that the damping ratio is $D_1=5\%$ constant in the Nonviscous model, while in other models $D_1$ is 5% only at the first peak and varies with the angular frequency $\omega=2\pi f$ as follows.

$$\text{Kelvin model:}\quad D_1=\frac{\tan\delta}{2}=\frac{\omega\xi_1}{2G_1}=\frac{\pi V_{s1}\xi_1}{4G_1H}\frac{\omega}{\omega_1}=\frac{\pi\xi_1}{4H\rho_1 V_{s1}}\frac{f}{f_1}=D_0\frac{f}{f_1} \tag{4.2.37}$$

$$\text{Maxwell model:}\quad D_1=\frac{\tan\delta}{2}=\frac{G_1}{2\omega\xi_1}=\frac{G_1H}{\pi V_{s1}\xi_1}\bigg/\frac{\omega}{\omega_1}=\frac{H\rho_1 V_{s1}}{\pi\xi_1}\bigg/\frac{f}{f_1}=D_0\bigg/\frac{f}{f_1} \tag{4.2.38}$$

Here, $\omega_1=2\pi f_1=(\pi V_{s1})/(2H)$ is the first resonant frequency in the surface layer and $\xi_1=$viscosity. $D_0=5\%$ is chosen in the above equation so that the peak values coincide at the first resonance among the three damping models as shown in Fig. 4.2.8. However, they widely diverge in the higher order peaks; staying constant in all the

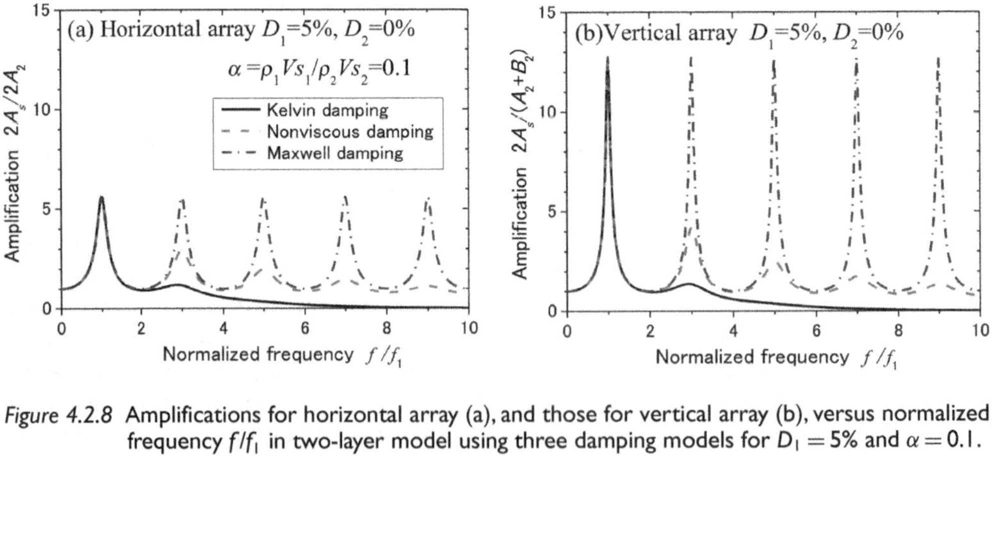


*Figure 4.2.8* Amplifications for horizontal array (a), and those for vertical array (b), versus normalized frequency $f/f_1$ in two-layer model using three damping models for $D_1=5\%$ and $\alpha=0.1$.

peaks in the Maxwell model, moderately reducing in the Nonviscous model and drastically reducing in the Kelvin model. According to Eq. (1.6.12) in Sec. 1.6.1, the wave attenuation coefficient by internal damping $\beta \approx \omega D/V_s$ is proportional to the damping ratio $D$, while $D$ is proportional to $\omega^1$, $\omega^0$, $\omega^{-1}$ in the Kelvin, Nonviscous and Maxwell models, respectively, as indicated in Eqs. (1.5.33)–(1.5.35). Hence, $\beta$ is proportional to $\omega^2$, $\omega^1$, $\omega^0$ and the peak amplifications depend on the frequency in the same order as clearly shown in Fig. 4.2.8. The relative effect of the damping models is very similar in the two types of amplification $|2A_s/2A_2|$ and $|2A_s/(A_2+B_2)|$, though the absolute amplification values are much larger in the vertical array than in the horizontal array.

As already mentioned, laboratory soil tests demonstrate that internal soil damping is mostly non-viscous or frequency-independent. In contrast, if site amplifications are investigated by means of earthquake observations, they sometimes look like the Maxwell type with nearly identical peak values rather than the Nonviscous type with monotonically decreasing peak values with increasing frequency. This difference in the damping mechanism will be discussed again in Sec. 4.3.3.

## 4.3 SITE AMPLIFICATION BY EARTHQUAKE OBSERVATION

As already shown in Fig. 4.1.2 conceptually, there are two types of earthquake observation system to investigate seismic amplification between the ground surface and the base layer (engineering bedrock); the horizontal array and vertical array. The vertical arrays have been increasingly deployed all over Japan (more than 700) after the 1995 Kobe earthquake in particular, accumulating a number of strong motion records to study the site amplifications during destructive earthquakes at various site conditions. As one of typical examples, records at 4 vertical array sites near the epicenter during the main shock and associated small shocks of the 1995 Kobe earthquake are addressed in the following to examine the site amplifications observed in the vertical arrays.

### 4.3.1 Amplification of maximum acceleration or maximum velocity

The site amplification between ground surface and base can be defined in various ways. The simplest is the ratio of maximum accelerations or velocities between the two levels. In order to consider the effect of frequency, the ratio of Fourier spectra or response spectra can also be used. The peak value of the spectra ratios or the average value of the ratios in frequency intervals relevant to structural designs may be chosen as the site amplification, too.

Fig. 4.3.1 illustrates the locations of 4 vertical array sites (PI, SGK, TKS and KNK with the arrows indicating their principal axes of maximum accelerations) near the earthquake fault (the strike slip fault) with the epicenters of the main shock and numerous aftershocks during the 1995 Kobe earthquake ($M_J = 7.2$: $M_J$ is the earthquake magnitude used in the Japanese Meteorological Agency and similar to the moment magnitude $M_W$). As depicted in soil profiles in Fig. 4.3.2, the maximum depths of the vertical arrays are around 80 to 100 m. From the deepest level to the ground surface, 3 to 4 three-dimensional accelerometers (EW, NS, UD-directions) are installed.

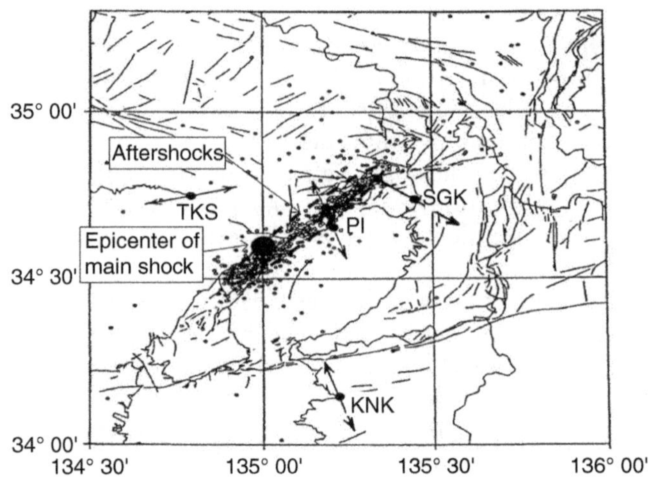


*Figure 4.3.1* Four vertical array sites PI, SGK, TKS, KNK near fault during 1995 Kobe earthquake.

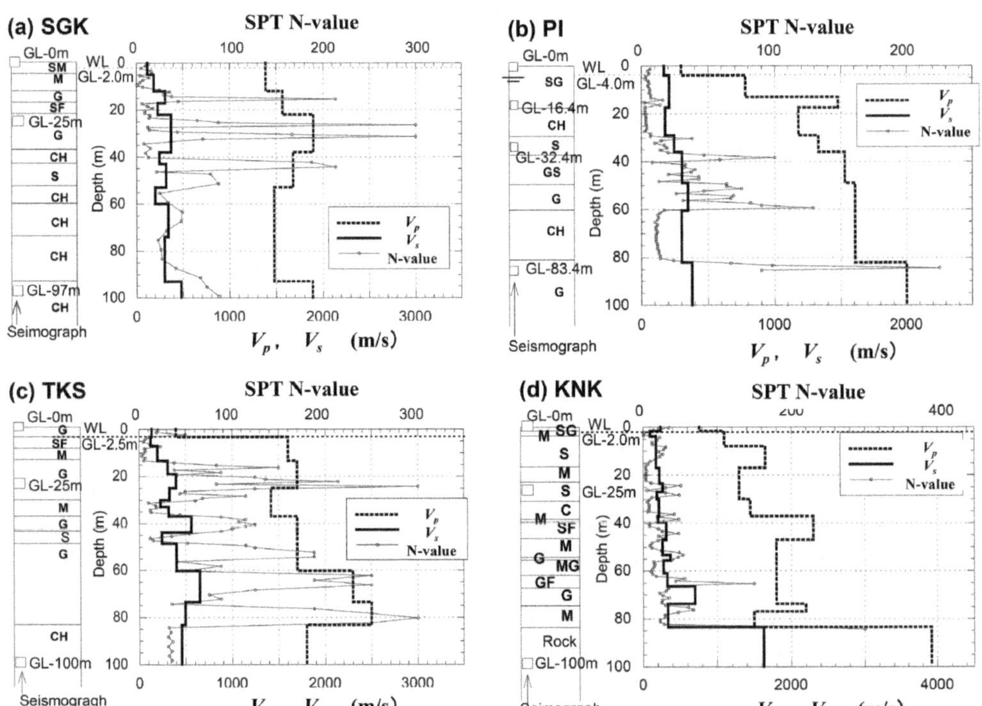


*Figure 4.3.2* Soil profiles at four vertical array sites with $V_p, V_s$, SPT *N*-values and seismograph depths: (a) PI, (b) SGK, (c) TKS and (d) KNK (Kokusho and Matsumoto 1998).

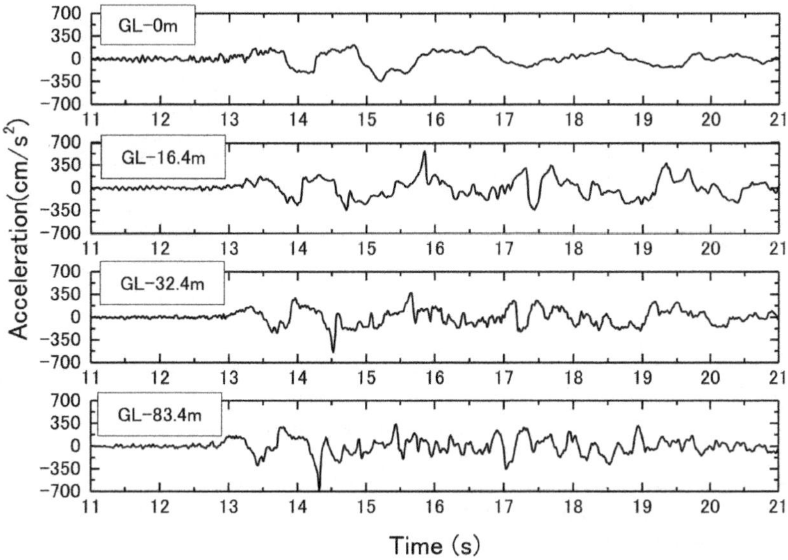


*Figure 4.3.3* Acceleration time histories of NS direction in PI vertical array during main shock of 1995 Kobe earthquake (Sato et al. 1996).

The geologies are manmade/Holocene near the surface followed by Pleistocene in deeper soils, except for Tertiary stiff rock in the base layer in the KNK-site (Sato et al. 1996, Kokusho et al. 2005a). Fig. 4.3.3 shows the acceleration time-histories in the NS direction at PI, the nearest to the fault. Because of the severe liquefaction in the manmade fill layer of about 15 m thick, the surface acceleration at GL.-0 m (Ground Level) evidently decreased compared to the lower level at GL.-16.4 m.

Fig. 4.3.4 illustrates variations of maximum acceleration along depth during the main shock in two horizontal and one vertical directions in the four vertical array sites. Unlike similar diagrams presented in another literature (Kokusho and Matsumoto 1998), the maximum response values in the following diagrams are corrected reconsidering the directional offsets of the downhole accelerometers. The maximum accelerations tend to increase monotonically with decreasing depth at KNK where the earthquake fault was remote and the maximum induced soil strain was of the order of $10^{-4}$ (Kokusho et al. 2005a). In contrast, maximum horizontal accelerations evidently decrease at the ground surface at PI the nearest site from the fault, and the acceleration tends to decrease in the middle depth in SGK. The deamplification in the surface acceleration in PI are due to extensive liquefaction in reclaimed decomposed granite soil, as already mentioned, leading to a typical example of the base-isolation as will be discussed in Sec. 5.11. Also note that no such liquefaction-induced deamplification but a sharp increase is visible in the vertical acceleration in the liquefied layer at PI.

Fig. 4.3.5 depicts amplifications defined as the ratios of the maximum accelerations between surface and deepest base level plotted versus the corresponding $V_s$-ratios between base and surface. The close and open symbols represent the main shock and aftershocks (for PI not the aftershocks but small shocks recorded before the

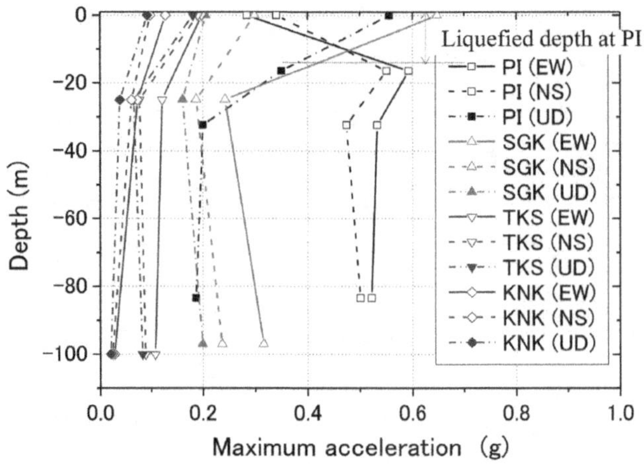


*Figure 4.3.4* Depth-dependent variations of maximum horizontal and vertical accelerations at four vertical array sites during main shock of 1995 Kobe earthquake (Kokusho et al. 2005a), with permission from ASCE.

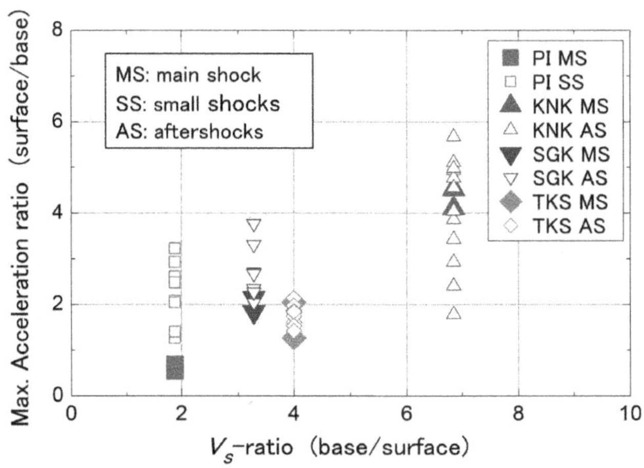


*Figure 4.3.5* Maximum horizontal acceleration ratio (surface/base) versus $V_s$-ratio (base/surface) at four vertical array sites during 1995 Kobe earthquake (Kokusho et al. 1998).

main shock), respectively. The amplifications are obviously increasing with increasing $V_S$-ratio for the main shocks and aftershocks in the four sites despite the data scatters, indicating that the $V_S$-ratio is one of the key parameters governing the amplification. Also visible is that the amplification tends to be smaller for the main shock than the aftershocks in those sites nearer to the fault in particular.

In order to see the effect of shaking intensity, the same maximum acceleration ratios are plotted for the main shock and aftershocks versus corresponding maximum accelerations at the deepest levels in the log scale in Fig. 4.3.6(a). Though the acceleration ratios vary considerably from site to site, their global decreasing trends with

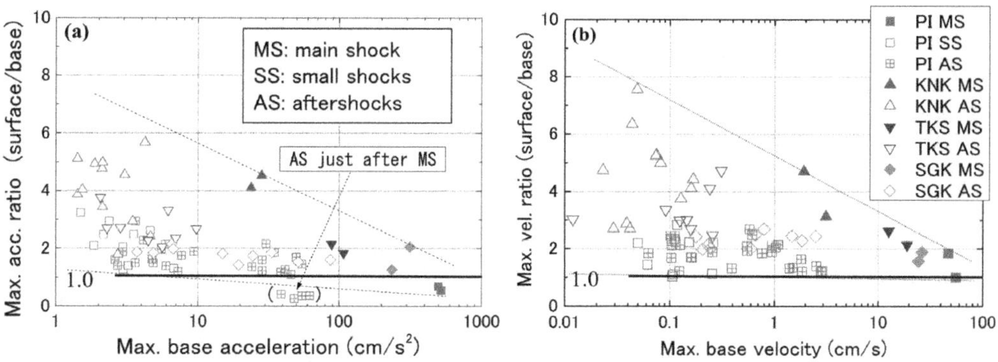


*Figure 4.3.6* Ratios of maximum accelerations (surface/base) versus max. base accelerations (a), and Ratios of maximum velocities (surface/base) versus base velocities (b) at four vertical array sites during 1995 Kobe earthquake (Kokusho et al. 1998).

increasing base accelerations are obviously seen as outlined by the pair of dashed lines and also in the individual sites. Some of the plots in PI located below the horizontal line 1.0 correspond to the main shock and aftershocks following immediately, indicating that the liquefaction-induced deamplification or base-isolation occurred during the main shock and sustained for some time after that (Kokusho and Matsumoto 1998). In Fig. 4.3.6(b), the ratio of maximum velocities between the surface and base are plotted versus the maximum velocities at the base, wherein the velocities are calculated from the acceleration records. The similar decreasing trends can be recognized also in the velocity ratios with increasing particle velocities at the base, though they do not seem to go lower than unity.

### 4.3.2 Spectrum amplification

The decreasing trends of the site amplification in terms of maximum accelerations/velocities in Fig. 4.3.6 seem to reflect the changes of soil properties due to increasing soil strain and pore-pressure buildup as well as the difference of frequency components in earthquake motions during the main shock and aftershocks. In order to single out the effect of nonlinear soil properties, it is necessary to investigate the spectrum amplifications between surface and base of the vertical array records.

Figs. 4.3.7(a), (b) and (c) exemplify Fourier spectrum ratios calculated from the surface and base records at GL.-83.4 m in EW, NS and UD directions, respectively, for five small shocks occurred before the main shock of 1995 Kobe earthquake. The dotted curves are for the individual small shocks and the solid curve is their average. The spectrum ratios thus obtained give frequency-dependent amplification of this site for stationary harmonic input motions. It is noted in Figs. 4.3.7(a) and (b) that the peak frequencies are almost consistent from one shock to another, and also in two orthogonal directions EW and NS. Also noted in Figs. 4.3.7(c) is that the amplification in vertical motions is considerably different from those in horizontal motions, so that

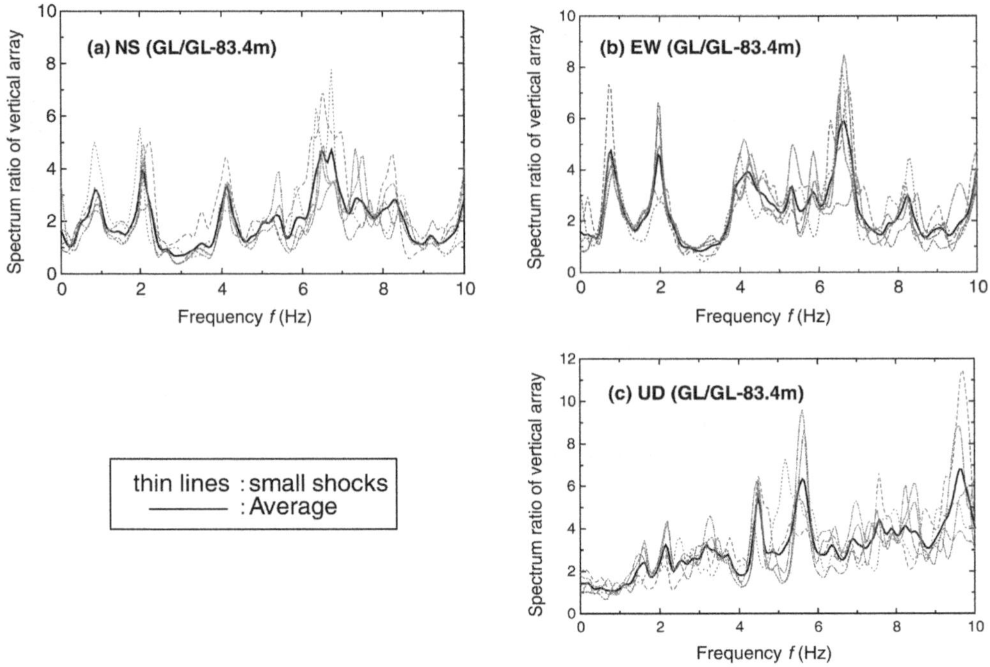


*Figure 4.3.7* Fourier spectrum ratios of PI vertical array between surface and base (GL.-83.4 m) for small shocks: (a) EW, (b) NS, (c) UD (Aoyagi 2000).

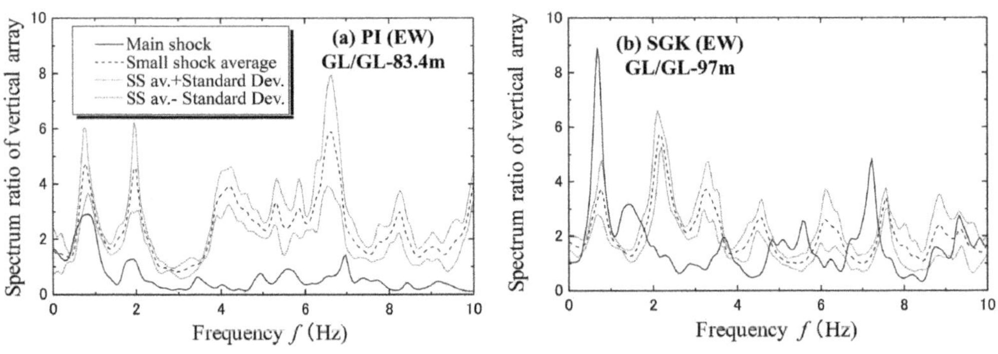


*Figure 4.3.8* Fourier spectrum ratios of horizontal motions in vertical array between surface and base compared between main shock and small shocks: (a) PI, (b) SGK (Aoyagi 2000).

the predominant frequencies tend to be higher. It is because the P-wave velocity much faster than the S-wave velocity is largely involved in the vertical motion amplifications.

Fig. 4.3.8 shows the Fourier spectrum ratios of horizontal motions between surface and base during the main shock of the Kobe earthquake at PI (a) and SGK(b) both located near the fault with the solid curves (Kokusho et al. 2005a). In the same diagrams, averages and standard deviations of the spectrum ratios for the small shocks

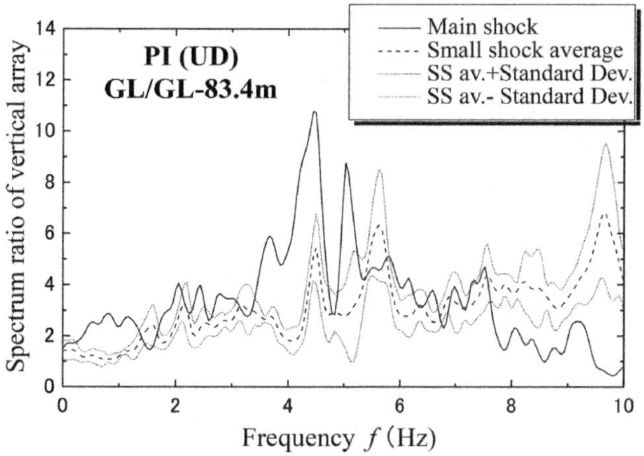


*Figure 4.3.9* Fourier spectrum ratios of vertical motions in vertical array between surface and base in PI compared between main shock and small shocks (Aoyagi 2000).

or aftershocks are superposed. The comparison with the small shocks reveals that peak frequencies during the main shock tend to shift to lower values. The shift looks to be greater for higher order peaks probably because soil nonlinearity tends to be manifested more in shallower soils associated with higher vibration modes. The peak spectrum amplitudes are mostly smaller in the main shock than in the small shocks reflecting strain-dependent increase in soil damping. This trend is particularly remarkable in PI where extensive liquefaction occurred in the top 15 m, so that the higher order peaks are very much shifted and damped. In SGK where no liquefaction was witnessed, the first peak frequency reflecting the global dynamic response of soils above the deepest seismometer is only marginally affected by strain-dependent property changes, while the higher-order peaks are obviously influenced in both frequency and amplitude.

Fig. 4.3.9 depicts the Fourier spectrum ratio of vertical motions between surface and base during the main shock compared with that during the small shocks at PI. The peak spectrum amplification during the main shock is almost comparable with those for small shocks, though peak frequencies are not agreeable between the main shock and small shocks. This is probably because the vertical motions are attributable to P-wave and its wave velocity and attenuation are insensitive to pore-pressure buildup and induced large strains in saturated soil layers, although some other factors such as low-saturated surface soils above the water table may have made the difference in peak frequencies.

### 4.3.3 Amplification reflecting frequency-dependent damping

One of the dominant factors governing site amplification is soil damping. As explained in Sec. 2.3.3, the damping ratio of soils is evaluated as nonviscous or frequency-independent in laboratory soil tests because soils are essentially frictional materials. For the nonviscous damping, the amplification for the vertical array of the two-layer

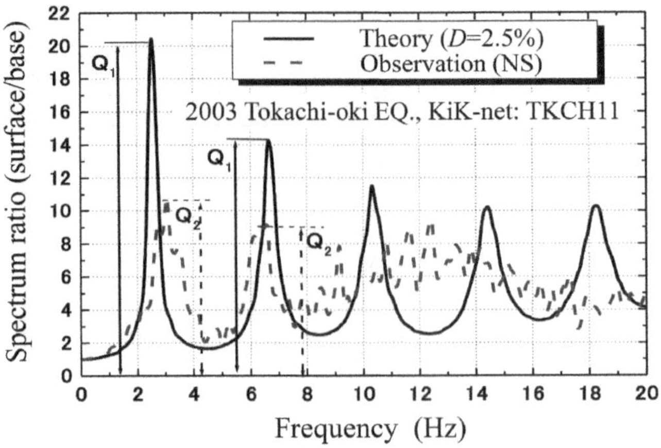


*Figure 4.3.10* Observed Fourier spectrum amplification in vertical array between surface and base compared with theoretical transfer function at KiK-net site (Kokusho 2013a).

model tend to gradually decrease with increasing frequency as shown with the dashed curve in Fig. 4.2.8(b). In contrast, Fourier spectrum amplifications calculated from vertical array records such as those in Figs. 4.3.7(a), (b) are not decreasing with increasing frequency and look more or less similar to the Maxwell model with the chain-dotted curve in Fig. 4.2.8(b). This trend seems to become clearer in smaller shocks than main shocks in the same PI site if Fig. 4.3.7 is compared with Fig. 4.3.8. It indicates that a damping mechanism other than frequency-independent nonviscous damping may also be involved in the field. In the following, some observations and the theoretical background of the in situ frequency-dependent damping are addressed.

#### 4.3.3.1 Damping in observed site amplification

Fig. 4.3.10 exemplifies a spectrum amplification between the surface and base (GL.-100 m) calculated from vertical array records at a KiK-net site (See Sec. 4.4) for the main shock during the 2003 Tokachi-oki earthquake in Hokkaido Japan and compares with the theoretical transfer function. The transfer function was calculated based on the multi-reflection theory of one-dimensional SH-wave using site-specific $V_S$-logging data assuming the non-viscous soil damping of $D = 2.5\%$ constant. A fair correspondence can be recognized in the first and second peak frequencies between observation and theory. In terms of the peak amplification values, however, no distinctly increasing or decreasing trend can be seen in the observed spectrum ratios whereas the theoretical peaks tend to decrease monotonically with increasing frequency. In order to have better matching between them, the damping ratio $D$ in calculating the transfer function may need to be adjusted different from the constant $D = 2.5\%$. For example, the theoretical peak value $Q_1$ in Fig. 4.3.10 is compared with the observational value $Q_2$ not only for the first peak but also for the higher-order peaks in cases where the peak frequencies

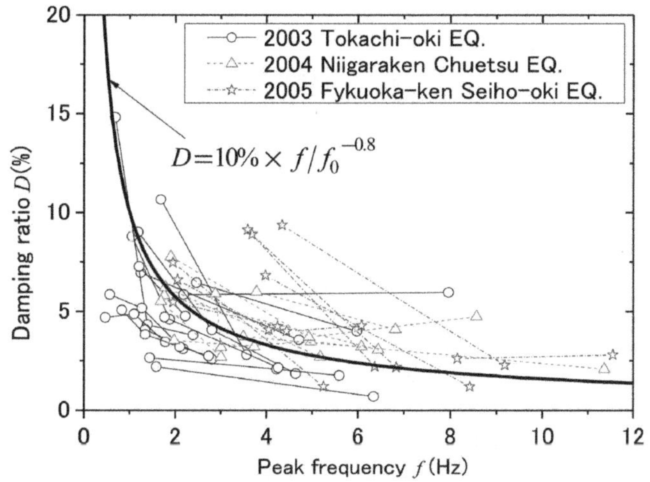


*Figure 4.3.11* Equivalent damping ratios determined from spectrum amplification peaks versus peak frequencies obtained from KiK-net records during three large earthquakes (Kokusho 2013a).

are similar. The $D$-value is adjusted peak by peak so that the corresponding two peak values agree. Hence, damping ratios for individual peaks can be determined as:

$$D = \frac{Q_1}{Q_2} \times 2.5\% \tag{4.3.1}$$

Fig. 4.3.11 summarizes the damping ratios $D$ of individual peaks thus determined versus corresponding peak frequencies for spectrum amplifications obtained in a number of vertical array sites during three strong earthquakes (the 2003 Tokachi-oki EQ., the 2004 Chuetsu EQ. and the 2005 Fukuoka-ken Seiho-oki EQ.) recently occurred in Japan (Kokusho 2013a). The same symbols in this diagram connected with lines are from the same sites but of different peaks. Despite data dispersions from site to site, a consistent trend is undeniable that the equivalent damping ratios $D$ thus determined in Eq. (4.3.1) are frequency-dependent so that the $D$-values tend to be lower as the peak frequencies become higher. In all these cases, the maximum surface accelerations exceeded 0.1~0.2 g and hence soils in shallow depths may have experienced a certain degree of strain-dependent nonlinearity though not so strongly nonlinear as to trigger liquefaction. Thus, not only for small seismic shocks with linear soil properties but also for relatively large shocks accompanying nonlinear soil response, the frequency-dependency of damping thus evaluated can be detected from ground motion records.

Two mechanisms are suspected to cause the frequency-dependency observed here. One is that already addressed in the equivalent linear analysis in Sec. 3.2.4, wherein the equivalent linear damping value, chosen to be compatible with the representative strain amplitude, tends to underestimate higher frequency components because they tend to induce smaller strains than lower frequency motions. This apparent frequency-dependency by strain-dependent effect seems to become more pronounced with increasing soil nonlinearity in larger strains. The other is wave scattering due to heterogeniety of in situ soils, which tends to be more dominant with decreasing strains as will be mentioned later.

The frequency dependency of damping ratio $D$ by wave scattering is sometimes formulated as:

$$D = D_0'\left(\frac{f}{f_0}\right)^{-m} \tag{4.3.2}$$

Here $D_0' = D$-value for $f$ = reference frequency $f_0$, and $m$ is a positive exponent controlling the frequency dependency of damping, wherein $m = -1$, 0, 1 correspond to the Kelvin, Nonviscous-Kelvin and Maxwell damping, respectively. A solid curve in Fig. 4.3.11 is drawn by Eq. (4.3.2) for $D_0' = 10\%$, $f_0 = 1.0$ Hz and $m = 0.8$, which may be able to roughly represent the plots in view of frequency dependency of damping in relatively large strains.

Though a rigorous interdependency between the frequency-dependent damping observed in site amplifications by wave scattering and the strain-dependent damping observed in laboratory soil tests is not theoretically clarified, the following formula is sometimes used to integrate them (Sato and Kawase 1992).

$$D = D_0 + D_0'\left(\frac{f}{f_0}\right)^{-m} \tag{4.3.3}$$

This indicates that in situ damping ratio is the addition of the strain-dependent part $D_0$ which is independent of frequency and the frequency-dependent part $D_0'(f/f_0)^{-m}$ independent of strain. The contribution of the latter part tends to lessen with increasing $D_0$-value as strain increases, leading to be overwhelmingly strain-dependent during destructive earthquakes. Actually, the amplification in PI in Fig. 4.3.8(a) during the main shock wherein strong soil nonlinearity with severe liquefaction occurred seems to be analogous to the transfer function of the Nonviscous damping shown with the dashed curve in Fig. 4.2.8(b). In contrast, the spectrum amplification for the small shocks in Figs. 4.3.7(a), (b) seems to be more like the chain dotted curve in Fig. 4.2.8(b) for the Maxwell damping.

Fig. 4.3.12 shows another research results on the frequency-dependency of soil damping in wave propagations investigated in situ by means of wave logging and small earthquake observations (Fukushima and Midorikawa 1994). In this diagram, $2D$ or the inverse of $Q$-values ($2D = 1/Q$) determined at different sites are plotted versus frequency on the full logarithmic diagram. Here, the $D$-values are modified from measured values in order to reduce site-specific differences in $D$-values due to different $V_s$-values by using a correlation between $V_s$ and $D$ as typically illustrated in Fig. 2.1.1. Also assumed is that the $D$-values are constant for $f > 5$ Hz based on observational data. This diagram indicates $m = 0.5$–$0.7$, intermediate between the Nonviscous ($m = 0$) and Maxwell model ($m = 1$).

In geotechnical problems, such as liquefactions or slope failures, high frequency components in seismic motions are not a big issue because they do not result in large strains compared to low frequency components. However, in designing machines and equipments with high resonant frequencies, high-frequency ground motions have to be evaluated with greater care. It is particularly important for medium to stiff soil

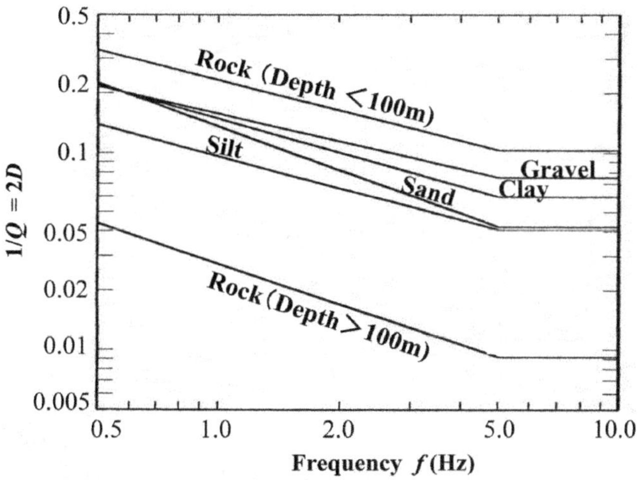


*Figure 4.3.12* Frequency-dependency of soil damping in wave propagations investigated in situ (Fukushima and Midorikawa 1994).

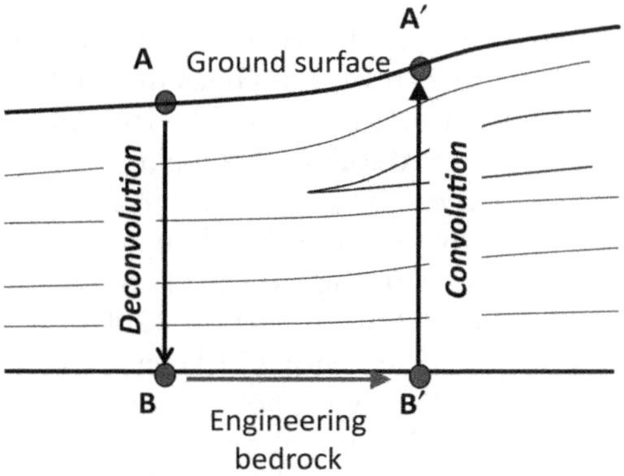


*Figure 4.3.13* Convolution and deconvolution to correlate design motions at two sites A and A′.

sites where the strain-dependency of damping is less dominant than the frequency-dependency.

In evaluating design seismic motions, one-dimensional site response analyses are carried out wherein the damping ratio together with its frequency dependency play a key role as well as the S-wave velocity profiles. If a design motion is prescribed at the ground surface, at Point A in Fig. 4.3.13 for example, the motion at Point A′ with a different soil profile is determined in the following procedure. First deconvoluting the motion at the engineering bedrock (Point B) by using the soil profile at A and then convoluting the motion at Point A′ from the deconvoluted motion at B′ (assuming the same as B on the same bedrock) by using the soil profile at A′. In this case, the assumption of frequency-dependency of damping will not significantly affect the calculated motion

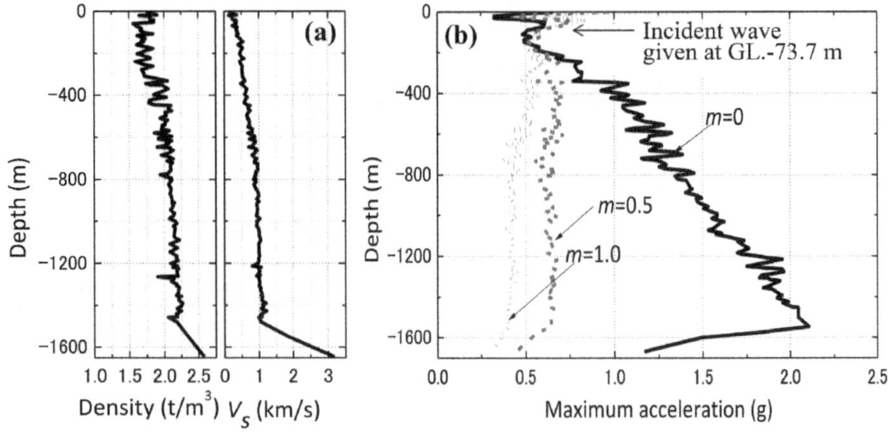


*Figure 4.3.14* Maximum acceleration down to GL.-1600 m for $m = 0, 0.5, 1.0$ (Mantani 2002).

at A′, because the frequency-dependency works oppositely in the deconvolution and convolution, cancelling the assumption employed in the procedures.

The frequency-dependency of damping tends to become explicit in calculating seismic motions in one-way from the bedrock to the ground surface or vice versa. Fig. 4.3.14(a) exemplifies a deep soil profile at Higashinada, Kobe, Japan, where a very deep borehole from the ground surface down to GL.-1691 m was drilled together with the suspension-type wave logging and soil density logging (Matsumoto et al. 1998). All the layers there consist of Holocene and Pleistocene soils down to 1500 m deep followed by tertiary stiff rocks. The S-wave velocity $V_s$ changes from less than $V_s = 200$ m/s near the ground surface as shown in Fig. 4.3.15(a) to $V_s = 500$ m/s at GL.-200 m, $V_s = 1000$ m/s at GL.-800 m, and followed by $V_s = 3200$ m/s at GL.-1600 m.

This deep soil profile was modeled with 172 layers of cohesive and non-cohesive soils underlain by the stiff rocks. The main shock record during the 1995 Kobe earthquake at the deepest level of the PI vertical array site (GL.-83.4 m) located nearby was given to the similar depth of the model (GL.-73.7 m, between Holocene and Pleistocene) to make an equivalent-linear analysis to the great depth (Kokusho and Mantani 2002, Mantani 2002). The strain-dependent change of modulus and damping used here followed the modified HD model in Eqs. (3.1.26), (3.1.27) wherein the reference strain $\gamma_r$ was assumed proportional to the square root of the effective overburden stress. The analysis revealed that $G/G_0$ was no less than 80% below the depth GL.-200 m, indicating that the soil nonlinearity was not significant below that level. As for the frequency-dependency of damping, the values $D$, $D_0$, $D_{\max}$ in Eq. (3.1.27) are all assumed proportional to $(f/f_0)^{-m}$ according to Eq. (4.3.2), wherein the reference frequency $f_0 = 0.78$ Hz was chosen as the predominant frequency of the input PI-motion. The minimum damping ratio $D_0$ was assumed 2% constant down to GL.-300 m, and decreasing in proportion to the power of 0.6 of the soil depth in the light of previous research by Yamamizu et al. (1983). The exponent $m$ in Eq. (4.3.2) was varied in three steps as $m = 0$ (Nonviscous damping), $m = 0.5$ and $m = 1.0$ (Maxwell damping) and its effect was compared.

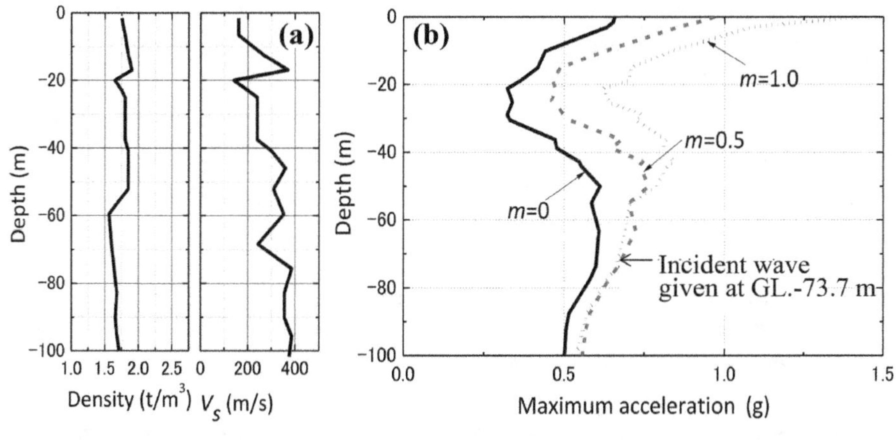


*Figure 4.3.15* Maximum acceleration down to GL.-100 m for $m = 0, 0.5, 1.0$ (Mantani 2002).

Fig. 4.3.14(b) shows the variation of maximum accelerations from the surface down to GL.-1691 m. For $m = 0$ (Nonviscous damping), the maximum acceleration tends to increase almost linearly with increasing depth down to about GL.-1550 m where $V_s$ jumps from 1000 m/s to 3200 m/s. In contrast to this strange trend, the cases with $m = 0.5$ and 1.0 yield more plausible results where the acceleration tends to gradually decrease with increasing depth.

Fig. 4.3.15(b) illustrates the variation of maximum accelerations along the depth for the top 100 m soil profile obtained in the same analysis. It indicates that for larger $m$ the acceleration tends to be amplified more near the surface because of smaller damping ratios prescribed for higher frequency motions. The frequency-dependent damping in the shallow depth with relatively large induced strain may largely reflect the strain-dependent damping which cannot be represented by a single value in the linear/equivalent linear analysis. Thus, the $m$-values for different damping models give the significant effect on the evaluation of surface acceleration, though for all the $m$-values the depth-dependent acceleration variations look to be plausible in the depth as shallow as 100 m.

In the following, the frequency-dependency of damping is discussed in stiff ground with small seismic strains or in deep ground, wherein frictional Nonviscous damping is not manifested even during strong earthquakes. In these conditions, the damping ratio may not be constant but frequency-dependent. This will give a considerable effect in evaluating acceleration response in deep/stiff soils particularly for design seismic motions for machines and equipments with high resonant frequencies.

#### 4.3.3.2 Outline of wave scattering theory

As a major reason why the frequency-dependency cannot be ignored in in situ damping unlike in laboratory damping by soil tests, the heterogeneity of in situ soils is responsible. As seismic wave propagates in heterogeneous soils, high-frequency scattering waves are generated. In view of energy, the wave scattering is interpreted that a part of the wave energy is imparted to the scattered high-frequency waves. In seismology,

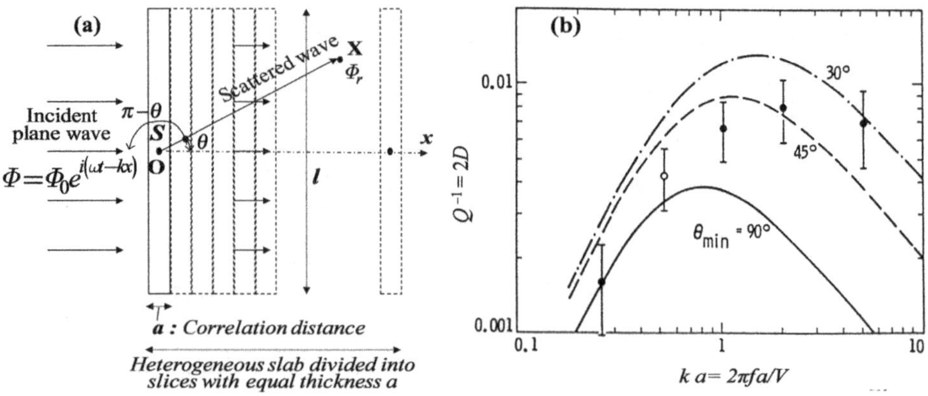


*Figure 4.3.16* Concept of incident wave scattering in heterogeneous medum (modified from Wu 1982) (a), and Theoretical relationships between $Q^{-1} = 2D$ versus *ka*-value (Frankel and Clayton 1986) (b).

the coda wave has been investigated for a long time as the seismic waves scattered by heterogeneities in the earth crust (Toksoz et al. 1988, Frankel and Clayton 1986). The similar scattering effects by soil heterogeneity may have a certain effect even on the site amplification in shallower depths.

A concept of weakly scattering theory (Wu 1982) of waves propagating in two-dimensional inhomogeneous medium may be illustrated as in Fig. 4.3.16(a), where a plane harmonic wave of the angular frequency $\omega$, the wave number $k$ and the displacement amplitude $\Phi_0$ travels in $x$-direction as:

$$\Phi = \Phi_0 e^{i(\omega t - kx)} \tag{4.3.4}$$

The inhomogeneous medium is assumed to randomly vary its wave velocity as $V(1 \pm \gamma_v)$ with the average $V$ and the coefficient of variation $\gamma_v$ and characterized by various correlation functions in space. If the exponential function is chosen, the correlation at a point with the offset distance $r$ and the correlation distance $a$ is expressed as:

$$N(r) = e^{-r/a} \tag{4.3.5}$$

The medium is modeled here as an infinitely large slab in Fig. 4.3.16(a) and divided into slices of an equal thickness $a$ where the incident wave $\Phi$ propagates normal to them. Let the amplitude of scattered wave at Point X be $\Phi_r$, when the plane wave propagates through the unit area by the unit distance. The energy ratio of scattered wave to the original plane wave is defined as $(\Phi_r/\Phi_0)^2$. By integrating the energy ratio in terms of one slab with thickness $a$ and height $l$, the ratio of loss energy by scattering, $\Delta E$, to the incident wave energy, $E$, can be expressed as:

$$\frac{\Delta E}{E} = \frac{1}{l}\int \left(\frac{\Phi_r}{\Phi_0}\right)^2 dS = \frac{1}{l}\int \left(\frac{\Phi_r}{\Phi_0}\right)^2 r d\theta \tag{4.3.6}$$

Here the integration is implemented along the arc length $S$ for scattering angle; from a minimum scattering angle ($\theta = \theta_{min}$) to fully backward angle ($\theta = \pi$). As explained in Sec. 1.6.1, the energy of the plane wave in Eq. (4.3.4) is expressed using the attenuation coefficient due to internal damping $\beta$ as:

$$E = \left|\rho V \omega^2 \Phi^2\right| = \left|\Phi_0^2 \rho V \omega^2 e^{-2\beta x} e^{2i\omega(t-x/V_s)}\right| = \Phi_0^2 \rho V \omega^2 e^{-2\beta x} \tag{4.3.7}$$

Its derivative in terms of the travel distance $x$ is written as:

$$\frac{\partial E}{\partial x} = -2\beta \Phi_0^2 \rho V \omega^2 e^{-2\beta x} \tag{4.3.8}$$

Hence the ratio of the derivative to the incident wave energy $E$ is expressed as:

$$\frac{1}{E}\left|\frac{\partial E}{\partial x}\right| = \frac{2\beta \Phi_0^2 \rho V \omega^2 e^{-2\beta x}}{\Phi_0^2 \rho V \omega^2 e^{-2\beta x}} = 2\beta = 2kD = kQ^{-1} \tag{4.3.9}$$

Here, $Q^{-1} = 2D$ is twice the damping ratio $D$. The left term of Eq. (4.3.9) is also written using the loss energy ratio for the slice of thickness $a$ in Eq. (4.3.6) as:

$$\frac{1}{E}\left|\frac{\partial E}{\partial x}\right| = \frac{1}{a}\frac{\Delta E}{E} \tag{4.3.10}$$

Thus, the following formula is derived from the above two equations.

$$Q^{-1} = \frac{1}{k}\frac{1}{E}\left|\frac{\partial E}{\partial x}\right| = \frac{1}{ak}\frac{\Delta E}{E} \tag{4.3.11}$$

Substituting Eq. (4.3.6) into Eq. (4.3.11), the value $Q^{-1}$ can be obtained as follows for the medium wherein the wave velocity varies randomly with the exponential function Eq. (4.3.5) with the correlation distance $a$ (Frankel and Clayton 1986).

$$Q^{-1} = 2k^2 a^2 \gamma^2 \int_{\theta_{min}}^{\pi} \left\{1 + a^2 [2k \sin(\theta/2)]^2\right\}^{-3/2} d\theta \tag{4.3.12}$$

Fig. 4.3.16(b) shows the variation of $Q^{-1}$ versus a non-dimensional number $ka$ for the minimum scattering angle $\theta_{min} = 30°$, $60°$, $90°$. It is obviously seen that the value $Q^{-1}$ or twice the damping ratio $2D$ takes a peak near $ka = 1.0$ and decreases as $ka$ increases or decreases. Considering $ka = 2\pi fa/V$, the damping ratio tends to increase with increasing frequency $f$ up to $f \approx V/2\pi a$ and decrease thereafter. Namely, the wave scattering tends to occur most efficiently around $f = V/2\pi a$ and the associated damping ratio in the incident wave becomes maximum there in accordance with the energy transfer from the incident wave to the scattering wave. With increasing or decreasing frequency from there, the scattering wave energy tends to decrease.

In the frequency-dependency formulated in Eq. (4.3.2), the damping ratio $D$ tends to decrease with increasing frequency. This indicates that the downslope only on the right side of the peak in Fig. 4.3.16(b) is observed in seismic records in surface layers (Kinoshita 1983). For example, if the peak frequency is assumed as $f = 1.0$ Hz, then

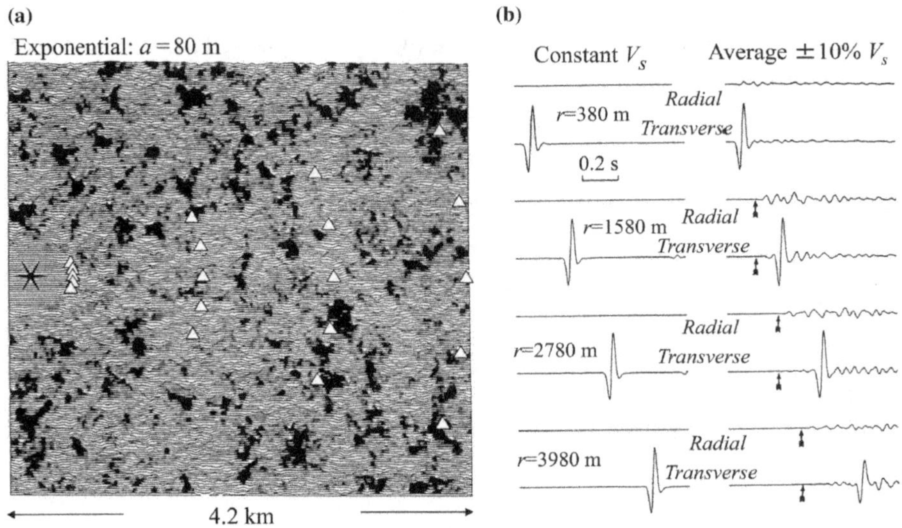


*Figure 4.3.17* Finite difference model of earth crust with heterogeneous $V_s$ (a), and Wave propagation simulation results for homogeneous and heterogeneous cases (arrows indicate arrival times of generated waves from SV to P conversion) (b) (modified from Frankel and Clayton 1986).

$a/V \approx 1/(2\pi)$ holds because $ka = 2\pi fa/V \approx 1.0$. Hence the correlation distance representing the soil heterogeneity should be $a \approx 35$ m for a soil layer with S-wave velocity $V_s = 200$ m/s to exhibit frequency-dependency such as Eq. (4.3.2) in the frequency range $f > 1.0$ Hz.

Different from the above-mentioned theoretical considerations, numerical calculations on wave propagations have been conducted using finite difference methods or finite element methods (Frankel and Clayton 1986, Sato and Kawase 1992). Fig. 4.3.17(a) exemplifies a finite difference model of earth crust of 4.2 km by 4.0 km surrounded by wave-absorbing boundaries (Frankel and Clayton 1986). The variations of $V_s$ for the inhomogeneous medium (Average $V_s \pm 10\%$ Standard Deviation) are expressed there with the waving lines wherein the dark patches indicate the values higher than the average velocity. The SV-wave starting from a source (the star mark) and received at a set of triangular points on the model with varying distances are calculated both for the radial and transverse components and illustrated in Fig. 4.3.17(b) to compare with a homogeneous medium with constant $V_s$ where no internal damping is considered. It is confirmed that, in the homogeneous medium, the transverse component (SV-wave) propagates without damping and no radial component is excited at all. The wave amplitudes here are already adjusted by considering the geometrical damping. In the inhomogeneous medium, the transverse component tends to attenuate apparently with distance, while the scattering waves are exited. It reflects the energy transfer from the original SV-wave to the scattering wave of various frequencies not only in the transverse but also radial direction. The $Q^{-1}$-values calculated from this numerical simulation are shown in Fig. 4.3.16(b) by the plots with error bars versus $ka$, showing the variation similar to the weakly scattering theory near the peak $ka = 1.0$. Besides this research, Wu (1982) found a fair agreement on the downslope right of the peak $ka > 1.0$ between earthquake observation results and the wave-scattering theory.

In other numerical analyses by FEM, the effect of heterogeneity in S-wave velocity in surface soils on site amplifications were studied (Sato and Kawase 1992), wherein the internal soil damping was also incorporated to know the effect on the wave scattering. It was shown in the analysis that high-frequency waves generated by scattering tend to attenuate in a short distance because of the soil damping, and hence the greater the soil damping is during strong earthquakes the less becomes the wave-scattering effect. Also indicated was that the damping ratio can be expressed as the sum of the two components of being strain-dependent and frequency-dependent as formulated in Eq. (4.3.3).

### 4.3.4 Microtremor H/V spectrum ratio

Microtremors are ambient ground vibrations with unnoticeably small amplitudes caused by natural phenomena such as ocean waves and winds and also by human activities such as traffics and machines. The properties of microtremors were first investigated by Kanai and Tanaka (1954). Since then, microtremor measurements have been increasingly implemented as a convenient tool to characterize seismic site response without deploying earthquake observation arrays. One of the goals of microtremor measurements is to know the site-specific predominant frequency of ground motions during earthquakes. As mentioned in Sec. 4.2, the predominant frequency in site amplification is theoretically determined from the multi-reflection of the SH-wave and calculated from the thicknesses of a set of soft soil layers overlying a stiff base layer and their S-wave velocities. Unlike the earthquake waves, however, microtremors consist of various source-specific body and surface waves which are sometimes very localized.

If microtremors are measured at two locations on the top of the surface layer and outcropping base layer as illustrated in Fig. 4.3.18(a) similar to the horizontal earthquake observation array, and their spectra $SH_s$ and $SH_b$, respectively, are compared, the spectrum ratio $SH_s/SH_b$ is expected to characterize site-specific earthquake amplification. However, the spectrum ratio thus obtained may not give the amplification directly compatible with that of the SH-wave in many cases. It is because the microtremor wave-field is consisting not necessarily of SH waves but possibly of surface waves of various origins and tends to be changeable in distance and time. Nogoshi and Igarashi (1971) actually found that the vertical component of microtremors is mostly composed of Rayleigh waves. The same authors also indicated by incorporating a basic theory of Rayleigh wave as mentioned in Sec. 1.4.1 that the spectrum ratio between horizontal to vertical components of microtremor records at a single location (H/V-spectrum ratio) may reflect the seismic site response reasonably well.

Nakamura (1989) applied the H/V-spectrum ratio of microtremor measurements at a single location to estimate the site-specific amplification characteristics corresponding to the SH-wave multi-reflection. In the simplified model shown in Fig. 4.3.18(a) for the surface soft soil with the thickness $H$ and impedance $\rho_1 V_{s1}$ underlain by the base layer with the impedance $\rho_2 V_{s2}$, the horizontal and vertical spectra at the surface of the soft layer, $SH_s$ and $SV_s$, respectively, may be expressed (Nakamura 2000) as:

$$SH_s = Amp_h \times SH_b + SRH_s, \quad SV_s = Amp_v \times SV_b + SRV_s \tag{4.3.13}$$

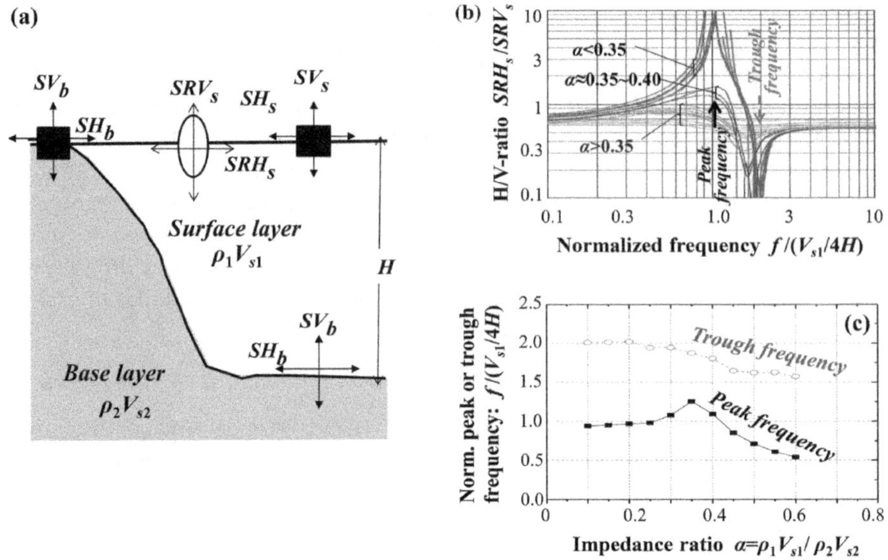


*Figure 4.3.18* Cross-section of surface soft layer and outcropping base layer with wave spectra at different points (a), Frequency-dependent H/V ratio for harmonic Rayleigh wave in a two-layer model (modified from Nakamura 2008) (b), and Its peak or trough frequency versus impedance ratio $\alpha$ (modified from Ohmachi et al. 1994) (c).

where, $Amp_h$, $Amp_v$ = spectrum amplifications in horizontal and vertical motions of vertically propagating body waves, $SH_b$, $SV_b$ = spectra of horizontal and vertical motions at the outcropping base layer, and $SRH_s$, $SRV_s$ = spectra of Rayleigh waves in the horizontal and vertical directions at the soft soil surface. Then, the next formulation may be possible.

$$\text{H/V-spectrum ratio} = SH_s/SV_s = \frac{Amp_h \times SH_b + SRH_s}{Amp_v \times SV_b + SRV_s} \tag{4.3.14}$$

If the contribution of the Rayleigh wave is ignored ($SRH_s = SRV_s = 0$), and also assuming that the vertical motions in the surface layer will not amplify near the first peak frequency of the horizontal motions ($Amp_v \approx 1.0$), then $SH_s/SV_s \approx Amp_h$, provided that the H/V-spectrum ratio at the outcropping base layer can be approximated as $SH_b/SV_b \approx 1.0$. Thus, the H/V-spectrum ratio at the soft soil surface may represent site-specific amplification characteristics corresponding to the SH-wave multi-reflection.

If the Rayleigh waves are dominant compared to the body waves, the H/V-ratio becomes $SH_s/SV_s \approx SRH_s/SRV_s$. For the simplified two-layer model in Fig. 4.3.18(a), it is shown that H/V-ratio ($SRH_s/SRV_s$) of a harmonic Rayleigh wave varies with normalized frequency, $f/(V_{s1}/4H)$ as typically calculated in Fig. 4.3.18(b), for the impedance ratio $\alpha = \rho_1 V_{s1}/\rho_2 V_{s2} = 1/4.5$–$1/1.2$, with a set of curves having peaks and troughs (Ohmachi et al. 1994, Kudo 1995). Note that the curves are quite dependent on the $\alpha$-value as indicated in the diagram, although the peak frequencies of the

H/V-ratio coincide with $f = V_{s1}/4H$ in most cases. Fig. 4.3.18(c) shows the peak and trough frequencies shown in Fig. 4.3.18(b) for the ratio $SRH_s/SRV_s$ plotted versus impedance ratio $\alpha = \rho_1 V_{s1}/\rho_2 V_{s2}$ (modified from Ohmachi et al. 1994). For smaller impedance ratios $\alpha \leq 0.35$, the peak frequency almost coincides $f = V_{s1}/4H$, indicating that the Rayleigh wave H/V-ratio can detect the site-dependent predominant frequency corresponding to the 1/4 SH-wave length, while it tends to be lower than that as $\alpha$ becomes larger than 0.4. This suggests that the H/V-ratio may be effective even for Rayleigh waves in detecting predominant frequency if a sharp impedance contrast (lower $\alpha$-values) exists between the surface and base layer.

Microtremors may consist of surface waves as well as body waves. In Fig. 4.3.19, site response spectra of earthquake records (in two directions NS and EW) are compared with microtremor measurements in the same locations in Mexico city (modified from Nakamura et al. 2003). In the upper three diagrams, the H/V-spectrum ratios for earthquake motions (thick curve) at a stiff hill surface (a) and soft soil surface (b), (c) are compared with H/H-spectrum ratios (thin curves) at the same sites for the same earthquake motion. The H/H-spectrum ratios were calculated by dividing the spectrum of horizontal motion at a site by that at a common reference site on an outcropping base layer as already defined in the horizontal array amplification in Fig. 4.1.2(a).

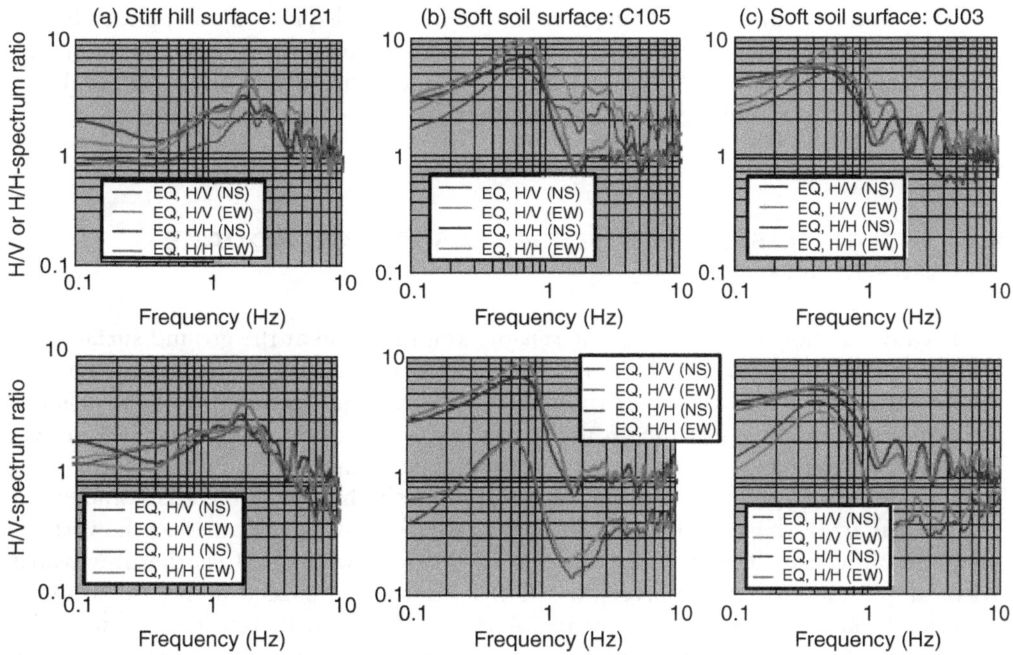


*Figure 4.3.19* H/V-ratios at single points compared with H/H-ratios relative to reference base point during earthquakes (upper diagrams), and comparisons of H/V-ratios during earthquakes and microtremors: (a) Outcropping base layer, (b), (c) Soft soil surface (modified from Nakamura et al. 2003).

It is shown that the H/V-ratios mostly coincide with the H/H-ratios in terms of the predominant frequency and amplification value as well, indicating that the H/V-scheme may be able to estimate seismic site amplifications if the wave-field is essentially composed of the SH-wave. In the lower diagram of Fig. 4.3.19, the same H/V-ratios for the earthquake (thick curves) are compared with microtremor H/V-ratios (thin curves) measured at the same sites. Though the peak frequencies show a good coincidence, the amplitude ratios tend to be smaller in the microtremors than in the earthquake in the soft soil sites. This is presumably because unlike the earthquake motion rich of body waves, the microtremors tend to be governed by Rayleigh waves in deep soft soil sites (Nakamura et al. 2003).

Many site investigations have been carried out to date using the microtremor H/V-spectrum ratios (e.g. Ohmachi et al. 1994, Horike et al. 2001, Nakamura et al. 2003). According to them, it is generally accepted that the predominant frequency in site amplification can be captured by the H/V-scheme at least. The reliability of frequency determined tends to be better as the impedance contrast between a set of surface layers and a base layer becomes sharper which tends to yield sharp H/V-spectrum peaks as indicated in Fig. 4.3.18(b). As for the spectrum amplification values during earthquakes, microtremor H/V-spectrum ratios in Kushuiro city in Japan was found to yield lower amplification than H/H-spectrum ratios evaluated from the earthquake motions at the same sites relative to a common reference motion (Horike et al. 2001). On the other hand, it was found to be able to roughly evaluate the amplification near the predominant frequency unless the effect of Rayleigh wave is dominant in Mexico city (Nakamura et al. 2003). Thus, there still remains a dispute on its applicability in evaluating earthquake site amplification characteristics in general, though it is unanimously accepted that the H/V-scheme is effective and cost-efficient method to evaluate the site-specific predominant frequency at least.

## 4.4 SITE AMPLIFICATION FORMULAS BY EARTHQUAKE OBSERVATION

As already mentioned in Sec. 4.1, the seismic amplification at the ground surface relative to the outcropping base layer nearby is expressed as $2A_s/2A_b$ and named as the horizontal array amplification or the incident wave amplification, wherein $A_s$ is the upward wave amplitude at the ground surface and $A_b$ is that at the base ($A_2$ in Sec. 4.1 is written here as $A_b$). On the other hand, if the surface motion is compared with the motion measured in the base layer at the same site by downhole seismometers in the vertical array, the amplification becomes $2A_s/(A_b+B_b)$ and named as the vertical array amplification or composite wave amplification, wherein $B_b$ is the downward wave amplitude in the base layer. In establishing seismic hazard maps, the amplification $2A_s/2A_b$ is needed rather than $2A_s/(A_b+B_b)$. It is therefore worthwhile to evaluate the horizontal array amplification $2A_s/2A_b$ from the vertical array database of $2A_s/(A_b+B_b)$ increasingly available recently. The KiK-net data in about 700 vertical array sites covering all over Japan are available for international researchers only a few hours after the occurrence of all earthquakes together with associated bore-hole and wave logging data from the website: http://www.kik.bosai. go.jp/kik/.

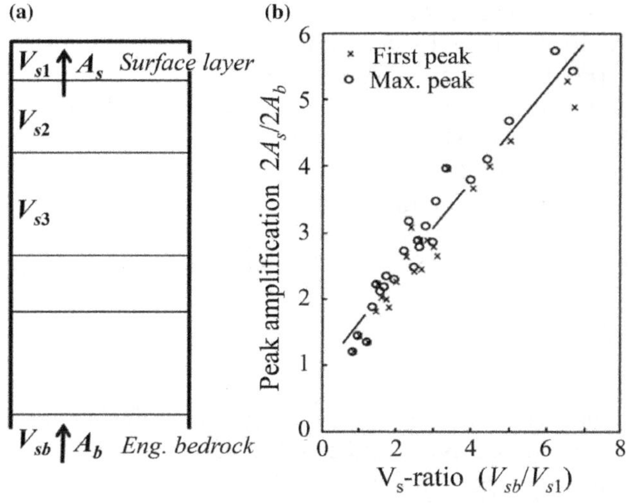


*Figure 4.4.1* Seismic zonation study: (a) Horizontally layered soil models, (b) Peak amplification $2A_s/2A_b$ versus $V_s$-ratio between base and surface (modified from Shima 1978).

As obviously seen in Eq. (4.2.35) for a simplified two-layer system, the peak amplification $2A_s/2A_b$ is written as a function of the impedance ratio $\alpha$ between the two layers and a variable $\delta_1 = \tan^{-1} 2D_1$ associated with damping ratio in the surface layer $D_1$. As the impedance ratio may be simplified as $\alpha = \rho_1 V_{s1}/\rho_2 V_{s2} \approx V_{s1}/V_{s2}$ using $\rho_1/\rho_2 \approx 1.0$ in many cases, the $V_s$-ratio $V_{s1}/V_{s2}$ serves as an important parameter for the amplification.

In actual site amplification evaluations too, $V_s$-ratios are considered as one of the key parameters. Shima (1978) did seismic zonation studies in Tokyo and nearby areas based on horizontally layered soil models as illustrated in Fig. 4.4.1(a) idealized from a number of bore-hole logging data available. Transfer functions of these models between ground surface and engineering bedrock as well as their first and higher-order peak values were calculated by the one-dimensional multi-reflection theory of SH-wave. In Fig. 4.4.1(b), the peak values $2A_s/2A_b$ are plotted versus $V_s$-ratios, $V_{sb}/V_{s1}$, between base velocities $V_{sb}$ and surface velocities $V_{s1}$. The plots are approximated here by a linear function of the $V_s$-ratio $V_{sb}/V_{s1}$. This indicates that the amplification between surface and base may be practically evaluated by the $V_s$-ratios between base and surface layers irrespective of the intermediate layers in realistic field soil profiles. This finding further implies that amplification at a site relative to another site may be evaluated from the ratios of surface S-wave velocities $V_{s1}$ at the two sites if they share the common base layer. This constitutes a basic concept of the seismic microzonation based on the near-surface S-wave velocity.

### 4.4.1 Site amplification formula using near-surface $V_s$

Because a surface layer is sometimes too thin to represent a site, a $V_s$-value not necessarily of the top surface layer but an average $V_s$-value of multiple layers to a certain depth is preferred in microzonation. Thus, empirical formulas were proposed using the $V_s$-values near ground surface.

The empirical formula by Joyner and Fumal (1984) evaluates maximum surface acceleration (PGA), maximum velocity (PGV) or 5% damping response spectrum using the average S-wave velocity in surface layers based on strong motion records in US. Here, the $V_s$-value is averaged over the soil thickness corresponding to a quarter-wave length from the ground surface (denoted here as $V_{s1/4}$) assuming seismic motions with the predominant period of $T = 1.0$ s. Amplification ratio $A$ in the maximum velocity at the ground surface relative to outcropping bedrock (assuming $V_s = 1060$ m/s) is given by the following formula using $V_{s1/4}$ (m/s in unit).

$$A = 23(V_{s,1/4})^{-0.45} \tag{4.4.1}$$

Another formula was proposed incorporating records during the 1989 Loma Prieta earthquake in California and associated small shocks, wherein the spectrum amplifications averaged over the period $T = 0.4$–$2.0$ s, denoted as $AHSA$, were correlated as follows with the S-wave velocity averaged over the surface layers of top 30 m, $V_{s30}$ (m/s) (Borcherdt et al. 1991).

$$AHSA = \frac{701}{V_{s30}}\text{: weak motions,} \quad AHSA = \frac{598}{V_{s30}}\text{: strong motions} \tag{4.4.2}$$

A similar empirical formula was proposed based on earthquake records in Japan (Midorikawa 1987), wherein the ratio of maximum velocity amplitudes, $A$ in the next equation between ground surface and base was correlated with $V_{s30}$, and compared with the seismic intensities observed during the 1923 Great Kanto earthquake (TCEGE 1999). The formula is expressed differently as follows depending on whether $V_{s30}$ is smaller than 1100 m/s or not.

$$\left.\begin{array}{ll} A = 68V_{s30}^{-0.60}; & V_{s30} < 1100\text{ m/s} \\ A = 1.0; & V_{s30} > 1100\text{ m/s} \end{array}\right\} \tag{4.4.3}$$

Fig. 4.4.2 summarizes these empirical formulas on $V_s$-dependent amplifications. Thus, the seismic site amplifications are practically evaluated using the S-wave velocities averaged over a certain depth in the surface soils relative to the base layer velocity. Among them, $V_{s30}$, average over the depth of 30 m is often used in practice for developing seismic zonation maps (e.g. BSSC 2003). Though the average S-wave velocity over a certain depth is conveniently used in combination with geomorphological maps, it inevitably ignores the effect of site-dependent soil profiles on seismic amplifications.

### 4.4.2 Amplification formula using average $V_s$ in equivalent surface layer

In order to improve the seismic amplification evaluations by taking site-dependent soil profiles into considerations, a concept of an equivalent surface layer was introduced by Kokusho and Sato (2008) based on the vertical array amplification database. In Fig. 4.4.3(a), spectrum ratios $2A_s/(A_b + B_b)$ observed at a KiK-net vertical array site

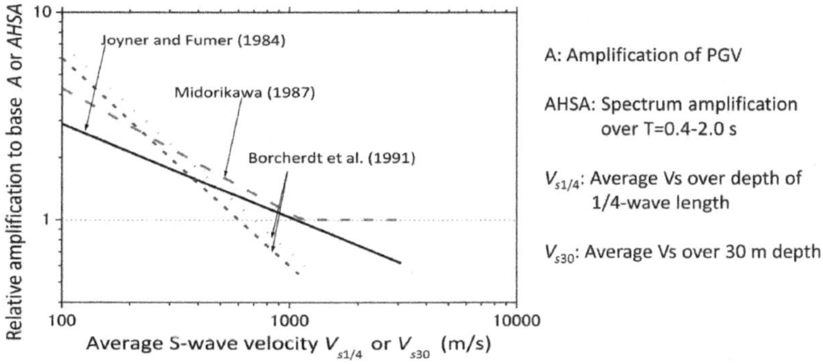


*Figure 4.4.2* Empirical formulas on site amplifications using average S-wave velocities near surface (TCEGE 1999).

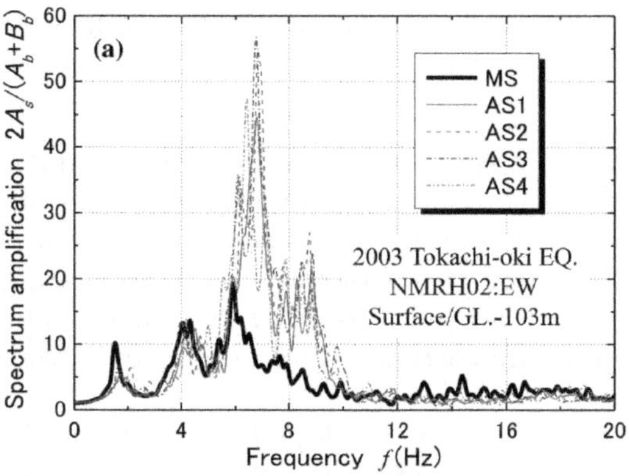


| (b) Layer i | Thick-ness $H_i$(m) | Depth (m) | Vs in layers $V_{si}$ (m/s) | Average $V_s$ from surface (m/s) | Fund. mode freq. by Eq.(1) $f$ (Hz) | Identifiy peaks by observed peaks |
|---|---|---|---|---|---|---|
| 1 | 4 | 4 | 110 | 110 | 6.88 | 3rd peak |
| 2 | 6 | 10 | 320 | 181 | 4.54 | 2nd peak |
| 3 | 30 | 40 | 500 | 347 | 2.17 | |
| 4 | 8 | 48 | 430 | 359 | 1.87 | 1st peak |
| 5 | 34 | 82 | 510 | 409 | 1.25 | |
| 6 | 12 | 94 | 650 | 430 | 1.14 | |
| 7 | 7 | 101 | 870 | 449 | 1.10 | |

*Figure 4.4.3* Typical spectrum amplification of vertical array and peak frequencies compared with 1/4-wave length formula (Kokusho and Sato 2008).

during 2003 Tokachi-oki earthquake are exemplified for the main shock and several aftershocks. A good reproducibility of site response during multiple aftershocks is observed as well as a clear difference in the spectrum ratios between the main shock and aftershocks reflecting strain-dependent soil properties in the higher-order peaks as already mentioned. In order to identify the soil layers generating peak frequencies in the spectrum ratio, the fundamental mode frequencies $f$ of the layered soil system were calculated by the following quarter-wave length formula (the same as Eq. (4.2.22)) based on $V_s$-logging data and tabulated in Fig. 4.4.3(b).

$$f = \frac{1}{4\sum_{i=1}^{n}(H_i/V_{si})} \tag{4.4.4}$$

Here, $H_i$ and $V_{si}$ are the thickness and S-wave velocity of the $i$-th layer numbered sequentially from the top, and $H_i/V_{si}$ is summed up layer by layer down to the base.

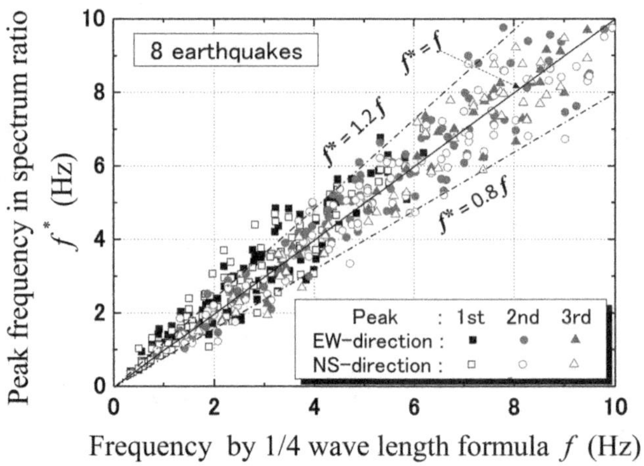


*Figure 4.4.4* Peak frequencies $f^*$ in observed spectrum ratios compare with $f$ calculated by 1/4 wave length formula using $V_s$-profile data (Kokusho 2013a).

This frequency $f$ corresponds to a set of layers with the thickness equal to the quarter wave length. The calculated frequency was compared one by one with the peak frequencies in the observed spectrum ratio in Fig. 4.4.3(a) to identify the equivalent surface layer of the thickness $H = \sum H_i$ consisting of one or more layers generating the fundamental mode frequency as tabulated in Fig. 4.4.3(b). Note that there can be multiple equivalent surface layers in the same site corresponding to individual peak frequencies.

In Fig. 4.4.4, the peak frequencies $f$ calculated by Eq. (4.4.4) using given $V_s$-profile data are taken in the horizontal axis to compare with the peak frequencies $f^*$ identified in the observed spectrum ratios in the vertical array for about 100 main shock records during eight strong earthquakes ($M_J = 6.4 \sim 8.0$) occurred in Japan from the year 2000 to 2008. There exists a satisfactory correspondence between $f$ and $f^*$ for not only the 1st but also higher-order peaks. Then, the average S-wave velocity $\overline{V}_s$ for each equivalent surface layer can be calculated from the fundamental mode frequency $f$ and the corresponding thickness $H = \sum H_i$ as:

$$\overline{V}_s = 4Hf \tag{4.4.5}$$

The average $V_s$-value thus determined for the equivalent surface layer serves as a key parameter to evaluate site-dependent amplifications for seismic zonation.

Next, the transfer function, $2A_s/(A_b + B_b)$ was calculated for each site using the S-wave velocity profile employed in calculating the average velocity. The damping ratio $D$ was tentatively assumed 2.5% in all the layers and also postulated to be frequency-independent. Fig. 4.4.5 exemplifies a typical transfer function $2A_s/(A_b + B_b)$ (thick dotted curve) which is compared with the observed spectrum ratios in EW and NS directions (thin solid curves) at a site. If a peak in the transfer function could be found at about the same frequency as that in the observational spectrum ratio as in this

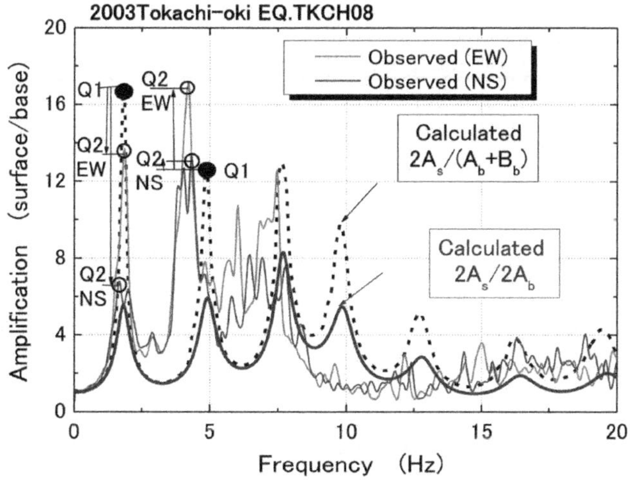


*Figure 4.4.5* Typical transfer functions $2A_s/(A_b + B_b)$ and $2A_s/2A_b$ compared with observation (Kokusho 2013a).

example, it was identified as the corresponding peak and the damping ratio previously assumed as $D = 2.5\%$ was modified by the next equation to have the same peak value

$$D = \frac{Q_1}{Q_2} \times 2.5(\%) \tag{4.4.6}$$

where $Q_1$ is the peak value of the transfer function using the soil data and $Q_2$ is that of the observed spectrum ratios both in EW and NS directions. Not only the 1st peak but also the higher order peaks were compared in this manner if possible, and the modified $D$-value was determined as the average in EW and NS directions. Then, the transfer function $2A_s/2A_b$ was calculated using the modified $D$-value using the same soil-layer model and the peak amplification was read off. In some cases, it was found that peak frequencies in the transfer functions $2A_s/2A_b$ and $2A_s/(A_b + B_b)$ significantly disagree. This is due to inappropriate downhole seismometer depths relative to the layer boundary of sharp impedance change, as will be discussed in Sec. 4.4.4, and these cases were discarded in the later data analysis. In this way, the peak amplifications in the spectrum ratio $2A_s/2A_b$ to be used for seismic zonation studies, were calculated based on the earthquake records at a number of vertical array sites (Kokusho and Sato 2008, Kokusho 2013a).

In the current practice of seismic zonation, the average S-wave velocity $V_{s30}$ is often used in making simplified site amplification evaluations (e.g. Midorikawa 1987, Borcherdt et al. 1991), where $V_{s30}$ is the averaged S-wave velocity over the top 30 m from the ground surface. In Fig. 4.4.6, the peak values in $2A_s/2A_b$ thus calculated from a number of vertical array data are plotted versus the velocity ratio, $V_{sb}/V_{s30}$, for main shock records during the eight strong earthquakes ($M_J = 6.4{\sim}8.0$) abbreviated here as EQ. 1~8. Here, $V_{sb}$ = S-wave velocity at stiff base layers where the downhole seismometers are installed (around 100 m deep), and $V_{s30} = 30/T_{30}$ where $T_{30}$ is the

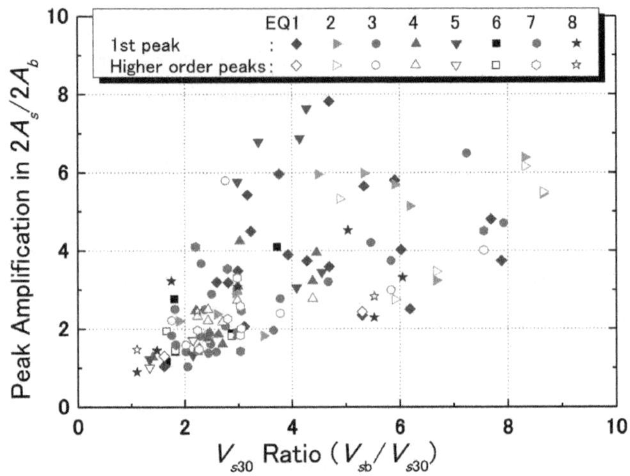


*Figure 4.4.6* Peak spectrum amplifications of $2A_s/2A_b$ for main shocks of 8 earthquakes versus average S-wave velocity ratios $V_{sb}/V_{s30}$ using $V_{s30}$ of top 30 m surface soils (Kokusho 2013a).

S-wave travel time (s) in the top 30 m. Obviously, a positive correlation can be recognized, though data are largely scattered. Some inconsistencies are evidently seen in the plots among different earthquakes and the 1st and higher order peaks, presumably because only the top 30 m is taken into account without considering site-specific soil profiles.

Fig. 4.4.7 shows the relationship of the peak amplifications in $2A_s/2A_b$ for the eight earthquakes plotted versus the $\overline{V}_s$-ratio defined here as the division of $V_{sb}$ by the average velocity $\overline{V}_s$ in Eq. (4.4.5). Note that a large number of plots are overlapping in the zone of $V_{sb}/\overline{V}_s \leqq 4.0$. In a good contrast to Fig. 4.4.6, the plots in Fig. 4.4.7 using $\overline{V}_s$ show a fairly good correlation for both 1st and higher order peaks despite differences in various influencing factors such as earthquake intensities, predominant frequencies, shaking durations and soil profiles. This indicates the importance to define the average S-wave velocity properly by identifying the site-specific equivalent surface layers wherein the individual amplification peaks are excited.

A simple linear correlation may be derived in Fig. 4.4.7 for the data-points of $V_{sb}/\overline{V}_s \leqq 10$ (the normally encountered condition), with the determination coefficient $R^2 = 0.79$ (Kokusho 2013a) as:

$$\frac{2A_s}{2A_b} = 0.369 + 0.626\left(\frac{V_{sb}}{\overline{V}_s}\right) \tag{4.4.7}$$

The equation is not so different from what was proposed based on 3 earthquakes (EQ3, EQ4 and EQ5) (Kokusho and Sato 2008): $2A_s/2A_b = 0.175 + 0.685(V_{sb}/\overline{V}_s)$ as superposed in the same diagram. Eq. (4.4.7) may be conveniently used because of its applicability to a wide variety of base layers with $V_{sb} = 400$ m/s to 3000 m/s for a number of the vertical array sites used here (Kokusho 2013a).

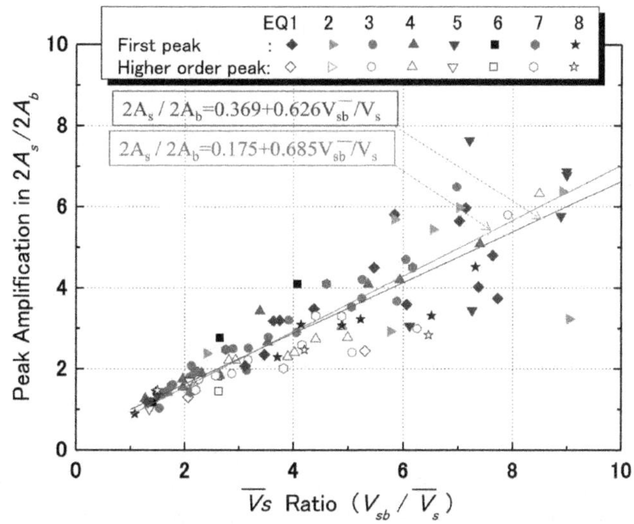


*Figure 4.4.7* Peak spectrum amplifications of $2A_s/2A_b$ for main shock of 8 earthquakes versus average S-wave velocity ratios $V_{sb}/\bar{V}_s$ and empirical equation for approximation (Kokusho 2013a).

Relative amplifications for the same seismic motions in an area underlain by a common base layer is evaluated as follows;

(i) Based on the microtremor measurements for example, decide the predominant frequency $f$ of a given site using the H/V spectrum ratios (Nakamura 1989).
(ii) Estimate the thickness of soft soil or Holocene layer $H$ where the fundamental frequency is exerted, which may be read off in geological maps or soil logging data available nearby in urban areas.
(iii) Then, the average S-wave velocity can be calculated by $\overline{V}_s = 4Hf$. If, a $V_s$-profile is obtained together with bedrock depth by surface wave methods or downhole methods, $\overline{V}_s$ is readily calculated by Eqs. (4.4.4) and (4.4.5).
(iv) Calculate the S-wave velocity ratio $V_{sb}/\overline{V}_s$ from $V_{sb}$ of the common base layer and $\overline{V}_s$ obtained above.
(v) Relative amplification between two different sites can be readily obtained from the $2A_s/2A_b$-values individually calculated in Eq. (4.4.7) from $V_{sb}/\overline{V}_s$. To be precise, the amplification thus obtained by using. (4.4.7) is slightly changeable depending on the value of $V_{sb}$ to be chosen among common base layers in the two sites, though its effect is ignorable for design purposes.

### 4.4.3 Effect of soil-nonlinearity

In order to examine the effect of strain-dependent soil nonlinearity on the site amplification, the first peak values of the spectrum ratios are compared in Fig. 4.4.8(a) between the main shocks (PGA $\approx 0.1$–$2.4$ g) in the vertical axis and the small aftershocks (mostly four aftershocks for PGA $\leq 0.1$ g versus one main shock) in the horizontal axis at

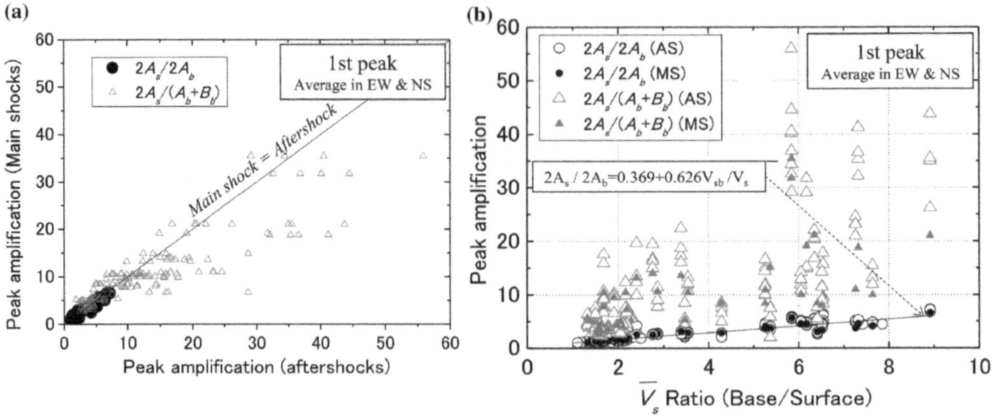


*Figure 4.4.8* First peak amplifications, $2A_s/2A_b$ and $2A_s/(A_b+B_b)$, for main shocks and aftershocks of 8 earthquakes: (a) Peak amplifications for main shocks versus for corresponding aftershocks, (b) Peak amplifications versus $V_{sb}/\overline{V}_s$ for main shocks and aftershocks (Kokusho 2013a).

individual sites for the eight earthquakes (EQ.1–EQ.8) in Japan. The two kinds of amplifications, $2A_s/(A_b+B_b)$ and $2A_s/2A_b$ obtained from the same vertical array records as described above are plotted with triangles and solid circles, respectively. The peak values of $2A_s/(A_b+B_b)$ are dispersed in a wide range, and the majority is plotted around or below the diagonal line of Main shock = Aftershock, indicating that during strong main shocks the site amplifications in terms of $2A_s/(A_b+B_b)$ tend to be lower than aftershocks due to the soil nonlinearity. In contrast, the peak values in $2A_s/2A_b$, despite smaller absolute values, concentrate near the diagonal line, indicating that the soil nonlinearity has only a marginal effect on the first peak value at least.

In Fig. 4.4.8(b), the same first peak values in the amplifications $2A_s/2A_b$ and $2A_s/(A_b+B_b)$ are plotted versus the $\overline{V}_s$-ratio, $V_{sb}/\overline{V}_s$, for not only the main shocks but also aftershocks of the eight earthquakes. Close and open circles are the peak values of $2A_s/2A_b$ while close and open triangles are those of $2A_s/(A_b+B_b)$, for the main shock and aftershocks, respectively. For $2A_s/2A_b$, a clear correlation can be recognized between the peak amplifications and the $\overline{V}_s$-ratios, which again can be approximated by Eq. (4.4.7) as shown in the diagram. Note that the differences in peak amplifications between the main shocks and corresponding aftershocks are almost invisible. In contrast, the peak values for $2A_s/(A_b+B_b)$ are not well correlated with the $\overline{V}_s$-ratios. Furthermore, the peak values for the main shocks are evidently smaller than those for the corresponding aftershocks in most sites. This indicates that the soil nonlinearity effect is very pronounced in $2A_s/(A_b+B_b)$ but almost invisible in $2A_s/2A_b$.

In order to account for this difference, a simple 2-layer system, a surface layer underlain by an infinitely thick base layer shown in Fig. 4.4.9(a) (the same as Fig. 4.2.1(b)) was studied (Kokusho and Sato 2008) assuming the impedance ratio between the two layers for small strain properties as $\alpha=\rho_1 Vs_1/\rho_2 Vs_2=0.3$. The transfer functions $2A_s/(A_2+B_2)$ and $2A_s/2A_2$ are already given in Eqs. (4.2.32) and (4.2.34), respectively. In order to take account the effect of strain-dependent

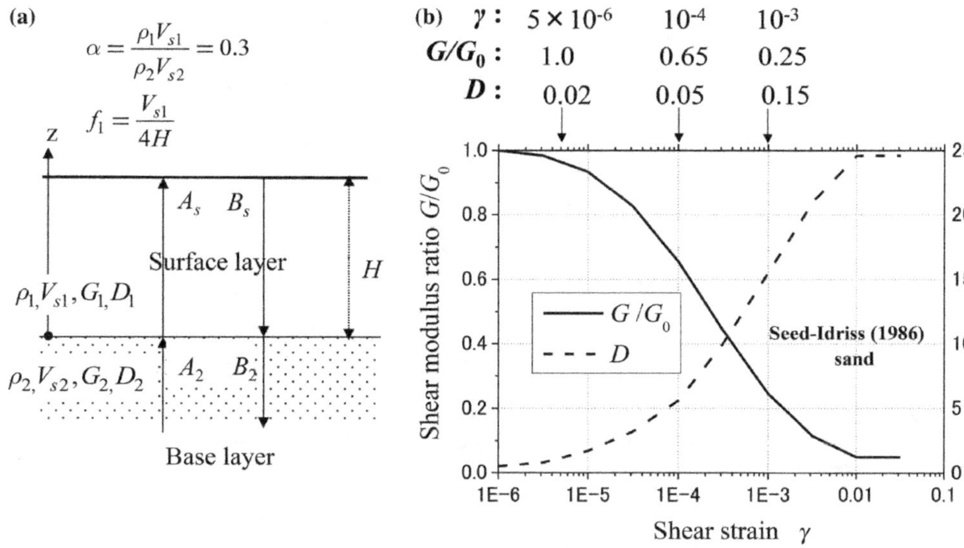


*Figure 4.4.9* Two-layer system (a), and Strain-dependent properties of surface layer (b).

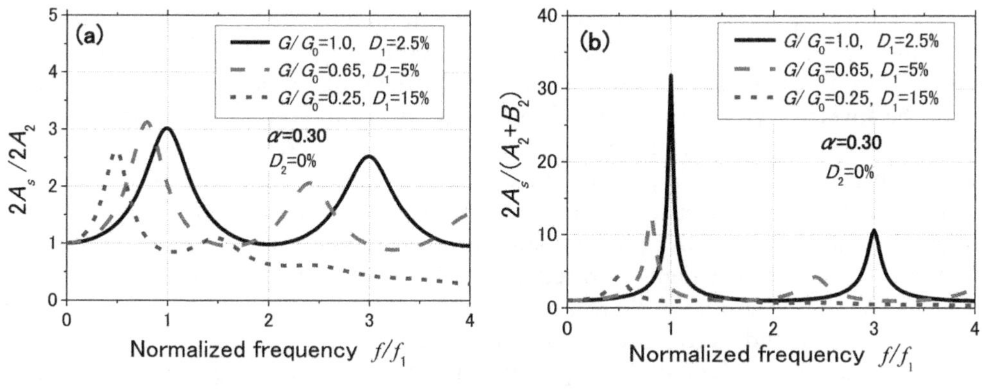


*Figure 4.4.10* Transfer functions of two-layer system: (a) $2A_s/2A_2$, (b) $2A_s/(A_2+B_2)$, for 3-step strain-dependent nonlinearity in surface layer: (Kokusho and Sato 2008).

soil properties on the amplifications, the shear modulus ratio $G/G_0$ and the damping ratio $D_1$ in the surface layer are parametrically changed; $G/G_0 = 1.0$, 0.65, 0.25 and $D_1 = 2.5$, 5, 15%, for the effective strain levels of $5 \times 10^{-6}$, $1 \times 10^{-4}$, $1 \times 10^{-3}$; respectively, assuming a typical degradation curve for clean sand (Seed et al. 1986) as indicated in Fig. 4.4.9(b), while in the base layer $D_2 = 0$. The calculated results of $2A_s/2A_2$ and $2A_s/(A_2 + B_2)$ are shown in Figs. 4.4.10(a) and (b), respectively. Here, the transfer functions between surface and base are shown versus the normalized frequency, $f/f_1$, where $f_1$ = first resonant frequency of the surface layer for small strain properties ($G/G_0 = 1.0$). The nonlinear soil properties have a clear effect on the peak

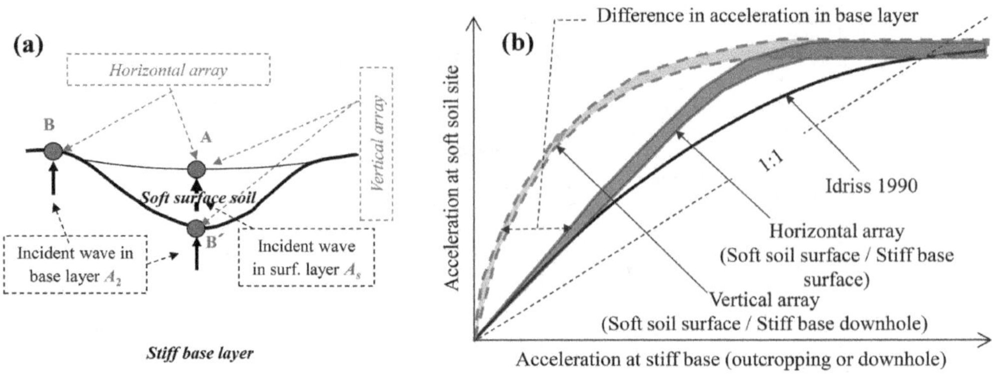


*Figure 4.4.11* Horizontal and vertical arrays (a), and Conceptual comparison of accelerations at stiff base site, outcropping (horizontal array) or downhole (vertical array), versus at soft soil sites (b).

frequency, due to degraded shear modulus both in $2A_s/2A_b$ and $2A_s/(A_b + B_b)$. However, in terms of the peak amplifications, the soil nonlinearity has much smaller impact on $2A_s/2A_b$ than on $2A_s/(A_b + B_b)$ for the 1st peak in particular. This is because in Eq. (4.2.32) the impedance ratio $\alpha^*$ affects $2A_s/2A_b$ as $|2A_s/2A_b| \approx |1/\alpha^*|$ at the peak. Under the paramount effect of radiation damping by the $\alpha^*$-value, the influence of strain-dependent properties becomes weaker. Furthermore, the impedance ratio $\alpha = \rho_1 V_{s1}/\rho_2 V_{s2}$, which becomes smaller with the degraded modulus or S-wave velocity in the surface layer, tends to give larger amplification, compensating the effect of increased soil damping during strong earthquakes in $2A_s/2A_b$. In contrast, the strain-dependent properties give a paramount effect on $2A_s/(A_b + B_b)$ with the absence of $\alpha^*$ in (4.2.34). Thus, the strain-dependent difference in soil properties between main shock and aftershocks tends to have smaller influence on the amplification in $2A_s/2A_b$ than in $2A_s/(A_b + B_b)$ as demonstrated in Fig. 4.4.8 by actual site amplification data.

Summarizing the above, there are two definitions of the site amplification; the horizontal array and vertical array as mentioned before and shown here again in Fig. 4.4.11(a). Fig. 4.4.11(b) conceptually shows the accelerations at the stiff base, the point B (outcropping) or B′ (downhole), taken in the horizontal axis versus those at the soft soil surface the point A in the vertical axis. The acceleration on the soft soil surface shows a strong nonlinear correlation with downhole base acceleration at B′ due to the soil nonlinearity effect as schematically illustrated with the dashed shaded belt in Fig. 4.4.11(b). In the horizontal array, the same acceleration on the soft soil surface tends to exhibit a remarkably linear correlation with the outcropping base acceleration at B as illustrated with the dark shaded belt in Fig. 4.4.11(b). This trend is actually substantiated in Fig. 4.4.8.

A similar research result of the horizontal array amplification available in United States is superposed on Fig. 4.4.11(b) with the solid line, wherein not peak spectrum amplifications but PGAs at soft soil sites and outcropping rock sites are compared (Idriss 1990). The curve was estimated from smaller earthquake records of similar epicenter distances during the 1985 Mexican earthquake and the 1989 Roma Prieta earthquake and also from numerical analyses for strong motions. The curve is obviously nonlinear reflecting the soil nonlinearity and the amplification between soil and rock tends to be lower than unity for PGA larger than around 0.4 g.

The above findings based on the actual records in Japan as well as the simple model analysis in Figs. 4.4.9 and 4.4.10 may indicate that soil nonlinearity is not so pronounced in the horizontal array unlike the vertical array. Nevertheless, it is obvious that in a certain point, where the soil moduli substantially decrease due to liquefaction, the amplification will be less than unity because the wave cannot arrive at the soft soil surface (base isolation effect in Sec. 5.11). In that case, the downward waves in the overburden soils diminish and the downhole accelerations at B′ become identical with the outcropping acceleration at B, merging the two curved belts in the horizontal and vertical arrays as illustrated in the diagram. Up to that point the site amplification defined by the horizontal array may possibly be more linear than normally anticipated.

### 4.4.4 Effect of downhole seismometer installation depth

In the vertical arrays monitoring site amplifications in exactly the same sites, the spectrum ratios $2A_s/(A_b+B_b)$ are obtained, and the spectrum ratios $2A_s/2A_b$ for the seismic zonation studies has to be calculated from them. Figs. 4.4.12 exemplifies the transfer functions $2A_s/(A_b+B_b)$ (thick dashed curves) at 6 vertical array sites (a)–(f) and compares with the corresponding transfer functions $2A_s/2A_b$ (thick solid curves). They are calculated using the soil profiles tabulated together. The installation depths of seismometers are indicated with the arrows, and the layer boundaries of major $V_s$-contrast are also pointed out with the triangle marks in the tabulated profiles. At all the sites, the peak frequencies of $2A_s/(A_b+B_b)$ are mostly compatible with those of the observed spectrum ratios in EW and NS directions.

If the two types of transfer functions $2A_s/(A_b+B_b)$ and $2A_s/2A_b$ are compared in each site, the coincidence in peak frequencies is perfect in (a) and good in (b) but tends to be poorer in (c), (d) and very poor in (e), (f). The reason may be accounted for by examining the soil profiles. In (a) and (b), the $V_s$-value at the depth of downhole seismometer is much larger than the upper layers, and the downhole seismometer is not so deep from the boundary of the major $V_s$-contrast. In (c) and (d), the $V_s$-value at the base layer is not very large compared to the upper layers though the seismometer depth is not so deep from a boundary of major $V_s$-contrast. In (e) and (f), though the $V_s$-value at the base layer is much larger than the upper layers, the depth of seismometer is too deep from the upper boundary of major $V_s$-contrast to properly detect the response of the upper layers. These observations suggest the significance of seimometer depth in deploying the downhole seismometer appropriately considering site specific soil profiles.

In order to find out a basic rule how the peak frequencies in the two transfer functions, $2A_s/(A_b+B_b)$ and $2A_s/2A_b$, are governed by soil profiles, a simplified three-layer model shown in Fig. 4.4.13(a) is analyzed. The model consists of a soft surface layer (1st layer: thickness $H_1$), an intermediate layer (2nd layer: $H_2$) and a stiff base layer (3rd layer) with infinite depth. The corresponding impedance ratios are $\alpha_{12}=(\rho V_s)_1/(\rho V_s)_2$ between the 1st and 2nd layer, $\alpha_{23}=(\rho V_s)_2/(\rho V_s)_3$ between the 2nd and 3rd layer, and the downhole seismometer is at the top of the 3rd layer. The two transfer functions, $2A_s/(A_b+B_b)$ and $2A_s/2A_b$, are calculated for the model and shown in Figs. 4.4.13(b), (c) against the normalized frequency $f/f_1$ where $f_1=V_{s1}/H_1$.

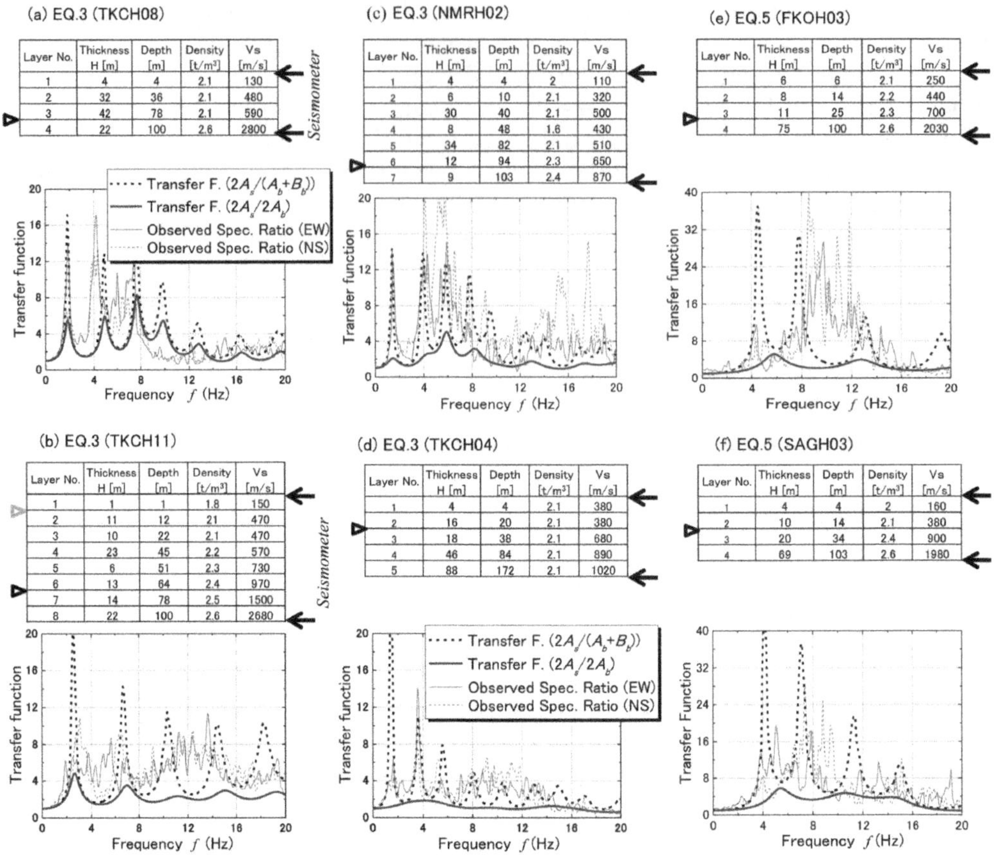


*Figure 4.4.12* Examples of transfer functions $2A_s/(A_b + B_b)$ compared with $2A_s/2A_b$ together with observed spectrum ratios, and tabulated soil profiles at 6 vertical array sites (a)–(f) (Kokusho 2013a).

In Fig. 4.4.13(b), wherein the thickness of intermediate layer $H_2$ is parametrically varied with respect to $H_1$ for $\alpha_{12} = 0.2$ and $\alpha_{23} = 0.8$ unchanged, the transfer function $2A_s/2A_b$ (solid curve) is stable with its peak frequency $f/f_1 \approx 1.0$ irrespective of the $H_2/H_1$-ratio because of the clear impedance contrast $\alpha_{12} = 0.2$ between the surface and intermediate layer. In contrast, the shapes and peak frequencies of the transfer function $2A_s/(A_b + B_b)$ (dashed curve) tend to change with increasing $H_2/H_1$-ratio for $H_2/H_1 \geqq 2.0$ in particular. This is because, in $2A_s/(A_b + B_b)$, the depth of seismometer serves as a virtual rigid boundary with no radiation of wave energy below as already mentioned in Sec. 3.2.3 even if it is in the midst of a uniform layer. In Figs. 4.4.13(c), where the impedance ratios $\alpha_{12}$, $\alpha_{23}$ are parametrically changing (keeping $\alpha_{12} \times \alpha_{23} = 0.16$ constant) with the constant thickness ratio $H_2/H_1 = 4.0$, the two transfer functions $2A_s/(A_b + B_b)$ and $2A_s/2A_b$ are quite different for $\alpha_{23} \approx 0.64$ or larger, though they coincide to each other if $\alpha_{23} \approx 0.4$ or smaller. This indicates that a sharp impedance contrast is preferred at the boundary near the downhole seismometer.

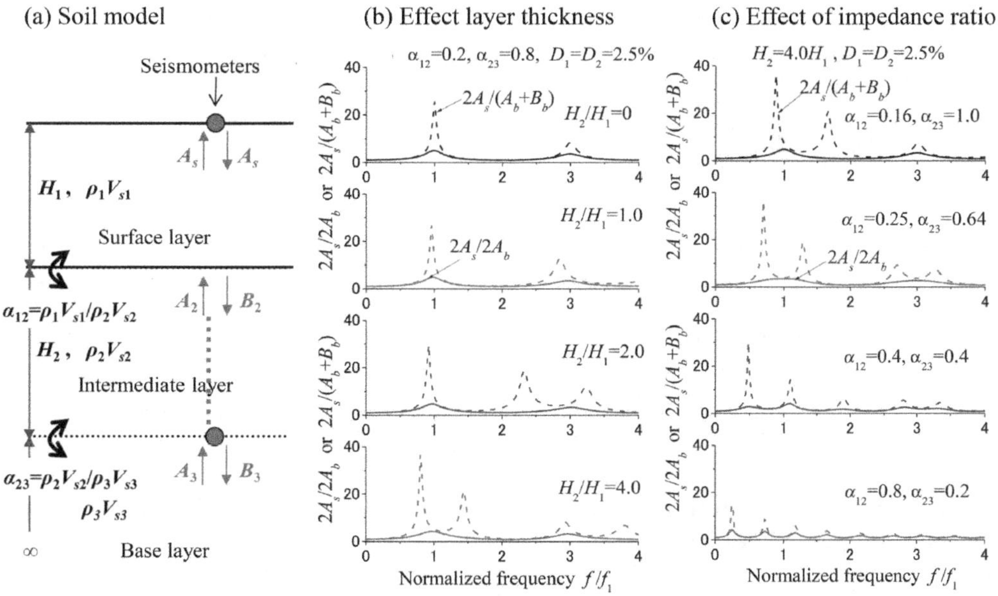


*Figure 4.4.13* 3-layer model with variable downhole seismometer depth relative to major layer boundary (a), Transfer functions for various layer thicknesses (b) and various impedance ratios (c) (Kokusho 2013a).

Consequently, in installing vertical array systems, the downhole seismometers should be in a stiff base layer not too deep from the layer boundary. A sharp impedance contrast at the base boundary is preferable. If the depth of the downhole seismometer is twice deeper than the surface layer thickness from the boundary with major impedance contrast, namely $H_2/H_1 > 2.0$, peak frequencies different from those in $2A_s/2A_b$ tend to appear in $2A_s/(A_b+B_b)$. Thus, it should be borne in mind that there are appropriate site-specific installation depths for the downhole seismometers of vertical arrays, so that the peak frequencies in the spectrum amplifications are in accordance with those of horizontal arrays. The depth should never be the deeper, the better.

## 4.5 SSI AND RADIATION DAMPING IN ONE-DIMENSIONAL WAVE PROPAGATION

In calculating the seismic response of superstructures, a lumped mass-spring system with the mass elements $m_i$ $(i=1-n)$ resting on the foundation ground such as in Fig. 4.5.1 is employed with horizontal acceleration given at the ground surface. Normally, the ground surface is assumed to be rigid as in (a), and the inertial load $-m_i(\ddot{u}_i+\ddot{z})$ is given to the $i$-th lumped mass $m_i$, where $\ddot{u}_i$ = acceleration of the mass relative to the ground and $\ddot{z}$ = absolute ground acceleration. When the structure is heavy, however, the ground acceleration nearby may be affected so as to be different from the far field ground acceleration $\ddot{z}$ by $\ddot{u}_G$, changing the inertial load to $-m_i(\ddot{u}_i+\ddot{z}-\ddot{u}_G)$

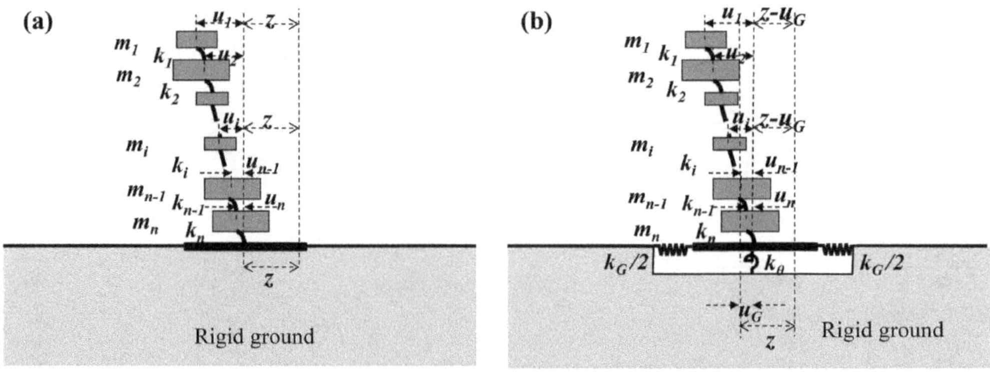


*Figure 4.5.1* Lumped mass-spring structure model resting on foundation ground: (a) Directly on rigid ground without SSI, (b) Connected via spring with SSI.

as shown in (b). This effect named as SSI (Soil-Structure-Interaction) is often modelled by a soil spring with a spring constant $k_G$ connecting the foundation and the far field ground in Fig. 4.5.1(b). Not only the horizontal springs for the horizontal motions but also the rotational springs $k_\theta$ are used to allow the rotation of the structure. These SSI springs tend to lower the resonant frequency of the system compared to that on the rigid ground.

Not only the resonant frequencies but also damping characteristics tend to change because of SSI. If the rigid ground is assumed as in Fig. 4.5.1(a), no vibration energy radiates into the ground. Thus, the vibration without SSI decays only by the internal damping in the structure. If the effect of radiation damping cannot be ignored in this case, the internal damping has to be added on purpose by a certain amount to compensate the rigid foundation.

The SSI effects on superstuctures are three-dimensional problems. Theoretically, vibrations of rigid foundations of different configurations resting on an ŭniform elastic half-space have been solved to yield closed form solutions on equivalent spring constants and radiation damping. Numerically, quite a few tools are available in direct analyses and indirect substructure analyses (e.g. NIST 2012). Apart from those specialized state of the art, very fundamental and simple characterizations of SSI and radiation damping are considered in the following in the realm of one-dimensional wave propagation.

### 4.5.1 Soil-structure interaction (SSI)

The degree of SSI may be quantified by comparing the ground motion beneath a superstructure with that in the far field. Fig. 4.5.2 shows (a) an elastic half-space in the far field, (b) a one-degree of freedom lumped mass-spring system on the elastic half-space and (c) a shear-vibration system on the elastic half-space. Horizontal displacement $u(t,z)$ of the SH-wave propagating in the half space with velocity $V_s$ and density $\rho$ in $z$-direction is expressed as:

$$u(t,z) = u_u(t - z/V_s) + u_d(t + z/V_s) \tag{4.5.1}$$

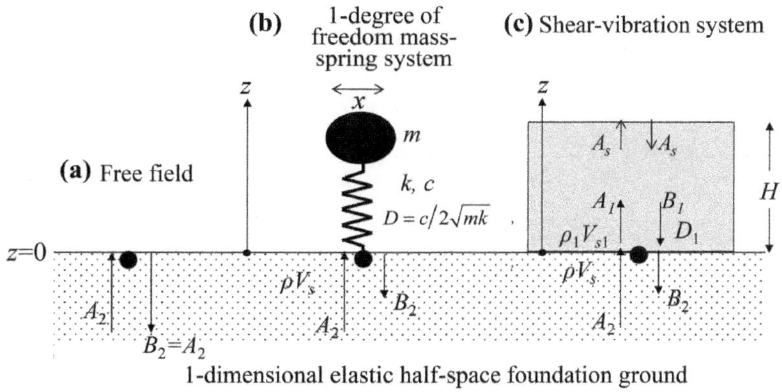


*Figure 4.5.2* SSI-effect comparing free field and simplified structures on foundation ground: (a) Free field, (b) 1D-freedom mass-spring system, (c) Shear-vibration system.

where $u_u(t-z/V_s)$ and $u_d(t+z/V_s)$ are upward and downward wave components. The vibration of the lumped mass-spring model is formulated as:

$$m\ddot{x}+c\dot{x}+kx=-m\ddot{u}(t,0) \tag{4.5.2}$$

Here, $m=$ mass, $k=$ spring constant, $c=$ dashpot constant (damping ratio $D=c/2\sqrt{mk}$: Kelvin damping), and $x=$ horizontal displacement of the mass relative to the ground. Because the reaction to the ground is $c\dot{x}+kx$, shear stress $\tau(t,0)$ working on the ground from the mass-spring system having the contact area $a$ is;

$$\tau(t,0)=\frac{c\dot{x}+kx}{a} \tag{4.5.3}$$

The shear stress is formulated also from Eq. (4.5.1) as:

$$\tau(t,0)=G\left.\frac{\partial u(t,z)}{\partial z}\right|_{z=0}=-\rho V_s[\dot{u}_u(t,0)-\dot{u}_d(t,0)] \tag{4.5.4}$$

Hence, the next formula is obtained from the above two equations.

$$\frac{c}{a\rho V_s}\dot{x}+\frac{k}{a\rho V_s}x=-\dot{u}_u(t,0)+\dot{u}_d(t,0) \tag{4.5.5}$$

The following equation modified from Eq. (4.5.2) is combined with what is obtained by differentiating Eq. (4.5.5) once in terms of time $t$.

$$\ddot{x}+\frac{c}{m}\dot{x}+\frac{k}{m}x=-\ddot{u}(t,0)=-\ddot{u}_u(t,0)-\ddot{u}_d(t,0) \tag{4.5.6}$$

Then, the next simultaneous equation is reached (Hoshiya and Yamazaki 1979).

$$\begin{aligned}&\left(1+\frac{c}{a\rho V_s}\right)\ddot{x}+\left(\frac{c}{m}+\frac{k}{a\rho V_s}\right)\dot{x}+\frac{k}{m}x=-2\ddot{u}_u(t,0)\\&\left(1-\frac{c}{a\rho V_s}\right)\ddot{x}+\left(\frac{c}{m}-\frac{k}{a\rho V_s}\right)\dot{x}+\frac{k}{m}x=-2\ddot{u}_d(t,0)\end{aligned}\tag{4.5.7}$$

If harmonic waves are assumed as:

$$u_u(t,0)=A_2e^{i\omega t},\quad u_d(t,0)=B_2e^{i\omega t},\quad x=Xe^{i\omega t}\tag{4.5.8}$$

then, Eq. (4.5.7) is written as:

$$\begin{aligned}&\left\{\left(1+\frac{c}{a\rho V_s}\right)\omega^2-i\left(\frac{c}{m}+\frac{k}{a\rho V_s}\right)\omega-\frac{k}{m}\right\}X=-2\omega^2A_2\\&\left\{\left(1-\frac{c}{a\rho V_s}\right)\omega^2-i\left(\frac{c}{m}-\frac{k}{a\rho V_s}\right)\omega-\frac{k}{m}\right\}X=-2\omega^2B_2\end{aligned}\tag{4.5.9}$$

From this, the next formula is obtained.

$$A_2+B_2=-\left(1-\frac{k}{m}\frac{1}{\omega^2}-i\frac{c}{m}\frac{1}{\omega}\right)X\tag{4.5.10}$$

Consequently, the amplitude of ground motion beneath the mass-spring system, $A_2+B_2$, shown in Fig. 4.5.2(b) relative to that in the free field, $2A_2$, in Fig. 4.5.2(a) can be written as:

$$\begin{aligned}\frac{A_2+B_2}{2A_2}&=\left(1-\frac{k}{m}\frac{1}{\omega^2}-i\frac{c}{m}\frac{1}{\omega}\right)\Big/\left\{\left(1+\frac{c}{a\rho V_s}\right)-\frac{k}{m}\frac{1}{\omega^2}-i\left(\frac{c}{m}+\frac{k}{a\rho V_s}\right)\frac{1}{\omega}\right\}\\&=\left(1-\left(\frac{\omega_0}{\omega}\right)^2-2iD\frac{\omega_0}{\omega}\right)\Big/\left\{(1+2D\alpha)-\left(\frac{\omega_0}{\omega}\right)^2-2i\left(D+\alpha/2\right)\frac{\omega_0}{\omega}\right\}\end{aligned}\tag{4.5.11}$$

Here, $\omega_0^2=k/m$, $D=c/2\sqrt{mk}$, and $\alpha=(\omega_0 m/a)/(\rho V_s)$ corresponds to the impedance ratio of mass-spring system because $\omega_0 m/a$ is equivalent to the seismic impedance in terms of the dimension.

In the case of the large-width shear-vibration system in Fig. 4.5.2(c), with the height $H$, the equivalent impedance $\rho_1 V_{s_1}$ and the internal damping ratio $D_1$, resting on the elastic half space, dynamic response immediately below the system may be approximated by the two-layer soil model discussed in Sec. 4.2.2 and the next equation is derived from Eqs. (4.2.28) and (4.2.29).

$$\frac{B_2}{A_2}=\frac{(1-\alpha^*)+(1+\alpha^*)e^{-2ik_1^*H}}{(1+\alpha^*)+(1-\alpha^*)e^{-2ik_1^*H}}\tag{4.5.12}$$

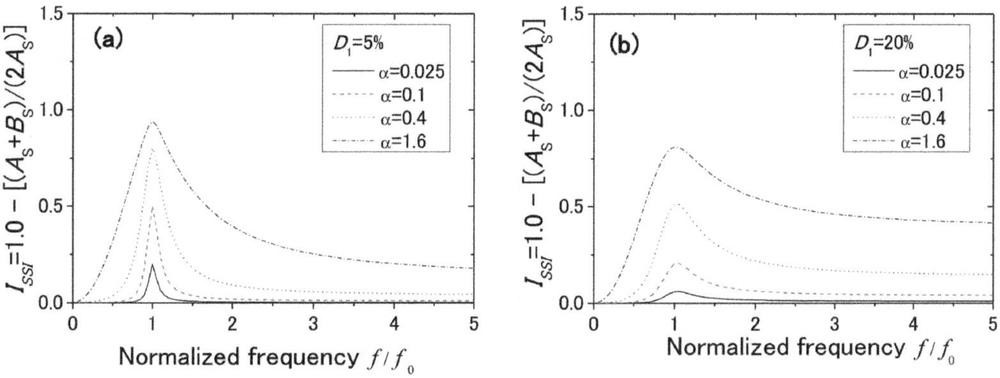


*Figure 4.5.3* $I_{SSI}$ versus $f/f_0$ corresponding to 1D-freedom mass-spring system for four stepwise impedance ratios: (a) $D_1 = 5\%$, (b) 20%.

The ratio of ground motion amplitude immediately below the model, $A_2 + B_2$ in Fig. 4.5.2(c), to that in the free field, $2A_2$ in (a), is written as:

$$\frac{A_2 + B_2}{2A_2} = \frac{1 + e^{-2ik_1^* H}}{(1 + \alpha^*) + (1 - \alpha^*)e^{-2ik_1^* H}} \tag{4.5.13}$$

The Index of SSI, $I_{SSI}$ may be reasonably defined here as:

$$I_{SSI} = 1 - \left| \frac{A_2 + B_2}{2A_2} \right| \tag{4.5.14}$$

so that $I_{SSI} = 0$ in the free field where $B_2 = A_2$, and $I_{SSI}$ beneath structures increases with increasing structural mass or rigidity up to $I_{SSI} = 1.0$ with $B_2 = -A_2$.

Let us calculate $I_{SSI}$ of the two simple structure models with typical parameters in Figs. 4.5.2(b) and (c) using Eqs. (4.5.11) and (4.5.13), respectively. As the key value of SSI, the impedance ratio $\alpha$ is chosen differently for different structures. For a light-weight wooden house with the smallest SSI effect, $\alpha = (\omega_0 m/a)/(\rho V_s) = 0.025$ assuming typically that the house with the horizontal area $a = 10\,\text{m} \times 10\,\text{m}$, the mass $m = 50\,\text{t}$ and the resonant frequency $f_0 = \omega_0/2\pi = 2\,\text{Hz}$ is resting on the foundation ground with $\rho = 1.8\,\text{t/m}^3$ and $V_s = 140\,\text{m/s}$. For a 5-story RC building, $\alpha = (\omega_0 m/a)/(\rho V_s) = 0.11$ assuming $a = 20\,\text{m} \times 20\,\text{m}$, $m = 2000\,\text{t}$ and $f_0 = \omega_0/2\pi = 4\,\text{Hz}$, while $\rho = 2.2\,\text{t/m}^3$ and $V_s = 500\,\text{m/s}$ in the foundation ground. For a massive concrete dam with the density $\rho_1 = 2.3\,\text{t/m}^3$ and S-wave velocity $V_{s1} = 2000\,\text{m/s}$ resting on the ground $\rho = 2.4\,\text{t/m}^3$ and $V_s = 1200\,\text{m/s}$, $\alpha = \rho_1 V_{s1}/\rho V_s = 1.6$. Thus, the impedance ratio is changed in four steps as $\alpha = 0.025$, 0.1, 0.4 and 1.6.

Fig. 4.5.3 shows the variations of $I_{SSI}$-value versus normalized frequency $f/f_0$ calculated by Eqs. (4.5.11) and (4.5.14) corresponding to the mass-spring system for the four values of $\alpha$ and the internal damping ratios of the structure (a) $D_1 = 5\%$ and (b) 20% for the Kelvin damping. This indicates that the SSI-effect becomes dominant particularly in and near the resonance. Its effect tends to reach in a broader frequency

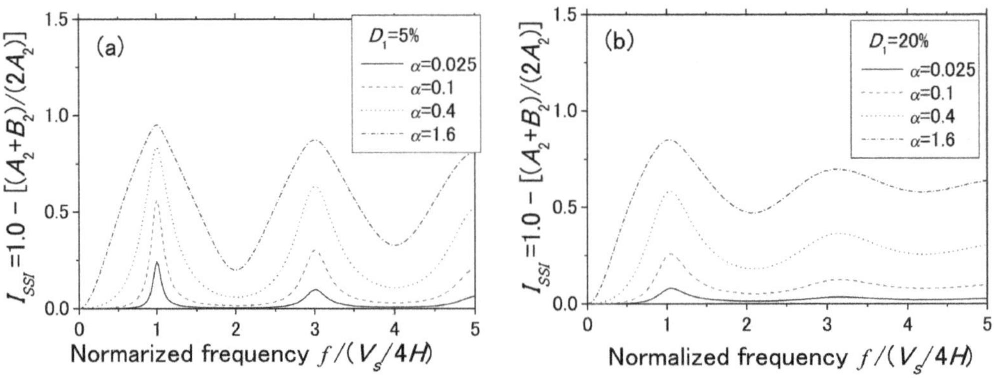


*Figure 4.5.4* $I_{SSI}$ versus $f/(V_s/4H)$ corresponding to shear-vibrating system for four stepwise impedance ratios: (a) $D_l = 5\%$, (b) 20%.

range nearby with increasing $\alpha$ and increasing internal damping ratio of the structure $D_1$. The similar results calculated by Eqs. (4.5.13) and (4.5.14) corresponding to the shear-vibration structure are shown in Fig. 4.5.4 for the four stepwise values of $\alpha$ and the internal damping ratios of the structure $D_1 = 5\%$ (a) and 20% (b) for the Nonviscous damping. Note that there are multiple peaks corresponding to multiple degrees of resonance in this case, wherein the first peak has almost the same peak amplitude as in Fig. 4.5.3. Thus, in the two different models, it is evident that the SSI-effect is highly variable with frequency relative to the resonant frequency. Also noted is that with increasing $D_1$-value as the structure approaches failure, the SSI-effect tends to be less-dependent on frequency and become larger between peak frequencies, too.

In the above evaluations, the structure is assumed to vibrate horizontally only in the shear mode, though it may vibrate not only in the shear mode but also in the bending or rocking mode. Despite this limitation, the above-mentioned findings may help the SSI-effect even in actual problems be understood qualitatively. In normal engineering practice, a structure is idealized by the lumped mass-spring model as in Fig. 4.5.1(a) and the free-field earthquake motions are directly given on a rigid ground surface. This indicates $I_{SSI} = 0$, which may be justified for a wooden house because $I_{SSI} < 0.2$ even in resonance as indicated in Fig. 4.5.3. However, for heavy structures with larger impedance ratio, $I_{SSI}$ tends to increase, necessitating the SSI-effect to be taken into account particularly near the resonance and near failures with large internal damping.

### 4.5.2 Radiation damping

The damping mechanisms of seismic vibrations in superstructures resting on foundation ground are attributed to internal damping and radiation damping as illustrated in Fig. 4.5.5(a). The internal damping occurs due to the loss of earthquake energy in structural members, while the radiation damping occurs due to the energy migration from a structure to foundation ground, though the gross energy is unchanged.

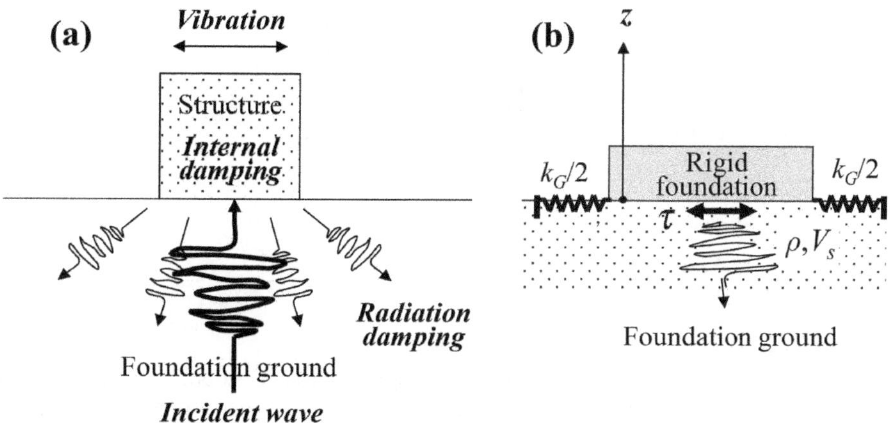


*Figure 4.5.5* Radiation damping versus internal damping (a), and Simplified evaluation for rigid foundation on elastic half-space (b).

Though the radiation damping occurs three-dimensionally because the structure has a limited dimension in contrast to the unlimited foundation ground expanding three-dimensionally, it is simplified here to understand its basic mechanism in the arena of one-dimensional SH-wave wave propagation.

#### 4.5.2.1 *Rigid foundation*

Imagine that a rigid foundation resting on an elastic half-space with density $\rho$ and S-wave velocity $V_s$ illustrated in Fig. 4.5.5(b) vibrates horizontally in harmonic motion with the displacement amplitude $u_0$ and angular frequency $\omega$.

$$u = u_0 e^{i\omega t} \tag{4.5.15}$$

If the foundation width is infinitely large, the vibration energy radiates into the half-space as one-dimensionally propagating SH-wave as:

$$u = u_0 e^{i(\omega t + kz)} \tag{4.5.16}$$

Here, $z$-axis is directing upward and $k = \omega / V_s$ is the wave number. Shear stress $\tau$ beneath the foundation is written using $G = \rho V_s^2$ as:

$$\tau = G \left. \frac{\partial u}{\partial z} \right|_{z=0} = iGku_0 e^{i\omega t} = iGku = i\rho V_s \omega u \tag{4.5.17}$$

The foundation is actually of finite width wherein the horizontal SSI-effect is modeled by a lateral spring with the spring constant $k_G$. Considering the spring reaction for the foundation displacement $u$ relative to free-field $k_G u$, which is assumed

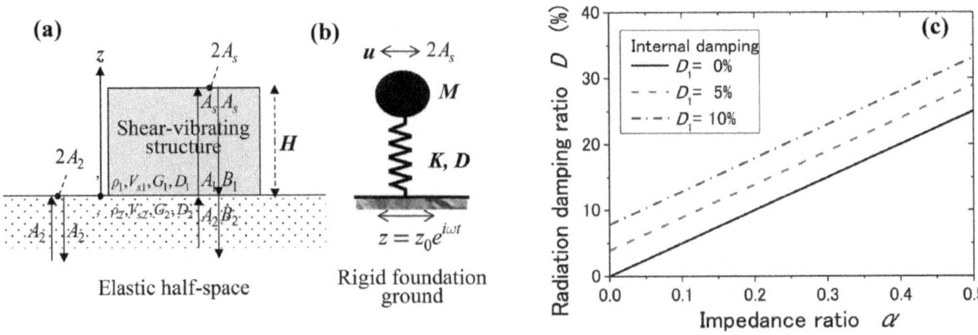


*Figure 4.5.6* Shear-vibration structure on elastic half-space (a), 1D-freedom mass-spring system on rigid ground (b), and Equivalent damping ratio $D$ versus $\alpha$ derived by comparing them (c).

to be frequency-independent, the shearing force $T$ is correlated with the relative displacement using the foundation area $A$ as (Tajimi 1965):

$$T = (k_G + i\rho V_s\omega)Au = (k_G^2 + \rho^2 V_s^2\omega^2)^{1/2} A u e^{i\delta}, \quad \tan\delta = \rho V_s\omega / k_G \tag{4.5.18}$$

The radiation damping of the foundation is expressed by the phase angle $\delta$ in the above equation based on the correlation in Eq. (1.5.32) as:

$$D = \frac{\tan\delta}{2} = \frac{\rho V_s\omega}{2k_G} \tag{4.5.19}$$

It is proportional to angular frequency $\omega$ and inversely proportional to the lateral spring constant $k_G$.

#### 4.5.2.2 *Shear-vibration structure*

In place of the rigid foundation in Fig. 4.5.5(b), imagine the shear-vibration structure as in Fig. 4.5.6(a), the dynamic response may be approximated by the two-layer system as addressed in Sec. 4.2.1. The peak amplification of the system in resonance between the displacement amplitude at the top of the structure $2A_s$ and that at the free ground surface $2A_2$ is given by Eq. (4.2.35). Then, the structure in Fig. 4.5.6(a) is compared with one-degree of freedom lumped mass-spring system on the rigid ground shown in Fig. 4.5.6(b) vibrated by the harmonic force $Fe^{i\omega t}$ as:

$$m\ddot{u} + c\dot{u} + ku = Fe^{i\omega t} \tag{4.5.20}$$

Using the resonant angular frequency $\omega_0 = \sqrt{k/m}$ and damping ratio $D = c/(2\sqrt{mk})$,

$$\ddot{u} + 2D\omega_0\dot{u} + \omega_0^2 u = \left(\frac{F}{m}\right)e^{i\omega t} \tag{4.5.21}$$

The absolute amplitude is written as:

$$|u| = \frac{F/k}{\{(-(\omega/\omega_0)^2 + 1)^2 + 4D^2(\omega/\omega_0)^2\}^{1/2}} \tag{4.5.22}$$

and the peak amplification $Q$ at the resonance frequency $\omega = \omega_0$ is obtained in the following (if $D$ is not too large) as the ratio of the absolute peak displacement $|u|_{\omega=\omega_0}$ to the spring displacement due to static force $F/k$.

$$Q = \frac{|u|_{\omega=\omega_0}}{(F/k)} = \frac{1}{2D} \tag{4.5.23}$$

On the other hand, if the ground displacement of the mass-spring system in Fig. 4.5.6(b) is written as $z = z_0 e^{i\omega t}$, Eq. (4.5.21) is rewritten as:

$$\ddot{u} + 2D\omega_0\dot{u} + \omega_0^2 u = -\ddot{z} = \omega^2 z_0 e^{i\omega t} = \left(\frac{F}{m}\right) e^{i\omega t} \tag{4.5.24}$$

Since $F = m\omega^2 z_0$, the following correlation holds:

$$\frac{F}{k} = \left(\frac{m}{k}\right)\omega^2 z_0 = \left(\frac{\omega}{\omega_0}\right)^2 z_0 \tag{4.5.25}$$

and $F/k = z_0$ at $\omega = \omega_0$. This indicates that the amplification $Q$ in Eq. (4.5.23) is also interpreted as a ratio of the peak absolute displacement $|u|_{\omega=\omega_0}$ to the ground displacement amplitude $z_0$ at $\omega = \omega_0$, which corresponds to the free surface displacement amplitude of the foundation ground $2A_2$ in Fig. 4.5.6(a). If the peak displacement amplitude $u$ of the lumped mass is assumed to correspond to the first resonance amplitude in Eq. (4.2.35) ($n = 1$) at the top of the shear-vibration structure $2A_s$, the next equations can be derived from Eq. (4.5.23).

$$Q = \left|\frac{2A_s}{2A_2}\right| = \frac{1}{\sinh(\pi\delta_1/4) + \alpha\cosh(\pi\delta_1/4)} = \frac{1}{2D} \tag{4.5.26}$$

$$D = \frac{[\sinh(\pi\delta_1/4) + \alpha\cosh(\pi\delta_1/4)]}{2} \tag{4.5.27}$$

where $\delta_1 = \tan^{-1} 2D_1$. Fig. 4.5.6(c) shows the equivalent damping ratio $D$ of the one-degree of freedom mass-spring system resting on the rigid ground in Fig. 4.5.6(b) calculated by Eq. (4.5.27) versus the impedance ratio of the shear-vibration structure on the elastic half-space in Fig. 4.5.6(a). Here the internal damping ratio of the ground is $D_2 = 0$ (which does not affect the result so much) and that of the structure is varied as $D_1 = 0$, 5, 10%. Note that the equivalent damping ratio $D$ reflecting the radiation damping is linearly correlated with the impedance ratio $\alpha$. The impedance ratio $\alpha$ has a strong impact on $D$; $D = 10\%$ for $\alpha = 0.2$ and $D = 20\%$ for $\alpha = 0.4$ even for $D_1 = 0$. If the internal damping $D_1$ increases from zero, it tends to increase the equivalent damping ratio $D$ by a constant increment as shown in Fig. 4.5.6(c). Though this evaluation ignores three-dimensional energy radiation in actual conditions, it indicates a significant role of the impedance ratio between structure and foundation ground to determine the radiation damping in one-dimensional wave propagation and how the effect of internal damping is superposed on it.

## 4.6 ENERGY FLOW IN WAVE PROPAGATION

Seismic design in practice is based on inertial forces given by accelerations (e.g. maximum or equivalent accelerations) or seismic coefficients. Though the force-based design method has long been used to date, it is recognized increasingly that acceleration alone may not be an appropriate parameter for seismic damage evaluations. More and more strong acceleration records with maximum values exceeding 1 g have been obtainxd in recent years. Fig. 4.6.1 summarizes horizontal peak ground accelerations (PGAs) observed during recent earthquakes. They are increasing almost monotonically as years go by, arriving at almost 3 g. This is presumably because the density of earthquake observation networks (for example K-NET and KiK-net in Japan) are getting denser in their deployments to be able to pick up localized higher PGAs than before. On the other hand, cases are increasing where no significant structural damage was reported despite the high observed PGAs; e.g. 1.8 g in Tarzana, California during the 1994 Northridge earthquake, USA, 1.7 g in Tokamachi during the 2004 Niigata-ken Chuetsu earthquake, Japan, and 2.8 g in Tsukidate during the 2011 Tohoku earthquake, Japan.

Actually, the acceleration is not relevant to determine structural damage by itself, because the seismic wave energy defined in Eq. (1.2.25), which seems to closely correlated with the damage, is dependent not only on the wave amplitude, acceleration or particle velocity, but also on the soil impedance where the ground motion is recorded. In United States, Arias Intensity (Arias 1970) proportional to the time-integral of squared acceleration time-histories is sometimes used in earthquake engineering to measure a sort of seismic intensity similar to the energy, though it is not actually the wave energy in the exact physical meaning because of non-involvement of soil impedance. The wave amplitude, acceleration or velocity, cannot be meaningful by itself in view of damage

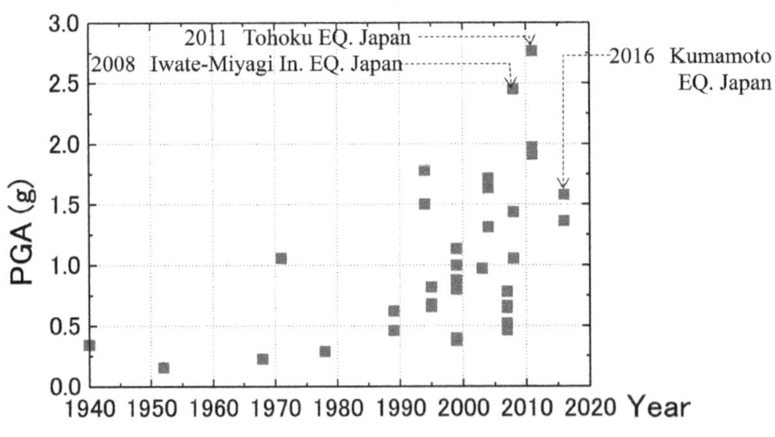


*Figure 4.6.1* PGAs observed during recent strong earthquakes in recent years (modified from Kokusho 2009).

potential without the associated impedance values where that value was recorded if the wave energy governs the damage. Hence, when a design motion with a given amplitude is discussed, it is essential to identify the soil condition or impedance where the motion was defined or recorded.

The seismic energy was investigated by seismologists (e.g. Gutenberg and Richter 1942, 1956) in order to evaluate the total energy released from a seismic source based on observed earthquake records. However, from the viewpoint of engineering design in particular, very few researchers tried to investigate earthquake motions in terms of energy. Among them, Sarma (1971) calculated site-specific seismic energies from velocity records and compared with spherically radiated energy from the source. An energy-based design method has been proposed for buildings (Akiyama, 1999), although it is still limited within the energy capacity of superstructures resting on rigid ground, without considering the energy demand from deformable ground to the superstructures. In the liquefaction evaluation, energy-based methods have been proposed, where the energy capacity for liquefaction triggering is compared with the input earthquake energy demand assuming the spherical energy radiation for body waves (Davis and Berrill 1982), though it is not used in practice.

In most of the energy-related investigations, the energy concept is limited in the energy capacities of soils and structures, while the energy demands in design earthquakes are not discussed. Kokusho and Motoyama (2002) performed a basic study on the energy demands of seismic waves in surface layers based on the one-dimensional multi-reflection theory of the SH-waves using vertical array records during the 1995 Kobe earthquake, which was followed by theoretical study on the same topic by Kokusho et al. (2007). Similar studies using a number of vertical array data was further carried out to understand general trends of the energy demand in surface layers (Kokusho and Suzuki 2011, 2012). In the following Sections, the energy demands in surface soil layers are discussed based on the calculations of wave energy flows in simplified models and actual soil models of vertical array sites using the multi-reflection theory of one-dimensional SH-wave propagations.

### 4.6.1 Energy flow at a boundary of infinite medium

In order to understand the mechanism controlling the energy flow in layered soil deposits, let us go back to simplified models. The first is an infinite medium consisting of two parts with the horizontal boundary as shown in Fig. 4.6.2(a), where $z$-axis is taken upward from the boundary. The S-wave velocities are $V_{s1}$ and $V_{s2}$, and the soil densities are $\rho_1$ and $\rho_2$ in the upper and lower parts, respectively. The wave displacements $u_1$, $u_2$ respectively are expressed as:

$$u_1 = A_1 e^{i(\omega t - k_1 z)} \tag{4.6.1}$$

$$u_2 = A_2 e^{i(\omega t - k_2 z)} + B_2 e^{i(\omega t + k_2 z)} \tag{4.6.2}$$

where, $\omega$ = angular frequency, $k_1$ and $k_2$ are the respective wave numbers defined by $k_1 = \omega/V_{s1}$, $k_2 = \omega/V_{s2}$. Here, $A_1$ is the amplitude for upward wave in the upper part, and $A_2$, $B_2$ are the amplitudes for upward and reflecting downward waves in the lower part, respectively. This corresponds to a modification of the two-layer model in Fig. 4.2.1(b), wherein downward wave $B_1 = 0$. Hence, the amplitude ratios among

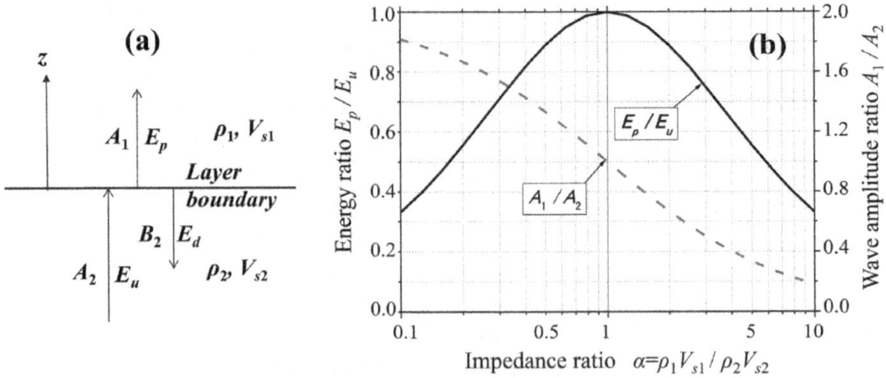


*Figure 4.6.2* Vertically propagating SH-wave in Infinite media consisting of two properties (a), and Amplitude ratio $A_1/A_2$ and energy ratio $E_u/E_p$ versus impedance ratio (b) (Kokusho et al. 2007).

$A_1, A_2, B_2$ can readily be obtained from Eqs. (4.2.9) and (4.2.10) using the impedance ratio $\alpha = \rho_1 V_{s1}/\rho_2 V_{s2}$.

$$\frac{A_1}{A_2} = \frac{2}{1+\alpha} \tag{4.6.3}$$

$$\frac{B_2}{A_2} = \frac{1-\alpha}{1+\alpha} \tag{4.6.4}$$

Because the energy of one-directionally propagating SH-wave passing through a unit horizontal area is proportional to the square of particle velocity amplitude times associated impedance ratio as defined in Eq. (1.2.25), the corresponding energy ratios are written as:

$$\frac{E_p}{E_u} = \frac{\rho_1 V_{s1} A_1^2}{\rho_2 V_{s2} A_2^2} = \frac{4\alpha}{(1+\alpha)^2} \tag{4.6.5}$$

$$\frac{E_d}{E_u} = \frac{\rho_2 V_{s2} B_2^2}{\rho_2 V_{s2} A_2^2} = \frac{(1-\alpha)^2}{(1+\alpha)^2} \tag{4.6.6}$$

where $E_u$, $E_p$ and $E_d$ are cumulative energies for the upward wave in the lower part, propagating wave in the upper part, and downward wave in the lower part, respectively. In Fig. 4.6.2(b), the amplitude ratio and the energy ratio of the waves are taken along the two vertical axes versus the logarithm of impedance ratio in the horizontal axis. Note that the energy ratio $E_p/E_u$ decreases symmetrically as the impedance ratio $\alpha$ is departing from unity, whereas the amplitude ratio monotonically decreases with increasing impedance ratio. This indicates that, if a layer boundary exists, the energy $E_p$ always decreases from $E_u$ because a part of the energy $E_d$ is inevitably transferred to the reflecting wave due to the different impedance between the two media. If the internal damping is considered, $\alpha$ in Eqs. (4.6.5) and (4.6.6) have to be replaced with $\alpha^* = \rho_1 V_{s1}^* / \rho_2 V_{s2}^*$ defined in Eq. (4.2.27), although this gives only a marginal difference in the energy ratios for the damping-values normally conceivable.

Such a condition only with upward energy and no downward energy as in Fig. 4.6.2(a) may indicate that the upper medium is infinitely large with no upper reflecting boundary. It may also be interpreted that the upward energy $E_p$ is completely dissipated in the upper medium and hence no downward energy comes down. In this interpretation, $E_p$ in Eq. (4.6.5) may be accounted for as the upper limit of seismic wave energy to be given to a destructing structure wherein all the energy is dissipated as a perfect energy absorber.

## 4.6.2 Energy flow of harmonic wave in two-layer system

Next, the energy flow is considered for harmonic waves propagating in the two-layer model shown in Fig. 4.6.3(a) consisting of the surface layer of the height $H$ underlain by the infinitely thick base layer. The displacement amplitude ratios of the upward and downward waves in the base layer, $A_2$, $B_2$ and those at the ground surface $A_s$ already available in Eqs. (4.2.32), (4.2.33) can be used to obtain the energy flow. For the stationary harmonic wave propagation, cumulative wave energies in a given time period is used here. The ratio of the upward energy at the ground surface $E_s$ to that in the base layer $E_u$ is written as:

$$\frac{E_s}{E_u} = |\alpha^*| \left| \frac{A_s}{A_2} \right|^2 = \frac{4|\alpha^*|}{|(1+\alpha^*)e^{ik_1^* H} + (1-\alpha^*)e^{-ik_1^* H}|^2} \tag{4.6.7}$$

while the ratio between the downward and upward energies at the top of base layer, $E_d$ and $E_u$ is written as:

$$\frac{E_d}{E_u} = \left| \frac{B_2}{A_2} \right|^2 = \left| \frac{(1-\alpha^*) + (1+\alpha^*)e^{-2ik_1^* H}}{(1+\alpha^*) + (1-\alpha^*)e^{-2ik_1^* H}} \right|^2 \tag{4.6.8}$$

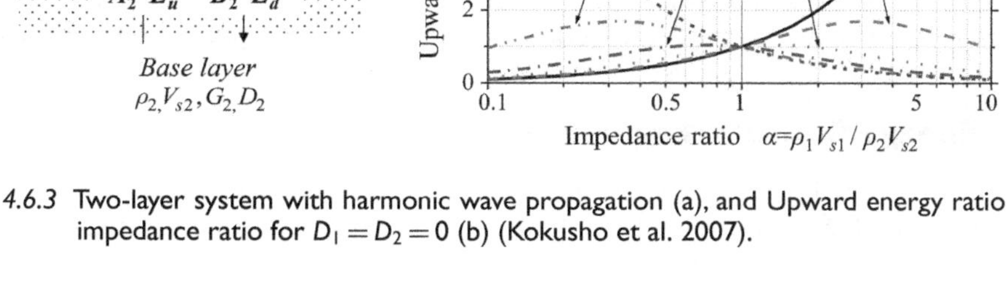


*Figure 4.6.3* Two-layer system with harmonic wave propagation (a), and Upward energy ratio versus impedance ratio for $D_1 = D_2 = 0$ (b) (Kokusho et al. 2007).

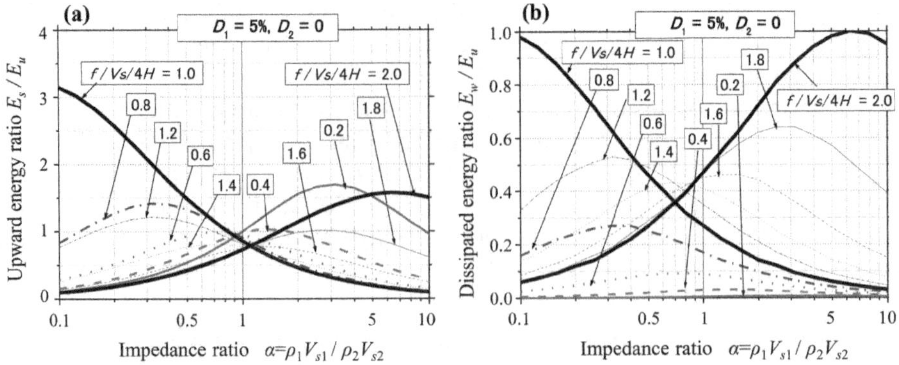


*Figure 4.6.4* Upward energy ratio versus impedance ratio (a), and Dissipated energy ratio versus impedance ratio (b), for a two-layer system of $D_1 = 5\%$, $D_2 = 0$ (Kokusho et al. 2007).

Furthermore, dissipated energy in the surface layer $E_w$ is expressed as $E_w = E_u - E_d$, because in the stationary vibration the difference between the upward and downward energies at the top of the base layer has to be balanced with the energy dissipated in the surface layer, hence,

$$\frac{E_w}{E_u} = 1 - \frac{E_d}{E_u} \tag{4.6.9}$$

Fig. 4.6.3(b) depicts the upward energy ratio $E_s/E_u$ in Eq. (4.6.7) for stepwise varying normalized frequencies, $f/(V_s/4H)$, versus the impedance ratio $\alpha$ in the log-scale for zero internal damping ($D_1 = D_2 = 0$). The energy ratio varies symmetrically with respect to the center axis $\alpha = 1.0$ where $E_s/E_u = 1.0$ quite reasonably. The maximum energy in the surface layer occurs in the first resonance, $f/(V_s/4H) = 1.0$, for $\alpha < 1.0$, whereas it occurs in the second resonance, $f/(V_s/4H) = 2.0$, for $\alpha > 1.0$, when $E_s$ is much larger than $E_u$ reflecting energy accumulation in the surface layer due to the resonance. By substituting $k_1 H = \omega H/V_{s1} = \pi/2$ into Eq. (4.6.7), $E_s/E_u = 1/\alpha$ is obtained as the maximum energy ratio, indicating that the softer the surface soil the larger is the ratio of energy in resonance accumulated in the surface layer. Also note that for off-resonance frequencies, such as $f/(V_s/4H) \leq 0.6$ or $f/(V_s/4H) \leq 1.4$, $E_s$ tends to be much lower than $E_u$ with decreasing $\alpha$ for $\alpha < 1.0$.

In Figs. 4.6.4(a), (b), the upward energy ratio $E_s/E_u$ and $E_w/E_u = 1 - E_d/E_u$ calculated by Eq. (4.6.7) and Eq. (4.6.9), respectively, for $D_1 = 5\%$ is shown versus $\alpha$. The energy $E_s$ tends to decrease evidently compared to the case of $D_1 = 0\%$, though the near-resonance energy accumulation effect in the surface layer can still be recognized. It is also noteworthy that a considerable energy $E_w$ out of $E_u$ tends to be dissipated in and near resonance due to the multi-reflection of the wave trapped in the surface layer.

### 4.6.3 Energy flow of irregular wave in two-layer system

Similar energy flow for irregular seismic wave with a limited duration in the two-layer system in Figs. 4.6.5(a) is studied here. In the surface layer with the thickness $H = 30$ m

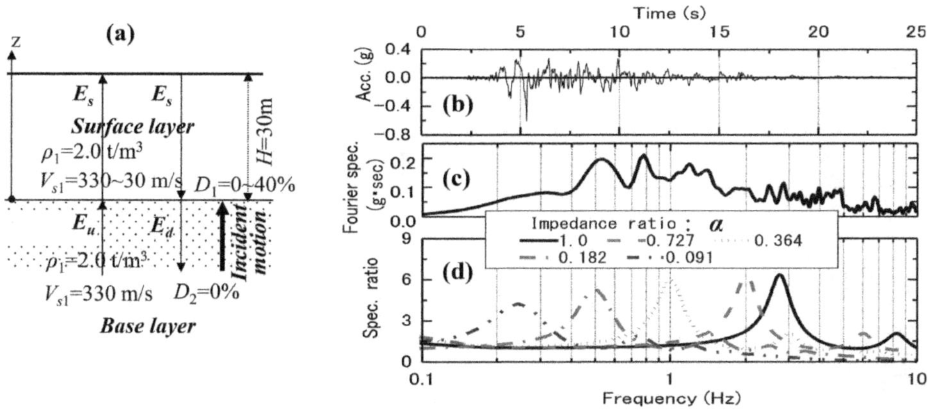


*Figure 4.6.5* Energy flow for irregular wave: (a) Two-layer model, (b) Time-history, (c) Fourier spectrum of input wave, (d) Transfer functions of two-layer model (Kokusho et al. 2007).

and density $\rho_1 = 2.0\,\text{t/m}^3$, S-wave velocity and damping ratio are parametrically varied as $V_{s1} = 330$–$30\,\text{m/s}$ (impedance ratio $\alpha = 1.0$–$0.091$) and $D_1 = 0$–$40\%$, respectively. It is underlain by a base layer with $\rho_2 = 2.0\,\text{t/m}^3$, $V_{s1} = 330\,\text{m/s}$, $D_2 = 0\%$, and the incident wave is given at the top of the base layer. The seismic wave is the incident acceleration motion at GL.-83.4 m in the PI vertical array (in the principal direction where maximum acceleration occurred) during the 1995 Kobe earthquake in Japan. Its time history and Fourier spectrum are depicted in Figs. 4.6.5(b) and (c). Fig. 4.6.5(d) shows transfer functions between ground surface and base layer for $\alpha = 1.0$–$0.091$ and $D_1 = 5\%$. Note that the two-layer system with $\alpha = 0.182$ tends to have the peak frequency similar to the input earthquake motion.

The wave energy $E$ of one-directionally propagating SH-wave through a unit horizontal area in a time interval $t = t_1$–$t_2$, and the associated energy flux $dE/dt$ can be written by the following formula as already explained in Eqs. (1.2.25), (1.2.26).

$$E = \rho V_s \int_{t_1}^{t_2} (\dot{u})^2 dt \tag{4.6.10}$$

$$\frac{dE}{dt} = \rho V_s (\dot{u})^2 \tag{4.6.11}$$

where, $\dot{u}$ = particle velocity and $\rho V_s$ = seismic impedance.

Fig. 4.6.6(a) depicts the time-histories of wave energy $E_s$ at the ground surface for $D_1 = 0$ and 5% in the upper and lower diagram, respectively. $E_s$ monotonically increases with time to the end of the seismic motion because it is the cumulative energy arriving at the ground surface. Obviously, the ultimate $E_s$-value which may have something to do with earthquake damage of structures on the surface is dependent not only on the damping ratio $D_1$ but also on the impedance ratio $\alpha$. In Fig. 4.6.6(b), the time-histories of the difference between upward and downward energies $E_u - E_d$ in the base layer are depicted in the similar manner. Here, the time-dependent increase and

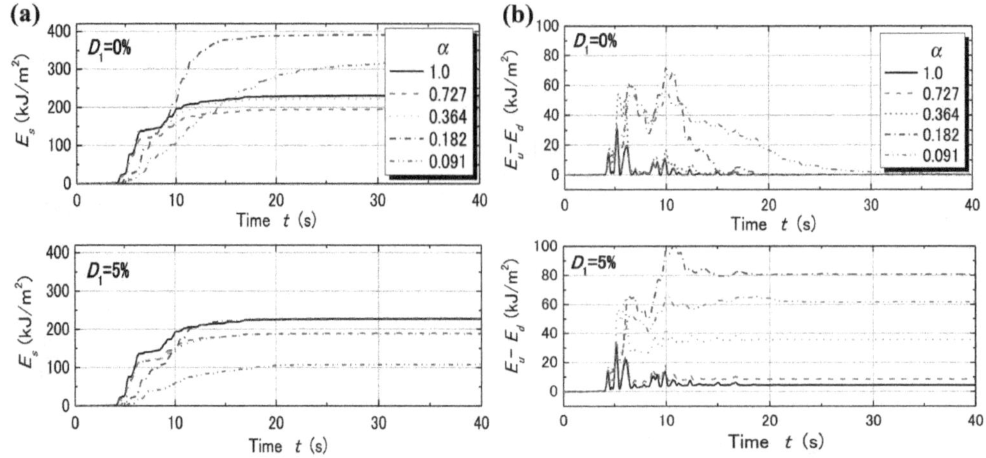


*Figure 4.6.6* Time-histories of irregular wave energies for $D_1 = 0\%$ (top) and 5% (bottom): (a) Upward energy at ground surface $E_s$, (b) Energy difference $E_u - E_d$ in base layer.

decrease are evidently seen, reflecting that the energy temporarily stored in the surface layer eventually returns to the base layer. For $D_1 = 0$, the values $E_u - E_d$ return to zero eventually, indicating that no energy dissipation occurs, while for $D_1 = 5\%$ they tend to converge to certain non-zero values reflecting the energy loss in the surface layer, which is also dependent on the impedance ratio.

Based on the series of similar calculations for different $D_1$-values, the ratios of surface energy $E_s$ to upward energy $E_u$ at the end of the earthquake motion are plotted versus the impedance ratio in Fig. 4.6.7(a). If $D_1 = 0$ or 2.5%, $E_s/E_u > 1.0$ at $\alpha = 0.182$ or nearby indicating that the energy is temporarily stored in the surface layer because the two-layer system with this $\alpha$-value is in/near resonance with the input motion. For $D_1 > 10\%$ however, $E_s/E_u$ tends to be smaller than unity and decrease monotonically with decreasing $\alpha$ (for softer soils). Fig. 4.6.7(b) shows the ratios of dissipated energy $E_w = E_u - E_d$ to $E_u$ at the end of the earthquake motion for different $D_1$-values. Needless to say, $E_w/E_u$ tends to be larger with increasing $D_1$. It takes the maximum at $\alpha = 0.182$ where the two-layer system of this $\alpha$-value is near resonance with the input motion. Thus, the energy storage effect in the surface layer near resonance is cancelled by the increasing dissipated energy $E_w$ with increasing $D_1$.

Apart from the cumulative energy calculated above, the energy flux $dE/dt$ defined in Eq. (1.2.26) is calculated in the surface layer as well as in the base layer, and the ratios of the maximum values $(dE_s/dt)_{\max}/(dE_u/dt)_{\max}$ are plotted versus the impedance ratios $\alpha$ in Fig. 4.6.7(c). For the damping ratio $D < 5\%$, the maximum energy flux takes a peak at an impedance ratio slightly higher than that for $E_s/E_u$ in Fig. 4.6.7(a). However, the ratios of energy flux tend to show similar $\alpha$-dependent variations to those of the cumulative energy in that they tend to monotonically decrease with decreasing $\alpha$ for $D > 10\%$.

Thus, Fig. 4.6.7 indicates that during strong earthquakes when a soft surface layer manifests larger damping value, the seismic wave energy $E_s$ or the maximum energy

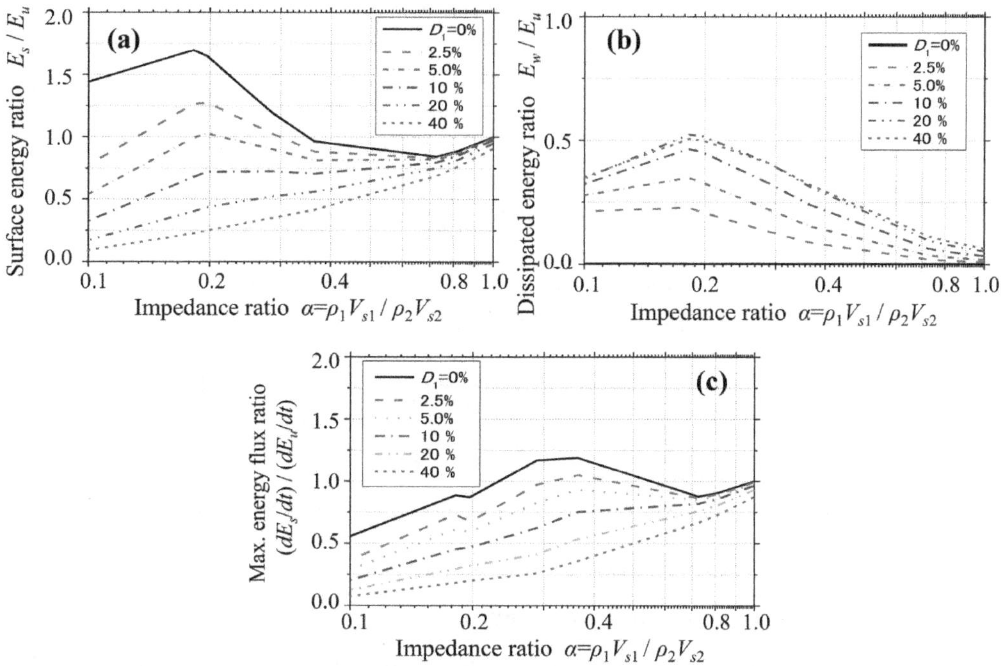


*Figure 4.6.7* Energy ratio in two-layer system for irregular seismic wave versus impedance ratio: (a) Surface energy versus base energy $E_s/E_u$, (b) Dissipated energy versus base energy $E_w/E_u$, (c) Max. surface energy flux versus max. base energy flux $(dE_s/dt)/(dE_u/dt)$ (Kokusho et al. 2007).

flux $(dE_s/dt)_{\max}$ arriving at the ground surface tends to decrease relative to $E_u$ or $(dE_u/dt)_{\max}$ in the base layer with decreasing impedance ratio $\alpha$ or S-wave velocity in the surface layer despite the energy storage effect near resonance, at least for this particular earthquake motion used here. Actually, this trend can be confirmed in a number of vertical array earthquake records as explained in the following.

### 4.6.4 Energy flow calculated by vertical array records

Research on the seismic wave energy for the energy demand is still limited in number, not only due to historical backgrounds that the energy concept was traditionally not so popular as the force-equilibrium in engineering design but also due to difficulties to have subsurface ground motion records. After the 1995 Kobe earthquake, however, a great number of vertical array strong motion observation stations including the KiK-net system were deployed all over Japan, which recorded earthquake motions at downhole depths as well as at the ground surface in the same site for various soil profiles. Here, the subsurface energy flows are calculated utilizing the records during nine strong earthquakes by assuming the vertical propagation of SH waves to know depth-dependent energy demands.

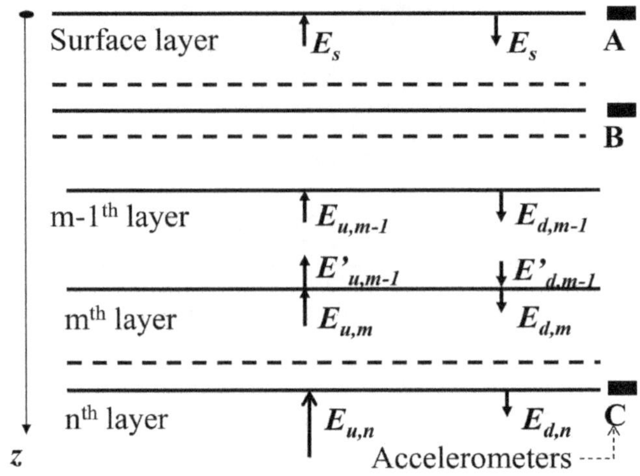


*Figure 4.6.8* Horizontal soil layers with vertical array seismometers (Kokusho and Suzuki 2011).

#### 4.6.4.1 *Energy flow calculation procedure*

The wave energy $E$ and energy flux $dE/dt$ are formulated in Eqs. (4.6.10) and (4.6.11), respectively, where $\dot{u}$ is the particle velocity not directly of observed downhole records but of traveling waves in either upward or downward direction. Therefore, it is essential to separate a measured subsurface motion into the upward and downward waves in order to evaluate the energy flow. In the multiple reflection theory of SH wave, a level ground is idealized by a set of horizontal soil layers as shown in Fig. 4.6.8. Let $E_{u,m}$, $E_{d,m}$ denote the upward and downward energies at the upper boundary of the $m$-th layer and $E_{u,m-1}$, $E_{d,m-1}$ the corresponding energies at the upper boundary of the $(m-1)$-th layer, respectively. Because of the internal damping, the upward and downward energies at the lower boundary of the $(m-1)$-th layer may be different from $E_{u,m-1}$, $E_{d,m-1}$ and denoted here as $E'_{u,m-1}$, $E'_{d,m-1}$. Then, it is easy to understand that the principle of energy conservation holds at the boundary between $m$-th and $(m-1)$-th layer as:

$$E_{u,m} + E'_{d,m-1} = E'_{u,m-1} + E_{d,m} \equiv E_t \tag{4.6.12}$$

If the wave energies are evaluated at the end of a given earthquake motion, the energy $E_t$ in Eq. (4.6.12) means the gross energy passing through the boundary during the earthquake. From this, the following equation is readily derived.

$$E_{u,m} - E_{d,m} = E'_{u,m-1} - E'_{d,m-1} \equiv E_w \tag{4.6.13}$$

Here, $E_w$ stands for the energy dissipated in soil layers above the layer boundary during the earthquake, because all the energy computed here is assumed to transmit vertically.

It is also clear that the dissipated energy $E_w$ can be calculated from $E_u$ and $E_d$ at any intermediate depth as:

$$E_w = E_u - E_d \tag{4.6.14}$$

Based on the multiple reflection theory, the upward and downward SH waves and corresponding wave energies at arbitrary levels can be evaluated from a single record at any level using the free surface boundary condition (e.g. Schnabel et al. 1972). If downhole records are available, however, they will considerably improve the energy flow evaluation which may not fully comply with the simple theory. If seismic records are obtained not only at the ground surface (Point A) but also at two subsurface levels, B and C as illustrated in Fig. 4.6.8 for example, then the energy flow between B and C can be calculated by using earthquake records at the two levels as already mentioned in Sec. 3.2.3. For the energy evaluation between the ground surface (Point A) and Point B on the other hand, two sets of energy flow can be calculated using the earthquake record either at A or B together with the boundary condition at the free ground surface. The two sets of energy are then averaged with the weight of relative proximity to the corresponding points to have the averaged energy flow.

Nine earthquakes (Kobe EQ. and EQ1 to EQ8) used here are with magnitudes from $M_J = 6.4$ to 8.0 and the focal distances from 9 to 227 km. The depths of the vertical arrays from the ground surface to the deepest accelerometer span from 83 m to 260 m, nearly 100 m in most sites, and the S-wave velocities at the base are widely diverged as $V_s = 380-2800$ m/s due to differences in geology (Kokusho 2013a). Four array sites for the Kobe EQ. and one site for EQ7 consist of accelerometers at three or more different levels including the ground surface, while all others belonging to the KiK-net consist of only two levels, surface and base. The scalar sum of the wave energies calculated from the two orthogonal horizontal acceleration records are used for the energy flow evaluations. Equivalent linear soil properties optimized for main shock records were incorporated in the evaluations wherein the damping ratio was assumed as Nonviscous (Kokusho and Suzuki 2011).

#### 4.6.4.2 Energy flow in two vertical array sites

Typical examples of the energy flow are shown below in two sites; (i) Port Island (PI) where multiple accelerometers were installed in soft deposits and strongly nonlinear response during the 1995 Kobe earthquake ($M_J = 7.2$) was recorded near the fault, and (ii) Taiki (TKCH08: KiK-net) in Hokkaido where a pair of accelerometers at soft soil surface and stiff bedrock recorded strong ground motions during the 2003 Tokachi-Oki earthquake (EQ3: $M_J = 8.0$). Table 4.6.1 shows soil profiles and properties in the two vertical array sites.

##### *(i) Port Island (PI)*

In PI, all the soils are Quaternary, and $V_s$ at the deepest level is smaller than 400 m/s as indicated in Table 4.6.1(a) (Kokusho and Motoyama 2002, Kokusho and Suzuki 2011). Extensive liquefaction occurred in the reclaimed soil down to around 15 m from the surface. Main shock records at 3 levels (GL.-0 m, −32.4 m and −83.4 m) were used for the energy evaluation. In the lower two panels of Fig. 4.6.9(a), the

*Table 4.6.1* Soil profiles and properties at two vertical array sites: (a) PI, (b) KiK-net Taiki.

| Layer No. | Depth (m) | Layer thickness (m) | Soil density $\rho$ (t/m$^3$) | Small-strain properties $V_s$ | Small-strain properties D (%) | Main-shock properties $V_s$ (m/s) | Main-shock properties D (%) | Seismometer depth (m) |
|---|---|---|---|---|---|---|---|---|
| (a) Port Island (PI) vertical array (1995 Kobe earthquake) | | | | | | | | |
| 1 | GL-0 | 4 | 1.7 | 170 | 2 | 79 | 40 | A: GL-0 |
| 2 | GL-4.0 | 12.4 | 2 | 210 | 2 | 47 | 42 | |
| 3 | GL-16.4 | 1.1 | 2 | 210 | 2 | 47 | 42 | |
| 4 | GL-17.5 | 11.5 | 1.7 | 180 | 1 | 135 | 30 | |
| 5 | GL-29.0 | 3.4 | 2 | 245 | 1 | 165 | 6.3 | |
| 6 | GL-32.4 | 3.6 | 2 | 245 | 1 | 165 | 6.3 | B: GL-32.4 |
| 7 | GL-36.0 | 13 | 2.2 | 305 | 1 | 245 | 6.3 | |
| 8 | GL-49.0 | 11.5 | 2.2 | 350 | 1 | 282 | 6.3 | |
| 9 | GL-60.5 | 21.5 | 1.8 | 303 | 1 | 253 | 6.3 | |
| 10 | GL-82.0 | 1.4 | 2.2 | 380 | 1 | 328 | 6.3 | |
| 11 | GL-83.4 | Base | 2.2 | 380 | 1 | 329 | 6.3 | C: GL-83.4 |
| (b) KiK-net Taiki vertical array (2003 Tokachi-oki earthquake: EQ3) | | | | | | | | |
| 1 | GL-0 | 4 | 1.8 | 130 | 2.5 | 86 | 6.8 | A: GL-0 |
| 2 | GL-4.0 | 32 | 2.1 | 480 | 2.5 | 398 | 4.8 | |
| 3 | GL-36.0 | 42 | 2.2 | 590 | 1 | 559 | 2.2 | |
| 4 | GL-78.0 | 22 | 2.6 | 2800 | 1 | 2800 | 1 | |
| 5 | GL-100 | Base | 2.6 | 2800 | 1 | 2800 | 1 | B: GL-100 |

particle velocity time histories at the surface (GL.0 m) are shown in two orthogonal horizontal axes (the principal axis with maximum acceleration and normal to that). In the top panel, the energy at the surface $E_s$ as a scalar sum of the two axes (calculated from the velocity time histories and the impedance of the surface layer) is shown. In the lower two panels of Fig. 4.6.9(b), the upward and downward velocity waves at the deepest level (GL.-83.4 m) are shown in the two axes. In the top panel, the time histories of the energies at the deepest level calculated from the velocities are shown. It is noted that the upward and downward energies, $E_u$ and $E_d$, show time-dependent monotonic increase because they are cumulative energies transmitted by one-directionally propagating waves. In contrast, the difference $E_u - E_d$ indicates the energy balance in the soil layers upper than the deepest level and hence shows both increase and decrease with timc.

In Fig. 4.6.10, the corresponding energy fluxes of upward waves $dE_s/dt$ or $dE_u/dt$ calculated in Eq. (4.6.11) are depicted at three depths (GL.-0 m, -32.4 m and -83.4 m) (Kokusho et al. 2007). Different from the cumulative upward energy $E_s$ or $E_u$, the energy fluxes tend to fluctuate significantly having several peaks along the time axis. These peaks appear at the moments for $E_s$ or $E_u$ undergoing steep time-dependent increases in Fig. 4.6.9(a) or (b) and are very much dependent on wave forms of the earthquake motions. Also noted is that the peaks tend to decrease in heights and occur later with decreasing ground depths.

Fig. 4.6.9(c) shows the distributions of the energies in PI, $E_u$, $E_d$, $E_w$ along the depth summed up in the two orthogonal directions. The energies between B and C

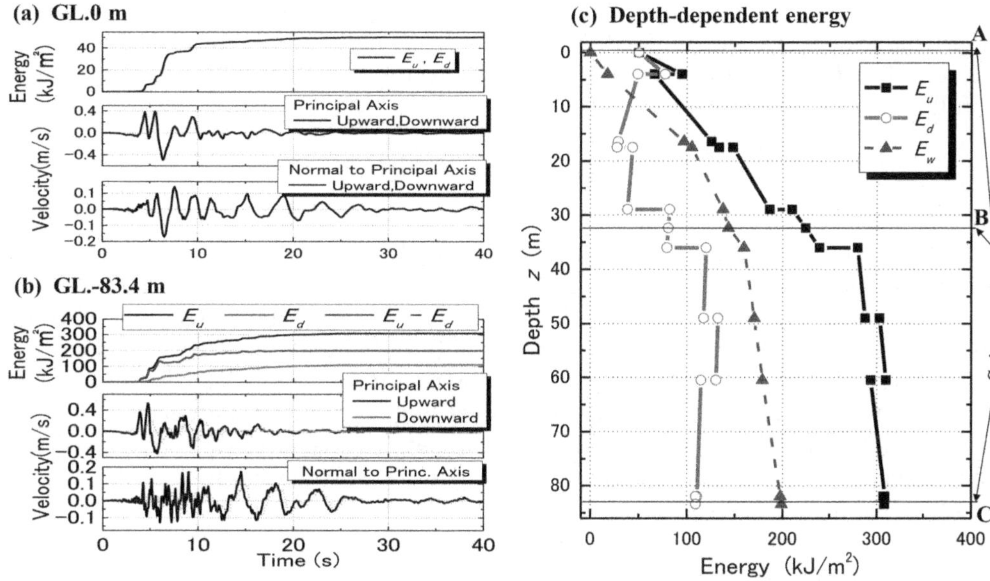


*Figure 4.6.9* Calculation of wave energies in PI: (a) Time-histories of energy and velocity at GL.0 m. (b) The same at GL.-83.4 m, (c) Depth-dependent energies (Kokusho and Suzuki 2011).

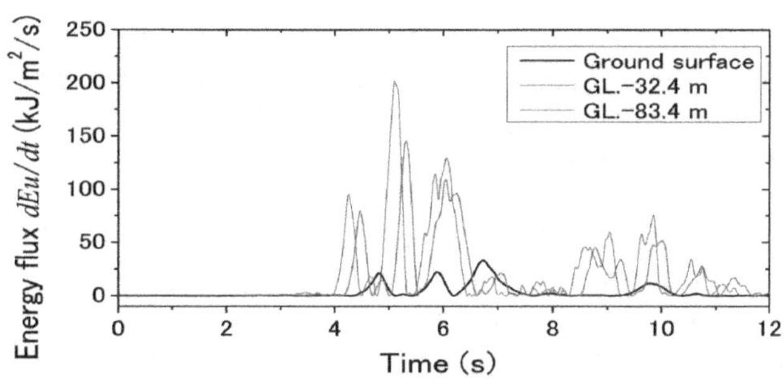


*Figure 4.6.10* Time-dependent changes of energy flux in PI at three depths (Kokusho and Motoyama 2002).

are uniquely determined from the combination of Record B and C (Kokusho and Motoyama 2002, Kokusho and Suzuki 2011). In contrast, either Record A at the surface or Record B is sufficient to calculate the distribution between A and B, because the free surface boundary condition is available there. In PI, where strong soil nonlinearity due to the extensive liquefaction occurred in the surface layer, Record B was exclusively used for the calculation between A and B because the soil properties there were likely to be less influenced by the soil nonlinearity near surface than Record A. Record A was used for computing the energy at the surface A only, which was 50 kJ/m$^2$ in contrast to 86 kJ/m$^2$ calculated from Record B. The energies at B obtained from the combination of Records B and C were $E_u = 236$ kJ/m$^2$ and $E_d = 80$ kJ/m$^2$ whereas those from Record B together with the free surface condition were $E_u = 212\,\text{kJ/m}^2$ and $E_d = 82\,\text{kJ/m}^2$. Though the differences were not large, the energies $E_u$ and $E_d$ at the intermediate depths were averaged. In order to avoid discontinuity near the intermediate point B in $E_w$ calculated by Eq. (4.6.13), the averaging operation was implemented as follows.

$$E_w = \frac{E_{u,m} - E_{d,m}}{2} + \frac{E'_{u,m-1} - E'_{d,m-1}}{2} \tag{4.6.15}$$

Fig. 4.6.9(c) shows an obvious decreasing trend of $E_u$ with decreasing depth particularly in the top 36 m. The downward energy $E_d$ is evidently smaller in the top 36 m than in the deeper part. As a result, the energy $E_w = E_u - E_d$ dissipated cumulatively above a given depth tends to increase considerably with increasing depth. The increasing rate is particularly large from the surface down to around 15 m deep, because of the extensive liquefaction occurred in the reclaimed soil. The energy loss per unit volume in the liquefied soil can be read off from the diagram as $100\,\text{kJ/m}^2/16.4\,\text{m} = 6\,\text{kJ/m}^3$ on average, which is comparable with the dissipated energy density for liquefaction to be addressed in Sec. 5.6.1.

*(ii) Taiki (TKCH08: KiK-net)*

Table 4.6.1(b) shows profiles and soil properties at one of the KiK-net sites, Taiki (TKCH08). Quite different from PI, the bedrock is very stiff ($V_s = 2800$ m/s) at the deepest point (GL.-100 m), while the small-strain $V_s$ in the surface layer is as low as $V_s = 130$ m, which further degraded during the main shock. Main shock records in two horizontal directions at the surface (Record A) and the deepest level at GL.-100 m (Record B) were used for the energy flow calculation (Kokusho and Suzuki 2011).

In the lower two panels of Fig. 4.6.11(a), the velocity time histories at the surface (GL.0 m), calculated from Record A are shown in NS and EW directions, while in the top panel the upward energy at the surface calculated from the velocity time histories at A are shown as the sum in the two orthogonal directions. In Fig. 4.6.11(b), the velocity time histories of upward and downward waves at the deepest level of GL.-100 m calculated from Record B in the two directions and the energy time histories at the same level are shown in the same manner. Both upward and downward energies, $E_u$ and $E_d$, show rapid time-dependent increase with a marginal difference to each other, resulting in a small value of $E_u - E_d$, indicating that energy dissipation in this site is very small, reflecting the very stiff soil condition in the deeper portion.

In Fig. 4.6.11(c), the energy flows along the depth are calculated either from Record A at the surface or from Record B at the base combined with the free surface condition, and plotted with different symbols. The solid thick lines with the close symbols are the averages of the two calculations with the weight of the proximity to the levels A and B. The two energy flows calculated are very similar to each other, indicating that the soil model is a good reproduction of the actual ground at this particular site. Thus, the averaging procedure tends to modify the depth-dependent energy variations to a certain degree, though the energy values at the base and at the surface are unaffected by this procedure.

In this site, despite almost the same upward energy as in PI, more than $300\,\text{kJ/m}^2$ at the deepest level, less than $100\,\text{kJ/m}^2$ passed through the boundary (GL.-78 m)

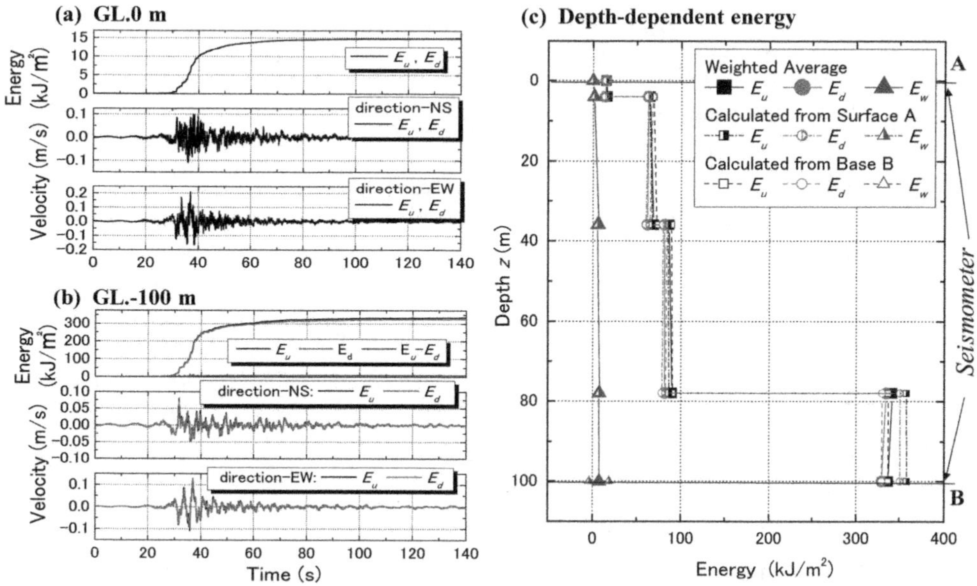


*Figure 4.6.11* Calculation of wave energies at KiK-net Taiki: (a) Time-histories of energy and velocity at GL.-0 m, (b) The same at GL.-100 m, (c) Depth-dependent energies (Kokusho and Suzuki 2011).

of the drastic impedance change and only 15 kJ/m$^2$ reached the ground surface. A small difference between $E_u$ and $E_d$ indicates that the considerable upward energy was reflected at the intermediate boundaries and returned to the deeper ground as the downward energy, before arriving at the soft soil layer near the surface. This also means that the dissipated energy $E_w$ could not be large because the most energy transmitted only in the deep and stiff layers with small energy loss.

#### 4.6.4.3 General trends of energy flow observed in vertical arrays

Fig. 4.6.12 depicts the variations of upward energy $E_u$ along the depth $z$ calculated for 9 earthquakes at 30 vertical array sites. On account of large differences in the energies depending on the individual earthquakes, the horizontal axis is taken in logarithm. Like PI and Taiki explained above in detail, the upward energies show obvious decreasing trends in most sites with decreasing depth irrespective of the differences in the absolute upward energy. In some sites, the $E_u$-value decreases to less than 1/10 from the base to the surface. The decreasing trend is more pronounced near the surface in contrast to that below the depth of 50 m–100 m.

There has been a traditional view employed not only in seismology (Gutenberg and Richter 1942) but also in earthquake engineering (Joyner and Fumal 1984) that the wave energy (= square of velocity amplitude × wave impedance) is kept constant as the seismic wave propagates underground. Hence the velocity amplitude has normally been considered inversely proportional to the square root of the impedance $\rho V_s$. Fig. 4.6.12 indicates that this traditional view may not hold at least in the ground of

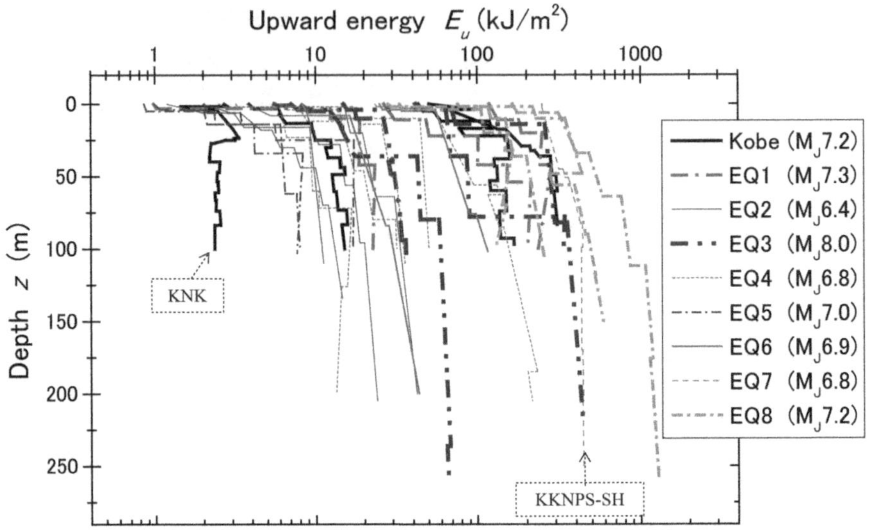


*Figure 4.6.12* Variations of upward energy $E_u$ along depth calculated for 9 earthquakes at 30 vertical array sites (Kokusho and Suzuki 2011).

shallow depths, though there are exceptional sites, KKNPS-SH and KNK indicated in the chart, where the energy tends to be almost constant or decreasing mildly up to the ground surface.

In order to examine how the decreasing trend is influenced by site conditions including those exceptional sites, Fig. 4.6.13(a) shows the ratios of upward energies between surface and base ($E_s/E_u$) plotted versus the inverse impedance ratios $(\rho V_s)_{\text{base}}/(\rho V_s)_{\text{surf.}}$ (Kokusho and Suzuki 2011). From the data points located in between the two dashed curves, it may be recognized that the energy ratio tends to decrease with the increasing inverse impedance ratio despite the data dispersions. The data point for KKNPS-SH may be explained in this way though KNK is far from the trend. Also noted is that, out of 30 sites, the energy ratio $E_s/E_u > 0.3$ holds in only 4 sites, among which $E_s/E_u$ is exceptionally large ($E_s/E_u > 0.8$) in KKNPS-SH and KNK. In all the other sites, only less than 10% to 30% of the upward energy at the deepest level arrived at the surface.

It may well be expected that not only the impedance ratio as mentioned above but also the damping ratios of the individual sites may affect the energy ratio $E_s/E_u$. Hence, the plots are differentiated with 4 symbols in Fig. 4.6.13(a) according to 4 steps of $D_{MA}$, average damping ratios modified for the ground thickness 100 m (Kokusho and Suzuki 2011). This differentiation however does not seem to highlight the effect of damping as obviously as that of the impedance ratio.

In Fig. 4.6.13(b), the ratios of the dissipated energy to the upward energy $E_w/E_u$ calculated at the deepest levels of the vertical arrays are taken in the vertical axis versus the modified average damping ratios $D_{MA}$ in the horizontal axis. It reveals that, in most sites, the dissipated energy is less than 30% to 40% of the upward energy. Despite the large data scatters, the plots are in between the pair of dashed curves shown in the diagram, indicating that the energy ratio $E_w/E_u$ tends to increase with increasing

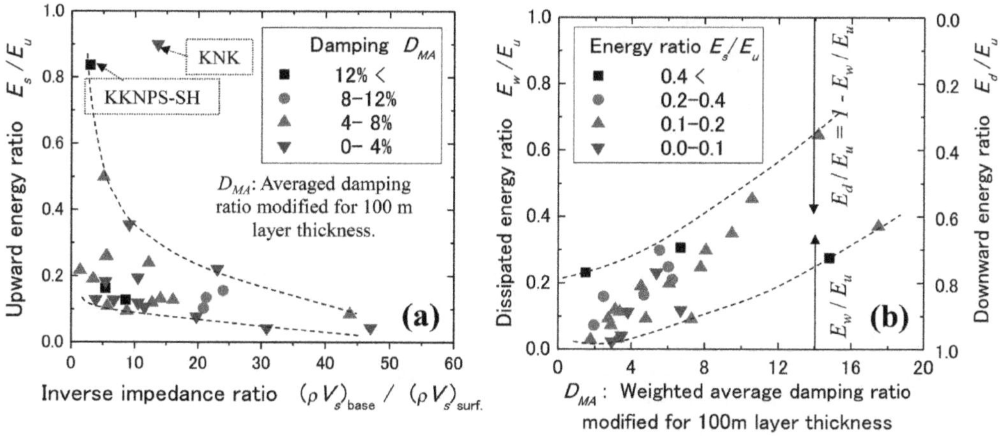


*Figure 4.6.13* Upward energy ratio between surface and base ($E_s/E_u$) plotted versus inverse impedance ratios $(\rho V_s)_{base}/(\rho V_s)_{surface}$ for all sites (a), and ratios of dissipated energy to upward energy at deepest level $E_w/E_u$ versus modified average damping ratios $D_{MA}$ for all sites (Kokusho and Suzuki 2011).

damping ratio $D_{MA}$. The plots are classified into 4 steps of the upward energy ratio $E_s/E_u$ with different symbols, revealing that the plots with higher $E_s/E_u$ tend to be at higher positions in the diagram compared to the others. This suggests that the more the energy reaches the ground surface, the larger energy loss can occur presumably in the shallow ground.

It is readily understood from Eq. (4.6.14) that the downward energy normalized by the upward energy, $E_d/E_u$, is expressed as $E_d/E_u = 1 - E_w/E_u$, meaning that the vertical axis in Fig. 4.6.13(b) can also represent $E_d/E_u$ in the opposite direction as shown in the right side of the diagram. Hence, it can be said that, in most sites during strong earthquakes, more than 60 to 70% of the upward energy at the deepest level goes back to the deeper ground without being dissipated in the upper soil layers. This further indicates that the major mechanism to make the surface energy ratio as low as $E_s/E_u < 0.1$–$0.3$ in most sites is not the soil damping absorbing the upward energy, because the energy ratio $E_w/E_u < 0.3$–$0.4$ is not large enough to account for the low $E_s/E_u$-value. Instead, the drastic energy decrease at the surface relative to the base should be largely attributed to wave reflections that occur at intermediate layer boundaries interrupting the energy to go up more or less.

#### 4.6.4.4 *Correlation of upward energy ratio with impedance ratio*

As already shown, the upward energy tends to decrease considerably in most sites as it goes up from the base (about 100 m deep) to the ground surface. In order to evaluate how the upward wave energy tends to decrease as it approaches to the ground surface, an empirical formula was developed (Kokusho and Suzuki 2012), wherein the ratios of upward energies between layers are correlated to the corresponding impedance ratios using the same data set of the vertical array records mentioned above. Out of the depth-dependent upward energy variations at 30 sites, 24 sites have been used, wherein the difference in the upward energies at the deepest level calculated from the

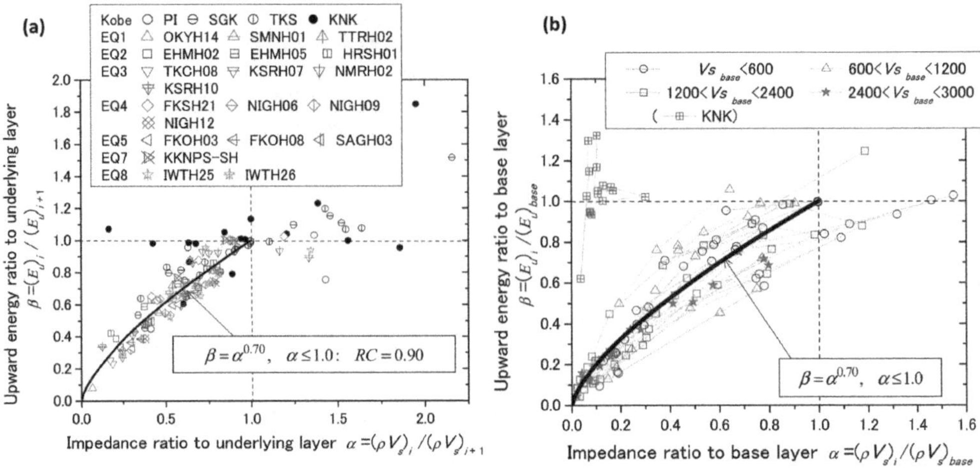


*Figure 4.6.14* Impedance ratios $\alpha$ versus corresponding upward energy ratios $\beta$ compared with empirical formula: (a) Between neighboring layers, (b) Between given layer and base layer (Kokusho and Suzuki 2012).

two measured motions, one at the ground surface and the other at the deepest level, were within about 25%. The impedance ratio $\alpha$ and the ratio of upward energy $\beta$ defined in the following equations between two neighboring layers, $i$ (upper) and $i+1$ (lower) were calculated for $i=1$ to $n-1$ from surface to base.

$$\alpha=\frac{(\rho V_s)_i}{(\rho V_s)_{i+1}},\quad \beta=\frac{(E_u)_i}{(E_u)_{i+1}} \tag{4.6.16}$$

The soil density $\rho$ was assumed depending on the S-wave velocity as; 1.6–2.0 t/m$^3$ for $V_s\leq300$ m/s, 2.0–2.2 t/m$^3$ for 300 m/s $\leq V_s\leq$ 700 m/s, 2.3–2.4 t/m$^3$ for 700 m/s $\leq V_s\leq$ 1000 m/s, and 2.5–2.7 t/m$^3$ for 1000 m/s $\leq V_s<$ 3000 m/s. The energy ratios $\beta$ are plotted versus the corresponding impedance ratios $\alpha$ in Fig. 4.6.14(a) for all layers above the deepest levels in the 24 vertical array sites with different symbols. For the majority of the data points $\alpha\leq1.0$, because the impedance ratio is normally less than unity ($\rho V_s$ is getting larger in deeper layers). For $0\leq\alpha\leq1.0$, it is quite reasonable to assume that $\beta=0$ for $\alpha=0$, and $\beta=1$ for $\alpha=1$ (a uniform ground). Hence, a simple power function $\beta=\alpha^n$ may be practical to approximate the plots (the KNK plots evidently biased from others probably due to some site-specific problem are omitted), and the power $n=0.70$ is obtained from the least mean-square method with the determination coefficient $R^2=0.81$.

$$\beta=\alpha^{0.70},\quad 0\leq\alpha\leq1.0 \tag{4.6.17}$$

Thus, Eq. (4.6.17), shown in Fig. 4.6.14(a) with the thick solid curve, represents the data points fairly well up to $\alpha=1.0$.

Next, the impedance ratio $\alpha$ and the upward energy ratio $\beta$ are redefined, different from Eq. (4.6.16), between an arbitrary layer $i$ ($i = 1$ to $n-1$) and the deepest layer (base layer) in the vertical arrays as follows.

$$\alpha = \frac{(\rho V_s)_i}{(\rho V_s)_{\text{base}}}, \quad \beta = \frac{(E_u)_i}{(E_u)_{\text{base}}} \tag{4.6.18}$$

Here, $(\rho V_s)_{\text{base}}$ and $(E_u)_{\text{base}}$ are the wave impedance and the upward energy in the base layer, respectively. In Fig. 4.6.14(b), the data points for all the layers at the 24 vertical array sites are plotted on the $\alpha$–$\beta$ diagram, wherein the symbols are connected with dashed lines for individual sites and differentiated according to 4 classes of $V_s$-values at the base layers. Although the plots are more dispersed than those in Fig. 4.6.14(a), the curve by the same Eq. (4.6.17), using $\alpha$ and $\beta$ redefined in Eq. (4.6.18) and superposed, seems to averagely represent the plots. This indicates that it may be possible from a practical point of view to use Eq. (4.6.17) in order to evaluate the upward energies in shallow soil layers from the upward energy at a base layer by considering the impedance ratios between the two corresponding layers. Also noted here is that, out of the 4 classes of $V_s$-values at the base layers, the plots corresponding to 2400 m/s $< V_s <$ 3000 m/s show good fitting with the empirical curve, indicating the applicability of this equation even to base layers as stiff as the seismological bedrocks.

#### 4.6.4.5 Upward energy at the deepest level of vertical array

The upward energies at the deepest levels (base layers) $(E_u)_{\text{base}}$ calculated from the upward waves (obtained by the multi-reflection analyses using the observed downhole records) and the impedance-values there are plotted versus hypocentral distances $R$ in Fig. 4.6.15 at all the vertical array sites with various close symbols corresponding to the 9 earthquakes. Though the plots are widely dispersed, the decreasing trends in $(E_u)_{\text{base}}$ with increasing $R$ for individual earthquakes is recognizable. Among the 9 earthquakes, the plots of EQ3 with $M_J = 8.0$ (2003 Tokachi-Oki earthquake) are reasonably located relatively higher on the right side of the diagram, while others with $M_J$ around 7 are lower on the left side.

The straight dashed lines drawn in the chart represent the following formula between incident energies $E_{IP}$ versus hypocentral distances $R$ calculated for individual earthquake magnitudes by assuming the spherical energy radiation of the body waves from the center of energy release (e.g. Sarma 1971, Davis and Berrill 1982).

$$E_{IP} = \frac{E_{\text{Total}}}{4\pi R^2} \tag{4.6.19}$$

where $E_{IP}$ is in kJ/m$^2$, $R$ in m, and $E_{\text{Total}}$, the total energy in kJ assumed to radiate from the hypocenter, is calculated as:

$$\log E_{\text{Total}} = 1.5M + 1.8 \tag{4.6.20}$$

The earthquake magnitude $M$ is Surface Wave Magnitude $M_S$ originally by Gutenberg (1956), but Japanese Magnitude $M_J$ is used here because the two magnitudes are not so different.

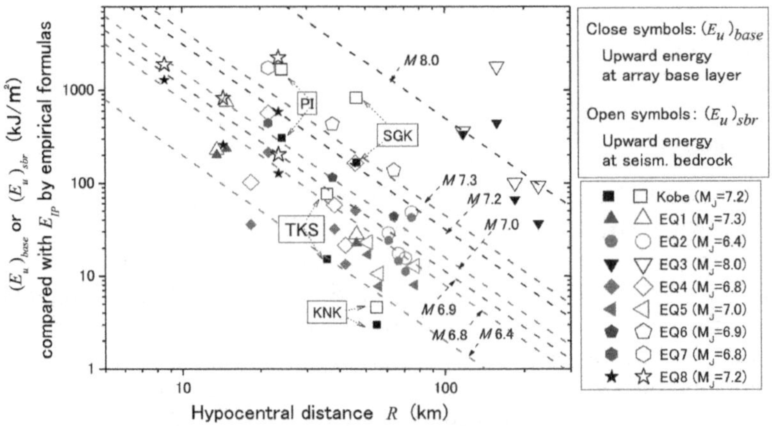


*Figure 4.6.15* Upward energies at array base layer $(E_u)_{base}$ or seismological bedrock $(E_u)_{sbr}$ versus hypocenteral distance $R$ compared with incident energies $E_u$ versus $R$ lines by simple formulas.

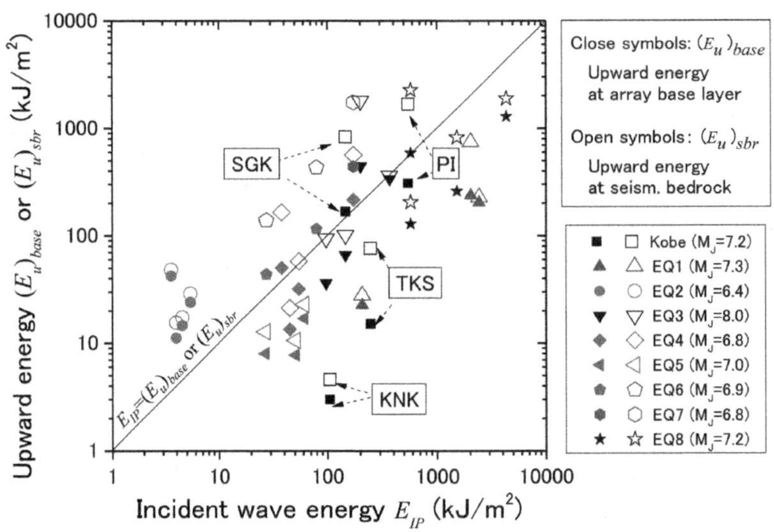


*Figure 4.6.16* Upward energies at seismological bedrock $(E_u)_{sbr}$ or at array base $(E_u)_{base}$ versus incident energy $E_{IP}$ by empirical formulas.

In Fig. 4.6.16, the energies $(E_u)_{\text{base}}$ at individual sites are directly plotted with solid symbols in the vertical axis versus the incident energies per unit area $E_{IP}$ calculated by Eqs. (4.6.19) and (4.6.20) for corresponding earthquake Magnitudes and hypocenteral distances in the horizontal axis on the log-log diagram. The majority of the solid symbols are located near the diagonal line of $E_{IP} = (E_u)_{\text{base}}$ or lower. In contrast, the plots for EQ2 (2001 Geiyo earthquake) are all high above the diagonal line. This may somehow reflect the path effects by this particular intra-plate earthquake occurred 46 km deep, while other earthquakes except EQ3 (a plate boundary subduction earthquake) are all crustal earthquakes.

Based on the finding that Eq. (4.6.17) may be useful to roughly evaluate the energy ratio between two layers including stiff rocks, the same equation is further applied to estimate the upward energies at the seismological bedrock $(E_u)_{sbr}$ from those at the base layers $(E_u)_{base}$ of the vertical arrays. The impedance for the seismological bedrock is determined here by assuming $V_s = 3000$ m/s and $\rho = 2.7$ t/m$^3$. In Figs. 4.6.15, 4.6.16, the energies $(E_u)_{sbr}$ thus calculated are superposed with corresponding open symbols for the 9 earthquakes. Though the estimated energies at the seismological bedrock $(E_u)_{sbr}$ are widely dispersed again, the decreasing trends of the energies with increasing $R$ for individual earthquakes are still recognizable in Figs. 4.6.15. It may be pointed out in Figs. 4.6.16 that $(E_u)_{sbr}$ tends to have better matching than $(E_u)_{base}$ with $E_{IP}$ as a whole though individually the matching becomes better in some sites and worse in other sites.

It may well be expected that the simple formulas Eqs. (4.6.19) and (4.6.20) cannot make good energy predictions because they completely neglect pertinent fault parameters, such as fault type, dimension, directivity, asperity, etc. For example, the observed energy in PI and SGK-site during the 1995 Kobe earthquake in Fig. 4.6.16 overshoot the calculated $E_{IP}$-values by 3–4 times. Forward directivity effect (Somerville 1996) may possibly have been involved to make the energy in these sites extraordinary large in contrast to TKS and KNK during the same earthquake. More detailed study will be needed to incorporate fault and path mechanisms of individual earthquakes. In the meantime, however, the simple formulas Eqs. (4.6.19) and (4.6.20) may be employed to estimate the incident energy for engineering purposes, though very crudely.

### 4.6.5 Design considerations in view of energy

Energy-based design considering earthquake wave energy has rarely been employed in engineering practice compared to force-based design using acceleration or seismic coefficients. Nevertheless, the energy concept seems to have a great advantage over the force-equilibrium concept in view of a great applicability of energy capacity to failures in engineering materials such as soil liquefaction under irregular seismic loading. In a straight forward energy-based design concept, the energy capacity of a structure may be compared directly with the energy demand provided by a given earthquake motion. This allows a designer to roughly capture the safety allowance of a structure against a given earthquake motion before implementing the detailed analysis. Hence, in developing the energy-based design, it is preferred that not only the energy capacity but also the energy demand be discussed from a viewpoint of structural performance. In the following, how the earthquake energy demand is correlated with structural behavior and site-dependent earthquake damage are discussed.

#### 4.6.5.1 Energy-based structure design

Obviously, the degree of structural damage is determined by induced strains in supporting members of superstructures relative to their threshold yield or ultimate strains. In the performance-based design framework, the structural performance and structural damage during strong earthquakes are evaluated in terms of induced strains or

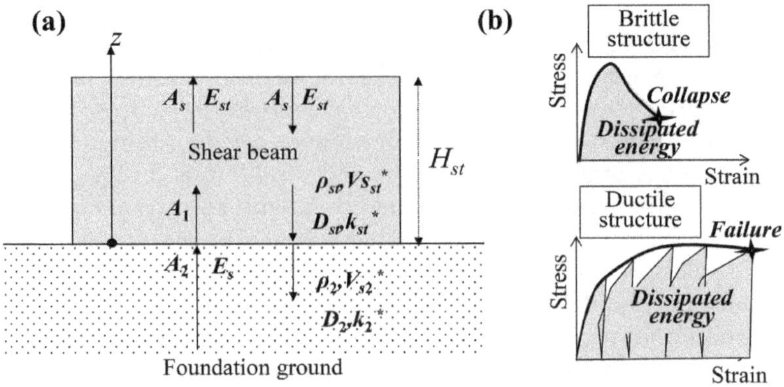


*Figure 4.6.17* Shear vibration structure on foundation ground approximated by 2-layer system (a) and Stress-strain relationships and dissipated energies for brittle and ductile structures (b).

deformation levels. Consequently, it may be meaningful to revisit to a basic relationship between the induced strain and seismic wave energy supplied to it from foundation ground by a simplified model.

If a superstructure is idealized as a shear vibration system of a large lateral dimension resting on uniform foundation ground, then the soil-structure interaction may be approximated by a 2-layer soil system as depicted in Fig. 4.6.17 which was already discussed in Sec. 4.2.2. Here, the notations have to be changed as $H \to H_{st}$, $\rho_1 \to \rho_{st}$, $V_{s1}^* \to V_{S_{st}}^*$, $\alpha^* \to \alpha_{st}^*$, $k_1^* \to k_{st}^*$, $E_s \to E_{st}$, $E_u \to E_s$, according to the replacement of the surface layer in Fig. 4.6.3(a) by the superstructure in Fig. 4.6.17(a), wherein $H_{st}$ = height of structure, $\rho_{st}$ = equivalent density of structure, $V_{S_{st}}^*$ = complex equivalent S-wave velocity of structure, $\alpha_{st}^*$ = complex impedance ratio, $k_{st}^*$ = complex wave number of structure, $E_s$ = upward wave energy at ground surface, $E_{st}$ = upward wave energy in superstructure. Then, the shear strain $\gamma$ in the superstructure is given by differentiating the displacement $u_1$ in Eq. (4.2.30) in terms of $z$ as:

$$\gamma = \frac{\partial u_1}{\partial z} = 2k_{st}^* A_s \sin k_{st}^*(H_{st} - z)e^{i(\omega t - k_{st}^* H_{st})} \tag{4.6.21}$$

The amplitude $A_s$ in this equation can be converted to $A_2$ by using Eq. (4.2.32). On the other hand, the energy flux at the foundation ground in Fig. 4.6.17(a), $dE_s/dt$, is correlated with the upward wave amplitude $A_2$ at the ground surface by using Eq. (4.6.11) as:

$$|A_2| = \left| \frac{(dE_s/dt)}{\omega^2 \rho_2 V_{s2}^*} \right|^{0.5} \tag{4.6.22}$$

Hence, the absolute shear strain of the superstructure in Eq. (4.6.21) can be expressed by the energy flux at the ground surface $dE_s/dt$ (Kokusho et al. 2007) as:

$$|\gamma| = \left| \frac{4 \sin k_{st}^*(H_{st} - z)}{(1 + \alpha_{st}^*)e^{ik_{st}^* H_{st}} + (1 - \alpha_{st}^*)e^{-ik_{st}^* H_{st}}} \right| \left| \frac{\alpha_{st}^*(dE_s/dt)}{\rho_{st} V_{S_{st}}^{*3}} \right|^{1/2} \tag{4.6.23}$$

The first absolute value on the right-hand side indicates the contribution of resonance effect on the induced strain of structure. In reality, superstructures are not so simple as uniform shear-vibration systems but more like complicated mass-spring systems with limited lateral dimensions and vibrate in shear-bending modes. However, it may be possible to find equivalent parameters for the idealization wherein Eq. (4.6.23) holds basically.

Under strong earthquake motions, the structure exceeds a certain yield strain and dissipates plastic strain energy. For brittle structures with small ductility, such as concrete buildings with insufficient reinforcements, masonry or brick buildings, the maximum shear strain or associated dissipated energy induced in a single loading cycle is decisive for the collapse of the structure as schematically illustrated at the top of Fig. 4.6.17(b). Consequently, the supplied energy flux $dE_s/dt$ in that cycle has to be compared with the dissipated energy as indicated in Eq. (4.6.23) for the performance based design. In contrast, for structures with higher ductility factors such as embankments, dams and soil retaining structures, the cumulative strain or associated dissipated strain energy by repeated loadings as schematically illustrated at the bottom of Fig. 4.6.17(b) is crucial for structural performance wherein the cumulative dissipated energy by cyclic loading is compared with the wave energy demand $E_s$ during a given earthquake motion. Thus, the performance of brittle structures with low damping ratios is very much dependent on the wave forms of individual earthquake motions or the energy flux, while that of ductile structures with high damping is not so much dependent on the wave forms but on the cumulative wave energies.

Based on the above considerations, it may be said that the wave energy or energy flux, the degree of resonance, and the impedance ratio between the structure and the ground are the three key factors influencing the induced strain or degree of damage in a given structure having a prescribed stiffness represented by $V^*_{s_{st}}$. Considering that the seismic wave energy is relatively small at the ground surface in soft soil sites as already demonstrated, it is unlikely that the structural damage by seismic shaking is always higher in soft soil sites than in stiff soil sites as generally perceived from old times. Instead, the site-dependent structural performance seems to be different from one earthquake to another depending on the wave energy and other seismic, structural and geotechnical factors including the degree of resonance (Kokusho et al. 2007).

#### 4.6.5.2 Earthquake damage versus upward wave energy

As already indicated, the seismic wave energy at ground surface tends to be smaller in soft soil sites than in stiff soil sites because, unlike the wave amplitude, it always decreases due to wave reflections in passing through layer boundaries and also because surface soft soils cannot store so much energy due to high energy dissipation during resonance. This finding may not be compatible with a widely accepted perception from old times that soft soil sites tend to undergo heavier earthquake damage than stiff soil sites.

During the 1923 Kanto earthquake, in Japan, a great number of wooden houses collapsed in down-town soft soil areas than Pleistocene stiff soil areas in Tokyo and triggered big fires killing about 100 thousand people. The same trend seems to hold in the 1987 Loma Prieta earthquake, when major damage of wooden houses and life

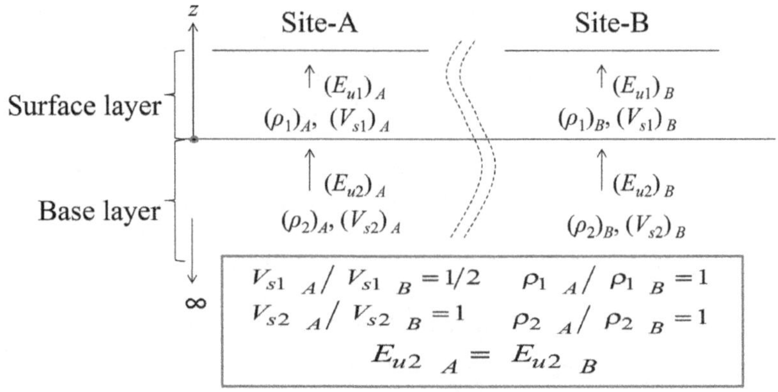


*Figure 4.6.18* Conceptual comparison of upward energy at two sites A and B by using two-layer model.

lines was concentrated in soft soil areas along the San Francisco Bay. However, during 1995 Kobe earthquake in Japan on the contrary, buildings and civil engineering superstructures were heavily damaged in areas of relatively competent soils. In contrast, structural damage directly due to seismic inertial effect was not so serious in soft soil areas along seashore where geotechnical damage was prevalent due to liquefaction, less than a few kilometers apart from the heavily damaged areas (e.g. Matsui and Oda 1996, Tokimatsu et al. 1996).

In discussing earthquake damage, one must be careful if it is structural damage by shaking effect such as failures of structural members or by geotechnical damage of foundations or bearing soils which may also deteriorate superstructures. With regard to the structural shaking damage, earthquake-induced strain in superstructures depends as indicated in Eq. (4.6.23) on the wave energy flux at the foundation ground, the degree of resonance, the impedance ratio between structure and ground, and of course the structural stiffness. In this regard, the wave energy or energy flux at the ground surface tends to be lower in softer soils as already mentioned, which seems to be inconsistent with the general perception that the earthquake damage becomes greater in soft soil sites.

In order to deal with this problem, let us classify the earthquake-related damage into the structural damage due to seismic shaking and the geotechnical damage caused by earthquake-induced soil deformations. With regard to the geotechnical damage, let us compare two different site conditions A and B shown in Fig. 4.6.18 by using the two-layer model (the impedance of the surface and base layers, $\rho_1 V_{s1}$ and $\rho_2 V_{s2}$, respectively). The two sites are of almost the same condition except that $V_s$ of the surface layer in Site-A is a half of that in Site-B. As indicated in Eq. (1.2.19), shear strain in the surface layer is given as $\gamma = -\dot{u}/V_{s1}$ using $\dot{u}$ = particle velocity, and hence the upward wave energy in the surface layer $E_{u1}$ can be written using Eq. (1.2.25) as:

$$E_{u1} = \rho_1 V_{s1} \int (\dot{u})^2 dt = \rho_1 V_{s1}^3 \int \gamma^2 dt \tag{4.6.24}$$

or it is modified as:

$$\int \gamma^2 dt = \frac{E_{u1}}{\rho_1 V_{s1}^3} \tag{4.6.25}$$

The term on the left in Eq. (4.6.25) can be regarded as a parameter of cumulative squared strain in terms of time. This parameter seems to represent geotechnical damage because, for soils behaving as ductile materials, the failure is determined not by a single strain amplitude but by some sort of cumulative strain parameter such as in Eq. (4.6.25).

If Site-A and B is compared in Fig. 4.6.18, assuming $(V_{s1})_A/(V_{s1})_B = 1/2$, $(V_{s2})_A/(V_{s2})_B = 1$, $(\rho_1)_A/(\rho_1)_B = 1$, $(\rho_2)_A/(\rho_2)_B = 1$, and also the upward energy in the base layer is identical $(E_{u2})_A = (E_{u2})_B$, the impedance ratios between surface and base $\alpha = \rho_1 V_{s1}/\rho_2 Vs_2$, $\alpha_A$ at Site-A and $\alpha_B$ at Site-B, are correlated as $\alpha_A/\alpha_B = 1/2$. Using the empirical equation $\beta = \alpha^{0.70}$ in Eq. (4.6.17) on the upward energy ratio $\beta$ and the corresponding impedance ratio $\alpha$ derived from a number of vertical array records, the following can be obtained.

$$\frac{\beta_A}{\beta_B} = \frac{(E_{u1}/E_{u2})_A}{(E_{u1}/E_{u2})_B} = \frac{(E_{u1})_A}{(E_{u1})_B} = \left(\frac{\alpha_A}{\alpha_B}\right)^{0.70} = \left(\frac{1}{2}\right)^{0.70} = 0.62 \tag{4.6.26}$$

Thus, the upward energy in the surface layer in Site-A becomes smaller, 62% of Site-B, whereas the cumulative strain parameter defined in Eq. (4.6.25) is 5 times larger in Site-A than Site-B as follows.

$$\frac{\left(\int \gamma^2 dt\right)_A}{\left(\int \gamma^2 dt\right)_B} = \frac{[(E_{u1})_A/(E_{u1})_B]}{[(\rho_1 V_{s1}^3)_A/(\rho_1 V_{s1}^3)_B]} = \frac{0.62}{(1/2)^3} = 5.0 \tag{4.6.27}$$

This indicates that, although the upward energy at the ground surface $E_u$ is smaller in soft soil sites than in stiff soil sites, the cumulative soil strain can be larger. This seems to be compatible with the generally accepted perception at least for geotechnical damage or geotechnically-induced structural damage that softer soil sites tend to suffer heavier earthquake damage.

On structural damage due to seismic shaking, the effect of resonance in low-damping structures has to be properly accounted for in order to discuss if soft soil sites are more prone to earthquake damage than stiff soil sites. Also noted here is that there are quite a few reports published recently denying a generally increasing trend of structural damage in soft soil sites, wherein structural damage by seismic inertial effect tends to decrease with increasing geotechnical damage in soft soil sites (Suetomi and Yoshida 1998, Trifunac and Todorovska 2004, Bakir et al. 2005). More investigations are certainly needed to understand this important topic properly, wherein the important first step is to carefully classify case histories of earthquake-induced damage if they are actually caused by structural shaking or geotechnical effects.

## 4.7 SUMMARY

1 Various parameters govern site amplification during strong earthquakes between ground surface and engineering bedrock. Besides the earthquake-related parameters, such as wave amplitudes, frequency contents and durations, which are strongly dependent on earthquake magnitudes and fault distances, the site-related parameters, such as S-wave velocity profiles, soil internal damping, strain-dependent nonlinear soil properties and local geological/topographical profiles are important. Among them, the amplification is firstly dependent on the S-wave impedance ratio between the surface layer and the engineering bedrock.

2 In defining site amplifications during earthquakes, two earthquake observation systems are available; (i) horizontal array observation on different surface geologies at nearby sites and (ii) vertical array observation in the same site at ground surface as well as downhole depths of different geologies. In order to make use of the vertical array data in microzonation studies, the downhole records have to be modified so that they are free from downward waves from overburden.

3 H/V-spectrum ratios (Horizontal motion divided by Vertical motion) at a single surface point for not only earthquake observations but also microtremor measurements may be able to substitute the horizontal array in order to predict site-specific seismic response, predominant frequency in particular.

4 A two-layer model consisting of a surface soft soil layer underlain by a stiff base layer can capture major dynamic response characteristics of actual soft soil sites underlain by engineering bedrock despite the simplification, as exemplified in the quarter wave-length formula which is applicable to realistic soil conditions to calculate predominant frequencies.

5 Transfer functions between ground surface and base for the vertical arrays are determined with no regard to the properties of the base layer, while those for the horizontal arrays are strongly influenced by the properties and the impedance ratio at the base boundary. This makes the amplification values in the former much higher than in the latter particularly under smaller damping ratios in the surface layers as actually confirmed in observed spectrum amplifications.

6 The Nonviscous damping model normally employed in soil dynamics calculates site amplifications between surface and base monotonically decreasing from lower to higher order peaks, while the Maxwell damping model gives constant amplifications for all the peaks. Observed spectrum site amplifications during strong earthquakes in soft soil sites tend to be similar to the former, while those during weak earthquakes in relatively stiff soil sites are similar to the latter. This is because frequency-independent frictional soil damping becomes dominant due to induced large strains in the former, while wave scattering in heterogeneous deposits can play a major role under small induced soil strains in stiff soils.

7 The site amplification is considerably different in horizontal and vertical motions because the S and P-wave velocities associated with them, respectively, are quite different. Furthermore, the amplifications for the horizontal and vertical motions tend to change quite differently from weak to strong earthquakes particularly in saturated liquefiable soils, because the S-wave velocity is sensitive to induced strains and pore-pressure buildup while the P-wave velocity is not.

8 In order to evaluate relative site amplifications in microzonation studies, average S-wave velocities of top 30 m of surface soils $V_{s30}$ are often used. Site amplification database during recent strong earthquakes indicate that the $V_{s30}$-values show a positive but poor correlation with spectrum peak amplifications between surface and base. Average velocities $\overline{V}_s$ in equivalent surface layers generating spectrum amplification peaks are correlated much better with the peak amplifications and recommended to be employed for microzonation studies in place of $V_{s30}$, wherein $\overline{V}_s$ may be readily estimated by microtremor measurement and soft soil thickness.

9 Observed peak values of spectrum amplifications in horizontal arrays do not change significantly due to soil nonlinearity between strong and weak earthquakes compared to those in vertical arrays, though the associated peak frequencies become lower during strong earthquakes. This observation by a number of site response records can readily be accounted for theoretically by a simplified two-layer model considering strain-dependent equivalent linear soil properties and radiation damping.

10 Spectrum amplifications in vertical arrays may sometimes differ significantly from those in horizontal arrays in peak frequencies and spectrum shapes, because the installation depths of downhole seismometers were not appropriately chosen relative to the layer boundaries of major impedance contrast. Hence, care is needed how to choose the appropriate depth according to site-specific soil profiles in deploying the downhole seismometer.

11 The SSI (Soil-structure-interaction) effect evaluated for a simplified shear-vibration structure resting on semi-infinite foundation ground is very frequency-dependent and particularly dominant near resonance for heavy and rigid structures with high resonant frequency but almost negligible for light-weight wooden houses. The off-resonance SSI-effect tends to be greater with increasing internal structural damping as the structures approach to failures. The damping ratio of a lumped-mass model resting on rigid ground should be increased considering the SSI effect in proportion to the impedance ratio between the structure and foundation ground in addition to the internal structural damping.

12 The seismic wave energy is dependent not only on the wave amplitude, acceleration or particle velocity, but also on the soil impedance where the ground motion is recorded. It is meaningless to define design motions, acceleration or velocity, without referring the associated impedance value. Hence, when a design motion with a given amplitude is discussed, it is essential to identify the soil condition where the motion is defined.

13 The energy flow in the upward and downward waves and the energy dissipation as their difference can be calculated assuming one-dimensional SH-wave propagation in a horizontally layered ground. The wave energy, unlike wave amplitude, always tends to decrease at layer boundaries of different soil properties depending on its wave impedance ratio because of the reflected waves generated there. SH-wave propagation through a boundary in a two-layer model with no upper boundary in the top layer can theoretically determine the upper limit of energy supply to an overlying perfect energy absorber.

14 A number of vertical array strong motion records indicate that the upward energy tends to decrease considerably near the ground surface in most cases, and the

energy ratio between arbitrary two layers in the same soil profile is approximately in proportion to the power of 0.7 of the corresponding impedance ratio. This trend indicates that the upward energy at the ground surface tends to be smaller in sites of softer soils because of the lower impedance ratios than stiffer soil sites. In addition, high material damping values during strong earthquakes tend to dissipate wave energy in soft soils to make the energy storage in resonance difficult to occur.

15 This finding seems to be incompatible with a perception widely accepted that soft soil sites tend to suffer heavier earthquake damage than stiff soil sites. However, the smaller upward energy still tends to induce larger soil strains and can account reasonably for greater geotechnical damage in softer soils among various earthquake damage. In this context, it is essential to differentiate earthquake damaging mechanisms into direct inertial effects on structures and geotechnical effects on foundations in statistically assessing earthquake damage.

16 Upward energies at the deepest levels calculated from vertical array records or those further extrapolated to the seismological bedrock show a certain degree of compatibility with the well-known empirical earthquake energy formula despite considerable data scatters. Hence, the simple energy formula may be used to determine the incident energy in energy-based design with some modifications depending on individual earthquakes reflecting specific fault rupture mechanisms.

17 In a straight forward energy-based design, the energy capacity closely connected with the induced strain of structural members may be compared directly with the energy demand associated with a given upward wave motion. In a simplified relationship, the induced strain of a given structure is governed by the incident earthquake energy flux from the foundation ground, the equivalent impedance ratio between the structure and the foundation ground, and the degree of resonance. If the structure is ductile and of high damping such as soil structures, the cumulative wave energy is more important than the energy flux.

Chapter 5

# Liquefaction

Liquefaction is a kind of ground failure which tends to occur during strong seismic shaking in loose, saturated non-cohesive soils. Though the liquefaction had often been witnessed from old days and documented in archives, it was after the year 1964, when the Niigata earthquake in Japan and the Alaskan earthquake in USA caused devastations by extensive soil liquefactions, the geotechnical research started to understand its mechanism and how to mitigate the effect. Since then, quite a few liquefaction cases have been observed during a number of earthquakes, and the significance of its effect on structural damage has increasingly been recognized.

The liquefaction research started from clean sands such as those involved in the Niigata earthquake, and the triggering mechanism and major influencing parameters were investigated. Liquefaction cases involving low-plastic fines with particle sizes smaller than 0.075 mm also drew attention since the 1964 Alaskan earthquake, such as the 1999 Kocaeli earthquake in Turkey and the 2011 Tohoku earthquake Japan. On the other hand, gravelly soils have been recognized to be liquefiable in several case histories. At present, almost all non-cohesive loose saturated soils are categorized as potentially liquefiable irrespective of their particle sizes.

In this chapter, the liquefaction triggering mechanism considering a variety of soils is first discussed in terms of the influencing factors and how to evaluate the liquefaction potential for engineering purposes, followed by post-liquefaction soil behaviors in terms of residual strengths and deformations. Mitigation measures for liquefaction as well as base-isolation effects of earthquake motions in liquefied deposits are also addressed.

## 5.1 TYPICAL LIQUEFACTION BEHAVIOR

In order to understand the essence of liquefaction behavior, two typical test results are shown here: a scaled model test of saturated sand deposit as a total system and an undrained cyclic loading test on a saturated sand specimen as a soil element from the deposit.

### 5.1.1 Scaled model test

A uniform sand 2 m high deposited in water in an acrylic transparent tube is hit by a spring-powered metal hammer as illustrated in Fig. 5.1.1(a). This causes instantaneous

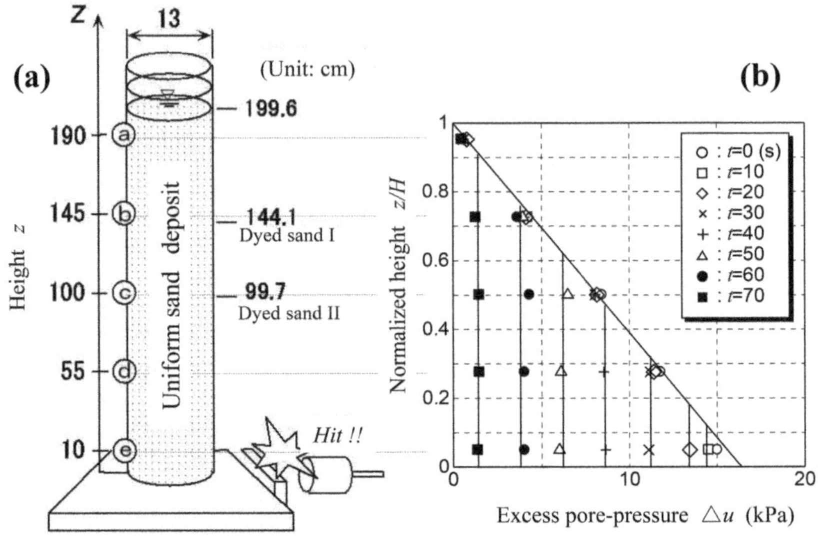


*Figure 5.1.1* Liquefaction model test for uniform sand deposit in acrylic tube (a), and Excess pore-pressure along height (b) (Kojima 2000).

liquefaction throughout the sand deposit with excess pore-pressure, measured at 5 depths, increasing linearly with depth as indicated by the line at $t=0$ in Fig. 5.1.1(b). The depth-dependent pressure gradient is equal to the critical hydraulic gradient, $i_{cr} = (\rho_{sat} - \rho_w)/\rho_w$, where $\rho_{sat}$, $\rho_w$ = densities of saturated sand and water, respectively. With elapsing time, the pore-pressure tends to decrease as the plane of re-sedimentation (wherein liquefied sand particles suspended in water reconsolidate and recover the effective stress) proceeds from the bottom to the top, and the excess pore-pressure below the plane becomes identical.

Fig. 5.1.2(a) depicts the time histories of excess pore-pressures measured by the five pore-pressure gages from the shallower to deeper depths. It is remarkable that the pressures stay constant until certain times shown with the arrows, which tend to appear earlier at greater depths. Then, it is followed by decreasing trends of the pore-pressures with time almost linearly down to zero. Obviously, the arrows correspond to the moments at which the re-sedimentation plane passes those particular depths. Correspondingly, Fig. 5.1.2(b) shows time-dependent sand settlements at three depths in the sand deposit (sand surface, dyed sand I, and dyed sand II indicated in Fig. 5.1.1(a)). The settlements occur almost linearly with time until $t_1$, $t_2$, $t_3$ also indicated in Fig. 5.1.2(b), corresponding to the three depths. It is clear that no visible settlement occurs after the sedimentation plane passes those particular depths.

These diagrams clearly indicate that sand particles at a depth in a uniform saturated deposit are being liquefied as long as the re-sedimentation plane is below that depth, and the entire liquefaction comes to the end when the plane reaches to the surface. In parallel with the re-sedimentation and re-consolidation, the pore-water is squeezed

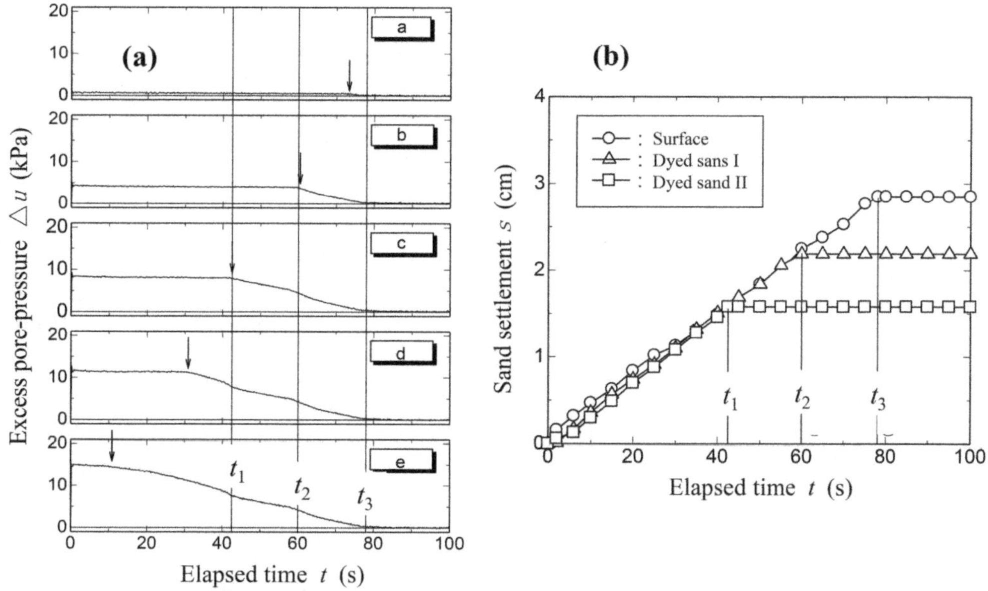


*Figure 5.1.2* Time-histories at different levels of sand deposit: (a) pressure variations, (b) Settlements (Kokusho 2003).

out. It is confirmed that the settlement velocity of sand particles is equal to the seepage velocity of pore water under the critical hydraulic gradient (Kokusho 2003).

Such a model test is quite effective for understanding the liquefaction mechanism qualitatively in a global soil deposit with various boundary conditions, though a wide gap exists between the model and prototype in quantitatively understanding the stress-strain behavior of liquefied sand in situ. Hence, the soil element tests are carried out to obtain quantitative design values from the stress-strain relationships of in situ soils under in situ stress conditions.

## 5.1.2 Undrained soil element test

Fig. 5.1.3(a) shows time histories of dynamic shear stress $\tau_d$, shear strain $\gamma$ and excess pore-pressure $u$, obtained by an undrained torsional simple shear test on saturated clean sand of relative density $D_r \fallingdotseq 50\%$, consolidated with isotropic stress $\sigma_c' = 98$ kPa corresponding to ground depth about 10 m and cyclically loaded in 30 cycles simulating earthquakes. When the cyclic shear stress is applied over 10 cycles and the excess pore-pressure approaches to the maximum value, the shear strain starts to grow rapidly up to 20% with further cyclic stress applications. The strain develops almost symmetrically in the positive and negative sides up to larger values, though the symmetry does not hold for larger strains due to imperfect mechanical performance of the test equipment.

The top panel in Fig. 5.1.3(b) illustrates the effective stress path in the same test. The effective stress $\sigma_c'$ tends to decrease from the initial point A to the left as the pore pressure builds up due to negative dilatancy of the soil skeleton in undrained loading

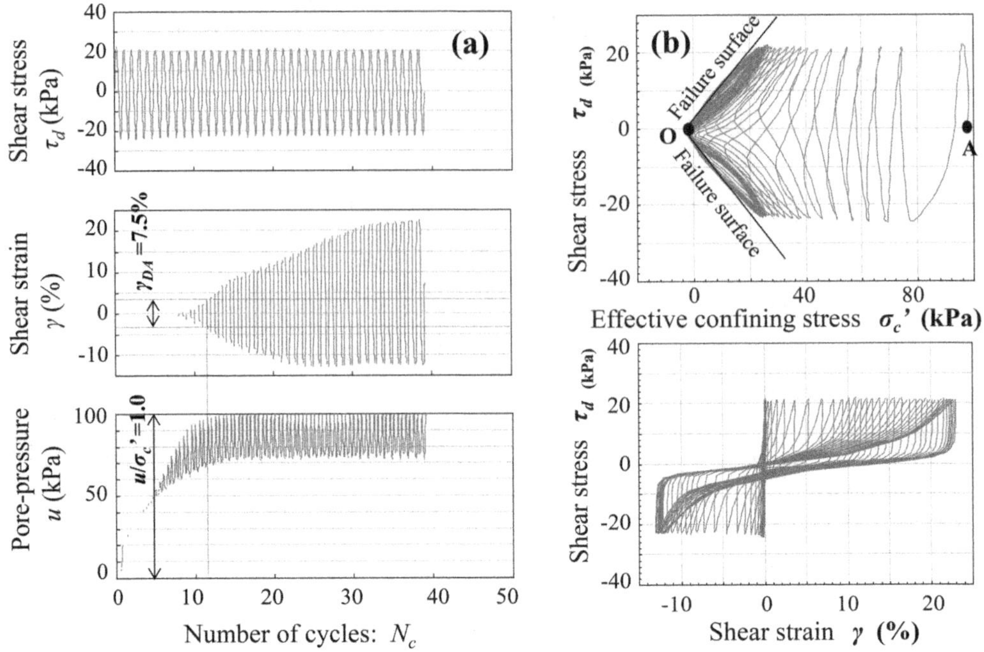


*Figure 5.1.3* Torsional simple shear test result on clean sand ($D_r \fallingdotseq 50\%$): (a) Shear stress, shear strain and pore-pressure, (b) Effective stress path and stress-strain curve (Kusaka 2012).

and eventually reaches to the origin O. There, the effective stress is completely lost because the excess pore pressure $u$ increases up to the initial effective stress $\sigma_c'$, causing sudden increase of induced shear strain amplitude $\gamma$ as indicated in Fig. 5.1.3(a). The first arrival of the stress path at the origin O is called the state of initial liquefaction or liquefaction onset. It is normally recognized that the induced double amplitude strain at the initial liquefaction is about $\gamma_{DA} = 7.5\%$ in shear strain for simple shear tests and $\varepsilon_{DA} = \gamma_{DA}/(1+\nu) = 5\%$ in axial strain for triaxial tests considering Poisson's ratio $\nu = 0.5$ in the undrained condition.

In the stress-strain curve shown in the bottom of Fig. 5.1.3(b), the secant modulus tends to become evidently lower, and the strain ranges with very low mobilized shear resistance is widening with increasing loading cycles. However, the modulus can never be so low for this sand of $D_r \fallingdotseq 50\%$ as to behave like liquid because the shear resistance starts to pick up at some strains in every cycle due to the positive dilatancy of soil skeleton. This type of behavior is named "cyclic mobility" and should be considered differently from liquefaction of very loose contractive soils which may undergo truly liquid-like flow failures.

Very loose soils, such as clean sands looser than $D_r \fallingdotseq 30\%$ or sands not so loose but containing a lot of non-plastic fines, tend to be contractive without positive dilatancy under high effective confining stresses in particular, developing fully or partially flow-type failures like liquid. In very contractive soils, soil specimens in test devices tend to be nonuniform during undrained shearing (loosening or forming the water film at the

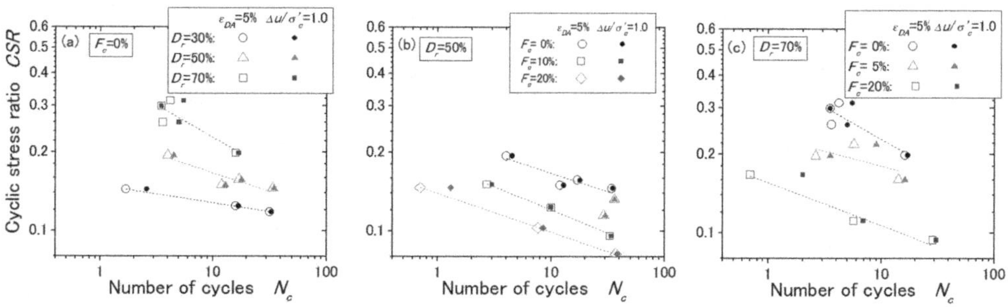

*Figure 5.1.4* *CSR* for initial liquefaction ($\varepsilon_{DA} = 5\%$ or $\Delta u/\sigma_c' = 1.0$) versus number of cycles $N_c$ for sands with different relative densities $D_r$ and fines contents $F_c$ (Kokusho et al. 2013a).

top of the specimen) due to "void redistribution" (NRC 1985, Kokusho et al. 2003), making soil element tests on uniform specimens almost meaningless.

For sands with $D_r \fallingdotseq 50\%$ or denser, the soil element liquefaction tests are easier to simulate in situ soil behavior during earthquakes, though complex in situ stresses cannot be fully reproduced. Low shear resistance is restricted within a certain strain interval (though widening with increasing load cycles), beyond that shear resistance revives as "cyclic mobility." Even in a very dense sand of $D_r \approx 100\%$, this type of behavior can be observed by a larger number of loading of large stress amplitude with nearly 100% pore-pressure buildup. Even in clayey soils with high plasticity, a similar type of shear behavior can be seen with about 90% pressure buildup and associated shear stiffness degradation as will be seen in Sec. 5.9.

### 5.1.3 How to interpret element test data

Stress-controlled element tests yield the time-histories of pore-pressure and strain as shown in Fig. 5.1.3. There may be multiple ways in reducing these time-history data to quantify how liquefiable the tested soils are under the cyclic loading. In the normal practice, a chart is drawn, where the vertical axis is the cyclic stress ratio (*CSR*) and the horizontal axis is the number of cycles $N_c$ for a soil specimen to attain a certain strain amplitude or 100% pressure buildup, $\Delta u/\sigma_c' = 100\%$.

Fig. 5.1.4 shows typical diagrams obtained from triaxial tests on Futtsu beach sand with $D_r \approx 30{\sim}70\%$ and $F_c = 0{\sim}20\%$. A clear trend is visible despite some data scatters that the *CSR* for the pore-pressure ratio $\Delta u/\sigma_c' = 100\%$ or the double amplitude axial strain $\varepsilon_{DA} = 5\%$ tends to decrease with increasing number of load cycles $N_c$. The *CSR* versus $N_c$ correlation may be approximated by a straight line in the log-log scale as indicated in Fig. 5.1.4, though the semi-log scale is also used wherein only $N_c$ is in the log axis. Then, cyclic resistance ratio (*CRR*) is determined from the correlation as the *CSR*-value corresponding to a given number of cycles $N_{eq}$ equivalent to a design earthquake motion. In the current liquefaction potential evaluation method (SBM: stress-based method), the *CRR*-value thus decided is compared with a *CSR*-value, a stress ratio of a sinusoidal motion with the number of cycles $N_{eq}$, which is equivalent to a design seismic motion in terms of damage level as explained in Sec. 3.1.6.

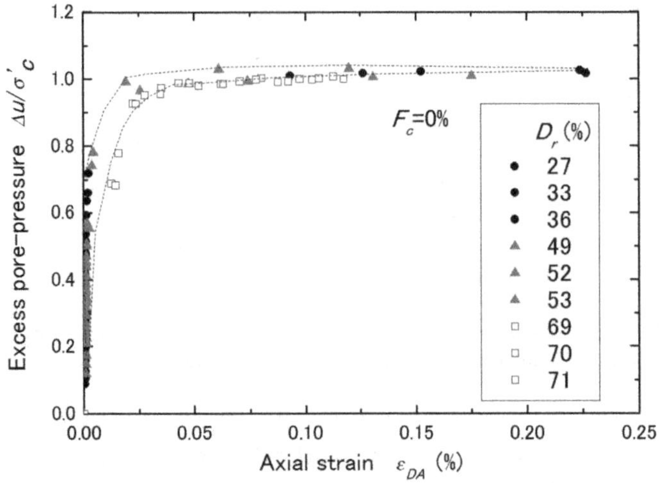


*Figure 5.1.5* Excess pore-pressure ratio versus double amplitude axial strain in triaxial tests on clean sands with varying density.

Fig. 5.1.5 depicts relationships between the axial strain $\varepsilon_{DA}$ versus excess pore-pressure ratio $\Delta u/\sigma'_c$ for the same tests on clean sands ($F_c = 0$) of $D_r = 27$–70% as shown in Fig. 5.1.4. The pressure buildup ratio $\Delta u/\sigma'_c$ at each cycle is plotted versus the corresponding double amplitude axial strain $\varepsilon_{DA}$ in the vertical and horizontal axis, respectively. Note that the plots are concentrated along a narrow band within a pair of dotted curves despite wide differences in $D_r$ and stepwise increments of $\Delta u/\sigma'_c$ or $\varepsilon_{DA}$ in individual tests. It is also pointed out that the pore-pressure builds up to 100% at around the double axial strain amplitude of $\varepsilon_{DA} = 5\%$. This almost unique relationship between $\Delta u/\sigma'_c$ and $\varepsilon_{DA}$ was recognized in previous research also in smaller strain levels, wherein triaxial tests on clean sands with $D_r = 45$–80% indicated not only the uniqueness in the $\Delta u/\sigma'_c$ versus $\varepsilon_{DA}$ correlation but also the existence of threshold strain for the initiation of pressure-buildup at around a single amplitude shear strain $\gamma = 1 \times 10^{-4}$ (NRC 1985).

It was also demonstrated experimentally that the energy dissipated in cyclically loaded sand serves as a decisive factor for pore-pressure buildup and liquefaction triggering (Towhata and Ishihara 1985, Yanagisawa and Sugano 1994, Figueroa et al. 1994). Fig. 5.1.6(a) exemplifies a typical stress versus strain curve of clean sand with $D_r \approx 50\%$ where the dissipated energy $\Delta W$ per unit volume in a given cycle is calculated from the hysteresis area ABCD, and the cumulative energy is summed up to a certain cycle as $\sum \Delta W$. In Fig. 5.1.6(b), time-histories are depicted for the cumulative energy in the top panel together with the axial stress, strain and excess pore-pressure. It is obviously seen that not only the strain but also the cumulative energy show drastic increase at the last stage of the pressure-buildup just before liquefaction is triggered.

Fig. 5.1.7 shows cyclic triaxial test results for a clean sand under the effective confining stress $\sigma'_c = 98$ kPa where the pore-pressure buildup ratios $\Delta u/\sigma'_c$ are plotted versus the cumulative dissipated energies $\sum \Delta W$ for a clean sand having different

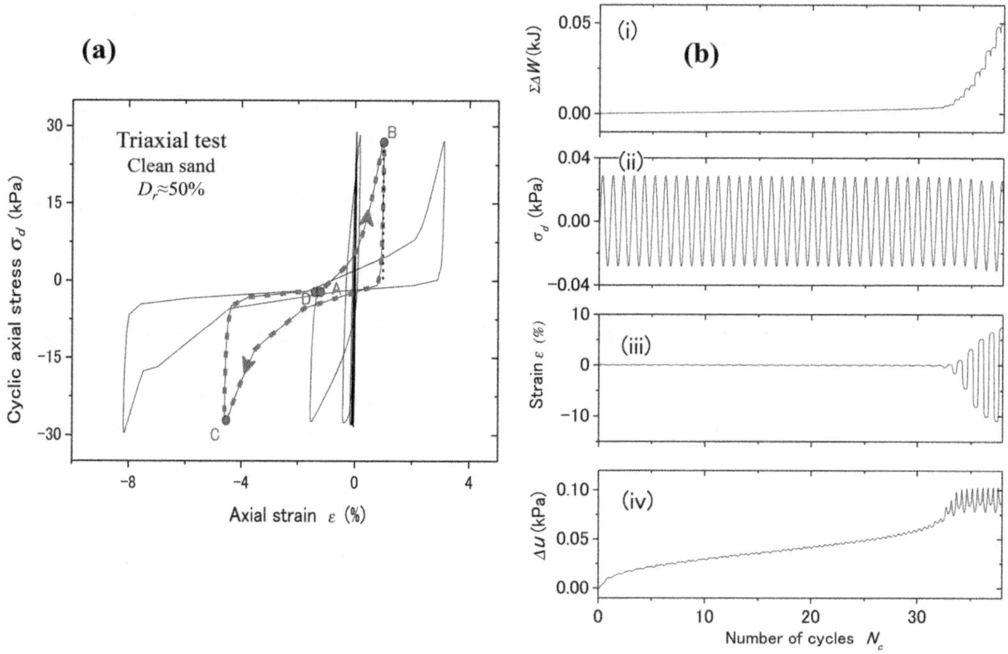


*Figure 5.1.6* Typical stress-strain hysteresis curve (a), and Time histories of (i) cumulative dissipated energy, (ii) shear stress, (iii) shear strain, (iv) excess pore-pressure (b) (Kokusho 2013b).

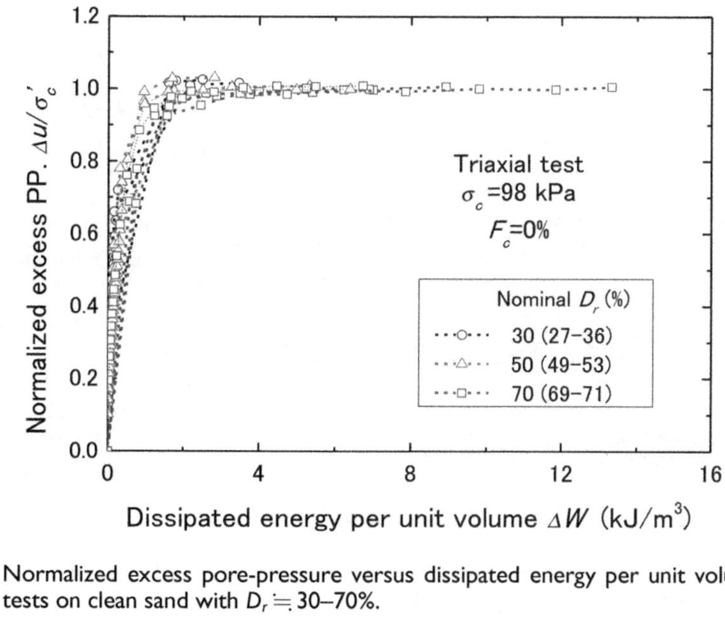


*Figure 5.1.7* Normalized excess pore-pressure versus dissipated energy per unit volume in triaxial tests on clean sand with $D_r \fallingdotseq$ 30–70%.

densities $D_r = 30–70\%$. The $\Delta u/\sigma'_c$-value is markedly well correlated with $\sum \Delta W$ and tends to attain 1.0 in the narrow range of the energy. Although this relationship can pave a way to an energy-based liquefaction potential evaluation as will be addressed in Sec. 5.6, the liquefaction potential is currently evaluated solely by the stress-based approach, where cyclic resistance ratios determined from the *CSR* versus $N_c$ relationships for particular sand such as in Fig. 5.1.4 are compared with seismically induced stress ratios (Seed and Idriss 1971). This is mainly because the liquefaction-related design principle shares a common historical background with other structural designs based on force-equilibrium.

## 5.2 GENERAL CONDITIONS FOR LIQUEFACTION TRIGGERING

Before starting in-depth discussions on liquefaction mechanisms, general conditions how liquefaction is triggered in situ soils are summarized in geotechnical and seismic aspects for a beginner in liquefaction studies.

### 5.2.1 Geotechnical conditions

*(i) Soil types*

Soil types potentially liquefiable are non/low-cohesive soils including SP, SW, SM, GS, GW, GM, ML, CL in the standard soil classification (ASTM 1985). Fig. 5.2.1 shows

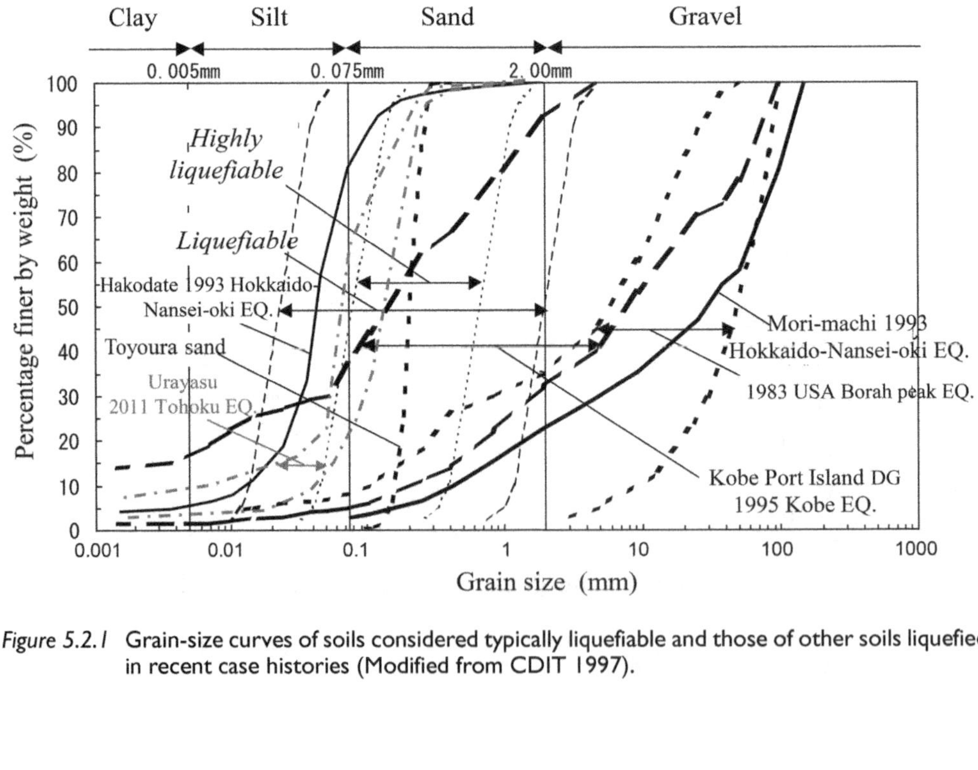


*Figure 5.2.1* Grain-size curves of soils considered typically liquefiable and those of other soils liquefied in recent case histories (Modified from CDIT 1997).

representative grain size curves of non/low-cohesive soils liquefied in previous earthquakes (modified from CDIT 1997). Among them, the ranges of highly liquefiable and liquefiable sands are illustrated with two pairs of curves, in the center of which Toyoura sand often used in Japan is found. These poorly-graded clean sands have long been considered to be typically liquefiable. However, more recent earthquakes (e.g. the 1999 Kocaeli earthquake Turkey, the 1999 Chi-Chi earthquake Taiwan, and the 2011 Tohoku earthquake Japan) increasingly incurred severe liquefactions not only in clean sands but also in sands containing a plenty of non/low-plastic fines. It should be emphasized that even sandy soils containing a lot of fines ($F_c = 50$–$80\%$) or even silty soils with little sand particles liquefy if the fines are non/low-plastic. Furthermore, well-graded gravelly soils are reported to have liquefied extensively during some recent earthquakes such as the 1983 Borah Peak earthquake in USA, the 1993 Hokkaido Nansei-oki earthquake and 1995 Kobe earthquake in Japan. Fig. 5.2.1 indicates that a variety of soils not necessarily typical poorly-graded sands are also liquefiable.

In this regard, the percentage in terms of dry weight of particles finer than 0.075 mm $W_f$ out of total soil weight $W_t$ is defined as fines content $F_c$ (%) such that:

$$F_c = \frac{W_f}{W_t} \times 100 \tag{5.2.1}$$

Similarly, the percentage of soil particles coarser than 2 mm $W_G$ is defined as gravel content $G_c$ (%) such that:

$$G_c = \frac{W_G}{W_t} \times 100 \tag{5.2.2}$$

Also note that gravelly soils liquefied so far are all well-graded, containing not only coarse particles but also sands and fines. It is sometimes considered that gravelly soils, distinctively more permeable than sandy soils, are difficult to develop pore-pressure buildup for liquefaction triggering. However, in situ gravels are well-graded in most cases with their permeability as low as sands, providing undrained conditions necessary for pressure buildup (JGS Committee 2001). It may also be pointed out that well-graded gravels tend to have higher absolute densities than poorly graded sands. Nevertheless, gravelly soils with very low relative densities actually liquefied during past earthquakes. The effect of particle gradations of fines-containing sandy and gravelly soils together with the effect of plasticity of fines on the liquefaction potentials will be discussed in Sec. 5.4.

#### *(ii) Saturation*

Liquefaction occurs basically in saturated soils under water table, because pore-pressure buildup in the undrained condition is the key mechanism in liquefaction triggering. However, liquefaction-like behavior can occur not only in perfectly saturated soils with saturation $S_r = 100\%$ but also in imperfectly saturated soils $S_r = 90$–$80\%$. Here, the saturation $S_r$ is defined as a volume ratio of pore-water to soil void in percentage. The effect of saturation on the liquefaction susceptibility will be discussed in Sec. 5.7.

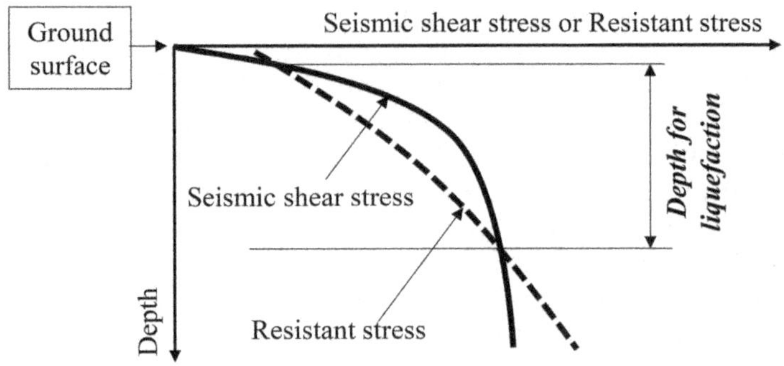


*Figure 5.2.2* Conceptual illustration on depth limitation for liquefaction occurrence.

*(iii) Relative density*

One of the governing parameters of liquefaction during cyclic loading is how dense soil is, though more subtle microscopic fabric of soil particles is also influential to liquefaction as will be addressed in Sec. 5.3. If a soil is densely packed, the excess pore-pressure becomes slow to build up because it tends to dilate beyond a certain shear strain limit. Because the absolute density of soil varies corresponding to its particle size, grading and shape, the relative density $D_r$ is often used as a common scale for all non-cohesive soils to measure how dense soil is packed in terms of 0 to 100% such as

$$D_r = \frac{e_{\max} - e}{e_{\max} - e_{\min}} = \frac{1/\rho_{\min} - 1/\rho}{1/\rho_{\min} - 1/\rho_{\max}} \tag{5.2.3}$$

where $e_{\max}, e_{\min}$ = maximum and minimum void ratios, and $\rho_{\min}, \rho_{\max}$ = corresponding minimum and maximum soil dry densities determined by test methods for the minimum and maximum density. The test methods are standardized differently in different countries, such as USA (ASTM 2001) and Japan (JGS 2008, 2009). The relative density is often used in laboratory tests, though it cannot be determined efficiently in situ. In normal engineering practice, instead, penetration resistance values by Standard Penetration Test (SPT) or Cone Penetration Test (CPT) are used, which are considered to reflect in situ relative densities.

*(iv) Soil depth*

It is easy to understand that shallower soils tend to liquefy more because the densities are lower and the ages are younger than deeper soils. Besides these factors, there exists a depth limitation for liquefaction as illustrated in Fig. 5.2.2. Namely, the liquefaction-resistant stress shown with the dashed line tends to increase monotonically with increasing effective overburden stress with a curve slightly nonlinear. On the other hand, the seismically-induced shear stress is obtained from the sum of seismic inertial forces down to given depths considering their phase differences. As the result, the seismic shear stress shown with the solid curve increases almost linearly with depth near the ground surface where the phase difference is still small but tends to converge to a certain value on account of a cancelling effect of the inertial forces due to growing phase differences with increasing depth. Hence, a depth-limit for liquefaction susceptibility surely appears, below which the seismic shear stress cannot be larger than the resistance. This limit may depend on soil conditions and seismic motions, though it is normally taken as 20 m in engineering practice in Japan (e.g. JRA 2002, AIJ 2001).

*(v) Aging effect*

It is generally accepted that younger Holocene soils have higher liquefaction susceptibility than older Pleistocene soils, and manmade fill soils are most susceptible to liquefaction. Chemical bonding between soil particles or mechanical effects on particle arrangements, named as soil fabric, developing in geological time may cause such aging effects. Recent earthquakes in Japan triggered liquefaction much more often in reclaimed soils or manmade fills than in young natural Alluvial sands, indicating that the aging effect is very important even among very young soils. However, this effect is not taken into account in current liquefaction evaluations in a quantitative manner (e.g. in terms of hundreds or thousands of years) and is still an important subject of ongoing research. This issue will be discussed in the later Sections.

*(vi) Initial stress*

Soil liquefaction incurs structural damage when it occurs near or beneath structures, where soil is under the influence of sustained initial shear stress by the overlying structures. Major liquefaction damage such as settlements, sliding and overturning of superstructures actually occurs under the influence of sustained initial shear stresses. In current engineering practice, the liquefaction susceptibility has mostly been evaluated in a level ground where structure-induced initial shear stresses do not work. It is thus significant to consider the effect of initial shear stress not only on liquefaction-triggering but also on post-liquefaction deformations as will be discussed in Sec. 5.8.

### 5.2.2 Seismic conditions

Past experiences indicate that seismically-induced liquefaction is triggered by earthquakes with the earthquake magnitude larger than 5 to 6. Liquefactions tend to occur more often for earthquakes with larger magnitudes even in very long hypocentral distances because it gives larger seismic stresses or larger seismic wave energies.

In simplified liquefaction evaluations currently employed in practice, ground surface acceleration or its maximum value, Peak Ground Acceleration (PGA), is used as a key seismic parameter. However, it should be noted that liquefaction is directly determined by seismically induced soil strain which is more closely correlated with particle velocity as indicated in Eq. (1.2.19). Hence, the particle velocity rather than the acceleration in seismic motion will play a significant role in triggering liquefaction, indicating that, even for the same accelerations, the motion with longer dominant period or larger particle velocity tends to cause liquefaction.

The duration of seismic motions is another important factor to cause soil liquefaction. A larger number of load cycles tends to build up higher excess pore-pressure, leading to higher liquefaction potential. Even after the onset of liquefaction, longer duration of motion gives greater seismic energy to the soil and makes induced strain and post-liquefaction settlement larger.

In the following, the geotechnical conditions will be discussed in terms of the above-mentioned pertinent parameters in each Section in detail. The seismic conditions will be addressed mainly in Sec. 5.5.5.

## 5.3 GEOTECHNICAL CONDITIONS FOR LIQUEFACTION TRIGGERING

Various soil parameters for triggering soil liquefaction (also named as initial liquefaction and liquefaction onset) are discussed here based mainly on laboratory element tests concerning the effects of effective confining stress, relative density, stress-strain history and soil fabric.

### 5.3.1 Effect of confining stress

It was one of the important subjects in early days of liquefaction research how to simplify in situ stress conditions properly in laboratory element tests. If a level ground is horizontally shaken by the SH-wave, the soil element ($K_0$-consolidated) is cyclically sheared by seismic shear stress in the undrained condition as illustrated in Fig. 5.3.1(a). This stress condition may be reproduced by a simple shear test in Fig. 5.3.1(b), where the soil specimen $K_0$-consolidated (vertical stress $\sigma'_v$ and horizontal stress $\sigma'_h = K_0\sigma'_v$) is sheared cyclically with a stress amplitude $\tau_d$. As already mentioned in Sec. 2.2.3, the simple shear device, despite its mechanical simplicity, has some deficiency in applying uniform shear stresses all around the specimen. The more sophisticated torsional simple shear test is also available using a hollow cylindrical specimen to have better reproducibility of in situ stresses.

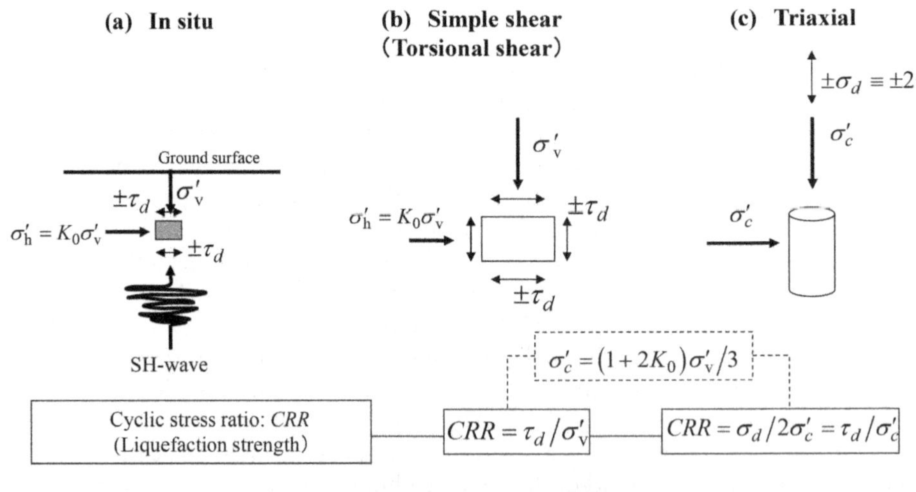


*Figure 5.3.1* Comparison of stress conditions in situ and laboratory: (a) In situ, (b) Simple shear test, (c) Triaxial test, and corresponding definitions of *CRR*.

The first endeavor to simulate liquefaction-triggering in cyclic loading element tests was undertaken by Seed and Lee (1966) not in the simple shear but in the triaxial condition as indicated in Fig. 5.3.1(c). This pioneering experiment demonstrated that 100% pore-pressure buildup due to the negative dilatancy in undrained cyclic loading is the key mechanism of seismic liquefaction. The soil specimen was isotropically consolidated with the effective stress $\sigma_c'$, and cyclically loaded with the axial stress of single amplitude $\sigma_d$ with the horizontal stresses unchanged. Since then, this triaxial test has been implemented as the standard undrained cyclic loading test for liquefaction to date, though the stress system seems quite different from in situ.

The effect of the difference in stress conditions was discussed by Seed and Peacock (1971) by comparing the simple shear test on $K_0$-consolidated specimens in (b) and the triaxial test on isotropically consolidated specimens in (c), and a relationship between the cyclic stresses for liquefaction triggering in the field $\tau_d$ and the triaxial test $\sigma_d$ was proposed as

$$(\tau_d/\sigma_v')_{\text{field}} = c_r \times (\sigma_d/2\sigma_c')_{\text{triax}} \tag{5.3.1}$$

where the coefficient $c_r$ connecting the two stress ratios was determined to be 0.55–0.70 for clean sands with the relative density $D_r = 40$–$85\%$.

A significant effect of the effective overburden $\sigma_v'$ or confining stress $\sigma_c'$ in Eq. (5.3.1) on the liquefaction resistance was also found in the experiment by Seed and Lee (1966). Considering this effect, the liquefaction resistance has normally been expressed as a ratio of the cyclic shear stress amplitude to the effective overburden or confining stress and named as the cyclic resistance ratio (*CRR*). The cyclic shear stress amplitude above is determined so that 100% pore-pressure buildup or double amplitude strain (in triaxial $\varepsilon_{DA} = 5\%$ axial strain or in simple shear $\gamma_{DA} = 7.5\%$ shear strain) is attained as a condition of the initial liquefaction in a given number of cycles.

In order to know the difference in the cyclic resistant ratios in the simple shear and triaxial shear tests, Ishihara and Yasuda (1975) compared the two test results on isotropically consolidated specimens of the same clean Fuji-river sand of $D_r = 55\%$ as depicted in Fig. 5.3.2. The cyclic resistance ratios (*CRR*) for 100% pore-pressure buildup plotted versus the number of loading cycles ($N_c$) show a good coincidence between the two tests, indicating that no clear difference arises in liquefaction resistance between the two testing methods if both specimens are isotropically consolidated.

In the field, $K_0$-consolidated soils are cyclically sheared while the lateral strain is constrained. In order to simplify this condition in laboratory tests by isotropically confined specimens, it is necessary to choose appropriate confining stress to obtain the same liquefaction resistance. Fig. 5.3.3 depicts cyclic stress ratios for 100% pore-pressure buildup versus $N_c$ relationships obtained by torsional simple shear tests on $K_0$-consolidated specimens for $K_0 = \sigma_h'/\sigma_v' = 0.5$, 1.0, 1.5 (Ishihara et al. 1977). If the cyclic stress amplitudes reaching the initial liquefaction in a certain number of cycles are normalized by the vertical effective stress as $\tau_d/\sigma_v'$ in (a), the results are plotted separately depending on the $K_0$-values. In contrast, if the mean effective confining stress $\sigma_c' = (1 + 2K_0)\sigma_v'/3$ is chosen in place of $\sigma_v'$ as shown in (b), the same datasets concentrate along the single unique line. Thus, the test results indicate that the simplified cyclic shear tests on isotropically consolidated specimens with $\sigma_c' = (1 + 2K_0)\sigma_v'/3$

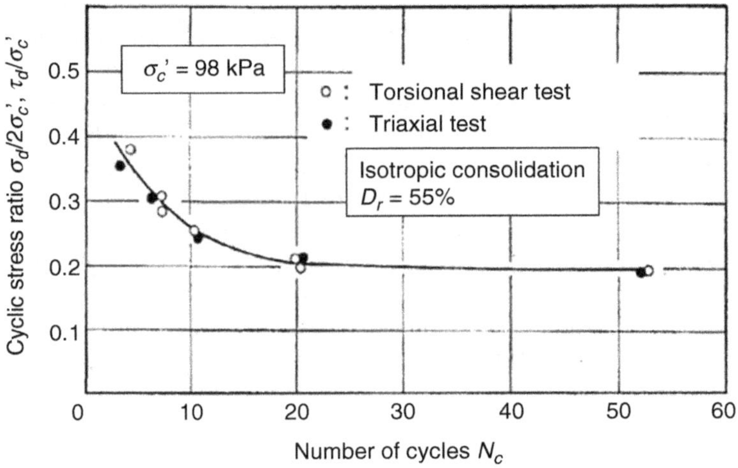


*Figure 5.3.2* Comparison of *CSR* for 100% pore-pressure buildup versus $N_c$ relationships between torsional simple shear and triaxial tests on isotropically consolidated specimens (Ishihara and Yasuda 1975).

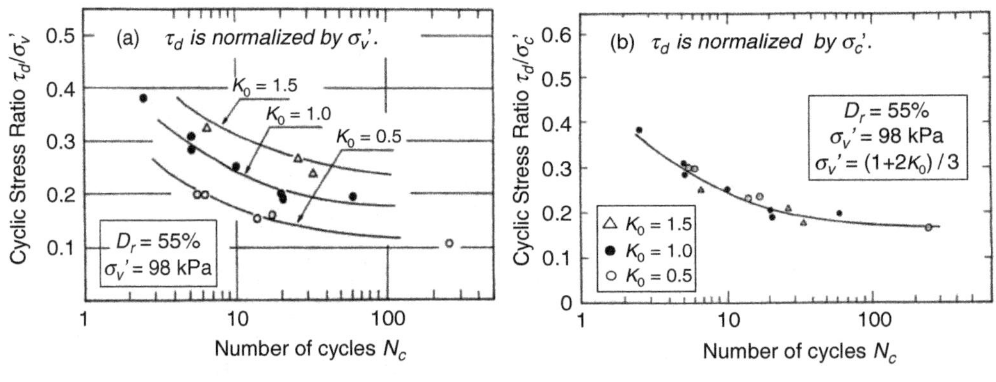


*Figure 5.3.3* $K_0$-consolidarted torsional simple shear test for $K_0 = 0.5$, 1.0, 1.5: (a) Vertical axis normalized by $\sigma_v'$ (b) Vertical axis normalized by $\sigma_c' = (\sigma_v' + 2\sigma_h')/3$ (Ishihara et al. 1977), by permission of Oxford University Press.

can yield the liquefaction resistance of $K_0$-consolidated specimens. This result seems to be compatible with what Seed and Peacock (1971) indicated in Eq. (5.3.1). There are however other test results that showed some difference, depending on different sample preparation methods and soil densities, in the cyclic resistance ratios between isotropically consolidated triaxial tests and $K_0$-consolidatated torsional shear tests normalized by $\sigma_c'$ (e.g. Tatsuoka et al. 1986a, Yamashita and Toki 1992). Though the isotropically

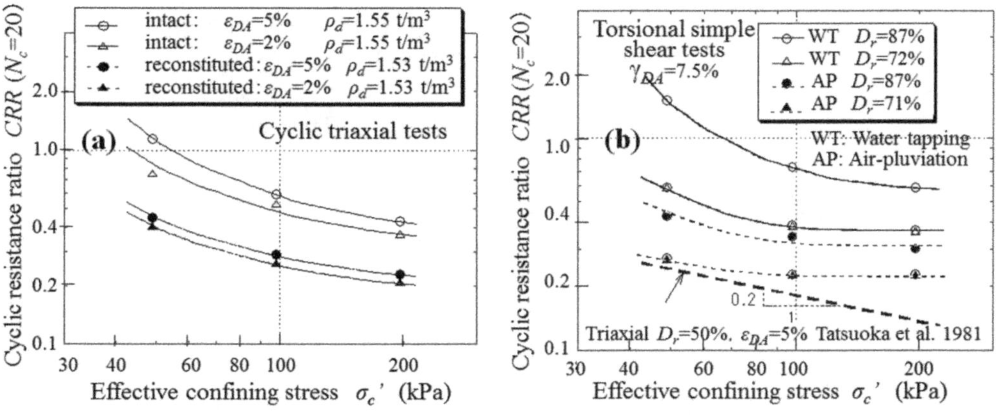


*Figure 5.3.4* Decreasing trend of *CRR* with increasing effective confining stress $\sigma'_c$: (a) *CRR* by triaxial tests on intact Pleistocene Narita sands, (b) *CRR* by torsional simple shear tests on reconstituted Toyoura sand by two sample preparation methods (Kokusho et al. 1983a).

consolidated triaxial test is often employed in practice as a convenient test method in evaluating liquefaction resistance of in situ soils recovered intact, there may still exist some uncertainties involved in this practice in evaluating in situ liquefaction resistance.

Cyclic resistance ratio (*CRR*) is defined as the cyclic shear stress normalized by the effective confining stress, as already mentioned. It was assumed in the early stage of liquefaction research (Lee and Seed 1967) that the normalization would make the value independent of the confining stress. However, *CRR* was found to be actually dependent on the confining stress (Tatsuoka et al. 1981, Kokusho et al. 1983a). Fig. 5.3.4(a) shows *CRR* for $\varepsilon_{DA} = 2$ or 5% and $N_c = 20$ obtained by triaxial tests on Pleistocene Narita sands, which were sampled intact by block from in situ, plotted versus associated effective confining stresses $\sigma'_c$ on the log-log scale. In the same diagram, a similar decreasing trend in *CRR* with increasing $\sigma'_c$ can also be seen for specimens reconstituted from the same sand with significantly lower absolute *CRR* values despite almost identical dry densities. The decreasing trend is pronounced in the low confining stress range in particular. Fig. 5.3.4(b) depicts a similar diagram obtained for reconstituted Toyoura sand by torsional simple shear tests. The specimens were prepared by two different methods as described later on; air-pluviation (AP) and water tapping (WT) to make dense specimens of $D_r \fallingdotseq 70$–90%. *CRR* for $\gamma_{DA} = 7.5\%$ in $N_c = 20$ tends to decrease with increasing $\sigma'_c$ again for the two types of sample preparation, though the AP specimens show smaller dependency on $\sigma'_c$ than the WT specimens. Soil fabric which seems to be introduced in dense sands by the WT method is likely to cause the stronger $\sigma'_c$-dependent variation of *CRR* than that by the AP method. Similar plots on the same Toyoura sand of $D_r \fallingdotseq 50\%$ obtained by triaxial tests (Tatsuoka et al. 1981) superposed in the chart indicate stronger $\sigma'_c$-dependency of *CRR* ($\varepsilon_{DA} = 5\%$, $N_c = 20$) with the gradient $\log(CRR)/\log\sigma'_c = 0.2$ up to higher $\sigma'_c$ of 100–200 kPa.

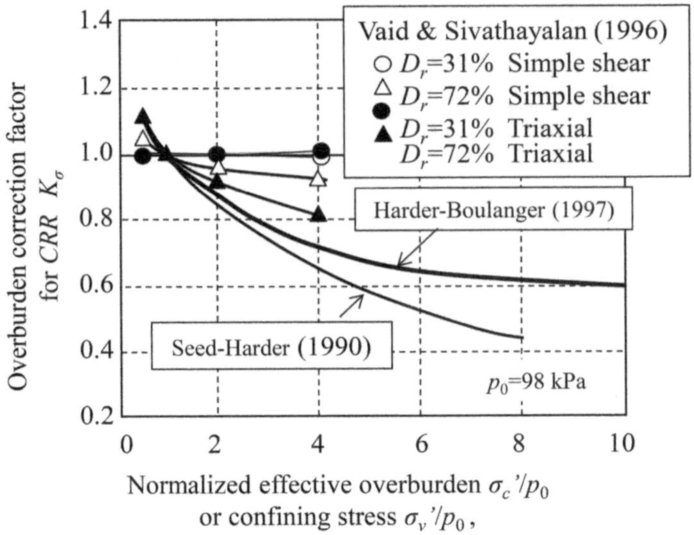


*Figure 5.3.5* Correction factor for *CRR*, $K_\sigma = CRR/CRR_{\sigma_c'=98\,\text{kPa}}$ for various sands by various tests (replotted from Idriss and Boulanger 2008) by permission of EERI.

In North-American practice, the effect of effective overburden on the *CRR*-value is taken into account in engineering design by introducing an overburden correction factor $K_\sigma$ defined (Idriss and Boulanger 2008) as:

$$K_\sigma = CRR/CRR_{\sigma_c'=98\,\text{kPa}} \tag{5.3.2}$$

The variations of $K_\sigma$ obtained in different tests versus normalized stresses $\sigma_v'/p_0$ for simple shear tests or $\sigma_c'/p_0$ for triaxial tests are summarized in Fig. 5.3.5 (Idriss and Boulanger 2008). It indicates that the stress-dependent variations of $K_\sigma$-values are very much different according to sands, densities and test methods. For example, in test results on Fraser Delta sand in Canada in the diagram using the NGI-type simple shear device (Vaid and Sivathayalan 1996), the decreasing trend of *CRR* for $\gamma_{DA} = 7.5\%$, $N_c = 10$ with increasing $\sigma_c'$-values is not so strong for sands with lower $D_r$ values in particular.

The *CRR*-variations appear to stem from the fact that the soil dilatancy varies depending on the effective confining stress. Fig. 5.3.6 schematically illustrates the difference in volume changes in drained shear tests. It is well known that, in monotonic shearing in (a), a soil of a given density tends to contract under high confining stresses versus dilate under low confining stresses with increasing shear strain. This mechanism has been discussed in Sec. 3.1.5, where the soil dilatancy is differentiated by the steady state line (SSL) on the state diagram, so that the soil if sheared tends to contract or dilate above or below the line (Casagrande 1971). The same soil, if sheared cyclically as shown in Fig. 5.3.6(b), contracts with no regard to the confining stress. However, the same soil under higher confining stresses tends to contract much more than that

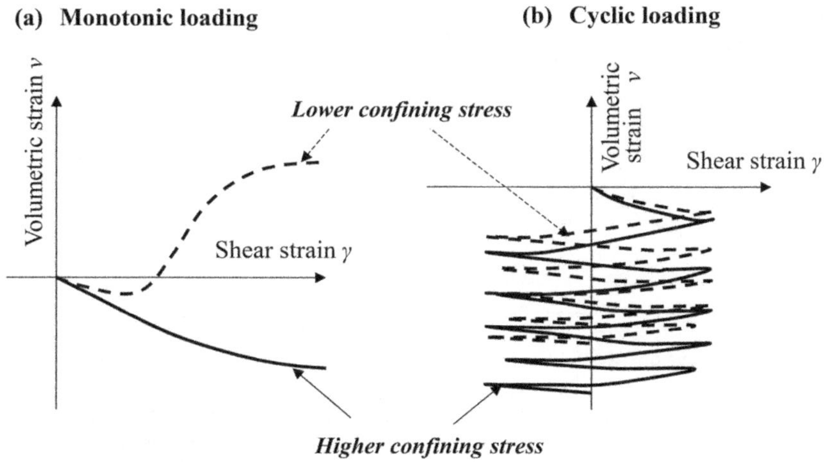


*Figure 5.3.6* Difference in conceptual dilatancy behavior of the same density sands due to smaller and larger confining stresses: (a) Drained monotonic loading, (b) Drained cyclic loading.

under low confining stresses for the identical void ratio. This explains why the same soil in undrained shearing tends to give a lower *CRR* under higher effective confining stresses.

In actual field conditions, sand deposits are seldom uniform along the depth and often very variable in densities, physical properties and geological backgrounds. Hence, care is needed if a sand deposit concerned belongs to the same sand so that the *CRR*-modification by $K_\sigma$-value is reasonable. In evaluating in situ *CRR* by laboratory tests on intact samples in major projects, it is recommended to conduct the liquefaction tests on samples recovered from multiple depths by employing the corresponding confining stresses.

## 5.3.2 Effect of relative density and soil fabric

### 5.3.2.1 Relative density versus CRR

From an early stage of the liquefaction research, the relative density $D_r$ has been considered one of the key parameters. Fig. 5.3.7(a) shows *CRR*–$D_r$ relationships for a clean sand obtained in a kind of undrained simple shear tests utilizing a shaking table (De Alba et al. 1976). The *CRR*-values is defined for pore-pressure buildup $\Delta u/\sigma'_c = 100\%$ or single-amplitude induced strain $\gamma = 5$–$25\%$ for the number of cycles $N_c = 10$. For $\gamma = 5\%$, *CRR* is almost proportional to $D_r$ up to $D_r = 80\%$, while it turns to be non-linear at higher $D_r$-values and drastically increasing with increasing $\gamma$. This is because the cyclic mobility effect aforementioned tends to be more pronounced with increasing $D_r$, cyclically mobilizing shear resistance and impeding strain developments. Thus, it is of utmost importance to determine the shear strain amplitude for liquefaction-induced

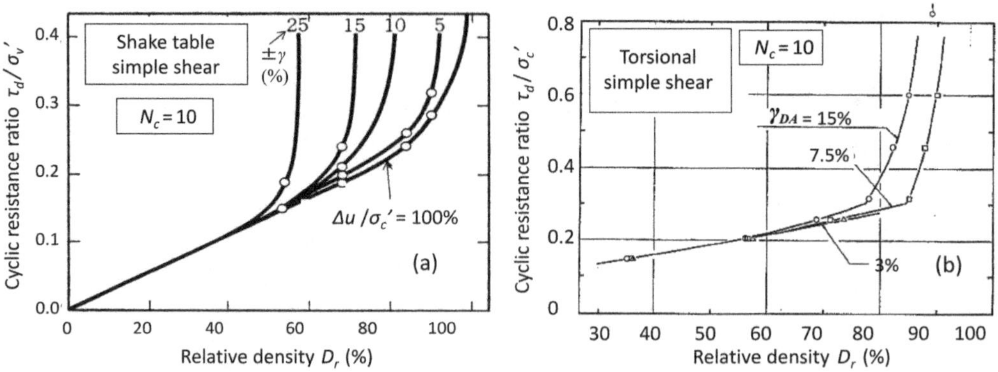


*Figure 5.3.7* CRR corresponding to various strain amplitudes $\gamma$ versus relative density $D_r$ obtained from undrained simple shear tests for clean sands: (a) Shake table test results (De Alba et al. 1976) with permission from ASCE, (b) Torsional simple shear test results (Tatsuoka et al. 1982).

structural failure. If the double amplitude strain $\gamma_{DA}=7.5\%$ or $\varepsilon_{DA}=5\%$ is considered corresponding to the initial liquefaction of 100% pore-pressure buildup, it may be said that *CRR* is proportional to $D_r$ up to $D_r=70$–$80\%$.

Fig. 5.3.7(b) shows a similar result on the *CRR* for $N_c=20$ versus $D_r$ relationship obtained by another research using a torsional simple shear device for clean Toyoura sand prepared by the air-pluviation (AP) method (Tatsuoka et al. 1982). It indicates again that the *CRR* versus $D_r$ relationships are almost linear up to $D_r=70$–$80\%$ for $\gamma_{DA}=3$–$7.5\%$. Reflecting the linear correlation, the following simple formula was typically used for the liquefaction evaluation in Japan (Ishihara 1977, Tatsuoka et al. 1978).

$$CRR=0.0042\times D_r(\%) \tag{5.3.3}$$

#### 5.3.2.2 Influence of soil fabric on CRR

Though *CRR* essentially depends on the relative density, it was also found experimentally to be highly dependent on microscopic fabric of soil samples prepared in different methods in the laboratory. Fig. 5.3.8 shows *CRR* for $\varepsilon_{DA}=5\%$ or $\Delta u/\sigma_c'=1.0$ plotted versus $N_c$ obtained by cyclic triaxial tests on same clean sand of $D_r \fallingdotseq 50\%$ (Mulilis et al. 1977). The samples were prepared in 6 different methods a–f described in the chart, among that Method-a (the moist sand is densified with high-frequency vibration) and Method-f (the dry sand is pluviated in air with a constant fall height and a constant flux named as AP) appear to make the strongest and weakest fabric, respectively, with their difference in *CRR* becoming almost 60–70% for the same $D_r=50\%$ sand.

Fig. 5.3.9(a) depicts *CRR*-values obtained in torsional simple shear tests for $\gamma_{DA}=3$ or 7.5% in $N_c=20$ plotted versus relative densities $D_r$ of specimens prepared by two methods: water tapping (WT: sand is rained in a water-filled mold and then densified by tapping from outside) and the air-pluviation (AP) (Kokusho et al. 1983a). As the *CRR*-values tend to increase drastically for $D_r \geq 80\%$ for both samples, the

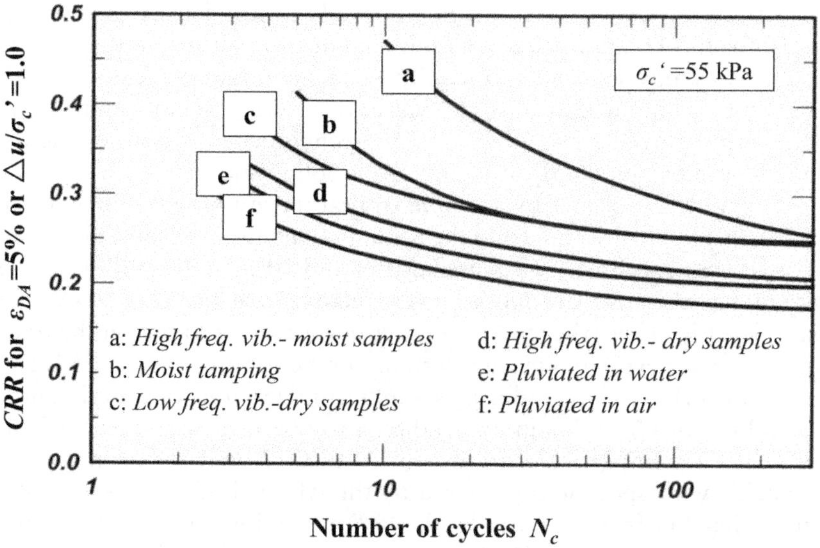


*Figure 5.3.8* *CRR* versus number of cycles $N_c$ for specimens prepared by different methods (modified from Mulilis et al. 1977) with permission from ASCE.

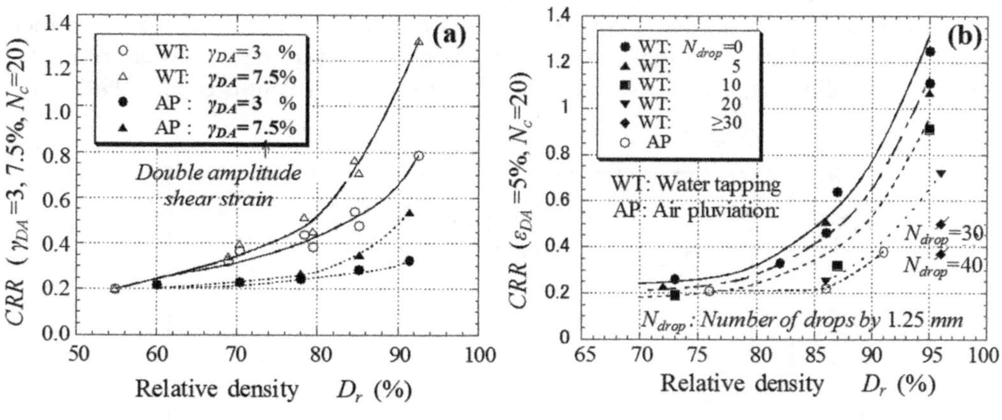


*Figure 5.3.9* Variations of *CRR* ($\varepsilon_{DA} = 5\%$, $N_c = 20$) versus relative density $D_r$ for specimens prepared by two different methods WT & AP (a), and Impact of a given number of small drops $N_{drop}$ on *CRR* versus $D_r$ plots (b) (Kokusho et al. 1983a).

difference between the two *CRR*-values is almost doubled there. This is probably because the WT method tends to introduce the soil fabric more resistant to cyclic loading in terms of particle orientations and contacts than the AP method particularly for denser sands.

These laboratory test results indicate that liquefaction resistance cannot be determined solely by the relative density but very much influenced by how soil fabric was formed when the soil was deposited in situ. This further suggests that liquefaction tests have to be done on samples recovered directly from in situ soils. However, how to recover sand samples without changing not only in situ density (neither densifying loose sands nor loosening dense sands) but also in situ delicate soil fabric may never be an easy task. Sandy soil tends to readily lose its soil fabric because of mechanical disturbance such as induced strains and vibrations during in situ sampling and laboratory handling. Conventional tube sampling techniques may not be able to recover sandy soils without significantly affecting subtle soil fabric.

Fig. 5.3.9(b) demonstrates by a simple laboratory test how significantly mechanical disturbance may deteriorate soil fabric and liquefaction resistance (Kokusho et al. 1985). Triaxial test specimens (5 cm in diameter and 10 cm in height) of clean Toyoura sand were prepared by the WT method with relative densities $D_r \fallingdotseq 70$–95% to build in strong soil fabric. Then, the specimen, after dewatered, was put on the brass-cup of the liquid-limit test device and given a certain number of drops $N_{\mathrm{drop}}$ of 1.25 mm representing the disturbance effects during sampling and laboratory handling procedures. The *CRR*-values for $\varepsilon_{DA} = 5\%$ in $N_c = 20$ tend to obviously decrease with increasing $N_{\mathrm{drop}}$ for $D_r \approx 95\%$ in particular (down to 1/4 of *CRR* without drop $N_{\mathrm{drop}} = 0$) and almost coincide with specimens prepared by the AP method. The test results vividly demonstrate that soil fabric built in by the WT method is very easy to deteriorate by a series of faint shocks, though $D_r$ does not noticeably change due to the dropping as indicated by the vertical alignment of the plots in the diagram.

Fig. 5.3.10 shows the *CRR*-values for $\varepsilon_{DA} = 5\%$ in $N_c = 15$ versus $D_r$ for intact samples recovered in place by an in-situ freezing technique. In this sampling, in situ soils are first frozen by circulating a coolant through underground vertical pipes and then drilled to recover intact to make laboratory tests in a test apparatus after trimming, setting, consolidating with in situ stresses and thawing (Yoshimi et al. 1994). This

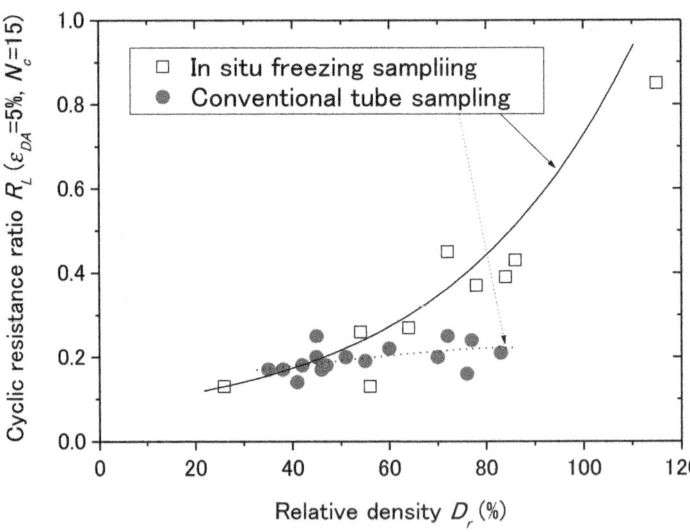


*Figure 5.3.10* *CRR* versus relative density $D_r$ by cyclic triaxial tests on soil specimens recovered by in situ freezing sampling and conventional tube sampling (Yoshimi et al. 1994).

sampling technique is generally believed to be effective to preserve in situ soil fabric, though very costly. The *CRR*-values by open symbols for the freezing sampling in the diagram tend to increase markedly for $D_r \geq 60\%$, while those of the conventional tube sampling methods with close symbols are obviously lower with no increasing tendency up to $D_r = 80\%$, indicating that in situ sand samples by conventional tube sampling may give distinctively lower *CRR* for $D_r \geq 60\%$.

Thus, *CRR*-values of in situ soils are basically dependent on their relative densities, though there seems to exist no unique correlation between the two, because soil fabric formed in soil sedimentation processes tend to have a great influence on *CRR*. The best way to evaluate the in situ *CRR*-values would be to have undisturbed samples by in situ freezing and subsequent drilling, or block sampling from dewatered trenches by hand very carefully.

In normal engineering practice, where undisturbed sampling plus laboratory tests are too costly, *CRR* is determined by in situ sounding such as Standard Penetration Tests (SPT) or Cone Penetration Tests (CPT). The penetration resistance of SPT or CPT is recognized to have a close correlation with relative density. However, the effect of soil fabric may not be fully reflected in the penetration tests.

### 5.3.3 Effect of stress/strain history

From the above findings, it may well be inferred that the *CRR*-values of natural sand deposits are strongly influenced by how deposits were stratified and what kind of mechanical and geochemical histories they experienced since then. Two typical long-term mechanical effects for in situ sands may be overconsolidation and preshearing.

It is easy to understand that the overconsolidation tends to increase the $K_0$-value and hence the effective confining stress $\sigma'_c$. Because *CRR* defined as $\tau_d/\sigma'_c$ can be uniquely correlated with $\sigma'_c = (1 + 2K_0)\sigma'_v$ as already mentioned in Sec. 5.3.1, *CRR*-values defined by $\tau_d/\sigma'_v$ become higher for overconsolidated soils under the same relative density. Besides that, the overconsolidation will raise *CRR* higher even for the same $\sigma'_c$ and $D_r$ due to the change of soil fabric. Fig. 5.3.11 summarizes increasing rates of *CRR* relative to that of $OCR = 1.0$ with increasing overconsolidation ratios ($OCR$) obtained by different investigators (Ishihara and Takatsu 1979, Kokusho et al. 1983a, Tatsuoka et al. 1988). Though the evaluations are largely diverted among researchers depending on different sands and test methods, the effect may be approximated by a simple power function as:

$$CRR/CRR_{OCR=1} = (OCR)^m \tag{5.3.4}$$

where the exponent $m$ varies 0.1~0.5. Combining the $K_0$-effect with this, *CRR* of overconsolidated clean sands $(\tau_d/\sigma'_v)_{OCR}$ may be formulated using *CRR* of isotropically and normally consolidated specimens $(\tau_d/\sigma'_c)_{OCR=1}$ as:

$$(\tau_d/\sigma'_v)_{OCR} = \frac{1 + 2K_0}{3}(OCR)^m(\tau_d/\sigma'_c)_{OCR=1} \tag{5.3.5}$$

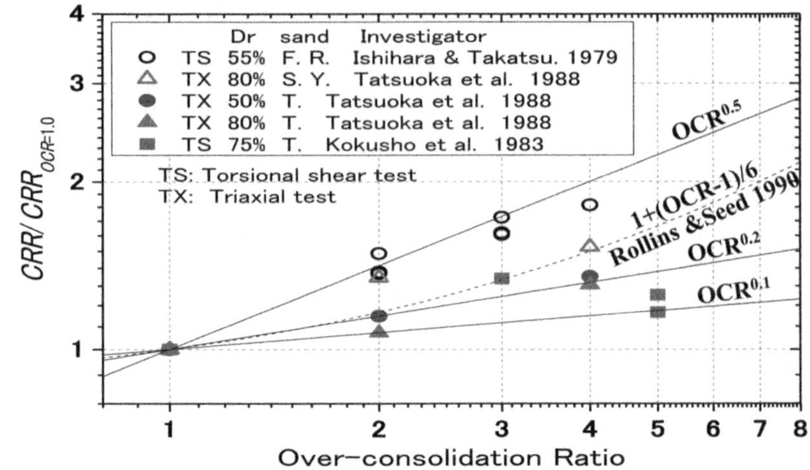


*Figure 5.3.11* Increasing rate of *CRR* with increasing *OCR* by different test methods on various sands (Ishihara and Takatsu 1979, Kokusho et al. 1983a, Tatsuoka et al. 1988, Rollins and Seed 1990).

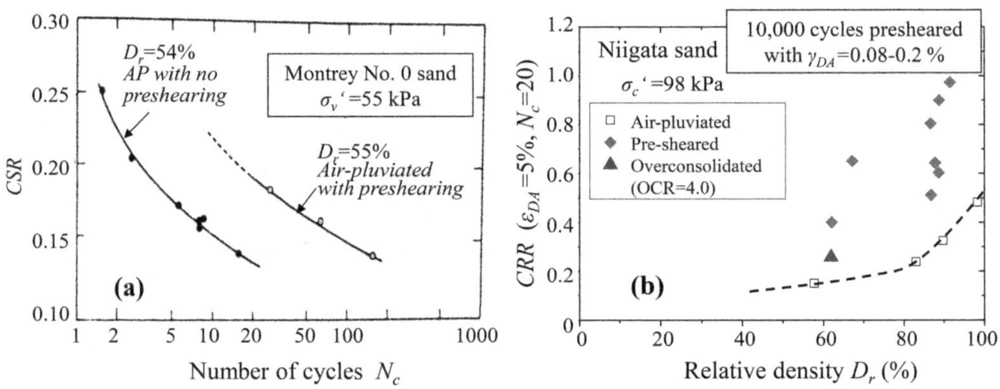


*Figure 5.3.12* Effect of small-strain preshearing on AP clean sands: (a) *CSR–$N_c$* curves without/with preshearing by shake-table tests (Seed et al. 1977) and (b) *CRR–$D_r$* plots with/without preshearing and with over-consolidation by cyclic triaxial tests (Tokimatsu et al. 1986).

Similar test results in USA indicate another formula also shown in the diagram assuming a linear function of OCR (Rollins and Seed 1990) as:

$$CRR/(CRR)_{OCR=1} = 1 + (OCR - 1)/6 \tag{5.3.6}$$

Another suspected long-term effect in situ is low-strain pre-shearing by a number of small seismic vibrations that soils may have experienced after deposition. Fig. 5.3.12(a) compares $CSR{\sim}N_c$ curves obtained by simple shear tests using a shaking table for the same clean sand prepared by the AP method, without or with adding a given number of low-strain preshearing (Seed et al. 1977). The sand with preshearing shows higher liquefaction resistance by about 50% than that without preshearing despite almost identical relative densities $D_r = 54\sim55\%$. The effect of preshearing was also investigated by Tokimatsu et al. (1986) in triaxial tests in which ten thousands cycles of 0.08~0.2% axial strain were applied to dense AP sand specimens before conducting undrained cyclic loading for liquefaction. Fig. 5.3.12(b) indicates the significant effect of preshearing, doubling or tripling the *CRR*-values of AP-sands without considerable changes in $D_r$. It also indicates that the effect of preshearing employed here is much more dominant than the overconsolidation effect of $OCR = 4.0$ also plotted in the diagram.

## 5.4 EFFECT OF GRAVELS AND FINES

### 5.4.1 Particle grading

Though clean sands are the most typical liquefiable soil, sands containing non/low-plastic silts and gravels are also liquefiable. In order to deal with soils containing various particle size, the uniformity coefficient $C_u$ is defined to represent the particle grading as:

$$C_u = \frac{D_{60}}{D_{10}} \tag{5.4.1}$$

where $D_{60}$ and $D_{10}$ are the particle sizes of 60% and 10% fines by weight, respectively.

In Fig. 5.4.1(a), grain size curves of granular soils, RS1, RS2 and RS3 are illustrated with solid lines to be addressed in the later discussions. These materials were reconstituted from fluvial sands and gravels of non-weathered hard particles to make the mean grain sizes $D_{50} = 0.14$, 0.40, 1.15 mm and the uniformity coefficients $C_u = 1.44$, 3.79, 13.1 for RS1, RS2, RS3 respectively. Soil names DG1, DG2, DG3 written on the

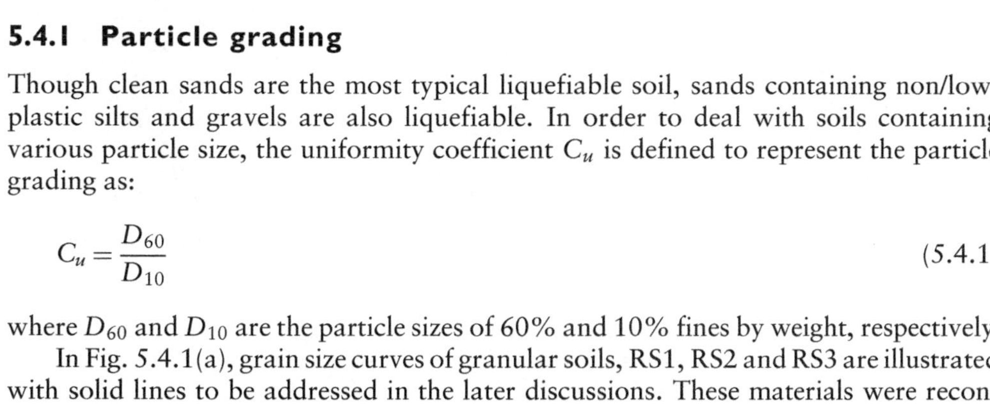


*Figure 5.4.1* Reconstituted sandy/gravelly soils RS1 (DG1), RS2 (DG2), RS3 (DG3): (a) Grain size curves, (b) Maximum or minimum density $\rho_{max}$ or $\rho_{minx}$ versus fines content $F_c$ (Kokusho 2007).

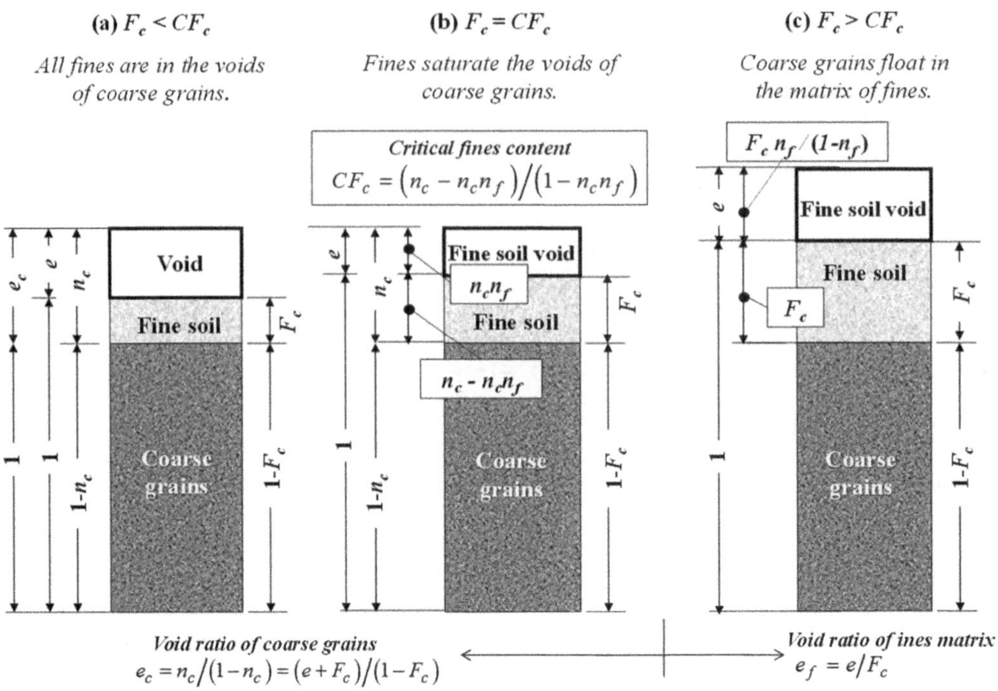


*Figure 5.4.2* Binary packing model for soils consisting of two-size particles changing soil structure with increasing fines content: (a) $F_c < CF_c$, (b) $F_c = CF_c$, (c) $F_c > CF_c$ where $CF_c$ = critical fines content.

same diagram are decomposed granite soils having the same grain size curves and will also be addressed later. For the RS1 and RS3, low-plasticity fines ($I_p = 6$) are mixed with varying fines content as shown by thin dashed curves. The maximum and minimum densities $\rho_{max}$, $\rho_{min}$ of these materials measured by a standardized test method (JGS 2009) are shown in Fig. 5.4.1(b). The densities $\rho_{max}$, $\rho_{min}$ increase from RS1 to RS3, indicating that well-graded soils tend to have larger density than poorly-graded soils. The same diagram also shows the variations of $\rho_{max}$ and $\rho_{min}$ with increasing fines content $F_c$ for the poorly-graded RS1 and well-graded RS3. The densities seem to be increasing or almost stationary with increasing $F_c$ up to certain $F_c$-values, then followed by downturns for both RS1 and RS3 at different $F_c$.

The $F_c$-dependent variations of $\rho_{max}$ and $\rho_{min}$ may be interpreted in the light of a simplified "binary packing" model illustrated in Figs. 5.4.2(a)–(c), where a soil assumed to consist of only two particle sizes, coarse grain and fines matrix, is packed in a container of a given volume (e.g. Skempton and Brogan 1994). If the fines content $F_c$ is low as in (a), the fines stay in the void of the coarse grains. In this condition, the void ratio of coarse grains $e_c$ (neglecting the fines) is formulated using $e$ = global void ratio as:

$$e_c = \frac{e + F_c}{1 - F_c} \tag{5.4.2}$$

With increasing $F_c$, there exists a threshold named here as "critical fines content", $CF_c$, at which the fines are just enough to saturate all the void of coarse grains as shown in (b). In this situation, the volume of fines including the internal voids is equal to the porosity of the coarse grains $n_c$ if the total soil volume is unity, and the solid volumes of fines and fines plus coarse grains are $n_c(1-n_f)$ and $1-n_c+n_c(1-n_f)=1-n_cn_f$, respectively, where $n_f$ = porosity of fines. Hence the critical fines content $CF_c$ can be expressed, assuming the same solid density for both particles, as:

$$CF_c = \frac{n_c(1-n_f)}{1-n_cn_f} \tag{5.4.3}$$

Beyond $CF_c$, the fines matrix overflows the coarse grain voids and interrupts the direct contacts between grain particles, drastically changing the soil structure from grain-supporting to matrix-supporting. Then, the soil voids are only in fines matrix as in (c), and the matrix void ratio $e_f$ can be correlated with the global void ratio $e$ as:

$$e_f = \frac{e}{F_c} \tag{5.4.4}$$

Correspondingly, the density increases as $F_c$ increases from 0 to $CF_c$ because the void is filled with fines without increasing the total soil volume and then start to decrease thereafter because the total volume increases with the increasing fines matrix of lower density. In Fig. 5.4.1(b), the values $CF_c$evaluated in Eq. (5.4.3) from $n_c$, $n_f$ corresponding to the maximum and minimum density tests are shown with the ranges in between parallel lines, indicating that $CF_c$ is evaluated smaller for the well-graded RS3 than the poorly-graded RS1 because of the smaller $n_c$ value. In the measured $F_c$-dependent density curves shown in Fig. 5.4.1(b), too, the density peaks tend to occur at lower $F_c$ in RS3 than RS1, though the absolute value is not agreeable due to the difference between the simplified binary packing model and the actual continuous particle size distributions.

Thus, the effect of particle gradation or uniformity coefficient on liquefaction resistance of gravelly soils and silty sands may be differentiated into two different mechanisms; the dependency of soil density on the uniformity coefficient $C_u$, and the change in the soil structure with increasing fines content $F_c$ from grain-supporting to matrix supporting. As for the fines content, the plasticity of fines is another important subject considering that fine soils involved in liquefaction so far are essentially non/low-plastic. In the following, the effect of particle grading is first discussed on liquefaction resistance of poorly-graded sands and well-graded gravelly soils. Then, the effect of non/low-plastic fines on the liquefaction resistance of sands and gravelly soils is dealt with after discussing the effect of the plasticity of fines in general.

## 5.4.2 Liquefaction resistance of gravelly soils

### *5.4.2.1 Gravelly soils actually liquefied*

Liquefactions of gravelly soils were reported way back in the 1948 Fukui earthquake in Japan, the 1964 Alaskan earthquake in USA and several other earthquakes including Chinese earthquakes. Detailed investigations on liquefied gravelly soils were conducted

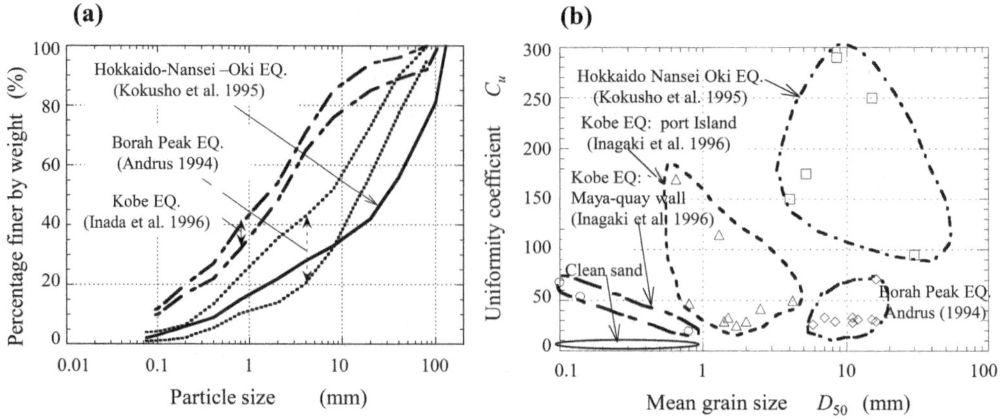


*Figure 5.4.3* Field investigation results on particle grading of previously liquefied gravelly soils: (a) Grain size curves, (b) $D_{50}$ versus $C_u$ plots (Kokusho 2007).

during the 1983 Borah Peak earthquake in USA (Andrus 1994), the 1993 Hokkaido-Nansei-Oki earthquake (Kokusho et al. 1995) and the 1995 Kobe earthquake in Japan (e.g. Inagaki et al. 1996). Fig. 5.4.3(a) shows the grain size curves of those liquefied gravelly soils. They are all well-graded, containing gravels up to 100 mm as well as sands and even some fines. They may actually have contained boulders still larger, too large for sieving tests. Fig. 5.4.3(b) plots the mean grain size $D_{50}$ versus the uniformity coefficient $C_u$ of previously liquefied gravelly soils in the horizontal and vertical axes. If compared with typical poorly graded clean sands also shown in the diagram, it is clear that very well-graded gravelly soils with large $C_u$ and $D_{50}$ actually liquefied. During the Hokkaido Nansei-Oki earthquake, rock debris exceeding a half meter were involved in the liquefied soils (Kokusho et al. 1995), indicating that there is literally no upper limit for the gravel size.

For such well-graded soils, one may imagine that sand particles are exclusively responsible for liquefaction, while larger gravelly particles are only floating in sand particles. This view may be justified in gap-graded gravelly sands where sand particles are overwhelming so that there is little direct contact between gravels as already mentioned in the binary packing model shown in Fig. 5.4.2. This seems unlikely however in the well-graded soils actually liquefied in previous earthquakes shown in Fig. 5.4.3(a), where the grading curves are smooth and large gravel particles, too, are considered to take part in liquefaction.

Absolute dry densities of these well-graded gravelly soils are much higher than poorly-graded sands. However, gravelly soils are sometimes very loosely deposited with a low relative density showing very low $N$-values and S-wave velocities. Fig. 5.4.4 shows SPT $N$-value versus S-wave velocity ($V_s$) plots in those gravels previously liquefied. This indicates very low values; $N = 5{\sim}16$ and $V_s = 60\,\text{m/s}{\sim}210\,\text{m/s}$ as low as poorly-graded loose sand despite the significant difference in the absolute density. For instance, the dry densities of liquefied gravelly soils actually measured were $\rho_d = 1.7{\sim}2.0\,\text{t/m}^3$ for the decomposed granite fill in Kobe and $2.0{\sim}2.1\,\text{t/m}^3$ for the

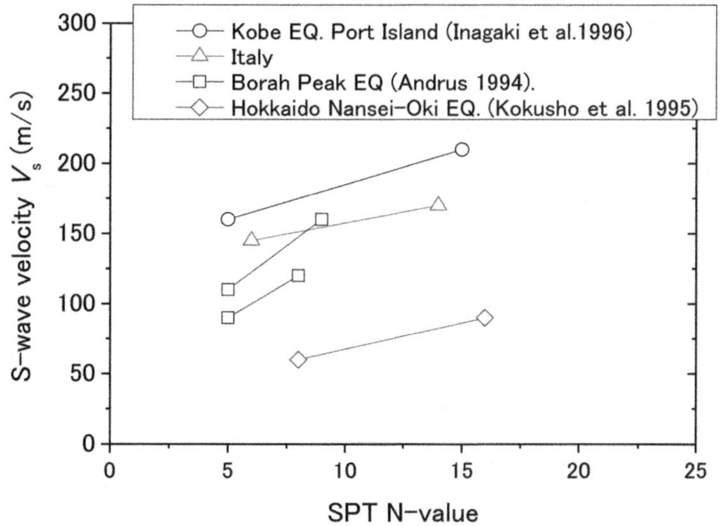


*Figure 5.4.4* SPT *N*-values versus S-wave velocities in liquefied gravelly soils (Kokusho 2007).

Hokkaido case, versus it is 1.4~1.5 t/m$^3$ for typically liquefiable clean sands. Despite this distinctive difference, well-graded gravelly soils may liquefy if the relative density is sufficiently low.

#### 5.4.2.2 *Liquefaction resistance by cyclic triaxial test*

Undrained cyclic loading test results in a systematic test program for sandy and gravelly soils are visited here to see the effect of grain size distributions on liquefaction resistance of sandy/gravelly soils. The tested three materials are RS1, RS2 and RS3 ($C_u = 1.44$, 3.79, 13.1, respectively) shown with the thick lines in Fig. 5.4.1(a) with parametrically changing relative densities (Kokusho et al. 2004). The specimen size was 200 mm in height and 100 mm in diameter about 5 times the maximum gravel size and all the tests were performed under the isotropic confining stress of $\sigma'_c = 98$ kPa. The open symbols in Fig. 5.4.5 show the cyclic stress ratios $CSR = \sigma_d/2\sigma'_c$ for the double amplitude axial strain $\varepsilon_{DA} = 5\%$ versus the number of loading cycles $N_c$ for RS1–RS3 having nominal relative density $D_r \approx 50\%$. All the plots are corrected for the MP-effect (Sec. 2.2.3.4) based on Tokimatsu and Nakamura (1986) and Tanaka et al. (1991). Though the *CSR*-values for RS1 (open circles) seem to be slightly lower than those for RS2 and RS3 (other open symbols), the differences may not be so clear compared to the degrees of data dispersions.

In Fig. 5.4.6(a), the *CRR*-values $\sigma_d/2\sigma'_c$ for $\varepsilon_{DA} = 5\%$, $N_c = 20$ are plotted versus relative densities varying from $D_r \approx 20\%$ to 90% for RS1–RS3. Despite some data dispersions, it may well be judged that *CRR* is almost uniquely correlated with the relative density $D_r$ for all RS1–RS3. The same *CRR*-values are plotted in Fig. 5.4.6(b) versus the uniformity coefficients $C_u$ of the three tested materials with solid symbols and tied to each other by solid lines for individual $D_r$ as a parameter. It may be said that the *CRR*-values are essentially unchanged with increasing $C_u$, though there are

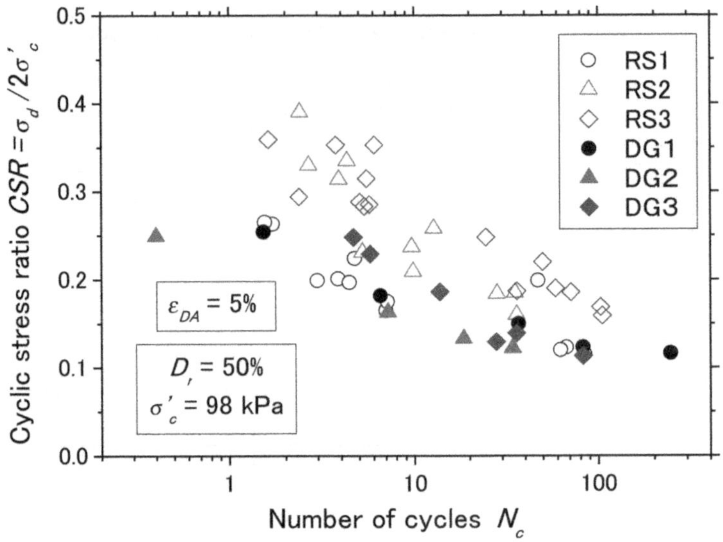


*Figure 5.4.5* Cyclic stress ratios $CSR = \sigma_d/2\sigma'_c$ for $\varepsilon_{DA} = 5\%$ versus the number of loading cycles $N_c$ for sands and gravelly soils (Kokusho et al. 2004) with permission from ASCE.

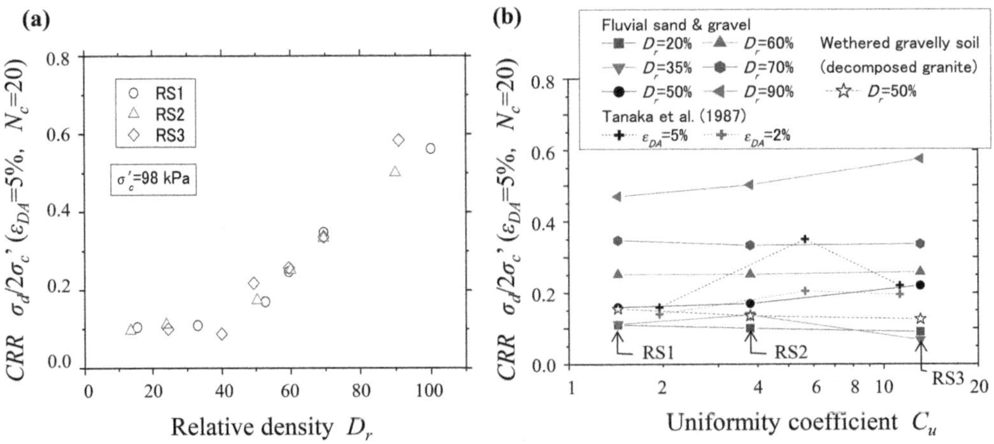


*Figure 5.4.6* *CRR* for RS1~RS3 with different uniformity coefficients $C_u$ and relative densities $D_r$: (a) *CRR* versus $D_r$, (b) *CRR* versus uniformity coefficients $C_u$ (Kokusho et al. 2004) with permission from ASCE.

$C_u$-dependent increasing trends for $D_r \approx 50\%$ and 90% by only 20–40% from RS1 ($C_u = 1.44$) to RS3 ($C_u = 13.1$). Other test data on gravelly soils by Tanaka et al. (1987) are also plotted on the same diagram, showing a nonsystematic $C_u$-dependent variation. On the other hand, Evans and Zhou (1995) conducted a series of undrained

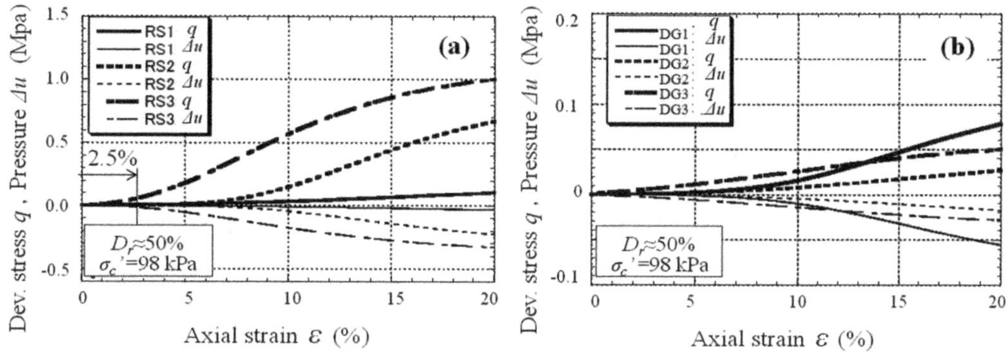


*Figure 5.4.7* Deviator stress $q = \sigma_1 - \sigma_3$ and pore-pressure $\Delta u$ versus the axial strain $\varepsilon$ during post-liquefaction undrained monotonic loading tests: (a) Fluvial sand and gravels RS1~RS3 of fresh and hard particles, (b) DG1~DG3 of weathered crushable particles (Kokusho 2007) with permission from ASCE.

cyclic triaxial tests on sand-gravel specimens with varying gravel contents, showing about 50% increase in liquefaction resistance with increasing gravel content for the same relative density, though the material was not continuously graded but gap-graded by sands and gravels. If the marginal $C_u$-dependency in Fig. 5.4.6(b) is compared with the considerable $D_r$-dependent increase of more than 4 times in the $CRR$ between $D_r = 40$ to 100% shown in Fig. 5.4.6(a), it may well be said that $CRR$ of gravelly soils defined by $\varepsilon_{DA} = 5\%$ is almost uniquely determined by the relative density irrespective of their particle gradations or uniformity coefficients.

### 5.4.2.3 Post-liquefaction behavior of gravelly soils

Post-liquefaction behavior of liquefied soils attracts a greater attention these days, because the performance-based design increasingly employed needs such soil performance. In order to simplify liquefaction behavior influenced by initial shear stresses, the following two-step test procedure may be employed; (i) a stress-controlled cyclic undrained test for isotropically consolidated specimens to attain initial liquefaction in a level ground without initial shear stress, and (ii) a post-liquefaction strain-controlled monotonic undrained test on the same specimen just after the preceding liquefaction test assuming that the initial stress starts to work after the liquefaction.

Fig. 5.4.7(a) shows the variations of the deviator stress $q = \sigma_1 - \sigma_3$ and the excess pore-pressure $\Delta u$ versus the axial strain $\varepsilon$ during the post-liquefaction undrained monotonic loading tests for RS1–RS3. Specimens of $D_r \approx 50\%$ were cyclically loaded under isotropic effective confining stress $\sigma_c' = 98$ kPa up to the double amplitude strain $\varepsilon_{DA} \approx 10\%$, then sheared monotonically in the undrained condition. The curves indicate that the soils behave almost like a liquid up to a certain axial strain without any mobilized stress $q$ or pressure change $\Delta u$. Then, $q$ and $\Delta u$ start to pick up due to positive dilatancy which tend to occur at smaller strain and with higher rate for well-graded RS3 than for poorly-graded RS1. The ultimate shear resistance of

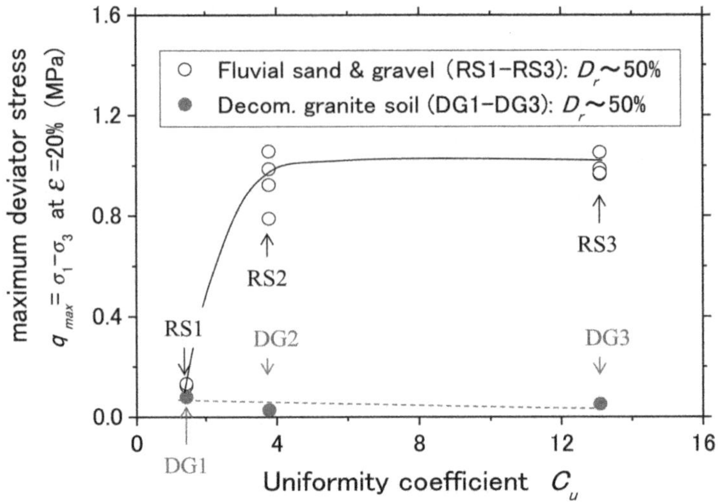


*Figure 5.4.8* Maximum deviator stress $q = q_{\max}$ versus uniformity coefficient $C_u$ for RS1~RS3 with identical relative density $D_r \approx 50\%$ (Kokusho et al. 2004) with permission from ASCE.

RS3 at $\varepsilon = 20\%$ becomes considerably larger than that of RS1 despite almost the same $D_r$.

The test results on the post-liquefaction shear resistance observed in Fig. 5.4.7(a) are in a sharp contrast to the liquefaction test results shown in Fig. 5.4.6(a). It should be emphasized here that the $CRR$-value is defined at the double amplitude axial strain $\varepsilon_{DA} = 5\%$ which corresponds to $\varepsilon = 2.5\%$ in the single amplitude strain in the monotonic loading test. The $q$-curves in Fig. 5.4.7(a) are almost identical up to $\varepsilon = 2.5\%$ for RS1, RS2, RS3 of the same relative densities, then become divergent for larger strains. This indicates that the soil strength is uniquely determined by the relative density $D_r$ in the smaller strain up to a single strain amplitude $\varepsilon \approx 2.5\%$ corresponding to the initial liquefaction, then the effect of particle grading or uniformity coefficient starts to dominate for strains larger than that. Namely, well-graded gravelly soils tend to exhibit larger positive dilatancy and higher shear resistance than poorly-graded sands having the same relative density if they are sheared up to higher strains, because their absolute densities are larger, although their strength for the initial liquefaction corresponding to the lower strain level ($\varepsilon = 2.5\%$) is mostly determined by the relative density.

In Fig. 5.4.8, maximum deviator stresses $q = q_{\max}$ at the axial strain $\varepsilon = 20\%$ are plotted (open circles) versus the uniformity coefficients $C_u$ for RS1–RS3 with identical relative density $D_r \approx 50\%$ based on the test results shown in Fig. 5.4.7(a). According to this, $q_{\max}$ tends to increase with increasing $C_u$ up to around $C_u = 4$, and $q_{\max}$ of well-graded soils is several times larger than that of poorly-graded soils. If this is compared with the results in Fig. 5.4.6, it may be stated that the well-graded gravelly soils are as prone to the initial liquefaction as poorly-graded sands, but less prone to large post-liquefaction deformations.

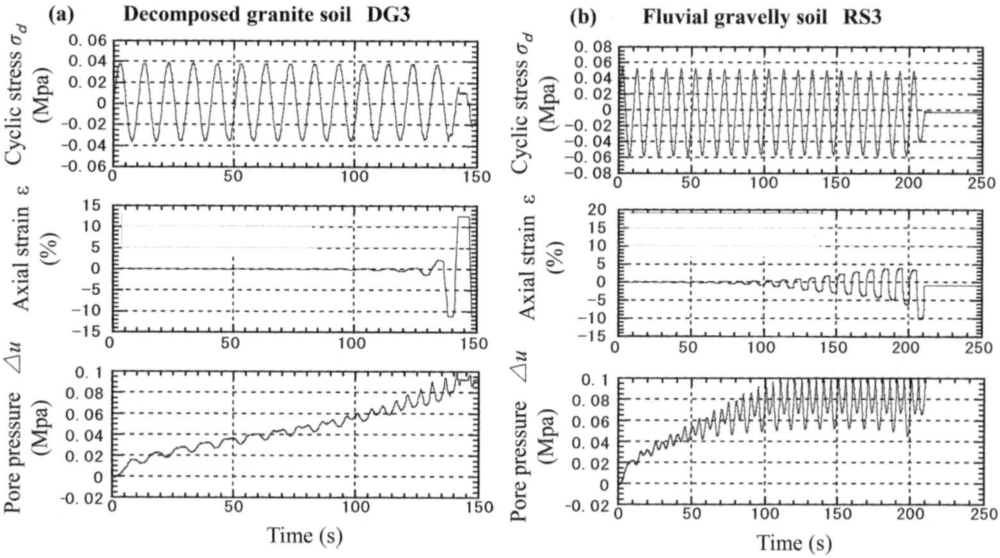


*Figure 5.4.9* Typical liquefaction test data of DG3 (a) compared with that of RS3 (b), having the same grain size and relative density $D_r \approx 50\%$ (Hiraoka 2000).

#### *5.4.2.4 Effect of particle crushability*

So far, the particles of granular soils were all fresh (non-weathered), hard in quality and difficult to crush. In nature, particles of some sort of granular soils are easy to crush. One of such crushable soils is decomposed granite, wherein gravel particles are strongly weathered and their angular ends are easy to break. Another typical crushable granular soils is calcareous sands consisting of porous and roundish particles originated from corals. There are still volcanic and other types of crushable granular soils in nature. In the following, some test results on decomposed granite will be addressed to see the effect of particle crushability on the liquefaction resistance in comparison with non-crushable fluvial granular soils of fresh grains.

Three materials, DG1, DG2, DG3, were reconstituted artificially using decomposed granite soil originated from Kobe city in Japan to have the same grain size curves as RS1, RS2, RS3, as shown in Fig. 5.4.1(a). *CSR* versus $N_c$ plots obtained for DG1, DG2, DG3 of $D_r \approx 50\%$ by a series of undrained cyclic triaxial tests are superposed in Fig. 5.4.5 with solid symbols. They are plotted lower than RS1, RS2, RS3 of the corresponding grain sizes, respectively, particularly for the gravelly DG3. Typical time histories for DG3, of cyclic axial stress $\sigma_d$, axial strain $\varepsilon$ and excess pore-pressure $\Delta u$, are compared in Fig. 5.4.9 with those for RS3 having the same grain size. Despite the same density $D_r \approx 50\%$, DG3 behaves like loose sand with small cyclic-mobility pore-pressure changes followed by abrupt increase in axial strain just before the initial liquefaction, while RS3 behaves like dense sand with large cyclic-mobility pore-pressure fluctuations and gradual strain increase.

*CRR*-values of the DG soils read off from liquefaction test results are also plotted in Fig. 5.4.6(b) with the open star-symbols. If they are compared with the solid circles for the fluvial soils of the same relative density $D_r \approx 50\%$, the *CRR*-values of the DG soils are almost the same at $C_u = 1.44$, but getting smaller with increasing $C_u$, indicating that coarser gravels are damaged by weathering more strongly than sands.

The DG soil specimens of $D_r \approx 50\%$, after undrained cyclic loading up to $\varepsilon_{DA} \approx 10\%$, were monotonically loaded in the undrained condition in the same manner as the fluvial soils to see their post-liquefaction behavior. Fig. 5.4.7(b) shows the variations of deviator stress $q = \sigma_1 - \sigma_3$ and excess pore-pressure $\Delta u$ versus the axial strain $\varepsilon$ for DG1~DG3. Note that the scale in vertical axis in Fig. 5.4.7(b) is about one tenth of that in Fig. 5.4.7(a), indicating that the shear resistance for larger strains is very much depressed in DG2 and DG3 in particular presumably because the coarser particles tend to reflect the weathering more strongly. The same observation can be made in Fig. 5.4.8, where the maximum deviator stress $q_{\max}$ at the axial strain $\varepsilon = 20\%$ is plotted with the close circles versus the uniformity coefficient $C_u$ for DG1~DG3. Unlike the fluvial soils, no increasing trend in $q_{\max}$ can be seen with increasing $C_u$ from the poorly-graded to well-graded DG soils.

Thus, the crushability of gravelly particles plays a significant role not only in the liquefaction potential but also the post-liquefaction undrained shear resistance. If gravel particles are non-weathered and hard in quality, well-graded soils exhibit strong shear resistance against post-liquefaction larger deformations, although their resistance against the liquefaction triggering by cyclic loading for the double-amplitude axial strain $\varepsilon_{DA} = 5\%$ is more or less the same as poorly-graded sands of the same relative density.

### 5.4.3 Liquefaction resistance of fines-containing soils

In liquefaction cases after the 1964 Niigata earthquake such as the 1999 Kocaeli earthquake Turkey, the 1999 Chi-Chi earthquake Taiwan and the 2011 Tohoku earthquake Japan, it was increasingly recognized that sands containing plenty of non/low-plastic fines are as liquefiable as clean sands. In this respect, it is readily conceivable that the plasticity or cohesion of fines has a lot to do with the liquefaction susceptibility of these soils.

#### 5.4.3.1 *Plasticity of fines*

Fig. 5.4.10(a) shows the triangular classification chart of liquefied sand before 1980 summarized by Tokimatsu and Yoshimi (1983). It indicates that, while soils with fines content $F_c$ as high as 60~70% have liquefied, none of the soils with the clay content $C_c > 20\%$ did liquefy in good agreement with a study in China (Seed and Idriss 1981). According to another study in China (Finn 1982), plasticity index $I_p = 10$ or lower seems to be the condition for liquefaction. Fig. 5.4.10(b) summarizes physical properties of sand boils erupted from reclaimed soils during 4 earthquakes (Mori et al. 1991), indicating that the erupted soils are mostly $C_c < 10$ despite very high fines content up to $F_c \fallingdotseq 100\%$. Though the ejecta may change its $F_c$ and $C_c$ from original soils during the sand boiling process, it indicates at least that there seems to be no upper limit of $F_c$ for liquefied sands to be erupted.

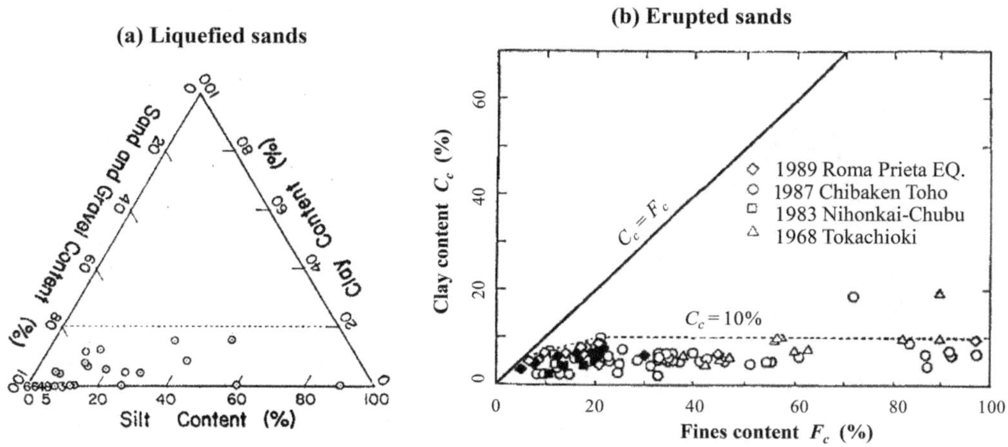


*Figure 5.4.10* Physical properties of sands containing fines liquefied in previous earthquakes: (a) Triangular classification of liquefied sands (Tokimatsu & Yoshimi 1983), (b) $F_c$ versus $C_c$ plots for erupted sands (Mori et al. 1991).

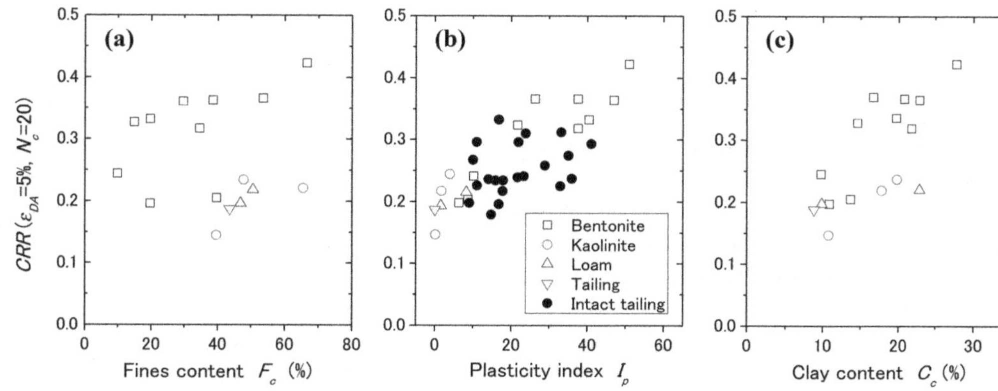


*Figure 5.4.11* Effects of properties of fines on liquefaction resistance of fines-containing soils: (a) $CRR \sim F_c$, (b) $CRR \sim I_p$, (c) $CRR \sim C_c$ (modified from Koseki et al. 1986).

Figs. 5.4.11(a)–(c) show the results from systematic undrained cyclic triaxial tests using sand specimens containing a variety of fine soils ($F_c = 10$–$67\%$, $I_p = 0$–$51$, $C_c = 9$–$28\%$) to see how their properties influence the $CRR$-values (Koseki et al. 1986). The figure obviously shows that the physical properties have definite effects on $CRR$, among which the plasticity index $I_p$ and the clay content $C_c$ show clearer positive correlations with $CRR$. Thus, these two variables representing the cohesion of fines can serve as relevant indices for screening the liquefaction potential. In contrast, $F_c$ is not so closely correlated with $CRR$, probably because $F_c$ is not directly representing soil plasticity compared to $I_p$ or $C_c$.

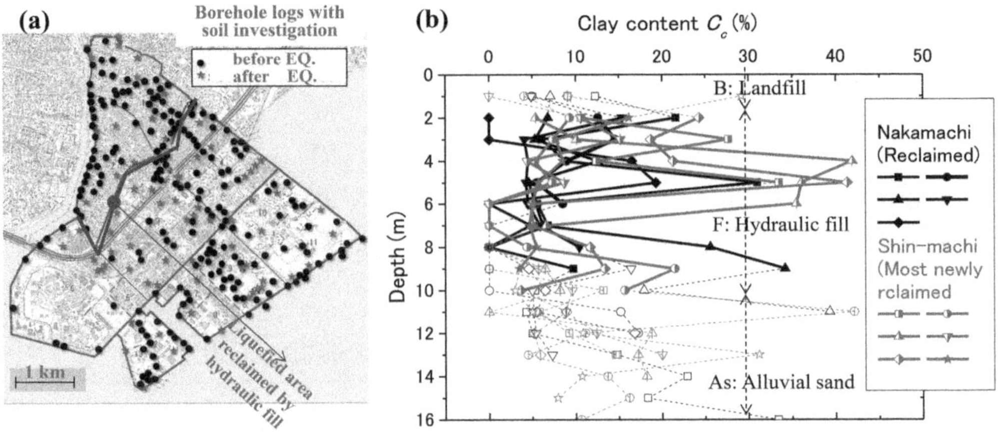


*Figure 5.4.12* Map of Urayasu city along the Tokyo bay and liquefied areas (a), and Depth-dependent $C_c$-variations along different boreholes in the liquefied areas (b).

In this regard, the following recent case history may be worth to be addressed here for further considerations on the effect of plasticity and also aging (Kokusho et al. 2014a). During the 2011 Tohoku earthquake in Japan ($M_J = 9.0$), extensive liquefaction occurred along the Tokyo bay area, more than 200 km far from the nearest edge of the earthquake fault. In Urayasu city in particular shown in Fig. 5.4.12(a), extensive liquefaction occurred in a wide area newly reclaimed after 1968 by hydraulically filled sea-bed soils. Huge amount of ejecta containing lots of fines, all non-plastic, covered ground surface, and the original ground surface subsided by 0.1–0.3 m at least. In a good contrast, another area of the same city existed before 1948 did not liquefy despite very similar soil profiles and properties. There were a plenty of borehole logs and SPT data available in this area as plotted in the figure, sharing similar soil profiles all over Urayasu, consisting of surface land fill (B-layer), alluvial sand layer (As-layer) and underlying Holocene soft clay on Pleistocene gravelly base. A big difference in the soil profile in the reclaimed area was the existence of hydraulic fill (F-layer) between B and As-layer, while it was missing in the non-reclaimed and non-liquefied area. Hence, the F-layer is considered to be responsible for the liquefaction during the earthquake.

In Figs. 5.4.13(a), (b), correlations for the F-layer between $F_c$–$I_p$ and $F_c$–$C_c$ are shown, respectively, with various sysmbols corresponding to stepwise vertical strains induced by soil subsidence as a key parameter. All the physical properties were obtained from soils sampled by the SPT split spoons after the earthquake. The properties are widely varied; $F_c = 0$–$100\%$, $I_p = 0$–$60$ and $C_c = 0$–$50\%$, indicating that soils presumably liquefied were very inhomogeneous and variable in plasticity. In each chart, horizontal and vertical dashed lines are drawn corresponding to Japanese design criteria (e.g. JRA 2002, AIJ 2001) with which soils are initially screened as potentially liquefiable by $F_c \leq 35\%$, $I_p \leq 15$ and $C_c \leq 10\%$. The earthquake-induced subsidence strains classified into four groups here were quantified

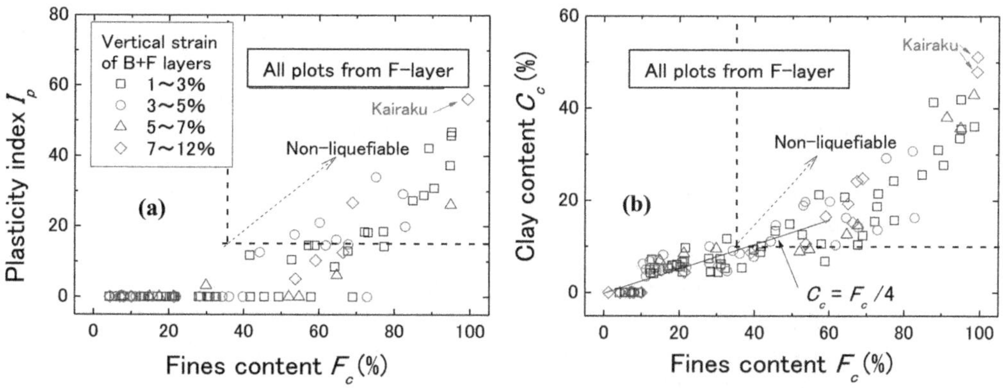


*Figure 5.4.13* Cross-correlations between physical properties of liquefied hydraulically-filled F-layer in Urayasu with stepwise subsidence strains: (a) $F_c$–$I_p$, (b) $F_c$–$C_c$, (c) $C_c$–$I_p$.

from ground surface subsidence divided by the total thickness of the F plus B layers, because the F-layer is considered to have liquefy by itself and presumably let the B-layer liquefy by the upward seegage flow. Despite the difficulty in interpreting the chart (because only a single strain value is available at each soil investigation point, while SPT-based physical properties have multiple values at individual depths of the same point), it may be said that the plots for larger subsidence strains (heavier liquefaction) tend to be located in zones of smaller $I_p$ or smaller $C_c$ except some abnormal plots belonging to a specific site (e.g. Kairaku). In contrast, a larger number of plots with larger strains are located for large $F_c$-values exceeding the threshold $F_c = 35\%$ indicating that $F_c$ may not be a better index than $I_p$ and $C_c$ for screening liquefiability (Kokusho et al. 2014a).

Fig. 5.4.12(b) shows depth-dependent variations of clay content $C_c$ at 11 boreholes in the liquefied area in Urayasu, where the soil profiles are composed of B-layer, F-layer and As-layer from the ground surface to 16 m deep. The plots are connected with either thick solid lines in the F-layer or thin dotted lines in the B and As-layers. The soils are very variable along the depth, with sublayers of low $C_c$ interbedded with those of high $C_c$ unlikely to liquefy. Nevertheless, widespread non-plastic ejecta covered the reclaimed area during the earthquake, and the soil subsidence strain in liquefied deposits exceeded 5% at many points as indicated in Fig. 5.4.13. A possibility may be that intensive liquefaction occurred initially in soils with small $C_c$, and then destabilized the soils with higher $C_c$, erupting considerable non-plastic ejecta to the ground surface. This indicates that sand deposits containing fines of higher plasticity than normally defined as liquefiable in design criteria may not be free from liquefaction if they are interbedded by liquefiable sands containing non/low-plastic fines as often observed in hydraulic filling soils as in Urayasu. This possibility was already suggested for the same reclaimed land before the earthquake in the discussion made by Yoshimi (1991) in conjunction with previous case histories on non-plastic ejecta in reclaimed deposits shown in Fig. 5.4.10(b).

#### 5.4.3.2 *Effect of non-plastic fines*

Liquefaction potential of soils, initially screened by design criteria in terms of physical properties, is then evaluated using SPT or CPT. If sands contain a measurable amount of fines, a boundary curve between liquefaction and non-liquefaction is modified in current engineering practice so that the liquefaction strength is raised for the same penetration resistance in accordance with fines content $F_c$ as will be explained in Sec. 5.5.2. The modification seems to be necessary because the penetration resistance cannot uniquely predict *CRR* by itself, necessitating another parameter such as fines content $F_c$. This $F_c$-dependent modification of liquefaction strength was originated from liquefaction case studies (Tokimatsu and Yoshimi 1983, Seed and De Alba 1984), wherein empirical boundary curves, developed in situ separating liquefaction/non-liquefaction on a *CRR* versus penetration resistance diagram, were found strongly dependent on fines content, though the mechanical basis of the finding was not clear. The same $F_c$-dependency of the boundary curves on the *CRR* versus penetration resistance diagram was also recognized when the *CRR*-values were determined by laboratory cyclic loading tests on intact samples (e.g. Suzuki et al. 1995).

In contrast to these findings, however, quite a few laboratory tests using reconstituted specimens showed that *CRR* clearly decreases with increasing $F_c$ of low plasticity fines for the same relative density. Thus, a lack of understanding of mechanism still remains in the current practice for liquefaction potential evaluation using the penetration resistance in relation with laboratory test results for sands containing fines of no/low-plasticity.

Fig. 5.4.14 shows triaxial test results by several investigators who measured the variations of *CRR*-values of sands by increasing contents of fines ($I_p \leq 5$) while keeping the relative densities unchanged (Huang et al. 1993, Sato et al. 1997, Polito and

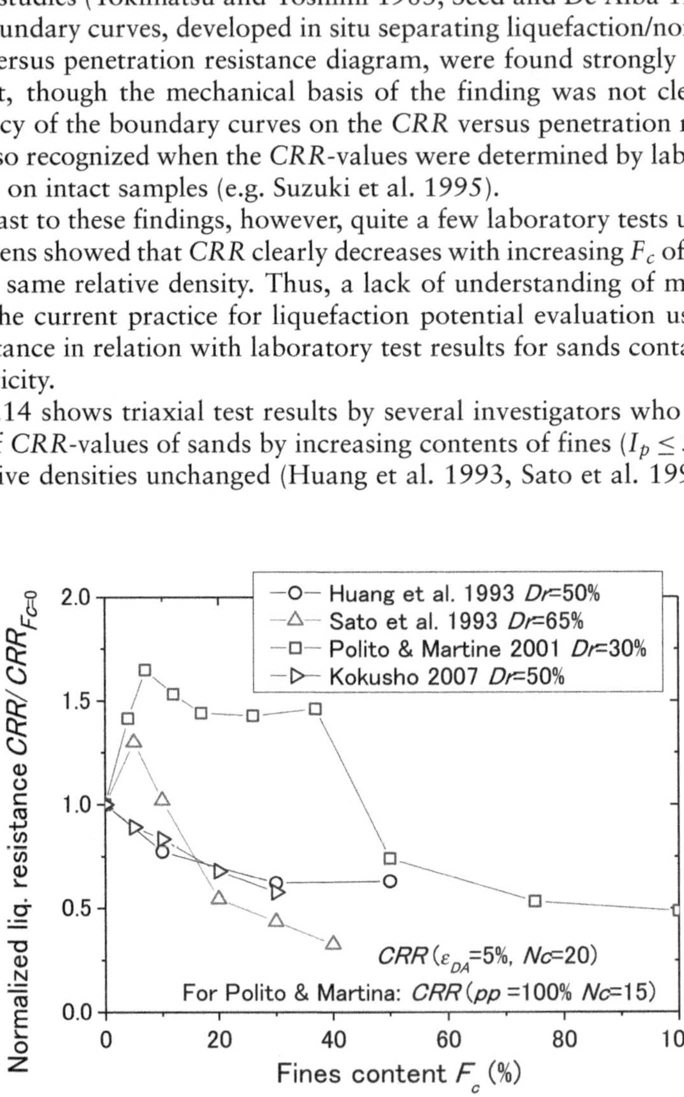


*Figure 5.4.14* Relative changes of *CRR*, $CRR/CRR_{Fc=0}$ with increasing fines content $F_c$ of sands in different previous investigations.

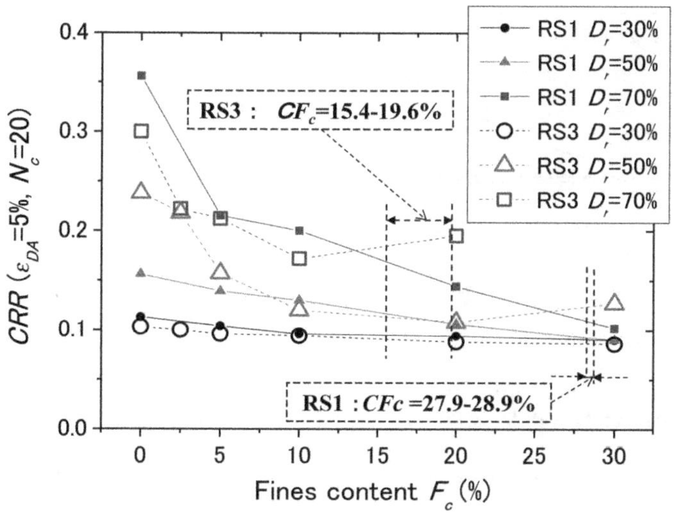


*Figure 5.4.15* Variations of *CRR* with increasing $F_c$ for poorly-graded RS1 and well-graded RS3 (Kokusho 2007).

Martina 2001, Kokusho 2007). The *CRR*-values in the vertical coordinate normalized by those for $F_c = 0$ tend to decrease from certain peak values with increasing $F_c$ despite detailed differences among the individual tests. Fig. 5.4.15 depicts the similar effect of $F_c$ on *CRR* for not only poorly-graded sands but also well-graded gravelly soils obtained by the test series using the soil materials explained in Fig. 5.4.1(a) (Kokusho 2007). The *CRR*-values for $\varepsilon_{DA} = 5\%$ and $N_c = 20$ were measured for RS1 and RS3 of relative density $D_r \approx 30{\sim}70\%$ containing low-plasticity fines of $I_p = 6$ increasing step by step from $F_c = 0$ to 30% to have the dashed grain size curves shown in Fig. 5.4.1(a). The liquefaction resistance obviously decreases with increasing $F_c$ particularly for larger $D_r$ not only for the poorly-graded RS1 but also for the well-graded RS3. Also noted in Fig. 5.4.15 is that the decrease occurs in the narrower range of $F_c$ for RS3 ($F_c = 0$–$10\%$) than for RS1 ($F_c = 0$–$30\%$). This difference between RS1 and RS3 may be explained by the binary packing model in Fig. 5.4.2, though qualitatively. Thus, these test results indicate that the mixture of non/low-plastic fines tend to significantly reduce the liquefaction resistance of granular soils of the same relative densities, presumably because fines tends to change soils from being dilative to contractive during shearing as to be discussed in the later Sections.

#### 5.4.3.3 *Effect of fines on post-liquefaction behavior*

In order to investigate the effect of fines content on the post-liquefaction behavior of granular soils, undrained monotonic loading tests were carried out for the same soils without drainage immediately after the prior undrained cyclic loading. Figs. 5.4.16(a) and (b) show the curves of deviator stress $q$ or pore-pressure $\Delta u$ versus axial strain $\varepsilon$ for RS1 and RS3, respectively, for the relative density $D_r \approx 50\%$ and the fines content

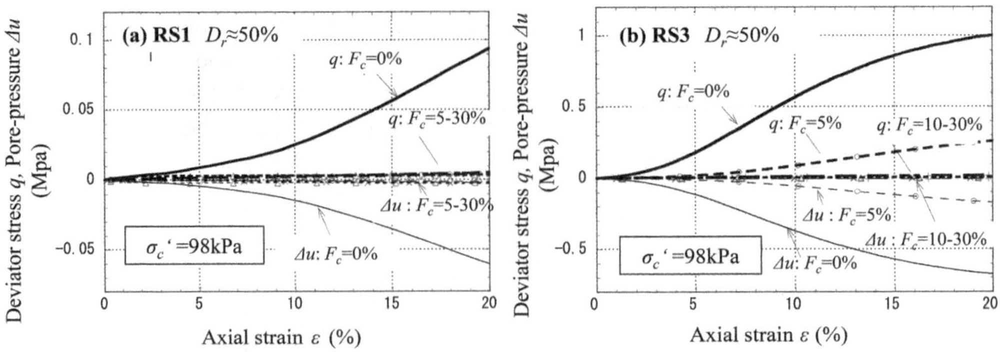


*Figure 5.4.16* Deviator stress/pore-pressure versus axial strain curves in post-liquefaction monotonic loading tests with fines: (a) Poorly-graded RS1, (b) Well-graded RS3 (Kokusho 2007).

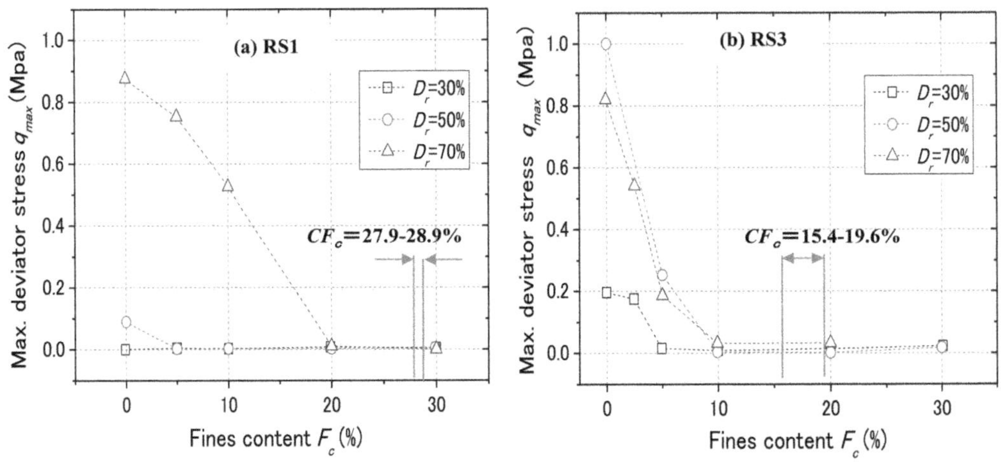


*Figure 5.4.17* Maximum deviator stress $q_{max}$ versus fines content $F_c$ in post-liquefaction monotonic loading tests: (a) Poorly-graded RS1, (b) Well-graded RS3 (Kokusho 2007).

$F_c$ parametrically changing. In the prior cyclic loading histories, all the specimens attained almost 100% pore pressure buildup and the axial strain $\varepsilon_{DA} \approx 10\%$. In the initial stage of monotonic loading, the stress increases gradually with increasing strain reflecting softened shear stiffness due to the preceding cyclic loading. The mobilized stress $q$ is much greater for RS3 than RS1 in the case of $F_c = 0$ due to the difference in the particle gradation as explained before (Note again that the scale in the vertical axis in Fig. 5.4.16(a) is about one tenth of that in Fig. 5.4.16(b)). In both RS1 and RS3, however, increasing fines content tends to dramatically reduce the stress remobilization after liquefaction.

In Figs. 5.4.17(a) and (b), the post-liquefaction maximum deviator stress $q = q_{max}$ defined at $\varepsilon = 20\%$ are plotted versus fines content for RS1 and RS3, respectively, with $D_r \approx 30$, 50 and 70%. Although the $q_{max}$-value decreases eventually to only less than a few percent of that for $F_c = 0$ with increasing $F_c$ both in RS1 and RS3, it tends to decrease more drastically in denser soils than looser soils. Also note that it tends to occur by a smaller increase of $F_c$ in RS3 than RS1 for the dense soil of $D_r = 70\%$ in particular. The reason may again be explained by the difference in the critical fines content $CF_c$ between RS1 and RS3 mentioned before. The residual strength reduction is much more considerable than the cyclic strength reduction indicated in Fig. 5.4.15. Thus, fines content in liquefiable granular soils has a more significant effect on the post-liquefaction shear resistance than the liquefaction resistance.

## 5.5 LIQUEFACTION POTENTIAL EVALUATION BY IN SITU TESTS

The relative density $D_r$ is one of the key parameters for liquefaction potential in laboratory tests. However, $D_r$ is difficult to measure and inconvenient to use in situ. Furthermore, the liquefaction resistance is dependent not only on $D_r$ but also largely on in situ stress conditions and the soil fabric induced during soil deposition, subsequent loading histories and aging. In engineering practice, in situ penetration tests are widely used for the liquefaction potential evaluation, in which the cyclic resistance ratio $CRR$ is directly determined from penetration resistance. The penetration resistance reflects not only the relative density but also the vertical and horizontal effective stresses. It may be expected to also reflect subtle effects of soil fabric to a certain extent.

### 5.5.1 Penetration tests and data normalizations

Among the penetration tests, Standard Penetration Test (SPT) has been the most popular method, although Cone Penetration Test (CPT) is increasingly employed recently. Swedish Weight Sounding (SWS) is another simple test sometimes used in Japan. S-wave velocity measurements are also pursued by a group of researchers to develop some correlations with in situ liquefaction potential reflecting soil fabric (e.g. Andrus and Stokoe 2000). In the following, penetration tests for liquefaction potential evaluations are addressed in terms of the testing methods, and the basic concepts and procedures for the $CRR$ evaluations using the penetration resistance are reviewed.

#### *5.5.1.1 Overview of penetration tests*

*(1) SPT*

The SPT setup is schematically illustrated in Fig. 5.5.1(a). A bore-hole is drilled in advance to a certain depth and the SPT probe is driven at its bottom by a hammer with a mass of 63.5 kg and drop height 75 cm. SPT blow-counts called as "$N$-value" are defined as the number of blows to penetrate the probe by 30 cm. The drilling and hammer-driving are conducted in turn in every 1 or 2 m depth interval to a required depth. In this test, the types of penetrated soil can be identified by a split spoon sampler at the tip of the probe. The physical properties of soils; fines content $F_c$, plasticity index $I_p$ and clay content $C_c$, necessary for screening liquefiable soils can be checked by the samples.

The SPT $N$-values are discontinuous variables at depths of 1 m interval at most, difficult to grasp continuous soil profiles. The test procedure is not so efficient because

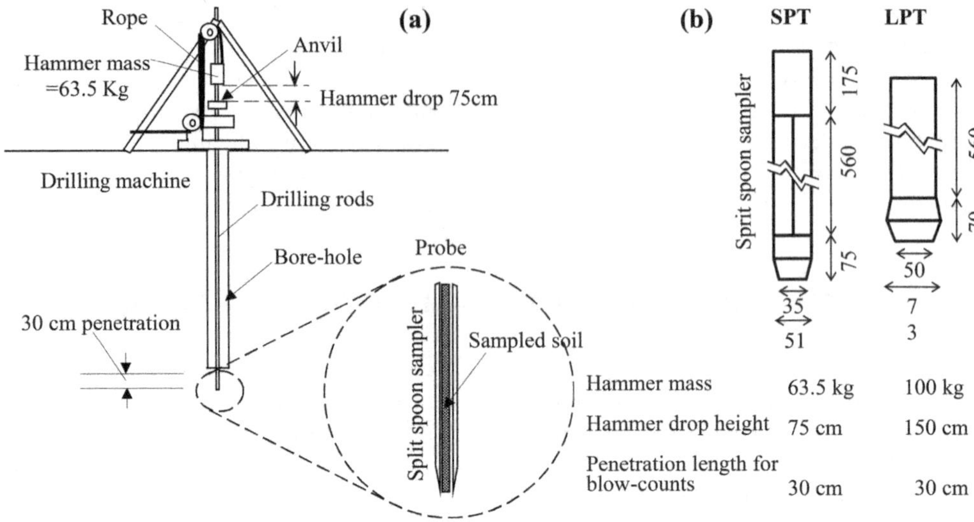


*Figure 5.5.1* Schematic illustration of SPT setup (a) and probes of SPT and LPT in Japan (b).

*Table 5.5.1* Summary of SPT rod energy ratio $ER_m$ in different countries (Seed et al. 1985) with permission from ASCE.

| *Country* | *Hammer type* | *Hammer release* | *Estimated rod energy $ER_m$ (%)* | *Correction factor for 60% rod energy* |
|---|---|---|---|---|
| Japan | Donut | Free-fall (Tonbi) | 78 | 78/60 = 1.30 |
| | Donut | Rope & Pulley with special throw release | 67 | 67/60 = 1.12 |
| USA | Safety | Rope & Pulley | 60 | 60/60 = 1.00 |
| | Donut | Rope & Pulley | 45 | 45/60 = 0.75 |
| Argentina | Donut | Rope & Pulley | 45 | 45/60 = 0.75 |
| China | Donut | Free-fall | 60 | 60/60 = 1.00 |
| | | Rope & pulley | 50 | 50/60 = 0.83 |

drilling and hammer-driving have to be repeated intermittently, when water level at the bore-hole fluctuates potentially disturbing the bottom soil by boiling failure. Another problem is that the test results are sensitive to technical details such as the dimensions of the drilling rod, hammer, rope and pulley, and how the hammer is dropped. These details may reduce the driving energy by the hammer actually transmitted to the rod down to 40% to 90% of the theoretical gravity energy according to worldwide investigations (Schmertmann and Palacios 1979, Kovacs et al. 1983).

Seed et al. (1985) summarized the rod energy ratio $ER_m$ as in Table 5.5.1 based on international research on the SPT practice and proposed $ER = 60\%$ to be adopted as the standard value. By using the estimated rod energy ratio $ER_m$ in the table (where hammer release methods; free-fall or rope and pulley, are focused), SPT-values

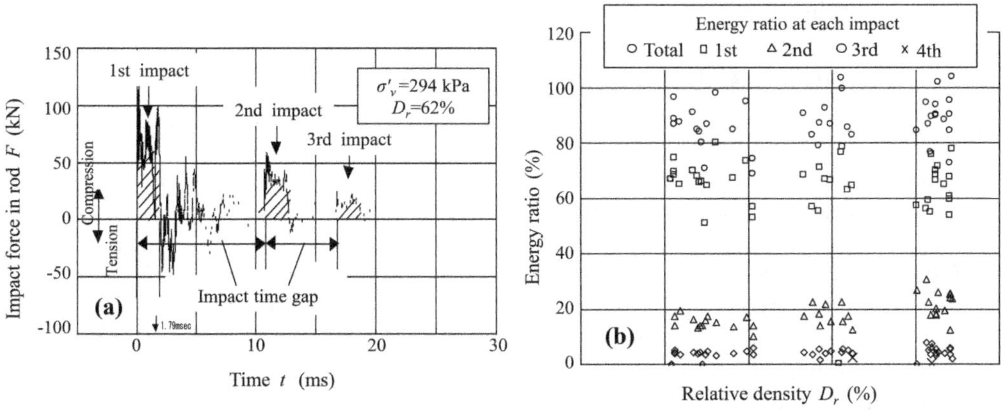


*Figure 5.5.2* Time-history of measured impact force in rod (a), and Correlation between energy ratio and relative density for short rod length (Suwa et al. 2000).

$N_m$ in individual tests may be converted to the standard $ER = 60\%$ in the following formula.

$$N_{60} = N_m \cdot \left( \frac{ER_m}{60} \right) \tag{5.5.1}$$

In US, the hammer energy measurement and energy correction using Eq. (5.5.1) is implemented in engineering practice.

In Japan, the energy correction is not normally done, probably because the SPT practice is believed more or less standardized. For example, similar investigations on the SPT procedures were conducted there in terms of the blow-counts instead of the energy measurement. It was found that the difference in $N$-value due to the hammer release methods was negligibly small for $N < 12$ (Yoshimi and Tokimatsu 1983), and in denser Pleistocene sands the free-fall (Tonbi) method gave $N$-values 86% of the rope-pulley method (Oh-oka 1984).

With regard to the hammer energy measurement practice, the energy is measured only for the first impact among multiple impacts actually occurring in a single hammer fall (Kovacs et al. 1983). Fig. 5.5.2(a) exemplifies the time-history of impact force transmitted in a short length SPT rod (4.93 m from tip to anvil out of the total length 6.43 m). It clearly shows that the hammer gives multiple impacts (the shaded time ranges), supplying energies intermittently for the rod to penetrate into the ground. Fig. 5.5.2(b) shows energy ratios (the rod wave energy divided by the theoretical hammer gravitational energy) in percentage at each impact versus various relative densities $D_r$ of the penetrated soil. It indicates that the energy ratio, 60–80% in the first impact tends to approach nearly 90–100% eventually irrespective of $D_r$, if the subsequent impact energies are added. In the short rod case, upward tension wave generated from downward compression wave at the probe tip returns to the anvil and separates the hammer-anvil contact, temporarily cutting the energy supply earlier than the case of a longer rod. Thus, the short rod length tends to reduce the first impact energy significantly but not the total energy. Thus, the correction for short rod length (Schmertmann and Palacios 1979) sometimes employed in the SPT practice may not be necessary (Idriss and Boulanger 2008) as used to be considered.

### *(2) LPT*

In order to investigate coarse gravelly soils for which SPT is not powerful enough to penetrate, there exist a penetration test similar to SPT but with the larger size and larger driving energy named as Large Penetration Test (LPT). However, the specification for LPT has not yet been standardized at all and is quite different internationally. The dimensions of the probe and the major specifications of LPT employed in Japan are available in Fig. 5.5.1(b) in comparison with SPT. In Fig. 5.5.3(a), correlations between SPT-blow counts and LPT-blow counts are shown for poorly-graded sands and well-graded sandy gravels, indicating that the blow-counts ratio, SPT to LPT, $N/N_d$ both for the 30 cm penetration, tends to increase from 1.5 to 2.0 with increasing gravel contents (Yoshida and Kokusho 1987). In Fig. 5.5.3(b), the blow-counts ratios $N/N_d$ are plotted versus the mean grain sizes $D_{50}$ based on a larger database, indicating again that $N/N_d$ increases from 1.5 to 2.5 with increasing $D_{50}$ from 0.1 to 10 mm (Tokimatsu 1988).

In Canada and USA, Becker Penetration Test (BPT) is often used as a powerful test method for gravelly soils (Harder and Seed 1986, Sy and Campanella 1993). It uses double action diesel pile-hammer driving double-wall casing tubes of 16.8 cm in diameter, 3 m long per piece, continuously into the ground. Some formulas are also proposed converting the BPT blow counts $N_b$ for 30 cm penetration to the SPT blow counts for liquefaction evaluations.

### *(3) CPT*

In cone penetration tests (CPT), a cone with cross sectional area $A_p = 10\,\text{cm}^2$ shown in Fig. 5.5.4(a) is penetrated statically by a hydraulic jack with a constant speed 1–2 cm/s

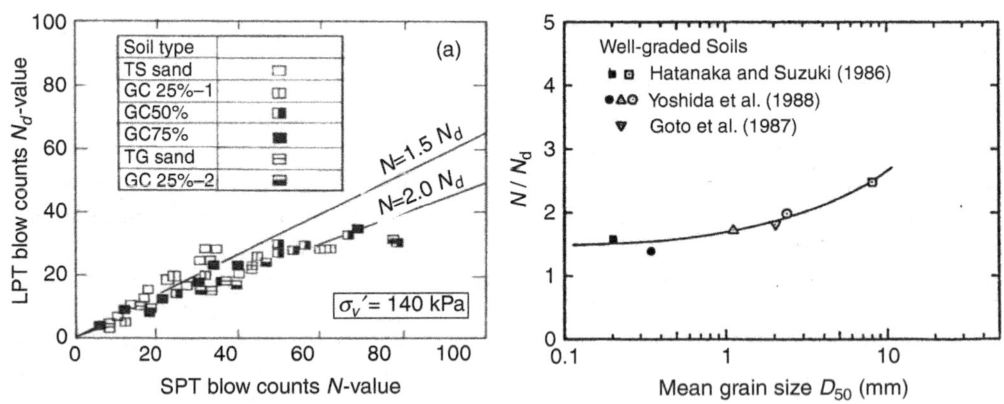


*Figure 5.5.3* LPT $N_d$-value versus SPT $N$-value for sands and gravelly soils: (a) obtained in calibration chamber tests (Yoshida and Kokusho 1987), (b) $N/N_d$-ratio versus mean grain size $D_{50}$ correlation (Tokimatsu 1988).

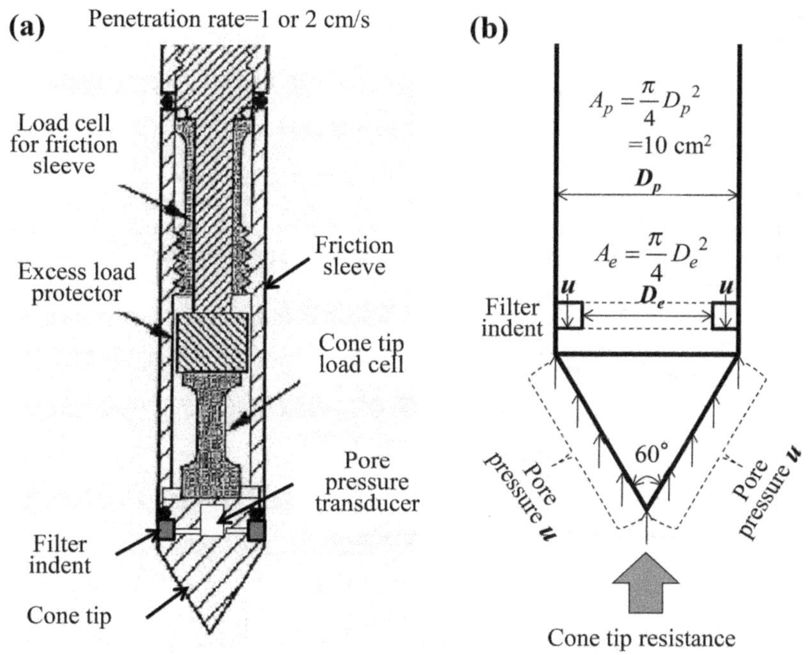


*Figure 5.5.4* Sketch of electronic CPT probe: (a) Side-sectional view, (b) cross-sectional areas considering indent for pressure transducer (JGS Soil Investigation Editing Committee 2004).

continuously. During the penetration, cone-tip resistance $q_c$, friction at a sleeve $f_s$ and pore-pressure $u$ behind the cone tip are continuously measured electronically. Unlike SPT, CPT can obtain seamless records of these variables during the penetration, giving detailed variations of soil profiles without disturbing soils by prior-drilling. The test procedures are simpler and easier to standardize to avoid the differences in technical details. CPT is applicable exclusively to soft clayey soils and loose sands, which are easy to penetrate, but still needs a heavy counterweight (e.g. a built-in CPT vehicle) to provide the reaction force.

When the cone penetrates into the ground below water table, the tip resistance $q_c$ is influenced by the pore-water pressure because of a filter indent for the pressure transducer behind the cone as shown in Fig. 5.5.4(b). Hence, the corrected tip resistance $q_t$ is obtained as,

$$q_t = \frac{P_m}{A_p} + \left(1 - \frac{A_e}{A_p}\right) u = q_c + \left(1 - \frac{A_e}{A_p}\right) u \qquad (5.5.2)$$

Here, $P_m$ = tip load transducer reading, $A_e$ = effective area excluding the indented area from the gross area $A_p$ (Robertson 1990).

A drawback in the CPT is a difficulty to directly see investigated soils unlike SPT. However, the measurements of sleeve friction and pore pressure can be utilized to

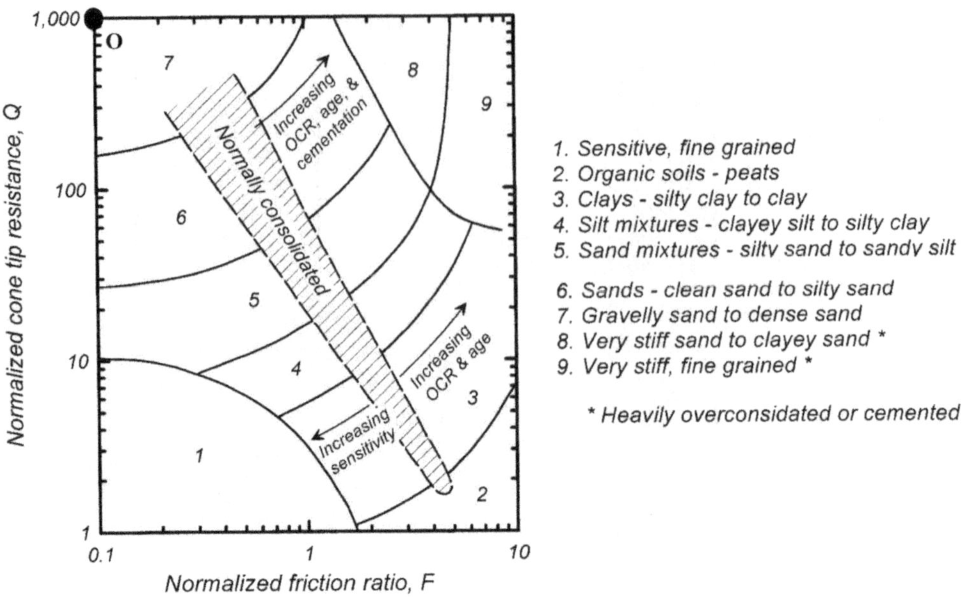


*Figure 5.5.5* Empirical chart categorizing soils into nine types based on two non-dimensional values, $Q$ and $F$ (modified from Robertson 1990).

overcome this difficulty. The sleeve friction $f_c$ in particular is combined with the tip resistance to categorize soil types without soil sampling (Robertson 1990). Fig. 5.5.5 depicts an empirical chart categorizing soils into the nine soil types described there on the full-logarithmic $Q$-$F$ diagram wherein non-dimensional values, $Q$ and $F$ are defined by the following equations (Robertson and Wride 1998).

$$Q = \frac{q_c - \sigma_{vc}}{p_0}\left(\frac{p_0}{\sigma'_{vc}}\right)^n \quad F = \frac{f_s}{q_c - \sigma_{vc}} \tag{5.5.3}$$

Here, $\sigma_{vc}$ = total vertical stress, $\sigma'_{vc}$ = effective vertical stress, $p_0$ = unit stress (98 kPa), and the exponent $n$ varies from 0.5 for sands to 1.0 for clays. Boundaries between the soil types 2 to 7 in Fig. 5.5.5 are approximated by concentric circles centered at O, the above left corner of the diagram. Hence, the radius from the point O can serve as an index $I_c$ representing the soil types 2–7 and $I_c$ is formulated by $Q$ and $F$ as,

$$I_c = [(3.45 - \log Q)^2 + (\log F + 1.22)^2]^{0.5} \tag{5.5.4}$$

This parameter $I_c$ is further used to estimate fines content $F_c$, the pertinent parameter in the current liquefaction potential evaluations practice. For example, Jeffery and Davies (1993) proposed

$$F_c = 0\ (I_c < 1.26), \quad F_c = 1.75 I_c^{3.25} - 3.7\ (1.26 < I_c < 3.5), \quad F_c = 100\ (3.5 < I_c) \tag{5.5.5}$$

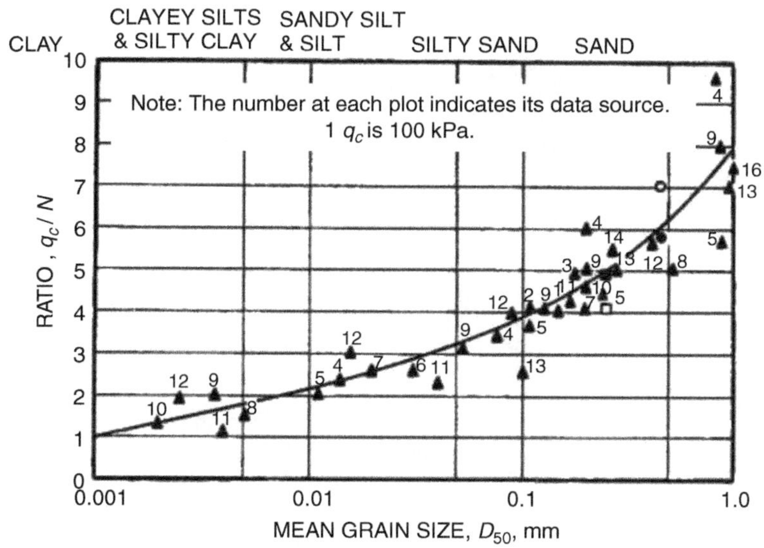


*Figure 5.5.6* Ratio of penetration resistances CPT to SPT, $q_c/N$, as function of mean grain size $D_{50}$ based on different data sources (modified from Robertson and Campanella 1985) with permission from ASCE.

and, Suzuki and Tokimatsu (2003) proposed

$$F_c = I_c^{4.2} \tag{5.5.6}$$

As CPT is getting popular in liquefaction evaluations, several empirical curves correlating the CPT tip resistance directly with in situ *CRR* have been proposed. However, the CPT resistance is sometimes used in SPT-based liquefaction evaluations by converting to the SPT *N*-values. Fig. 5.5.6 shows such a chart summarizing the ratio of CPT-resistance to SPT blow counts, $q_c/N$, as a function of mean grain size $D_{50}$ derived from different data sources (Robertson and Campanella 1985). The approximation curve in the chart indicates that the $q_c/N$-value takes 1.0 to 7.0 as the soils shift from clay to coarse sand, where $1q_c$ is identical to 100 kPa. Suzuki and Tokimatsu (2003) proposed empirical formula based on a great number of test data in Japan as:

$$N = 0.341 \cdot I_c^{1.94}(q_t - 0.2)^{(1.34 - 0.0927 \cdot I_c)} \tag{5.5.7}$$

### *(4) SWS*

Swedish Weight Sounding (SWS) is a simple and handy in situ test shown in Fig. 5.5.7, possible to conduct only by human power down around 10 m deep. The tip is a metal screw point, 20 cm long and 50 N in weight, with dimensions written in the figure. At the start of the test, the tip is statically driven into the ground by a set of weights uploaded step by step on the top of the rod up to the total weight $W_{sw} = 1000$ N. If the penetration is shorter than 25 cm, then a rotation is given to the rod by a handle

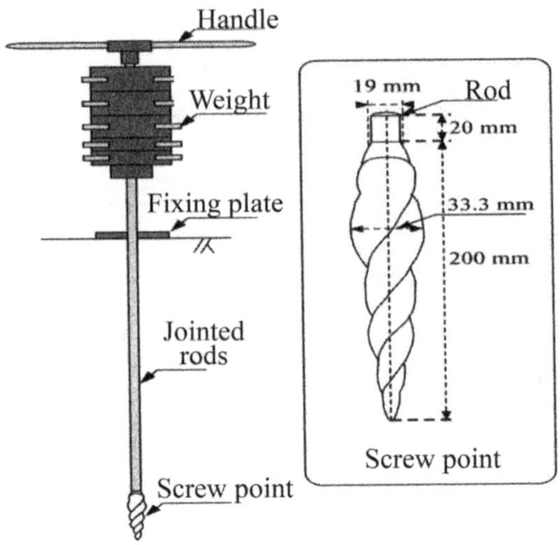


*Figure 5.5.7* Sketch of SWS and screw point (Tsukamoto et al. 2004).

and the number of half turns $N_a$ to reach 25 cm in the total penetration is counted and $N_a$ is converted to $N_{sw}$ using the penetration $L_{sws}$ (cm) such as $N_{sw} = (100/L_{sws}) \times N_a$ indicating that $N_{sw} = 4N_a$ if $L_{sws} = 25$ cm. The SWS penetration resistance values are normally converted to the SPT $N$-value (Inada 1960) as:

$$N = 0.002W_{sw} + 0.067N_{sw} \tag{5.5.8}$$

Another formula proposed for liquefaction evaluations for sand is (Tsukamoto et al. 2004):

$$N = \frac{\sqrt{e_{\max} - e_{\min}}}{10}(N_{sw} + 40) \tag{5.5.9}$$

where, $e_{\max}$ and $e_{\min}$ are the maximum and minimum void ratios.

#### 5.5.1.2 Correction of penetration resistance by overburden

It is easy to understand the penetration resistance tends to increase with increasing ground depth even in a uniform soil deposit. In order to have a depth-independent unique value for the same relative density in different depths, a correction is made for SPT $N$-values or CPT tip resistance values $q_t$ to have corrected $N_1$-values or $q_{t1}$-values, respectively, as,

$$N_1 = C_N N, \quad q_{t1} = C_N q_t \tag{5.5.10}$$

As the correction factor $C_N$ in the above, either of the following two formulas are normally employed in practice, where $\sigma'_v$ = effective overburden stress, $a$ = a constant and $n$ = an exponent.

$$C_N = \frac{a + 1.0}{a + \sigma'_v/p_0} \tag{5.5.11}$$

$$C_N = \left(\frac{p_0}{\sigma'_v}\right)^n \tag{5.5.12}$$

In both equations, $C_N = 1.0$ if $\sigma'_v = p_0 \equiv 98\,\text{kPa}$ and tends to decrease with increasing $\sigma'_v$. These equations were drawn from calibration chamber tests of dry clean sands (Gibbs and Holtz 1957). Actually the next formula substituting $a = 0.7$ into Eq. (5.5.11) proposed by Meyerhof (1957) is often used.

$$N_1 = \frac{1.7}{0.7 + \sigma'_v/p_0} N \tag{5.5.13}$$

On the other hand, Kokusho et al. (1985) found better applicability of Eq. (5.5.11) of $a = 1.5$ to different clean sand by a series of calibration chamber tests. As for Eq. (5.5.12), the next equation with $n = 0.5$ is often used (Liao and Whitman 1986).

$$N_1 = \left(\frac{p_0}{\sigma'_v}\right)^{0.5} N \tag{5.5.14}$$

In such equations, only the vertical stress $\sigma'_v$ is considered, while the effect of horizontal effective stress $\sigma'_h = K_0\sigma'_v$ is not explicitly included. In overconsolidated soils, the earth-pressure coefficient at rest $K_0$ tends to be larger from around 0.5 to over 1.0, which will inevitably affect the correction factor. The influence of $K_0$ on $N$-value and its depth-dependent correction needs special attentions in evaluating a liquefaction potential in overconsolidated deposits.

#### *5.5.1.3 SPT N-value versus relative density*

Correlations between $N$-value and $D_r$ were investigated and proposed by many investigators for a variety of soils. Among those, a correlation proposed in the same paper by Gibbs and Holtz (1957) and almost exclusively employed in liquefaction evaluation practice is

$$D_r = 21\sqrt{\frac{N}{(\sigma'_v/p_0) + 0.7}} \tag{5.5.15}$$

where $D_r$ is in %. This equation if combined with Eq. (5.5.13) yields:

$$D_r = 16.1\sqrt{N_1} \tag{5.5.16}$$

All these equations above were obtained actually in the calibration chamber tests on dry clean poorly graded sands. Nevertheless, they are used for almost all granular soils

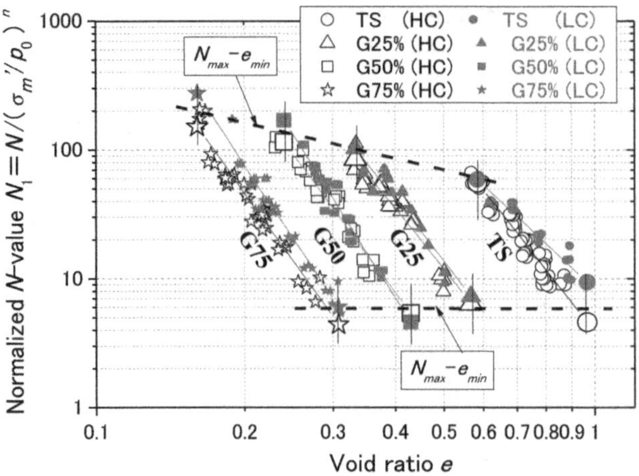


| Soil type | $D_{50}$ | $C_u$ | $\rho_s$ | $e_{min}$ | $e_{max}$ |
|---|---|---|---|---|---|
| TS | 0.34 | 1.95 | 2.701 | 0.584 | 0.996 |
| G25 | 1.13 | 5.67 | 2.674 | 0.334 | 0.567 |
| G50 | 2.28 | 11.3 | 2.668 | 0.240 | 0.429 |
| G75 | 7.30 | 31.1 | 2.653 | 0.161 | 0.308 |

*Figure 5.5.8* Normalized SPT *N*-values versus void ratios e for 4 granular materials in calibration chamber tests and tabulated physical soil properties (Kokusho 2007).

irrespective of differences in particle gradations. In the following, a series of calibration chamber tests, where not only poorly graded sands but also well-graded sandy gravels were tested, are addressed to better understand SPT-*N* versus $D_r$ correlations of various granular soils with different particle gradations (Kokusho and Yoshida 1997, Kokusho 2007).

Artificial soil layers were placed with various relative densities in a large calibration chamber, 2.0 m inside diameter and 1.5 m height, as illustrated in Fig. 2.3.2(a). The soils were saturated and vertically pressed with given overburden stresses by an overlying rubber bag beneath the cap in the $K_0$-stress condition without horizontal displacement in the pressure chamber. The overburden was initially set as 50 kPa and then increased step by step either to the maximum of 200 kPa in the first series of test (LC test) or to the maximum of 1 MPa in the second series (HC test). The stress condition in the soil layer was monitored vertically and horizontally by pressure cells installed at the bottom and side wall of the container. At every step of the overburden stress, SPT was carried out through openings of the cap and rubber bag into the soils.

Four soils with grain size curves shown in Fig. 2.3.2(b) were used in the tests; one river sand (TS) and three gravelly soils (G25, G50 and G75 gravels) with their uniformity coefficients $C_u$ also shown in the diagram. The measured *N*-values were normalized according to Eq. (5.5.12) using the mean stress $\sigma'_m = (\sigma'_v + 2\sigma'_h)/3$ in place of $\sigma'_v$ as:

$$N_1 = N/\{(\sigma'_v + 2\sigma'_h)/3p_0\}^n = N/(\sigma'_m/p_0)^n \qquad (5.5.17)$$

The $N_1$-values thus normalized are plotted versus the void ratios in Fig. 5.5.8. On the full logarithmic chart, the data points may be approximated by parallel lines located differently depending on different particle gradations despite the slight data separations recognized between the HC and LC test series. Maximum and minimum *N*-values;

$N_{1\,max}$ and $N_{1\,min}$ defined as the intersections of these parallel lines with vertical straight lines corresponding to $e = e_{min}$ and $e_{max}$ (tabulated in the figure for individual soils) are marked with large solid symbols of the same kinds in Fig. 5.5.8. $N_{1\,max}$ tends to increase with increasing uniformity coefficient $C_u$, while $N_{1\,min}$ may be judged to be almost unchanged despite the data dispersions as approximated by the thick dashed lines. $N_{1\,min}$ and $N_{1\,max}$ may be correlated with the uniformity coefficients $C_u$ (Kokusho 2007) as:

$$N_{1\,min} = 5.8, \quad N_{1\,max} = 42.6C_u^{0.46} \tag{5.5.18}$$

This implies that $N$-values of well-graded gravelly soils of large $C_u$-values can be as small as poorly-graded loose sand if the soils are loose enough despite tremendous differences in the void ratio, whereas the $N$-values can be considerably larger than that of dense sand if they are dense.

In Fig. 5.5.9, the normalized $N$-values $N_1$ are plotted versus the relative densities $D_r$, indicating that well-graded soils can take a wider range of $N$-values than poorly-graded soils and the difference between them tends to widen for $D_r$ larger than around 50%. A solid curve drawn in the figure corresponding to Eq. (5.5.16) for a clean sand (Meyerhof 1957) can be expressed as $(N_1)_{\sigma'_m=98\,\text{kPa}} = 1.23(N_1)_{\sigma'_v=98\,\text{kPa}} = 47.3 \times (D_r/100)^2$ for $n = 0.5$ and $\sigma'_m = (1 + 2K_0)\sigma'_v/3$ with $K_0 = \sigma'_h/\sigma'_v = 0.5$, because $N_1$-value in Eqs. (5.5.16) and (5.5.17) are defined differently in terms of $\sigma'_m = 98\,\text{kPa}$ (effective mean stress) and $\sigma'_v = 98\,\text{kPa}$ (effective vertical stress), respectively. Obviously, this equation matches the test data for the clean sand (TS) well, indicating a good compatibility for the poorly-graded sands between the well-known Meyerhof's equation and the test results by Kokusho and Yoshida (1997).

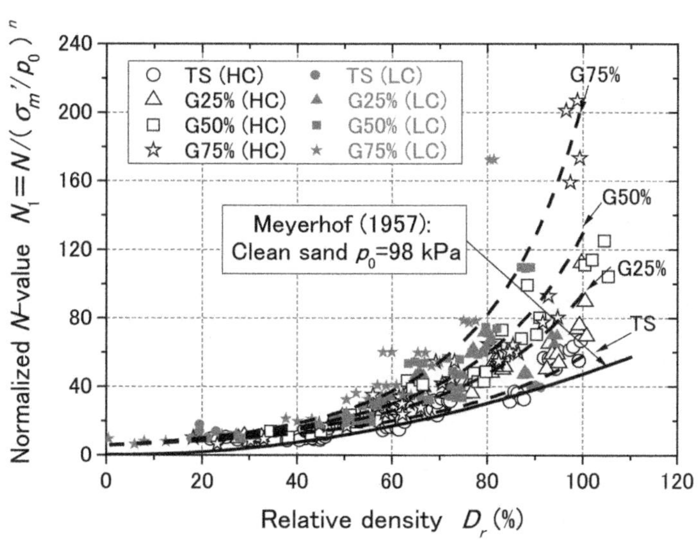


*Figure 5.5.9* Normalized SPT $N$-values versus relative density plots for 4 materials compared with proposed empirical formula by dashed curves (Kokusho 2007).

Based on the linearity of the normalized $N$-value versus void ratio relationships in the log-log scale in Fig. 5.5.8, $N$-value may be formulated by the next equation,

$$N = N_{1\min}(N_{1\max}/N_{1\min})^{D_r^*}(\sigma'_m/p_0)^n \tag{5.5.19}$$

where, $D_r^*$ is a logarithmic relative density defined here as

$$D_r^* = \frac{\log(e_{\max}/e)}{\log(e_{\max}/e_{\min})} \tag{5.5.20}$$

The exponent $n$ in Eq. (5.5.19) plotted against $D_r^*$ in Fig. 5.5.10(a) may be approximated by the following simple function irrespective of the different soils used in the tests (Kokusho and Yoshida 1997).

$$n(D_r^*) = 0.27(D_r^*)^{-0.4} \tag{5.5.21}$$

In Fig. 5.5.10(b), $n$ versus $D_r$ correlations in Eq. (5.5.21) for the four soils (almost identical by using $e_{\max}$ and $e_{\min}$ tabulated in Fig. 5.5.8) are compared with $n = 0.5$ in Eq. (5.5.12) or $n = 0.784 - 0.521D_r$ proposed by Idriss and Boulanger (2008). This indicates that $n$ is inherently $D_r$-dependent, although it may be approximated to be nearly 0.5 (Liao and Whitman 1986) for $D_r = 30$–$60\%$ wherein the liquefaction is a major concern.

Thus, the $N$-values for granular soils with various grain size curves can be expressed by the uniformity coefficient $C_u$, the logarithmic relative density $D_r^*$, and the confining pressure $\sigma'_m$ as

$$N = 5.8(7.3C_u^{0.46})^{Dr^*}\left(\frac{\sigma'_m}{p_0}\right)^{0.27(D_r^*)^{-0.4}} \tag{5.5.22}$$

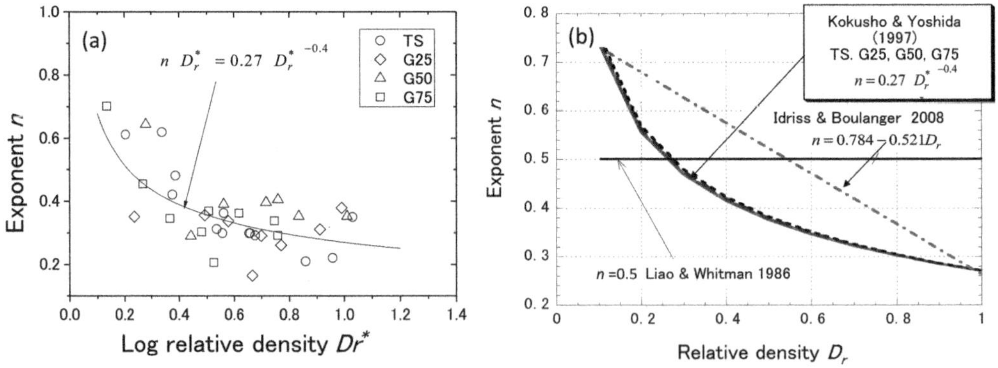


*Figure 5.5.10* Exponent $n$ versus relative density (Kokusho and Yoshida 1997) (a), and Comparisons with other proposals (b).

This empirical equation if superposed with the dashed curves in Fig. 5.5.9 shows a fairly good coincidence with the test results. Thus, the SPT $N$-value versus $D_r$ correlation is shown to be highly dependent on the particle gradation.

### 5.5.2 Liquefaction resistance versus penetration resistance

Two different approaches have been employed in establishing correlations between the liquefaction resistance and the in situ penetration tests to be used in design practice. In the first approach, laboratory-based liquefaction tests on intact samples recovered in the field were implemented to correlate with corresponding in situ penetration tests. In the second approach, a number of liquefaction case histories during previous earthquakes are revisited, and cyclic shear stress ratios ($CSR$) estimated from maximum ground surface accelerations (PGA) during those events are directly correlated to corresponding in situ penetration tests. In the following, the two approaches are outlined and compared in terms of their features and compatibility.

#### 5.5.2.1 *Evaluation using laboratory tests*

In this approach employed mostly in Japan, sand samples are taken out from in situ deposits without disturbance and tested in undrained cyclic triaxial tests to have cyclic resistance ratios $CRR$ under isotropic effective confining stresses $\sigma_c'$ corresponding to in situ stresses. In parallel with this, penetration tests are implemented in the same soil deposits and the normalized penetration resistances are correlated with the corresponding $CRR$-values. This approach can apply to soils irrespective of their depths and densities, although some uncertainties are left if in situ liquefaction triggering mechanism can exactly be reproduced by the cyclic triaxial tests in the isotropically consolidated specimens as already discussed.

Another important concern in this approach is that not only the density but also the fabric of in situ soils have to be preserved for the laboratory tests because only faint vibration or shearing during sampling and lab handling procedures may change in situ soil fabric and hence liquefaction resistance significantly as already mentioned. In situ freezing sampling though very costly may be employed to have intact samples of uncemented sandy soils. Block sampling by hands from dewatered trenches in shallow depth may be another option if carefully implemented. Recently, a more advanced and economical tube sampling technology is emerging called "Gel-Push Sampling" in which uncemented sands or gravelly soils may possibly be recovered without disturbance by utilizing a kind of gel in the sampling tube (Tani et al. 2007).

The in situ freezing sampling has often been employed in Japan in establishing the $CRR$ versus $N_1$ correlations for major design codes. The pioneering research using this sampling was carried out by Yoshimi (1994) and Yoshimi et al. (1994). Fig. 5.5.11 shows $CRR$–$(N_1)_{78}$ correlation for clean sands, wherein $CRR$ in the vertical axis is defined as $\sigma_d/2\sigma_c'$ ($\varepsilon_{DA} = 5\%$, $N_c = 15$) and $(N_1)_{78}$ in the horizontal axis means the SPT $N_1$-value with the hammer energy 78% of the theoretical gravitational energy (nearly the average SPT energy ratio in Japan). It is remarkable that the $CRR$-value by the in situ freezing (solid dots) tends to increase drastically for $(N_1)_{78}$ larger than 20–25. In contrast, open symbols in the same figure for specimens recovered from in situ by conventional tube sampling show no clear increasing trend with increasing

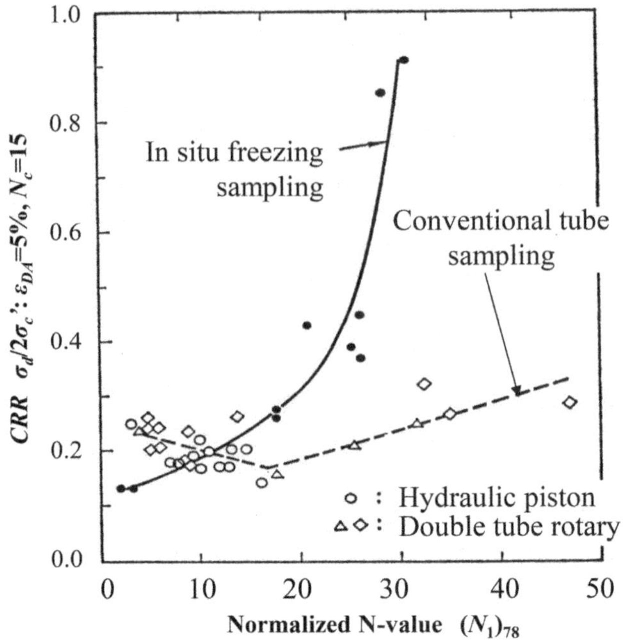


*Figure 5.5.11* *CRR* versus normalized SPT $(N_1)_{78}$ correlation based on triaxial tests on intact sands by in situ freezing sampling (Yoshimi et al. 1994).

$N_1$-value, resulting in significantly lower $CRR$-values for $N_1 \approx 10$ or higher, whereas $CRR$-values for $N_1 < 10$ are obtained higher presumably due to artificial densification.

Fig. 5.5.12 shows $CRR$–$N_1$ plots obtained similarly from cyclic triaxial tests on clean sands taken out by block sampling from a sand layer artificially placed in the soil container shown in Fig. 2.3.2(a) already mentioned and combined with SPT $N_1$-values measured in the same container test (Kokusho et al. 1983b, Kokusho et al. 1985). Although the $CRR$-value here is defined differently for the number of cycles $N_c = 10$, and the $N_1$-values are normalized from $N$-values with $a = 1.5$ in place of $a = 0.7$ in Eq. (5.5.11), the curve for $\varepsilon_{DA} = 5\%$ shows fairly good coincidence with that obtained by the in situ freezing sampling, and the drastic increase of $CRR$ for $N_1 > 20$–$25$ is also visible. This indicates a possibility that careful manual sampling of clean sands in block from dewatered trenches may be able to substitute very expensive in situ freezing sampling.

After the 1995 Kobe earthquake, the liquefaction resistance evaluation in road bridge designs in Japan (JRA 2002) was revised by conducting a comprehensive research program using in situ freezing sampling. Fig. 5.5.13(a) depicts the $CRR$ versus $N_1$ plots of intact specimens for $F_c < 5\%$ (clean sand) sampled from man-made fills and Holocene/Pleistocene sands (Matsuo 1997). The solid curve approximating the plots for clean Holocene sands are almost agreeable to those in the previous two figures, showing a sharp increase of $CRR$ for $N_1 > 20$–$25$ again. The dashed curve drawn there, which corresponds to an old version of the design code used before the

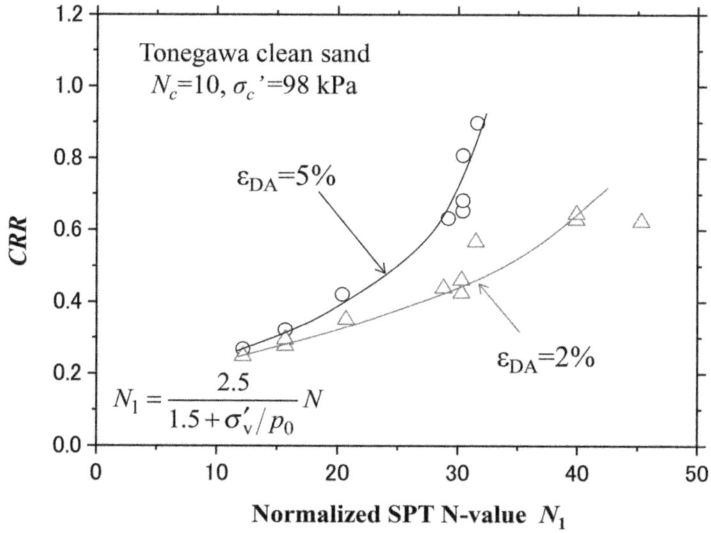


*Figure 5.5.12* *CRR* versus normalized SPT $N_1$ correlation based on triaxial tests on intact sands by block sampling from pressure chamber (Kokusho et al. 1983b).

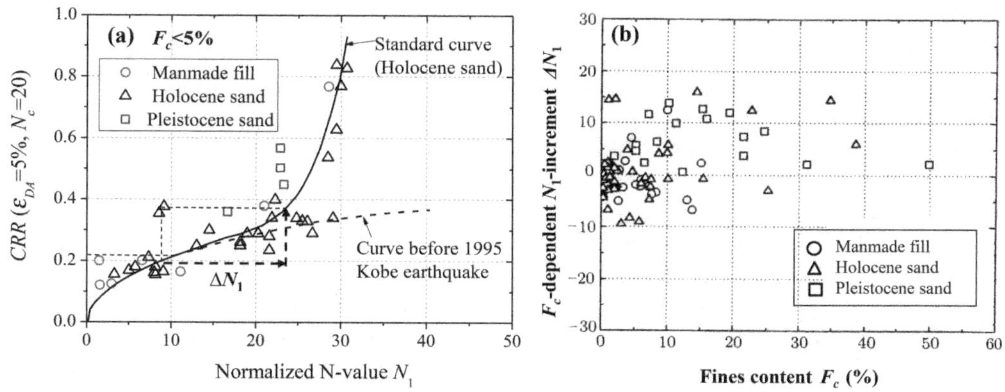


*Figure 5.5.13* *CRR* based on triaxial tests on intact sands by in situ freezing sampling in various soils versus SPT $N_1$-value (a), and $F_c$-dependent $N_1$-increment chart (b) (modified from Matsuo 1997).

1995 Kobe earthquake based on sand samples by conventional tube sampling, showing no increasing trend of *CRR* with increasing $N_1 > 20{\sim}25$, demonstrates again how important the sample quality is in evaluating the in situ liquefaction resistance.

On the other hand, as pointed out based on liquefaction case histories by Tokimatsu and Yoshimi (1983), the *CRR*–$N_1$ plots in Fig. 5.5.13(a) may reflect a significant effect of fines content. In order to take this effect into account, a *CRR*–$N_1$

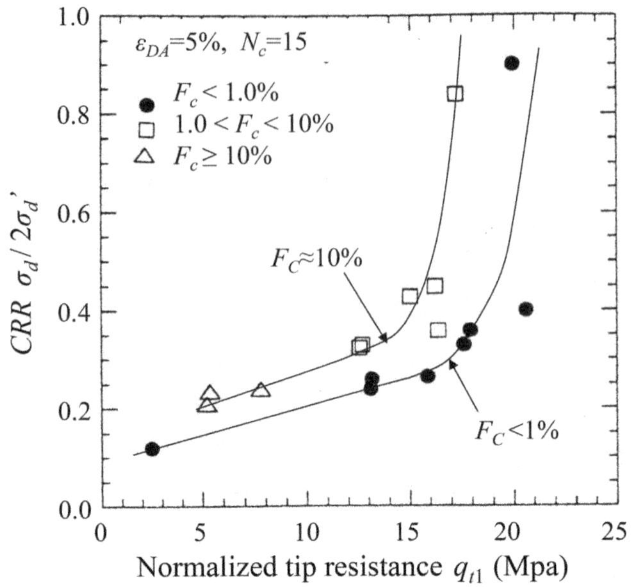


*Figure 5.5.14* *CRR* versus normalized CPT $q_{t1}$-value based on triaxial tests on intact sands by in situ freezing sampling (modified from Suzuki et al. 1994).

plot of various $F_c$-values in the diagram is compared with a coordinate on the standard clean sand curve (the solid curve) having the same *CRR* to read off the difference in $N_1$-value, $\Delta N_1$ as indicated with the dashed arrow. Fig. 5.5.13(b) shows a diagram of $\Delta N_1$ thus obtained versus corresponding $F_c$-values. This diagram indicates despite considerable data scatters that $\Delta N_1$ tends to increase with increasing $F_c$ for different types of soils. By using the $\Delta N_1$ versus $F_c$ relationship thus obtained, *CRR* for sands with given $F_c$-values may be determined from modified $N_1$ (=original $N_1 + \Delta N_1$) using the standard clean sand curve.

In Fig. 5.5.14, a similar relationship developed by Suzuki et al. (1994, 1995) to evaluate *CRR* from in situ CPT resistance by using the in situ freezing sampling is shown for sands with varying fines content. The *CRR*-value in the vertical axis is for $\varepsilon_{DA} = 5\%$, $N_c = 15$, and normalized CPT tip resistance $q_{t1}$ is defined as $q_{t1} = q_t/(\sigma_v'/p_0)^{0.5}$: $p_0 = 98$ kPa. As indicated by the approximation curves drawn on the chart, *CRR* tends to increase sharply for $q_t > 10$–20. Also noted here again is the evident effect of fines content on *CRR* versus $q_{t1}$ curves, so that *CRR* tends to be larger for higher $F_c$ under the same penetration resistance.

#### 5.5.2.2 *Evaluation using case histories*

In this approach, a number of liquefaction case histories during previous earthquakes are studied, wherein the maximum ground surface accelerations (PGA) at various sites are estimated from earthquake magnitudes and hypocentral distances using empirically derived attenuation curves. Then the *CSR*-values of given soil depths at a site

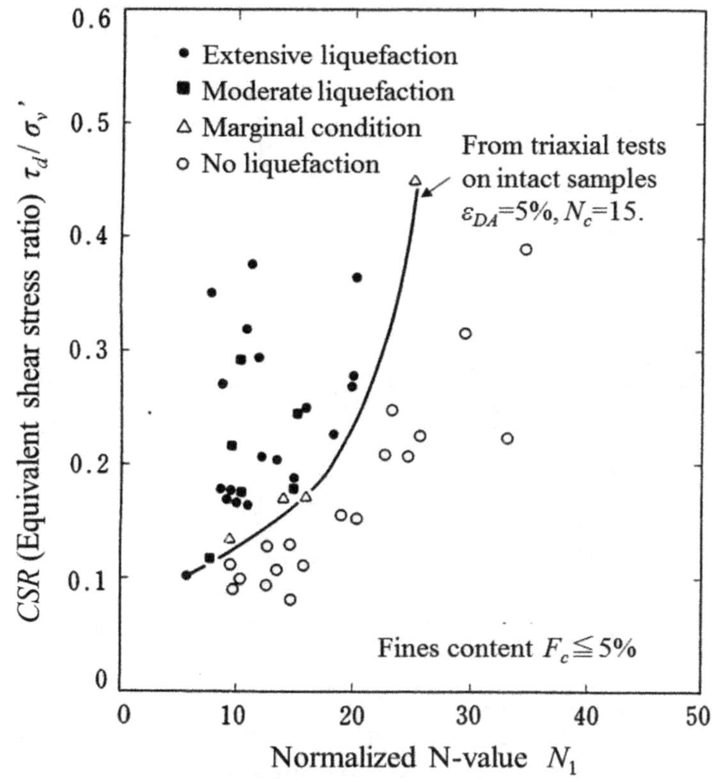


*Figure 5.5.15* *CSR* versus normalized SPT $N_1$-value based on liquefaction case histories mainly in Japan (Tokimatsu and Yoshimi 1983).

are evaluated in a simplified method according to PGA-values and depths. How to determine the seismic stress amplitudes along the depth and the number of equivalent cycles in evaluating *CSR* is addressed in Sec. 5.5.5.

On the other hand, the associated penetration test data available, in most cases SPT $N$-values, are collected, and normalized by the effective overburden stresses $\sigma_v'$ to have the $N_1$-values as already explained. Then, the data are plotted on the $CSR$–$N_1$ diagram as shown in Fig. 5.5.15 with discriminatory symbols according to surface manifestations of liquefaction occurrence site by site; for example, close symbols if evidences of liquefaction such as sand boils, ground settlements, fissures are observed versus open symbols if no evidence was there or even other symbols for intermediate observations. Then a boundary curve most appropriately discriminating the symbols is drawn as a condition of liquefaction triggering in terms of $N_1$.

Fig. 5.5.15 shows such a diagram developed by Tokimatsu and Yoshimi (1983) for sands with fines content $F_c \leq 5\%$ based on case histories mainly in Japan. Here, the equivalent shear stress amplitude $\tau_d$ is calculated from the maximum stress $\tau_{max}$ by $\tau_d = r_n \tau_{max}$, using a stress reduction coefficient $r_n$ to be explained in Sec. 5.5.5,

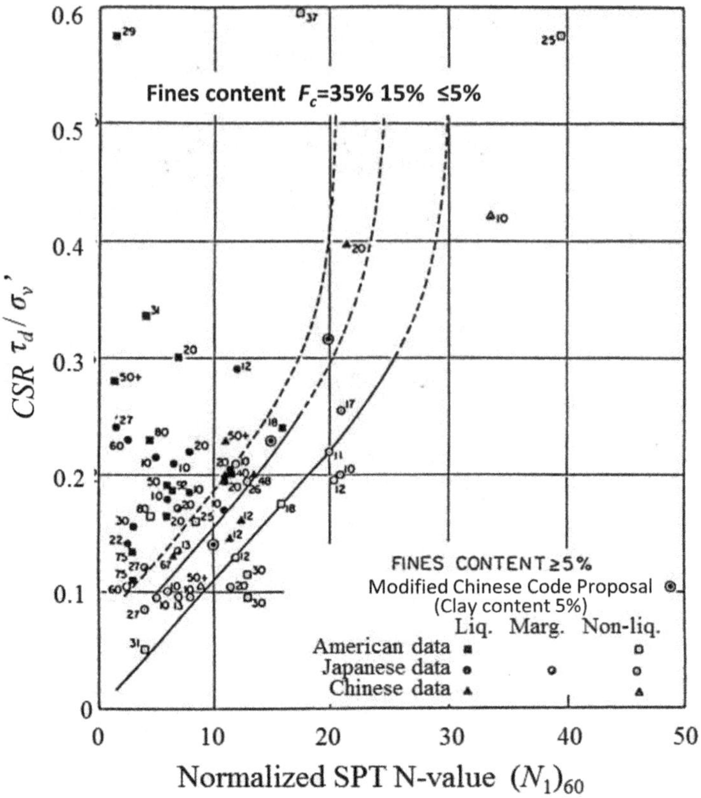


*Figure 5.5.16* *CSR* versus normalized SPT $(N_1)_{60}$-value based on liquefaction case histories with parametrically changing fines content and boundary curves separating liquefaction/non-liquefaction (Seed and De Alba 1984) with permission from ASCE.

normalized by the effective overburden stress as $\tau_d/\sigma_v'$, and taken as *CSR* under the $K_0$-anisotropic stress condition in the vertical axis. A boundary may be drawn without difficulty between the close and open symbols very similar to a solid curve in the chart, which was actually obtained by the laboratory tests on intact soils already explained in Fig. 5.5.11. The proximity of the two curves obtained by the two different approaches as already acknowledged by Tokimatsu and yoshimi (1983) helps to increase their credibility.

Fig. 5.5.16 shows the similar diagram developed by Seed and De Alba (1984) based on international case history data with the fines content as a key parameter again. It shows that the boundary curve of $F_c \leq 5\%$ almost coincides with the corresponding curve in Fig. 5.5.15, and the curve tends to shift left and upward with increasing $F_c$. Thus, the significant effect of fines content on the $CRR$ versus $N_1$ relationships is recognized internationally as an important parameter in liquefaction potential evaluations. These boundary curves proposed in 1980s have been revised by later investigations in USA but not so significantly changed since then (Idriss and Boulanger 2008).

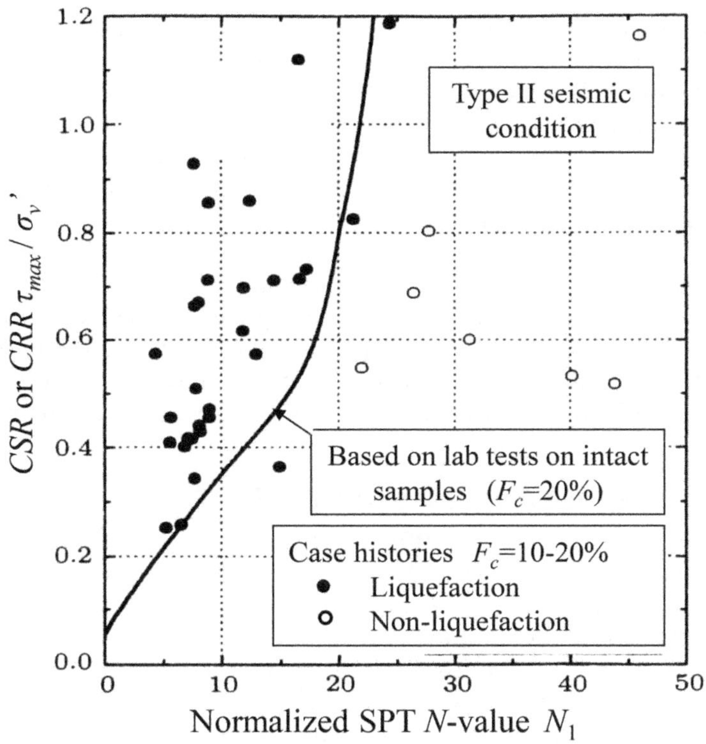


*Figure 5.5.17* CRR versus normalized SPT $N_1$-value curve based on lab tests in Japanese design code compared with liq./non-liq. plots based on corresponding case history data (Matsuo 1997).

This approach using actual field liquefaction performance are directly substantiated by the case histories and hence persuasive in shallow sand deposits in particular, though there seems to be some limitations. The evidences of liquefaction occurrence such as sand boils or fissures tend to become difficult to appear as the liquefied layer becomes deeper. Liquefaction in medium dense to dense sands or gravelly sands which tend to dilate with growing shear strain may not provide surface evidences clear enough to be detected as in loose sands. These limitations may be cleared to some extent by combining with the evaluation method based on lab tests, if the agreement between the two approaches can be confirmed. In general, they seem to be agreeable as already indicated in Fig. 5.5.15. Fig. 5.5.17 shows another comparison of the lab test-based curve (solid line) with the close/open plots for the case history studies conducted in the same sites (Matsuo 1997). Note that *CRR* or *CSR* in the vertical axis is defined in terms of the maximum seismic stress in place of the equivalent stress amplitude. This indicates again that under a particular seismic condition (some particular type seismic motion (Type II) in the Japanese design code (JRA 2002)), the lab-test curve for $F_c = 20\%$ is mostly compatible with the liquefaction/non-liquefaction boundaries for sands of $F_c = 10\text{–}20\%$ based on case histories in the same sites.

### 5.5.3 $F_c$-dependency of *CRR* – penetration resistance curve

Previous liquefaction case histories have demonstrated that sands containing a plenty of non/low-plasticity fines liquefied as frequently as clean sands. As already shown in the previous figures, *CRR*-values corresponding to an identical penetration resistance are evaluated higher if the associated $F_c$-values are higher based on both laboratory soil tests and case histories. Consequently, the $F_c$-dependent increase of *CRR* is considered as an important requirement in liquefaction potential evaluations in current design codes. On the other hand, liquefaction resistance of sands tends to decrease with increasing $F_c$ of low/non-plastic fines under the same relative density in laboratory tests as addressed in Sec. 5.4.3.2. Thus, the mechanical basis of the $F_c$-dependency in the *CRR* versus penetration resistance relationship has not been clarified yet. In the following, research results by a series of undrained cyclic triaxial tests coupled with miniature cone tests in the same specimens having various contents of low-plastic fines are addressed to directly compare the *CRR*-values with the penetration resistances.

#### 5.5.3.1 *Mini-cone triaxial tests for $F_c$-dependency*

In a series of tests (Kokusho et al. 2005b, 2011b, 2012 a,b), mini-cone penetrations were carried out for soil specimens 100 mm in diameter and 200 mm in height prior to triaxial liquefaction tests. The pedestal below the soil specimen was modified as shown in Fig. 5.5.18, so that a mini-cone built in the test device can penetrate into the specimen from the bottom. To realize that, the pedestal consists of two parts, a circular base to which the cone rod is fixed and a movable metal cap, through which the cone rod penetrates into the overlying specimen. By opening a valve, water in a reservoir inside the pedestal between the movable cap and the base is squeezed

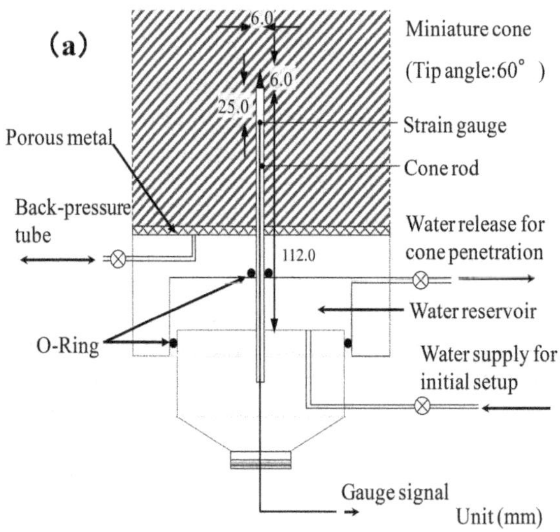


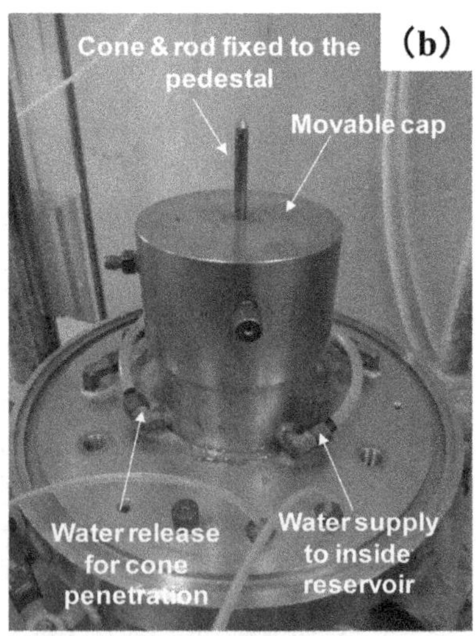


*Figure 5.5.18* Lower part of modified pedestal in mini-cone triaxial apparatus (Kokusho et al. 2012a): (a) Cross section, (b) Photograph, with permission from ASCE.

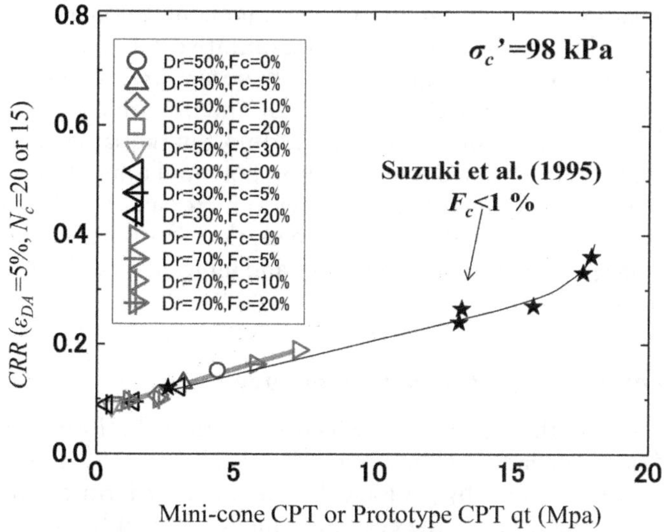


*Figure 5.5.19* $CRR \sim q_t$ plots by mini-cone triaxial tests on specimens with varying $F_c$ and $D_r$ (Kokusho et al. 2012a,b) compared with plots by prototype CPT (Suzuki et al. 1995), with permission from ASCE.

by the cell pressure, resulting in the settlement of whole specimen at the top of the pedestal, and thereby the cone penetrates into the specimen by 25 mm (from the initial to final projection in the specimen, 45 to 70 mm). The mini-cone is 6 mm diameter and 60 degrees tip angle, about 1/6 times smaller in size and the penetration speed is about 2 mm/s, 1/10 times slower than the prototype CPT. In the test sequence, the penetration test was first conducted in the undrained condition, and after reconsolidating the liquefaction test was carried out on the same specimen. It was confirmed in advance (Kokusho et al. 2005b) that the liquefaction resistance *CRR* was almost unaffected by the preceding mini-cone test and reconsolidation. Also shown by additional test series was that identical mini-cone tests in the drained conditions increased the penetration resistance for the same *CRR* by only $q_t = 1$ MPa, marginal within the range of data scatters (Kokusho et al. 2005b).

In the test, the relative density $D_r$, and fines content $F_c$ of the specimens were parametrically varied to investigate their effects on the *CRR* versus $q_t$ correlations. In Fig. 5.5.19, the *CRR*-values for $\varepsilon_{DA} = 5\%$, $N_c = 20$ are plotted versus the mini-cone resistances $q_t$. The resistance $q_t$ was determined from the average of maximum cone resistances during penetrations in 3~4 test specimens for one set of liquefaction tests. All symbols are located along the thick straight line in the diagram despite the wide varieties in $D_r$ (30–70%) and $F_c$ (0–30%), indicating that the liquefaction resistance *CRR* is uniquely correlated to $q_t$ irrespective of the difference in $D_r$ and $F_c$. The solid star symbols in the diagram, by the way, are *CRR* versus $q_{t1}$ plots based on in situ CPT and associated triaxial tests on intact clean sands sampled by in situ freezing by Suzuki et al. (1995) (see Fig. 5.5.14). The two research results, quite different in many ways, coincide surprisingly well at least in the linear part of the globally nonlinear correlation, indicating the applicability of the mini-cone triaxial test results to field conditions not only in a qualitative but also quantitative manner.

The uniqueness of the *CRR* versus $q_{t1}$ plots occurs presumably because the difference in $D_r$ or $F_c$ results in the changes in *CRR* and $q_t$ with a certain proportion so that the plots for different $D_r$ or $F_c$ are aligned almost in the same line. This finding is quite contradictory with the current liquefaction potential evaluation practice already mentioned, where *CRR* is to increase according to increasing $F_c$. Hence, there may be a significant difference in the *CRR* versus penetration resistance correlation between reconstituted fresh specimens in the laboratory and intact aged soils in situ with respect to the $F_c$-dependency.

### 5.5.3.2 Cementation effect in $F_c$-dependency

In order to investigate the aging effect which may affect the $F_c$-dependency, a small quantity of ordinary Portland cement up to CC = 1.0% was added to fines-containing sand to simulate cementation by geological aging in short-term accelerated tests (24 hours curing time). Here, CC is the cement content, defined by the dry weight ratio of cement to total soil. In the *CRR* versus $q_t$ diagrams in Fig. 5.5.20(a), the open symbol plots are without cement (CC = 0) identical with those in the previous figure and aligning in the line. By adding the cement, the plots tend to move up to the half-close (CC = 0.5%) further to the full-close symbols (CC = 1.0%) as indicated by the dashed arrows for $F_c$ = 0%, 5%, 10%, 20% and 30% individually in the diagram. Thus, it is clearly seen that *CRR* increases with a larger rate than $q_t$ so that each plot tends to go up to a higher position than the solid line for CC = 0 by the cementation effect. In the diagram, *CRR* tends to be larger with increasing $F_c$ for the same CC = 1% up to $F_c$ = 20%, indicating not only the cement content but also the fines content plays a significant role in these changes. However, *CRR* for $F_c$ = 30% tends to decrease from

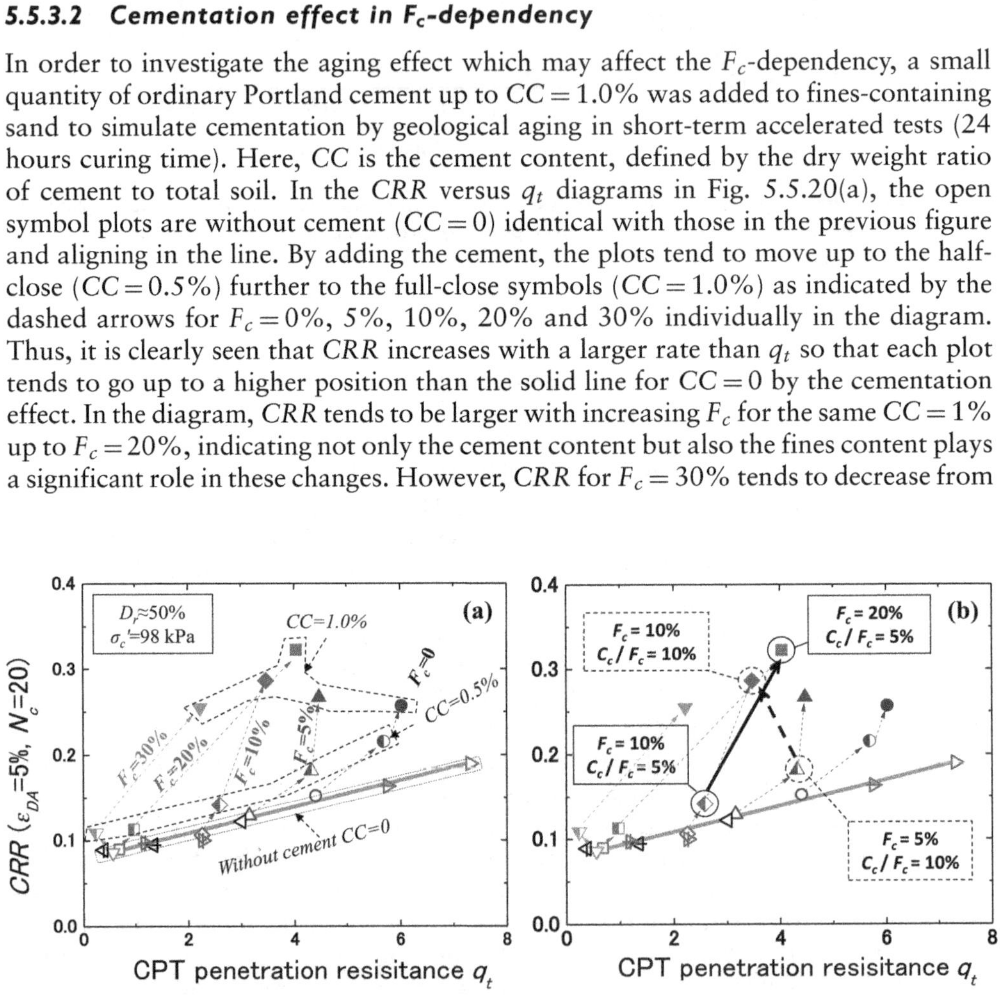


*Figure 5.5.20* *CRR*~$q_t$ plots in mini-cone triaxial tests on specimens with small amount of cement simulating cementation by geological aging in short-term accelerated tests (Kokusho et al. 2011b, 2012a,b) with permission from ASCE.

$F_c = 20\%$, presumably because the change in soil structure may occur between the two $F_c$-values from grain-supporting to matrix-supporting already discussed by using the binary packing model in Sec. 5.4.1.

Here, the ratio $CC/F_c$ may represent the chemical activity of fine particles, which plays an important role in long-term geochemical reactions in cementation because of their surface area much larger than coarser grains. The specimens with higher value of $CC/F_c$ in this accelerated test may be considered older in the geological age because of stronger cementation exhibited in natural soils. In Fig. 5.5.20(b) showing exactly the same plots as (a), those with $CC/F_c = 5\%$ or 10% are encircled and connected with the thick arrows, indicating that, for the same $CC/F_c$-value, $CRR$ tends to increase with increasing $F_c$ from 5 to 10% or 10 to 20%. It implies that, given a certain fine soil with a particular chemical activity ($CC/F_c$), the $CRR$–$q_t$ line tends to be higher with increasing $F_c$. One may take this result for granted because in the accelerated test, the increase in fines means the increase in the total amount of cement. However, if the accelerated test can reproduce an essential mechanical process of geological long-term cementation, this qualitative result does actually simulate the nature; a larger volume of fines with the same chemical activity results in clear increase in $CRR$ in the same aged soil.

The similar trend was actually observed in intact sands of Holocene and Pleistocene ages, wherein intact samples with a given $F_c$ tend to be plotted higher on the $CRR$ versus $q_t$ diagram than corresponding disturbed samples (Kokusho et al. 2012b). The difference between the intact and disturbed samples seems to be larger with increasing soil age, though more research on in situ soils of various ages are needed. Thus, it may be concluded that not the fines content itself but the aging effect by cementation, which becomes more pronounced in sands with higher $F_c$, can facilitate the mechanical basis why liquefaction strength should be raised depending on fines content under the same penetration resistance in the current practice.

### 5.5.4 Evaluation on gravelly soils

As already shown in Fig. 5.4.3, very well-graded gravelly soils with large $C_u$ and $D_{50}$ actually liquefied in previous earthquakes, though their absolute densities are much higher than poorly-graded sands. In order to determine the $CRR$ versus penetration resistance correlation for gravelly soils, two approaches are available again; cyclic triaxial tests on intact gravelly soils from in situ and case history studies in liquefied gravelly soils.

In Fig. 5.5.21(a), the cyclic resistant ratios ($CRR$) of well-graded gravelly soils in triaxial tests are plotted versus SPT $N_1$-values, where the close symbols are Holocene soils actually liquefied in Japan ($N_1 =$ around 25 or lower), while open circles and triangles are Pleistocene dense gravels and Holocene gravelly soils without liquefaction case histories (Tanaka et al. 1992, Kokusho et al. 1995, Inagaki et al. 1996, Matsuo and Murata 1997). All the tests were carried out using large cyclic triaxial apparatuses on intact specimens (20–30 cm in diameter and 40–60 cm in height) recovered by in situ freezing sampling. $CRR$ in the vertical axis is defined here for $\varepsilon_{DA} = 2$–$2.5\%$ in the dense Pleistocene gravels or $\varepsilon_{DA} = 5\%$ in other gravelly soils, and the number of cycles $N_c = 20$. The dashed curve in the diagram is the $CRR$ versus $N_1$ correlation for poorly-graded clean sands by Yoshimi et al. (1994) already explained in Fig. 5.5.11

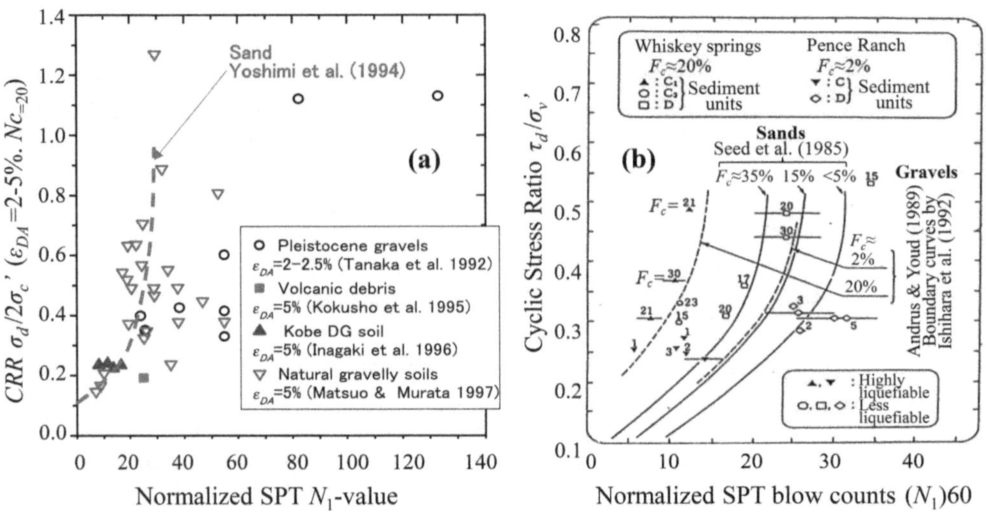

*Figure 5.5.21* *CSR* or *CRR* versus SPT $N_1$-values for gravelly soils obtained by two different approaches: (a) Triaxial tests on intact samples, (b) Case study during 1983 Borah Peak earthquake (Original data by Andrus and Youd 1989, modified by Ishihara et al. 1992).

where *CRR* is defined for $\varepsilon_{DA} = 5\%$, $N_c = 15$. Though the plots of gravelly soils seem to be compatible with the curve of sand up to $N_1 \approx 20$, *CRR* is evidently lower than sands with the same penetration resistance for a number of plots of $N_1 > 20$ among the dispersed data points. Assuming that the differences in the detailed definition of *CRR* among the various data are not large enough to completely change the trend, this finding may be difficult to accept for engineers who intuitively consider the gravelly soils are stronger and difficult to liquefy than sands. In the design code in Japan (JRA 2002), *CRR* of gravelly soils tends to become lower than sands actually under the same SPT $N_1$-value.

In order to properly understand this trend, it is necessary to revisit the SPT *N*-value versus relative density correlations for sandy and gravelly soils in Fig. 5.5.9. This figure indicates that under the same relative density, well-graded gravelly soils yield higher $N_1$-values mainly because of the larger absolute density than poorly-graded sands. This trend becomes clearer for denser states $D_r \geq 50\%$ or $N_1 \approx 20$ or larger. On the other hand, Fig. 5.4.6(a) indicates that poorly-graded sands and well-graded gravelly soils exhibit comparable liquefaction resistance if their relative density is the same. Combining the above two test results, it may be explained why the *CRR*-values of well-graded gravels were obtained much lower than those of poorly-graded sands under the same SPT $N_1$-value for $N_1 \approx 20$ or larger in Fig. 5.5.21(a).

On the other hand, Fig. 5.5.21(b) shows *CSR* versus modified SPT $(N_1)_{60}$ plots, obtained by Andrus and Youd (1989) in a liquefaction case study during the 1983 Borah Peak earthquake, where the occurrence of liquefaction in gravelly soils was evidenced by sand boils, fissures and lateral spreading observed during the earthquake (Andrus 1994). The dashed boundary curves were drawn in the diagram so that it could separate highly liquefiable plots from less liquefiable ones (Ishihara et al. 1992). If it is compared with similar curves for sands (solid curves) proposed by Seed et al. (1985), the *CRR*-value of gravels seems to be higher than that of sands. This trend is completely opposite to that shown in Fig. 5.5.21(a), wherein many gravelly soils tend to exhibit lower *CRR* than sandy soils in the laboratory tests on intact gravelly soils for the $N_1$-value larger than around 20.

As already observed in the monotonic loading test results in Fig. 5.4.7(a), it is remarkable that despite almost the same $D_r$-value, mobilized shear resistance is quite high in well-graded gravels in comparison to poorly-graded sands. If the shear resistance at double amplitude strain much larger than 5% (equivalent to 2.5% single amplitude strain) is concerned, the relative density can no more serve as a common scale. Instead, the particle gradation or uniformity coefficient $C_u$ makes a big difference even for soils of the same relative density. This implies that well-graded clean gravelly soils are less prone to post-liquefaction large deformations in the field such as cracks, differential settlements and lateral spreading to identify the occurrence of liquefaction. This observation, also considering that the $N_1$ versus $D_r$ correlation is not so much different between gravels and sands for $N_1 < 20$ as shown in Fig. 5.5.9, may be able to account for why the $CSR$–$N_1$ boundary curves in the gravel case histories tend to be higher than those of sands for the same SPT $N_1$-value in Fig. 5.5.21(b). The shear resistance for those large-strain ground deformations can become significantly higher for well-graded gravels than sands as demonstrated in Fig. 5.4.7(a).

## 5.5.5 Overview of current practice of liquefaction potential evaluation in SBM

In an earlier stage of liquefaction research, a stress-based method (SBM), comparing undrained cyclic strength with seismically induced shear stress, was proposed (Seed and Idriss 1971) and standardized for liquefaction potential evaluations in design codes in many countries. Though there seems to be some differences in SBM procedures in different design criteria in different countries, they are basically composed of common steps as outlined in Fig. 5.5.22. In the following, the major steps in the SBM liquefaction evaluation are outlined.

### 5.5.5.1 *Basic evaluation steps*

The SBM consists of Step-1 to 3, described in the following.

Step-1: Soils are initially screened in terms of physical properties, such as $F_c \leq 35\%$, $I_p \leq 15$, $C_c \leq 10{\sim}15\%$ in the case of Japanese design codes (JRA 2002, AIJ 2001). These thresholds were determined from quite a few liquefaction case histories worldwide wherein soils containing fines with plasticity were involved. Although there can be other criteria concerning the particle sizes such as mean grain size $D_{50}$ or fines content $F_c$, $I_p$ or $C_c$ seems more appropriate because the plasticity serves as a key parameter for liquefaction susceptibility as already discussed. Though these thresholds have mostly been compatible with previous liquefaction cases, some exceptional cases seem to have occurred as suggested in Fig. 5.4.12 during the 2011 Tohoku earthquake in Japan, where soils with the plasticity higher than the threshold values seem to

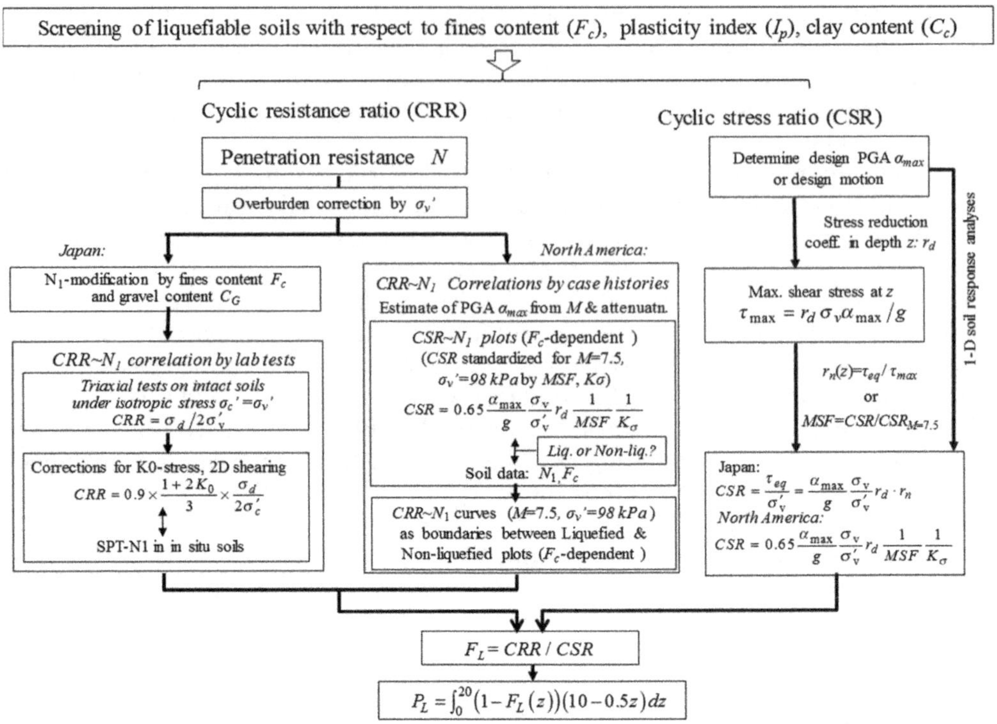


*Figure 5.5.22* Flow chart on major steps in stress-based liquefaction potential evaluation procedure.

have liquefied in young-age hydraulically-filled deposits where low and high-plasticity sublayers were interbedded.

Step-2: For potentially liquefiable soils screened above, a $F_L$-value (Factor for liquefiability) is calculated by dividing the cyclic resistance ratio ($CRR$) of the soil by the cyclic stress ratio ($CSR$) for a given seismic motion.

$$F_L = \frac{CRR}{CSR} \tag{5.5.23}$$

If $F_L$ is lower than 1.0, the soil is judged to liquefy.

Step-3: The $F_L$-values are dependent on depth $z$ in a given soil profile. In order to judge a depth-dependent cumulative effect of liquefaction on structures on/near the ground surface, a $P_L$-value defined by the next equation is used.

$$P_L = \int_0^{20} (1 - F_L(z))(10 - 0.5z)dz \tag{5.5.24}$$

where $F_L(z)$ is the $F_L$-values at depth $z$, and $F_L(z) = 1.0$ if $F_L(z) > 1.0$. The $P_L$-value was first proposed by Iwasaki et al. (1978b) in Japan and employed in USA by the name *LSI*

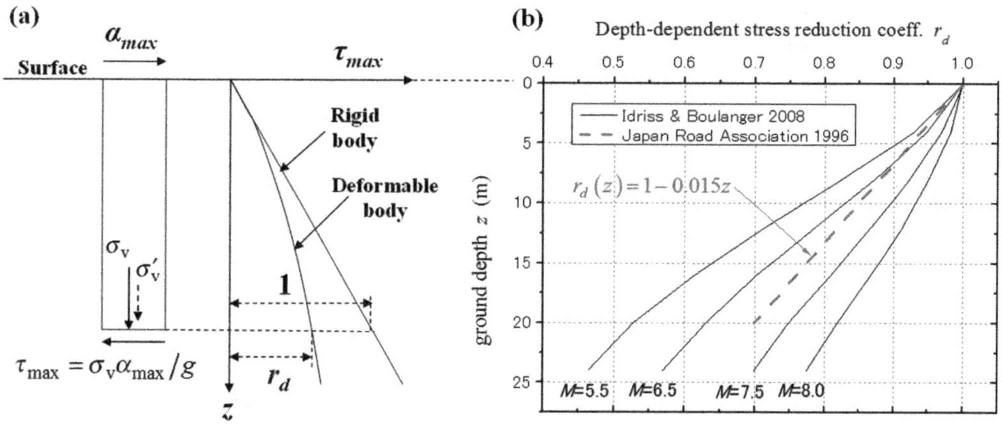


*Figure 5.5.23* Seismic shear stress induced in a level ground (a), and Depth-dependent stress-reduction coefficient $r_d(z)$ in USA and in Japan (b).

(Liquefaction Severity Index). It is the integration of $1 - F_L(z)$ from the surface $z = 0$ to the depth of $z = 20$ m with the weight of $10 - 0.5z$, and gives the maximum value $P_L = 100$ for $F_L(z) = 0$ (the severest liquefaction) and the minimum value $P_L = 0$ for $F_L(z) > 1.0$ (no liquefaction) in all the depth. Normally, $P_L > 15$ is considered to result in significant liquefaction damage near ground surface according to case histories in Japan. In the following, the evaluation procedures in Step-2, the major portion on SBM, will be explained in detail.

#### 5.5.5.2 How to decide CSR

Basically, the seismically induced shear stress can be evaluated by one-dimensional SH-wave propagation in a level ground using a design seismic motion. However, in a simpler method sometimes followed in practice, the peak ground acceleration $\alpha_{max}$ (PGA), at the surface is first decided from given earthquake magnitude and the acceleration attenuation curve. Then the peak shear stress $\tau_{max}$ at a depth $z$ is given, by depth-dependent total overburden stress $\sigma_v(z)$ and the acceleration of gravity $g$, as $\tau_{max} = \sigma_v(z) \cdot \alpha_{max}/g$ for the soil column postulated as a rigid body as illustrated in Fig. 5.5.23(a). Considering soil deformability causing a phase difference in the dynamic response, $\tau_{max}$ tends to vary as shown in the figure and may be formulated using the depth-dependent stress-reduction coefficient in terms of ground depth $r_d(z)$ as:

$$\tau_{max} = r_d(z) \cdot \sigma_v(z) \cdot \frac{\alpha_{max}}{g} \qquad (5.5.25)$$

Needless to say, $r_d(z)$ is quite dependent on the frequency content of seismic motions, the S-wave velocity profiles of particular sites, and the strain-dependent changes in soil properties as well. For the convenience in design, $r_d(z)$ may be simplified based

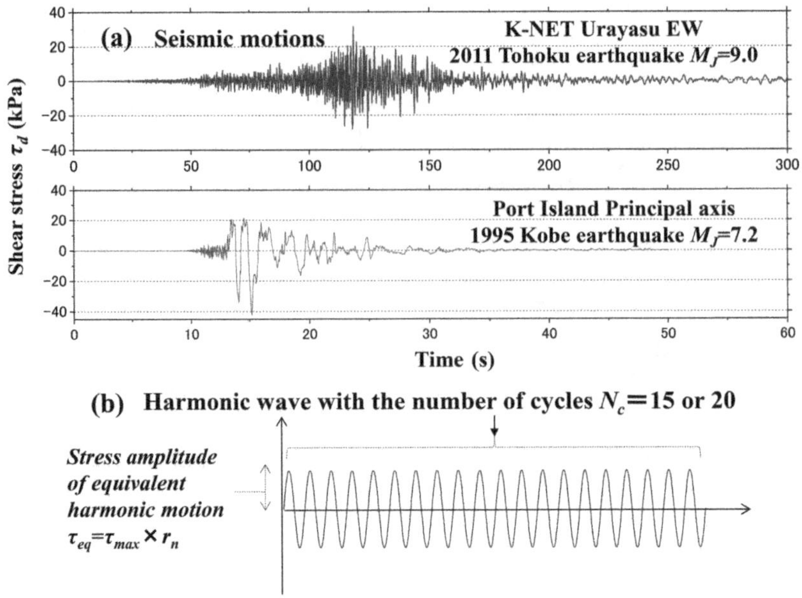


*Figure 5.5.24* Conversion of irregular stress time histories to equivalent harmonic motions with stress amplitude and number of cycles.

on a number of 1-D soil response analyses. In USA, $r_d(z)$ is chosen depending on the earthquake magnitude as shown in Fig. 5.5.23(b) (Idriss and Boulanger 2008), because longer period motions dominant in larger magnitude earthquakes result in milder depth-dependent decay. In Japan, a single linear function $r_d(z) = 1 - 0.015z$ is used in design criteria (JRA 2002, AIJ 2001) independent of earthquakes. Because the depth is limited $z = 20$ m maximum for liquefaction potential evaluation in Japan, the minimum $r_d$ becomes 0.70 there.

In order to compare the seismic stress with the liquefaction resistance of soils, in terms of shear stress amplitude in the harmonic wave to trigger liquefaction in a given number of cycles, an irregular stress time history with the maximum stress $\tau_{\max}$ is converted to a harmonic wave. In Fig. 5.5.24(a), two extreme stress time histories of strong earthquake motions, the 2011 Tohoku earthquake ($M_J = 9.0$) with a very long duration of major shaking of 3 minutes, and the 1995 Kobe earthquake ($M_J = 7.2$) of only 20 seconds, are exemplified. In the SBM, such different motions are converted to harmonic motions as illustrated in (b) with equivalent stress amplitudes $\tau_{eq}$ and equivalent numbers of cycles $N_{eq}$ so that the harmonic motions have the equivalent effects on the liquefaction triggering.

Though both $\tau_{eq}$ and $N_{eq}$ may be adjustable to have the equivalent harmonic motions, if $N_{eq}$ is first fixed as $N_{eq} = 15$, then $\tau_{eq}$ is modified accordingly using the coefficient $r_n$ (Tokimatsu and Yoshimi 1983) as:

$$\tau_{eq} = r_n \cdot \tau_{\max} \qquad (5.5.26)$$

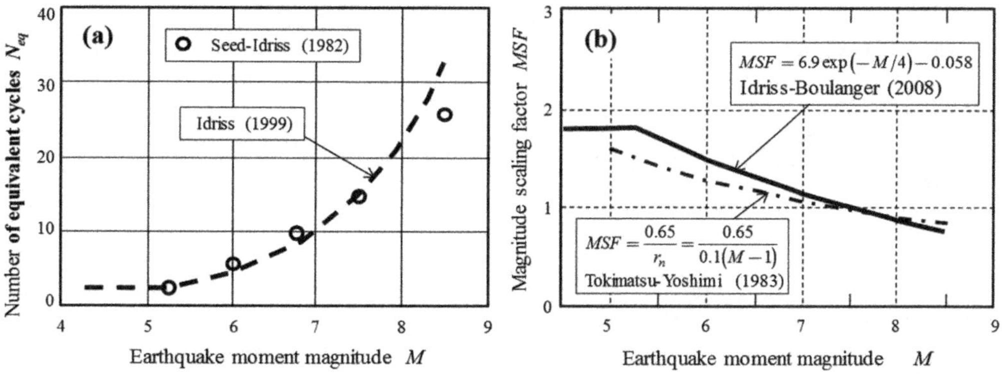


*Figure 5.5.25* Number of equivalent cycles $N_{eq}$ at 65% of peak stress versus earthquake magnitude $M$ (a), and Magnitude scaling factor $MSF$ versus $M$ compared with coefficient $r_n$ (b), (replotted from Idriss and Boulanger 2008), by permission by EERI.

where, $r_n$ is a stress reduction coefficient in terms of seismic wave irregularity given by the following empirical formula of earthquake magnitude $M$ (Tokimatsu and Yoshimi 1983) and employed in the Japanese design code (AIJ 2001).

$$r_n = \frac{\tau_{eq}}{\tau_{\max}} = 0.1(M - 1) \tag{5.5.27}$$

This formula was developed in accordance with empirical relationships proposed by Seed et al. (1975a) between earthquake magnitudes and the number of equivalent cycles. The $r_n$-value tends to increase with increasing earthquake magnitude.

Using Eqs. (5.5.25) and (5.5.26), the $CSR$-value is defined as the ratio of $\tau_{eq}$ to the effective overburden stress $\sigma'_v$ as:

$$CSR = \frac{\tau_{eq}}{\sigma'_v} = \frac{r_n \tau_{\max}}{\sigma'_v} = \frac{\alpha_{\max}}{g}\frac{\sigma_v(z)}{\sigma'_v(z)} \cdot r_n \cdot r_d(z) \tag{5.5.28}$$

In North American practice, $CSR$ in Eq. (5.5.28) with $r_n = 0.65$ is used after Seed and Idriss (1971). In that case, $N_{eq}$ can be determined as a function of the earthquake magnitude $M$ as illustrated in Fig. 5.5.25(a) (Idriss and Boulanger 2008) based on studies using many different earthquake records using the concept of fatigue theory explained in Sec. 3.1.6.1. Thus, the combination, $r_n = 0.65$ and $N_{eq} = 15$ for $M$ 7.5 earthquakes, is used as default values. In Japan, another combination, $r_n = 0.65$ and $N_{eq} = 20$ (corresponds to $M$ 8.0 earthquakes according to Fig. 5.5.25(a)) is used as the default value in the design code of road bridges (JRA 2002).

#### 5.5.5.3 How to decide CRR

The $CRR$-values are normally determined from in situ penetration tests using the $CRR$ versus penetration resistance correlations already prescribed. In Japan, the correlations have been constructed by combining triaxial test results on intact soils by in situ freezing sampling with penetration tests in the same soils as already explained in Sec. 5.5.2.1. In this case, the $CRR$-values by triaxial tests $CRR_{tx} = \sigma_d / 2\sigma'_c$ on isotropically consolidated specimens are converted to $\tau_{eq}/\sigma'_v$ corresponding to $K_0$-consolidated in situ soils as,

$$\frac{\tau_{eq}}{\sigma'_v} = 0.9 \cdot \frac{1 + 2K_0}{3} \cdot CRR_{tx} \tag{5.5.29}$$

Here, the constant 0.9 reflects the multi-directional shaking effect in situ, which cannot be considered in triaxial tests. Namely, soils are sheared in arbitrary horizontal directions during earthquakes, which can be expressed as the combined shearing in two orthogonal directions with a phase difference. It was shown in laboratory tests that the shearing in two horizontal directions reduces the liquefaction resistance to 80–90% of that in one direction (Seed et al. 1978). Tokimatsu and Yoshimi (1982) also showed that the fatigue theory can explain the $CRR$-value in the two-directional loading reducing to the similar percentage based on Eq. (3.1.47).

In North America, the $CRR$ versus penetration resistance correlations have been constructed based on numerous case history studies during previous earthquakes as the boundary curve segregating the plots of liquefaction and non-liquefaction on the chart, where the earthquake magnitudes and the effective overburden stresses are modified to the default values, $M = 7.5$ and $\sigma'_v = 98$ kPa, respectively. Unlike Eq. (5.5.29), the $CRR$-values $\tau_{eq}/\sigma'_v$ can be directly compared with $CSR$ given by design seismic motions without considering the effects of $K_0$ and multi-directional shearing because these effects may already be included to a certain extent in the field-based $CRR$-values. However, the standardized $CRR$-values for the $M$ 7.5 earthquake and the overburden $\sigma'_v = 98$ kPa, $CRR_{M=7.5,\sigma'_v=98\,\mathrm{kPa}}$, given in the design chart has to be modified to $CRR$ for particular earthquake magnitude $M$ and particular overburden $\sigma'_v$ (Idriss and Boulanger 2008) as:

$$CRR = MSF \cdot K_\sigma \cdot CRR_{M=7.5,\sigma'_v=98\,\mathrm{kPa}} \tag{5.5.30}$$

Here, the coefficient $K_\sigma$ is already defined in Eq. (5.3.2) as the overburden correction factor of $CRR$ by $\sigma'_v$, and $MSF$ is Magnitude Scaling Factor to adjust the liquefaction resistance (under the number of equivalent cycles $N_{eq} = 15$ constant) in accordance with earthquake magnitudes $M$.

$$MSF = \frac{CRR}{CRR_{M=7.5}} \tag{5.5.31}$$

The $MSF$-value (Seed et al. 1975a, Idriss and Boulanger 2008) are shown in Fig. 5.5.25(b), which tends to decrease monotonically with increasing $M$, and takes $MSF = 1.0$ at $M = 7.5$. $MSF$ is correlated with the stress reduction coefficient $r_n$ in Eq. (5.5.27) as,

$$MSF = \frac{0.65}{r_n} = \frac{0.65}{0.1(M-1)} \tag{5.5.32}$$

and the two coefficients are compatible to a certain extent as indicated in the same diagram. This *CRR*-evaluation method was developed very empirically, employing quite a few postulates in evaluating *CRR*. It is therefore critical to follow the same postulates as far as possible in applying this evaluation to individual sites in order to have appropriate results (Idriss and Boulanger 2008).

Thus in the SBM currently used for liquefaction evaluation in engineering practice, uncertainties are involved in converting widely varying irregular seismic motions to uniform motions of appropriate amplitudes and number of cycles. The constants, $r_n$ or *MSF* and $N_{eq}$, are firstly dependent on earthquake magnitudes $M$ but also found to be dependent on hypocentral distances and soil depths (Green and Terri 2005). Another uncertainty is in the calculation by the fatigue theory with which the correlations $N_{eq}{\sim}M$ and $r_n$ $(MSF){\sim}M$ have been developed. The constants $\alpha$ and $b$ defining the fatigue curve mentioned in Sec. 3.1.6.1 tend to widely vary among researchers (Tokimatsu and Yoshimi 1982) and may affect the SBM results considerably.

## 5.6 ENERGY-BASED LIQUEFACTION POTENTIAL EVALUATION

There exists another liquefaction potential evaluation based on energy, first proposed a few decades ago but not yet employed in practice. Unlike the stress-based method (SBM), the energy-based method (EBM) can directly deal with irregular seismic motions without converting to harmonic motions. In the following, after reviewing previous research, its outlines are addressed in terms of pertinent experimental data on dissipated energy or liquefaction capacity, evaluation procedures how to compare the capacity with the demand, and some typical EBM examples to compare with SBM.

### 5.6.1 Review on Energy-Based Method

The energy-based liquefaction evaluation method (EBM) was first proposed by Davis and Berrill (1982) and Berrill and Davis (1985), following a theoretical paper by Nemat-Nasser and Shokooh (1979) that the pore-pressure buildup is directly related to the amount of energy dissipated in the unit volume of soil (dissipated energy density). Undrained cyclic loading tests focusing on the dissipated energy in soil specimens were conducted using a torsional simple shear apparatus by Towhata and Ishihara (1985), in which a unique relationship was found between shear work (the dissipated energy) and excess pore-pressure being independent of the shear stress history. Yanagisawa and Sugano (1994) conducted similar cyclic simple shear tests on the effect of irregularity of cyclic stress on the dissipated energy to find a unique relationship, again. Laboratory soil tests were also conducted by Figueroa et al. (1994) using a strain-controlled torsional shear device, which demonstrated that the dissipated energy per unit volume during cyclic loading was closely connected to pore-pressure buildup under different confining stresses. Green et al. (2000) proposed an energy-based pore-pressure generation model and showed that it can approximate test results on sands and silt-sand mixtures of various densities.

In the first liquefaction potential evaluation by EBM proposed by Davis and Berrill (1982), the dissipated energy in liquefiable sands (capacity) was directly correlated

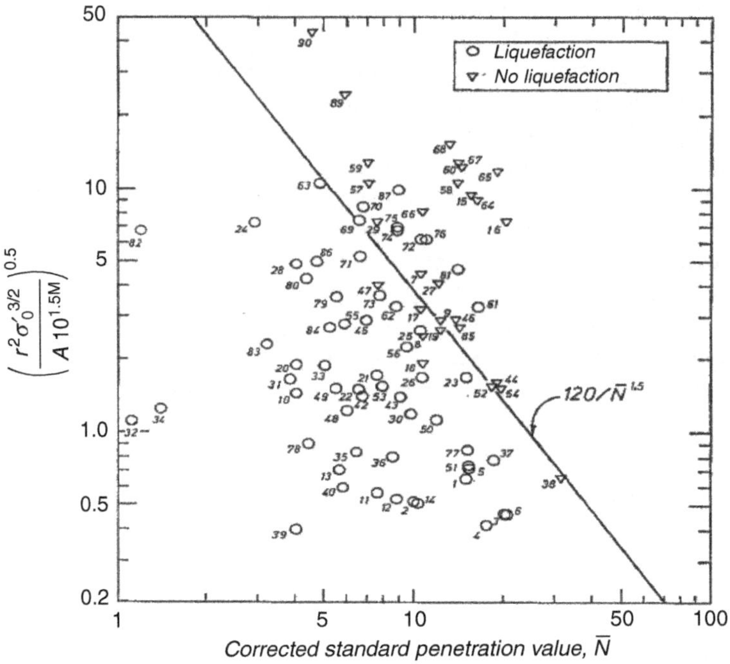


*Figure 5.6.1* Energy-based method directly comparing seismic energy calculated from empirical equations with previous liquefaction case histories (Berrill and Davis 1985).

with seismic energy (demand). The energy arriving at a site was calculated by essentially the same formulas as Eqs. (4.6.19) and (4.6.20). In their method, it was not addressed, however, at which depth the incident energy is given at a site, or how it transmits upward to liquefiable sand layers. Instead, variables consisting of earthquake magnitude, source distance and other pertinent parameters were calculated in liquefied/non-liquefied sites individually during previous earthquakes. They were then directly plotted versus corrected SPT blow-counts in the diagram as in Fig. 5.6.1 and compared with liquefaction case histories to empirically obtain a boundary curve discriminating liquefaction/non-liquefaction (Berrill and Davis 1985) by an analogous approach as explained in Sec. 5.5.2.2 in the stress-based method.

On the other hand, Kazama et al. (1999) proposed an energy-based scheme to evaluate liquefaction potential, in which cumulative dissipated energy in soil layers during a given seismic motion was evaluated in one-dimensional equivalent linear analysis and compared with the energy capacity for the soil layers to liquefy. The evaluation was carried out in the following steps:

(i) Elastic strain energy in each soil element is evaluated by the equivalent linear analysis using the following formula, wherein $G_{eq}$ is the equivalent linear shear modulus obtained from the analysis.

$$W_E(t) = \frac{1}{2} G_{eq} \{\gamma(t)\}^2 \tag{5.6.1}$$

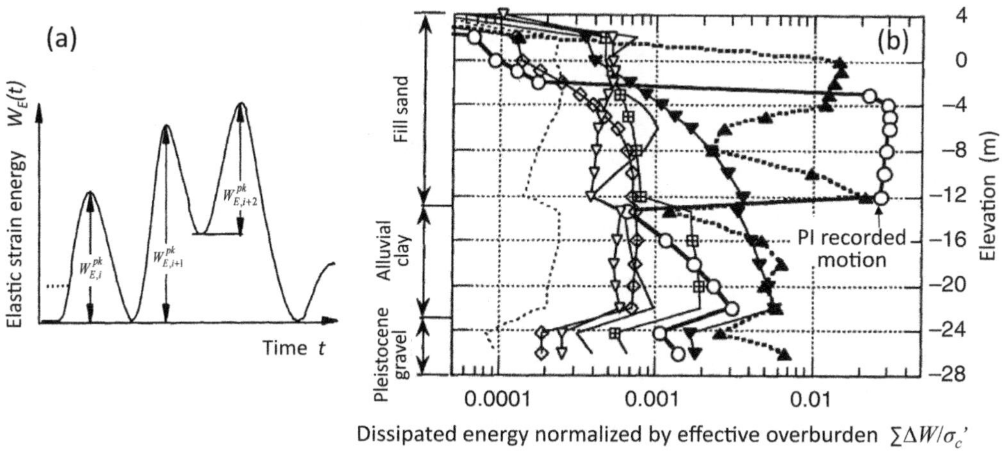


*Figure 5.6.2* Energy-based method using equivalent-linear soil response analysis (Kazama et al. 1999): (a) How to calculate peak elastic strain energies $W^{pk}_{E,i}$, (b) Cumulative dissipated energies along depth at Port Island site calculated for 8 strong motion records.

(ii) The time history of $W_E(t)$ is plotted as schematically illustrated in Fig. 5.6.2(a), and from individual peak values $W^{pk}_{E,i}$, corresponding dissipated energies are calculated in the next equation using equivalent damping ratios $D_{eq}$, wherein the multiplier 1/2 indicates that $\Delta W_i$ corresponds to the dissipated energy in every half cycle of loading.

$$\Delta W_i = 4\pi D_{eq} W^{pk}_{E,i} \times 1/2 \tag{5.6.2}$$

(iii) Then the $\Delta W_i$-values are summed up in the time sequence to have time-dependent cumulative dissipated energy to compare with a threshold energy capacity of each layer for various liquefaction behaviors not only the initial liquefaction.

This energy-scheme was applied to Port-Island (PI), Kobe Japan where extensive liquefaction occurred during the 1995 Kobe earthquake. Eight different earthquake motions were given to the soil profile and depth-dependent variations of the cumulative dissipated energy $\sum \Delta W$ at the end of the individual motions normalized by effective overburden stresses $\sigma_v'$ were evaluated as indicated in Fig. 5.6.2(b). The maximum value of $\sum \Delta W/\sigma_v'$ attained 0.02–0.03 for the motion recorded at PI (open circles) in the filled soil where extensive liquefaction occurred during the earthquake. The same authors proposed a procedure wherein the pore-pressure buildup ratio $u/\sigma_v'$ and the modulus degradation $G/G_1$ ($G_1$ = shear modulus at first cycle of undrained test) evaluated from the $\sum \Delta W$-values in each layer are incorporated in design.

Kokusho (2013b) proposed EBM to evaluate in situ liquefaction potential by directly calculating upward seismic wave energy $E_u$ (energy demand) and comparing

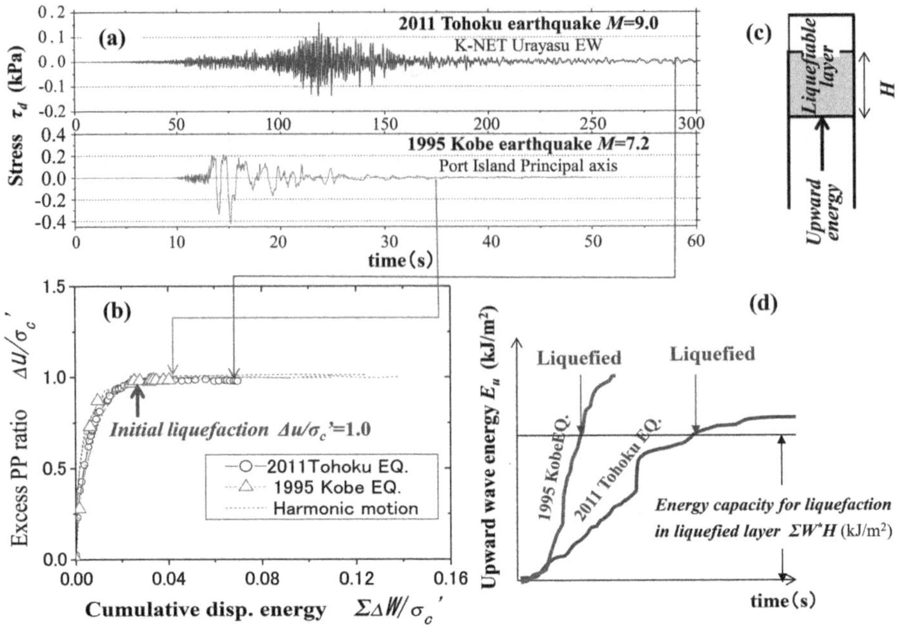


*Figure 5.6.3* Schematic illustration on basic idea of energy-based method by Kokusho (2013b).

it with the energy capacities in liquefiable layers as illustrated in Fig. 5.6.3. Based on experimental studies already referred, the energy capacity is almost uniquely determined from the energy dissipated in sand until liquefaction occurs irrespective of earthquake durations, dominant periods and wave forms. Fig. 5.6.3(a) shows torsional simple shear test results of $D_r = 50\%$ clean sand for the two extreme earthquake motions; the long duration 2011 Tohoku earthquake and the short duration Kobe earthquake. The excess pore-pressures in Fig. 5.6.3(b) are almost uniquely correlated with cumulative dissipated energy $\sum \Delta W$ until the initial liquefaction (Kaneko 2015). The energy capacity for liquefaction $\sum W^*H$ is correspondingly determined from the cumulative dissipated energy $\sum \Delta W$ where $W^*$ = strain energy associated with the dissipated energy $\Delta W$ determined in laboratory liquefaction tests and $H$ = layer unit thickness as will be explained later.

On the other hand, the energy demand can be calculated as the upward seismic wave energy $E_u$ (already discussed in Sec. 4.6.4) transmitted to liquefiable layers during particular earthquakes to compare with the unique liquefaction energy capacity $\sum W^*H$ irrespective of the difference of the motions as schematically depicted in Figs. 5.6.3(c) and (d). It should be noted here that the wave energy $E_u$ has to be compared not directly with $\sum W^*$ but with $\sum W^*H$ because the dimensions of $E_u$ and $W^*$ are energy/area and energy/volume, respectively. One of the advantages of this EBM is that the energy demand of a given earthquake motion can be grasped at a glance to compare with the energy capacity as illustrated in Fig. 5.6.3.

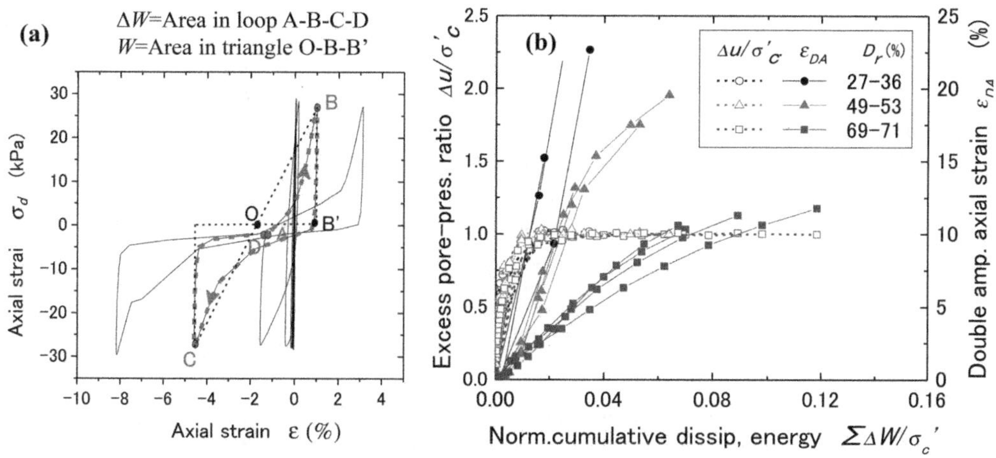


*Figure 5.6.4* Cyclic triaxial test results: (a) Dissipated and maximum elastic strain energies, ΔW and W, in stress-strain curves, (b) Normalized cumulative dissipated energy versus excess pore-pressure ratio or double amplitude axial strain obtained from a series of cyclic triaxial tests (Kokusho 2013b).

## 5.6.2 Dissipated energy for liquefaction in lab tests

Fig. 5.6.4(a) typically shows how the dissipated energy $\Delta W$ in a single loading cycle is defined as the area of the stress-strain hysteresis loop A-B-C-D in cyclic triaxial tests. The triangular area OBB′ means the maximum elastic strain energy in the cyclic loading and denoted as $W$. Accumulated dissipated energy per unit volume is obtained by adding $\Delta W$ in each cycle of loading to $k$-th cycle as,

$$\sum \Delta W = \sum_k \left( \int_A^D \sigma_d d\varepsilon \right)_k \tag{5.6.3}$$

In Fig. 5.6.4(b), the pore-pressure normalized by the initial effective stress $\Delta u/\sigma_c'$ and the strain amplitude $\varepsilon_{DA}$ in the vertical axes are plotted against the cumulative dissipated energy $\sum \Delta W/\sigma_c'$ in the horizontal axis with different symbols for $D_r \approx 30$, 50 and 70%. Here, the dissipated energy per unit volume $\Delta W$ is normalized by the effective confining stress $\sigma_c'$, where $\Delta W$ has the dimension of stress. This normalization is meaningful also because the cumulative dissipated energy $\sum \Delta W$ for pore-pressure buildup or given induced strains was found to increase almost in proportion to the confining stress. For example, Fig. 5.6.5(a) shows $\sum \Delta W$ versus $\sigma_c'$ plots based on previous test data by Figueroa et al. (1994) by strain-controlled cyclic torsional shear tests for the pore-pressure buildup $\Delta u/\sigma_c' = 100\%$. Obviously, $\sum \Delta W$ tends to increase almost in proportion to the confining stress in most of the data. Fig. 5.6.5(b) depicts similar test data by stress-controlled torsional shear tests wherein $\sum \Delta W$ for the induced strain $\gamma_{DA} = 7.5\%$ seems to be proportional to $\sigma_c'$ despite some data scatters. Hence, it is

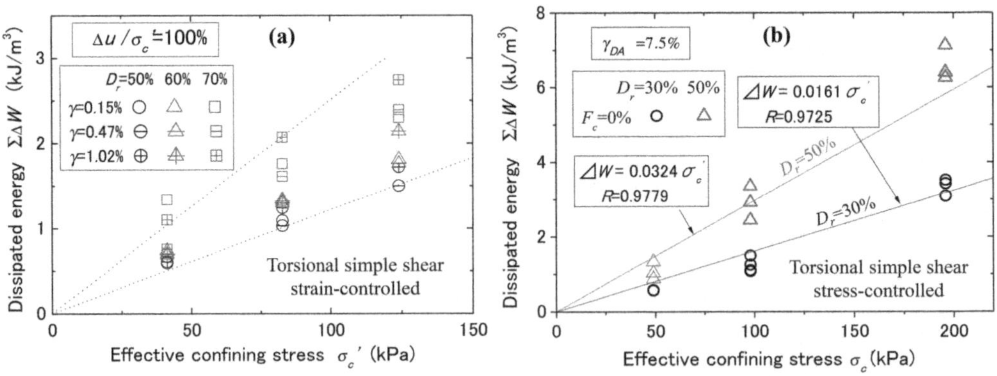

*Figure 5.6.5* Effect of confining stress on cumulative dissipated energy by torsional shear tests: (a) Strain-controlled test for 100% pressure buildup (plotted from original data by Figueroa et al. 1994), (b) Stress-controlled test for $\gamma_{DA}=7.5\%$ (Kaneko 2015).

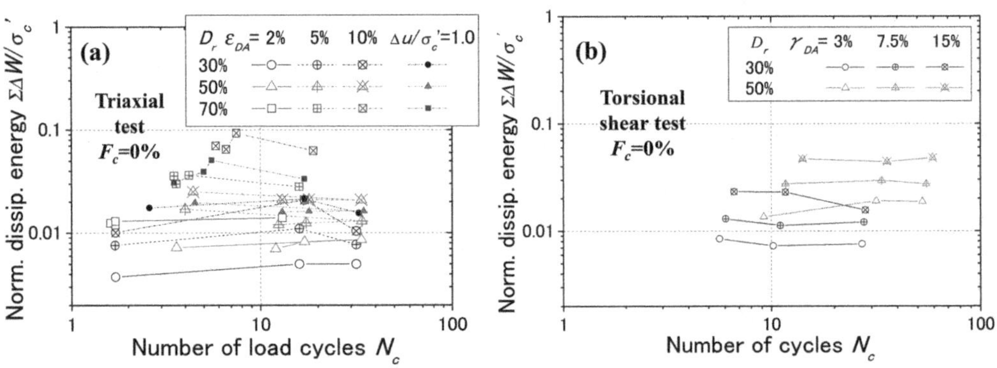

*Figure 5.6.6* Normalized cumulative dissipated energy for given strains or pressure buildup versus number of cycles: (a) Cyclic triaxial test (Kokusho 2013b), (b) Torsional shear test (Kaneko 2015).

meaningful to express the cumulative dissipated energy density in the normalized form as $\sum \Delta W/\sigma_c'$ from the mechanical point of view, too. In Fig. 5.6.4(b), the pore-pressure buildup correlates very well with the dissipated energy, and becomes $\Delta u/\sigma_c'=1.0$ at around $\sum \Delta W/\sigma_c'=0.02$. It is remarkable that the difference in the $\Delta u/\sigma_c' \sim \sum \Delta W/\sigma_c'$ correlations for different $D_r$ is small, while the $\varepsilon_{DA} \sim \sum \Delta W/\sigma_c'$ correlations are very much dependent on $D_r$. For the individual $D_r$-values, the cumulative dissipated energy $\sum \Delta W/\sigma_c'$ can be correlated almost consistently with the strain $\varepsilon_{DA}$ not only up to the initial liquefaction ($\varepsilon_{DA}=5\%$) but also even after that and serves as an indicator for the severity of liquefaction.

In Fig. 5.6.6(a), the cumulative dissipated energies $\sum \Delta W/\sigma_c'$ are plotted in the vertical axis of log-log charts versus the number of cycles $N_c$ in the horizontal axis to attain specific values of the strain amplitudes, $\varepsilon_{DA}=2$, 5, 10%, or the pressure buildup, $\Delta u/\sigma_c'=1.0$ in the triaxial tests on clean sands of $D_r\approx 30$–70%. There are groups of 2 to 4 data-points with identical symbols in the charts corresponding to the same specific strains $\varepsilon_{DA}$ or $\Delta u/\sigma_c'=1.0$ having different number of the loading cycles $N_c$. The lines connecting the same symbols do not show consistently increasing or decreasing trend of $\sum\Delta W/\sigma_c'$-values with increasing $N_c$, despite those for the dense sands showing non-systematic up-down trends particularly in higher strains. Fig. 5.6.6(b) shows similar plots to (a) obtained by torsional simple shear tests using the same clean sand of $D_r\approx 30\sim 50\%$. The plots are for $\sum\Delta W/\sigma_c'$-values attaining the strain amplitudes, $\gamma_{DA}=3$, 7.5, 15%. From the two diagrams (a) and (b), the lines connecting the same symbols may be judged to be essentially flat for easily liquefiable loose sands with lower $\sum\Delta W/\sigma_c'$-values. This indicates that the cumulative dissipated energy $\sum\Delta W/\sigma_c'$ determines particular strain amplitudes or pore-pressure buildup almost uniquely irrespective of $N_c$ and $CSR$. This further indicates that the $CSR$-$N_c$ lines corresponding to particular strains or pore-pressure buildup as shown in Fig. 5.1.4 for example, which is normally considered as a basis for the stress-based approach of liquefaction potential evaluation and also interpreted as the lines of equal damage in the fatigue theory (Annaki and Lee 1977, Green and Terri 2005), also represents the lines of equal dissipated energy. This observation paves a way to EBM using soil test data in SBM.

From the $CSR\sim N_c$ chart in Fig. 5.1.4 obtained by cyclic loading tests, $CRR$ for $N_c=20$ for example can be determined for $\varepsilon_{DA}=2$, 5, 10%, and $\Delta u/\sigma_c'=1.0$. The $CRR$-values are directly correlated with corresponding dissipated energy $\sum\Delta W/\sigma_c'$ calculated from the same test data to develop a $CRR\sim\sum\Delta W/\sigma_c'$ chart shown in Fig. 5.6.7 (Kokusho 2013b). Note that the values $\sum\Delta W/\sigma_c'$ in the vertical axis correspond to

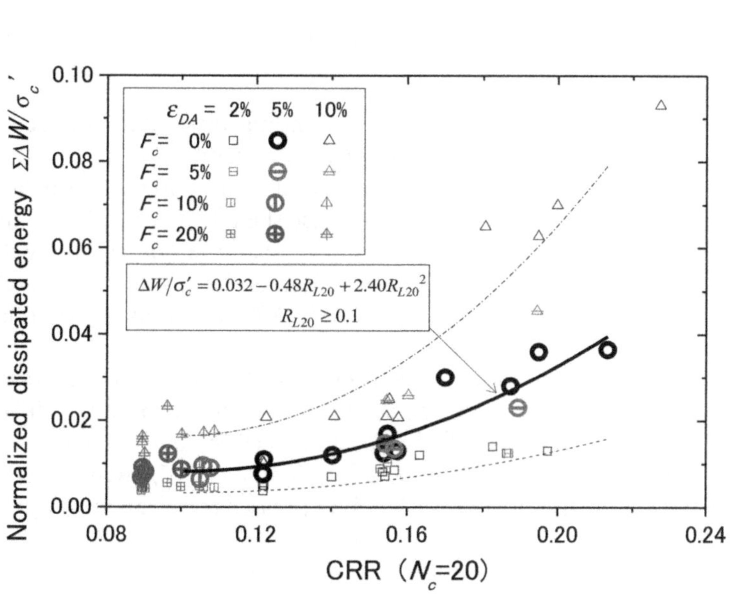


*Figure 5.6.7* $CRR$ ($N_c=20$) versus $\sum\Delta W/\sigma_c'$ plots for various $D_r$ and $F_c$ approximated by a parabolic function (Kokusho et al. 2012).

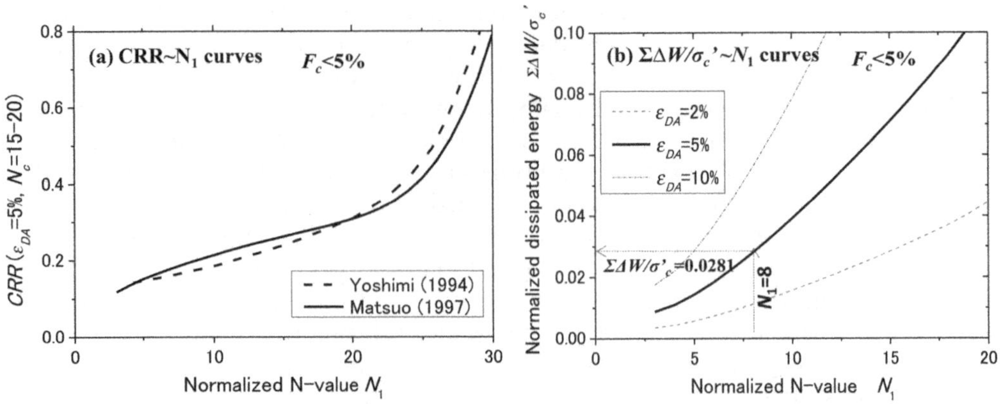


*Figure 5.6.8* CRR versus $N_1$ curves by two researchers (a), and $\sum \Delta W/\sigma_c'$ versus $N_1$ curves derived by combining with CRR versus $\sum \Delta W/\sigma_c'$ curve (b).

the dissipated energies to attain the axial strain $\varepsilon_{DA} = 5\%$ by arbitrary stress amplitudes and numbers of cycles, while the *CRR*-values in the horizontal axis represent the stress amplitudes at $N_c = 20$. Despite some data scatters, the *CRR*-value for the strain level $\varepsilon_{DA} = 5\%$ (open circles) seems to be uniquely correlated with $\sum \Delta W/\sigma_c'$ for sands with different relative densities and fines contents and approximated by the following parabolic function for $CRR \geq 0.1$ with the determination coefficient $R^2 = 0.86$.

$$\sum \Delta W/\sigma_c' = 0.032 - 0.48 \cdot CRR + 2.40 \cdot CRR^2 \tag{5.6.4}$$

If this relationship between *CRR* and corresponding dissipated energy $\sum \Delta W/\sigma_c'$ in Fig. 5.6.7 holds uniquely for sands with various densities and fines content, it may well be assumed to be applicable to natural sands with different soil fabric such as those formed in long geological histories, because the effect of soil fabric may possibly affect both *CRR* and $\sum \Delta W/\sigma_c'$ in such a way that the correlation will not differ considerably. This indicates that *CRR* versus $N_1$ correlations, already established and used in SBM, may easily be transformed into $\sum \Delta W/\sigma_c'$ versus $N_1$ correlations to be used in EBM.

For example, Fig. 5.6.8(a) shows empirical curves drawn between *CRR* ($\varepsilon_{DA} = 5\%$, $N_c = 20$) and $N_1$, developed for sands with a small amount of fines from several different sites independently by Yoshimi (1994) and Matsuo (1997), being almost coincidental to each other. The curve by Matsuo (1997), giving *CRR* ($\varepsilon_{DA} = 5\%$, $N_c = 20$, isotropic consolidation) from $N_1$, is expressed for clean sand by the following formula (JRA 2002).

$$CRR = \begin{cases} 0.0882\sqrt{N_1/1.7} & : N_1 < 14 \\ 0.0882\sqrt{N_1/1.7} + 1.6 \times 10^{-6}(N_1 - 14)^{4.5} & : 14 \leq N_1 \end{cases} \tag{5.6.5}$$

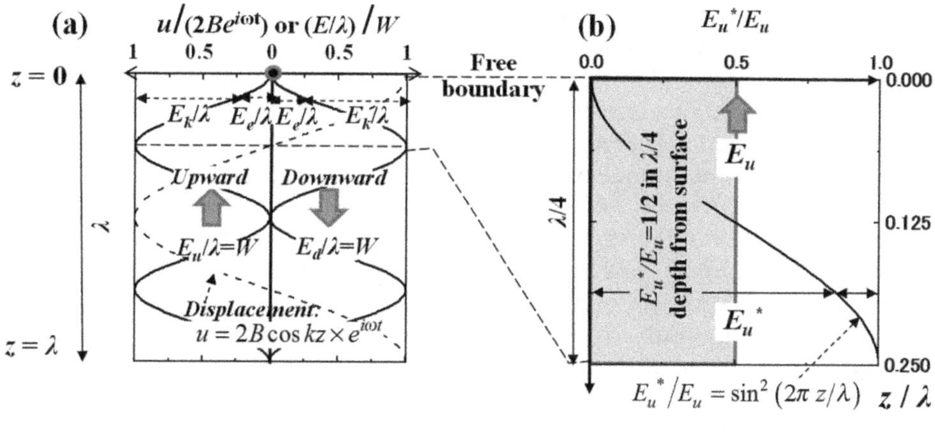


*Figure 5.6.9* Wave energy versus depth near free ground surface: (a) Upward and reflected downward waves, (b) Upward wave energy postulated in 1/4-wave length depth.

Fig. 5.6.8(b) shows a direct relationship between $N_1$ and $\sum \Delta W/\sigma_c'$ by combining Eqs. (5.6.4) and (5.6.5). The curves are drawn not only for $\varepsilon_{DA} = 5\%$ (initial liquefaction) but also for $\varepsilon_{DA} = 2\%$ and 10%. For sands with a given fines content, direct relationships between $N_1$ and $\sum \Delta W/\sigma_c'$ can be constructed in the same way by combining Eq. (5.6.4) with empirical $CRR \sim N_1$ relationship for particular fines content (e.g. JRA 2002).

### 5.6.3 How to compare capacity and demand

In EBM by Kokusho (2013b), the energy capacity for liquefaction is directly compared with the energy demand for a given earthquake motion in liquefiable layers. The upward SH-wave energy is considered here as the energy demand, because the wave energy causing liquefaction is the cumulative value and the associated downward energy also contributing the liquefaction originally constitutes a part of the upward energy. Some considerations incorporated in comparing the energy capacity with the upward energy are explained in the following.

There is an important issue about the energy demand for liquefaction that not all the upward energy is available in developing liquefaction in the shallow ground. It was already stated in Sec. 1.2 that the upward wave energy $E_u$ is shared evenly by kinetic energy $E_k$ and strain energy $E_e$, 50% each, both of which can supply the dissipated energy for liquefaction. However, if a stationary harmonic response of a soil column near the free ground surface is considered as illustrated in Fig. 5.6.9(a), the stationary vibrating displacement occurs as shown with the dashed curve with nodes and antinodes due to the upward and reflecting downward waves. Correspondingly, the strain energy $E_e$ is zero at the surface or any other antinodes and 100% at the nodes, while the kinetic energy $E_k$ is vice versa as illustrated with the solid curves. Thus, the two kinds of energy are distributed with a fixed rate of 0 to 100% depending on the position and is not convertible to one another, quite different from the one-directionally propagating wave. This is, however, an extreme case in the standing harmonic motion of a particular frequency and may not represent a realistic seismic response by transient irregular short-duration motion, wherein nodes and antinodes are difficult to appear regularly along the depth. However again, the free ground surface consistently serves as an antinode of the displacement with zero strain energy for all frequencies and its effect cannot be ignored even in the irregular seismic motions. Consequently, it is postulated here that the surface boundary effect, though fading away with increasing depth, can reach down to 1/4 of the wave length of $\lambda = V_s T$ for a representative period of seismic motion $T$ and the average wave velocity $V_s$, but not beyond that.

In Fig. 5.6.9(b) the upward SH-wave in the shallow depth of $\lambda/4$ from the surface is zoomed in. Considering the effect of the free surface boundary, the depth-dependent variation of an energy ratio for the harmonic standing wave with the representative wave length $\lambda = V_s T$ may be formulated (Kokusho 2016) as:

$$\frac{E_u^*}{E_u} = \sin^2\left(\frac{2\pi z}{\lambda}\right) \tag{5.6.6}$$

wherein $E_u^*$ stands for the upward energy which can compensate dissipated energy, and $E_u$ is the total upward energy at the same depth $z$. Namely, at the depth $z = \lambda/4$, the upward energy $E_u$ consists of 50% $E_k$ and 50% $E_e$, both of which can compensate the dissipated energy as in the one-directionally propagating wave energy ($E_u^*/E_u = 100\%$), while at the surface $z = 0$, $E_u$ is fixed to be $E_u^*/E_u = 0\%$ with no compensation allowed.

For earthquake motions, however, the application of Eq. (5.6.6) is obviously unrealistic because of non-harmonic irregularity and nonlinear soil properties during strong earthquakes. Hence, the energy ratio in Eq. (5.6.6) may well be simplified further and assumed as shaded in Fig. 5.6.9(b) to take the average value 1/2 down to the depth of $\lambda/4$ as:

$$\frac{E_u^*}{E_u} = \frac{1}{2} \tag{5.6.7}$$

Considering that dominant periods of earthquake motions $T$ in most liquefiable site conditions may be $T > 0.5$–$1.0$ s for average wave velocities $V_s > 160$ m/s in surface soil deposits, the minimum depth of $\lambda/4$ means around 20 m from the surface. This indicates that in normal liquefaction evaluation practice within the depth of $z = 20$ m, the upward energy should be halved to compare with the liquefaction energy capacity.

The next issue to consider is how to define the energy capacity to compare with the energy demand. It was shown in Sec. 1.6.2 that dissipated energy $\Delta E$ relative to the wave energy $E$ for SH-wave propagating in one wave length $\lambda$ is written as $\Delta E/E = 1 - e^{-4\pi D}$ in Eq. (1.6.36) using the damping ratio $D$. The dissipated energy $\Delta E$ for liquefaction has to be supplied by the wave energy $E$ in the field. It may well be assumed that the dissipated energy density for liquefaction per unit wave length $\Delta E/\lambda$ in in situ soil is identical to the dissipated energy density $\Delta W$ for liquefaction measured in laboratory cyclic loading tests on the same soil. As already mentioned in Sec. 1.5, the maximum elastic strain energy density $W$ defined in Eq. (1.5.4) is given in a half loading cycle, and the energy density $\Delta W$ in Eq. (1.5.3) is dissipated in one cycle of loading in an ideal viscoelastic material. As already observed in Fig. 1.6.2(a), in situ wave energy dissipation mechanism formulated as $\Delta E/E = 1 - e^{-4\pi D}$ may be approximated by $\Delta W/2W=2\pi D$ for larger $D$-values associated with liquefaction behavior.

This observation in one-cycle loading may be extended to the similar relationship as follows for the cumulative energies $\sum\Delta W$ and $\sum 2W$ if the damping ratio $D$ can be assumed constant during the liquefaction process.

$$\frac{\sum\Delta W}{\sum 2W}=2\pi D \tag{5.6.8}$$

A systematic test program actually shows that the damping ratio of sand is around $D=0.1$ to $0.2$ with the average 0.15 during liquefaction tests as shown in Fig. 5.6.10(b). Because the dissipated energy density for liquefaction $\sum\Delta W$ is supposed to be identical both in situ and in the laboratory, the upward wave energy density $E/\lambda$ should be compared with twice the cumulative elastic strain energy density $\sum 2W$ based on Eq. (5.6.8). As stated above, the upward energy should be halved to compare with the liquefaction energy capacity in the liquefiable shallow depth of 20 m. This means that $\sum 2W$ correlated with dissipated energy density $\sum\Delta W$ for liquefaction should be compare with the energy demand $E_u^*=E_u/2$ instead of $E_u$. If the upward wave energy $E_u$ is defined as the energy demand in the present EBM, the wave energy density $E/\lambda$ should be compared with four times the cumulative maximum strain energy, $\sum 4W$.

Apart from the ideal viscoelastic material, Fig. 5.6.10(a) exemplifies a typical stress-strain relationship obtained in undrained cyclic loading triaxial tests on saturated sands. In Fig. 5.6.10(b), the energy calculation results obtained from such stress-strain curves are shown cycle by cycle (Kokusho 2013b). In the vertical axis, the elastic maximum strain energies $W$ (*Area*(ODD′)) multiplied by 4 because of the above-mentioned reason and summed up in the loading sequence as $\sum W^*\equiv\sum 4W$ are plotted with open symbols versus the cumulative dissipated energies $\sum\Delta W$ (*Area*(ABCDEA)) in the horizontal axis. The same test data in Fig. 5.6.4(b) is used here again in the plots, which

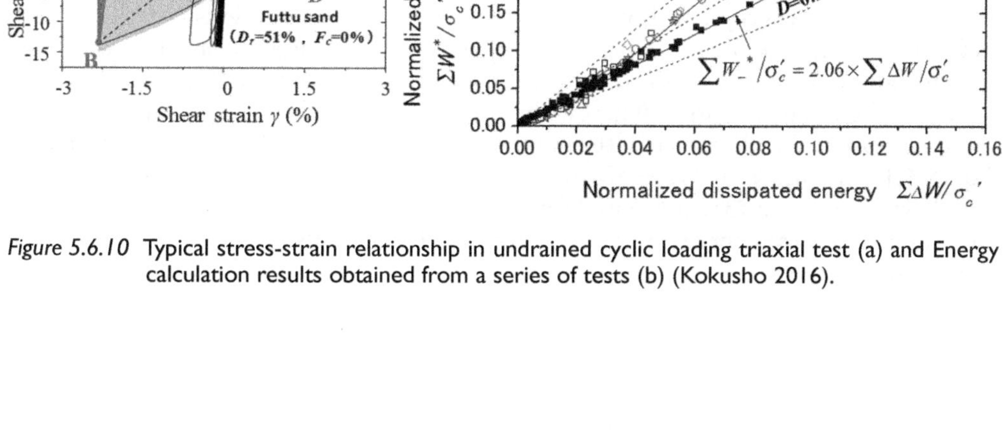


*Figure 5.6.10* Typical stress-strain relationship in undrained cyclic loading triaxial test (a) and Energy calculation results obtained from a series of tests (b) (Kokusho 2016).

may be approximated by the next equation to determine $\sum W^*$ from the cumulative dissipated energy for liquefaction $\sum \Delta W$ (Kokusho 2013b).

$$\frac{\sum W^*}{\sigma'_c} = 5.4 \times 10^{1.25 \times \log(\sum \Delta W/\sigma'_c)} \tag{5.6.9}$$

These $\sum \Delta W \sim \sum W^*$ plots may be compared with Eq. (1.5.7) $\Delta W/W = 4\pi D$ using some representative values of $D$, in order to know if the damping ratio $D$ can be represented by a constant value during cyclic loading liquefaction tests. The correlation $\sum \Delta W/\sum W^* = \sum \Delta W/\sum 4W = \pi D$ for $D = 0.1$, 0.15, 0.20 is shown with a set of dashed lines in Fig. 5.6.10(b) to compare with the open symbol plots. Obviously, nearly all the plots for different relative density $D_r$ and fines content $F_c$ are in between $D = 0.10$ and 0.20 throughout the cyclic loading liquefaction tests and may be approximated by $D \approx 0.15$ as the average, confirming the assumption to draw $\sum \Delta W/\sum 2W = 2\pi D$ in Eq. (5.6.8) from $\Delta W/2W = 2\pi D$.

Apart from using the elastic maximum strain energy $W$ (*Area*(ODD′) in a half cycle as in normal engineering practice, the strain energy actually needed in one cycle denoted here as $2W_-$ considering the partial energy recycling, corresponds to *Area*(ABB′CDD′EA) minus *Area* (BB′C) in Fig. 5.6.10(a) and can be evaluated in the same way as Eq. (1.6.40).

$$2W_- = \frac{\Delta W + Area(ABB'CDD'EA)}{2} \tag{5.6.10}$$

In order to compare this with the cumulative dissipated energy in the same manner as $\sum W^* \equiv \sum 4W$ versus $\sum \Delta W$, $2W_-$ is doubled here and summed up to individual cycles as $\sum W^*_- \equiv \sum (2 \times 2W_-)$ and plotted versus $\sum \Delta W$ with close symbols in Fig. 5.6.10(b) (Kokusho 2016). The relationship $\sum W^*_- \sim \sum \Delta W$ is not so different from $\sum W^* \sim \sum \Delta W$ for $\sum \Delta W$ up to 0.02~0.04, which corresponds to initial liquefaction (Kokusho 2013b). Beyond that energy, $\sum W^*$ obviously gives higher energy than $\sum W^*_-$, while $\sum W^*_-$ tends to be almost proportional to $\sum \Delta W$ all the way from zero to $\sum \Delta W = 0.12$. It is approximated by the next equation with a high coefficient of determination $R^2 = 0.997$.

$$\sum \frac{W^*_-}{\sigma'_c} = 2.06 \times \sum \frac{\Delta W}{\sigma'_c} \tag{5.6.11}$$

The difference between Eqs. (5.6.9) and (5.6.11) is partially attributed to that twice the elastic strain energy $2W$ given to a soil specimen in one cycle is correlated with $\Delta W$ in the former while recycling of a part of the strain energy from the first to the second half cycle loading is considered in the latter. The effect of nonlinear stress-strain curve on the calculated strain energy, the cyclic mobility effect in particular, may be another cause of the difference. If the notation $\sum W^*_- \equiv \sum (2 \times 2W_-)$ is reminded here, Eq. (5.6.11) implies $\sum W^*_-/2 = \sum 2W_- = 1.03 \times \sum \Delta W \approx \sum \Delta W$. This allows a very simple interpretation that the cumulative dissipated energy $\sum \Delta W$ is almost equal to the cumulative strain energy $\sum 2W_-$ in Eq. (5.6.10) actually supplied all through the liquefaction process.

### 5.6.4 Evaluation steps in EBM

If a design earthquake motion is given at the ground surface, the particle velocity $\dot{u}$ of the upward SH wave in any layer can be calculated by the one-dimensional equivalent linear response analysis. The associated upward energy (the energy demand) can be calculated from $\dot{u}$ for a time interval $t = 0 \sim t_1$ for the major shaking duration as already formulated in Eq. (1.2.25) as:

$$E_u = \rho V_s \int_0^{t_1} (\dot{u})^2 dt \tag{5.6.12}$$

where, $\rho$ = soil density, and $V_s$ = S-wave velocity of the layer.

In order to determine the upward energy, there may be another simplified method based on the well-known empirical formulas on incident seismic energy at a base layer, Eqs. (4.6.19) and (4.6.20), using earthquake magnitudes and hypocentral distance. From the incident energy, the upward energy $E_u$ at a given layer in a shallow depth may be evaluated by Eq. (4.6.17) using the seismic impedance ratio between the layer and the base where the incident energy is defined as explained in Sec. 4.6.4.

The evaluation steps for the EBM are illustrated in Fig. 5.6.11. Hereafter, the notation of summation $\sum$ in terms of loading cycles will be abbreviated for simplicity, so that $\sum \Delta W \rightarrow \Delta W$ and $\sum W^* \rightarrow W^*$.

a) At a given site, a soil model consisting of different layers is divided into "layer units" of a constant thickness $H = 1$ or 2 m in accordance with the availability of penetration test data with sequential numbers $i = 1$–$n$. The normalized dissipated

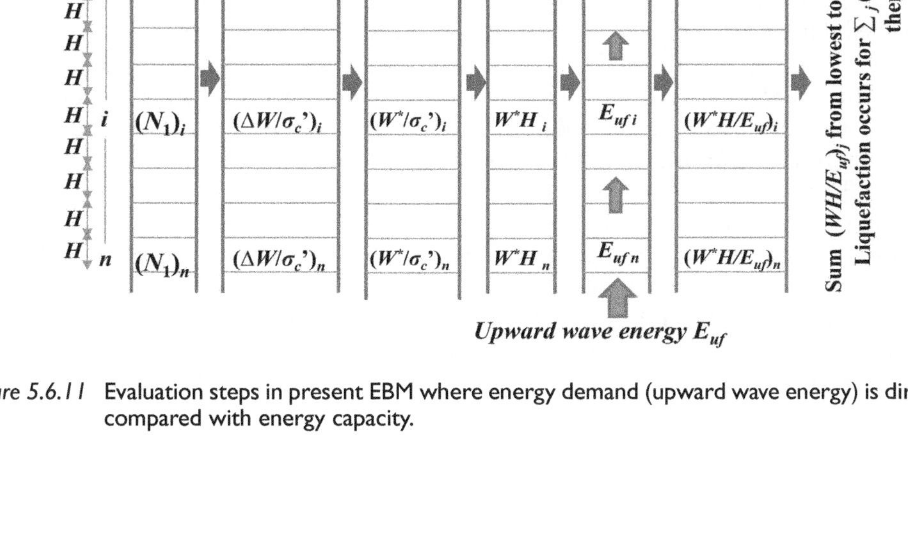


*Figure 5.6.11* Evaluation steps in present EBM where energy demand (upward wave energy) is directly compared with energy capacity.

energy density $\Delta W/\sigma_c'$ for liquefaction is determined for each layer unit from penetration test results, using Eqs. (5.6.4) and (5.6.5) for example.

b) The normalized strain energy density $W^*/\sigma_c'$ corresponding to $\Delta W/\sigma_c'$ for liquefaction is evaluated by Eq. (5.6.9), though Eq. (5.6.11) might also be used if $W_-^*$ is employed as the strain energy density in place of $W^*$.

c) Then the strain energy $W^*H$ for the unit with the thickness $H$ to liquefy is calculated as the energy capacity of the unit. In calculating $W^*$ from $W^*/\sigma_c'$, the effective confining stress $\sigma_c'$ is determined from the effective overburden stress $\sigma_v'$ as $\sigma_c' = (1+2K_0)\sigma_v'/3$.

d) The upward energy $E_u$ is calculated in Eq. (5.6.12) using the one-dimensional response analysis of the soil model, and the ultimate energy at the end of shaking of a given earthquake motion $E_{uf}$ is determined for each unit as the energy demand.

e) The liquefaction energy capacity $W^*H$ in each unit is directly compared with the energy demand $E_{uf}$ by calculating an energy ratio $W^*H/E_{uf}$. A unit with a smaller value of the energy ratio $W^*H/E_{uf}$ has higher and earlier liquefaction potential than other units in the same soil profile, although the overall liquefaction potential will be decided in g) below.

f) The energy ratios of individual units over the soil profile are arranged and numbered in sequence starting from the lowest energy ratio ($j$=1) toward higher ones and summed up as $\sum_j (W^*H/E_{uf})_j$ in terms of the sequence $j$, denoted here as $AER$, "accumulated energy ratio".

g) Liquefaction is considered to occur at most in those units where $AER \equiv \sum_j (W^*H/E_{uf})_j < 1.0$, because the upward energy can liquefy individual units in the above-mentioned sequence $j$ until it is all consumed by the dissipated energies in those units.

Thus, in the present EBM, the energy demand $E_{uf}$ is explicitly given, and liquefaction is judged to occur only in those layers where their total energy capacities $\sum_j (W^*H)_j$ are within the energy demand of the upward wave.

### 5.6.5 Typical EBM results compared with SBM

In order to compare the above-mentioned EBM with SBM, the first soil model addressed here is a hypothetical uniform sand deposit 10 m thick underlain by a stiff base shown in Fig. 5.6.12 (Kokusho 2013b). The sand deposit $K_0$-consolidated with its normalized SPT $N$-value $N_1 = 8$, the effective overburden and the S-wave velocity shown in the figure is divided into 5 layer units of $H = 2$ m thick each (L1 to L5), wherein L1 is unsaturated (the density $\rho_t = 1.8$ t/m$^3$) and L2 to L5 are saturated ($\rho_{sat} = 1.9$ t/m$^3$).

A horizontal acceleration motion (K-NET Urayasu EW) during the 2011 Tohoku earthquake ($M_J = 9.0$) is given at the ground surface either in the real time scale (RT: duration 236 s) or in a compressed half time scale (RT/2: duration 118 s). In Fig. 5.6.13, the two time histories (a) RT and (b) RT/2 given at the ground surface are shown at the top together with the upward energies calculated in the individual units at the bottom. Note that the upward energy dramatically decreases down to about 1/8 if the time scale is halved (RT/2).

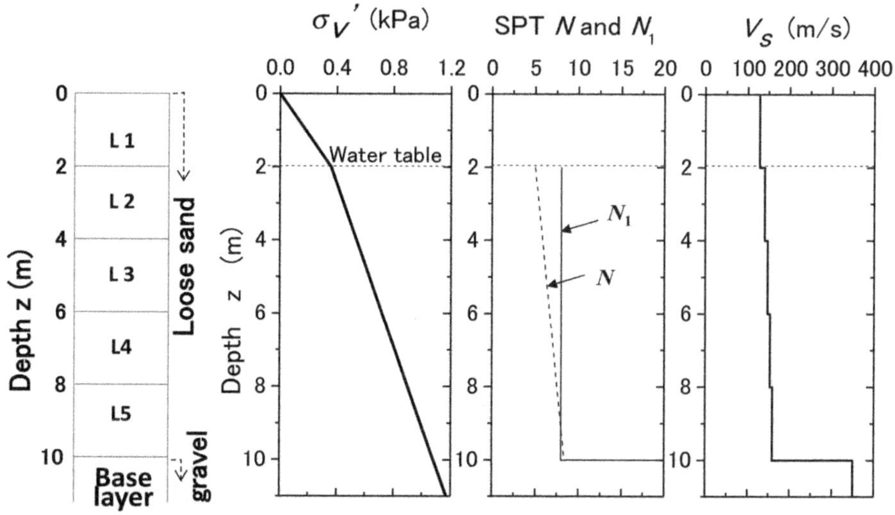


*Figure 5.6.12* Uniform soil model for liquefaction evaluated by EBM and SBM (Kokusho 2013b).

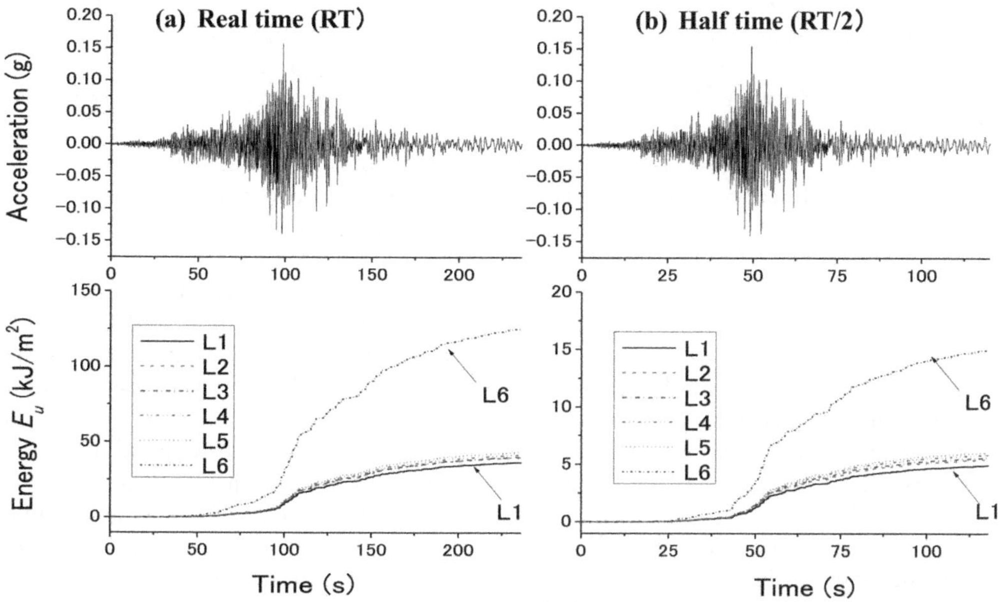


*Figure 5.6.13* Time histories of acceleration (top) and upward wave energy (bottom) given to the soil model: (a) Real-time motion (RT), (b) Compressed half-time motion (RT/2) (Kokusho 2013).

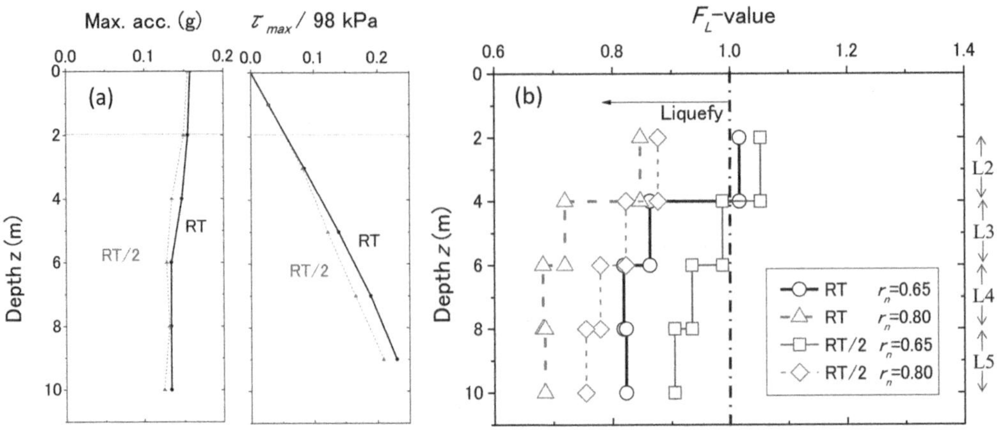


*Figure 5.6.14* Liquefaction evaluation results by SBM for RT and RT/2 motions: (a) Depth-dependent Max. acceleration and $\tau_{max}$, (b) Depth-dependent $F_L$ (Kokusho 2013b).

As shown in Fig. 5.6.14(a), the accelerations in the model are more or less the same, while the stress, identical at the shallow depth, is getting slightly smaller in RT/2 than in RT with increasing depth because of the shorter wave length in the RT/2-motion. In the SBM evaluation, the cyclic stress ratio is obtained from the maximum seismic shear stress $\tau_{max}$ and the effective overburden stress $\sigma'_v$ as $CSR = r_n \tau_{max}/\sigma'_v$, where the stress reduction coefficient $r_n$ introduced in Eq. (5.5.28) is determined as $r_n = 0.80$ for the $M = 9.0$ earthquake (Tokimatsu and Yoshimi 1983) and also $r_n = 0.65$ for the default value. In Fig. 5.6.14(b), the $F_L$-value thus evaluated is illustrated along the depth for the RT and RT/2-motions. The choice of $r_n = 0.65$ or $r_n = 0.80$ tends to have a greater effect on the $F_L$-value than the difference of input motions, RT or RT/2, despite the considerable difference of energy in the two motions. It is also noted that the deeper the unit, the lower its $F_L$-value and the higher its liquefaction potential for both RT and RT/2-motions in the uniform soil model.

In the EBM evaluation, the normalized dissipated energy per unit volume to liquefy the sand layer of $N_1 = 8$ can be calculated from Eqs. (5.6.4) and (5.6.5) as $\Delta W/\sigma'_c = 0.0281$ (Note all the summation signs $\sum$ for loading cycles in terms of the energies $\Delta W$ and $W^*$ are abbreviated). Then, the corresponding strain energy per unit volume for liquefaction is given as $W^*/\sigma'_c = 0.0621$ from Eq. (5.6.9). The liquefaction energy capacities $W^* H$ for the units $H = 2$ m thick to liquefy are calculated using the corresponding average confining stresses $\sigma'_c = \sigma'_v(1 + 2K_0)/3$, with $K_0 = 0.5$.

In Fig. 5.6.15(a), the liquefaction energy capacities $W^*H$ (thick lines) are shown together with the upward energies $E_{uf}$ along the depth. As mentioned, $E_{uf}$ is quite different between the two input motions, RT (medium thin line) and RT/2 (thin line). In Fig. 5.6.15(b), the energy ratio $W^*H/E_{uf}$ calculated in individual units is shown along the depth with the thin lines plus small solid symbols. Because the energy ratio $W^*H/E_{uf}$ is obviously smaller for the units in shallower depths both for RT and RT/2 motions, liquefaction tends to occur first in L2 and descend in sequence to the deeper

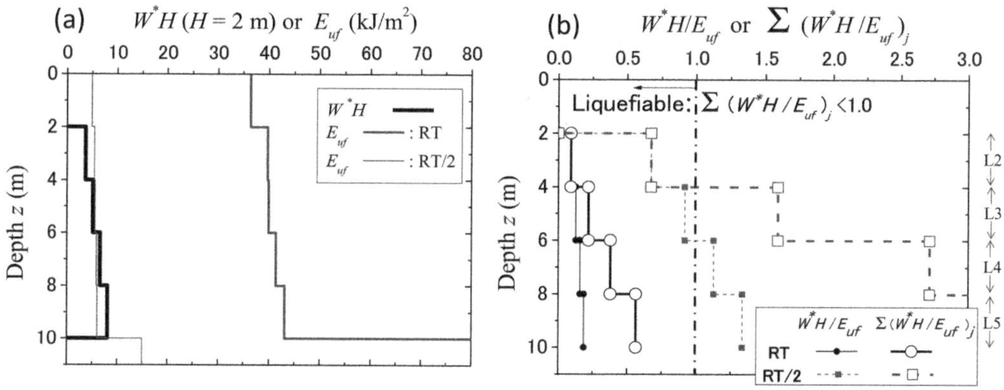


*Figure 5.6.15* EBM results for RT and RT/2 motions along depth: (a) Energy capacity $W^*H$ or upward wave energy $E_{uf}$, (b) Energy ratio $W^*H/E_{uf}$ or accumulated energy ratio $\sum W^*H/E_{uf}$ (Kokusho 2013b).

units. The thick lines with large open symbols in the same diagram are the values $AER = \sum_j (W^*H/E_{uf})_j$ calculated in the EBM step (f) explained above. For the RT-motion shown by the thick solid lines, $AER < 1.0$ holds from L2 to L5, indicating that the upward energy is enough to liquefy all the saturated units. In contrast, for the RT/2-motion shown by the thick dashed lines, $AER < 1.0$ only for L2, indicating that the upward energy is not enough to liquefy all but the unit L2. Thus, there exists a clear difference in liquefaction potential between the two input motions, reflecting the tremendous energy demand reduction in the RT/2-motion.

The results by EBM in Fig. 5.6.15(b) can be compared with those by SBM in Fig. 5.6.14(b). The results by SBM and EBM appear to be essentially consistent in that all the saturated units are to liquefy by the RT-motion. This consistency gets better if the stress reduction coefficient in SBM is taken as $r_n = 0.80$ considering the $M = 9.0$ earthquake, while the effect of input motions is inherently included in EBM. The effect of the half-time scale in the RT/2-motion is far more evident in EBM than in SBM. Another qualitative difference between the two methods is that the liquefaction potential is higher in the shallower units than in the deeper units in the uniform sand deposit in EBM, whereas it is vice versa in SBM.

The next soil model addressed here is a filled farmland which liquefied and fluidized during the 2003 Tokachi-oki earthquake ($M_J = 8.0$) in Hokkaido, Japan (Kokusho and Mimori 2015). The site was 230 km far from the epicenter of the off-shore plate-boundary earthquake, and the maximum acceleration recorded nearby was only 0.055 g as indicated in the acceleration time history of the long duration of about a minute in Fig. 5.6.16(a). The soil models with each unit thickness $H = 1$ m were developed consisting of the L1–L7 units where the upward energies were calculated as shown in (b). The SPT $N$-values for individual units were determined from SWS (See Sec. 5.5.1.1) sounding data at eight investigation points as shown in Fig. 5.6.16(c) using an empirical formula developed in Japan (Inada 1960). The thickness of the soft sandy fill was variable (4–7 m) depending on the SWS investigation points and

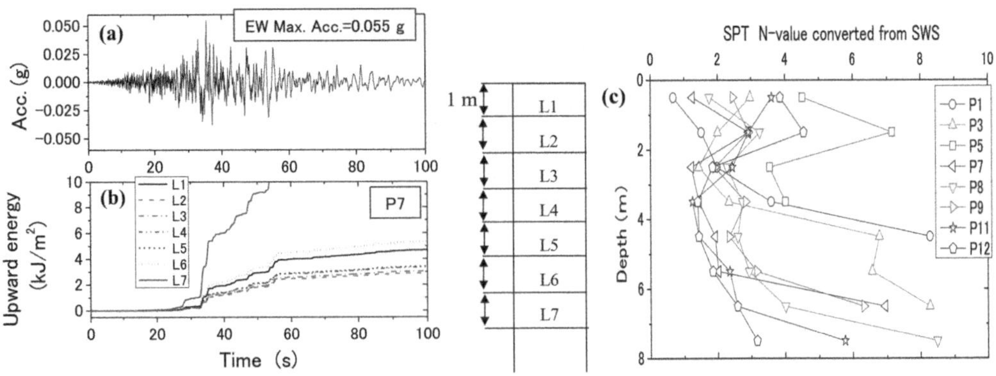


*Figure 5.6.16* Acceleration time history given to sand fill liquefied during a far-field *M* 8.0 earthquake with max. acc. about 0.05 g (a), Associated upward energies (b), and SWS-converted *N*-values versus soil depths at investigation points in liquefied site (c) (Kokusho and Mimori 2015).

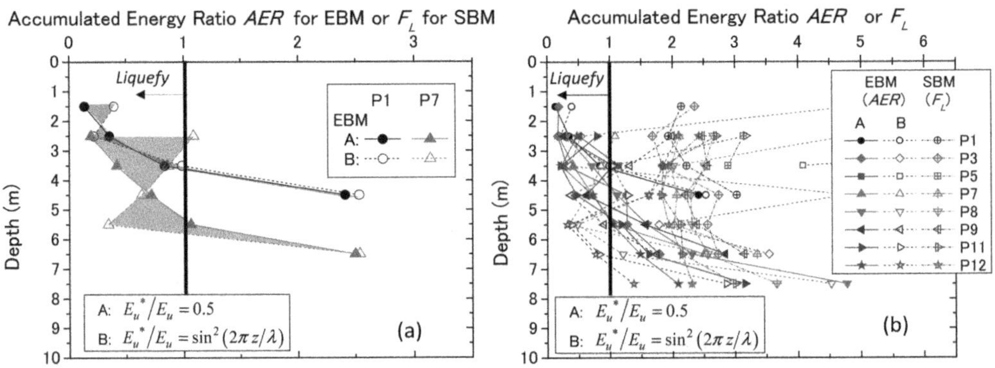


*Figure 5.6.17* *AER*-values by Method-A and B by EBM plotted versus soil depth: (a) At P1 and P7, (b) At 8 investigation points compared with $F_L$-values by SBM (Kokusho 2016).

the water table was 1–2 m below the ground surface (Tsukamoto et al. 2009). The normalized dissipated energy densities $\Delta W/\sigma_c'$ in individual units were evaluated from the $N$-values and average fines content ($F_c = 33\%$) in the same manner as mentioned above using Eq. (5.6.4) and a formula similar to Eq. (5.6.5) considering the fines content in the design code in Japan (JRA 2002). The seismic shear stress $\tau_{max}$ and the upward energy $E_u$ were calculated using the 1D response analysis with the input motion, K-NET Kitami in Fig. 5.6.16(a), given at the surface.

In Fig. 5.6.17(a), liquefaction potentials evaluated by EBM ($AER = \sum_j (W^*H/E_{uf})_j$) for individual depths are plotted with the close symbols connected with solid line at two representative soil investigation points, P1 and P7. Because $AER < 1.0$ is the condition for liquefaction, the shallower portion will liquefy both at P1 and P7 according to this EBM. In obtaining this result, a significant simplification by

Eq. (5.6.7) has been employed in the present EBM as already mentioned so that the upward energy $E_u^*$ to be able to compensate the dissipated energy is assumed constant as $E_u^*/E_u = 1/2$ within the depth of a quarter wave length for a dominant wave period $T$ from the ground surface (named as Method-A here). However, there may be more or less a certain depth-dependency in the energy ratio $E_u^*/E_u$, actually. Hence, a comparative study has been conducted to take into account the depth-dependent variation of $E_u^*$, wherein $E_u^*/E_u = \sin^2(2\pi z/\lambda) = \sin^2(2\pi t/T)$ in Eq. (5.6.6) is used (named as Method-B) in place of Eq. (5.6.7) to consider the extreme depth-dependent effect. Here, $t$ is the travel time of the SH wave from the ground surface to a particular depth $z$ using strain-dependent degraded S-wave velocities in individual layers, $T$ is the dominant period of seismic motion, and otherwise the same EBM procedure is followed here (Kokusho 2016).

In Fig. 5.6.17(a), the $AER$-values obtained by Method-B (open symbols connected with dashed lines) are superposed at two representative points P1 and P7 to compare with those by Method-A. Though the liquefied depths tend to be slightly deeper in Method-B than in A, the difference is not so significant. The liquefaction may probably occur in the shaded area on the diagram in between the two lines of Method-A and B, because they seem to represent the two most extreme cases. In Fig. 5.6.17(b) the same results for all eight points are shown for EBM and SBM. Again, all the points except P5 are evaluated to be liquefiable also in Method-B as in Method-A. In a clear contrast, the SBM-evaluation results superposed on the same diagram indicates no possibility of liquefaction at all because $F_L$-values are far beyond 1.0 despite that the effect of the earthquake magnitude $M$ 8 is taken into account by choosing the stress reduction coefficient $r_n = 0.70$ in accordance with Eq. (5.5.27). In EBM, Method-A may well be recommended as a simplified and practical tool on a safer side in evaluating liquefaction potential in shallow depths where the liquefaction potential can be too low by Method B (Kokusho 2016).

Thus, the EBM can predict liquefaction behavior very simply just by comparing upward wave energy (the energy demand) with the energy capacity correlated with the dissipated energy for liquefaction. It may be able to readily take account of various aspects of input seismic motions (dominant period, duration, number of wave cycles and irregularity) only in terms of energy, and hence can be of a great help to examine the reliability of conventional SBM liquefaction evaluations for a variety of earthquakes motions. It is still necessary, however, to apply this EBM to more case histories to demonstrate its reliability in much more practical conditions.

## 5.7 EFFECT OF INCOMPLETE SATURATION

Liquefaction occurs under ground water table, though soils there may not always be fully saturated particularly in manmade deposits. In many slope failures, soils are suspected to have liquefied during earthquakes even in lowly-saturated conditions. In recent years, innovative liquefaction mitigation measures are being developed, in which the soil saturation is deliberately lowered in situ by injecting air-bubbles or by other measures. Thus, liquefaction behaviors of unsaturated or imperfectly saturated soils are concerned in quite a few engineering problems.

### 5.7.1 Evaluation by laboratory tests

The degree of saturation or simply named as "saturation" is defined as $S_r = V_w/V_v$, where $V_v$ = volume of void and $V_w$ = volume of pore-water. Fig. 5.7.1 exemplifies undrained cyclic triaxial test results on clean sand ($F_c$ = 0%, $D_r \approx 50\%$) under the isotropic confining stress $\sigma'_c = 50$ kPa to see how the saturations $S_r$ = 100, 90, 80% influence the liquefaction behavior (Kochi 2008). Unlike fully saturated specimens, the triaxial test here controls the lateral stress, too, in addition to the axial stress to keep constant the effective mean stress which otherwise varies due to imperfect saturation. Namely, the lateral stress is cyclically regulated as $\Delta\sigma_3 = -\Delta\sigma_1/2$ in conjunction with the cyclic axial stress $\Delta\sigma_1$ so that the mean stress increment $\Delta\sigma_c = (\Delta\sigma_1 + 2\Delta\sigma_3)/3$ is always zero. The time-histories of the axial strain in Fig. 5.7.1 indicate the changes from abrupt increase to gradual increase together with increasing *CSR*-values necessary to liquefy as the saturation $S_r$ decreases. The pore-pressure buildup, 100% for $S_r$ = 100%, tends to be difficult to occur with decreasing $S_r$, indicating that the typical liquefaction failure with zero effective stress shifts to the cyclic softening failure with decreasing shear stiffness.

In Fig. 5.7.2(a), a ratio $CRR/CRR_{Sr=100\%}$ (*CRR* for $S_r$ < 100% divided by *CRR* for $S_r$ = 100%) is plotted versus $S_r$ or *B*-value (the pore-pressure coefficient) based on torsional simple shear tests (Yoshimi et al. 1989) or triaxial tests (Nakazawa et al. 2001) using Toyoura clean sand with $D_r$ = 40–70%. Despite the differences in the test device and the definition of *CRR* for liquefaction onset ($\gamma_{DA}$ = 7.5% for $N_c$ = 15 or $\varepsilon_{DA}$ = 5% for $N_c$ = 20, respectively), the two test results coincide fairly well. In terms of the *B*-value, *CRR* is influenced only marginally if it changes from 1.0 to 0.8, indicating that the *B*-value does not have to be more than 0.95 as normally specified in

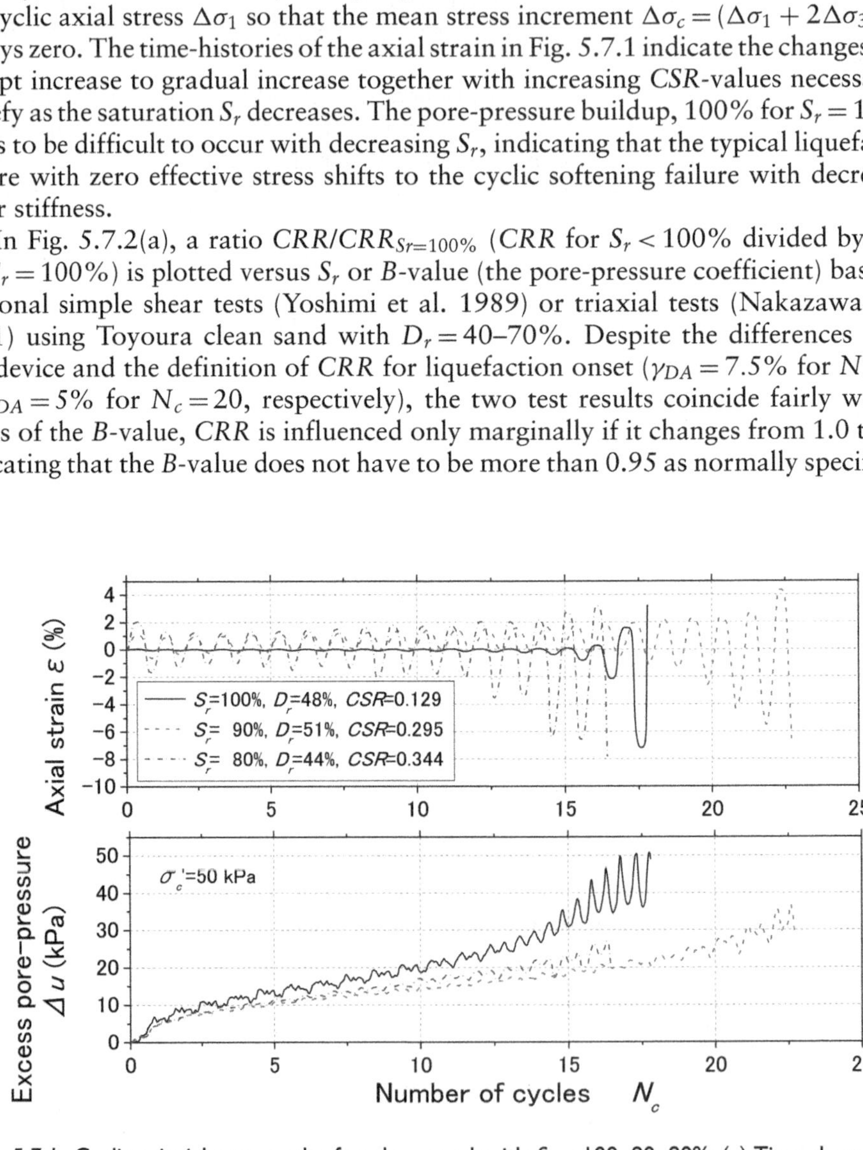


*Figure 5.7.1* Cyclic triaxial test results for clean sand with $S_r$ = 100, 90, 80%: (a) Time-depend axial strain, (b) Time-depend pore-pressure buildup (Kochi 2008).

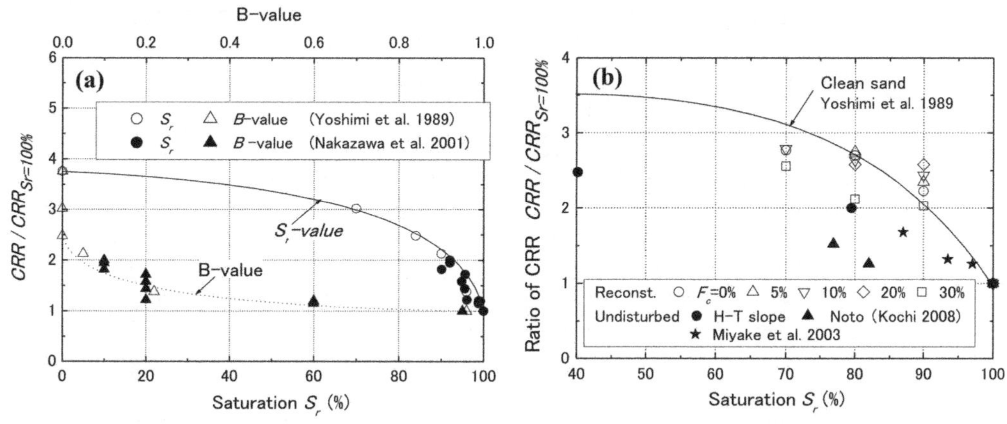


*Figure 5.7.2* Ratio of *CRR* due to saturation plotted versus $S_r$ or *B*-value by laboratory tests: (a) Reconstituted clean sands, (b) Reconstituted clean sands with non-plastic fines plus intact soils from in situ.

undrained cyclic liquefaction tests. On the other hand, the decrease in $S_r$ from 100% to 90% almost doubles the *CRR*-value, though the increasing rate of *CRR* for further decrease in $S_r$ tends to be smaller.

In Fig. 5.7.2(b), triaxial test results on reconstituted river sands mixed with non-plastic fines of $F_c = 0$–30% are plotted with open symbols (Kochi 2008, Hara et al. 2009). The fines-containing sand shows $S_r$-dependent change of *CRR* similar to the clean sands, though the increasing rate becomes smaller as $S_r$ gets smaller. In the same figure, the results on intact soils recovered from in situ are also plotted with close symbols. Those intact soils include samples from H-T slope ($F_c \approx 5 \sim 25\%$, $I_p = 0$, $C_u \approx 2 \sim 10$) failed during the 2004 Niigataken-Chuetsu earthquake, Noto road embankment ($F_c \approx 30$–40%, $I_p \approx 20$, $C_u \approx 40$) failed during the 2006 Noto-earthquake (Kochi 2008) in addition to intact samples by other researchers (Miyake et al. 2003). Those intact soils show lower increasing rates of *CRR* than the clean sands with decreasing $S_r$, presumably because they are affected by considerable amounts of fines or soil fabric different from reconstituted specimens.

## 5.7.2 Theoretical background

In order to deal with liquefaction in unsaturated soils, the effective confining stress $\sigma'_c$ is defined from the total stress $\sigma_c$ considering the effects of pore-air pressure $u_a$ and pore-water pressure $u_w$ as:

$$\sigma'_c = (\sigma_c - u_a) + \chi(u_a - u_w) = (\sigma_c - u_a) + \chi \cdot s \tag{5.7.1}$$

where $s = u_a - u_w$ is the suction ($u_a \geq u_w$) and $\chi$ is a material parameter for the contribution of the suction to the effective stress (Bishop and Blight 1963).

Fig. 5.7.3 exemplifies time histories of the pore-air and pore-water pressure together with those of the deviatoric stress and axial strain obtained in a undrained

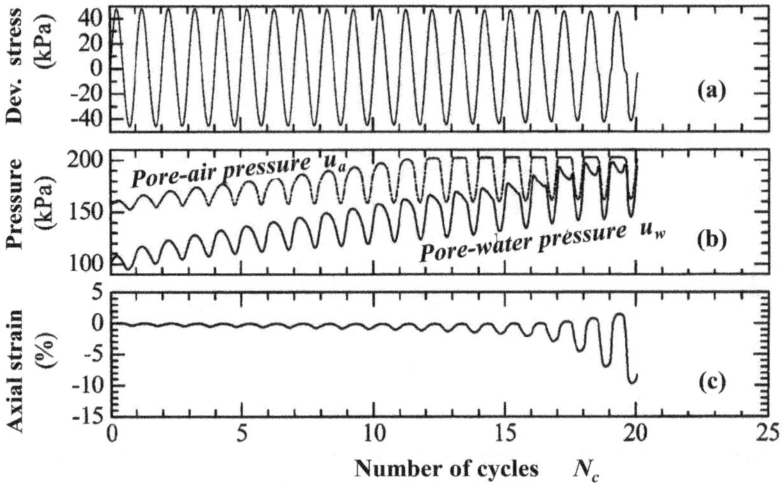


*Figure 5.7.3* Time histories of pore-air and pore-water pressures $u_a$, $u_w$, deviatoric stress and axial strain in undrained cyclic triaxial test on non-plastic silt (Okamura and Noguchi 2009).

cyclic triaxial test on non-plastic silt (Okamura and Noguchi 2009). The specimen is isotropically consolidated by $\sigma'_c = 196$ kPa with air and water-pressures, $u_{a0} = 160$ kPa and $u_{w0} = 98$ kPa respectively, inducing the suction $s = u_{a0} - u_{w0} = 62$ kPa in the initial condition. As cyclic loading goes on, both $u_a$ and $u_w$ build up so that the suction $s = u_a - u_w$ is kept constant as 62 kPa, raising the air-pressure $u_a$ up to 100% at $N_c = 12$. At this moment, the liquefaction-induced rapid strain increase does not occur because the suction and associated effective stress formulated in Eq. (5.7.1) is still sustained by the $u_w$-value. After that, $u_w$ further goes on increasing up to 100% so that $s = u_a - u_w = 0$ at $N_c = 18$, when liquefaction occurs together with complete loss of effective stress and sudden strain increase.

In unsaturated soils, the compressibility of pore-air has a significant effect on the pressure buildup. The air-compressibility largely depends on backpressure (hydrostatic pressure) in the ground, hence the liquefiability of unsaturated soils is governed by the backpressure. Because the pore-air volume in unit volume of soil with void ratio $e$ and saturation $S_r$ is $(1 - S_r)e/(1 + e)$, the volumetric strain $\varepsilon_v$ due to the change in the backpressure from $p_0$ to $\Delta p$ is formulated, according to the Boyle's law (Okamura and Soga 2006) as:

$$\varepsilon_v = \frac{\Delta p}{p_0 + \Delta p}(1 - S_r)\frac{e}{1+e} \tag{5.7.2}$$

Here, $p_0$ = initial back pressure (absolute pressure from vacuum) and $\Delta p$ = pore-pressure buildup in the process of liquefaction. If the maximum pressure buildup due to liquefaction $\Delta p = \sigma'_c$ is substituted in Eq. (5.7.2), then

$$\varepsilon_v^* = \frac{\sigma'_c}{p_0 + \sigma'_c}(1 - S_r)\frac{e}{1+e} \tag{5.7.3}$$

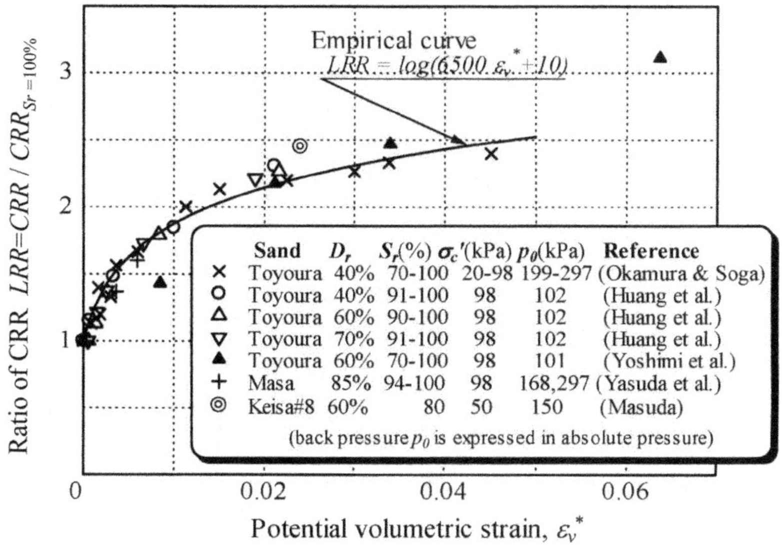


*Figure 5.7.4* Ratios of *CRR* between unsaturated and saturated (*LRR*) versus the volumetric strain $\varepsilon_v^*$ for various soils under different backpressures (Okamura and Noguchi 2009).

The value $\varepsilon_v^*$ in the equation corresponds to the highest potential volumetric strain for 100% pressure buildup. In Fig. 5.7.4, the ratios of *CRR* between unsaturated and saturated soils (denoted here as *LRR*) are plotted versus the volumetric strain $\varepsilon_v^*$ based on a number of triaxial liquefaction tests on clean sands carried out under different back pressures. Despite some data scatters, the plots can be approximated by a unique curve and the increasing rate in the *CRR* is expressed as a simple function of strain $\varepsilon_v^*$ (Okamura and Noguchi 2009) as:

$$LRR = \log(6500\varepsilon_v^* + 10) \tag{5.7.4}$$

Namely, the highest potential volumetric strain $\varepsilon_v^*$ during liquefaction considering the absolute initial backpressure serves as a key parameter for evaluating the liquefaction resistance. According to Eq. (5.7.3), the strain $\varepsilon_v^*$ in soil deposits having the same $S_r$ and $e$ is proportional to $1/(p_0/\sigma'_c + 1)$, indicating that $\varepsilon_v^*$ becomes larger with increasing $\sigma'_c$. In the light of Fig. 5.7.4, the larger $\varepsilon_v^*$ results in larger increase in *CRR* with decreasing saturation by the same amount.

Because $p_0$ defined by the absolute pressure is 98 kPa (atmospheric pressure) at the smallest, $\varepsilon_v^*$ stays small for a small confining stress $\sigma'_c$, resulting in a marginal *CRR* increment for an imperfectly saturated soil at a shallow depth. The extreme case of this is a 1 g scaled model shake table test for liquefaction where the effective stress is very low and a low saturated model is not a big issue to simulate a fully saturated prototype (Okamura and Soga 2006). The ratio of *CRR* for a soil of $S_r = 90\%$ with respect to $S_r = 100\%$ is calculated using Eqs. (5.7.3) and (5.7.4), and depicted along the depth in

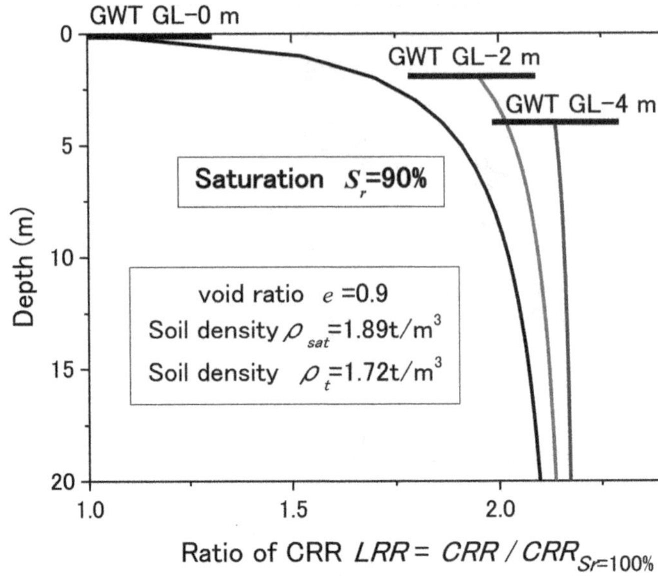


*Figure 5.7.5* Depth-dependent change in *CRR*-ratio of unsaturated sands of $S_r = 90\%$ calculated for 3 different ground water tables.

Fig. 5.7.5. If the ground water table (GWT) is at the surface, *CRR* is almost the same immediately below the water table despite the saturation of $S_r = 90\%$. However, as the water table is lowered to GL-2 m or deeper, the effect of the poor saturation becomes dominant to make the *CRR* almost doubled below the water table. This indicates that lowering the saturation may be promising as one of the countermeasures to mitigate liquefaction of soils except immediately at the ground surface.

The data points in Fig. 5.7.4 are all from clean sands with small matric suction. With increasing $F_c$ and suction, higher pore-pressure has to be built up until the suction and the effective stress become zero. This seems to indicate that the increasing rate of *CRR* for unsaturated soils tends to be larger in sands with fines than in clean sands (Okamura and Soga 2009). On the other hand, Unno et al. (2008) after conducting a series of triaxial tests on Toyoura clean sand found that the amount of volume change required to reach the zero effective stress state depends on the volume compressibility of soil skeleton, the saturation and the initial confining pressure. According to this finding, the increase of fines may increase the soil skeleton compressibility and cause lower increasing rate of *CRR* of unsaturated soils even for the same degree of saturation, as observed in test results in Fig. 5.7.2(b).

### 5.7.3 Effect on B-value and P-wave velocity

Soils are sometimes dealt with as porous elastic materials with fluid using the poroelasticity theory by Biot (1956). According to the theory, the equilibrium equations in three-dimensional axes $x$, $y$, $z$ using the summation convention are expressed

(Zienkiewicz et al. 1982) as:

$$\sigma_{ij,j} + \rho g_i = \rho \ddot{u}_i + \rho_f \ddot{w}_i \tag{5.7.5}$$

$$-p_{,i} + \rho_f g_i = \rho_f g k^{-1} \dot{w}_i + \rho_f \ddot{u}_i + n^{-1} \rho_f \ddot{w}_i \tag{5.7.6}$$

Here, $\sigma_{ij}$ = total stress (tension positive), $p$ = pore-pressure (compression negative), $g_i$ = gravity acceleration, $g$ = $i$-direction component of $g_i$, $u_i$ = solid phase displacement, $w_i$ = fluid phase displacement, $\rho$ = total density, $\rho_f$ = fluid density, $k$ = permeability coefficient, $n$ = porosity. Eq. (5.7.5) corresponds to the equilibrium of the solid phase considering the inertial force of the fluid in the second term on the right-hand side. Eq. (5.7.6) is the equilibrium of the pore fluid considering the viscous resisting force by the Darcy flow in the first term on the right. On the other hand, the equation of continuity can be written using $K_f$ = bulk modulus of the fluid (water + air) as:

$$\dot{\varepsilon}_{ii} + \dot{w}_{i,i} = -\frac{\dot{p}n}{K_f} \tag{5.7.7}$$

If the bulk modulus of solid soil particles is $K_s$, then the next equation can be derived using the volumetric strain $\varepsilon_v = \varepsilon_{ii}$ and the effective mean stress $\sigma'_m = \sigma'_{ii}/3$.

$$\dot{w}_{i,i} = -\frac{\partial \varepsilon_v}{\partial t} - \frac{(1-n)}{K_s}\frac{\partial p}{\partial t} + \frac{1}{K_s}\frac{\partial \sigma'_m}{\partial t} - \frac{n}{K_f}\frac{\partial p}{\partial t} \tag{5.7.8}$$

For the undrained condition $\dot{w}_{i,i} = 0$, and the incremental equation may be expressed as:

$$-\Delta\varepsilon_v - \frac{(1-n)}{K_s}\Delta p + \frac{1}{K_s}\Delta\sigma'_m - \frac{n}{K_f}\Delta p = 0 \tag{5.7.9}$$

Using the bulk modulus of the soil skeleton $K$ and substituting $\Delta\sigma'_m = K\Delta\varepsilon_v$ into the above equation, then:

$$\frac{\Delta p}{\Delta\varepsilon_v} = -\left(1 - \frac{K}{K_s}\right) \Big/ \left(\frac{1-n}{K_s} + \frac{n}{K_f}\right) \tag{5.7.10}$$

Furthermore, by substituting $\Delta\sigma'_m = \Delta\sigma_m + \Delta p$ and $\Delta\varepsilon_v = \Delta\sigma'_m/K$ into Eq. (5.7.9), the next equation for the pore-pressure coefficient $B$-value is obtained.

$$B \equiv -\frac{\Delta p}{\Delta\sigma_m} = \left(\frac{1}{K} - \frac{1}{K_s}\right) \Big/ \left(\frac{1}{K} - \frac{n}{K_s} + \frac{n}{K_f}\right) \tag{5.7.11}$$

If air-bubbles are mixed in the pore fluid, it tends to reduce the fluid bulk modulus $K_f$ drastically. In response to a change in pore pressure $\Delta p$, the volume change of the fluid is $nV_0\Delta p/K_f$ where $V_0$ = total soil volume. If the degree of saturation $S_r$ is assumed to be constant for a small $\Delta p$, the fluid volume change may be obtained by adding the

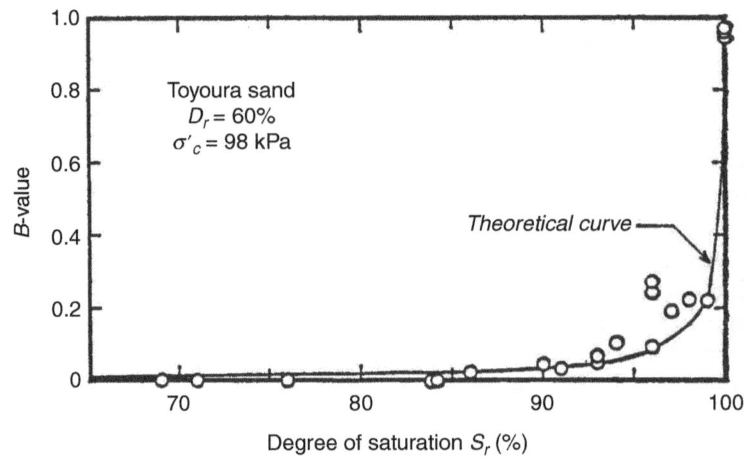


*Figure 5.7.6* Plots of B-value versus saturation $S_r$ by torsional simple shear tests on clean sand compared with theoretical curve (Yoshimi et al. 1989).

volume change of water $\Delta V_w$ and that of air $\Delta V_a$. The former is written as $\Delta V_w = nV_0S_r\Delta p/K_w$ where $K_w$ = bulk modulus of water, and the latter can be expressed as $\Delta V_a = (\Delta p/p)V_a = nV_0(1-S_r)(\Delta p/p)$ by using the Boyle's law where $V_a$ = air volume and $p$= absolute pore pressure from vacuum in positive hereafter (Lade and Hernandez 1977). Thus, the next equations are obtained.

$$nV_0\frac{\Delta p}{K_f} = \Delta V_w + \Delta V_a = nV_0S_r\frac{\Delta p}{K_w} + nV_0(1-S_r)\frac{\Delta p}{p} \tag{5.7.12}$$

Hence, $K_f$ can be written as:

$$K_f = \frac{K_w}{S_r + (1-S_r)(K_w/p)} \tag{5.7.13}$$

Substituting this into Eq. (5.7.11) and considering $K_s \gg K$, then the $B$-value is expressed (Lade and Hernandez 1977) as;

$$B \equiv \frac{\Delta p}{\Delta\sigma_m} = \frac{1}{1 + nK[S_r/K_w + (1-S_r)/p]} \tag{5.7.14}$$

In Fig. 5.7.6, $B$-values measured in specimens in torsional shear tests of the Toyoura clean sand $D_r = 60\%$ and isotropically consolidated with $\sigma'_c = 98$ kPa are plotted against the degrees of saturation $S_r$ (Yoshimi et al. 1989). The $S_r$-values were calculated from water contents and dry densities of individual specimens. The solid curve in the diagram represents the theoretical relationship in Eq. (5.7.14) where $n = 0.43$, $K = 6.7 \times 10^4$ kPa, $K_w = 2.23 \times 10^6$ kPa, and $p$ = absolute pressure in pore-fluid ($p = 98$ kPa and 343 kPa in unsaturated and saturated specimen, respectively).

Thus the theoretical formula can explain the test results fairly well. The curve shows that the $B$-value is a very sensitive index reflecting a marginal drop of saturation.

In order to know how much in situ soils are saturated, it is convenient to use the P-wave velocity $V_p$. The relationship between $V_p$ and the $B$-value can be derived theoretically as follows (Ishihara 1971). When soil is in vibration in the undrained condition, Eq. (5.7.5) reduces to the next equation because $\ddot{w}_i = 0$ and $\rho g_i = 0$, neglecting the effect of gravity.

$$\sigma_{ij,j} = \rho \ddot{u}_i \tag{5.7.15}$$

Substituting $\sigma_{ij} = \sigma'_{ij} - p\delta_{ij}$ into Eq. (5.7.15) yields

$$\sigma'_{ij,j} - p_{,i} = \rho \ddot{u}_i \tag{5.7.16}$$

Using Eq. (5.7.10), $p_{,i}$ can be written as:

$$p_{,i} = \frac{\partial p}{\partial \varepsilon_v} \varepsilon_{v,i} = -\left(1 - \frac{K}{K_s}\right) \varepsilon_{v,i} \Big/ \left(\frac{1-n}{K_s} + \frac{n}{K_f}\right) \tag{5.7.17}$$

and using the Hooke's law in terms of the effective stresses,

$$\sigma'_{ij} = K\varepsilon_v + 2G(u_{i,j} - \varepsilon_v/3) \ : \ i = j, \quad \sigma'_{ij} = G(u_{i,j} + u_{j,i}) \ : \ i \neq j \tag{5.7.18}$$

By Substituting Eqs. (5.7.17) and (5.7.18) into Eq. (5.7.16), the next is obtained.

$$\rho \ddot{u}_i = \left[K + \frac{1 - K/K_s}{(1-n)/K_s + n/K_f} + \frac{G}{3}\right] \varepsilon_{v,i} + G\nabla^2 u_i \tag{5.7.19}$$

If this equation is compared with the normal three-dimensional wave equation in Sec. 1.3,

$$\rho \ddot{u}_i = (\lambda + G)\varepsilon_{v,i} + G\nabla^2 u_i \tag{5.7.20}$$

then, the Lame's constant $\lambda$ is determined as:

$$\lambda = K + \frac{1 - K/K_s}{(1-n)/K_s + n/K_f} - \frac{2G}{3} \tag{5.7.21}$$

On the other hand, the P-wave equation in Sec. 1.3 is written as:

$$\rho \ddot{u}_i = (\lambda + 2G)\nabla^2 u_i \tag{5.7.22}$$

indicating that

$$V_p = \sqrt{(\lambda + 2G)/\rho} = \sqrt{\frac{1}{\rho}\left[K + \frac{1 - K/K_s}{(1-n)/K_s + n/K_f} + \frac{4G}{3}\right]} \tag{5.7.23}$$

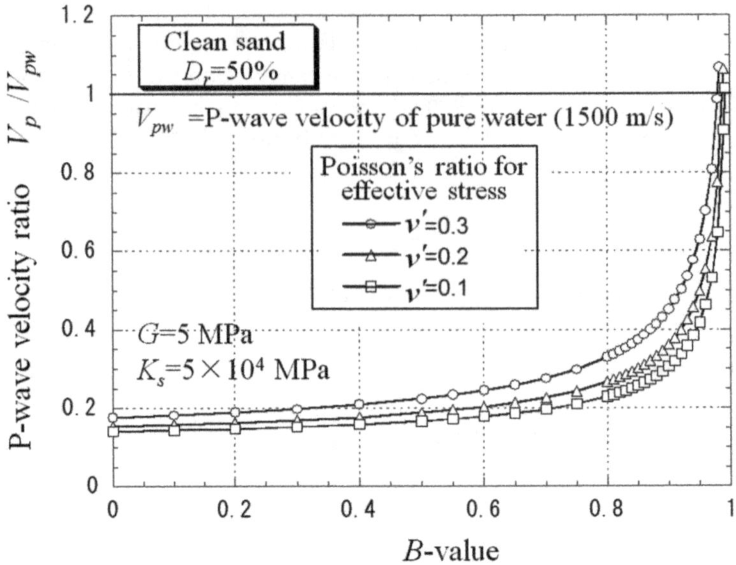


*Figure 5.7.7* P-wave velocity ratio $V_p/V_{pw}$ versus *B*-value for imperfectly saturated sand calculated for Poisson's ratio $\nu' = 0.1–0.3$ (Kokusho 2000a).

$V_p$ in Eq. (5.7.23) is normalized by the P-wave velocity of pure water $V_{pw}$ as:

$$\frac{V_p}{V_{pw}} = \sqrt{\frac{\rho_f}{\rho}\left[\frac{K}{K_w} + \frac{1 - K/K_s}{(1-n)K_w/K_s + nK_w/K_f} + \frac{4G}{3K_w}\right]} \tag{5.7.24}$$

The bulk modulus of soil skeleton is written using Poisson's ratio $\nu'$ in terms of the effective stress and $G = \rho V_s^2$ as:

$$K = \frac{2(1+\nu')}{3(1-2\nu')}G = \frac{2(1+\nu')}{3(1-2\nu')}\rho V_s^2 \tag{5.7.25}$$

Combining Eqs. (5.7.11), (5.7.23) and (5.7.25), and also using $K_s \gg K$, $K_f$, the next equation correlating the *B*-value with the velocity ratio $V_p/V_s$ can be obtained.

$$B = 1 - \frac{2(1+\nu')}{3(1-2\nu')}\frac{1}{(V_p/V_s)^2 - 4/3} \tag{5.7.26}$$

Fig. 5.7.7 illustrates a relationship between the *B*-value and $V_p/V_{pw}$ (Kokusho 2000a) calculated from Eqs. (5.7.24) and (5.7.26), respectively, for clean sand of $D_r = 50\%$ and $\nu' = 0.1{\sim}0.3$ assuming $K_w = 2.22 \times 10^3$ MPa (corresponding to P-wave velocity in water $V_{pw} = 1500$ m/s), $G = 5$ MPa corresponding to the similar research by Ishihara (1971). It shows a drastic drop of $V_p/V_{pw}$ from 1.0 to 0.20~0.33 (from $V_p = 1500$ m/s to 300~500 m/s) with decreasing *B*-value from 1.0 to 0.8. According

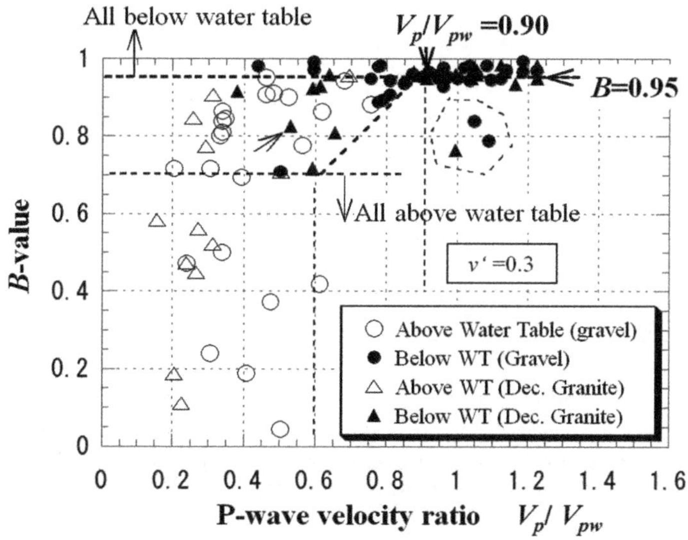


*Figure 5.7.8* In situ *B*-value evaluated from measured velocity ratio $V_p/V_s$ and P-wave velocity ratio $V_p/V_{pw}$ based on field wave-logging data in different sites (Kokusho 2000a).

to Fig. 5.7.6, the corresponding decrease in the degree of saturation $S_r$ is around 1% or less, indicating that the *B*-value is much more sensitive than $S_r$. Thus the P-wave velocity can conveniently be used to detect in situ *B*-values. The P-wave velocity is actually used in scaled model tests of saturated sand layers to quantify the degree of saturation. In these occasions, care is needed to consider a possibility that the measured $V_p$ may be influenced by a spatial variability of air-bubbles relative to the travelling P-wave length and can be higher than the real value (e.g. Naesgaard et al. 2007).

In Fig. 5.7.8, *B*-values calculated from Eq. (5.7.26), using the wave velocity ratio $V_p/V_s$ measured in situ in different sites and assuming $\nu' = 0.3$, are plotted versus the normalized P-wave velocity, $V_p/V_{pw}$ (Kokusho 2000a). It shows that except a few abnormal plots, the *B*-values thus obtained are greater than 0.90~0.95 if $V_p/V_{pw} = 0.8$–$0.9$ or larger. The close symbols all located under water tables satisfy $B > 0.7$, indicating that the increase in *CRR* due to the imperfect saturation is only a few percent according to Fig. 5.7.2(a). Thus, it seems to be difficult that imperfect saturation conditions under water tables in natural deposits can increase liquefaction resistance to a measurable degree.

### 5.7.4 Effect on residual strength

A simple evaluation of the seismic stability of unsaturated slopes may be possible by comparing post-liquefaction residual shear strengths of unsaturated soils with working stresses. Because slopes are not fully saturated in many cases, it is significant to know how the imperfect saturation affects the stress-strain behavior including the residual shear strength of unsaturated slopes after cyclic loading. For this goal, it

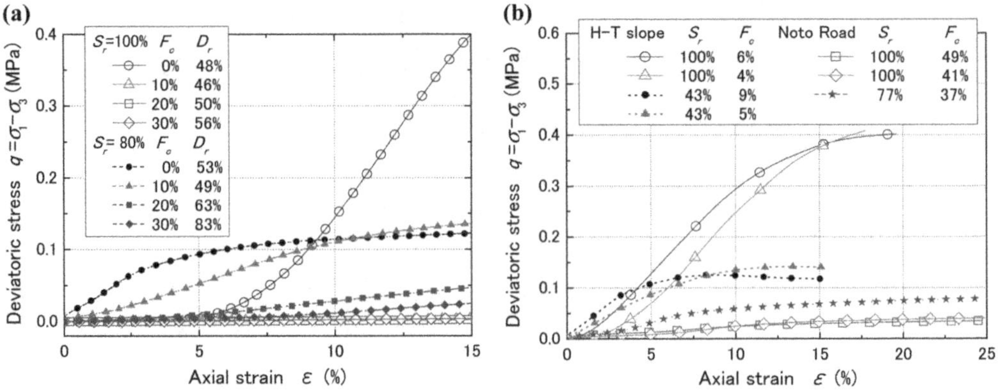


*Figure 5.7.9* Deviatoric stress versus axial strain in post-liquefaction undrained monotonic loading triaxial tests of saturated and unsaturated soils: (a) Reconstituted soils with varying $F_c$, (b) Intact soils from failed slopes with varying $F_c$ (Kochi 2008).

may be possible to conduct undrained strain-controlled monotonic loading tests on unsaturated specimens after undrained cyclic loading tests. Fig. 5.7.9(a) shows such triaxial test results on reconstituted isotropically consolidated specimens after cyclically loaded up to double amplitude axial strain of $\varepsilon_{DA}=10\%$. It depicts stress-strain curves up to the axial strain $\varepsilon=15\%$ obtained in a series of tests on sand specimens of $S_r=100\%$ or 80% without and with non-plastic fines. For the clean sand $F_c=0$, the post-liquefaction shear resistance picks up considerably from a post-liquefaction near-zero initial value with increasing strain for the medium density of $D_r\approx 50\%$ in the fully saturated condition $S_r=100\%$. In a good contrast, in the unsaturated clean sand $S_r=80\%$, the shear resistance, larger than the saturated sand in the post-liquefaction initial stage, will not grow largely but converge to a depressed value. If fines are contained as $F_c\geq 10\%$, the shear resistance develops very little in the fully saturated specimens of $S_r=100\%$, indicating a significant effect of fines on the undrained shear on granular soils, while the impact of fines are less significant for $S_r=80\%$. This is because the increase of fines changes the soil to be contractive, and the decrease of saturation causes smaller development of positive pore-pressure for $S_r=80\%$ than for $S_r=100\%$. Thus the above test results may indicate that the post-cyclic residual strength is larger in fully saturated clean sands of $S_r=100\%$ than in partially saturated ones, while in less dilative sands containing fines, the residual strength tends to be larger for partially saturated soils.

In Fig. 5.7.9(b), similar test results obtained for intact soils sampled in block from in situ slopes actually failed during previous earthquakes in Japan (already addressed in Fig. 5.7.2(b)) are shown for the axial strain up to $\varepsilon=15\sim25\%$ (Kochi 2008). The triaxial tests were conducted on specimens, either unsaturated in natural water contents or artificially fully-saturated, sampled from the H-T slope ($S_r=43\sim44\%$, $F_c=4\sim9\%$, $C_u\approx 2$, $I_p=\mathrm{NP}$) and the Noto slope ($S_r=77\%$, $F_c=37$–$49\%$, $C_u\approx 40$, $I_p\approx 20$). For the former soil with particle gradations similar to clean sands with low $F_c$, the post-liquefaction residual shear resistance becomes evidently larger in fully saturated specimens than in the specimens of natural water contents. For the latter soil with high $F_c$, the residual strength tends to be larger in the unsaturated specimens than in the fully saturated ones, though the absolute value is obviously lower than the former soils from the H-T slope with low $F_c$-values.

Thus, a very clear trend can be recognized in both reconstituted and intact soils that the post-liquefaction residual shear resistance is higher in the fully saturated condition than in the unsaturated condition in clean sands with small $F_c$, whereas it is vice versa for soils with high $F_c$ because the soil tends to be less dilative with increasing fines content.

## 5.8 EFFECT OF INITIAL SHEAR STRESS

As already mentioned, the liquefaction triggering mechanism in engineering practice is considered in a free field $K_0$-consolidated level ground without nearby structures. However, most liquefaction-induced damage actually occurs near structural foundations, embankments or sloping grounds where initial shear stresses are working and the $K_0$-consolidated condition with lateral constraint does not hold. Unlike normal structural damage caused by the seismic inertial force, the typical liquefaction damage is characterized by uneven settlement, residual displacement, flow and sliding failure caused by sustained initial shear stresses coming from structural loads due to the reductions of shear strength and stiffness, which may occur even after earthquake shaking. Thus the mechanisms of liquefaction triggering and post-liquefaction deformation under the working initial stresses should correctly be understood in the liquefaction evaluation in comparison with the liquefaction in the free-field level ground.

In order to deal with the liquefaction behavior considering the initial shear stress, the simplified two-step lab test procedure has sometimes been employed as already mentioned. First, the normal liquefaction test on isotropically consolidated specimen is carried out with no initial stress, then followed by the undrained monotonically loading strain-controlled test simulating the effect of the initial shear stress. This procedure is simple and convenient, though both effects of seismic and initial stresses should have been taken into account at the same time throughout the process to be precise. In the following, the exact liquefaction mechanisms during and post cyclic loading under the influence of sustained initial stresses are considered.

### 5.8.1 Laboratory tests considering initial shear stress

A soil element in the level ground is in the stress condition shown in Fig. 5.8.1(a), where shear stress is working there initially due to the anisotropic $K_0$-consolidation though it disappears, changing to the isotropic stress state with 100% pore-pressure buildup. In contrast, the elements near slopes and shallow foundations shown in (b) and (c) are under the influence of sustained initial shear stresses, before, during and after seismic shaking. This indicates that the soil may behave quite differently in liquefaction-triggering and post-liquefaction behavior from the $K_0$-consolidation. It is almost impossible to reproduce two or three-dimensional in situ initial stress conditions exactly in laboratory tests, and some simplifications portrayed in Figs. 5.8.2(a)~(c) have to be introduced according to the types of laboratory tests.

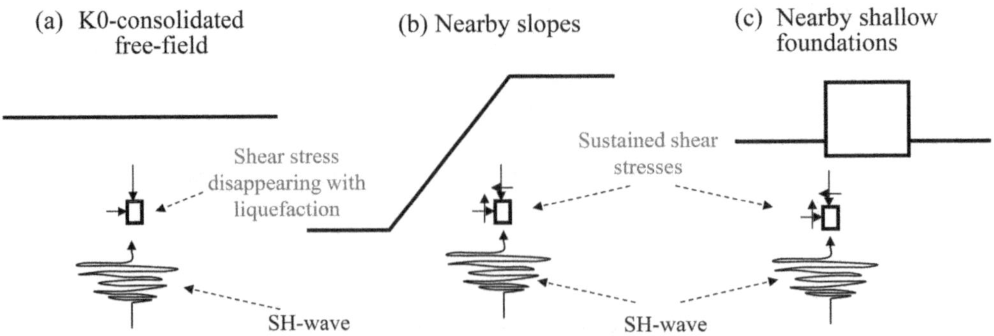


*Figure 5.8.1* Stress-conditions in soil elements: (a) In free-field, (b) Nearby slopes, (c) Nearby shallow foundations.

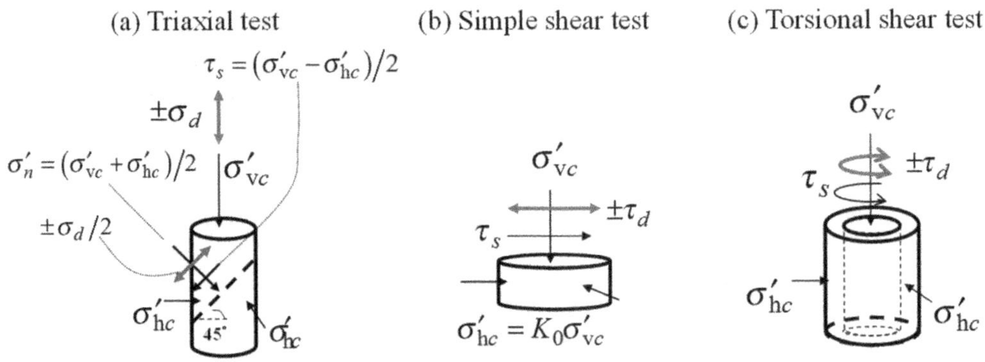


*Figure 5.8.2* Stress-states with initial shear stresses in soil elements for cyclic loading tests: (a) Triaxial test, (b) Simple shear test, and (c) Torsional shear test.

(a) In the cyclic triaxial test, the specimen is anisotropically consolidated with the vertical, horizontal, and mean stresses, $\sigma'_{vc}$, $\sigma'_{hc}$ and $\sigma'_c = (\sigma'_{vc} + 2\sigma'_{hc})/3$, respectively, applying the initial shear stress $\tau_s = (\sigma'_{vc} - \sigma'_{hc})/2$ on the 45° plane from the horizontal plane. Then, the cyclic axial stress $\pm\sigma_d$ is loaded in the undrained condition to give a cyclic shear stress $\tau_d = \sigma_d/2$ on the 45° plane.

(b) In the simple shear test, the specimen is initially consolidated in the $K_0$-condition with the lateral constraint; $\sigma'_{vc}$ vertically, $\sigma'_{hc} = K_0\sigma'_{vc}$ horizontally with the mean stress $\sigma'_c = (\sigma'_{vc} + 2\sigma'_{hc})/3$. Next, the initial shear stress $\tau_s$ is imposed on the horizontal plane in the drained condition, though it does not seem logical to consider $\tau_s$ in the $K_0$-stress condition. Then, the cyclic shear stress $\pm\tau_d$ is applied on the horizontal plane in the undrained condition.

(c) In the torsional simple shear test, the specimen is initially consolidated either isotropically with $\sigma'_c$ or anisotropically with $\sigma'_{vc}$ and $\sigma'_{hc}$ to impose the initial shear stress $\tau_s = (\sigma'_{vc} - \sigma'_{hc})/2$ on the 45° plane. Next, the initial shear stress $\tau_s$

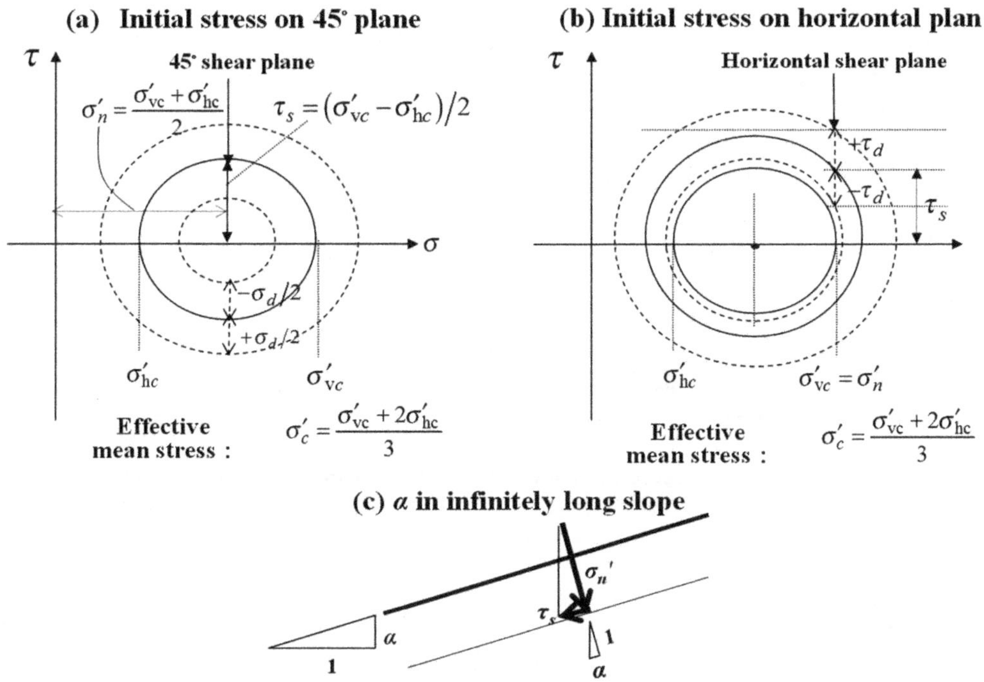


*Figure 5.8.3* Mohr's circle diagram: (a) Initial stress on 45° plane, (b) Initial stress on horizontal plane, (c) Initial shear stress ratio $\alpha$ compared with infinitely long slope.

is imposed on the horizontal plane in the drained condition. Then, the cyclic torsional shear stress $\pm\tau_d$ is superposed on the horizontal plane in the undrained condition.

In (a) and (b), the initial stress $\tau_s$ and cyclic stress $\pm\tau_d$ are working on the same plane, whereas in (c) $\tau_s$ can be imposed either on the 45° or horizontal planes, or the two planes at the same time if necessary, to have better reproducibility of in situ stress states.

The stress-states in the above soil element tests are illustrated in the two-dimensional Mohr's circle diagram in Fig. 5.8.3. In those elements where the initial shear stress $\tau_s = (\sigma'_{vc} - \sigma'_{hc})/2$ works on the 45° plane, the effective normal stress on the same plane is $\sigma'_n = (\sigma'_{vc} + \sigma'_{hc})/2$ as shown in (a), and the initial shear stress ratio $\alpha$ is defined as the ratio of the two stresses as $\alpha = \tau_s/\sigma'_n$. For those where the shear stress $\tau_s$ is imposed on the horizontal plane together with the normal stress $\sigma'_{vc}$ (if $\sigma'_{vc} = \sigma'_{hc}$ then $\sigma'_{vc} = \sigma'_c$) as shown in (b), $\alpha$ is defined $\alpha = \tau_s/\sigma'_{vc}$. It is easy to understand in (c) that the value of $\alpha$ thus defined corresponds to the ratio of shear stress to normal stress working on a plane parallel with the slope gradient $\alpha$.

In those tests where the cyclic shear stress $\tau_d$ is working on the same plane as the initial shear stress $\tau_s$, there can be two cases; "stress reversal" $\tau_d > \tau_s$ and "stress non-reversal" $\tau_d \leq \tau_s$. Fig. 5.8.4 exemplifies undrained triaxial test results on Futtsu sand,

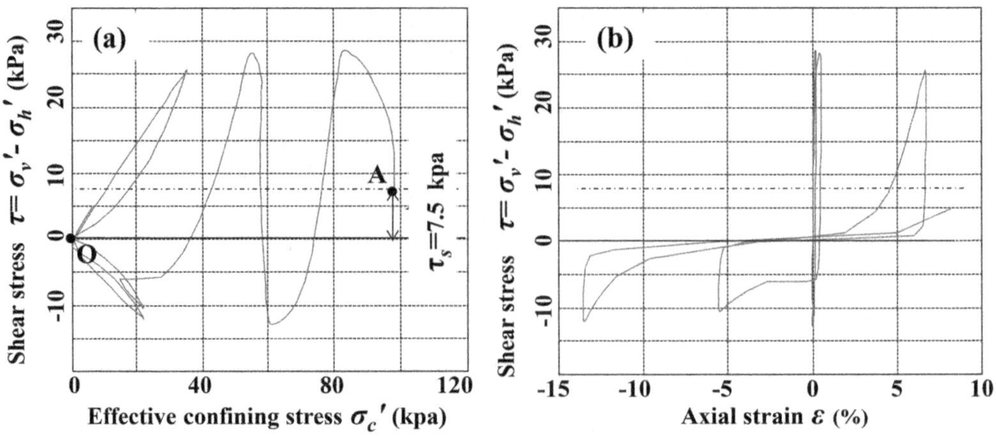


*Figure 5.8.4* Undrained triaxial test results anisotropically consolidated and cyclically loaded in stress reversal condition: (a) Effective stress path, (b) Stress-strain curve (Kato 2011).

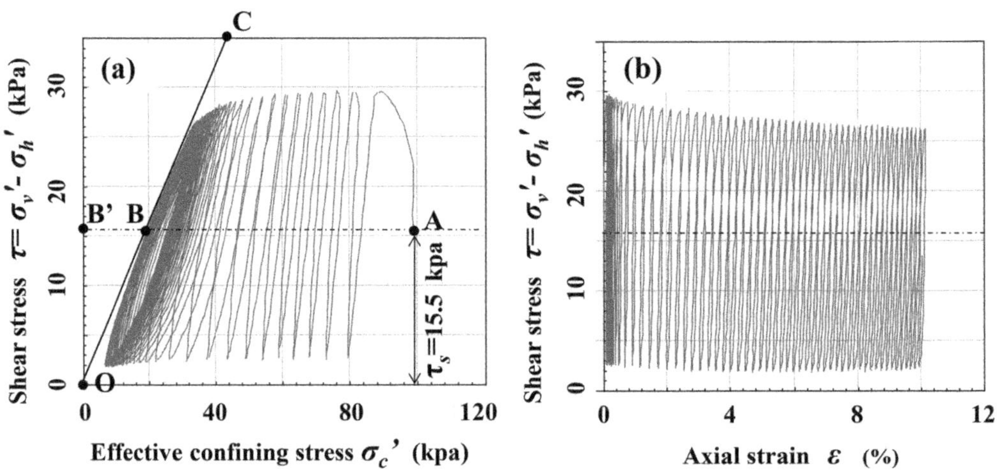


*Figure 5.8.5* Undrained triaxial test results, anisotropically consolidated and cyclically loaded in stress non-reversal condition: (a) Effective stress path, (b) Stress-strain curve (Kato 2011).

$D_r \approx 30\%$, $F_c = 0\%$, anisotropically consolidated with $\sigma'_{vc} = 108$ kPa, $\sigma'_{hc} = 93$ kPa, $\sigma'_c = (\sigma'_{vc} + 2\sigma'_{hc})/3 = 98$ kPa, $\tau_s = (\sigma'_{vc} - \sigma'_{hc})/2 = 7.5$ kPa, and loaded by the cyclic shear stress $\tau_d = \sigma_d/2 = 20.8$ kPa ($CSR = 0.212$). In the effective stress diagram (a), wherein $\tau = (\sigma'_v - \sigma'_h)/2$ and $\sigma'_c = (\sigma'_v + 2\sigma'_h)/3$ are taken along the vertical and horizontal axes, the stress-strain curve is shown in the diagram (b). Because $\tau_d \geq \tau_s$, it is the stress-reversal condition and the stress path starting from the point A can arrive at the origin O, where 100% pressure buildup and zero effective stress is attained, followed by drastic strain increase, as obviously seen in the figure.

Fig. 5.8.5 depicts a similar example in terms of the stress path (a) and the stress-strain curve (b) for the same sand specimen, $D_r \approx 30\%$, $F_c = 0\%$, with $\sigma'_{vc} = 118.7$ kPa,

$\sigma'_{hc} = 87.7$ kPa, $\tau_s = 15.5$ kPa and $\sigma'_c = 98$ kPa, sheared by the cyclic shear stress $\tau_d = 13.4$ kPa ($CSR = 0.137$). In this case, because of the stress non-reversal condition $\tau_d \leq \tau_s$ the stress path cannot reach the origin O, imposing the limitation in the excess pore-pressure buildup and associated strain development. The pressure buildup is limited at B by the line OC in Fig. 5.8.5(a), and the Mohr-Coulomb failure envelope is expressed as $\sin\phi' = (\sigma'_v - \sigma'_h)/(\sigma'_v + \sigma'_h)$ or in the following formula.

$$\frac{\tau}{\sigma'_c} = \frac{3\sin\phi'}{3 - \sin\phi'} \tag{5.8.1}$$

Here, $\phi'$ = internal friction angle in terms of the effective stress, and the point B is the intersection of the line OC and line AB′ for $\tau = \tau_s = (\sigma'_{vc} - \sigma'_{hc})/2$ = constant. The residual effective stress at B is expressed from Eq. (5.8.1) as,

$$\sigma'_c = \frac{3 - \sin\phi'}{6\sin\phi'}(\sigma'_{vc} - \sigma'_{hc}) \tag{5.8.2}$$

and the maximum pressure buildup $\Delta u_{\max}$, defined by the difference in confining stress at A ($\sigma'_{c0} = (\sigma'_{vc} + 2\sigma'_{hc})/3$) minus at B in Fig. 5.8.5(a), and normalized by the effective horizontal stress $\sigma'_{hc}$ is expressed, using a coefficient $K_c = \sigma'_{vc}/\sigma'_{hc}$, as

$$\frac{\Delta u_{\max}}{\sigma'_{hc}} = \frac{\sigma'_{c0} - \sigma'_c}{\sigma'_{hc}} = 1 - (K_c - 1)\frac{(1 - \sin\phi')}{2\sin\phi'} \tag{5.8.3}$$

Fig. 5.8.6 compares triaxial test results in terms of $\Delta u_{\max}/\sigma'_{hc}$ versus $K_c$ with the solid line by Eq. (5.8.3) assuming $\phi' = 36.6°$ and shows a fair agreement between them (Vaid and Chern 1983).

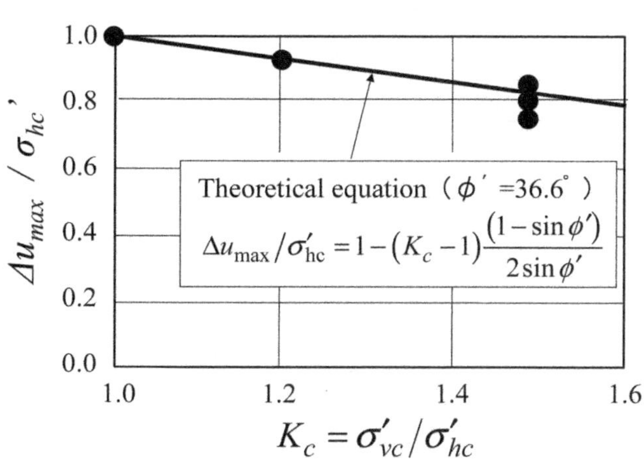


*Figure 5.8.6* Normalized maximum pressure buildup $\Delta u_{\max}/\sigma'_{hc}$ versus $K_c = \sigma'_{vc}/\sigma'_{hc}$ by triaxial tests with initial shear stress compared with theory (Vaid and Chern 1983).

### 5.8.2 Effect on liquefaction failure

A significant effect of the initial shear stress on the liquefaction triggering and associated strain development depending on the stress reversal or non-reversal during cyclic shearing was first pointed out by Yoshimi and Oh-oka (1975) using a ring torsional shear test device. Fig. 5.8.7 shows similar test results by Vaid and Finn (1979) using a constant-volume simple shear device, where the specimen $K_0$-consolidated and then loaded with initial shear stress on the horizontal plane in advance was cyclically sheared on the same plane with no volume change. In the case of stress reversal (No. 22 and 67), the pore pressure in (a) builds up swiftly to $\Delta u/\sigma_v' = 1.0$ and the shear strain in (b) grows up to $\gamma = 10\%$ accordingly, whereas in the non-reversal (No. 91) or intermediate case (No. 16), the pressure cannot buildup 100% and the strain are slow to increase.

In the above, in situ stress states were very much simplified in test specimens so that the initial shear stress $\tau_s$ and cyclic shear stress $\tau_d$ were working on the same plane. Fig. 5.8.8 shows the *CSR* versus $N_c$ relationship using a torsional simple shear apparatus for dense clean sand specimens of $D_r \approx 100\%$ to investigate how the soil behavior changes according to the planes of initial shear stress application (Kokusho et al. 1981). Three options are chosen here; (a) the initial shear stress $\tau_s$ is applied on the horizontal plane in isotropically consolidated specimens, (b) $\tau_s = (\sigma_v' - \sigma_h')/2$ is induced on the 45° plane by the anisotropic consolidation by the vertical and horizontal stress $\sigma_v'$ and $\sigma_h'$, (c) Initial shear stresses $\tau_s$ are working on the horizontal and 45° planes at the same time. In all the options the cyclic shear stress $\pm\tau_d$ is working on the same

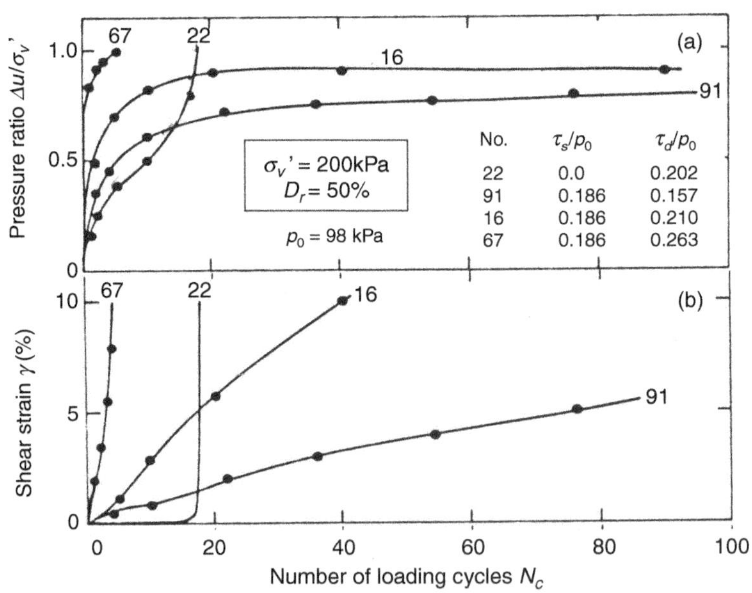


*Figure 5.8.7* Pressure buildup and strain increase versus number of cycles in simple shear tests in stress reversal/non-reversal conditions (Vaid and Finn 1979) with permission from ASCE.

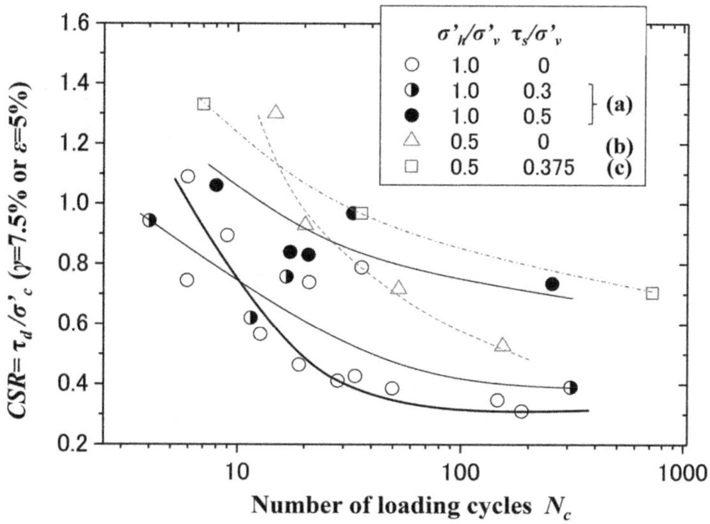


*Figure 5.8.8* *CSR* versus number of cycles $N_c$ in torsional undrained cyclic shear tests on dense clean sand with initial shear stresses working on different planes (Kokusho et al. 1981).

horizontal plane in the undrained condition. The *CSR*-value is defined corresponding to the induced torsional shear strain $\gamma = 7.5\%$ or axial strain $\varepsilon = 5\%$ with respect to the initial zero strain or in the cyclic double amplitude. If compared with the open circle plots which are isotropically consolidated and free from any initial shear stress, the plots with initial stresses $\tau_s$ tend to show higher *CSR* values no matter if $\tau_s$ works on the horizontal plane or on the 45° plane, though the trends in the $CSR{\sim}N_c$ curve seem to be different depending on the planes. Furthermore, if the two initial shear stresses are imposed on the 45° and horizontal planes simultaneously, *CSR* tends to increase further. This implies that in in situ soils loaded by the initial shear stresses on arbitrary planes 3-dimensionally different from the planes of cyclic stress applications are still influenced by them in their liquefaction behavior.

However, there exists another significant effect of the initial shear stress overwhelming the stress reversal/non-reversal effect mentioned above. Figs. 5.8.9(a)~(c) depicts $CRR = \tau_d/\sigma'_n$ corresponding to induced axial strain $\varepsilon = 1, 2.5, 5\%$ in the number of cycles $N_c = 10$ in the vertical axis versus the initial shear stress ratio $\alpha = \tau_s/\sigma_n{}'$ in the horizontal axis obtained by triaxial tests on clean sand specimens for three step of relative density $D_r = 45, 55, 65\%$, respectively (Vaid and Chern 1983). The plots above and below the diagonal line $\tau_s = \tau_d$ on each diagram correspond to the conditions of stress reversal and non-reversal, respectively. *CRR* tends to increase with increasing $\tau_s/\sigma'_n$ in the stress reversal zone $\tau_s < \tau_d$ for all the densities. This means that the cyclic strength of clean sand essentially tends to increase with increasing initial stress. However for the looser densities in (a) $D_r = 45\%$ and (b) 55%, $CRR = \tau_d/\sigma'_n$ for the prescribed strain starts to decrease with increasing $\tau_s/\sigma'_n$ in the stress non-reversal

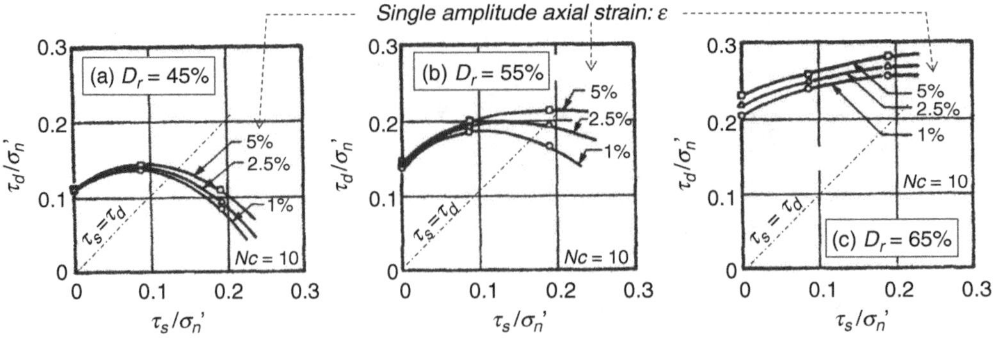


*Figure 5.8.9* $CRR = \tau_d/\sigma_n'$ for $\varepsilon = 1, 2.5, 5\%$ in $N_c = 10$ versus initial shear stress ratio $\tau_s/\sigma_n'$ in triaxial tests in stress reversal/non-reversal cases (Vaid and Chern 1983).

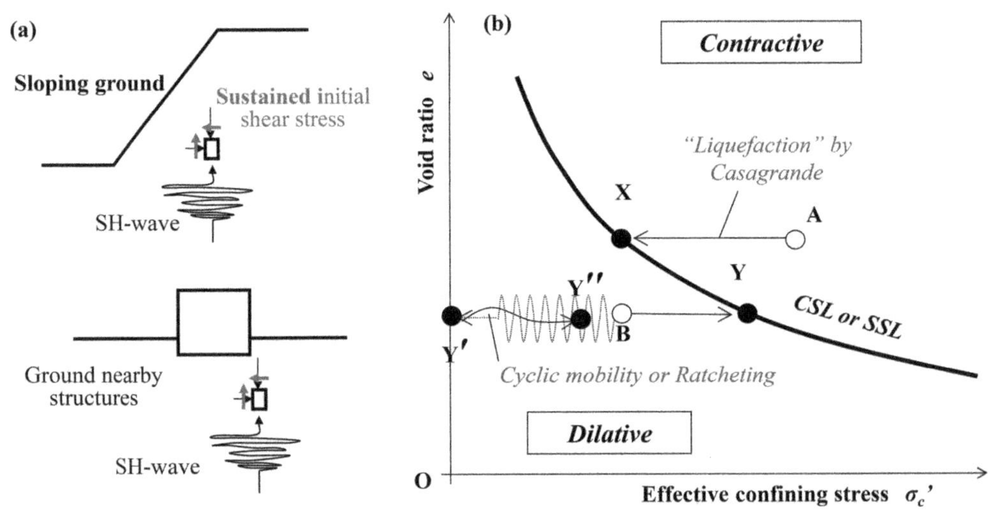


*Figure 5.8.10* Soils prone to lateral spreading under sustained initial shear stress (a), and $\sigma_c'$–e state diagram and CSL or SSL (Steady State Line) dividing into contractive and dilative states.

zone $\tau_s \geq \tau_d$. This indicates that the loose soil tends to fail mainly due to the increasing initial shear stress and the cyclic loading serves only as a trigger in the condition of pore-pressure buildup less than 100%.

### 5.8.3 Effect on failure mode

In order to discuss on the effect of initial shear stress on *CRR* either increasing or decreasing depending on the stress states, it is necessary to go back to a fundamental shear mechanism associated with the state-diagram addressed in Fig. 3.1.7. Though, soil liquefaction in level ground has been the basis in evaluating liquefaction triggering (Seed and Lee 1966), the sustained initial shear stress absent there is actually responsible for typical liquefaction-induced failures such as tilting, lateral spreading and sliding. In this respect, Casagrande (1971) provided a completely different view on the liquefaction mechanism focusing on the role of the sustained initial shear stress in near slopes and superstructures as illustrated in Fig. 5.8.10(a). He proposed use of the term "liquefaction" for a phenomenon in which the contractive sand loses its shear strength, not necessarily by cyclic loading but by monotonic loading as well. Actually, saturated slopes of contractive soils have sometimes undergone static flow-type failures without seismic effects when the undrained peak strength is lower than the static shear stress, called as "spontaneous liquefaction". For cyclic loading, the term "cyclic mobility" was proposed for the temporary zero effective stress condition in the dilative sand, accordingly. The same author, followed by Castro (1975) and several other researchers, utilized the concept of the Steady State Line (SSL) or Critical State Line (CSL) already addressed in Sec. 3.1.5 to interpret the mechanism as shown in Fig. 5.8.10(b). In the state diagram, liquefaction is interpreted as the result of undrained failure of saturated loose contractive sand. For example, a soil element starting at point **A** on the contractive side of SSL and eventually ending up with steady-state flow at constant volume at **X** on SSL. If a sand is monotonically loaded starting at **B** on the dilative side of SSL in the undrained condition, the point moves right toward **Y** on SSL with increasing effective confining stress $\sigma'_c$. If the same sand is loaded cyclically, the point moves from **B** to the left due to the negative dilatancy during cyclic loading (different from the positive dilatancy during monotonic loading as discussed in Sec. 3.1.5). It eventually reaches zero-effective stress at **Y′** under the zero-initial shear stress condition (in a level ground), which was demonstrated in the first triaxial liquefaction test by Seed and Lee (1966). However, subsequent undrained monotonic loading translates the point to the right to **Y″** and the resistance of the specimen revives again.

In order to merge the two different views of Seed and Casagrande, Vaid and Chern (1985) systematically performed triaxial tests to illustrate a unified picture of

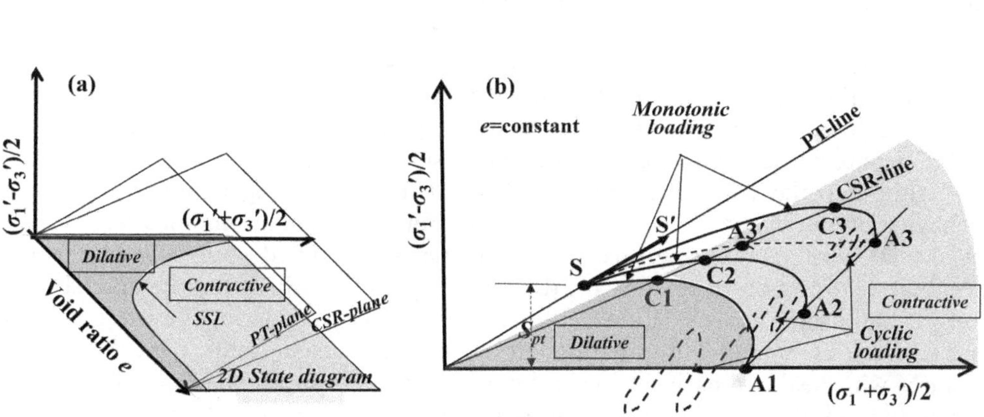


*Figure 5.8.11* Three-dimensional effective stress diagram (e~$(\sigma'_1 + \sigma'_3)/2$~$(\sigma'_1 - \sigma'_3)/2$) (a), and a typical section ($(\sigma'_1 + \sigma'_3)/2$~$(\sigma'_1 - \sigma'_3)/2$) at constant void ratio e (b) (Redrawn from Vaid and Chern 1985) with permission from ASCE.

the undrained monotonic and cyclic loading response of saturated sands including the effect of initial shear stress in cyclic loading tests. Fig. 5.8.11(a) shows a three dimensional effective stress diagram in $e \sim (\sigma_1' + \sigma_3')/2 \sim (\sigma_1' - \sigma_3')/2$ space, which is an extension of the state diagram shown in Fig. 5.8.10(b). Fig. 5.8.11(b) shows a typical vertical section of (a) at a constant void ratio $e$, where stress paths for monotonic as well as cyclic loading shown with solid and dashed curves, respectively, are drawn starting from the points A1, A2, A3 with different initial shear stresses $(\sigma_1' - \sigma_3')/2$, with $\sigma_3' =$ constant. For the monotonic loading, a contractive flow-type failure (liquefaction or limited liquefaction in the Casagrande's definition) is triggered at Points C1, C2, C3 on the straight line (called the *CSR*-line) shown in the diagram, only if the starting points A1, A2, A3 are on the contractive side of the 3-dimensional state diagram. The *CSR*-line is uniquely defined for stress paths of a given sand to have peak values and initiate strain softening thereafter. The similar concept was also proposed by Sladen et al. (1985) by the name of the collapse line. During the strain softening, the sand undergoes flow deformation, and if the flow is a limited type it is followed by subsequent strain hardening at Point S toward S′. The point S is on the PT (phase transformation)-line defined by Ishihara et al. (1975) shown in the diagram. Another condition for the contractive sand to undergo flow type failure is that the shear stress $(\sigma_1' - \sigma_3')/2$ on the *CSR*-line (at Point C3 for example) should be larger than the shear stress at S ($S_{PT}$) so that the stress path can undergo strain softening after reaching the peak (Vaid and Chern 1985).

In the case of the cyclic loading under sustained initial stress, Vaid and Chern (1985) demonstrated that the condition for the occurrence of flow-type failure is essentially the same as for the monotonic loading. Namely, the cyclic loading stress-path starting from the points A3 for example comes across the *CSR*-line at a point (A3′) with the shear stress higher than the stress $S_{PT}$ of Point S, so that the sand undergoes the flow-type or limited flow-type failure. One significant difference from the monotonic loading is that the cyclic loading builds up the pore-pressure, which translates the effective stress to the left (from A3 to A3′) on the *CSR* line and enables the flow-type failure to occur with the shear stress (initial shear stress + cyclic stress amplitude) smaller than C3 in the monotonic loading starting from A3. If any of the conditions necessary for triggering the flow-type or limited flow-type failure mentioned above is not met, then sand exhibits cyclic mobility in which strains tend to develop gradually in a ductile failure mode (Vaid and Chern 1985). In contrast, the flow-type failures may develop infinite or very large but limited strain quite abruptly leading to a dangerous brittle failure mode in liquefied ground due to sustained initial stresses.

Although other researches on the mechanical models merging monotonic and cyclic behavior of sands have been performed and obtained similar findings (e.g. Sladen et al. 1985, Alarcon-Guzman et al. 1988), issues associated with how to evaluate residual deformation depending on the failure types are still left to be agreed upon before establishing a unified liquefaction design practice considering the effect of initial shear stress. In many previous undrained shear tests on clean sands, typical flow type failures with peak strengths followed by infinitely large strain were not observed so often. Instead, limited flow-type failures are more prevalent, which are characterized by temporary quasi-steady state strengths followed by regained shear resistance for further straining.

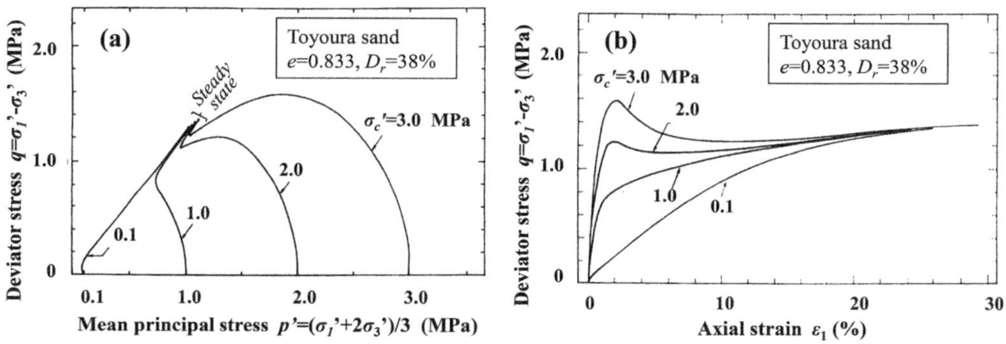


*Figure 5.8.12* Undrained monotonic loading test results of $D_r = 38\%$ Toyoura sand under different initial confining stress: (a) Effective stress paths, (b) Stress versus strain curves (Ishihara 1993) by permission of ICE Publishing.

For example, Ishihara (1993) made a comprehensive dataset of sand behavior in undrained monotonic loading from laboratory tests. Figs. 5.8.12(a) and (b) depict the undrained behavior of clean Toyoura sand with $D_r = 38\%$ under various effective confining stresses. The relatively loose sand is obviously dilative for the confining stress $\sigma_c' = 0.1$ MPa and starts to be slightly contractive for $\sigma_c' = 2.0$ MPa or higher in showing a temporary reduction after a peak in the deviator stress $q = \sigma_1' - \sigma_3'$ and an increase again as the limited flow-type failure. In contrast to a steady state flow with infinitely large strain, this response was called limited liquefaction by Casagrande (1971) and the temporary minimum value was termed as the quasi-steady state strength (Alarcon-Guzman 1988, Ishihara 1993). In normal liquefaction problems for shallow depths less than 10 m, the effective confining stress is $\sigma_c' = 0.1$ MPa or lower and the relative density of loose sand deposits would be around $D_r = 30{\sim}40\%$ in the loosest case in nature. Actually, the in situ soil density of clean sand in Niigata city, where extensive liquefaction occurred during the 1964 earthquake, was $D_r \approx 40\%$ as depicted in Fig. 5.10.16, although it may have been densified by several percent on average due to the liquefaction. This indicates that clean sands are normally on the dilative side and the flow-type failure (even the limited flow-type) seems to be difficult to occur according to laboratory tests. However, the presence of low/non-plastic fines mixed with clean sands may change the behavior significantly.

The significant role of fines in reducing the shear-induced dilatancy of clean sands was pointed out by several researchers (e.g. Ishihara 1993). Figs. 5.8.13(a) and (b) show typical effective stress paths and stress-strain curves, respectively, of undrained monotonic loading torsional shear tests on loose sand specimens with parametrically increasing non-plastic fines under the isotropic effective confining pressure $\sigma_c' = 98$ kPa. For the same low relative density $D_r \approx 30\%$, the clean sand of $F_c = 0$ is clearly dilative, while it becomes contractive with limited flow for $F_c = 10\%$ and undergoes perfect flow with almost zero residual strength for $F_c = 20\%$. This is presumably because the steady state line is significantly influenced by $F_c$, and accordingly the contractive

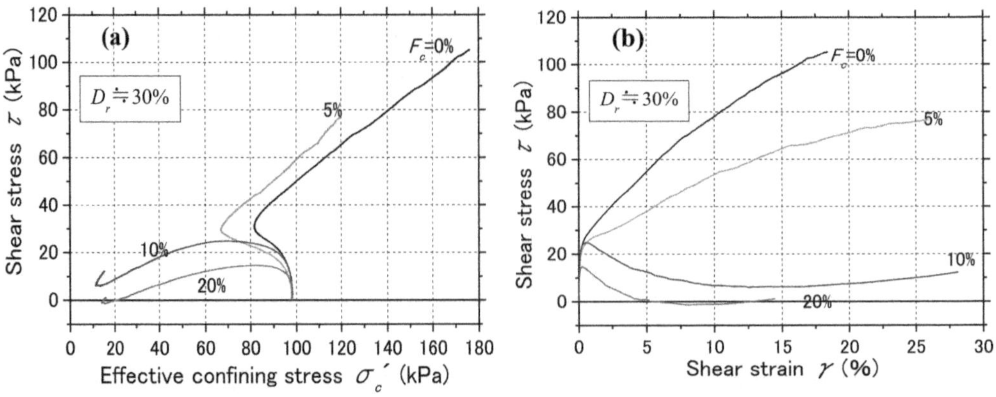


*Figure 5.8.13* Increasing contractive behavior of Futtu sand with increasing non-plastic fines, $F_c = 0 \sim 20\%$ by torsional shear tests: (a) $\tau \sim \sigma_c'$, (b) $\tau \sim \gamma$ (Kusaka et al. 2013).

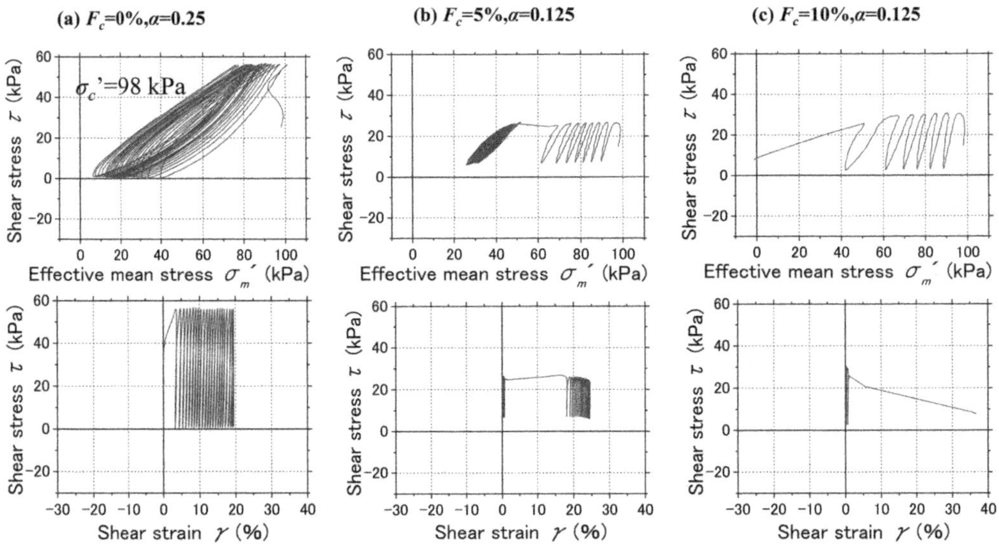


*Figure 5.8.14* Torsional cyclic loading test results of $D_r \approx 30\%$ under initial shear stress: (a) $F_c = 5\%$ with gradual failure, (b) $F_c = 5\%$ with limited flow failure, and (c) $F_c = 10\%$ with unlimited flow failure (Arai et al. 2015).

and dilative zones on the state diagram change drastically (e.g. Papadopoulou and Tika 2008).

In accordance with the monotonic loading tests, Fig. 5.8.14(a)~(c) compare test results of the undrained cyclic torsional shear tests on the same sand as in Fig. 5.8.13 with varying $F_c$ under the initial shear stress ratio $\alpha = \tau_s/\sigma_c' = 0.125$ or $0.25$ and

$\sigma_c' = 98$ kPa. For the clean sand with $\alpha = 0.25$ (a), the pore-pressure and associated strain tends to increase only gradually up to limited values due to the non-stress reversal condition making the failure-mode very ductile, because the clean sand is dilative as indicated by monotonic loading in Fig. 5.8.13. For the case of $F_c = 5\%$ with $\alpha = 0.125$ (b), a gradual increase of strain in the initial stage is followed by the limited flow failure and then the convergence to an ultimate value. For the case of $F_c = 10\%$ with $\alpha = 0.125$ (c), large strain occurs in the flow-type failure making the failure very brittle, because the sand with this $F_c$-value responds very contractively as observed in the monotonic loading. Though the effect of $F_c$ appears to be slightly different quantitatively, its enormous influence is observed in both monotonic and cyclic loading.

In current engineering practice, the effect of initial shear stress is represented by one of the influencing factors on the resistance to liquefaction triggering. For example in North American practice, a correction factor for initial shear stress $K_\alpha = CRR_\alpha / CRR_{\alpha=0}$ is used (Idriss and Boulanger 2008), where $CRR_\alpha$ is the cyclic resistance ratio under the initial shear stress ratio $\alpha = \tau_s / \sigma_v'$ ($\tau_s$ = initial shear stress, $\sigma_v'$ = effective normal stress). However, this effect should be focused not only on the $CRR$-values for liquefaction triggering but also more on how large the post-liquefaction strain develops and how suddenly it occurs.

Figs. 5.8.15(a), (b), and (c) summarize torsional shear test results on sands of $D_r \approx 30\%$ and 50% without/with non-plastic fines $F_c = 0{\sim}30\%$. The specimens were isotropically consolidated under $\sigma_c' = 98$ kPa and sheared in the drained condition by parametrically varied sustained initial shear stresses $\tau_s$ on the horizontal plane ($\theta_0 = 0°$) in (a), (b), whereas in (c) the specimens were anisotropically consolidated introducing

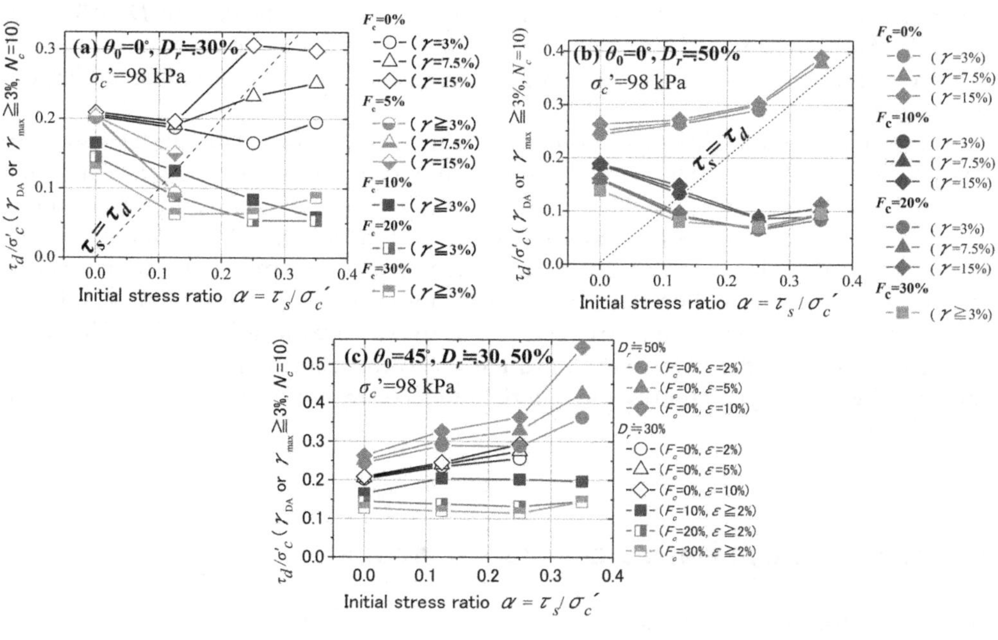


*Figure 5.8.15* Variations of $CRR$ with increasing initial shear stress ratio $\alpha$ for sands of various fines content $F_c = 0$–30%: (a) $D_r \approx 30\%$, (b) $D_r \approx 50\%$ (Kusaka et al. 2013, Arai et al. 2015).

the initial stress $\tau_s = (\sigma'_{vc} - \sigma'_{hc})/2$ on the $\theta_0 = 45°$ plane. The *CRR*-values defined for shear strains $\gamma = 3$, 7.5, 15% in (a), (b) and for axial strains $\varepsilon = 2$, 5, 10% in (c), respectively, in the number of cycles $N_c = 10$ are plotted versus the initial shear stress ratios $\alpha = \tau_s/\sigma'_c$ with different symbols and connected with lines to each other. For the case $\theta_0 = 0°$ in Figs. 5.8.15(a), (b), the enormous effect of $F_c$ on the $\alpha$-dependent *CRR* variations is observed for $D_r \approx 30\%$ and 50% as well, because the fines tend to change the soil from dilative to contractive as already seen. The diagonal dashed line corresponding to $\tau_s = \tau_d$ in the chart also influences liquefaction behavior; above the lines ($\tau_d > \tau_s$) the stress reversal tends to make pore-pressure build up easier than below the line. For contractive sands with $F_c \geq 5{\sim}10\%$, the *CRR*-values tend to decrease with increasing $\alpha$ up to $\alpha = 0.25$ despite the non-stress reversal conditions, showing the same trend as what was shown in Fig. 5.8.9 by Vaid and Chern (1983). For the case $\theta_0 = 45°$ in Figs. 5.8.15(c), the trend is similar, though the same soil seems to behave slightly dilative than the case $\theta_0 = 0°$, presumably because the plane of the initial shear stress is different by 45° from the plane of cyclic shearing.

All the plots are overlapped for the strains $\gamma = 3$, 7.5, 15% or $\varepsilon = 2$, 5, 10% in the case $D_r \approx 30\%$, $F_c \geq 10\%$, and $D_r \approx 50\%$, $F_c \geq 30\%$ both in $\theta_0 = 0°$ and 45° in Fig. 5.8.15. They represent brittle mode flow-type failures causing large strains as soon as $\gamma$ exceeds 3% or $\varepsilon$ exceeds 2%. If $\alpha = 0$ in these cases for a level ground, although the plots are also overlapped, the failures are not the brittle flow-type because of the absence of the initial shear stress, and only large cyclic strains tend to grow during shaking.

For the dilative clean sand $F_c = 0\%$, the *CRR*-value tends to increase with increasing $\alpha$. In the case of $D_r \approx 50\%$ in Fig. 5.8.15(b), all the plots for $F_c = 0\%$ are in the stress reversal condition, wherein the cyclic strain and the unilateral strain due to the initial shear stress develop together after the initial liquefaction. In the case of $D_r \approx 30\%$ in Fig. 5.8.15(a), the plots for $F_c = 0\%$ are crossing the line $\tau_s = \tau_d$ and separating to each other depending on the strain values $\gamma$, indicating that the ductile failure undergoes quite gradually.

How to evaluate the lateral deformation under the influence of initial shear stress to compare it with design thresholds is an essential part in the performance-based design (PBD). In the non-flow cyclic ductile failure, a designer's concern is how to evaluate the cyclic strain accumulation and compare it with a design value. In contrast, the flow-type brittle failures accompanying a sudden increase of limited or unlimited large strains are far serious and need greater care than non-flow type ductile failures. In the limited flow-type failures, it may still be possible to evaluate the deformation. However, it will change from the deformation evaluation to the critical stability evaluation in the unlimited flow-type failures wherein the induced strain is difficult to predict (Vaid and Chern 1985).

## 5.9 CYCLIC SOFTENING OF CLAYEY SOILS

As already mentioned, fine soils are as prone as sands to liquefaction if they are non/lowly-plastic. If their plasticity is high, they are exempt from liquefiable soils

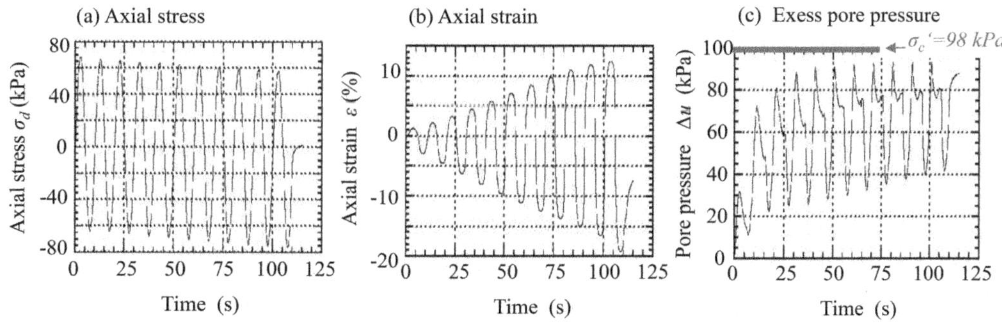


*Figure 5.9.1* Undrained cyclic triaxial test results of kaolin clay ($e = 0.987$, $I_p = 23$, OCR = 3) in time histories: (a) Axial stress, (b) Axial strain, (c) Excess pore-pressure (Yoshio 2002).

in liquefaction potential evaluation criteria. It is known however that cyclic loading will have a certain impact on post-seismic strength and stiffness in highly plastic fine soils, too, though they do not reduce shear resistance as drastically as liquefaction in non/low-plastic soils. One of the serious case histories known to date is the collapse of buildings due to shearing failure of highly plastic foundation clays during the 1985 Mexican earthquake (Mendoza 1987). This type of phenomenon may be categorized differently from liquefaction of non-cohesive soils and sometimes named as "cyclic softening" of clays. Though not so typical as liquefaction damage, one should be aware of its impact in engineering design and the associated mechanism in comparison with liquefaction.

### 5.9.1 Typical cyclic softening behavior

Fig. 5.9.1 depicts the time histories of (a) axial stress, (b) axial strain and (c) excess pore-pressure in an undrained cyclic triaxial test of kaolin clay ($e = 0.987$, $I_p = 23$, $OCR = 3$) $K_0$-consolidated with the vertical stress 294 kPa in a mold. Then, the specimen (5 cm in diameter, 10 cm in height) is isotropically consolidated with $\sigma_c' = 98$ kPa, and cyclically sheared by a harmonic motion ($f = 0.1$ Hz) with the cyclic stress ratio $CSR = \sigma_d/2\sigma_c' = 0.33$. Similar to sands in the process to liquefaction, the pore-pressure builds up to around 90% of the initial effective confining stress associated with stepwise development in strain. Note here that, in cyclic loading tests on clayey soils with the low water conductivities, the loading frequency is normally taken around 0.1 Hz or even lower to overcome a time-delay in the pore pressure response, though the pressure seldom comes up to 100%. The strain developments associated with the pressure buildup are not as drastic as in liquefaction despite larger void ratios than sands.

Fig. 5.9.2(a) exemplifies an effective stress path on the $p'$–$q$ diagram and the stress versus strain relationship on the $q$–$\varepsilon$ diagram based on the same test result on the kaolin clay. Here $p' = \sigma_1' + 2\sigma_3'$ and $q = \sigma_1' - \sigma_3'$. It exhibits a clear cyclic mobility response more like dense sands than loose liquefiable sands. The pressure increase is the largest in the first cycle followed by gradual stepwise increments, and the secant shear modulus

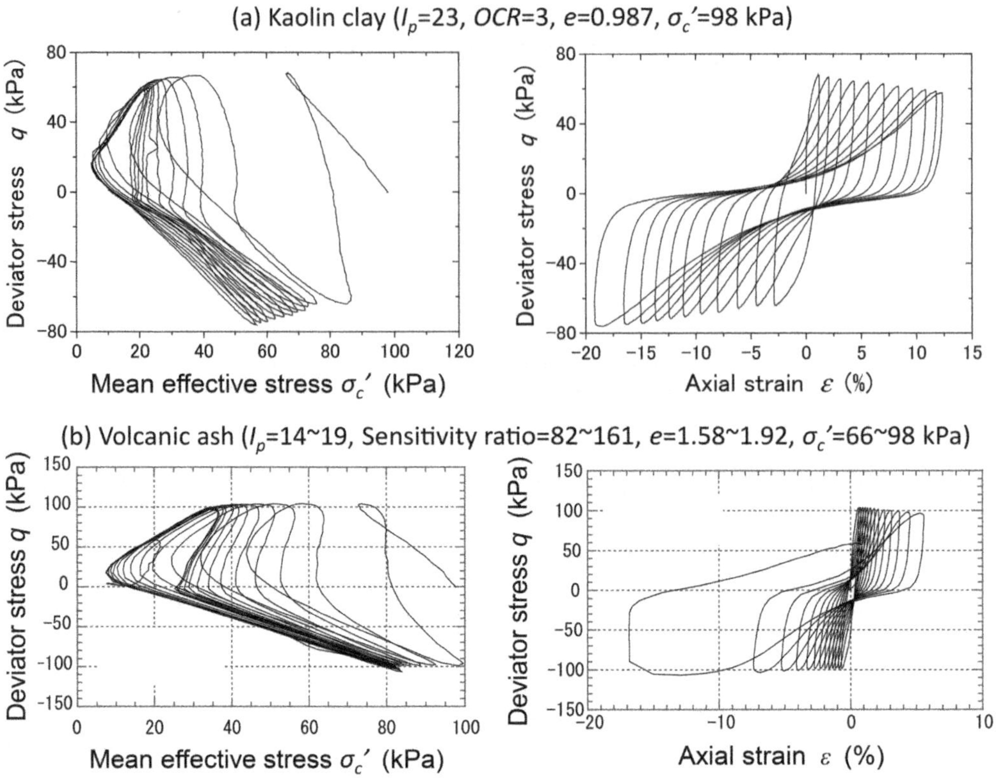


*Figure 5.9.2* Effective stress path (left) and stress-strain curve (right) in undrained cyclic triaxial tests: (a) Reconstituted kaolin clay (Yoshio 2002), (b) Intact volcanic ash (Ohkawa 2003).

connecting upper/lower tips of individual $q$–$\varepsilon$ hysteresis loops tends to degrade with the number of cycles reflecting the cyclic softening behavior.

In order to compare the cyclic loading response with that of natural intact clays, the same test was carried out on volcanic ash ($I_p = 14\sim19$, $e = 1.58\sim1.92$) several tens of thousands years old isotropically consolidated with the in situ confining stresses $\sigma_c' = 66\sim98$ kPa. The $p'\sim q$ and $q\sim\varepsilon$ relationships are exemplified in Figs. 5.9.2(b). Although this clay is highly sensitive (the sensitivity ratio $= 82\sim161$), cycle-wise modulus degradations corresponding to a large cyclic stress ratio ($CSR = 0.502$) is not considerable, not so different from the non-sensitive kaolin clay.

In Fig. 5.9.3, pore-pressure buildup ratios $\Delta u/\sigma_c'$ at the 11th loading cycle ($N_c = 11$) are plotted versus cyclic stress ratios $CSR = \sigma_d/2\sigma_c'$ for the kaolin clay of $I_p = 23$ and the intact highly sensitive volcanic ash of $I_p = 14\sim19$. It indicates that $CSR$ to attain about 90% pressure buildup is 0.33 for the former and 0.71 for the latter, which seems much higher than typically liquefiable loose sand despite void ratios of $e \approx 1.0$ or greater because of the plasticity. Despite the very high sensitivity, the intact volcanic ash exhibits larger $CSR$ to attain the 90% pressure buildup than the reconstituted kaolin clay. This seems to indicate that cyclic loading, despite the high

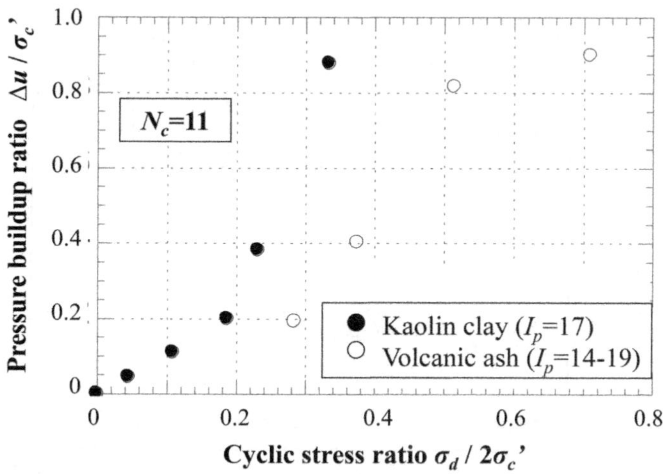


*Figure 5.9.3* Pore-pressure buildup ratio versus *CSR* ($N_c = 11$) for kaolin clay and highly sensitive volcanic ash (Ohkawa 2003).

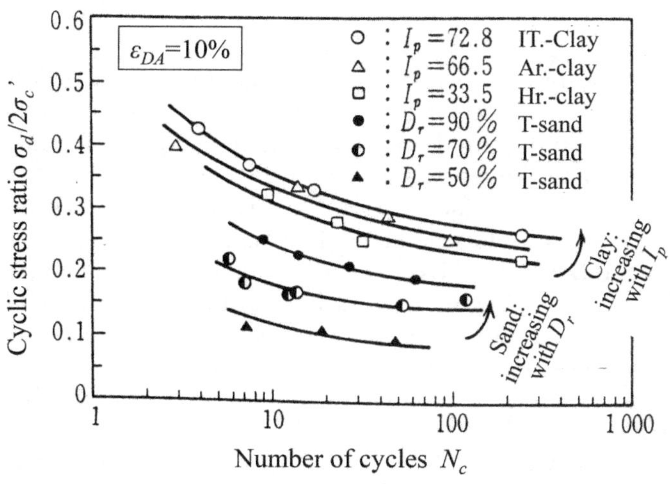


*Figure 5.9.4* *CSR* by triaxial tests ($\varepsilon_{DA} = 10\%$) versus number of cycles $N_c$ for intact clays compared with clean sand (Hyodo and Uchida 1998).

*CSR*, is not as damaging to the soil skeleton of this particular sensitive clay as the sensitivity test wherein soil particles are completely remolded by hand.

Thus, *CSR* to attain a certain pore-pressure buildup is likely to be greater in cohesive soils than in liquefiable sands. Fig. 5.9.4 shows *CSR*-values for the double amplitude axial strain $\varepsilon_{DA} = 10\%$ versus the number of cycles $N_c$ obtained in triaxial tests on three natural intact Holocene clays with the plasticity indexes $I_p = 34{\sim}73$

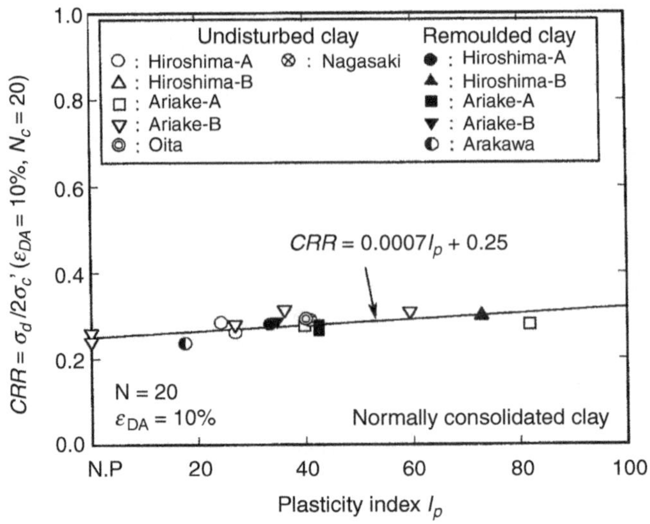


*Figure 5.9.5* CRR by triaxial tests ($\varepsilon_{DA} = 10\%$, $N_c = 20$) versus plasticity index $I_p$ for normally consolidated clays (Hyodo et al. 1999).

compared with those for the Toyoura sand (Hyodo and Uchida 1998). According to this result, *CSR* for $\varepsilon_{DA} = 10\%$ for the clays of $I_p \approx 30$–70 are much higher than that of the Toyoura clean sand (T-sand) of $D_r = 90\%$, implying the significant effect of the plasticity or cohesion on the resistance to cyclic loading.

In Fig. 5.9.5, the *CRR*-values, $\tau_d/\sigma'_c$ ($\varepsilon_{DA} = 10\%$, $N_c = 20$) of normally consolidated Holocene clays are plotted versus the plasticity indexes, $I_p = \text{NP}$ to over 70 obtained in cyclic triaxial tests (Hyodo et al. 1999). The soils here were sampled from various sites by tube-sampling and tested in undisturbed and remolded conditions. It can be seen that the effect of $I_p$ on *CRR* is not so significant in contrast to the effect of relative density $D_r$ on the liquefaction resistance of sands. According to the data, the *CRR*-values of both intact and remolded soils are almost uniquely correlated with $I_p$ as $CRR = 0.0007I_p + 0.25$.

All the above tests were conducted to obtain the resistance to cyclic softening of clays in level ground without sustained initial shear stresses. In the same context as the liquefaction problem considering the influence of initial shear stress, the effect of cyclic softening on soil stabilities near slopes and shallow foundations during cyclic loading in clayey soils needs special attention. Fig. 5.9.6 shows triaxial test results of specimens of two Holocene clays and Toyoura sand as well, anisotropically consolidated with $\sigma'_v = \sigma'_c + \sigma'_s$ vertically and $\sigma'_h = \sigma'_c$ horizontally, introducing the initial shear stress $\tau_s = \sigma'_s/2$ on the 45° plane, and cyclically loaded in the undrained condition (Hyodo and Uchida 1998). The *CRR*-values ($\sigma_d/2\sigma'_c$ for $\varepsilon_{DA} = 10\%$ in $N_c = 20$) are plotted in the vertical axis versus the initial shear stress ratio defined here as $\sigma'_s/2\sigma'_c$ in the horizontal axis. As already observed for clean sands, the *CRR* of clean sands here ($D_r = 50$ and 70%) tends to increase with increasing initial shear stress ratio $\sigma'_s/2\sigma'_c$ essentially in the

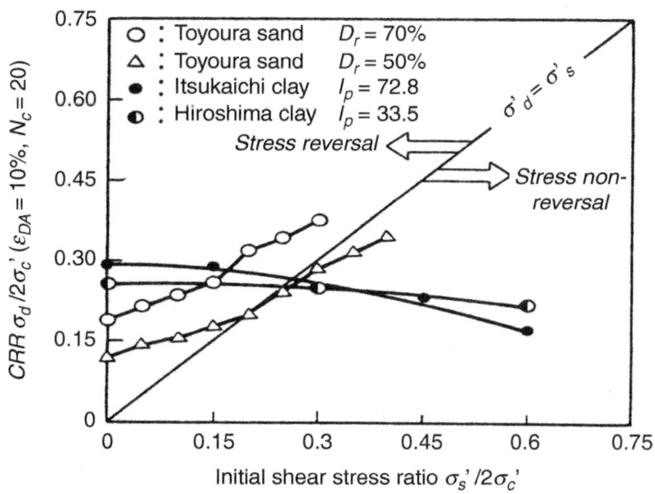


*Figure 5.9.6* CRR ($\varepsilon_{DA}$ = 10%, $N_c$ = 20) versus initial shear stress ratio $\sigma_s'/2\sigma_c'$ for Holocene clays compared with clean sand by triaxial tests (Hyodo and Uchida 1998).

stress reversal zone above the solid diagonal line in Fig. 5.9.6. In a good contrast, the *CRR*-values of clays tend to decrease gradually with increasing $\sigma_s'/2\sigma_c'$ crossing the diagonal line ($\sigma_d' = \sigma_s'$) into the stress non-reversal zone. This trend is analogous to what can be observed in sands containing non-plastic fines in Fig. 5.8.15, indicating that cohesive soils with much higher void ratio than sands are inherently contractive and tend to decrease dynamic shear strength in the non-reversal stress condition with increasing initial shear stress.

Thus, it may be said in general that cohesive soils are more resistant to cyclic loading, and failures like sand liquefaction are hard to occur in a level ground. However, large deformations may occur near slopes and foundations due to cyclic loading, because the absolute value of *CRR* tends to decrease with increasing initial shear stress.

### 5.9.2 Post-cyclic loading strength and deformation

In order to grasp the cyclic loading effect on the post cyclic shear strength and shear stiffness in cohesive soils, the simplified two-step laboratory test procedure can be employed as already mentioned; a normal undrained cyclic loading test on an isotropically consolidated specimen followed by a post-cyclic undrained monotonic loading test. In interpreting these test results, pore-pressure buildup ratios at the end of the cyclic loading with a certain stress amplitude and a given number of cycles are focused. Then, the pressure ratios are correlated with residual shear strength or deformation modulus in the stress-strain curve obtained in the strain-controlled monotonic loading tests conducted after the stress-controlled cyclic loading tests. Fig 5.9.7 summarizes such test results for cohesive soils with $I_p$ = 16–53 on the diagram where residual shear strength ratios $\tau_f/\tau_{f0}$ in the vertical axis are plotted versus the residual pore-pressure

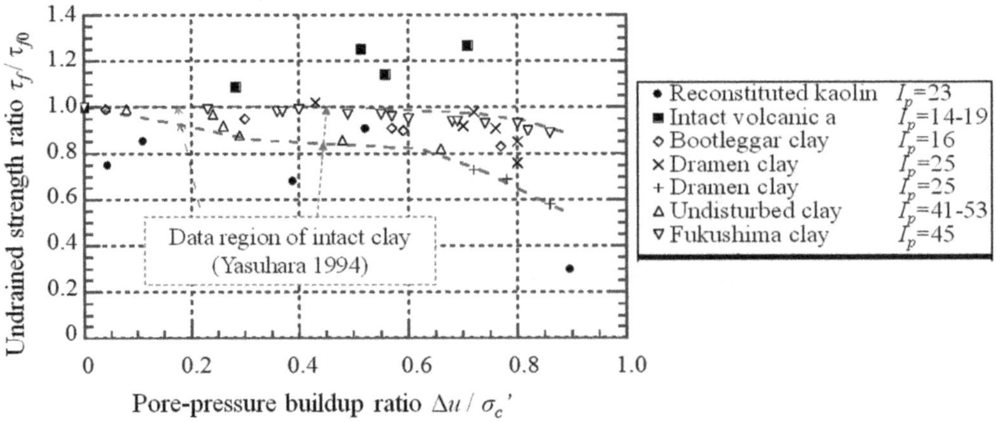


*Figure 5.9.7* Post-cyclic reduction in undrained strength ratio versus pore-pressure ratio in cohesive soils with various plasticity indexes $I_p$.

ratios $\Delta u/\sigma_c'$ in the horizontal axis. Here, $\tau_f$ and $\tau_{f0}$ are the residual shear strength with and without preceding cyclic loading, respectively. The kaolin clay ($I_p = 23$, $OCR = 3$) shown with the solid circles exhibit large drops of the strength ratio down to $\tau_f/\tau_{f0} \approx 0.3$ presumably because it was reconstituted in the laboratory without any aging effect. The highly sensitive volcanic ash ($I_p = 14{\sim}19$, $e = 1.58{\sim}1.92$) mentioned before shown with solid squares does not decrease but rather increase with increasing pore-pressure ratio presumably due to inhomogeneity of intact samples. It may well be judged though quite unexpectedly that the highly sensitive clay is not so much sensitive to seismic cyclic loading, indicating that the cyclic loading process may not be destructive enough to manually disturb the soil skeleton. For other intact clays, data points measured by either triaxial tests or simple shear tests on normally or overconsolidated clays are located between a pair of dashed lines (Yasuhara 1994), indicating the mild reduction of $\tau_f/\tau_{f0}$ down to 0.6–0.9 at $\Delta u/\sigma_c' = 0.9$. The strength reduction ratio does not seem to be clearly dependent on the plasticity index $I_p$, too. Thus, it may be said that unlike very contractive non-cohesive sandy soils where the residual strength may drastically decrease to develop flow-type failures, more than 60% of the shear strength is still left in cohesive soils after cyclic softening.

Fig. 5.9.8 shows the variations of deformation modulus $(E_{50})/(E_{50})_0$ in the vertical axis versus the pore-pressure ratio $\Delta u/\sigma_c'$ measured in the same cyclic triaxial tests. Here $(E_{50})$ and $(E_{50})_0$ are the secant deformation moduli (axial stress divided by axial strain in a test specimen) at the stress level 50% of the peak stress obtained from the monotonic loading stress-strain curves with and without preceding cyclic loading. The modulus tends to reduce considerably in contrast to the moderate reductions in the residual strength in Fig. 5.9.7 for all the soils and decrease down to less than $(E_{50})/(E_{50})_0 = 0.1$ at $\Delta u/\sigma_c' = 0.9$ for intact clays (Yasuhara and Hyde 1997) located in between the dashed lines. One of the symbols in the diagram represent ultra-soft clay from Mexico city with $I_p = 240$, that inflicted considerable building damage during

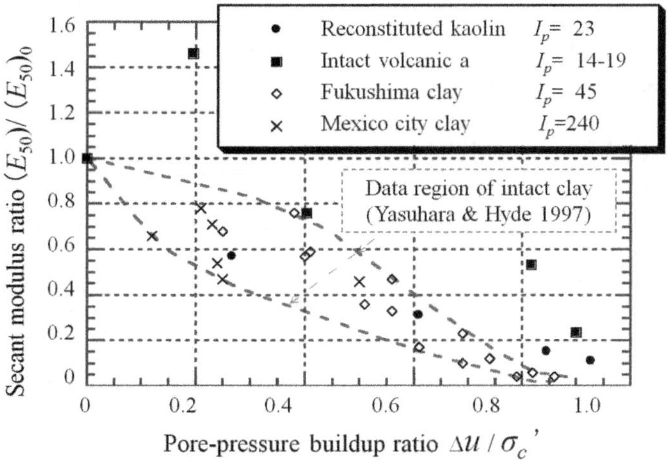


*Figure 5.9.8* Post-cyclic reduction in secant modulus ratio versus pore-pressure ratio in cohesive soils with various plasticity indexes $I_p$.

the 1985 Mexican earthquake. This data indicate that the big difference in $I_p$ does not seem to greatly differ the reducing trend of $(E_{50})/(E_{50})_0$ with increasing pore-pressure ratio.

## 5.10 LIQUEFACTION-INDUCED FAILURES AND ASSOCIATED MECHANISMS

### 5.10.1 Failure modes

Though liquefaction has occurred time and again historically, liquefaction-induced damage is getting large in scale and number, and diversifying in terms of the failure modes due to recent worldwide urbanization in poor geotechnical conditions. Fig. 5.10.1 illustrates schematically typical liquefaction-induced failure modes of structures so far experienced in categories (a) to (h); (a) Failures of pile foundations during shaking due to kinematic effects of degraded shear stiffness of liquefied soils and inertial effects of superstructures, (b) Free-field vertical ground settlements causing gaps relative to structures supported by piles, (c) Differential settlements and tilting of structures on shallow foundations and associated foundation failures, (d) Failures of retaining structures and deformation in backfills, (e) Sliding and lateral spreading of embankments, river levees and earth dams due to liquefaction in foundation soils or their own soils, (f) Uplifting and associated failures of buried structures, (g) Lifeline failures due to lateral flow or lateral spreading of liquefied ground, (h) Failures of pile foundations due to lateral flow of liquefied ground and their effects on superstructures.

Unlike normal structural damage during earthquakes, seismic inertial force has no direct impact on the most failures except (a). In liquefied sites, the inertial force acting on a superstructure tends to decrease due to a base-isolation effect to be discussed

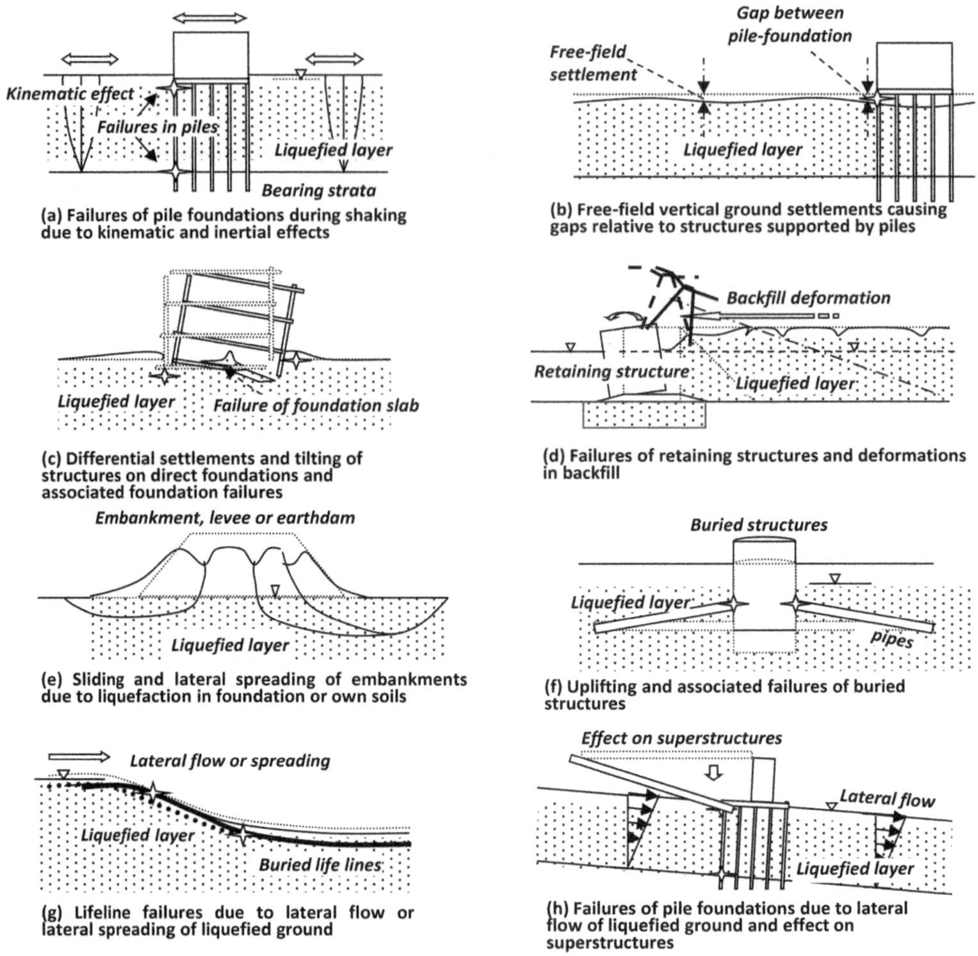


*Figure 5.10.1* Schematic illustrations of liquefaction-induced failure modes of structures experienced during previous earthquakes.

in Sec. 5.11. Most of the liquefaction-induced damage is characterized by the loss of bearing capacity, shear stiffness and associated residual deformations in foundation soils caused by dead loads of structures or slopes. Therefore, the damage may occur with time-delays not necessarily during earthquake shaking but after that, too. In the mode (b), the soil settles one-dimensionally by its own weight not by the effect of sustained initial shear stress. In the other modes (c)–(h), soils are always influenced by initial shear stresses loaded by structures, slopes, embankments and uneven ground surfaces. Almost all liquefaction soil failures occur in the undrained condition except the settlement failure (b), but in some cases the effect of void redistribution slightly different from the undrained condition plays an important role as will be discussed later.

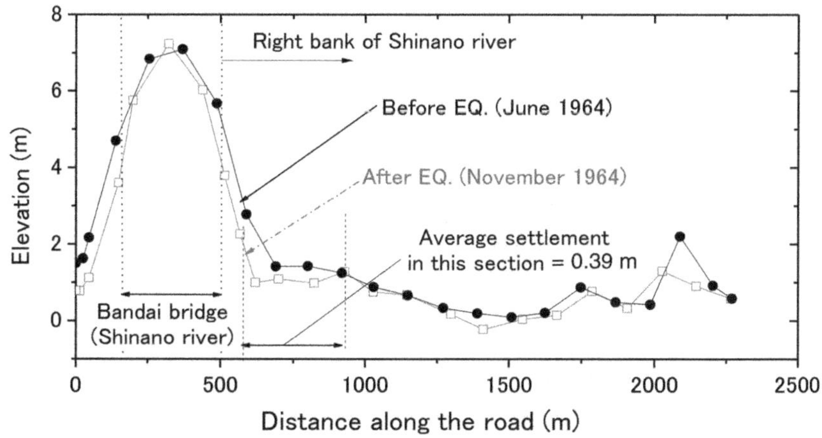


*Figure 5.10.2* Survey results along a national road passing through liquefied area before and after 1964 Niigata earthquake (Kokusho and Fujita 2002) with permission from ASCE.

In this Section, two major residual soil deformations induced by liquefaction; soil settlement in a level ground and lateral flow or lateral spreading in a gently sloping ground, will be focused to discuss their mechanisms. Then, the effects of residual soil deformations on structural foundations and soil-foundation interactions in liquefied deposits will be addressed, followed by countermeasures to mitigate the liquefaction damage.

## 5.10.2 Post-liquefaction settlement

### *5.10.2.1 Case histories*

One of the methods to measure liquefaction induced soil settlements is to compare survey data before and after earthquakes after subtracting tectonic effects. Fig. 5.10.2 shows an example data along a national road (from distance 0 to 2.3 km) passing through an area in Niigata city extensively liquefied during the 1964 Niigata earthquake (Kokusho and Fujita 2002). The difference in elevations before and after the earthquake were very variable from less than zero to more than 1 m, partly because the soil in this area not only subsided but also flowed laterally as will be explained later. The average settlement calculated from the changes of elevations for all the survey points in the right bank of Shinano river for the distance 560 to 2300 m was 15 cm and that for the distance 560 to 1000 m was 40 cm. The soil thickness liquefied during the earthquake estimated by the Japanese design code (JRA 2002) based on soil profiles and SPT $N$-values varied from 5 to 15 m in the same distance (Fujita 2001).

On the other hand, soil subsidence relative to foundations supported by tip-bearing piles can serve as a simple scale to evaluate liquefaction-induced vertical settlements by assuming the vertical rigidity of pile foundations. Fig. 5.10.3 exemplifies settlement histograms measured relative to a number of piles in two manmade islands filled by

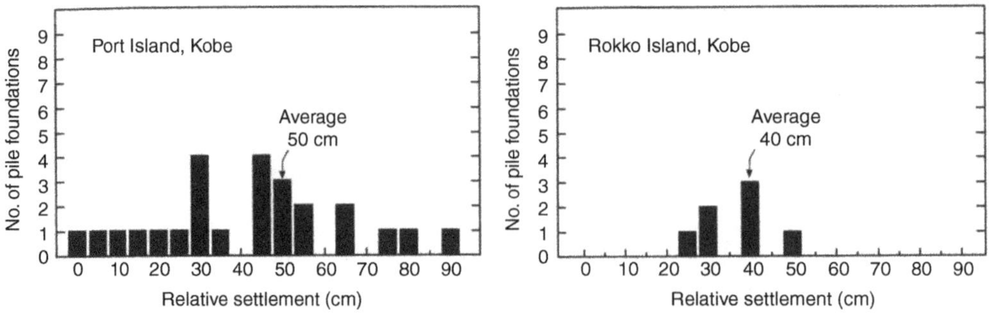


*Figure 5.10.3* Histograms of soil settlements relative to nearby pile foundations in liquefied manmade islands during 1995 Kobe earthquake (Ishihara et al. 1996)

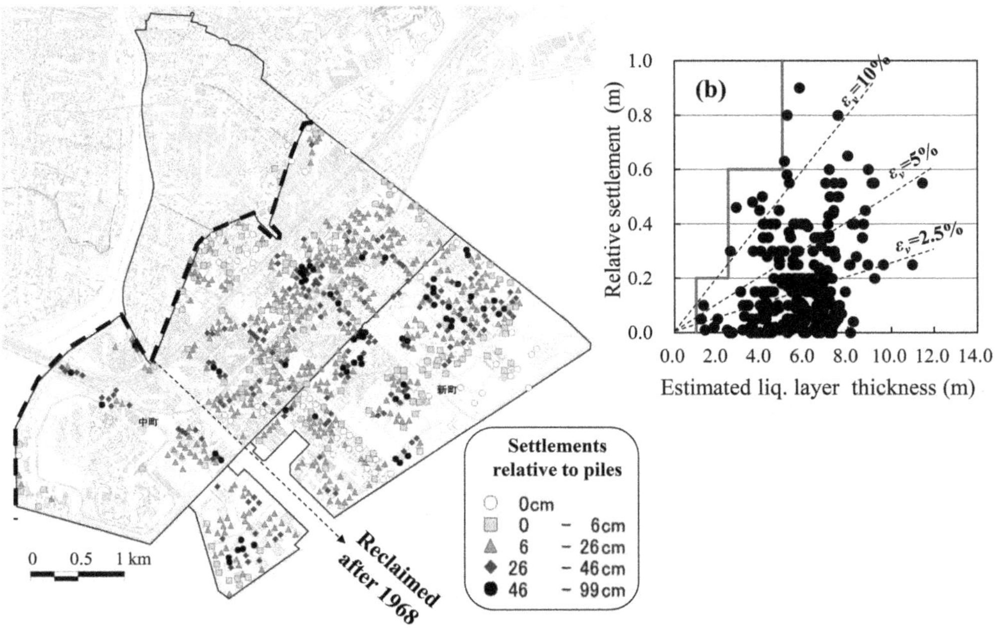


*Figure 5.10.4* Soil settlements relative to nearby pile foundations in Urayasu reclaimed areas during 2011 Tohoku earthquake (a), and Plots of settlement versus estimated liquefied layer thickness (b) (Urayasu city office 2012).

decomposed granite soils liquefied during the 1995 Kobe earthquake. The average settlement is 40–50 cm, though the data largely disperse to the maximum 90 cm (Ishihara et al. 1996). In Fig. 5.10.4(a), soil settlements relative to bearing pile foundations are summarized during the 2011 Tohoku earthquake on the map of Urayasu city along the Tokyo bay where extensive liquefaction occurred in hydraulically filled areas (Urayasu city office 2012). It clearly indicates that the relative settlements occurred exclusively

*Table 5.10.1* Typical liquefaction-induced soil settlements and vertical strains.

| *Earthquake & Site* | *Soil conditions* | *Estimated liq. thickness* | *Settlement* | *Vertical strain* | *Reference* |
|---|---|---|---|---|---|
| 1944 Tonankai EQ. Nagoya city | Loose Holocene sand | 5 m | 40 cm | 8% | Kishida 1969 |
| 1948 Fukui EQ. Maruoka | Loose Holoc. fine sand | 4 m | 30 cm | 7% | Kishida 1969 |
| 1964 Niigata EQ. Right bank Shinano river | Loose Holoc. sand | 5–15 m | 15–40 cm | 1–8% Aver.: 3% | Kokusho and Fujita 2002 |
| 1968 Tokachi-oki EQ. Aomori school yard | Very loose backfill sand | 5 m | 50 cm | 10% | Yoshimi 1970 |
| 1995 Kobe EQ. Port Island | Decomposed granite | 15 m | 0–90 cm Aver.: 50 cm | 0–6% Aver.: 3% | Ishihara et al. 1996 |
| 2011 Tohoku EQ. Urayasu reclaimed areas | Loose hydr. fill with fines | 5–8 m | 0–90 cm Aver.: 30 cm | 0–>10% Aver.: 5% | Urayasu city Office 2012 |

in the areas newly reclaimed after 1968 with the maximum settlement far exceeding 46 cm. In this area, the spatial distribution of ground surface settlements was also analyzed from air-photographs taken before and after the earthquake, which also indicated that the maximum settlement exceeded 50 cm in the reclaimed areas (Urayasu city office 2012). By assuming that the hydraulically-filled layer was responsible for the liquefaction, the settlements were correlated in Fig. 5.10.4(b) with liquefiable layer thicknesses estimated from numerous bore-hole data available in those areas. There seems to be a trend shown with the steps of solid line that the maximum settlements increase with increasing layer thickness (Urayasu city office 2012). The dashed lines in the diagram indicate proportional relationships between relative settlement and liquefiable layer thickness for the vertical soil strain $\varepsilon_v = 2.5$, 5, 10%. It is found that a lot of plots exceed $\varepsilon_v = 5\%$ and even $\varepsilon_v = 10\%$. This large settlement may be partially attributable that considerable volume of ejecta erupted and covered wide areas due to intensive liquefaction during this earthquake.

Table 5.10.1 summarizes typical values of liquefaction-induced soil settlements during previous earthquakes. Here, the vertical strain in liquefied layer was calculated by dividing the ground surface settlement by the corresponding thickness of liquefiable layer (Kishida 1969, Yoshimi 1970, Kokusho and Fujita 2002, Ishihara et al. 1996, Urayasu city office 2012). From the table and other data above, it may be said that the liquefaction induced settlement strains, though very variable presumably due to spatial variations of soils and associated lateral displacements, tend to be more than a few percent on average and exceed well over 5% up to 10% at the maximum.

#### *5.10.2.2 Post-liquefaction settlement by element tests*

Post-liquefaction soil settlements in a level ground may be reproduced in simple shear tests in the laboratory. $K_0$-consolidated soil specimens are liquefied by cyclic loading in the undrained condition and then reconsolidated to measure vertical settlement strains. A series of such stress-controlled cyclic simple shear tests were conducted

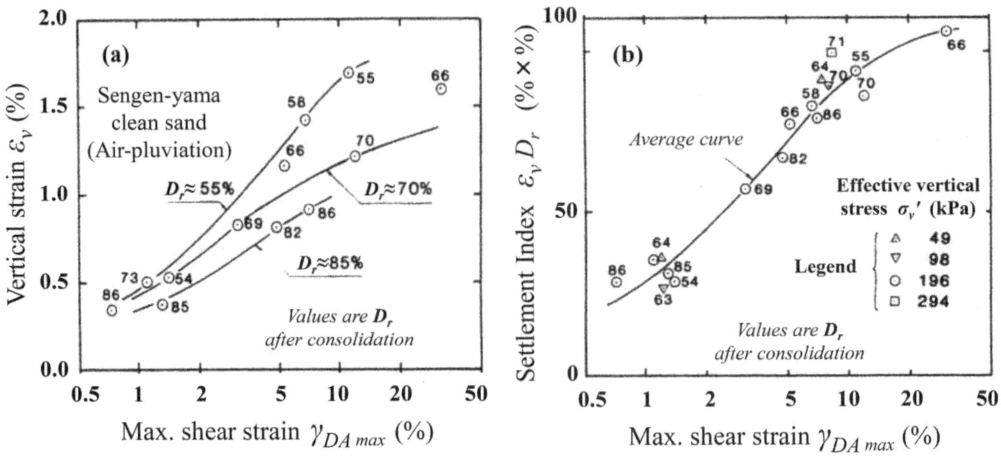


*Figure 5.10.5* Torsional cyclic shear test results on $K_{0\text{-consolidated}}$ clean sand specimens: (a) Vertical strain $\varepsilon_v$ versus maximum shear strain $\gamma_{DAmax}$, (b) Settlement index $\varepsilon_v D_r$ versus $\gamma_{DAmax}$ (Sasaki et al. 1982).

(Sasaki et al. 1982, Tatsuoka et al. 1984) using the torsional simple shear device on clean sand specimens (Sengen-yama sand) consolidated vertically by $\sigma_v' = 49{\sim}294$ kPa with lateral constraint by laminar metal rings. Fig. 5.10.5(a) shows that the vertical strains $\varepsilon_v$ plotted in the vertical axis have positive correlation with maximum double amplitude shear strains $\gamma_{DAmax}$ exerted during undrained cyclic loading (the last loading cycle) taken in the logarithmic horizontal axis depending on stepwise relative densities $D_r \approx 55$, 70, 85%. In Fig. 5.10.5(b), a settlement index $\varepsilon_v D_r$ introduced from the same dataset is plotted in the vertical axis versus $\gamma_{DAmax}$ almost uniquely for different $D_r$-values. This indicates that the vertical strain $\varepsilon_v$ tends to be inversely proportional to $D_r$. It also shows that the uniqueness of the relationship roughly holds for the wide range of overburden stress $\sigma_v' = 49{\sim}294$ kPa.

Similar simple shear tests were conducted on clean sand cyclically sheared one/two-directionally using several irregular seismic waves to simulate in situ seismic loading conditions more realistically (Nagase and Ishihara 1988). The sand specimen was isotropically-consolidated initially with $\sigma_c' = 196$ kPa while its lateral displacement was constrained by stacked annular plates. It was found that the vertical settlement strain is within 1% at the initial liquefaction (100% pressure buildup) and almost proportional to the excess pore-pressure until that moment as previously pointed out by Lee and Albeisa (1974). Also found was that the post-liquefaction settlement strain $\varepsilon_v$ after the 100% pressure buildup tends to increase almost uniquely with increasing maximum shear strain $\gamma_{max}$ induced by irregular loading irrespective of wave forms and one/two-directional loadings. Fig. 5.10.6 shows the results from the above-mentioned tests, wherein $\varepsilon_v$ tends to increase almost in proportion with single-amplitude maximum shear strain up to $\gamma_{max} = 8{\sim}10\%$ and then converge to a certain value depending on the relative density $D_r$. Based on the test results together with empirical correlations between $D_r$ and SPT $N_1$-values or CPT $q_c$-values, a design chart shown in

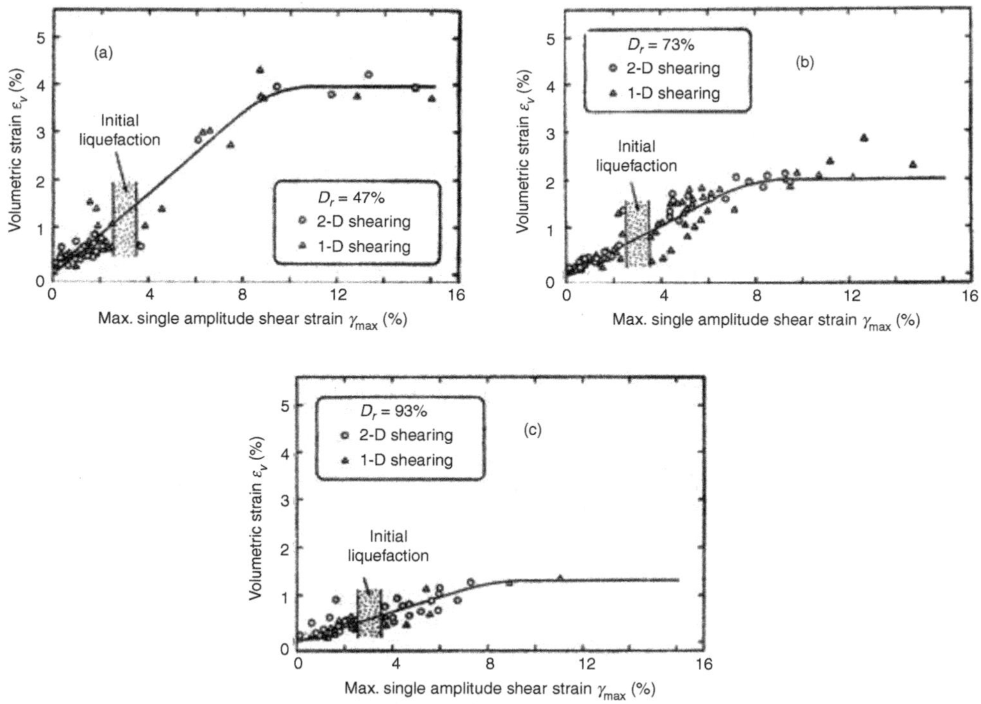


*Figure 5.10.6* Volumetric strains $\varepsilon_v$ versus max. shear strains $\gamma_{max}$ in one/two-directional cyclic simple shear tests on clean sand using irregular seismic waves: (a) $D_r = 47\%$, (b) $D_r = 73\%$, (c) $D_r = 93\%$ (Ishihara and Yoshimine 1992).

Fig. 5.10.7 was proposed by Ishihara and Yoshimine (1992) to evaluate the settlement strain $\varepsilon_v$ from $N_1$ or $q_{c1}$ and $F_L = CRR/CSR$ by assuming that $F_L \times (\gamma_{max}/3.5\%) = 1.0$, where $\gamma_{max}$ is in %, and 3.5% corresponds to nearly a half of the double amplitude shear strain $\gamma_{DAmax} = 7.5\%$. Herein, $F_L = 2.0$ was considered to correspond to the zero pore-pressure or $\varepsilon_v = 0$ condition. The design chart, though being based on laboratory tests on reconstituted clean sands, was applied to actual case histories of liquefaction-induced soil settlements during the 1964 Niigata earthquake and found to give reasonable evaluations within the range of observed values (Ishihara and Yoshimine 1992).

A similar research on the volumetric strain was conducted by triaxial tests in which sand specimens with $D_r = 70{\sim}100\%$ isotropically consolidated with $\sigma'_c = 49$–$196$ kPa were cyclically loaded in the undrained condition to certain axial strains followed by the drainage of pore-water to measure the volumetric strains $\varepsilon_v$ (Kokusho et al. 1983a). In Fig. 5.10.8, the volumetric strains $\varepsilon_v$ are plotted in the vertical axis versus the double amplitude maximum shear strains $\gamma_{DAmax}$ determined from the double amplitude maximum axial strain $\varepsilon_{DAmax}$ as $\gamma_{DAmax} = 1.5\varepsilon_{DAmax}$ and normalized by $D_r$ in the horizontal axis. The graph (a) is based on the same test data as already explained in Fig. 5.3.9(b) on reconstituted specimens of Toyoura clean sand prepared by WT and

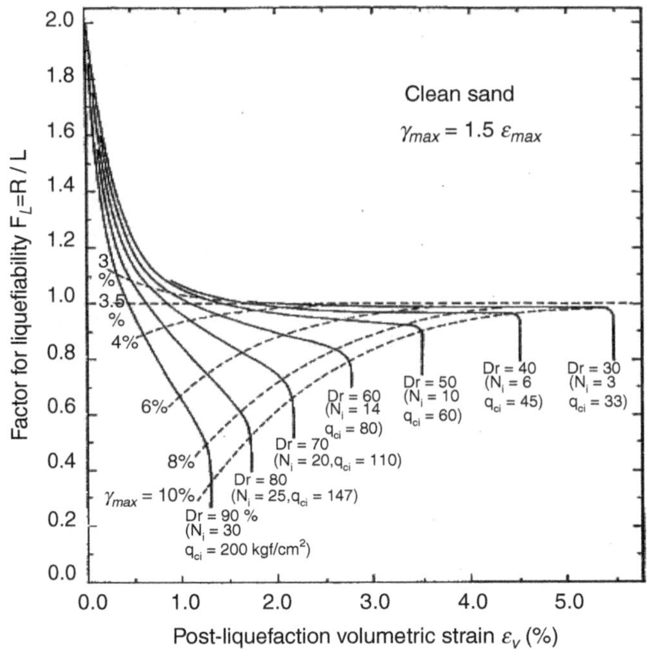


*Figure 5.10.7* Chart for determination of post-liquefaction volumetric strain as a function of factor of liquefiability $F_L$ by SPT $N_1$-value (Ishihara and Yoshimine 1992).

AP methods and given with various types of stress/strain histories. The plots may be approximated by the straight line through the origin, $\varepsilon_v(\%) = 8.5(\gamma_{\max}/D_r)$, despite data scatters, indicating that $\varepsilon_v$ is roughly proportional to maximum shear strain $\gamma_{\max}$ and inversely proportional to $D_r$ for the same $\gamma_{\max}$ as indicated by Tatsuoka et al. (1984).

It is noteworthy that all the plots are almost uniquely correlated with $\gamma_{\max}/D_r$ without systematic trends of deviations, implying that soil fabric depending on sample preparation methods, mechanical disturbances and overconsolidation have little to do with the post-cyclic loading volumetric strains. Also noted is that the plots from the triaxial test seem to be compatible with those from the torsional simple shear test for $K_0$-consolidated specimens using another poorly-graded clean sand (the cross symbol) by Tatsuoka et al. (1984) already addressed in Fig. 5.10.5, suggesting that the two test methods tend to yield similar results despite quite different test conditions.

On the other hand, Fig. 5.10.8(b) shows similar triaxial test results on intact sands ($F_c = 10\%$, $C_u = 4.8$) taken from Narita Pleistocene layer near Tokyo by block sampling. The intact sands, weakly cemented, were first tested and then the same specimens, remolded and reconstituted to be the same density, were tested again, under three confining stresses, $\sigma_c' = 49$, 98, 196 kPa. Again, the plots may be approximated by a straight line through the origin but with a quite different gradient,

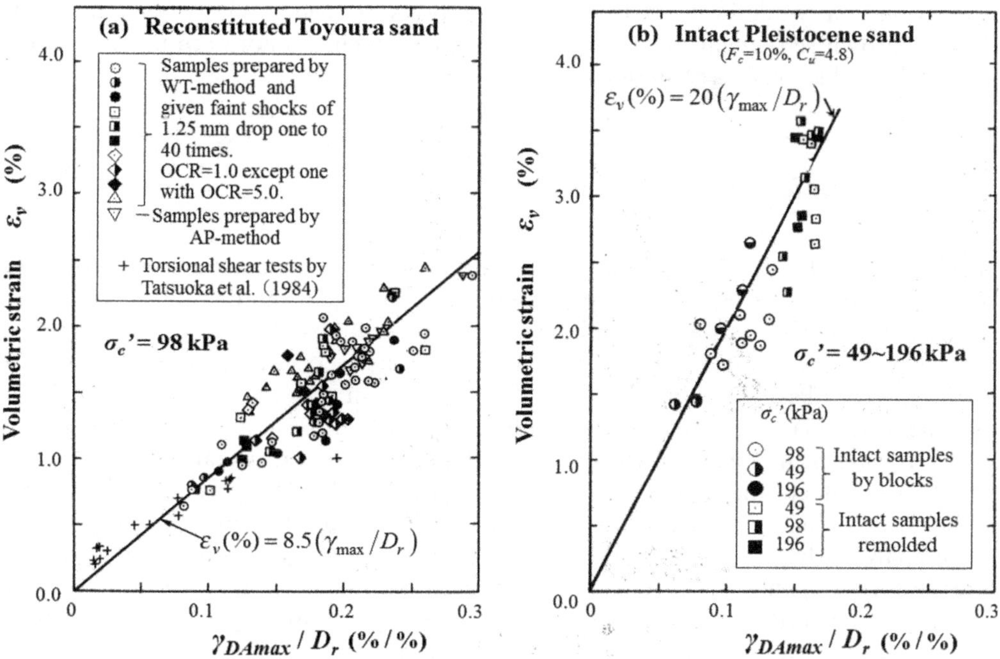


*Figure 5.10.8* Volumetric strain $\varepsilon_v$ versus max. shear strain $\gamma_{DAmax}$ divided by $D_r$: (a) Reconstituted Toyoura clean sand with various soil fabric, (b) Intact Pleistocene sand under different confining stresses (Kokusho et al. 1983a).

$\varepsilon_v(\%)=20(\gamma_{max}/D_r)$, indicating that the relationship is not unique but quite soil-specific. Presumably, larger $F_c$ and $C_u$ in Narita sand in Fig. 5.10.8(b) than Toyoura clean sand in Fig. 5.10.8(a) may be responsible for the larger gradient. Hence it is recommended in evaluating liquefaction-induced settlements to test site-specific soils or refer appropriate test data using soils with similar physical properties. Another important finding is that in situ soil fabric seems to have little impact on the relationship, presumably because the soil behavior becomes almost independent of subtle soil fabric once large strain develops after initial liquefaction. It is also interesting to see that the different confining stresses result in little notable differences in the approximation line, implying that the liquefaction-induced vertical strain may be calculated from seismically-induced maximum shear strain and relative density regardless of soil depths and the degree of soil disturbance once the soil-specific $\varepsilon_v$ versus $\gamma_{max}/D_r$ correlation becomes available.

Thus, the negative dilatancy due to cyclic loading is the major mechanism for the liquefaction-induced settlement. However, some of recent earthquakes have demonstrated that sand boiling during extensive liquefaction may increase soil settlement to a measurable degree. During the 2011 Tohoku earthquake near Tokyo and the series of 2011–2012 seismic events near Christchurch, a huge amount of ejecta was erupted widespread and covered the ground surface 10–20 cm thick. In the former case, the long duration of motion due to the $M_J = 9.0$ earthquake is suspected to have a certain impact on that, while in the latter the artesian pressure seems to have been involved. The additional settlement due to the sand boiling, which has not been taken account so far, has to be considered in design if necessary.

### 5.10.3 Liquefaction-induced lateral flow

The terminologies, lateral spreading and lateral flow, tend to be used without clear distinctions in their definition, though the former seems to include both non-flow and flow type failure. The lateral flow displacement in liquefied ground is immensely larger than the free surface settlement and may exceed several meters.

As typical flow failures that occurred in gently inclined slopes, large lateral deformations were observed along the Shinano river bank in Niigata city during the 1964 Niigata earthquake in Japan (Kawakami and Asada 1966), and along the beach and riverside near Anchorage during the 1964 Alaskan earthquake in USA (Seed 1968, McCulloch and Bonilla 1970). During the 1964 earthquakes and the 1983 Nihonkai-Chubu earthquake, large lateral flows of very gentle slopes occurred, which were analyzed by air-photographs (Hamada 1992). Liquefaction-induced flow failures occurred in the Lower San Fernando Dam with delayed time during the 1973 San Fernando earthquake (Seed et al. 1975b, Seed 1979) highlighted the peculiarity of this failure type. It may occur not only during but also after earthquake shaking. For example, the girders of Showa-Ohashi Bridge crossing the Shinano river fell down due to the flow failures of liquefied river bed during the 1964 Niigata earthquake. A taxi driver incidentally crossing the bridge during shaking witnessed that the failure occurred a few minutes after the end of the shaking. During the 1987 Edgecumbe earthquake in New Zealand, a bridge became out of service on account of lateral flow of the foundation ground about one hour after the cease of the earthquake shaking (Berrill et al. 1997). Hence, the lateral flow is considered to have actually been driven by the gravity force although it was initially triggered by the seismic effect. More recently, lateral spreading deformations occurred during 2010 and 2011 earthquakes near Christchurch New Zealand along river channels. They were surveyed by means of modern surveying techniques to yield a large database on permanent ground displacements (Robinson et al. 2014).

As another type of flow failures in flat lands behind retaining structures, liquefied manmade deposits behind quay walls or retaining walls underwent large lateral deformations triggered by seismically-induced wall movements toward the sea during the 1995 Kobe earthquake in Japan (Inagaki et al. 1996, Ishihara et al. 1996). It translated bridge piers or building foundations that were located near the retaining structures causing the falls of bridge girders, the deformations in superstructures or the ruptures of pile foundations. Similar damage occurred in the Akita Harbor during the 1983 Nihonkai-Chubu earthquake, in the Kushiro Harbor during the 1993 Kushiro-oki earthquake in Japan, and also in the Taichung Harbor during the 1999 Chi-Chi earthquake in Taiwan (Lee et al. 2000).

In the following, the two types of flow-failure are overviewed in typical case histories. Then the void-redistribution mechanism involved in many liquefaction-induced flow failures in gentle slopes will be discussed in more details.

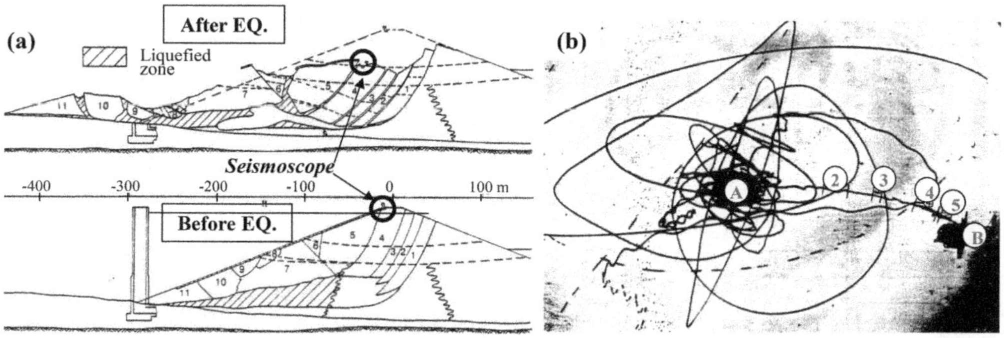


*Figure 5.10.9* Cross-sectional view of Lower San Fernando dam before and after 1973 San Fernando earthquake (Seed et al. 1975b) with permission from ASCE (a), and Two-dimensional seismometer records installed at dam crest (Seed 1979) by permission of ICE Publishing (b).

### *5.10.3.1 Case histories of lateral flow in gentle slopes*

Fig. 5.10.9(a) shows a cross-sectional view of the failure in the Lower San Fernando dam occurred during the 1973 San Fernando earthquake. The flow failure occurred in the upstream slope constructed by hydraulic filling. It was recorded by a seismoscope (a two-dimensional seismometer pen-recording ground motions on the horizontal plane) installed at the dam crest as shown in Fig. 5.10.9(b). This evidenced that the flow failure started 30 seconds after the end of shaking at Point A and continued to incline from 0 to 26° at Point B (Seed 1979) in the period of 50 seconds. The upstream slope of the dam was actually not so gentle (about 24°), but sublayers in the lower dam body formed by the hydraulic filling seem to have been gently inclined as suggested in the sketch of liquefied zones in (a).

The flow failure in liquefied soil deposits seems to have been involved also in submarine slides on-shore or off-shore triggered seismically. For example, Valdez and Seward, port cities in Alaska, USA, suffered great loss of human lives and properties by large scale submarine slides involving coastal areas (Coulter et al. 1966, Lemke et al. 1967). The inclination of the sea bed was 5° or less on average in the long slip surface of more than 2 km offshore from the beach, which is considerably less than the internal friction angle of the soil. The analogous on-shore failure occurred during the 1999 Kocaeli earthquake along the southern coast of the Izmit bay in Turkey (JGS 2000). The slope of the sea-bed originally about 5° consisting of coarse sand and gravel slid together with buildings near the beach. In another submarine slide which occurred 60 km off California coast during a 1980 medium magnitude earthquake (Field et al. 1982), a sea floor 2 by 20 km consisting of interbedded sand and mud with the seabed inclination only 0.25 degrees slid and became further flat with evidences of liquefaction such as sand boils on the sea floor.

During the 1964 Niigata earthquake, large lateral flows occurred in liquefied areas despite the apparent flat ground surface. Figs. 5.10.10(a) and (b) show the maps of Area-1 and 2, respectively, where the maximum displacement exceeded 4 m despite

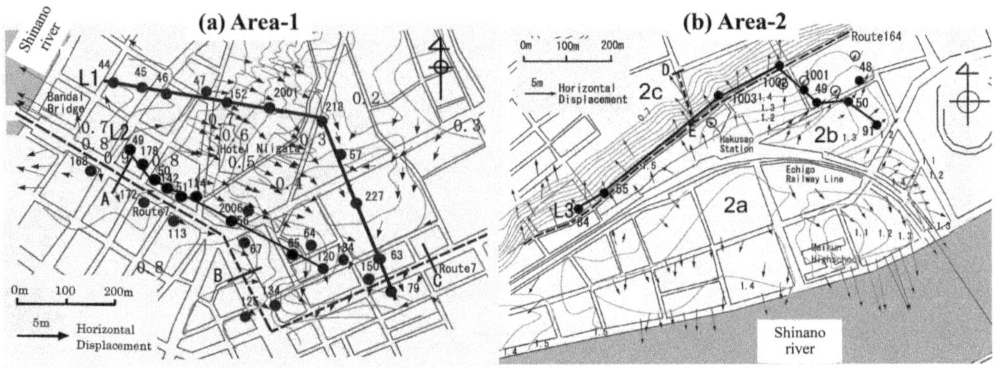


*Figure 5.10.10* Maps of Area-1 (a) and Area-2 (b) in Niigata city with elevation contours of 0.1 m (Kokusho & Fujita 2002) with displacement vectors of the ground surface (Hamada 1992) with permission from ASCE.

the very gentle slope of 1% or less (Kokusho and Fujita 2002). The elevation contours of 0.1 m pitch are superimposed on the maps together with liquefaction-induced displacement vectors of the ground surface calculated from air-photographs before and after the earthquake (Hamada 1992). It is remarkable that the displacement vectors, except nearby the Shinano river bank, are directing down-slope almost normal to the contours despite the very gentle slope of 1° or lower.

Borehole logging data in these areas were compiled as two-dimensional soil profiles shown in Figs. 5.10.11(a)~(c); along the lines L1, L2 in Area-1, and L3 in Area-2 drawn in Fig. 5.10.10. The soils essentially consisted of loose sands down to a depth of about 10 m with $N$-values less than 10 and the water table was located mostly around GL.-2 m from the ground surface. Soils in the liquefied areas were clean sands with fines content less than 5% as actually measured and plotted in Fig. 5.10.16(b). However, one or more sublayers with fine soils of either silt or clay were interbedded in the most soil profiles. The sublayer thicknesses were very variable, and in a few boreholes no distinctive fine soil sublayers were identified possibly because there exists a spatial discontinuity or they were too thin to be detected by normal borehole logging. The liquefaction susceptibility of the sandy soil was evaluated using the Japanese design code (JRA 2002) along the lines L1, L2 and L3 based on peak ground acceleration (PGA) 0.17 g recorded during the earthquake. The layers thus judged to be liquefiable were shaded in Fig. 5.10.11.

The down-slope flow displacements $D_{fn}$ normal to the contours were calculated from the displacement vectors in Fig. 5.10.10 and plotted versus the maximum surface inclinations $\beta_{\max}$ in Fig. 5.10.12(a). Here, the open and close symbols correspond Area-1 and Area-2 (2b, 2c), respectively. Obviously, the plots in the enclosed zones seem to indicate linear correlations despite large data scatters. Consequently, even slight surface gradients of less than 1% seem to have had a great influence on the lateral displacements in these areas, though the two linear correlations are distinctively different. This difference may be attributable to the fact that the average $N_1$-value in Area-1 was smaller than Area-2b,2c as shown in the $N_1$-histogram of Fig. 5.10.12(b),

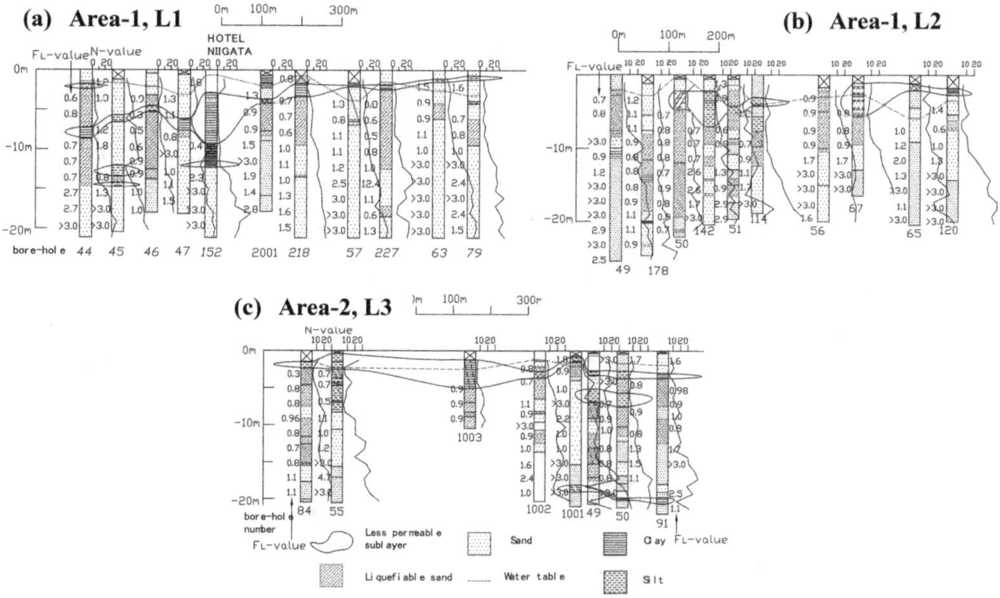

*Figure 5.10.11* Borehole logging data and two-dimensional soil profiles in Areas-1 and 2: (a) Area-1, Line L1, (b) Area-1, Line L2, (c) Area-2, Line L3 (Kokusho and Fujita 2002) with permission from ASCE.

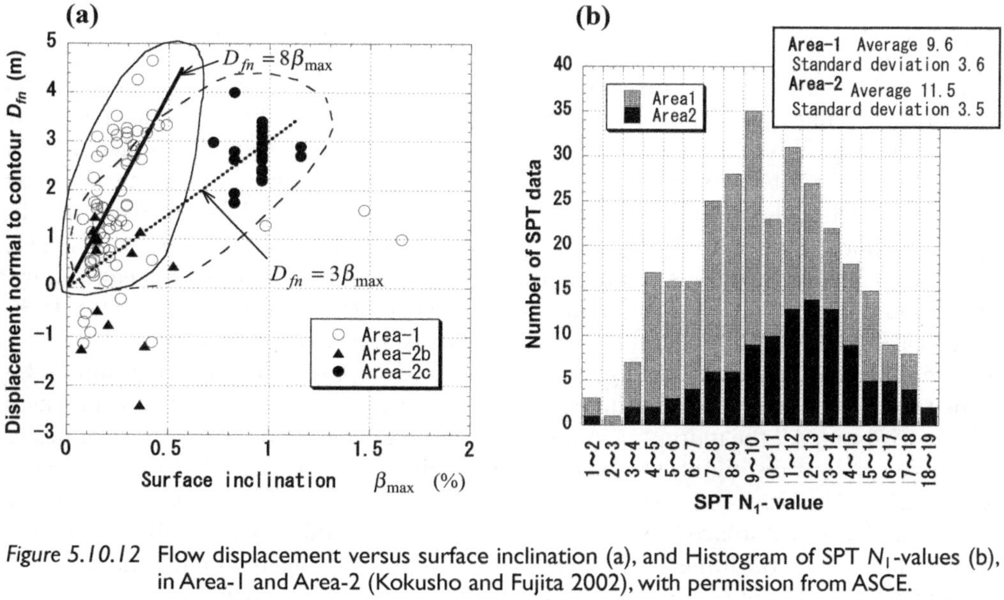

*Figure 5.10.12* Flow displacement versus surface inclination (a), and Histogram of SPT $N_1$-values (b), in Area-1 and Area-2 (Kokusho and Fujita 2002), with permission from ASCE.

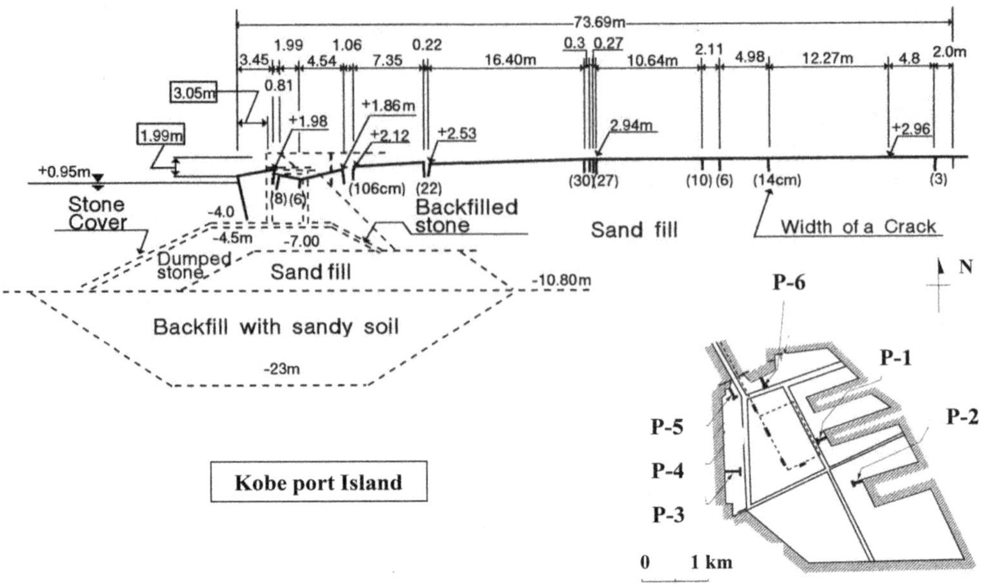


*Figure 5.10.13* Ground deformations in Port Island along a line normal to P-6 (Ishihara et al. 1996).

indicating that Area-1 was more prone to severe liquefaction than in Area-2 (Kokusho and Fujita 2002).

### 5.10.3.2 Case histories of lateral flow behind retaining walls

Unlike the Niigata city case mentioned above, manmade lands behind retaining walls are normally flat and free from sustained initial shear stress before liquefaction. If the retaining wall is translated or tilted forward during an earthquake, liquefied soils behind it tend to flow toward the displaced walls due to newly-induced shear stress by gravity. Thus, the mechanism for lateral flow in this case is different from that in gently inclined slope without retaining structures. As a typical example of this case, Fig. 5.10.13 shows the cross-section along a line normal to one of the displaced quay walls and the manmade ground behind in Kobe Port Island before and after the 1995 Kobe earthquake (Ishihara et al. 1996). The Port Island was constructed directly on a soft marine clay seabed of 10–15 m water depth by dumping decomposed granite (DG) soils transported from quarries in nearby mountains to make the total soil thickness about 15–17 m. All the margins around the island were retained by a number of heavy concrete caissons resting on the DG soils backfilled in seabed trenches to replace the soft marine clay. During the earthquake, a caisson quay wall P-6, shown in the figure for example, was displaced by about 3 m horizontally and 2 m vertically due to the seismic inertia and the liquefaction of the DG soils placed beneath and behind the wall. This inflicted horizontal displacements as well as vertical settlements in the manmade ground behind the wall together with a set of cracks parallel to the sea-line not only near the wall but far behind as sketched in the figure (Ishihara et al. 1996).

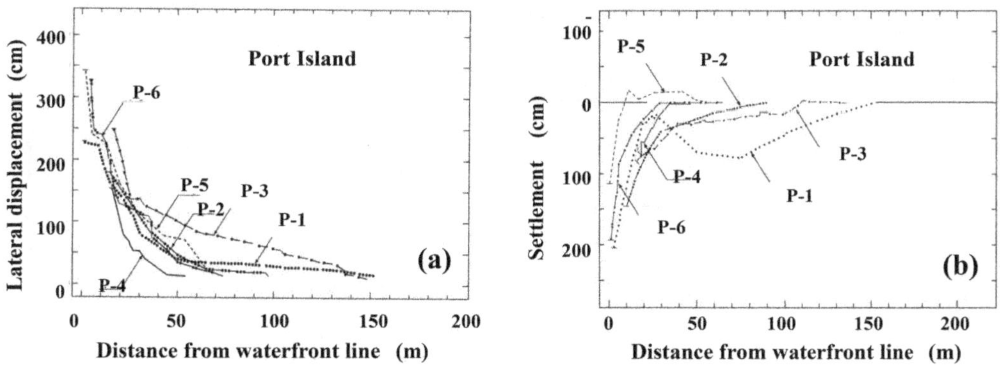


*Figure 5.10.14* Ground deformations versus distance from waterfront normal to quay walls: (a) Lateral displacement, (b) Settlement (Ishihara et al. 1996).

Fig. 5.10.14 shows the horizontal displacements and vertical settlements versus the horizontal distance from the sea-line along six survey lines (P-1 to P-6) normal to the quay walls around the Port Island. The displacements are highly concentrated just behind the walls and tend to decay rapidly with the distance from the walls, though minor influences tend to reach as far as 50 to 150 m. It is also observed that if the horizontal displacements are larger in a certain survey line, then the corresponding vertical displacements are also larger than in other lines, implying a certain correlation may exist between them.

Thus, the displacements concentrated behind displaced retaining structures seem to be quite different from the more or less widespread displacements observed in gently inclined liquefied ground in Niigata city. It may be explained by the different distributions of induced shear stress; namely in Niigata city, initial shear stress works all along the gentle slopes, while in Kobe PI, the shear stress is stronger near the displaced wall and tends to decay with distance. However, it may be pointed out that the displacement decays in Fig. 5.10.14 may have been different if the PI manmade ground was constructed by the hydraulic filling rather than the dumping method. In the hydraulic filling method, soils tend to be much more stratified and likely to develop water films due to the void redistribution to be discussed next. If this had happened, the horizontal displacements would have reached in a longer distance from the sea-line.

### 5.10.3.3 Void redistribution mechanism

The mechanism that leads to large liquefaction-induced lateral flow displacements in gentle slopes is still only poorly understood. While lateral displacements employed in current design methods may be assigned based on experiences in previous earthquakes, it lacks a theoretical basis from the mechanical point of view. As already discussed in Sec. 5.8 on the undrained shear mechanism in saturated sands, clean sands in shallow depths such as in Niigata city are rarely on the contractive side of SSL even for loose conditions with $D_r$ around 30%, and a mixture of non/low-plasticity fines is necessary for sands to be contractive and to occur flow-type failures under very low driving

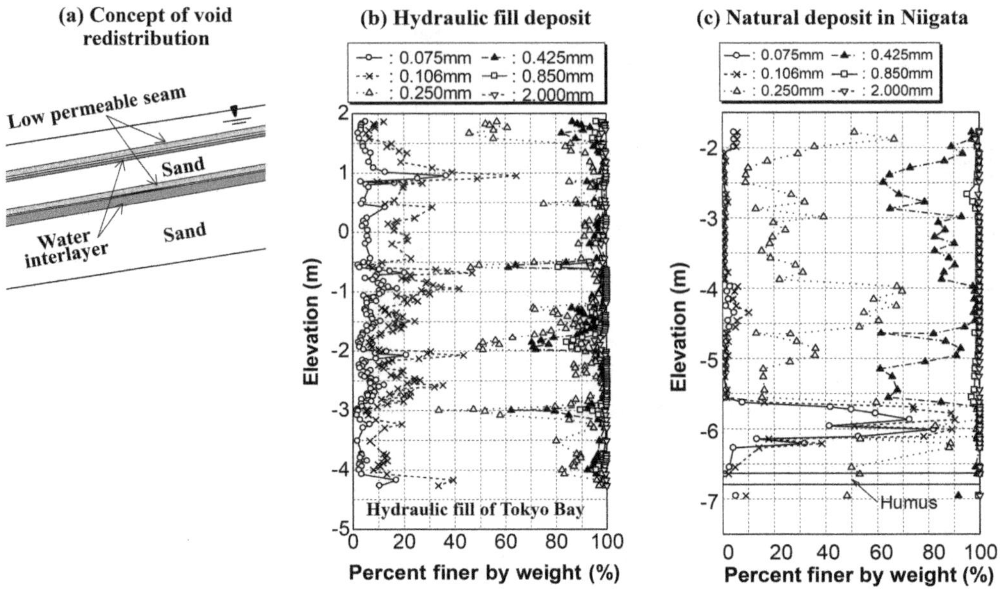


*Figure 5.10.15* Concept of void redistribution (a), and Sieving test results at two sites; hydraulic fill deposits (b) and natural sand deposits (c). (Kokusho and Kojima 2002) with permission from ASCE.

stresses. For example, centrifuge model tests on gently inclined saturated clean sand deposits under the application of sinusoidal motions in the sloping direction (e.g. Dobry et al. 1995) exhibit the dilative cyclic mobility response. Thus, liquefied clean sands under the influence of initial shear stress will not undergo flow-type displacements unless the dilative response is diminished. In this respect, it was shown that the dilative response of clean sand may reduce due to superimposed complementary cyclic shear stresses representing small aftershocks persisting after strong main shocks (Meneses et al. 1998). However, it is unlikely that this effect can be a major mechanism for large flow displacements in general.

In this regard, a concept of void redistribution as illustrated in Fig. 5.10.15(a) was addressed in a committee report in US (NRC 1985) and also discussed by Seed (1987) by using a special term, "water interlayer". Namely, silt seams sandwiched in sand deposits or silty sublayers capping sand deposits may cause the void redistribution wherein the liquefied sands beneath the silts tend to settle in exchange for upcoming pore-water and generate the water interlayers or "water film" eventually. This will considerably lessen the soil stability even in very gently-inclined slopes and may trigger a flow-type failure along the water films.

A sand layer, though represented by a single uniform layer, may comprise multiple sublayers with different grain sizes and silt seams which are too thin to detect in normal soil investigations. Figs. 5.10.15(b), (c) show soil stratifications investigated in situ by sieving tests in two sand deposits (Kokusho and Kojima, 2002): a hydraulically filled deposit along Tokyo Bay in (b), and a natural sand deposit in Niigata city

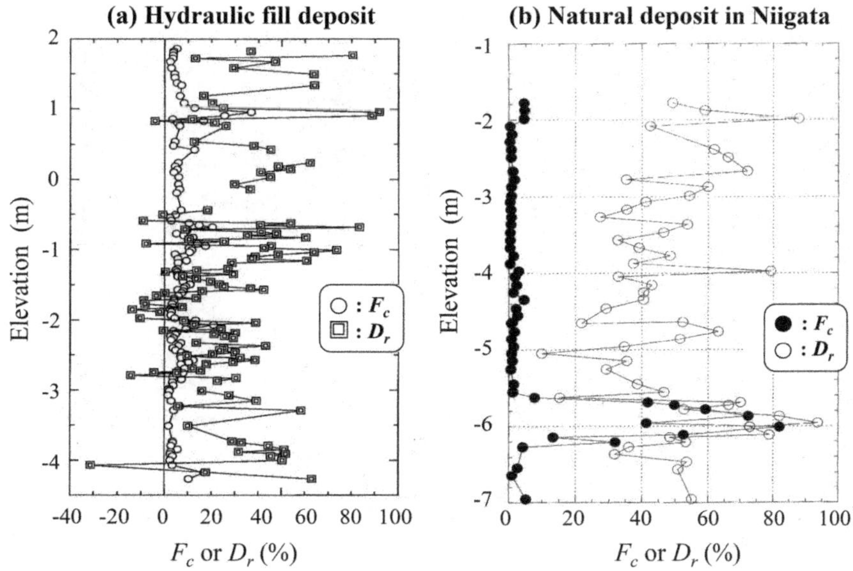


*Figure 5.10.16* Variation of relative density $D_r$ and fines content $F_c$ along elevation: (a) Hydraulic fill deposits, (b) Natural sand deposits (Kamikawa 2004).

where extensive liquefaction occurred during the 1964 Niigata earthquake in (c). The percentage finer by weight at each mesh size is plotted for the soil slice of 2 cm thick each versus its elevation. In the hydraulic fill deposit, the soil is highly variable and the fines content $F_c$ corresponding to the mesh size of 0.075 mm (#200 sieve) is fluctuating almost periodically by an interval shorter than 2 m together with other particle sizes changing accordingly. Normal borehole logging is likely to overlook such alternating thin fine layers and misinterpret the soil as uniform silty sand. In Niigata city, the soil is rather uniform, consisting of clean sand down to elevation GL.-5.6 m, and below that a silty or clayey layer about 0.6 m thick and a humus layer 15 cm thick appear. These low-permeability layers were confirmed to be continuous in the horizontal direction to at least 20 m. Figs. 5.10.16(a), (b) show relative densities $D_r$ and fines contents $F_c$ along the elevation in the same hydraulic fill and natural sand, respectively. In the fill, $D_r$ is around 30% on average though very fluctuating because of the large content of fragmented seashells, and sand sublayers with $F_c \approx 5\%$ sandwich many silty seams with $F_c$ up to 40%. In Niigata, $D_r$ is 40% on average except near the surface for the very clean sands with $F_c$ less than a few percent sandwiching a silt/clay layer of 1 m thickness with $F_c$ up to 80%.

It may be expected that the interbedded silty sublayers or seams result in differences in permeability in liquefied deposits, causing void redistribution in the form of water interlayers or water films. The formation of water films during liquefaction beneath low-permeability seams was observed in a number of model tests (e.g. Scott and Zuckerman 1972, Elgamal et al. 1989, Dobry et al. 1995, Kokusho 1999). The test results clearly indicated that the water films are readily formed after the onset of

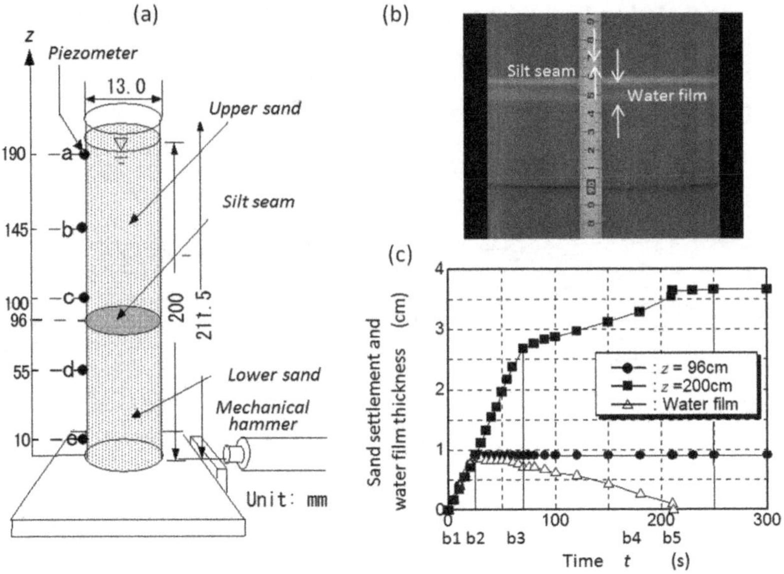

*Figure 5.10.17* One-dimensional liquefaction test of saturated sand layer sandwiching a silt seam: (a) Test equipment of lucite tube, (b) Photograph of water film beneath silt seam, (c) Time history of sand settlement and water film thickness. (Kokusho 2000b), with permission from ASCE.

liquefaction in loose sands beneath sandwiched seams of lower permeability because of pore-water migration or void redistribution and stay there much longer than the re-sedimentation of liquefied sand particles. As an example of such tests, Fig. 5.10.17 shows typical post-liquefaction behavior in saturated sand in one-dimensional liquefaction tests (Kokusho 1999, Kokusho and Kojima 2002). A saturated loose sand layer 200 cm high was prepared in a lucite tube in the same method already explained in Fig. 5.1.1. A non-plastic silt seam of 4 mm thick was sandwiched in the middle of the sand (at $z = 96$ cm of the soil column) as illustrated in Fig. 5.10.17(a). The relative densities of the upper and lower sand layers $D_r = 14\%$ and 39%, respectively, was instantaneously liquefied by a hammer impact. Fig. 5.10.17(b) shows the photograph of the water film formed beneath the seam (Kokusho 2000b, Kokusho and Kojima 2002). In Fig. 5.10.17(c), the variation of water film thickness beneath the silt seam as well as the settlement at the top of the upper and lower layers (at $z = 200$ cm and $z = 96$ cm respectively) are plotted against the elapsed time from the impact. The water film starts to show up just after the complete liquefaction triggered by the shock and stays there for sometime.

Time-histories of excess pore-pressures measured at points a~e in Fig. 5.10.17(a) in the same test are depicted in Fig. 5.10.18(a). The same excess pore-pressure data are plotted again along the depth at multiple time steps in Fig. 5.10.18(b). As soon as the excess pressure builds up 100%, it starts to decrease from the bottom in the upper and lower layers concurrently. Liquefaction in the lower sand layer ends at b2, followed

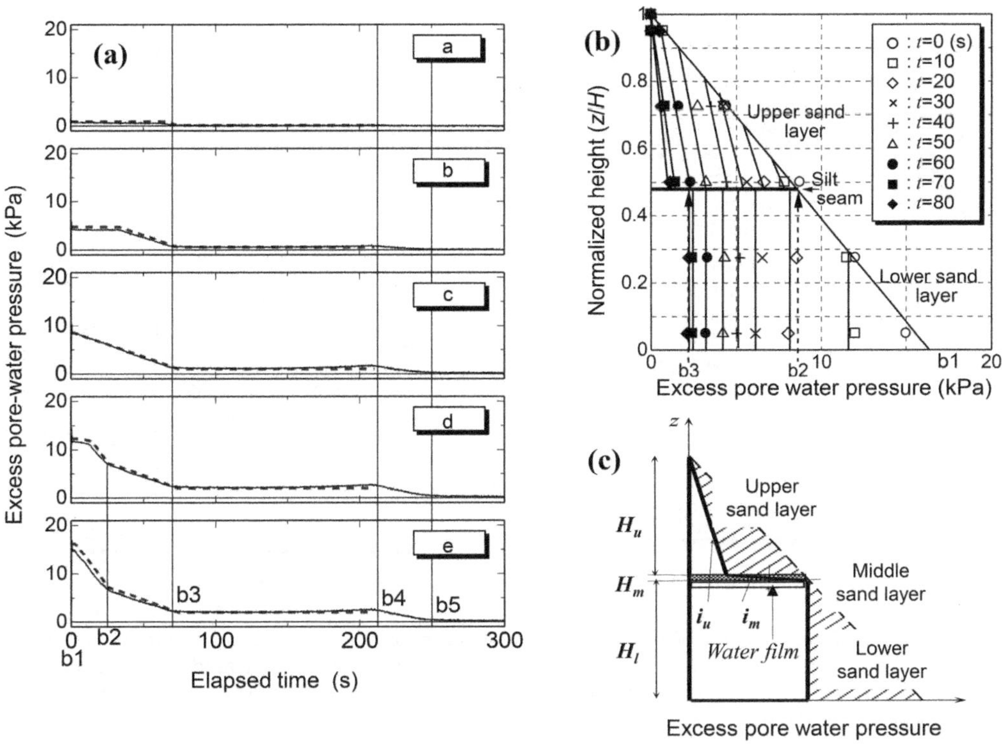


*Figure 5.10.18* Liquefaction test of sand layer sandwiching silt seam: (a) Time-histories of excess pore-pressure at 5 points, (b) Pore-pressure at time steps, (c) Schematic pressure distribution for water film generation. (Kokusho 2000b, Kokusho and Kojima 2002), with permission from ASCE.

by constant excess pore-pressure distribution throughout the lower layer. By extrapolating the pressure gradient from the upper and lower layers, there exists a distinctive discontinuity in pressure at the silt seam, introducing high hydraulic gradient in it. The water film continues to exist from b1 to b4 when all the water disappears there. Thus, a simple mechanism as schematically illustrated in Fig. 5.10.18(c) is working in generating a stable water film. The hydraulic gradient introduced in the middle silt seam, $i_m$, is expressed by the equation;

$$i_m = \frac{(\gamma'_u H_u + \gamma'_m H_m) - (\gamma'_u H_u - \sigma'_v)}{\gamma_w H_m} = i_{cr} + \frac{\sigma'_v}{\gamma_w H_m} \tag{5.10.1}$$

where $i_{cr} = \gamma'_m / \gamma_w$ is the critical hydraulic gradient, $\sigma'_v$ = effective vertical stress at the bottom of the upper layer, $i_m$, $i_u$ = hydraulic gradients, $H_m$, $H_u$ = thickness of those layers, $\gamma'_m$, $\gamma'_u$ = buoyant unit weights in the middle and upper layer, respectively, and $\gamma_w$ = unit weight of water.

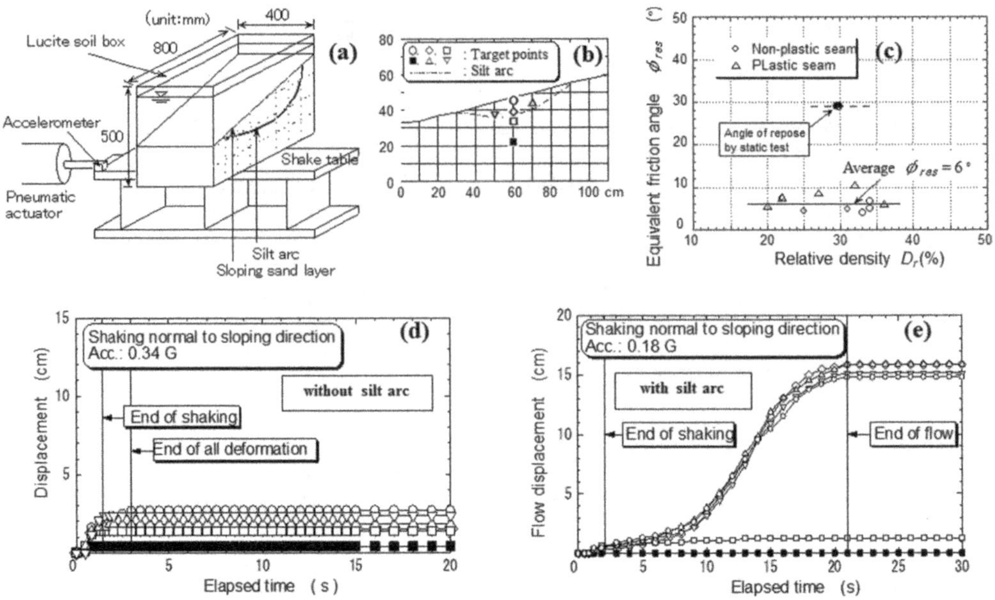


*Figure 5.10.19* Shake table test of 2-dimensional submerged sand slope without/with silt arc: (a) Model slope on shake table, (b) Target points of model, (c) Back-calculated friction angles, (d) Displacement time histories without silt arc (e) Displacement time histories with silt arc (Kokusho 2003).

Thus, a water film is readily formed just after the onset of liquefaction beneath a sandwiched sublayer with smaller permeability and stays there much longer than the re-sedimentation of liquefied sand particles. This indicates that the liquefied sand is actually in the drained condition locally, allowing the void redistribution to occur. Soil sublayers providing this mechanism seem to be abundant in the field, introducing multiple water films at different depths with different scales and durations. Kokusho and Kojima (2002) conducted the model tests on layered sands composed of different profiles, where stable water films almost always appeared beneath sublayers of lower permeability. Even if stable films did not appear, transient turbulence occurred near the boundary of sublayers leading to temporary instability. Thus, the void redistribution effect, stable water films or transient turbulence, will no doubt serve as slip planes if the soil has a sliding potential. In this regard, Naesgaard and Byrne (2005) suspected that another mechanism, called "soil mixing", may also be involved, leading to significant strength reduction due to volume contraction along a silt seam if the grain-size ratio between sand and silt satisfies a certain condition of mixing.

Fig. 5.10.19(a) demonstrates the effect of water film on slope instability in a 2-dimensional shaking table tests (Kokusho 2000b, Kokusho and Kojima 2002, Kokusho 2003, Kokusho 2006). Clean fine sand was rained in water to make a submerged loose sand slope in a rectangular lucite soil box. The sand slope was either entirely uniform, or sandwiching an arc-shaped silt seam near the surface. The model was subjected to 3 cycles of sinusoidal shaking perpendicular to the sloping direction and the displacement was monitored through the transparent side wall.

Figs. 5.10.19(d), (e) depict time-dependent displacements of the sand slopes without and with the silt arc, respectively, at representative target points shown in Fig. 5.10.19(b) with various symbols. The displacement in the homogeneous slope without the silt arc is limited and occurs mostly during shaking of the acceleration amplitude of 0.34 g. In the slope with the silt arc, the large flow displacement occurs after the end of shaking despite the smaller input acceleration of 0.18 g. The post-shaking delayed flow displacement occurs quite discontinuously along the silt arc with no displacement below that.

A basic question is why the clean sand on the dilative side of the Steady State Line under the low confining stress will not absorb ambient excess pore-water, blocking the flow displacement. The comparative observation of Figs. 5.10.19 (d) and (e) indicates that a water film, if formed beneath the seam, serves as a shear stress isolator which shields the deeper soil from the initial shear stress of the sloping ground, impeding the development of shear strain and positive dilatancy there (Kokusho, 2000b).

If the sliding occurs along the arc all through continuous water film, the residual strength would be zero. The residual friction angle actually back-calculated during the flow in the model test was 6° on average as plotted in Fig. 5.10.19(c) and almost independent of the sand density, plasticity of the silt seam and other parameters (Kokusho 2006). This is probably because the sliding can start as soon as the resistance becomes lower than the initial shear stress before the water film is fully developed, keeping the sand beneath the water film slightly dilative because of imperfect shielding from the shear stress due to imperfect development of the water film.

Fig. 5.10.20(a) shows an innovative undrained cyclic loading soil element test to simulate the void redistribution in a liquefied sand specimen using a hollow cylindrical torsional simple shear apparatus (Kokusho et al. 2003). The hollow specimen (100 mm/60 mm in the outer/inner diameters, and 200 mm in height) is supposed to represent a soil element extracted from a horizontal or gently-inclined sand layer beneath a low permeable seam as schematically illustrated in Fig. 5.10.20(b). The movement of the top loading plate above the specimen, representing the silt seam capping the liquefiable sand, was vertically restricted to reproduce the low permeability of the overlying seam. A light emitting diode (LED) was installed inside the hollow specimen so that the appearance of the water film can easily be identified through an opaque rubber membrane. Clean sand specimens with parametrically varying relative density $D_r$ were first isotropically consolidated with $\sigma'_c = 98$ kPa and then loaded by initial shear stress $\tau_s$ on the horizontal plane in the drained condition. Then, undrained cyclic loading tests were performed with $\tau_d/\sigma'_c \approx 0.2$.

As a typical example, Fig. 5.10.20(c) shows the time-histories of applied stress, measured strain, excess pore pressure, and water film thickness for the case $D_r = 28\%$, the initial stress ratio $\tau_s/\sigma'_c = 0.19$. Pore pressure builds up to 100% in the middle of the 3rd cycle, and the water film becomes visible 1.3 cycles later, increasing its thickness rapidly up to 5 mm eventually. It can also be pointed out that about 0.3 cycle earlier than water film being visible, the dilative cyclic response disappears in the pore-pressure measurement, which seems to serve as a better indicator for the full generation of water film than the LED observation (Kokusho et al. 2003).

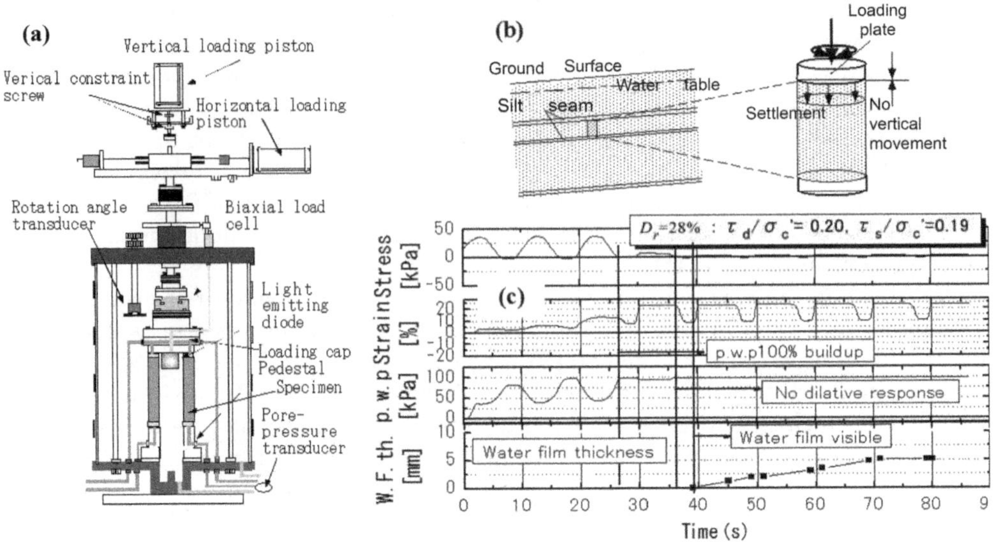


*Figure 5.10.20* Torsional shear test on water film generation: (a) Test equipment including LED, (b) Schematic view of test specimen simulating in situ layered soil, (c) Typical test results with initial stress showing cyclic shear stress, shear strain and water film thickness (Kokusho 2003).

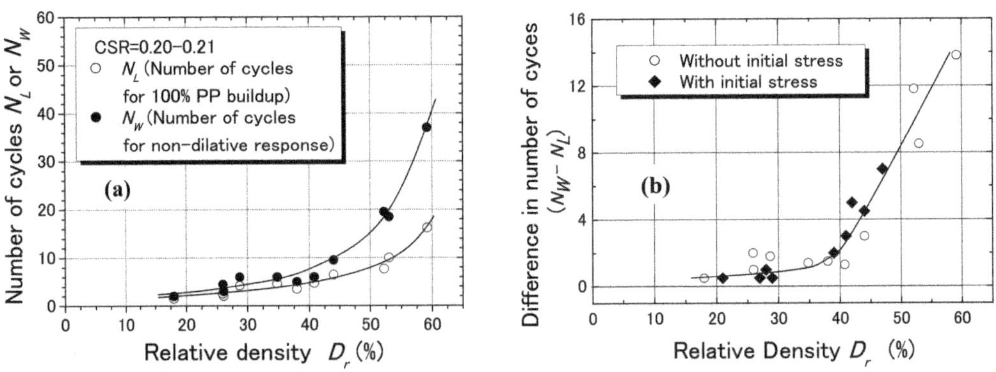


*Figure 5.10.21* Number of cycles for 100% pore-pressure buildup $N_L$ or non-dilative response $N_W$ versus relative density $D_r$: (a) $N_L$ or $N_W$ for tests without initial shear stress, (b) $(N_W-N_L)$ for tests without/with initial stress (Kokusho 2003).

In Fig. 5.10.21(a), the number of loading cycles for the 100% pore-pressure buildup (initial liquefaction) $N_L$ and that for the non-dilative response $N_w$ are plotted versus the relative densities $D_r$ of specimens from the series of tests without the initial shear stress. Obviously, loose sands with $D_r$ of about 40% or looser tend to show the non-dilative response due to the water film generation in less than a few cycles after the 100% pore-pressure buildup. For larger relative densities, $N_w$ tends to increase much more than $N_L$. In Fig. 5.10.21(b), the numbers of additional loading cycles $(N_w - N_L)$ from the onset of liquefaction to the non-dilative response are plotted versus the relative densities for the two test series without and with the initial shear stress. It indicates that the additional loading of one or two cycles is sufficient for water film generation and hence high flow potential in loose sands of $D_r$ around 40% or smaller. Also note that no large difference seems to exist between the cases without and with the initial shear stress, if the initial stress is smaller than the cyclic stress in the stress reversal condition, that is the case in gentle slopes. Thus, it may be said that the void redistribution or water film in layered sand deposits can serve as a significant mechanism for flow-type failures to occur in gentle slopes even if the sands sandwiching silt seams are clean and on the dilative side of SSL (Kokusho 2003).

The post-liquefaction void-redistribution mechanism in a sloping sand layer was investigated in volumetric-strain-controlled triaxial tests with constant shear stress (Boulanger and Truman 1996). A submerged infinite slope of liquefied sand overlain by a low-permeability cap layer is schematically illustrated in Fig. 5.10.22(a), wherein liquefaction-induced excess pore water comes up to dilate the sand at the top with the thickness $h_d$ and raise the pore pressure to a maximum value. The soil with the thickness $h_c$ below this zone contracts as there is a net outflow of water upward towards the dilating zone. As indicated on the $p'-q$ chart in Fig. 5.10.22(b), the corresponding friction angle initially $\phi'_{\text{mob}}$ for undrained condition (Point A) increases up to the peak friction angle $\phi'_p$ for the drained condition (Point B). This allows the

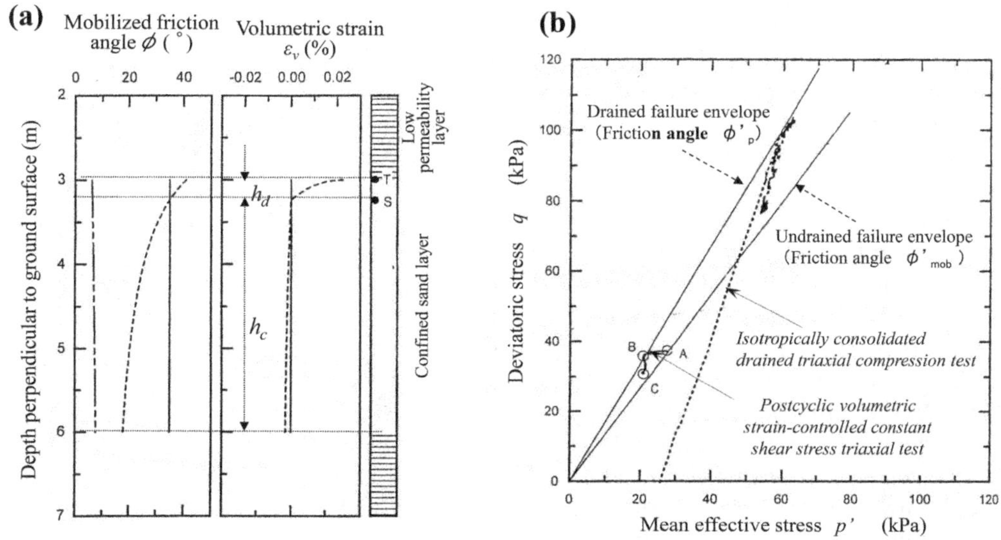


*Figure 5.10.22* Void redistribution mechanism simulated in triaxial tests: (a) Infinite sand slope capped with low permeability layer, (b) Effective stress path in volumetric strain-controlled constant shear stress test (modified from Boulanger and Truman 1996).

dilating zone to absorb the incoming water to a certain extent additionally. Further continued inflow of water would then reduce the pore-pressure toward the value corresponding to the steady state (Point C). When the element in the dilating zone has reached the steady state, its strength, if loaded undrained, would be equal to the driving shear stress acting parallel to the ground surface. Any further inflow of water to the dilating zone would cause instability of the slope, because with no further dilation being possible the water film will show up leading to instability of the slope as discussed above. Based on the analysis, Boulanger and Truman (1996) also indicated a procedure to evaluate the thickness of dilating zone $h_d$ and the threshold thickness of the contracting zone $h_c$ for triggering instability by comparing the maximum volume increase in the dilative zone, $V_{dil}$, with the volume decrease due to consolidation in the contractive zone, $V_{con}$. If the thickness of the contractive zone $h_c$ is larger than the threshold, then $V_{dil} < V_{con}$, and the instability is likely to occur in such a way that excess pore water concentrates at the top of the dilating zone and forms the water film.

The void redistribution effect has drawn increasing attention in recent years, though it may not be easy to integrate it into actual design methodologies due to in situ complex soil stratifications. More quantitative research on detailed case histories, sophisticated model tests and analytical efforts are needed to evaluate flow deformation for a variety of soil conditions and to predict the flow displacement, though some analytical research has already been conducted (Naesgaard 2011).

In this respect, Seed (1987) summarized case history data of lateral flows and proposed an empirical relationship between residual shear strengths back-analyzed from case histories and corresponding in situ penetration resistances. Comparing with similar relationships based on undrained laboratory tests, it was found that the residual strengths estimated from the case studies gave significantly lower strengths presumably due to the void redistribution and other in situ effects. Similar back-calculations from actual case histories of lateral flow failures have been implemented by quite a few investigators since then in North America. Fig. 5.10.23(a) shows one of them (Olson and Stark 2002), wherein post-liquefaction residual shear strengths $\tau_{res}$ normalized by pre-failure vertical effective stresses $\sigma'_v$ are correlated with normalized SPT blow counts $(N_1)_{60}$ for totally 33 case studies with different data reliabilities. Considerable scatters in the data points are visible, reflecting the complexity involved in in situ flow failure mechanisms including the void redistribution. It may be pointed out that $(N_1)_{60}$ values in the majority of the previous flow failure cases are lower than 10–15 (Seed 1987, Olson and Stark 2002), indicating that this type of failure seldom occurred in soils denser than that. In Fig. 5.10.23(b), equivalent residual friction angles $\phi_{res}$, calculated by $\phi_{res} \approx \tan^{-1}(\tau_{res}/\sigma'_v)$ from $\tau_{res}/\sigma'_v$-values in (a) are plotted versus $(N_1)_{60}$ with open circles. On the same diagram, the residual equivalent friction angles $\phi_{res}$ back-calculated from the model tests (Kokusho 2003) and plotted in Fig. 5.10.19(c) are superposed with close symbols against $N_1$, wherein $N_1$-values are converted from relative densities $D_r$ using the empirical formula Eq. (5.5.16). Though the residual friction angles $\phi_{res}$ evaluated from the case studies and model tests are not agreeable so well, they share distinctively low values of less than 10° with their majority overlapped between $\phi_{res} = 4 \sim 7°$.

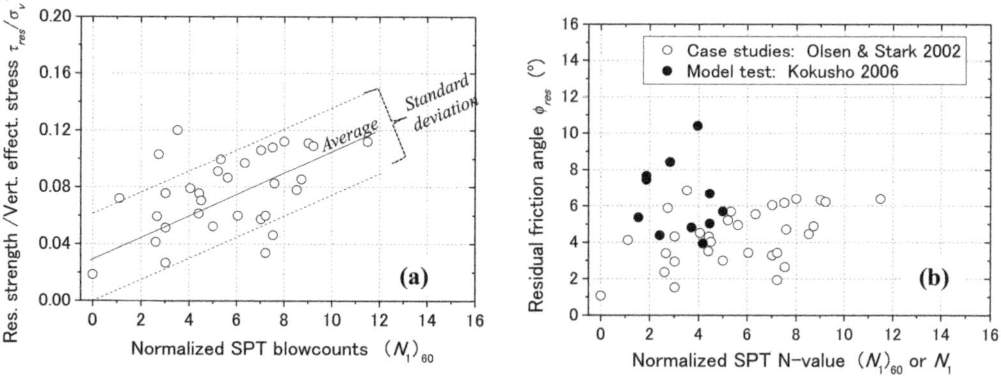


*Figure 5.10.23* Post-liquefaction residual shear strength divided by prefailure vertical effective stress, $\tau_{res}/\sigma_v'$, versus normalized SPT *N*-value $(N_1)_{60}$ back-calculated from case histories (replotted from Olsen and Stark 2002) (a), and Residual friction angle $\phi_{res}$ versus normalized SPT *N*-value $(N_1)_{60}$ from case studies (Olsen and Stark 2002) and model tests (Kokusho 2003) (b).

## 5.10.4 Liquefaction-induced effects on foundations

Settlements and lateral displacements in liquefied ground have significant effects on foundations of superstructures, buried structures and lifelines there. Hereafter, three representative cases are addressed; shallow foundations, buried structures and pile foundations.

### *5.10.4.1 Shallow foundations*

#### *(1) Relative settlement and tilting in case histories*

Free surface of liquefied ground settles according to the severity of liquefaction as discussed in Sec. 5.10.2. Shallow foundations resting on the liquefied ground tend to settle relative to the surrounding ground surface depending on the dead weight of superstructures. They often settle unevenly and tilt due to non-homogeneity of soils and non-symmetry of structures. Fig. 5.10.24 depicts relative settlements of building foundations versus tilting angles or gradients during previous earthquakes (Kokusho 2006). In the 1964 Niigata earthquakes, both settlements and tilting angles were considerable because the heavy 2~5 story RC buildings were directly on shallow foundations without/with friction piles resting on loose liquefiable clean sands. The maximum values of settlements and titling angles were 2.5 m and 8°, respectively (except the famous Kawagishi-cho apartment building which laid down nearly horizontally). RC buildings constructed on loose sandy soils after the Niigata earthquake are supported by bearing pile foundations in Japan. In the 2011 Tohoku earthquake, more than twenty thousand private houses on shallow foundations (sprit or slab foundations) were damaged mostly in hydraulically-filled reclaimed ground. The settlements and tilting were smaller than those during the 1964 Niigata earthquake mainly because most

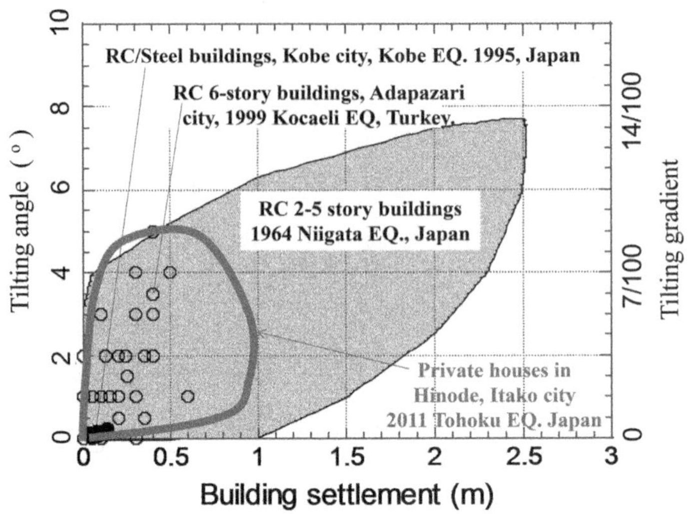


*Figure 5.10.24* Relative settlements of building foundations versus tilting angles or gradients during previous earthquakes.

of them in the recent earthquake were 2-story wooden light houses. During the 1999 Turkish Kocaeli earthquake, 6-story RC+brick buildings on shallow foundations resting on liquefied non-plastic silts settled and tilted considerably in Adapazari city. During the 1995 Kobe earthquake, buildings of shallow foundations settled and tilted only slightly (relative settlement around 10 cm and tilting angle 0.2°) despite that the underlying manmade fill (15 m thick) liquefied extensively and settled 50 cm maximum. The small relative settlements may presumably have something to do with the fill material of well-graded decomposed granite soils containing a lot of gravels.

For residential buildings, the tilting gradients have to be lower than some threshold for dwellers to live without health problems. It is said that normal people start to feel something strange in the gradient larger than 1/200, the value difficult to satisfy as the maximum limit once liquefaction occurs beneath the foundation. After the 2011 Tohoku earthquake, thresholds of tilting gradients for the liquefaction damage assessment were introduced by the Japanese government as follows; >1/20: severe, 1/20–1/60: medium, and 1/60–1/100: light.

### *(2) Effect of unliquefied surface layer*

Normally, there exist a layer at the top of soil profiles where liquefaction is difficult to occur because water table is lower than the ground surface and the surface soils tend to be densified by human and natural effects. If shallow foundations are resting on a unliquefiable layer overlying a liquefiable layer, it may well be postulated that the damage of superstructures resting on shallow foundations depends not only on the thickness of liquefiable layer but also on that of overlying unliquefiable layer. Ishihara (1985) investigated the manifestation of liquefaction at the ground surface during the

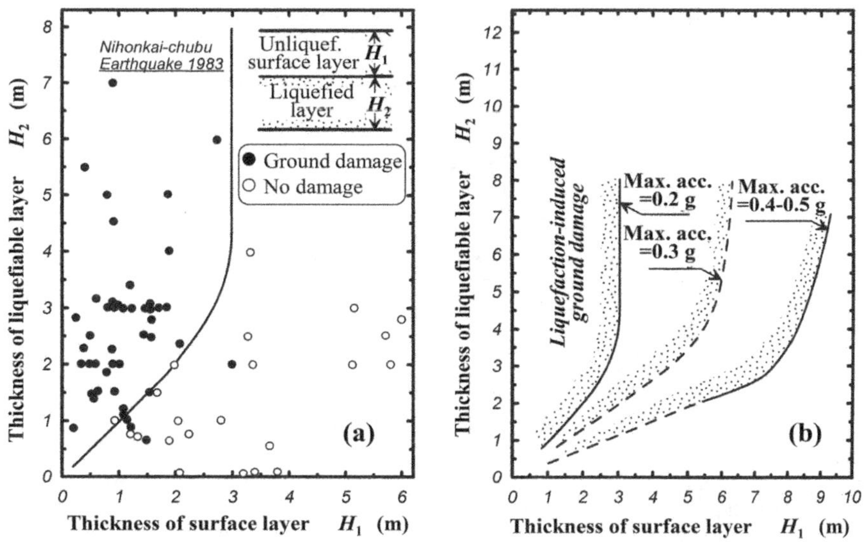


*Figure 5.10.25* Thickness of unliquefiable surface layer $H_1$ versus that of liquefiable layer $H_2$: (a) Boundary curves segregating plots of ground damage and no damage during earthquake with max. acc. 0.2 g, (b) Boundary curves with stepwise max. acc. (Ishihara 1985).

1983 Nihonkai-Chubu earthquake ($M_J = 7.7$). The thickness of surface unliquefiable layer $H_1$ and that of underlying liquefied layer $H_2$ were judged from many boring logs, and plotted in the horizontal and vertical axes in Fig. 5.10.25(a), respectively, with close/open circles if the liquefaction manifestations were visible/invisible at individual sites. Then, a boundary curve was drawn as indicated in the diagram segregating the close and open symbols for that particular earthquake of maximum surface acceleration (PGA) 0.2 g. It suggests that the liquefaction damage at the ground surface may be avoided on the right side of the curve, particularly if $H_1 > 3$ m regardless of $H_2$. By extending the database further to include case histories during the 1976 Tangshan earthquake, China, with PGA 0.4–0.5 g, the same author also proposed a diagram shown in Fig. 5.10.25(b) for parametrically changing maximum accelerations. Its applicability was also examined in case studies on ground settlements in City of Dagupan during the 1990 Philippine Luzon earthquake (Ishihara et al. 1993).

A similar diagram was also developed by Kokusho and Tsutsumi (2002) utilizing settlement data of RC buildings in Niigata city during the 1964 Niigata earthquake (BRI 1965) as illustrated in Fig. 5.10.26. The maximum acceleration in this case was 0.17 g according to the record in the Kawagishi-cho apartment. The boundary curve in this case seems to shift rightward so that no relative settlement occurred for $H_1 > 4$ m, 1 m thicker than that in Fig. 5.10.25(a) probably due to heavy RC buildings. Though some engineering judgement is involved in drawing the boundary curves depending on soil conditions and superstructures, the above diagrams may well suggest a mechanism that the surface unliquefiable layer serves as a sort of stiff slab to ease the relative settlements and tilting due to liquefaction in the underlying layer.

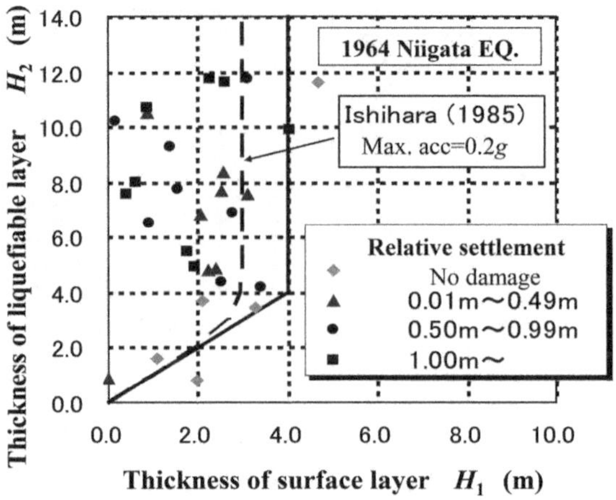


*Figure 5.10.26* Thickness of unliquefied surface layer $H_1$ versus liquefied layer $H_2$ during 1964 Niigata earthquake of max. acc. 0.17 g for relative settlement of RC building (Kokusho and Tsutsumi 2002).

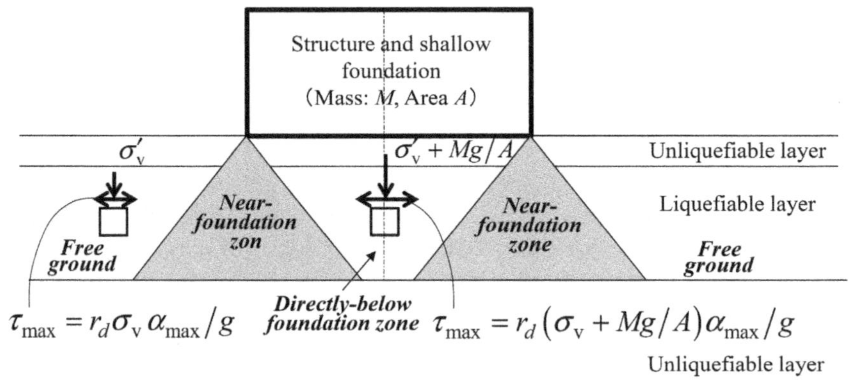


*Figure 5.10.27* Simplified model of shallow foundation resting on liquefied ground for foundation performance.

### *(3) Shallow foundation on liquefied soil*

Shallow foundations and associated working stresses will affect the performance of liquefiable soils near or below it in comparison to the free field and hence the foundation performance. In order to simplify the problem, let us consider the shallow foundation supporting a structure (the total mass $M$ and horizontal area $A$) resting on a unliquefiable surface layer underlain by a liquefiable layer illustrated in Fig. 5.10.27. Here, the simplified liquefaction potential evaluation practice explained in Sec. 5.5.5 is employed to examine the effect of the foundation. The cyclic stress ratio in the free ground $CSR_{free}$ is expressed in Eq. (5.5.28) using the total and effective overburden stresses, $\sigma_v$, $\sigma'_v$ and the PGA $\alpha_{\max}$ as:

$$CSR_{free} = \frac{\alpha_{\max}}{g} \frac{\sigma_v}{\sigma'_v} r_d r_n \tag{5.10.2}$$

where, $r_d$ and $r_n$ are the stress reduction coefficients in terms of ground depth and seismic motion irregularity. In contrast, the *CSR* below the foundation may be approximated as

$$CSR_{below} = \frac{\alpha_{\max}}{g} \frac{\sigma_v + Mg/A}{\sigma'_v + Mg/A} r_d r_n \tag{5.10.3}$$

because the overburden stresses are added by the structural dead load $Mg/A$. Here, the maximum acceleration of the foundation is assumed to be the same, $\alpha_{\max}$. Eq. (5.10.2) compared with Eq. (5.10.3) yields the following, if $\sigma_v/\sigma'_v > 1.0$ and the coefficients $r_d$ and $r_n$ are postulated to be identical in the two locations.

$$\frac{CSR_{free}}{CSR_{below}} = \frac{1 + (Mg/A)/\sigma'_v}{1 + (Mg/A)/\sigma_v} > 1.0 \tag{5.10.4}$$

Thus, the soil in the free ground are essentially more liquefiable than those directly below shallow foundations for the same soils having the same *CRR*. This trend becomes more dominant as $\sigma_v/\sigma'_v$ and $Mg/A$ are larger according to Eq. (5.10.4). Although the actual conditions are more complex because the depth-dependent coefficients $r_d$ may vary due to the structure and *CRR* may decrease below the foundation due to the larger overburden, model shaking table tests actually demonstrated that excess pore-pressure is easier to build up in free fields than immediately below foundations (e.g. Yoshimi and Tokimatsu 1977).

In the near-foundation ground shown in Fig. 5.10.27, large initial shear stress is working due to the foundation stress. Cyclic shear stresses are added there due to the rocking vibration of the structure. If the sand is on the dilative side of SSL, it tends to be more resistant to liquefaction than in the free ground, though it is still liquefiable particularly in the stress-reversal condition as discussed in Sec. 5.8. If the sand is on the contractive side, the brittle failure with flow deformation may tilt the structure.

It may be summarized in general that liquefaction occurs first in the free ground more easily and earlier, and tends to expand to zones near and directly below the foundations. If liquefaction occurs directly below the foundations, the foundation settles relative to the surrounding ground surface due to larger overburden by the structures than in the free field. The settlement occurs almost in one-dimensionally in the foundation center and 2 or 3-dimensionally in the margin involving local shear failure. This liquefaction mechanism seems to have a certain impact on liquefaction-induced foundation performance. Because larger settlement occurs in the foundation margin and heavier structural loads are normally carried there than in the center, the foundation slab tends to deflect in a convex shape sometimes inflicting tension-failures in the upper face near the center of the slab after liquefaction.

Fig. 5.10.28(a) shows a diagram of maximum liquefaction depths $H_{\max}$ from ground surface versus average foundation settlements $S_{av}$ of RC-buildings supported

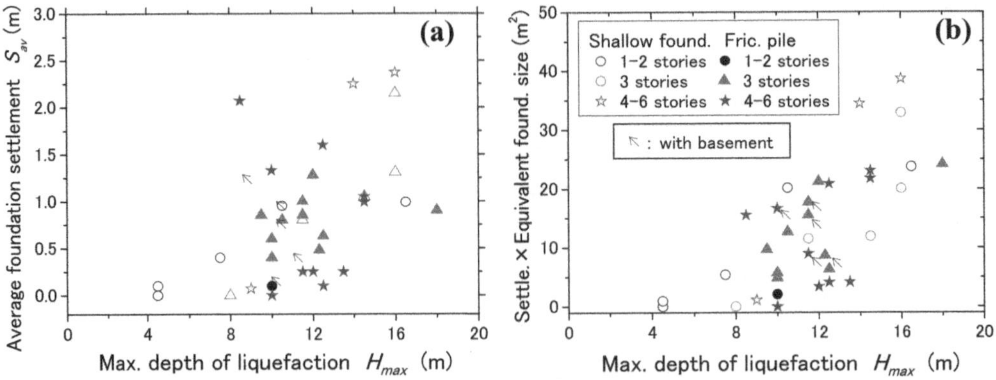


*Figure 5.10.28* RC-building settlements $S_{av}$ supported by shallow foundations (a), and Settlements multiplied by equivalent foundation size (b), versus max. liquefaction depths $H_{max}$ based on case histories during 1964 Niigata earthquake (replotted from data by Yoshimi and Tokimatsu 1977).

by shallow foundations (individual/continuous footings, mat foundations, without or with friction piles, and without or with basements) based on case history data during the 1964 Niigata earthquake on liquefaction-induced settlements (Yoshimi and Tokimatsu 1977, BRI 1965). Despite the large data scatters, it may be said that the settlement tends to increase with increasing maximum liquefaction depth $H_{max}$ and no settlement occurs for $H_{max} < 4$ m. In Fig. 5.10.28(b), the foundation settlement multiplied by an equivalent foundation size is plotted versus $H_{max}$ again. Here, the equivalent size is defined as the size of a square foundation having the same area. The plots are less scattered in (b) than in (a), suggesting that the settlements tend to be inversely proportional to the foundation size as pointed out by Yoshimi and Tokimatsu (1977). This seems to reflect that soils beneath larger foundations are less prone to severe liquefaction and large settlements. What was unexpected for the plots of Niigata earthquake shown in Fig. 5.10.28 is that different story buildings and the existence of basement or friction piles seem to have only negligible effect on the settlements. Unlike the Niigata case, building settlements during the 1999 Kocaeli earthquake (Yoshida et al. 2001) and the 1990 Luson earthquake (Acacio et al. 2001) are reported to have reflected the number of stories of buildings.

#### 5.10.4.2 Uplift of buried structures

Because the bulk densities of buried structures or pipes (normally around 1.0 t/m$^3$) are lower than the saturated sand density (around 1.8 t/m$^3$), they tend to lift up relative to surrounding soils due to the buoyant force working in liquefied soils. However, the liquefied soils are not perfect liquid in reality but preserve granular properties with dilatant behavior reviving in large shear deformations except for very loose contractive soils. Furthermore, the overlying unliquefied layer tends to block the uplifting. Thus, buried structures seldom uplift to such an extreme as their dead weights balance the buoyant forces.

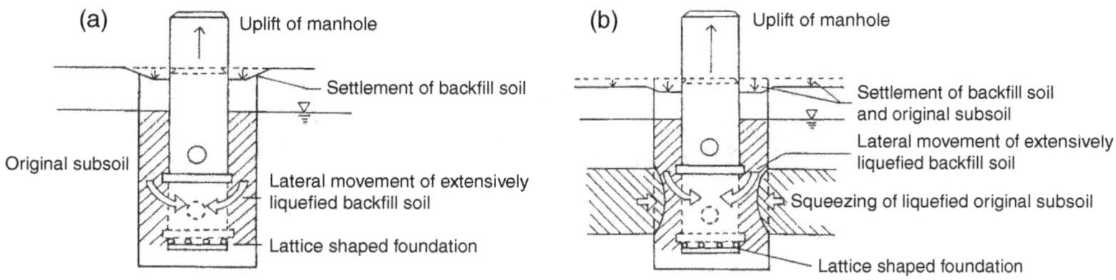


*Figure 5.10.29* Conceptual mechanism of manhole uplifted during 1993 Kushiro-oki earthquake: (a) Lateral movement of backfill soil, (b) Lateral movement of backfill and original soils (Koseki 1997).

As early cases of liquefaction-induced damage in buried structures, underground RC sewage tanks and gasoline tanks uplifted by maximum 2 m during the 1964 Niigata earthquake (BRI 1965). During the 1993 Kushiro-oki earthquake in Japan, a number of sewage manholes lifted up by more than 1 m (Koseki 1997). Similar manhole uplifts have been witnessed by several other earthquakes in Japan since then. It should be noted that in many of these cases what liquefied actually was not natural sandy soils but artificial sands backfilling the sewage facilities. Fig. 5.10.29 shows the mechanism of manhole uplifting illustrated schematically by Koseki (1997) based on excavation studies of damaged manholes after the earthquake. It was found that the uplift was caused by extensive liquefaction of backfill sands which moved laterally to fill the cavity beneath the ejected manhole as shown in (a). Also suspected was that sandy sublayers in the original ground liquefied and squeezed the backfill soil as shown in (b). A similar field investigation by Yasuda and Kiku (2006) on about 1600 manholes during the 2004 Niigataken Chuetsu earthquake in Japan revealed that the uplift displacements became larger for shallower water tables. Also suggested by the same authors was that the uplift of manholes buried by backfill sand tends to be larger in clayey ground than those in sandy ground, probably because the hydraulic conductivity of the ambient original ground has a strong impact on the intensity of liquefaction in the backfill.

There are several countermeasures to mitigate the uplift of buried structures, in addition to liquefaction mitigation measures of backfill sand, that are sometimes difficult to employ in existing structures. One of the effective ways is to install vertical walls around the periphery of a buried structure to the depth of an unliquefiable layer, preventing the lateral movement of liquefied sand toward the bottom of the structures (Yoshimi 1998). Other measures are also proposed and implemented such as to add counterweights inside the buried structures so that their bulk density becomes almost identical to liquefied soil density or to install vertical drains along the structures.

### *5.10.4.3 Pile foundations in liquefied soils*

If the lateral flow occurs, pile foundations are greatly displaced, leading to severe structural damage such as the collapse of bridge decks or the deformation of superstructures.

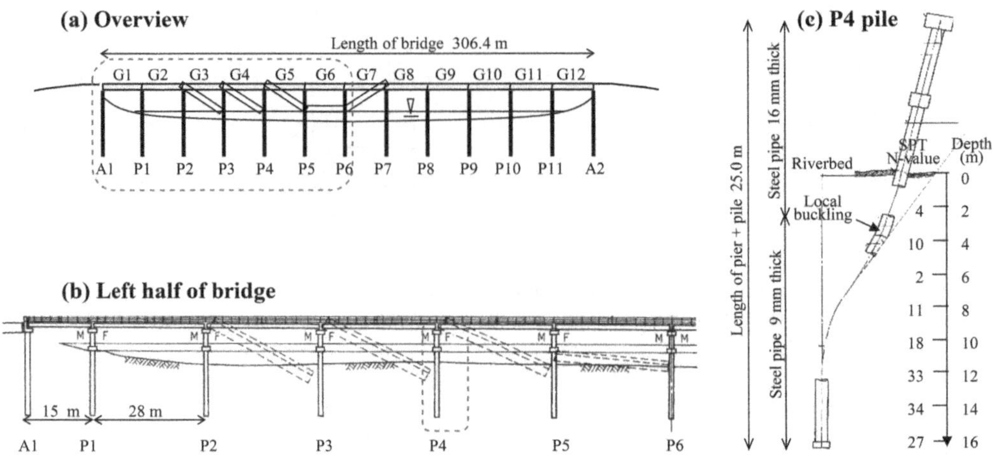

*Figure 5.10.30* Showa Oh-hashi bridge failed during 1964 Niigata earthquake: (a) Overview, (b) Left half of bridge, (c) Pile P4 pulled out after failure (modified from Fukuoka 1966).

The piles themselves are structurally damaged as demonstrated in the foundations of bridges and buildings during the 1964 Niigata earthquakes and the 1995 Kobe earthquake. In Figs. 5.10.30(a) and (b), the overview and the left half, respectively, of Showa Oh-hashi bridge which failed during the earthquake are illustrated (Fukuoka 1966). The bridge was composed of 12-span simple girders (G1–G12) supported by 11 piers (P1–P11) and two abutment piers (A1, A2) having the total length 306 m with the end spans 15 m and intermediate spans 28 m each. During the earthquake, there were several witnesses on the bridge who experienced the bridge failure; a taxi driver, a truck driver, two bicycle riders, a repair worker and a few pedestrians. All of them told that the bridge girders started to fall down a few minutes after they were shaken on the bridge by the major shaking, and it was just when most of them could evacuate safely to river banks (JSCE committee 1966). According to them, the girders G6, G7 shown in (a) fell down first and then G5, G4 and G3 in sequence. One of the steel pipes comprising P4 serving as the piers/piles (pile-in-group) welded in one piece supporting the girders (diameter 60 cm and total length 25 m) was pulled out to investigate after the earthquake. It was found to have been deformed largely in the ground at the 10 m depth from the ground surface and locally buckled in the embedded portion as shown in Fig. 5.10.30(c). The SPT $N$-values shown in the same diagram indicate that loose sand with $\mathrm{N}=10$ or smaller was down to 10 m from the surface presumably sandwiching a silty soil sublayer with the $N$-value $=2$ at 6 m deep. These observations suggest that the sand layer of about 10 m deep liquefied and flowed laterally possibly after water films were generated due to the elevation difference in the river cross-section. The flow failure may have propagated with a time-delay from the river center toward the left bank and displaced the bridge piles, moved the pier heads and dropped the girders in that sequence. Another witness account also reported that the middle part of the river flow was temporarily heaved, making a sandbar above the water level during that event.

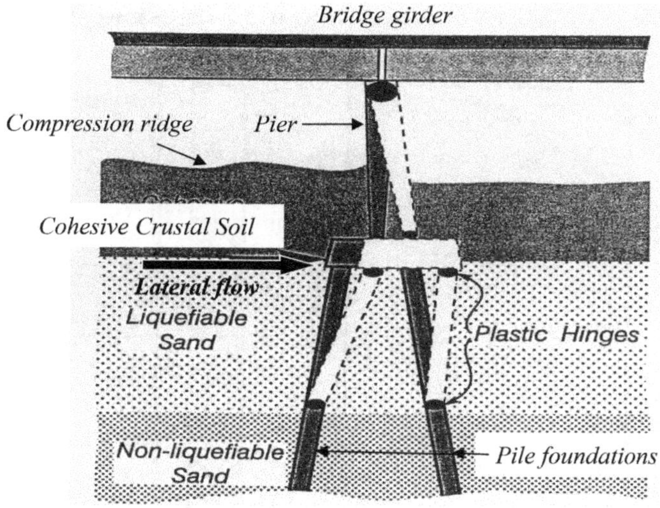


*Figure 5.10.31* Bridge foundation affected by lateral flow of crustal soils in New Zealand (Modified from Berrill et al. 1997).

During the 1987 Edgecumbe earthquake in New Zealand, a bridge became unserviceable due to lateral flow of foundation ground about one hour after the earthquake shaking ended (Berrill et al. 1997). It was observed that the foundations on piles passing through the liquefied soil to firm ground attracted large thrust caused by the lateral movement. Soil failure was observed at the ground surface behind the supporting piles. In a trench excavated there, a passive failure profile was found in an unliquefied crust overlying the liquefied soil at the bridge piers as sketched in Fig. 5.10.31 evidencing that the passive earth pressure was exerted in the crustal soil. This case history demonstrated the significant effect of drag force exerted on the piles by the overlying unliquefied soil in contrast to a smaller effect from the underlying liquefied layer. A possible remedy for existing bridges was also proposed to put in a crushing zone of weak material behind piers likely to be affected by the lateral flow.

After the 1995 Kobe earthquake, the damage of group piles which supported viaducts of an elevated highway system was investigated systematically by means of several in situ tests, such as drilling at pile heads, inspections of pile cracks by borehole cameras and non-destructive sonic tests using impulse elastic waves (Hamada et al. 2009). The piles were mostly of bored cast-in-place concrete types of 1–1.5 m in diameter and 27–40 m in length. It was found that earthquake-induced residual displacements both at the ground surface and pile heads tend to clearly increase in waterfront areas (within 100 m from quay walls) compared to inland areas (beyond 100 m from quay walls). Fig. 5.10.32 shows crack densities (the number of cracks per one meter pile length) for a number of pile foundations investigated along the route of the highway versus the soil depth. The data are shown in the two separate diagrams; (a) the piles near the waterfront (totally 14 investigated piles) and (b) the inland piles (totally 105 investigated piles), for the cracks 0.5 mm wide or larger (thick curves)

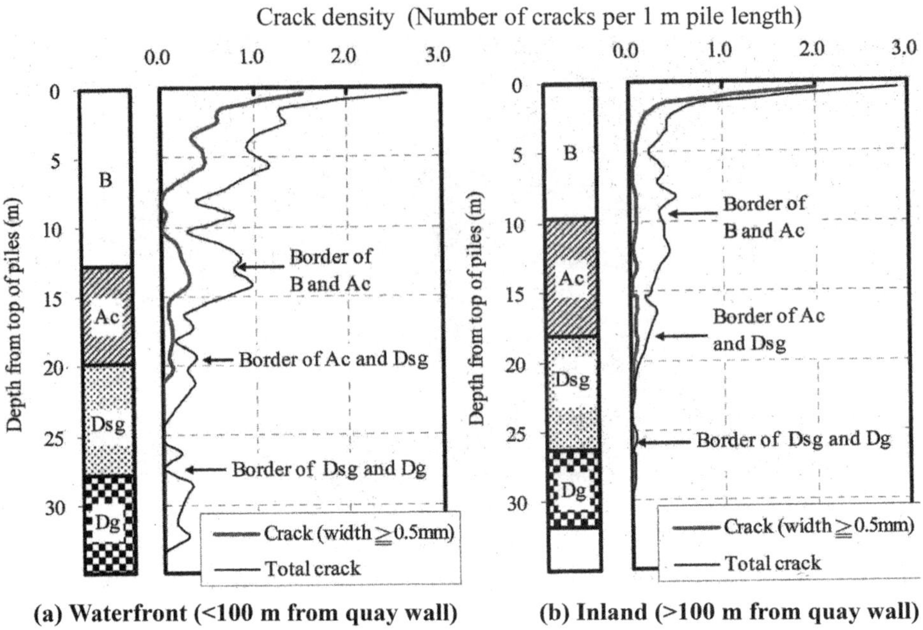


*Figure 5.10.32* Average crack densities of piles investigated versus soil depth for cracks wider than 0.5 mm and for all cracks: (a) Piles near waterfront, (b) Inland piles (Hamada et al. 2009).

and for all the cracks detectable (thin curves). The soil profiles in all this area were relatively similar as indicated in the same figure, comprising loose decomposed granite fills (B) at the top (around 10 m thick and the water table at GL.-2~-4 m, where liquefaction occurred), underlain by soft Holocene marine clay (Ac) and further by stiff Pleistocene sandy gravels (Dsg) and gravels (Dg). All the piles were supported by the stiff Dsg or Dg layer. Near the waterfront in (a), the density of cracks, no matter whether being wider than 0.5 mm or not, were the highest near the pile heads and still high in the upper half of the B-layer, and near the boundary between the B and Ac layers, whereas it became lower in the layers Dsg and Dg. In the inlands shown in (b), the crack densities were almost identically high near the pile head but definitely lower in larger depths, presumably reflecting the difference in the flow displacements. Namely, the dynamic effects of superstructures were dominant in bending moments in the pile heads both inland and water front, creating cracks of the highest density near the top. In addition, the residual soil displacements behind displaced quay walls gave large kinematic effects on the bending moments and the associated cracks in piles near the waterfront at the upper part of the liquefied B-layer presumably due to the effect of the unliquefied crust. The crack density tends to increase also near the lower boundary of the liquefied layer because of the higher bending moment exerted there due to the kinematic effect caused by the lateral flow. In contrast, the inland piles suffered much less cracks in the deeper portions, wherein the cracks were presumably inflicted by the dynamic effects of superstructures and the dynamic response of level

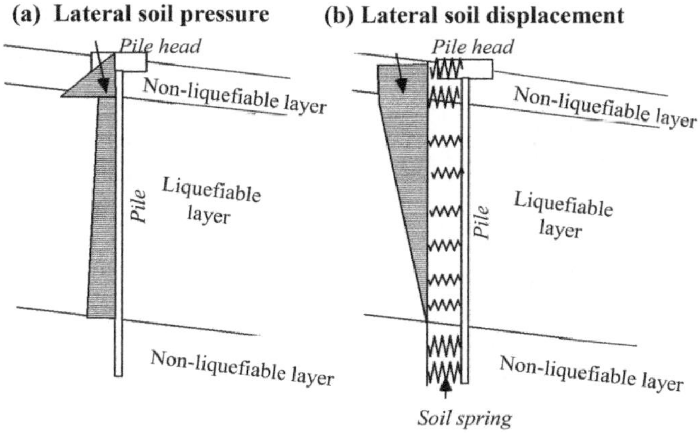


*Figure 5.10.33* Two different pile design methods considering lateral flow displacement in liquefied ground: (a) Applying lateral pressure directly on piles, (b) Applying kinematic soil displacement via soil springs.

ground. Field performance of various piles for building foundations in areas liquefied and flowed due to the displaced quay walls during the same earthquake were investigated also by Tokimatsu and Asada (1998). It was confirmed that the failures of piles concentrated at the interface between liquefied and unliquefied layers as well as near pile heads, indicating the significant effects of lateral ground displacements on pile damage particularly near waterfront.

As for designing pile foundations considering the lateral flow displacement in liquefied ground, two different methods may be available as illustrated in Fig. 5.10.33(a) applying lateral pressures directly on piles or (b) applying kinematic soil displacements via soil springs. The former method, though simpler than the latter, has a drawback that the pile reaction force highly dependent on the stiffness of the pile relative to that of the surrounding soil may not be properly taken into account. Hence, the latter method is recommended in most cases, wherein the soil-pile interaction is considered pseudo-statically. A basic equation for a pile connected with soil springs having Winkler's subgrade coefficients is employed here as follows (Chang 1937).

$$EI(d^4y/dz^4) = -kBy \qquad (5.10.5)$$

where, $y$ = horizontal pile displacement, $E$ = Young's modulus, $I$ = moment of inertia of piles, $z$ = soil depth, $B$ = pile diameter and $k$ = coefficient of nonlinear subgrade reaction or soil spring constant.

Fig. 5.10.34 shows how to idealize the soil-pile-superstructure interaction in the pseudo-static analyses for various soil displacements. As for the nonlinear subgrade reaction of the soil springs, the soil pressure $p$ on the pile is correlated with the horizontal pile displacement $y$ as illustrated in (a) using a nonlinear $p$–$y$ curve. The curve is defined by the initial soil spring constant $k_0$, the secant spring constant $k$ and the upper bound $p_{max}$ associated with passive failure of the soil. In the case shown in Fig. 5.10.34(b), the soil in the free-field is assumed unaffected by earthquakes and the

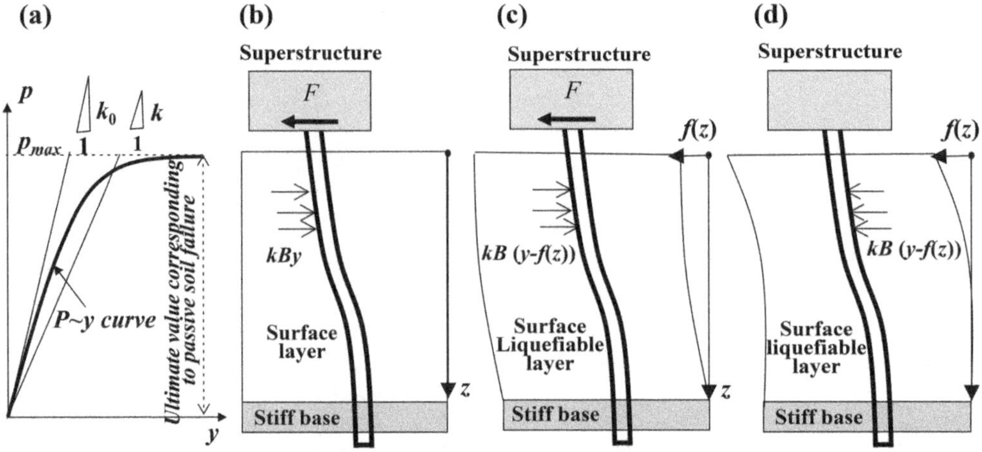


*Figure 5.10.34* Idealization for soil-pile-superstructure interaction: (a) *p–y* subgrade reaction curve, (b) Pile in non-deformed soil, (c) Pile in cyclically deformed soil, (d) Pile in laterally spreading soil (Modified from Tokimatsu and Suzuki 2009).

pile tends to deform by itself due to the inertial force $F$ from the superstructure. If the soil displacement becomes larger due to the dynamic response of liquefied ground as illustrated in (c), Eq. (5.10.5) is modified as:

$$EI(d^4y/dz^4) = kB[f(z) - y] \tag{5.10.6}$$

Here, the free-field ground displacement $f(z)$ is taken into account as the kinematic effect together with the inertial effects of piles and superstructures. If liquefaction-induced large lateral flow displacement occurs as shown in (d), $f(z)$ in Eq. (5.10.6) represents the residual soil displacement in the free-field, when the dynamic soil response is already ceased.

There are important issues to consider as follows in designing piles in laterally displaced soils in Fig. 5.10.34(d) by applying the kinematic soil displacement via soil springs:

(i) Free-field ground displacement: It is by no means easy to estimate the flow-induced lateral displacement of free-field, which is highly dependent on site, soil and seismic conditions. As already discussed, there are two types of lateral flows; i) in gentle slopes such as that occurred in Niigata city during the 1964 earthquake, and ii) behind displaced retaining structures such as that occurred in manmade lands during the 1995 Kobe earthquake. The basic mechanisms and how to estimate the magnitudes and spatial variations of the residual displacements may differ between the two. In the latter type, the experience during the Kobe earthquake (Ishihara et al. 1996) is largely referred in Japan as an important case history. For the former type, the case history data in the 1964 Niigata earthquake and many other earthquakes inflicting similar flow failures (e.g. Robinson et al. 2014) may serve as a data base. In addition, post-liquefaction ground displacements due to lateral spreading of embankments and slopes affect pile foundations nearby. Numerical analyses may be largely relied on to estimate the ground displacements in such cases.

(ii) Soil spring *p–y* curve in liquefied layer: The influence of liquefaction on the subgrade reaction coefficient $k$ is normally accounted for by a scaling factor or $p$-multipliers ($m_p$) to modify the $p$-value of the *p–y* curve in unliquefied soils. Normally, $m_p$ is taken as around 0.1 (e.g. Tokimatsu and Asada 1998), although it may depend on many factors such as relative density, physical soil properties, permeability of liquefied soil, severity of liquefaction, pile diameter and pile rigidity. Despite the considerable uncertainties involved in choosing the subgrade reaction, it is known that pile performance is rather insensitive to the $p$-multiplier in liquefied layer because the effect of unliquefied crust is overwhelming in many cases (Ashford et al. 2011).

(iii) Effect of unliquefiable crust: It should be noted again that the unliquefiable surface layer which normally caps liquefied soils has enormous effects on the pile behavior in terms of lateral displacements, bending moments, etc. as actually observed in previous case histories. Because large relative displacements tend to occur between the unliquefied crustal soil and piles, if the piles are rigid in particular, it is critical in design how to properly evaluate not only the *p–y* curve but also the ultimate lateral loads represented by the passive earthpressure imposed by the unliquefied crust against the pile foundation. The conventional Rankine earth pressure theory may be employed in estimating the passive earth-pressure, though the pressure may reduce to a certain extent because the upward seepage flow from the underlying liquefied sand tends to degrade the crustal soil.

In order to rationalize this very complicated problem on the pile design against lateral flow considering the effect of the unliquefied crustal soil layer, quite a few model tests have been conducted either in situ by using actual piles (e.g. Ashford and Juirnarongrit 2002), or in the laboratory using 1G and centrifuge shake tables (e.g. Tokimatsu, and Suzuki 2009). Fig. 5.10.35 depicts one of the examples of 1G model tests (Suda et al. 2007a), wherein a uniform clean saturated sand layer with an unsaturated (unliquefiable) portion at the top in a large-scale laminar shear box was first liquefied by shaking and then forcefully sheared monotonically from right to left with piles in it. During the monotonic shearing, a small vibration continued to preserve low shear resistance in the liquefied layer. Two 4.8 m long pile models with different flexural rigidities, a high-rigidity steel pile of diameter 32 cm and a low-rigidity PHC (pre-stressed concrete) pile of diameter 30 cm, were rigidly fixed at the base of the shear box.

In Fig. 5.10.36, the test results with liquefiable and unliquefiable layers, 2.4 m thick each, are shown in terms of (a) pile displacements and (b) bending moments of the high and low rigidity piles for the soil surface displacement 10 cm. The low-rigidity pile tends to deform largely almost in concert with the soil, while the high-rigidity pile deforms only a little, incurring the passive failure in the unliquefied crust in the right side of the pile due to the large difference in relative displacement. The failure mode in the crustal soil behind the rigid pile can be partitioned into typical passive failure near the surface and local shear failure below that. For the low-rigidity pile in contrast, the passive failure zone appears on the left side, because the pile head tends to move forward farther than the soil and push the surface portion of the crust. The bending

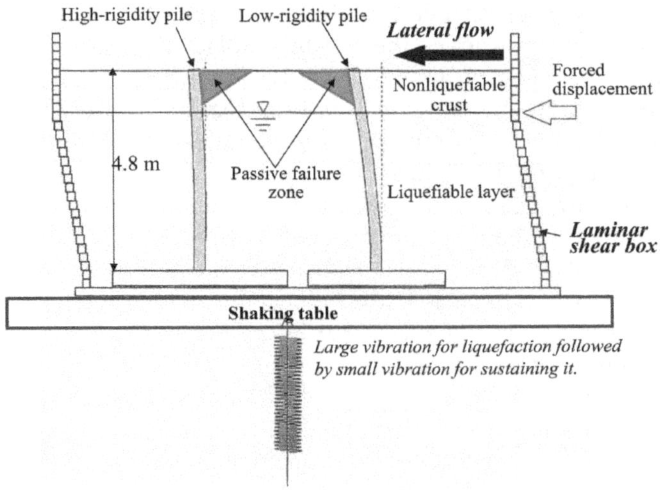


*Figure 5.10.35* 1G model tests of piles of different rigidities embedded in sand in a large-scale laminar shear box, liquefied by shaking and monotonic shearing.

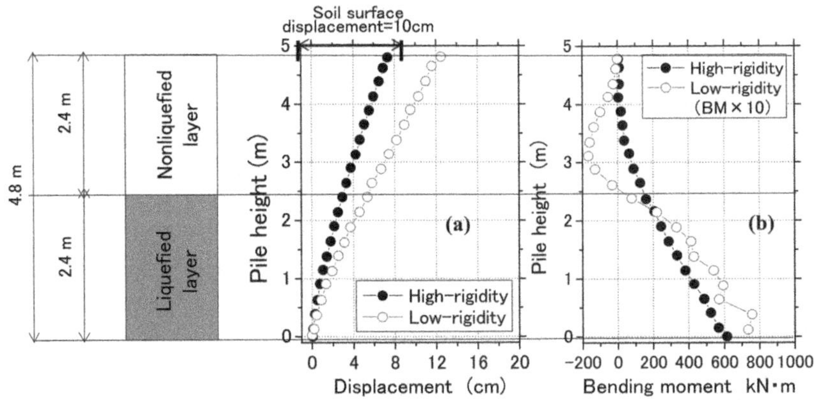


*Figure 5.10.36* Results of 1G model test on high and low rigidity piles in laterally spreading soil for soil surface displacement 10 cm: (a) Pile displacements, (b) Bending moments.

moment of the high-rigidity pile increases almost monotonically with depth, whereas that of the low-rigidity pile tends to show a negative peak near the bottom of the unliquefiable layer and then tends to increase with depth in positive values. Based on these test results, a design procedure was proposed incorporating the significant resistance in the crustal soil by using the subgrade reaction coefficient 10 times larger than in liquefied soil (Suda et al. 2007b).

## 5.10.5 Mitigation measures

As illustrated in Fig. 5.10.1, uneven settlements, tilting settlements, lateral flow and uplifting tend to occur as the consequence of liquefaction. The mitigation measures to take in advance may be classified as (i) to opt structural designs well-prepared for potential ground deformations, and (ii) to improve the ground more resistant to liquefaction against potential earthquakes. The details are available in other literatures including practical designs (e.g. Editing Committee of JGS 1998), and only the outline of the mitigation measures is briefly addressed in the following.

#### 5.10.5.1 Counter measures for shallow foundations and superstructures

Shallow foundations are directly affected by uneven ground settlements and other residual displacements caused by liquefaction. If the foundation is of high flexural rigidity, it can easily be remediated by leveling it by jacking up after suffering uneven settlement. Not only the foundation but also the superstructure is preferred to have a higher rigidity for better performance against the uneven ground settlements. If a widespread shallow foundation covers various soil conditions where different liquefaction-induced deformations are anticipated, joints prepared for the relative displacement may be introduced in advance at appropriate boundaries.

Foundations supported by tip-bearing piles are almost immune to the liquefaction-induced settlements, though the settlement gaps occur between the surrounding liquefied soils. This may affect the connections of buried lifelines and human access as well. The settlement gaps can be avoided if the surrounding soils are improved against liquefaction, as actually demonstrated for many condominium RC buildings during the 2011 Tohoku earthquake near Tokyo. Another significant problem of the bearing pile foundations is the mitigation against lateral spreading/flow. In addition to the design of piles themselves considering the effects of liquefied layers and unliquefied cap layers using the $p$–$y$ curves as mentioned earlier, the design consideration on superstructures against the lateral residual displacement of liquefiable ground is of extreme importance for bridges and towers. For example, girders of simple-beam bridges should be tied by strong wires to supporting piers as actually implemented after the Kobe earthquake in Japan. In the case of power transmission towers, independent pile caps of the four feet are sometimes tied with rigid RC-beams to make them deform together.

Another possible choice of the foundation on liquefiable deposits for better economy is a raft foundation with friction piles stopping in the middle of liquefiable deposits as an intermediate foundation-type between the two. In this type, the superstructure is supposed to settle together with the surrounding soil without intolerable tilting. A shallow foundation combined with lattice-type embedded walls underneath may be considered as a modification of the above foundation-type. In this foundation, the sand surrounded by lattice-walls is expected to be less liquefiable due to the confinement by the rigid walls, leading better foundation performance as demonstrated in a high RC building of the Kobe Harbor facilities during the 1995 Kobe earthquake. The combination with basements may be another modification of this type.

#### 5.10.5.2 Soil improvements

Soil improvement methods to increase the liquefaction resistance may be classified into the following four types; (i) densification, (ii) solidification, (iii) pore-pressure dissipation, and (iv) dewatering/desaturation. The soil improvement technologies are advancing very fast recently due to burning needs after severe liquefaction damage in many buildings and private houses during recent strong earthquakes such as the 1995 Kobe and 2011 Tohoku earthquakes in Japan and the 2010 Darfield earthquake and other 2011 earthquakes in New Zealand. One has to be aware of the most recent developments in the soil improvement technologies outlined below.

(i) Densification: This is very effective means of improving non/low-plasticity soils. In soils with high fines content, however, its effect tends to be small and slow to manifest. There are many different densification methods available with their own merits; (a) Vibratory tamper/rollers, (b) Vibro-rod/floatation, (c) Sand compaction pile, and (d) Heavy tamping.

  (a) The vibratory tamper/roller is used for improving bearing capacity in shallow ground and sometimes combined with replacements and solidifications by first removing surface soils by excavation and then refilling with other soils of better quality or further solidifying with some cement or lime before densifying. A big advantage of the shallow densification is visibility and hence better quality assurance of the improved soil.
  (b) In the vibro-rod method, a tube/H-shaped steel rod is penetrated into sandy soils by pushing and vibrating. In the vibro-floatation method, a cylindrical vibrator is penetrated together with water jet and vibration. In both methods, the rod and vibrator move up and down while additional sands, gravels, iron slags, etc. are fed along the peripheries of the penetrating rod to have greater compaction.
  (c) In the sand compaction pile (SCP) method, a steel tubular casing with a lid at the rod tip first penetrates down to some depths by vibration and then clean sand is fed to the casing. Then the casing moves up to discharge the sand below and move down to compact it by vibration with the closing lid. This movement generates compacted sand columns in the ground with the diameter much larger than the casing. Thus, this method compacts the soil by means of lateral compression as well as vibratory compaction. There are design charts available to evaluate improved SPT $N$-values as a function of the replacement area ratio (Editing Committee of JGS 1998). Such charts clearly indicate that high fines content tends to considerably reduce the compaction effect. In the gravel compaction pile method similar to the SCP-method, gravels with controlled grading are used in place of clean sands as the feeding material.
  (d) The heavy tamping method utilizes the potential energy of a heavy hammer of tens of tons falling from the height of tens of meters. The tamping can effectively compact non/low-plasticity soils such as sands, gravels and even waste materials such as fly ash (e.g. Kokusho et al. 2012c). The compaction effect normally reaches down around 10 m from the ground surface. In order to have proper compaction effects, non-cohesive granular soils are placed about one-meter thick before tamping so that the water table is about 1.5 m lower from the compaction surface. The effectiveness of this method largely depends on how frequently the hammer can fall for better work efficiency. The shocks and vibrations in falling the hammer may restrict its use only in sites far from houses and already existing facilities.

(ii) Solidification: It is divided into (a) Shallow soil stabilization, (b) Grouting, (c) Admixture in deeper ground. Because chemical agents are added to in situ soils in this method, their side-effects on natural environments through contamination of ground water, e.g. the change in PH, should be examined carefully in advance particularly in densely populated urban areas.

(a) The shallow soil stabilization, originally developed for cohesive soils to improve the trafficability, slope stability and resistivity against erosion, can be applied in shallow depth to have better performance of shallow foundations against liquefaction by making the unliquefiable and stable cap layer. The stabilization is normally implemented by mixing cement or quick/slaked lime powder with the original soil and compacting it.

(b) The grouting technology has been recently applied to soil improvements by grouting the mixtures of water, cement and fine soils to mitigate liquefaction nearby or below existing facilities in particular. It is categorized into compaction grouting, permeation grouting, and jet grouting. The compaction grouting is used for improving soils by forming column/sphere-shaped blocks under structures already in service, wherein the soil blocks compress and densify surrounding soils. The permeation grouting injects grout fluid of lower viscosity than the compaction grouting so that the fluid can penetrate in between sand particles to stabilize them. The jet grouting uses high-pressure fluid jetting out from nozzles of a rotating drilling rod moving vertically to cut in situ soils and form cement-mixing soil columns.

(c) The admixture solidifies liquefiable soils by mechanically mixing with cement in slurry or dry powder by rotating a large auger with mixing bars to form large-diameter (up to 3 m) soil-cement columns in the form of walls, grids, blocks, etc. It is sometimes used not only liquefaction mitigation but also for larger bearing capacity of foundations. For these purposes, it is sometimes required to make laboratory tests for evaluating properties of improved soils in advance. The design strength is normally taken as 1/3 to 1/5 of the laboratory strength, because in situ strength is so variable due to the spatial non-uniformity of admixture.

(iii) Pore-pressure dissipation: In this method normally named as "Gravel drains", the excess pore-pressure is dissipated through highly-permeable gravel drains installed in liquefiable sand deposits so that they drain excess pore-water during seismic loading to the ground surface before the pore-pressure builds up to 100%. The columns are installed by first penetrating a casing tube into the ground, filling it with gravels of regulated grain-size distributions and pulling up the tube while constructing the gravel column without compacting it. They are placed in grids of equal prescribed spacing or aligned along foundation margins. In order to determine the grid spacing properly, the consolidation theory developed for the vertical drains for consolidation of clay deposits can be utilized considering the time-dependent excess-pressure generation (Yoshikuni and Nakanodo 1974, Seed and Booker 1977). There is another method popular in US similar to the gravel drain named as "Stone column", wherein compacted

**Counter measures by foundations and superstructures**
- **Shallow foundations, Pile foundations**

**Soil improvements**
- **Densification:**
  - **Vibratory tamper/roller, Vibro-rod**
  - **Sand compaction pile, Heavy tamping**
  - **Stone columns**
- **Solidification:**
  - **Shallow soil stabilization**
  - **Grouting: Compaction, Permeation, Jet**
  - **Admixture in deeper ground**
- **Pore-pressure dissipation:** **Gravel drain,** **Stone columns**
- **Dewatering/desaturation:** **Air-injection**

*Figure 5.10.37* Various mitigation measures for liquefaction.

gravel columns not only drain pore-water but also densify the surrounding soils (e.g. Adalier et al. 2003).

(iv) Dewatering/desaturation: Dewatering is sometimes employed as an effective and economic measure for liquefaction mitigation. It works because, by lowering the water table, the unliquefiable cap layer becomes thicker and the effect of underlying liquefied layer on structures at the ground surface is suppressed. It also upgrades the liquefaction resistance because *CSR* of a given earthquake $CSR = (\alpha_{\max}/g)(\sigma_v/\sigma_v')r_d r_n$ in the free field in Eq. (5.10.2) tends to decrease because the effective overburden $\sigma_v'$ increases while the total overburden $\sigma_v$ does not change so much. However, the ground settlement associated with dewatering may become a problem in site conditions where highly compressible soil layers are underlying. As for the desaturation, liquefaction resistance *CRR* tends to rise considerably (maximum about twice) by lowering the saturation from $S_r = 100\%$ to 90% as mentioned in Sec. 5.7.1. Hence, economical liquefaction mitigation measures by injecting air bubbles into liquefiable soil layers have recently been developed and laboratory/in situ demonstration tests have been carried out (Okamura et al. 2009). The sustainability of the low saturation by air-bubbles for several tens of years as well as technical details on non-localized air-bubble injection are to be demonstrated in situ for its actual use.

## 5.11 BASE-ISOLATION DURING LIQUEFACTION

### 5.11.1 Base-isolation case histories

During past earthquakes having triggered liquefaction, seismic base-isolation effects were sometimes observed on structures resting on liquefied ground. One of the first known cases is apartment buildings in Kawagishi-cho, Niigata city, during the 1964 Niigata earthquake. The four-story RC buildings of shallow foundations resting on very loose liquefiable sand settled and tilted considerably. There were seismometers at the base and roof on one of the buildings recording the earthquake motions in

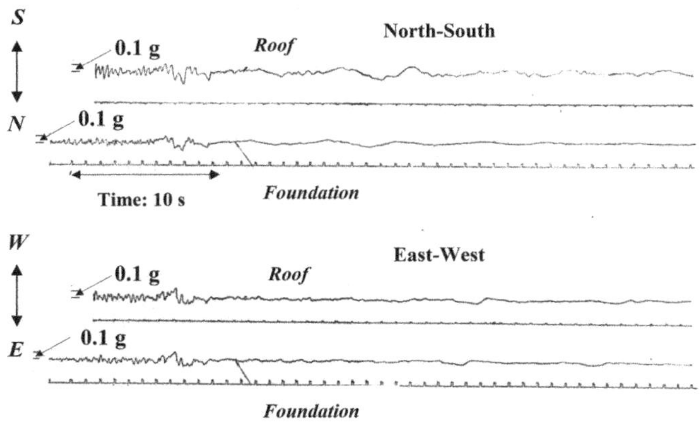


*Figure 5.11.1* Acceleration records reflecting liquefaction-induced base-isolation at Kawagishi-cho apartment in Niigata city during 1964 Niigata earthquake, Japan (Kanai 1966).

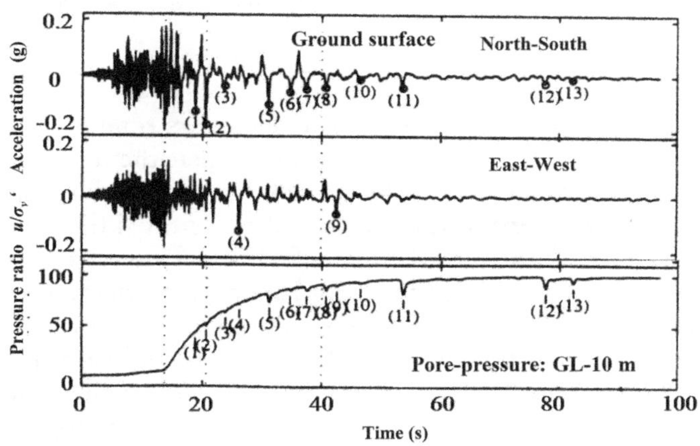


*Figure 5.11.2* Acceleration and pore-pressure records reflecting liquefaction-induced based isolation at Wildlife site during 1987 Imperial Valley earthquake, in USA (Adalier et al. 1997).

Fig. 5.11.1, that demonstrated the clear base-isolation wherein only a low-amplitude long-period motion sustained after some seconds of S-wave shaking (Kanai 1966).

Such seismic records reflecting the liquefaction-induced base-isolation have increased in number since then. Wildlife site records shown in Fig. 5.11.2 during the 1987 Imperial Valley earthquake in California, USA, consist of not only ground motions but also in situ excess pore-water pressure 10 m deep from the ground surface (Adalier et al. 1997). The horizontal acceleration time histories at the surface are peculiar reflecting the cyclic mobility of liquefying sand. Another cyclic-mobility type acceleration record was observed in the vertical array in Kushiro harbor, in Hokkaido, Japan during the 1993 Kushiro-Oki earthquake (Iai et al. 1995).

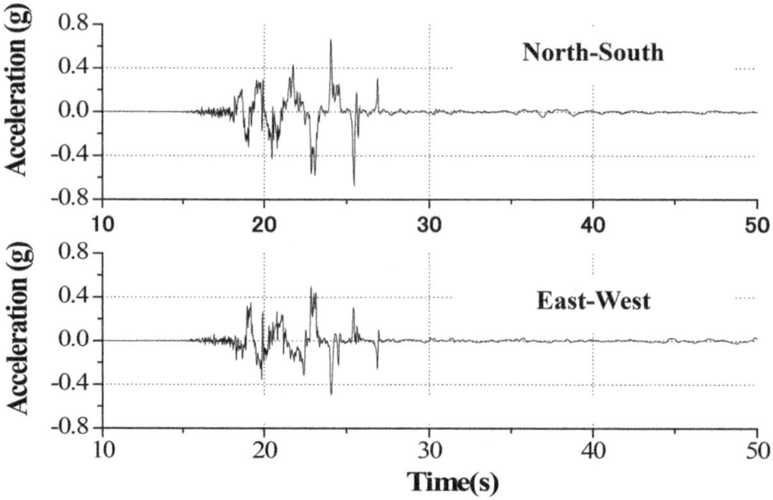


*Figure 5.11.3* Acceleration records reflecting liquefaction-induced based isolation at K-NET Kashiwazaki during 2007 Niigata-ken Chuetsu-oki earthquake, in Japan (plotted from K-NET data).

During the 1995 Kobe earthquake, downhole array strong motion records were obtained as shown in Fig. 4.3.3 in Kobe Port-Island, where the surface reclaimed layer of decomposed granite soils of about 16 m thick liquefied extensively. The records at 4 different levels in the vertical array clearly demonstrated the deamplification in horizontal accelerations on the surface of liquefied layer (Sato et al. 1996). The wave energy flow was calculated based on the records, indicating that considerable energy dissipation occurred in the liquefied layer (Kokusho and Motoyama 2002). The cyclic-mobility type acceleration motion in Fig. 5.11.3 was recorded also at Kashiwazaki city during the 2007 Niigataken Chuetsu-oki earthquake (JGS Reconnaissance Committee 2009). During the 2011 Tohoku earthquake in Japan, cyclic-mobility type acceleration response of young-aged man-made soils in Inage near Tokyo was obtained as shown in Fig. 5.11.4 (Tokimatsu et al. 2012). Thus, there is no doubt that liquefaction makes a significant difference in the seismic response of liquefied ground.

The effect of base isolation was also demonstrated by clear reductions in structural damage of buildings resting on liquefied ground. During the 1964 Niigata earthquake, the 4-story RC apartment buildings in Kawagishi-cho, previously mentioned, stayed perfectly intact despite considerable settlement and tilting, suffering no structural damage such as wall cracks or window-glass breakage. During the 1995 Kobe earthquake, a viaduct highway route running through coastal liquefied man-made lands experienced little damage in the superstructures directly by shaking, while another similar highway route passing through inland unliquefied areas only a kilometer apart, suffered severe damage in RC columns by strong shaking (Matsui and Oda 1996). During the 1999 Kocaeli earthquake in Turkey, apartment buildings in Adapazari city were damaged either by inertial effects or by liquefaction and the two types of damage did not overlap in the same area (Yoshida et al. 2001). It was also pointed out during

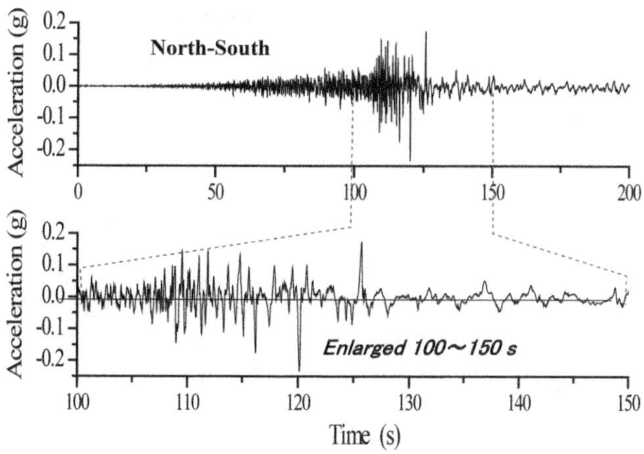


*Figure 5.11.4* Acceleration records reflecting liquefaction-induced based isolation at K-NET Urayasu during 2011 Tohoku earthquake, in Japan (plotted from K-NET data).

the 2004 Niigataken Chuetsu earthquake in Japan that heavy roof tiles of traditional Japanese wooden houses serve as an indicator if their foundation ground liquefied or not; no roof tile damage if liquefied. More recently, almost no structural damage of tens of thousands wooden houses including roof tiles by seismic inertial force in superstructures has been observed during the 2011 Tohoku earthquake in heavily-liquefied areas near Tokyo in Japan.

### 5.11.2 Base-isolation in terms of energy

Thus, the liquefaction-induced base-isolation is widely recognized not only in observed seismic records but also in actual performance of structures. However, the degree of base isolation seems to differ depending on the intensity of liquefaction and other conditions. In order to quantitatively understand the liquefaction-induced base-isolation mechanism in a simplified setting, let us consider a uniform fully-saturated sand layer illustrated in Fig. 5.11.5, wherein $\rho$ = soil density, $G_1$ = equivalent shear modulus, $V_{s1} = \sqrt{G_1/\rho}$ = associated S-wave velocity, and $D_1$ = internal damping ratio, all corresponding to the initial strain level of seismic loading. It is assumed that, during a given harmonic motion representing the seismic loading, liquefaction occurs in the upper portion with the thickness $H$ but not in the lower part due to some reasons such as aging effect, transforming the initially uniform layer into a two-layer system of liquefied and unliquefied layers. In the liquefied portion, $G_1$, $D_1$ and $V_{s1}$ change to $G$, $D$ and $V_s = \sqrt{G/\rho}$ cycle by cycle with increasing induced shear strain amplitude, whereas the properties in the unliquefied part are assumed unchanged for simplicity (Kokusho 2014).

The displacement amplitude and associated wave energy of upward harmonic SH wave in the unliquefied layer and liquefied layer at the layer boundary are $A_2$, $E_{u2}$ and $A_1$, $E_{u1}$, respectively, and those at the ground surface are $A_s$, $E_s$. Those of the downward wave are $B_2$, $E_{d2}$ and $B_1$, $E_{d1}$, correspondingly. Utilizing the equations on the wave amplitude ratios already discussed in Sec. 4.6.2, the ratios of the wave

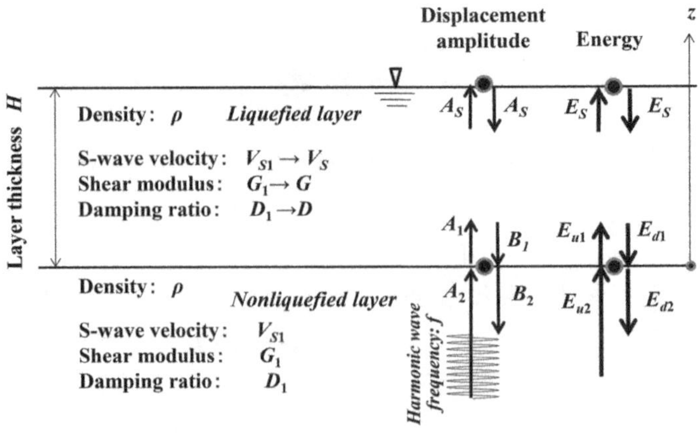


*Figure 5.11.5* A simplified model on liquefaction-induced base isolation in terms of energy (Kokusho 2014).

energies can be written in the following forms, where the complex wave number and the complex impedance ratio can be expressed as $k^* = \omega/V_s^* = \omega/[V_s(1+2iD)]$ and $\alpha^* = \rho V_s^*/\rho_1 V_{s1}^* = \alpha[(1+2iD)/(1+2iD_1)]^{1/2}$, respectively.

$$\frac{E_{u1}}{E_{u2}} = \left|\alpha^* \frac{A_1^2}{A_2^2}\right| = \left|\alpha^* \left[\frac{2}{(1+\alpha^*)+(1-\alpha^*)e^{-2ik^*H}}\right]^2\right| \tag{5.11.1}$$

$$\frac{E_s}{E_{u1}} = \left|\frac{A_s^2}{A_1^2}\right| = \left|e^{-2ik^*H}\right| \tag{5.11.2}$$

$$\frac{E_s}{E_{u2}} = \left|\alpha^* \frac{A_s^2}{A_2^2}\right| = \left|\alpha^* \left[\frac{2}{(1+\alpha^*)e^{ik^*H}+(1-\alpha^*)e^{-ik^*H}}\right]^2\right| \tag{5.11.3}$$

The upward energy at the bottom of the liquefied layer $E_{u1}$ decreases to $E_s$ at the ground surface. The ratio $E_s/E_{u1}$ is expressed by Eq. (5.11.2), which is further approximated for the condition $D \ll 1.0$ by using $k^* = \omega/V_s^* = k/(1+2iD)^{0.5} \approx k(1-2iD)^{0.5}$ as:

$$\frac{E_s}{E_{u1}} = \left|e^{-2ik^*H}\right| \approx e^{-2\frac{\omega D}{V_s}H} = e^{-2\beta H} \tag{5.11.4}$$

Here, $\beta = \omega D/V_s$, the wave attenuation coefficient by internal damping, is a key parameter to determine the energy dissipation during the wave propagation as already discussed in Sec. 1.6.2.

Thus, the base-isolation in liquefied layer splits into the two mechanisms, so that the energy transmission ratio at the ground surface compared to the underlying unliquefied layer $E_s/E_{u2}$ is expressed as the product of the two energy ratios.

$$\frac{E_s}{E_{u2}} = \frac{E_{u1}}{E_{u2}} \times \frac{E_s}{E_{u1}} \tag{5.11.5}$$

Here, $E_{u1}/E_{u2}$ corresponds to the reduction of energy transmission at the boundary as indicated in Eq. (5.11.1) due to the drastic drop of $V_s$ in the liquefied layer. In contrast, $E_s/E_{u1}$ corresponds to the reduction of energy transmission from the bottom to the surface in the liquefied layer by the wave attenuation in hysteretic soil damping as indicated in Eq. (5.11.2) or Eq. (5.11.4) due to the liquefaction-induced drastic change in $V_s$ and $D$ in the liquefied layer.

### 5.11.3 Soil properties by triaxial liquefaction tests

A set of results by stress-controlled undrained triaxial tests may be used to quantify sand properties varying with increasing induced strain for relative density $D_r = 30\sim50\%$ and fines content $F_c = 0\sim20\%$ (Kokusho et al. 2012a,b). In Fig. 5.11.6(a), a typical stress-strain relationship, shear stress $\tau = 0.5\sigma_d$ versus shear strain $\gamma = 1.5\varepsilon$, is exemplified, wherein $\sigma_d$ and $\varepsilon$ are the axial stress and the axial strain. The secant modulus calculated from a straight line connecting the plus and minus peaks of the $\tau\sim\gamma$ curve is denoted as $G_1$ in the first cycle of loading and $G$ in a given cycle. The damping ratio is calculated as $D = \Delta W/(4\pi W)$ from the dissipated energy per one cycle $\Delta W$ (the area ABB′CD) and corresponding maximum elastic strain energy $W$ (the area OBB′).

In Fig. 5.11.6(b), the values of $G/G_1$ and $D$ obtained from the tests are plotted cycle by cycle versus double amplitude strain $\gamma_{DA}$ in the semi-log graph for sands of different $D_r$ and $F_c$. The strain $\gamma_{DA}$ spans from about 0.1% in the first load cycle to the maximum 20~50% in the last cycle. The modulus ratios monotonically decrease from $G/G_1 = 1.0$ to nearly zero with increasing strain amplitude. Damping ratios start from $D_1 \approx 10\%$ in the first cycle and tend to increase to a maximum $D \approx 25\sim30\%$ at around $\gamma_{DA} = 1\%$, eventually converging to $D \approx 10\sim20\%$. For specimens $D_r = 50\%$ and $F_c = 0$, the decreasing trends in damping ratio $D$ after taking the maximum value is more conspicuous than other soils, reflecting the cyclic mobility appearing in the latter part of cyclic loading ($\gamma_{DA} \approx 5\%$ or larger), due to the dilative behavior of medium dense clean sands.

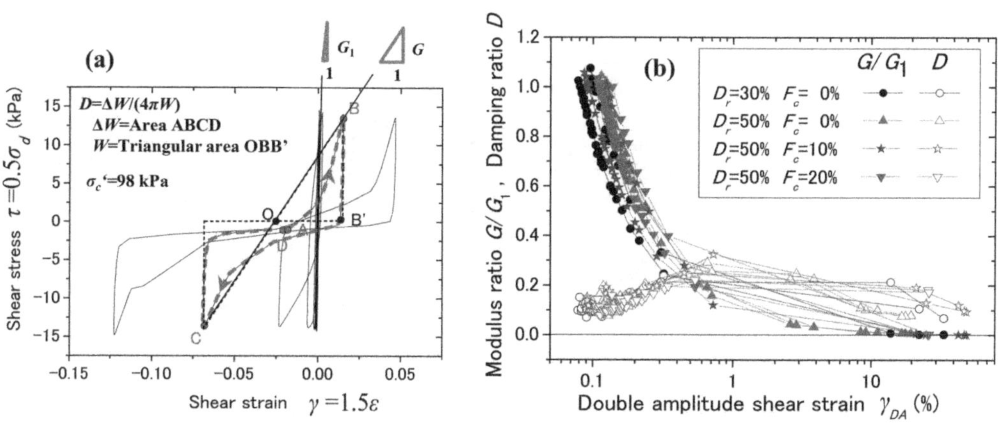


*Figure 5.11.6* Undrained triaxial test results to calculate base isolation effect: (a) Typical stress-strain curve of sands, (b) Modulus/Damping ratios versus double-amplitude shear strain for sands of various $D_r$ and $F_c$ (Kokusho 2014).

### 5.11.4 Energy calculation for base-isolation

Based on the triaxial test results, let us calculate the base-isolation effect in terms of energy for the simplified soil model with a liquefied layer thickness $H = 10$ m shown in Fig. 5.11.5 shaken by a harmonic wave of frequency $f = 1.0$ Hz. It is assumed that the soil properties by the cyclic triaxial test results can be applicable, starting from the initial values $G_1$, $D_1$, $V_{s1}$, and changing to $G$, $D$, $V_s = \sqrt{G/\rho}$ with increasing strain shown in Fig. 5.11.6(b). The shear modulus at the first cycle $G_1$ is determined using the equivalent linear modulus degradation (Hardin-Drnevich 1972b) considering the initial effective overburden stress in the soil model (Kokusho 2014). As for the damping ratio, the measured value in the triaxial test is directly used in the calculation assuming the hysteretic or nonviscous damping. The soil properties thus determined at the mid-depth of the liquefied layer and at the top of the unliquefied layer are used as representative values to compute the energy transmission ratios $E_{u1}/E_{u2}$ in Eq. (5.11.1) and $E_s/E_{u1}$ in Eq. (5.11.2).

In Fig. 5.11.7(a), the two energy ratio, $E_{u1}/E_{u2}$ and $E_s/E_{u1}$, thus calculated are plotted versus double amplitude shear strain $\gamma_{DA}$ in a semi-log diagram. Here, the stress-controlled cyclic triaxial test results in terms of $G$ and $D$ are directly applied cycle by cycle to evaluate the variations in the energy transmission ratio in the liquefied layer with increasing double amplitude shear strain, while the properties, $G1$ and $D1$, in the unliquefied layer are assumed unchanged. The plots with identical symbols are connected with thin dotted lines to show the sequential variation of the two energy ratios, $E_{u1}/E_{u2}$ and $E_s/E_{u1}$, with increasing $\gamma_{DA}$. Almost unique strain-dependent energy ratios hold both for $E_{u1}/E_{u2}$ and $E_s/E_{u1}$ despite the differences in $D_r$ and $F_c$. For $E_s/E_{u1}$, the plots for $F_c = 0\%$ tend to diverge from the thick approximation curve for larger $\gamma_{DA}$-values, reflecting the cyclic-mobility response due to the positive dilatancy of clean sand. Except for this diverging trend, $E_s/E_{u1}$ is smaller (the base-isolation effect is greater) than $E_{u1}/E_{u2}$ for all values of $\gamma_{DA}$, indicating that the base-isolation mechanism due to the wave attenuation in the liquefied layer is more dominant than the mechanism due to the impedance ratio at the boundary

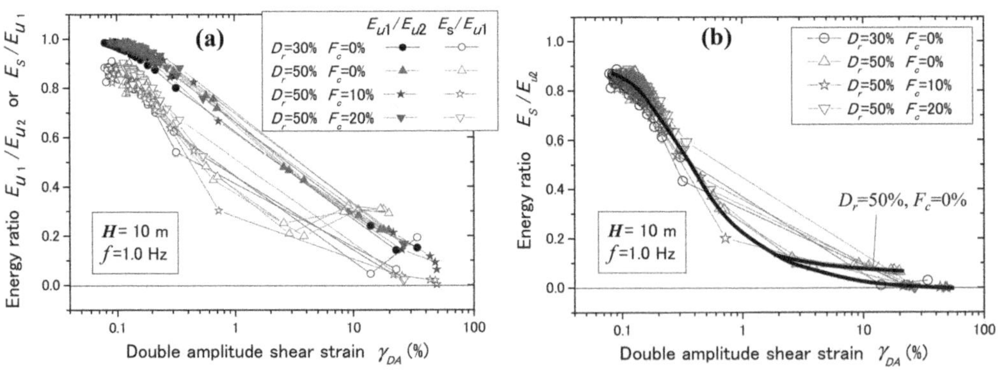


*Figure 5.11.7* Energy ratio calculated from soil properties by triaxial tests: (a) $E_{u1}/E_{u2}$ or $E_s/E_{u1}$ versus $\gamma_{DA}$, (b) $E_s/E_{u2}$ versus $\gamma_{DA}$ (Kokusho 2014).

for the condition of $H = 10$ m and $f = 1.0$ Hz. The damping ratio $D$ does not change so much with increasing $\gamma_{DA}$ as shown in Fig. 5.11.6(b). Hence, this base-isolation mechanism may be mostly attributed to the increase of wave attenuation coefficient $\beta = \omega D/V_s = 2\pi D/\lambda$ mainly because the wave length $\lambda = V_s/f$ is very much shortened with decreasing $V_s$ as liquefaction develops.

In Fig. 5.11.7(b), the global base isolation effect $E_s/E_{u2}$ calculated in Eq. (5.11.5) is plotted versus $\gamma_{DA}$. Again, the trend does not differ so much irrespective of $D_r$ and $F_c$ and may be approximated by the thick solid curve in the diagram. It starts from $E_s/E_{u2} = 0.8{\sim}0.9$ due to the wave attenuation in the pre-liquefaction sand layer, and drastically reduce with increasing $\gamma_{DA}$ to almost zero in contractive soils containing fines. For dilative clean sands, the curve tends to converge to a slightly higher value ($E_s/E_{u2} = 0.05{\sim}0.1$). Thus, perfect base-isolation may be difficult to occur in dilative clean sands compared to contractive sands containing low-plasticity fines, and some minor energy tends to arrive at the ground surface even during very severe liquefaction as observed in some of the actual recorded motions.

In the above, the thickness of liquefied layer was assumed as $H = 10$ m. The energy ratio $E_s/E_{u1}$ tends to increase (the base isolation effect tends to decrease) with decreasing thickness $H$, and become comparable with $E_{u1}/E_{u2}$ for around $H = 2.5$ m, though for a liquefied layer thicker than $H = 5$ m, the former tends to dominate the latter (Kokusho 2014). The base-isolation due to the impedance ratio acts at the boundary only, while that by the wave attenuation works all through wave propagation distance in the liquefied layer. Hence the latter mechanism tends to be more dominant than the former as the liquefied layer becomes thicker. As for the input frequency representing irregular seismic motions, if it is assumed higher than the present value $f = 1$ Hz, the effect of $E_s/E_{u1}$ tends be more dominant with shortening wave length in the liquefied layer, while the effect of $E_{u1}/E_{u2}$ at the boundary is unchanged. Thus, in major liquefaction cases, the wave attenuation effect within the liquefied layer seems to be more dominant than the impedance effect at the layer boundary.

However, the exception of this trend may potentially occur when very loose layered sand liquefies and continuous horizontal water films are formed in an earlier stage of shaking beneath low-permeable layers by the void-redistribution mechanism already discussed in Sec. 5.10.3 (Kokusho and Kojima 2002). If this actually happens, the energy transmission of the SH-wave becomes almost impossible at horizontally continuous water films, realizing near-perfect base-isolation of the SH-wave by that mechanism as presumably observed in the Kawagishi-cho seismic records during the 1964 Niigata earthquake as shown in Fig. 5.11.1.

## 5.12 SUMMARY

1 Cyclic triaxial tests of isotropically consolidated soil specimens are often implemented to evaluate liquefaction resistance of $K_0$-consolidated in situ soils. The cyclic resistance ratio ($CRR$) thus obtained in the laboratory test may be essentially applicable to in situ $CRR$ if the isotropic confining stress is equalized to the in situ effective mean stress. The $CRR$-value is not the same for the same sand, however, but subject to changing effective confining stress or soil depth.

2 The in situ $CRR$ is strongly dependent not only on the relative density but also on soil fabric which reflects how to deposit, stress/strain histories and aging effects. The aging effect is particularly important which is not fully comprehended in the present liquefaction evaluation criteria. In situ soil fabric is difficult to preserve in normal soil sampling and handling for laboratory soil liquefaction tests. Laboratory tests for in situ $CRR$ should be carried out on samples as intact as possible under in situ confining stress.

3 Previous case histories indicate that, in addition to typical clean sands, a variety of non/low-plastic soils are liquefiable, from fine soils to gravelly soils. Plasticity index $I_p$ or clay content $C_c$ is more appropriate for initial screening of soil liquefiability than grain size parameters such as the mean grain size $D_{50}$ or the fines content $F_c$ often used in practice.

4 There are two different views on liquefaction; "cyclic mobility" and "flow-type" which are not yet unified in present liquefaction evaluation practice, though their mechanisms can be clearly differentiated by the Steady State Line (SSL) on the State Diagram. The difference stems from that granular soils, if sheared in the undrained condition, always build up positive excess pore-pressure in cyclic loading in level ground without initial shear stress, while they develop either positive or negative pressure depending on the contractive or dilative side of SSL, respectively, by initial shear stresses near structural loads or sloping grounds.

5 Liquefaction potential is evaluated in present practice in a horizontal free field without sustained initial shear stress $\tau_s$. However, typical liquefaction damage occurs nearby structures or slopes wherein $\tau_s$ is working. If sand is on the dilative side of SSL, it gradually deforms with increasing loading cycles and $CRR$ tends to increase with increasing $\tau_s$, leading the liquefaction evaluation without $\tau_s$ to the safer side. The condition of stress reversal ($\tau_d > \tau_s$) or non-reversal ($\tau_d < \tau_s$) with respect to cyclic stress amplitude $\tau_d$ is another factor influencing the liquefaction potential of dilative sand by cyclic loading.

6 The increasing content of low/non-plastic fines drastically changes the shear-induced volumetric response from dilative to contractive for granular soils of the same relative density. Not only the $CRR$-values but also this effect of fines content should be borne in mind in evaluating liquefaction. If the soils are on the contractive side of SSL under the influence of initial shear stresses, brittle flow-type failures tend to occur and the $CRR$-values become smaller with increasing $\tau_s$. Thus, the significant difference in the failure modes in the dilative and contractive sides of SSL under the initial shear stress needs to be focused in design.

7 Imperfectly saturated sands are also liquefiable, though the $CRR$-values may become doubled at most if the saturation $S_r$ decreases from 100% to 90%. With regard to liquefaction-triggered slope failures, post-liquefaction undrained residual strength tends to be smaller with decreasing $S_r$ for dilative clean sands, though it is vice versa for contractive fines-containing sands.

8 In situ $CRR$-values for liquefaction evaluations are normally determined from penetration tests. In developing the penetration resistance versus $CRR$ correlations used there, the $CRR$-values have been back-calculated from previous liquefaction case studies or measured in laboratory undrained cyclic loading tests on intact soil samples and correlated with corresponding penetration test results in the same soil deposits. The above correlations developed by the

two different methods are confirmed to be almost consistent in some previous researches.

9 In the *CRR* versus penetration resistance correlations, the fines content $F_c$ is recognized to be an important parameter as substantiated in the case history studies and the laboratory tests as well. Hence the *CRR*-values are modified to be higher for higher $F_c$ under the same penetration resistance in liquefaction evaluation criteria used in current engineering practice worldwide, though its mechanical basis is not yet clarified. A series of mini-cone triaxial tests indicate that this $F_c$-dependency of the correlation may be attributable not to the fines content itself but to the aging effect by cementation. Sands with higher $F_c$ tend to develop stronger cementation and higher *CRR* in the same geological age, because finer soil particles with larger surface areas are more active in geochemical reactions.

10 Gravelly soils liquefied so far are well-graded and of much higher dry density than poorly-graded sands, though their relative densities $D_r$ and permeability constants are not much different from liquefiable sands. In characterizing liquefaction behaviors of well-graded granular soils containing fines, a concept of critical fines content by "Binary Packing Model" may be conveniently incorporated though qualitatively, wherein the soil structure changes from grain-supporting to matrix-supporting with increasing fines content.

11 For well-graded gravelly soils, *CRR*-values corresponding to the initial liquefaction, for double amplitude axial strain $\varepsilon_{DA}=5\%$, are not much different from poorly-graded sands of the same $D_r$. However, their post-liquefaction residual undrained strengths for larger strains are far greater, about 10 times larger than sands at $\varepsilon_{DA}=20\%$ for example. Because gravelly soils give much higher SPT $N_1$-values than sands for the same $D_r$ particularly for $D_r > 50\%$, *CRR* of gravelly soils tends to be lower than sands for the same $N_1$-values, as far as the initial liquefaction corresponding to $\varepsilon_{DA}=5\%$ is concerned. However, gravelly soils are much more resistant to post-liquefaction large-strain failures, if gravel particles are non-crushable and contain little fines.

12 In liquefaction potential evaluations, the stress-based method (SBM) is exclusively used in current engineering practice. In SBM, *CRR* is compared with *CSR* during design earthquakes, wherein the key issue is how irregular seismic motions are represented by harmonic motions with the equivalent amplitude and number of cycles. The energy-based method (EBM) is based on experimental facts that liquefaction behavior is uniquely dependent on the energy dissipated in sands almost irrespective of earthquake wave parameters. Hence, the liquefaction energy capacity can simply be compared with the wave energy demand of earthquake motions irrespective of earthquake wave motions. Comparative studies demonstrate that EBM and SBM predict almost comparable liquefaction behavior for some typical earthquake motions, while they tend to diverge for peculiar motions with too large/small accelerations in contrast to too small/large wave energies. It is hence recommended to employ EBM in supplementing SBM for various types of earthquake motions.

13 Soft cohesive soils tend to increase excess pore-pressure by cyclic loading, although they do not liquefy because of the plasticity or cohesion but do soften with degraded shear stiffness. Residual shear strength after cyclic loading tends to

go down to around 60~70% of the strength without cyclic loading. Near slopes and structural loads, the *CRR*-values tend to decrease with increasing initial shear stress, and hence the seismic instability under the initial shear stress may become a critical issue in very soft cohesive soils.

14 One-dimensional soil settlement due to liquefaction is correlated well with induced maximum shear strain during a given seismic motion. This correlation seems to be different for soils with different physical properties but insensitive to confining stresses, mechanical disturbance and soil fabric unlike the liquefaction resistance *CRR*. Recent case histories indicate that sand boiling may become large enough during heavy liquefactions to give additional settlement to be considered in design.

15 Previous case history studies indicate that liquefaction manifestations at the ground surface such as uneven settlements of shallow foundations tend to be eased if there is an unliquefied cap layer of a certain thickness overlying the liquefied layer. Also indicated is that liquefaction occurs first in the free ground and tends to expand to zones near and below shallow foundations, and hence soils beneath shallow foundations of wide dimensions are less prone to severer liquefaction and larger settlements.

16 Lateral flow or lateral spreading of liquefied ground nearby structural loads or slopes may occur in either of the following two mechanisms; undrained shear failure of fines-containing sands on the contractive side of SSL or void redistribution of layered sands on the contractive and even dilative side of SSL. Gently sloping ground in particular may flow in delayed failures due to the void redistribution mechanism if the sand is loose; $(N_1)_{60}$ lower than around 10–15, and sandwiches sublayers or seams with low permeability. Numerical modeling of detailed soil profiles for this mechanism is by no means easy, and the associated residual shear strengths for design are back-calculated from case histories. Besides that, lateral flows of liquefied level ground have occurred several times during recent earthquakes behind retaining walls displaced by earthquake effects.

17 Lateral flows in liquefied soils have significant effects on pile foundations to be considered in design. The effect of unliquefied cap layers is dominant in particular in determining the bending moments, displacements and failures of the piles, in contrast to the underlying liquefied layer. In order to consider the performance of piles with various flexural rigidities relative to lateral soil displacements, the design scheme wherein kinematic soil displacement by the lateral flow is given to the piles via soil springs is recommended rather than the scheme wherein lateral soil pressures are applied directly on the piles. How to evaluate $p{\sim}y$ curves of soil springs and earth-pressures particularly for the unliquefied cap layer is critical in appropriate design.

18 There are enough evidence that the base-isolation effect works in liquefied sites to considerably reduce the superstructural damage caused by inertial effects. This effect is caused by the two different mechanisms; a drastic impedance change between liquefied and non-liquefied layer and increasing energy dissipation during wave propagation in the liquefied layer. The latter mechanism tends to be greater than the former with increasing liquefied layer thickness and increasing predominant frequency.

19 Liquefaction mitigation measures may be opted from countermeasures in superstructure-design and soil improvements. Soil improvements are categorized into densifications, solidifications, pore-pressure dissipations and dewatering/desaturation, all of that have further detailed options. One has to be aware of fast technological advances in soil improvements associated with remediation works after recent earthquakes in Japan (the 1995 Kobe and 2011 Tohoku earthquakes) and in New Zealand (the 2010 Darfield and other earthquakes).

# Chapter 6

# Earthquake-induced slope failures

Slope failures induced by earthquakes occurred many times in the world and inflicted a large number of casualties in history. Some of major failures occurred before 1980 were summarized by Keefer (1984). One of the most devastating slope failure in recent history was 1920 Haiyuan earthquake ($M = 8.5$) in Ningxua Province, China. A number of slides were triggered during the earthquake in the hilly loess plateau area, and the debris, though essentially dry, flowed long distance and covered villages along valleys, killing about two-hundred thousands people (Close and McCormick 1922). Similar slope failures of wet loess slopes occurred in Tajik in Central Asia during 1989 small earthquake ($M = 5.5$), wherein landslides turned to be mud flows due to liquefaction and buried more than 100 houses under 5-meter thick mud with 220 villagers (Ishihara et al. 1990). During 1964 Alaskan earthquake ($M = 9.2$), a number of landslides occurred in the anchorage area, Alaska, USA, such as at L-Street and Fourth Avenue in downtown Anchorage. The largest slide occurred at Turnagain Heights in the suburb with the area of 3 km by 360 m maximum along sea bluffs and the soil volume 9.5 million $m^3$, destroying 75 homes. The soils sliding toward the bluff lines consisted mainly of slightly overconsolidated clays interbedding sandy silts or silty sands, implying possibly a partial involvement of seismic liquefaction of sandy soils (Committee on the Alaskan Earthquake 1971, Seed 1968).

Another devastating slope disaster occurred during 1970 Peruvian earthquake, where debris avalanche originated from a very high mountain top destabilized by a subduction earthquake ($M = 7.7$) attacked villages along valleys, killing at least 18 thousand people. The debris volume was 50~100 million $m^3$ of rock, ice, snow and soil that traveled 14.5 km from the source to the villages at an average velocity 280~335 km/hour (Plafker et al. 1971). During the 1999 Chi-Chi earthquake ($M = 7.3$) in Taiwan, a number of slope failures occurred in the mountainous central Taiwan. Among them, Chiufenerhshan landslide was one of the most devastating, which occurred very near (12–14 km) from the earthquake fault. The soil/rock mass of 1100 m wide and 2000 m long with the initial slope angle 21° slid along a dip plane of 5 m deep and buried 19 houses, killing 41 people (Lin et al. 2009a). Another devastating and larger slope failure occurred during the same earthquake was Tsaoling dip-slope slide, 3800 m wide by 2800 m long and the total volume of about 120 million $m^3$. Residents and their houses sitting on the sliding mass near the crest flew together and landed through a distance 3100 m away and 800 m downward, killing 32 people and 6 survived (Lin et al. 2009b). In El Salvador, a devastating landslide occurred in

Las Colinas during 2001 subduction earthquake ($M = 7.7$), where sliding pyroclastic soils of 200 thousand $m^3$ starting from the top of a slope 175 m high travelled as far as 700 m horizontally and covered houses killing 500 people (Konagai et al. 2009). During 2008 Wenchuan earthquake in China ($M_W = 7.7$), more than 150,000~180,000 landslides were triggered killing 20,000~30,000 people, among which 112 large landslides were with the areas greater than 50,000 $m^2$. These large landslides were markedly located close to the fault rupture in the narrow belt and mainly on the hanging wall side. More than a half of large landslides occurred in rock slopes (Chen et al. 2012, Guo et al. 2015).

In the earthquake country Japan, co-seismic slope failures have occurred very often; even after 1847 Zenkoji earthquake for example they occurred about once in every 15 years on average. One of the most devastating failure among them occurred at a volcanic mountain Unzen-Mayuyama in the Kyushu island triggered by volcanic shallow earthquake ($M_J = 6.4$) in 1792. Totally 440 million $m^3$ pyroclastic debris travelled 4.4 km and triggered tsunami in an inland sea, killing about 15 thousand people. During 1847 Zenko-ji earthquake ($M_J = 7.4$), more than 40 thousand slope failures occurred in central part of the main island of Japan. Among the failures, the sliding mass from Iwakura-yama, 30 million $m^3$ in volume, stopped a river and the breach of the natural dam caused considerable damage. More recently after the 1995 Kobe earthquake, there occurred several earthquakes which triggered significant number of slope failures in many parts of Japan. These damaged slopes in recent history include not only natural but also manmade slopes too. The most recent 2016 Kumamoto earthquake ($M_J = 7.3$) triggered about one thousand slope failures in volcanic slopes in Kyushu island. Before that, 2004 Niigata-ken Chuetsu earthquake ($M_J = 6.8$) in the central Japan and 2008 Iwate-Miyagi Inland earthquake ($M_J = 7.2$) in the northern Japan inflicted thousands of slope failures each, that will be discussed in detail in Sec. 6.5. These case histories indicate that slope failures caused by strong earthquakes can be very large in terms of number, volume as well as travel distance and hence very hazardous. It is also known that the significance of earthquake-induced slope failures is closely correlated with rainfalls before or after the earthquake.

Mechanical impacts of earthquakes on slope failures may be classified into (a) an inertial effect to drive the soil mass, and (b) a cyclic loading effect to weaken the shear resistance of the slope materials by pore-pressure buildup and disturbance of soil structures. After the initiation of sliding, the shear resistance of the soil mass may be further weakened during sliding.

In order to evaluate the potential of slope failures, slip-surface analyses are usually carried out in engineering practice taking account of the earthquake effect by the seismic coefficient. In this method, the possibility of slope failure is evaluated in terms of a safety factor $F_s$ wherein how large the post-failure deformation is out of the scope. For evaluating the post-failure slope displacements, a Newmark-method or its modified version is used assuming a rigid-block slide on a slip plane, though its practical application may be limited within a small displacement of around 1 m. On the other hand, another approximate evaluation method on slope deformations has been used by numerically calculating slope deformations due to self-soil weight using a two-dimensional FEM model considering the change of secant shear moduli before and after an earthquake, though large displacements are out of the scope, again.

In the following Sections, basic theories and methodologies employed in the current practice for slope failure evaluations during earthquakes will be addressed. After that, an energy principle governing the slope failures will be discussed based on a simple energy-based theory and shaking table model tests to evaluate post-failure runout distance. On the other hand, a number of case history data on slope failures during recent earthquakes in Japan are statistically analyzed to understand the actual behavior of failed slopes including failed volumes and runout-distances. The energy-based evaluation method is applied to the case histories to have an insight of the seismically-induced slope failures and to back-calculate mobilized friction coefficients for the evaluation of runout-distances.

## 6.1 SLIP-SURFACE ANALYSIS BY SEISMIC COEFFICIENT

In practice, a potential of seismically induced slope failure is calculated by a slip-surface analysis as illustrated in Fig. 6.1.1 essentially the same as the analysis for non-seismic slope failure. Namely, the driving force is compared with the resistant force along a potential slip surface under the influences of the horizontal seismic inertial force as well as the gravitational force. Here, the seismic inertial force is given as; $F = mgk$, using the mass of sliding soil $m$ and the horizontal seismic coefficient;

$$k = \frac{\text{horizontal acceleration}}{\text{gravitational acceleration}} \tag{6.1.1}$$

The comparison is made in the force balance along the slip surface or the moment balance around the center of the circular slip surface considering the arm length in terms of the safety factor $F_s$ as:

$$F_s = \frac{\text{resistant force}}{\text{driving force}} \quad \text{or} \quad F_s = \frac{\text{resistant moment}}{\text{driving moment}} \tag{6.1.2}$$

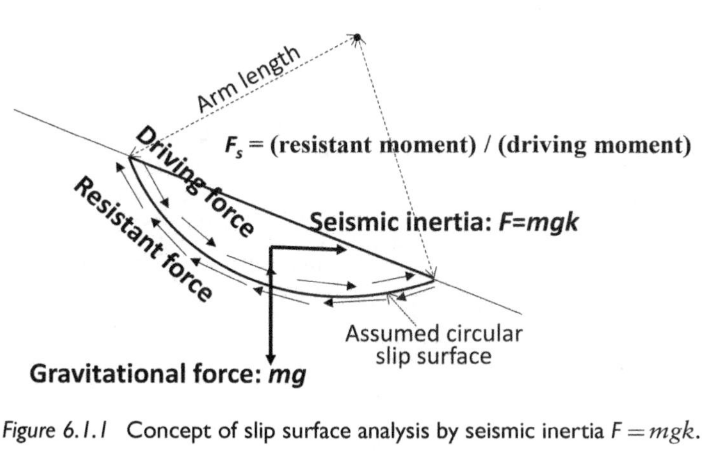


Figure 6.1.1 Concept of slip surface analysis by seismic inertia $F = mgk$.

If $F_s < 1.0$, then the slope is judged to fail. However, the safety factor larger than unity is sometimes assigned ($F_s > 1.0$–$1.2$ for example) considering a safety allowance during earthquakes in design practice. A circular slip surface is most commonly employed, though linear and other curves can be partially used. Several computational schemes are available for the slip surface analyses such as the Swedish method and Bishop method as explained in introductory text books in geotechnical engineering.

In the following, being different from such practical methods using the finite slip surface, a linear slip surface for an infinitely-long linear slope is considered in order to basically understand the force balance in seismically-induced slope failures.

### 6.1.1 Unsaturated slip plane

If the infinitely-long slope with its density $\rho$ and gradient $\beta = \tan\theta$ shown in Fig. 6.1.2(a) is loaded with seismic coefficient $k$, and the sliding occurs along a linear slip plane at the depth $D$, the forces $N$ and $T$ of a slope section (horizontal length $L$ and mass $M = \rho DL$) working normally and tangentially to the slip plane, respectively, are written as

$$N = Mg(\cos\theta - k\sin\theta), \quad T = Mg(\sin\theta + k\cos\theta) \tag{6.1.3}$$

Then, the safety factor $F_s$ for the slope to be stable is written using the friction coefficient $\mu = \tan\phi$ and the cohesion $c$ as:

$$F_s = \frac{N\tan\phi + cL/\cos\theta}{T} \tag{6.1.4}$$

$F_s \geq 1.0$ for the slope stability gives the following equation using Eq. (6.1.3).

$$\rho D(\cos\theta - k\sin\theta)\tan\phi + c/\cos\theta \geq \rho D(\sin\theta + k\cos\theta) \tag{6.1.5}$$

and, the critical seismic coefficient for the slope to slide at $F_s = 1.0$ is obtained as:

$$k_{cr} = \frac{(\tan\phi - \tan\theta) + c/(\rho D\cos^2\theta)}{(1 + \tan\phi\tan\theta)} \tag{6.1.6}$$

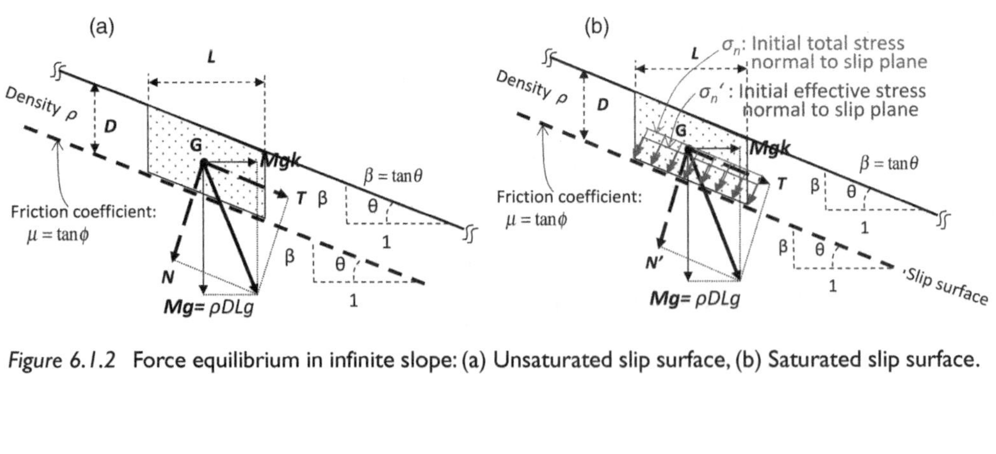


*Figure 6.1.2* Force equilibrium in infinite slope: (a) Unsaturated slip surface, (b) Saturated slip surface.

If the cohesion $c=0$, then:

$$k_{cr}=\frac{(\tan\phi-\tan\theta)}{(1+\tan\phi\tan\theta)}=\tan(\phi-\theta)=\frac{\mu-\beta}{1+\mu\beta} \tag{6.1.7}$$

indicating that without the seismic effect $k_{cr}=0$, the slope is stable for $\phi\geq\theta$ quite reasonably. If the seismic coefficient $k$ is prescribed, the critical friction coefficient $\mu_{cr}$ is written as:

$$\mu_{cr}=\tan\phi_{cr}=\frac{(\tan\theta+k)-c/\rho gD\,cos^2\theta}{(1-k\tan\theta)} \tag{6.1.8}$$

### 6.1.2 Saturated slip plane

Assume here that the slip surface is saturated and undrained during seismic loading below the groundwater table as indicated in Fig. 6.1.2(b) and consists of not stiff rocks but softer geo-materials wherein the principle of effective stress holds (the material properties are controlled by the effective stress). Then the seismic force normal to the slip plane is temporarily carried by the pore-pressure changes so that the effective stress stays unaffected. Hence, the normal force $N$ in Eq. (6.1.3) is replaced by $N'$ independent of seismic coefficient $k$ while $T$ is the same as in Eq. (6.1.3) as:

$$N'=\frac{\sigma'_{n0}L}{\cos\theta},\quad T=\frac{\sigma_{n0}L(\sin\theta+k\cos\theta)}{\cos^2\theta} \tag{6.1.9}$$

Here, $\sigma_{n0}$ and $\sigma'_{n0}$ are the total and effective normal stresses, respectively, working on the slip plane before an earthquake and expressed using the initial pore-water pressure $u_0$ as:

$$\sigma_{n0}=\rho gD\cos^2\theta,\quad \sigma'_{n0}=\sigma_{n0}-u_0 \tag{6.1.10}$$

The force-equilibrium equation is written as:

$$T=N'\tan\phi+\frac{cL}{\cos\theta} \tag{6.1.11}$$

Then, substituting $N'$ and $T$ in Eq. (6.1.9) to (6.1.11) makes the next equation for the saturated slip plane.

$$k_{cr}=\frac{\sigma'_{n0}}{\sigma_{n0}}\tan\phi-\tan\theta+\frac{c}{\sigma_{n0}} \tag{6.1.12}$$

If $c=0$, and modified friction coefficient $\mu^*$ and friction angle $\phi^*$ for the saturated slip plane are defined as:

$$\mu^*=\tan\phi^*=\left(\frac{\sigma'_{n0}}{\sigma_{n0}}\right)\tan\phi \tag{6.1.13}$$

then Eq. (6.1.12) for $c=0$ is simplified as:

$$k_{cr} = \tan\phi^* - \tan\theta \tag{6.1.14}$$

Here, a special care is needed how to determine the strength parameters $\mu = \tan\phi$ and $c$. Generally, seismically-triggered shear failure tends to occur in the undrained condition in saturated materials because the strain rate during the failure is high enough to prohibit the water migration. Hence, the constants $\phi_{cu}$ and $c_{cu}$ for the consolidated undrained conditions are normally chosen in design as $\phi$ and $c$ in Eqs. (6.1.12)–(6.1.14).

## 6.2 NEWMARK-METHOD

The Newmark model was originally started from a rigid block sliding on a horizontal plane (Newmark 1965) and extended to a block sliding on an inclined plane (Sarma 1975) and further to a soil mass sliding on a curved surface combined with dynamic FEM response analysis of the sloping body (Makdisi and Seed 1978). In the following, the theoretical backgrounds for calculating the slope displacement using these models are discussed.

### 6.2.1 Newmark-method for a rigid block on a straight slip plane

A force equilibrium of a block sliding along a straight slope shown in Fig. 6.2.1 can be formulated as follows (Sarma 1975). Let $\ddot{z}$ be the absolute horizontal ground acceleration, $s$ and $u = s\cos\theta$ the relative displacement of the block along the slope and its horizontal component, respectively. When the acceleration $\ddot{z}$ works in the upslope direction, the inertial force of the block parallel with the slope in the downslope direction is $M\ddot{z}\cos\theta$ and the normal component is $M\ddot{z}\sin\theta$. If the block slides downslope and also using $\mu = \tan\phi$ and $c=0$, the force equilibrium can be written as:

$$\ddot{z}\cos\theta - \ddot{u}/\cos\theta + g\sin\theta = \mu(-\ddot{z}\sin\theta + g\cos\theta) \tag{6.2.1}$$

Then, the relative horizontal acceleration can be expressed as:

$$\begin{aligned}\ddot{u} &= [\ddot{z}(\cos\theta + \mu\sin\theta) - g(\mu\cos\theta - \sin\theta)]\cos\theta \\ &= \left(\ddot{z} - g\frac{\mu\cos\theta - \sin\theta}{\cos\theta + \mu\sin\theta}\right)(\cos\theta + \mu\sin\theta)\cos\theta \\ &= [\ddot{z} - g\tan(\phi - \theta)]\cos(\phi - \theta)\frac{\cos\theta}{\cos\phi}\end{aligned} \tag{6.2.2}$$

The relative acceleration is zero ($\ddot{u}=0$) if $k \leq \tan(\phi - \theta)$, and takes a non-zero value only when

$$\ddot{z} - g\tan(\phi - \theta) = g[k - \tan(\phi - \theta)] > 0 \tag{6.2.3}$$

because the block will slip only if the seismic coefficient satisfies $k > k_{cr} = \tan(\phi - \theta)$ as already indicated in Eq. (6.1.7). Including the other case where the inertial force

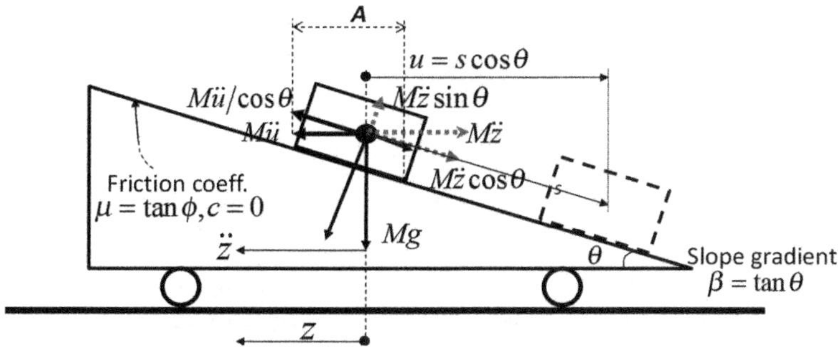


*Figure 6.2.1* Force equilibrium of a block sliding along a straight slope in Newmark model.

works in the upslope direction and the block slides upslope, Eq. (6.2.2) turns out to be

$$\ddot{u} = [\ddot{z} \mp g\tan(\phi \mp \theta)]\cos(\phi \mp \theta)\frac{\cos\theta}{\cos\phi} \tag{6.2.4}$$

with the threshold seismic coefficient as:

$$k_{cr}^{\pm} = \frac{\ddot{z}}{g} = \tan(\phi \mp \theta) \tag{6.2.5}$$

Here, the upper and lower signs correspond to downslope and upslope slide, respectively, and quite reasonably $k_{cr}^{+} \le k_{cr}^{-}$ because $\phi \ge \theta$.

In the saturated condition, Eqs. (6.2.4) and (6.2.5), using Eq. (6.1.13), turns to be:

$$\ddot{u} = [\ddot{z} \mp g(\tan\phi^* \mp \tan\theta)]\cos^2\theta \tag{6.2.6}$$

$$k_{cr}^{\pm} = \frac{\ddot{z}}{g} = \tan\phi^* \mp \tan\theta \tag{6.2.7}$$

Here, let us consider a very basic example where a block at a standstill initially at time $t = 0$ is given a rectangle-shaped acceleration time-history $\ddot{z} = kg$ only for a time $t = 0 \sim t_0$ and slides downslope along the unsaturated slip plane. Eq. (6.2.4) is integrated for the downslope slide to have the relative velocity $\dot{u}_0$ and relative displacement $u_0$ at $t = t_0$ as:

$$\dot{u}_0 = t_0[\ddot{z}_0 - g\tan(\phi - \theta)]\cos(\phi - \theta)\frac{\cos\theta}{\cos\phi} \tag{6.2.8}$$

$$u_0 = \frac{t_0^2}{2}[\ddot{z}_0 - g\tan(\phi - \theta)]\cos(\phi - \theta)\frac{\cos\theta}{\cos\phi} \tag{6.2.9}$$

After $t = t_0$ when $\ddot{z} = 0$, Eq. (6.2.4) for the downslope slide becomes

$$\ddot{u} = -g\tan(\phi - \theta)\cos(\phi - \theta)\frac{\cos\theta}{\cos\phi} \tag{6.2.10}$$

By integrating this using the velocity $\dot{u}_0$ and displacement $u_0$ at $t = t_0$ in Eqs. (6.2.8) and (6.2.9), the velocity and displacement after $t = t_0$ is written as:

$$\dot{u} = [\ddot{z}_0 t_0 - gt\tan(\phi - \theta)]\cos(\phi - \theta)\frac{\cos\theta}{\cos\phi} \tag{6.2.11}$$

$$u = \left[-\frac{1}{2}\ddot{z}_0 t_0^2 + \ddot{z}_0 t_0 t - \frac{1}{2}gt^2\tan(\phi - \theta)\right]\cos(\phi - \theta)\frac{\cos\theta}{\cos\phi} \tag{6.2.12}$$

Hence, the time $t_1$ for the block to stop is obtained by making $\dot{u} = 0$ in Eq. (6.2.11) and using the threshold seismic coefficient $k_{cr} = \tan(\phi - \theta)$ as

$$t_1 = \frac{\ddot{z}_0}{g\tan(\phi - \theta)} t_0 = \frac{k}{k_{cr}} t_0 \tag{6.2.13}$$

If $t_1$ is substituted in Eq. (6.2.12), the final horizontal displacement is obtained.

$$u_1 = \frac{\ddot{z}_0}{g\tan(\phi - \theta)}\frac{t_0^2}{2}[\ddot{z}_0 - g\tan(\phi - \theta)] \times \cos(\phi - \theta)\frac{\cos\theta}{\cos\phi} = \frac{k}{\tan(\phi - \theta)}u_0 = \frac{k}{k_{cr}}u_0 \tag{6.2.14}$$

If the slip plane is saturated, the displacements $u_0$ and $u_1$ in Eqs. (6.2.9) and (6.2.14) respectively are replaced as follows.

$$u_0 = \frac{t_0^2}{2}[\ddot{z}_0 - g(\tan\phi^* - \tan\theta)]\cos^2\theta \tag{6.2.15}$$

$$u_1 = \frac{k}{(\tan\phi^* - \tan\theta)}u_0 = \frac{k}{k_{cr}}u_0 \tag{6.2.16}$$

Fig. 6.2.2(a) shows an example calculation in the unsaturated case for $\ddot{z}_0 = g = 9.8\ \text{m/s}^2$, $t = 2$ s and $\beta = \tan\theta = 0.5$ ($\theta = 26.6°$), $\mu = \tan\phi = 0.8$ ($\phi = 38.7°$). The relative acceleration, which is non-zero because $\ddot{z}_0 = 9.8\ \text{m/s}^2$ exceeds $g\tan(\phi - \theta) = 2.1\ \text{m/s}^2$ only for $t = 0$–2 s as indicated in the top chart, is integrated to have the time histories of velocity in the middle and displacement in the bottom, respectively. The velocity and displacement at $t = 2$ s are $\dot{u}_0 = 17.6$ m/s, $u_0 = 17.6$ m and the time for the block to stop is $t_1 = 9.36$ s with the final displacement $u_1 = 82.4$ m.

If an arbitrary input acceleration motion is given, it is sliced in time increments into a series of rectangle-shaped motions, where the acceleration in each slice is assumed constant. Then velocities or displacements following individual rectangular slices already obtained in Eqs. (6.2.8)~(6.2.14) are superposed with the time-delays to have the global sliding behavior. As an example Fig. 6.2.2(b) shows a harmonic motion of five cycles with maximum acceleration 1.0 g, frequency 0.5 Hz given to a unsaturated slope of $\beta = \tan\theta = 0.1$ ($\theta = 5.7°$), $\mu = \tan\phi = 0.15$ ($\phi = 8.5°$). The threshold accelerations are $g\tan(\phi - \theta) = 0.049$ g downslope and $g\tan(\phi + \theta) = 0.254$ g upslope. In the calculation, the block starts to slide downslope as soon as $\ddot{z}$ exceeds 0.049 g and keeps sliding in the same direction even when $\ddot{z}$ changes directions and exceeds 0.245 g until the downslope velocity turns negative. This indicates that the upslope slide, which is

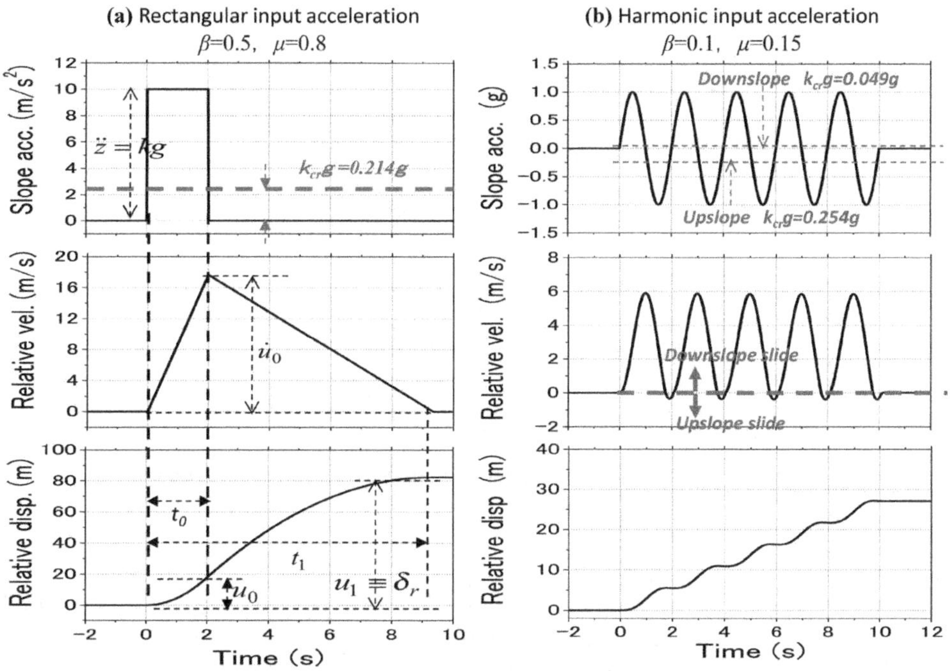


*Figure 6.2.2* Example calculations by Newmark model: (a) Rectangle-shaped acceleration given with $\beta = 0.5$ ($\theta = 26.6°$), $\mu = 0.8$ ($\phi = 38.7°$), (b) Harmonic acceleration given with $\beta = 0.10$ ($\theta = 5.7°$), $\mu = 0.15$ ($\phi = 8.56°$).

realized by a combination of the particular values $\mu$, $\beta$ and $\ddot{z}$ in this example, may not be easy to occur, because the downslope sliding velocity is overwhelming the upslope value normally.

### 6.2.2 Newmark method along a circular slip plane

A similar idea of Newmark model for the sliding block on the straight slope is applicable to curved slopes such as a circular slip surface. The same model as in the normal circular slip surface analysis, wherein the soil mass above a potential slip circle is divided into slices to calculate the force equilibrium of sliding mass as in Fig. 6.2.3, can be used to calculate the slope displacement. When the earthquake inertial force works in the downslope direction, the driving moment $M_D$ around the circle center O can be expressed as:

$$M_D = \sum W_i x_i + \sum k W_i y_i \tag{6.2.17}$$

The corresponding resisting moment if the slip surface is unsaturated is expressed as:

$$M_R = R\left[\sum W_i \left(\cos\alpha_i - k\sin\alpha_i\right)\tan\phi_i + \sum c_i l_i\right] \tag{6.2.18}$$

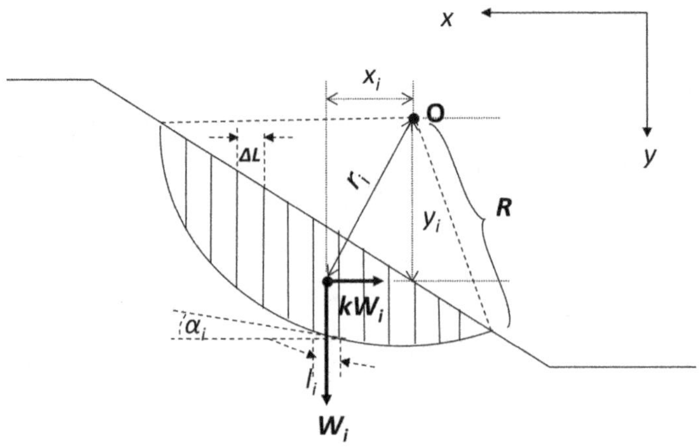


*Figure 6.2.3* Circular slip surface analysis using Newmark method.

Here, $W_i$ = weight, $r_i$ = radius, $x_i$, $y_i = x$, $y$ coordinates at the slice center, $l_i$ = slice width, $\alpha_i$ = angle of slip plane with respect to the horizontal plane, $\phi_i$ = friction angle, $c_i$ = cohesion of the $i$-th slice, and $R$ = radius of the circular slip plane as indicated in Fig. 6.2.3. The sliding is to occur when $F_s = M_R/M_D = 1.0$, hence;

$$F_s = \frac{M_R}{M_D} = \frac{R\left[\sum W_i\left(\cos\alpha_i - k\sin\alpha_i\right)\tan\phi_i + \sum c_i l_i\right]}{\sum W_i x_i + \sum k W_i y_i} = 1.0 \tag{6.2.19}$$

Hence, the threshold seismic coefficient downslope $k_{cr}^{+}$ is obtained as:

$$k_{cr}^{+} = \frac{R\sum W_i\cos\alpha_i\tan\phi_i + R\sum c_i l_i - \sum W_i x_i}{R\sum W_i\sin\alpha_i\tan\phi_i + \sum W_i y_i} \tag{6.2.20}$$

Then, the equilibrium of turning moments around Point O gives the following equation.

$$\begin{aligned}\ddot{\theta} &= \frac{(M_D - M_R)}{J}\\ &= \frac{\left\{\sum W_i x_i + k\sum W_i y_i - R\sum W_i\cos\alpha_i\tan\phi_i + Rk\sum W_i\sin\alpha_i\tan\phi_i - R\sum c_i l_i\right\}}{J}\\ &= \frac{\left\{k\left(R\sum W_i\sin\alpha_i\tan\phi_i + \sum W_i y_i\right) - \left(R\sum W_i\cos\alpha_i\tan\phi_i + R\sum c_i l_i - \sum W_i x_i\right)\right\}}{J}\\ &= \frac{\left(\ddot{z}/g - k_{cr}^{+}\right)\left(R\sum W_i\sin\alpha_i\tan\phi_i + \sum W_i y_i\right)}{J}\end{aligned} \tag{6.2.21}$$

Here, $\ddot{\theta}$ = rotational acceleration around O, $J$ = rotational moment of the sliding block around O, $\ddot{z} = kg$ = horizontal ground acceleration. If the inertial force in the upslope direction is also considered, the equilibrium equation and threshold seismic coefficient are written as:

$$\ddot{\theta} = \left(\frac{\ddot{z}}{g} \mp k_{cr}^{\pm}\right) \frac{(\pm R \sum W_i \sin\alpha_i \tan\phi_i + \sum W_i y_i)}{J} \tag{6.2.22}$$

$$k_{cr}^{\pm} = \frac{R\sum W_i \cos\alpha_i \tan\phi_i + R\sum c_i l_i \mp \sum W_i x_i}{\pm R\sum W_i \sin\alpha_i \tan\phi_i + \sum W_i y_i} \tag{6.2.23}$$

The upper and lower sign correspond to the downslope and upslope slide, respectively. Eqs. (6.2.22) and (6.2.23) have the similar forms as Eqs. (6.2.4) and (6.2.5), indicating that the Newmark method for slip circle failure is equivalent to the sliding block on the straight plane. The moment of inertia $J$ of the sliding mass in the above equation for Fig. 6.2.3 is expressed as the sum of those of individual slices $J_i$ as:

$$\begin{aligned} J = \sum_i J_i &= \rho g \Delta L \sum_i \int_{y_{ib}}^{y_{it}} (y^2 + x_i^2) dy = \rho g \Delta L \sum_i \left| \frac{1}{3} y^3 + x_i^2 y \right|_{y_{ib}}^{y_{it}} \\ &= \rho g \frac{\Delta L}{3} \sum_i (y_{it} - y_{ib})(y_{it}^2 + y_{it} y_{ib} + y_{ib}^2 + 3x_i^2) \end{aligned} \tag{6.2.24}$$

using $\Delta L$ = slice width, $y_{it}$, $y_{ib}$ = $y$-coordinates at the top and bottom of each slice at $x = x_i$. Using Eqs. (6.2.22) and (6.2.23), the time history of sliding angle $\theta$ can be calculated under a given ground acceleration, and the relative circular displacement along the slip plane can be obtained as $s = R\theta$ in the same manner as in the linear slip plane. Unlike the linear sliding, however, coordinates $x_i$, $y_i$ and $\alpha_i$ have to be revised with increasing circular displacement in order to apply these equations to large displacement problems.

### 6.2.3 Newmark-method combined with dynamic response analysis

So far, the slope and sliding block on it are assumed to be a rigid body and the uniform acceleration works throughout the slope including the sliding block before sliding. In order to take the effect of seismic amplification which may occur in actual slopes, dynamic response analyses may be conducted before the Newmark-type analyses to obtain slope acceleration reflecting the dynamic amplification (Makdisi and Seed 1978, Watanabe et al. 1984). The procedures are as follows; (i) a FEM analysis is implemented on the two dimensional discrete model of a slope using a design seismic motion, (ii) a slip plane (a circular plane normally) with the lowest safety factor is determined by comparing the resistance and driving force along potential slip planes based on the stress distribution by the FEM analysis, (iii) the global acceleration of soil mass above the slip plane is calculated at each time increment from accelerations at individual elements as their weighted average to represent the ground acceleration in Eq. (6.2.22), (iv) The Newmark-type analysis is implemented using Eq. (6.2.22) to

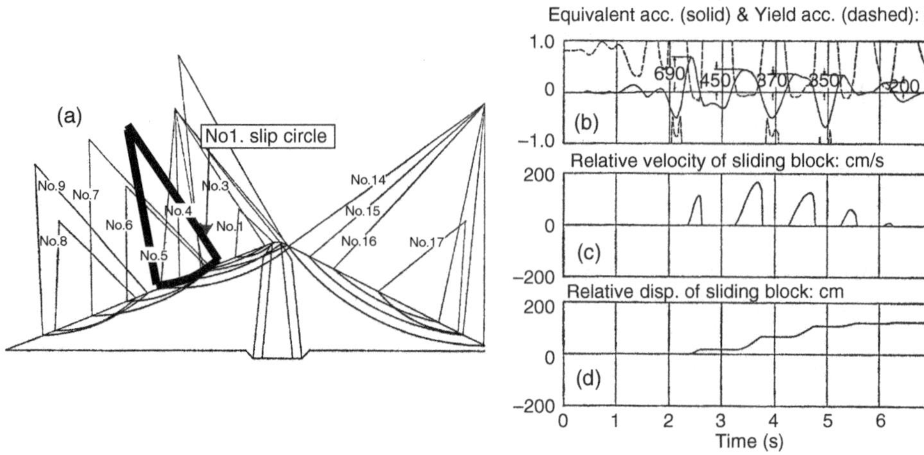


*Figure 6.2.4* Newmark-method combined with dynamic response analyses: (a) FEM model of embankment dam, (b) Equivalent acceleration compared with yield acceleration, (c) Relative velocity, (d) Relative displacement of sliding soil mass (Watanabe et al. 1984).

have the time history of slope displacement, considering the time-dependent change in the threshold seismic coefficient $k_{cr}^{\pm}$.

Figs. 6.2.4(a)~(d) exemplifies one of such studies carried out for an embankment dam (Watanabe et al. 1984). Out of several potential slip circles, the calculated results on the No.1 slip circle in (a) with the thick line are focused here. In (b), the global acceleration $\ddot{z}$ of the soil mass above the slip surface is compared with the threshold values $k_{cr}^{+}g$ (downslope yield acceleration), which is also changing in time due to the effect of vertical acceleration considered in this calculation, and only if the former exceeds the latter, the relative velocity is calculated as in (c) and the residual displacement is integrated from the velocity as in (d). More elaborations may be able to be incorporated in this type of analysis in terms of sliding conditions of the soil mass.

It is noteworthy however that the equivalent ground acceleration obtained as explained above is calculated from the dynamic response of a continuous slope model by the FEM analysis without considering the discontinuity of motions along the slip plane, and considerabe changes in the dynamic soil response that may occur by sliding is not incorporated. This indicates that a significant approximation is involved in determining the global acceleration after the sliding starts to develop large displacement. Thus, one should be aware that the application of this type of analysis is limited to a relatively small displacement, normally considered to be around 1 m.

## 6.3 SELF-WEIGHT DEFORMATION ANALYSIS USING DEGRADED MODULI

A simplified slope deformation analysis using FEM slope models was devised and implemented by comparing self-weight deformations of slopes before and after earthquakes (Lee 1974, Lee and Roth 1977). Unlike the slip surface analysis, the residual displacement of slope is assumed continuous in this analysis without a discontinuous

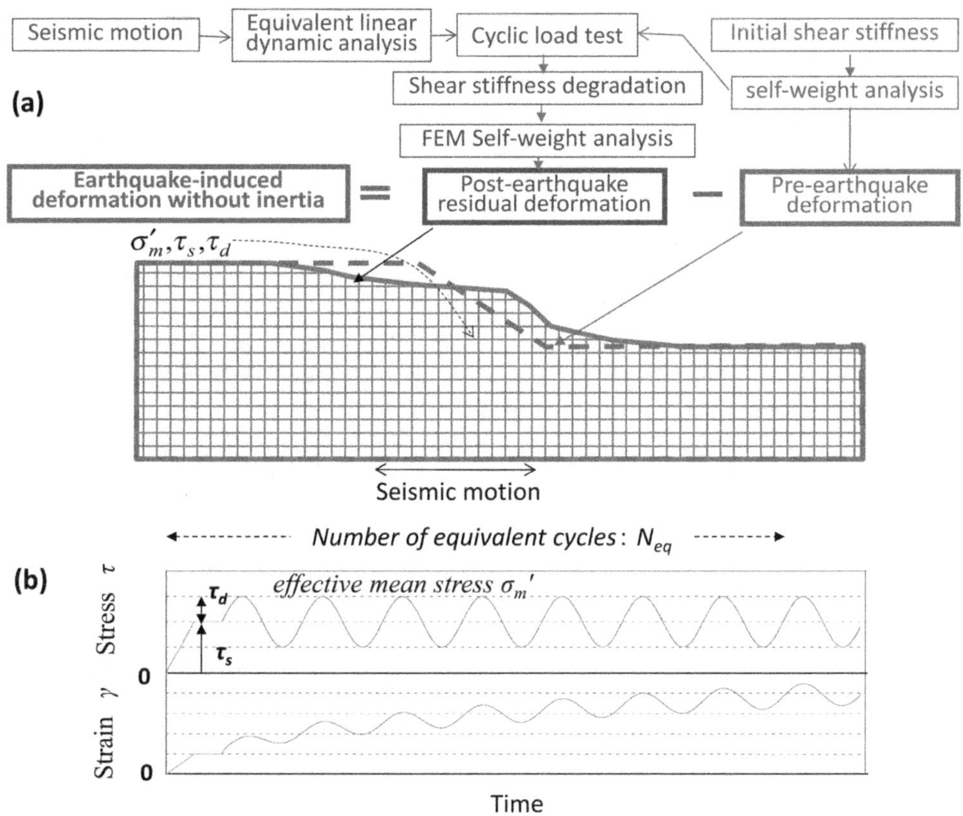


*Figure 6.3.1* Outline of self-weight deformation analysis of slope by degraded shear moduli: (a) Steps of analysis using FEM slope model, (b) Schematic stress history in soil tests for individual elements.

slip plane. Instead, the degradation of shear stiffness due to seismic effects is incorporated in the numerical analysis to calculate the post-earthquake slope deformation by the self-weight (the gravity-turn-on analysis) to compare with the pre-earthquake deformation. Namely, the seismic effect is taken into account not directly by inertial force but indirectly by degrading soil stiffness as the result of earthquake shaking.

### 6.3.1 Outline of analysis

The outline of the analysis is explained using a schematic chart in Fig. 6.3.1(a);

(i) The two-dimensional self-weight static analysis of a slope or embankment is first carried out using initial shear moduli to calculate initial stresses and deformations at various parts in the model. This will yield initial effective mean stress $\sigma'_m$, initial shear stress $\tau_s$ and initial deformation of slopes before an earthquake.

(ii) Then, the dynamic response analysis of the same model is conducted using the strain-dependent equivalent linear soil properties.

(iii) Consulting with the stress states obtained from the static and dynamic analyses above in representative elements, soil element tests are conducted as shown in Fig. 6.3.1(b) using the soil materials as follows;

a) The soil specimens are initially consolidated with effective mean stresses $\sigma'_m$ and the initial shear stresses $\tau_s$ corresponding to different elements.
b) A set of cyclic loading tests are conducted by the shear stress amplitudes $\tau_d$ and the number of cycles $N_{eq}$ equivalent to the seismic shear stresses calculated in those elements.
c) Accumulated shear strains after the cyclic loading in the soil tests are measured, from which degraded secant shear moduli are determined for soil elements at different portions.

(iv) Then, post-earthquake deformations are calculated by the self-weight analysis on the same model using the degraded secant moduli determined element by element from the cyclic loading tests.

(v) Finally, the earthquake-induced residual deformation can be obtained as the difference of the post-earthquake and initial deformations.

Thus, this method tries to capture the essential mechanism of seismically-induced slope deformation exclusively as the result of the soil modulus degradations by ignoring the seismic inertial effect. It is simple and easy to understand intuitively, though its applicability is limited to relatively small deformation without flow failure because the deformation is calculated by the continuum model without discontinuous slip planes. The reliability of calculated deformation largely hinges on how to properly determine the degraded equivalent moduli of various parts of the slope depending on the initial and earthquake stress conditions.

### 6.3.2 Equivalent moduli for residual deformation

Let us consider a soil element in a slope of the angle $\theta$ shown in Fig. 6.3.2(a), wherein the maximum and minimum principal stresses $\sigma'_1$, $\sigma'_3$ are acting on the principal planes which are inclined by the angle $\theta$ from the horizontal plane. The associated stress condition is expressed with the dashed circle ABC in the Mohr's stress diagram in Fig. 6.3.2 (c) before earthquake. If the seismic stress is superposed on it, the shear stress as well as the normal stress works cyclically on the former principal planes. The associated stress variations correspond to the arrows starting from Point A to A′ and A″, which is too complicated to reproduce in laboratory soil tests. Instead, for a soil element rotating by the angle $\theta$ anticlockwise as shown in Fig. 6.3.2(b), the initial stress state on the plane is represented by Point B with the normal and shear stresses $\sigma'_v$ and $\tau_s$ as indicated in Fig. 6.3.2(c). If the earthquake effect is approximated by vertically propagating SH wave even in the slope, the stress changes along the arrows from B to B′ and B″. This stress condition can be reproduced in torsional simple shear tests and even in triaxial tests though approximately using anisotropically consolidated specimens as already discussed in Sec. 5.8.1.

Fig. 6.3.3 shows schematic stress versus strain curves in the soil element tests. In case (a) without initial shear stress, the secant moduli of hysteresis loops, almost symmetric with the origin, tend to decease with the number of cycles $N_c$. If the maximum strain attained by cyclic loading with an equivalent stress amplitude $\tau_d$ and an

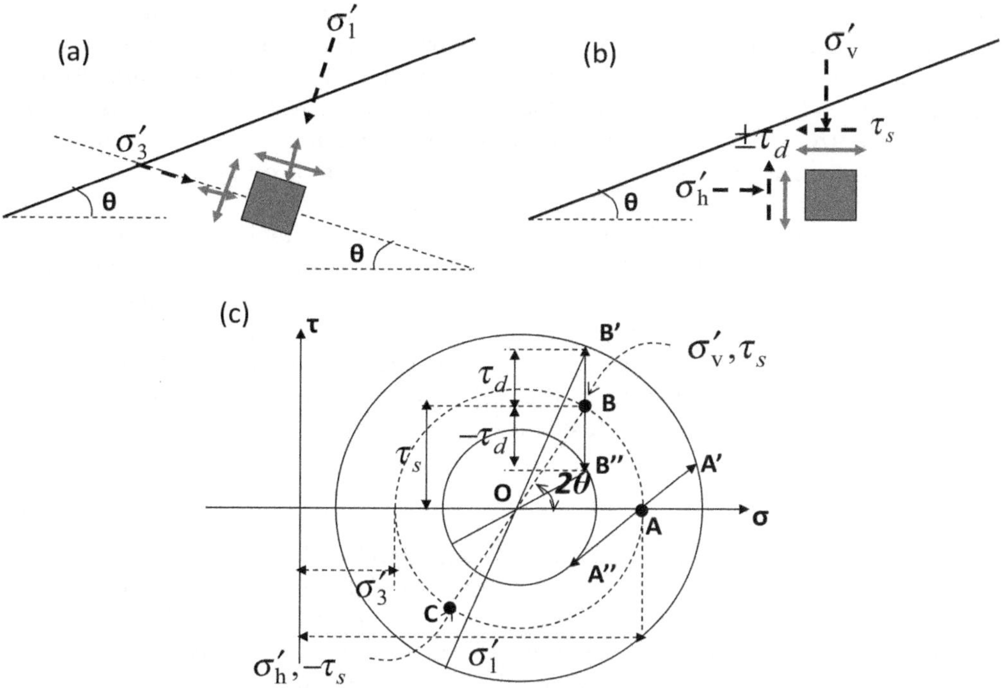


*Figure 6.3.2* Soil elements and associated Mohr's stress circle: (a) Soil element with inclined principal planes, (b) Soil element with horizontal/vertical non-principal planes, (c) Mohr's circle for (a) and (b).

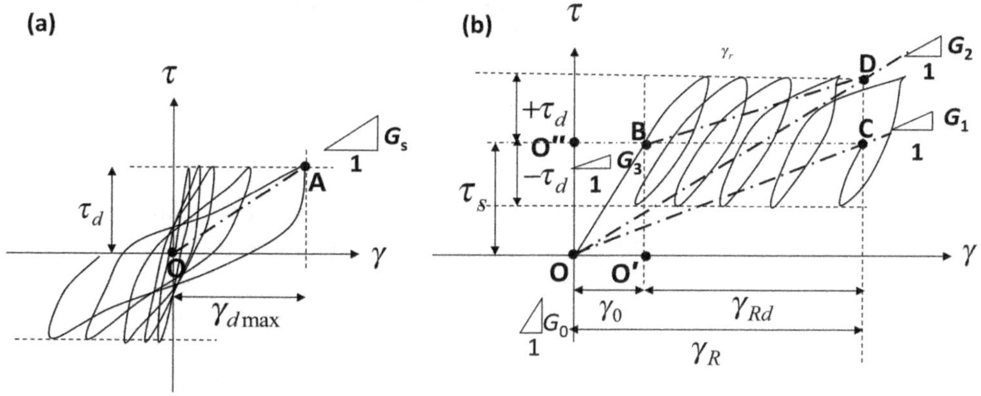


*Figure 6.3.3* Schematic illustration of stress-strain curves in cyclic loading test: (a) Without initial shear stress, (b) With initial shear stress.

equivalent number of cycle $N_{eq}$ is $\gamma_{d\max}$, the line connecting the origin O and the peak point A in the last cycle gives the secant modulus $G_s$ as:

$$G_s = \frac{\tau_d}{\gamma_{d\max}} \tag{6.3.1}$$

This modulus may be used if the soil is assumed to deform initially only under the seismic loading and eventually by the static self-weight as the post-cyclic additional monotonic loading along the degraded stress-strain curve. This assumption is sometimes employed in laboratory tests to obtain the post-liquefaction stress-strain response as mentioned in Chapter 5 by first conducting cyclic loading tests on specimens without initial shear stress followed by undrained monotonic loading. In this way, the secant moduli by the same stress-strain curve in the last cycle can be obtained corresponding to the self-weight static loading. In reality, the sustained initial shear stresses by the dead load will change the soil response under cyclic loading inevitably. If the soil is in the dilative zone as already discussed in Sec. 5.8, the shear moduli thus obtained in the absence of the initial shear stress tend to be smaller and lead to the conservative side in evaluating residual slope displacements.

In the case of Fig. 6.3.3(b), the initial shear stress $\tau_s$ by the self-weight of slope applied in the drained condition to the soil element gives the initial strain $\gamma_0$. The corresponding modulus $G_0$ may be defined as:

$$G_0 = \frac{\tau_s}{\gamma_0} \tag{6.3.2}$$

After that, the earthquake-induced residual strain develops by $\gamma_{Rd}$ from Point B during the undrained cyclic loading with the stress amplitude $\pm\tau_d$ eventually to Point C. If the total residual strain from the origin $\gamma_R = \gamma_0 + \gamma_{Rd}$ is induced only by the static stress $\tau_s$ with the implicit contribution of cyclic loading as actually assumed in the self-weight static analysis, the equivalent modulus $G_{eq}$ from Point O to C may be defined as:

$$G_{eq} \equiv \frac{\tau_s}{\gamma_R} = G_1 \tag{6.3.3}$$

If the contribution of the cyclic loading is taken account explicitly by employing cyclic stress amplitude $\tau_d$ in evaluating the modulus, the equivalent modulus $G_{eq}$ from Point O to D may be written as:

$$G_{eq} \equiv \frac{(\tau_s + \tau_d)}{\gamma_R} = G_2 \tag{6.3.4}$$

In the pioneering application of this type of analysis to an earth dam (Lee 1974), the modulus corresponding only to cyclic loading from Point B to D was chosen as:

$$G_3 = \frac{\tau_d}{\gamma_{Rd}} \tag{6.3.5}$$

and the equivalent modulus $G_{eq}$ from Point O to D was calculated as:

$$G_{eq} = \frac{1}{1/G_0 + 1/G_3} \tag{6.3.6}$$

as the composite modulus of $G_0$ and $G_3$ connected in series.

In any case, the earthquake-induced slope deformation $\delta_p$ is obtained as:

$$\delta_p = \delta_{ip} - \delta_i \tag{6.3.7}$$

Here, deformations $\delta_{ip}$ and $\delta_i$ are calculated from the gravity-turn-on analysis by self-weight using the post-earthquake equivalent modulus $G_{eq}$ and pre-earthquake modulus $G_0$ in Eq. (6.3.2), respectively. The problem herein is which equivalent modulus $G_{eq}$ among Eqs. (6.3.3), (6.3.4), (6.3.6) is proper to be employed in the analysis. The deformation analysis is conducted by applying the self-weight static load only, though the residual strain $\gamma_R$ obtained in the soil tests fully reflect the effects of cyclic loading under the self-weight loading. Considering that slope deformation or strain is calculated not directly by dynamic stress $\tau_d$ but by static shear stress $\tau_s$ due to self-weight using the seismically-degraded soil moduli, the definition of $G_{eq} = G_1$ in Eq. (6.3.3) seems reasonable.

This method, as a kind of hybrid methods, essentially depends on the FEM analysis while the moduli incorporated for different portions of the slope are determined from laboratory element tests. A large amount of engineering judgment has to be incorporated to properly determine the equivalent shear moduli considering the stress conditions in individual model elements, where a lot of ambiguities are included.

## 6.4 ENERGY-BASED SLOPE FAILURE EVALUATION

As already mentioned, slope failures induced by earthquakes occurred many times in the world and inflicted a huge number of casualties in history. As shown by Keefer (1984) in statistical analyses of seismically-induced slope failures worldwide, the numbers and affected areas of failed slopes are closely correlated to earthquake magnitudes and fault distances. This indicates that the earthquake energy which is directly related to the magnitude and distance may be a key parameter governing the seismic slope performance. On the other hand, it will be shown later that the earthquake inertial effect is only a trigger and the gravity is a major player in devastating slides wherein the failed debris travel long distance. The Newmark-type analyses using an earthquake acceleration time-history may be applicable to residual displacements of around a meter but irrelevant in evaluating longer travel distance. Instead, an innovative energy-based evaluation in which the gravitational potential energy is focused based on a simple energy principle may be applicable to evaluate long runout-distance slope failures.

A basic energy balance involved in earthquake-induced slope failures is first discussed in this Section, followed by shaking table tests to demonstrate the uniqueness of energy in determining the travel distance of failed soil mass. An energy-based evaluation for the travel distance of failed slopes is then explained using a simplified graphical method. How to determine the earthquake energy in situ for the evaluation is also discussed.

### 6.4.1 Energy balance in earthquake-induced slope failure

A basic equation on the energy balance in earthquake-induced slope failures is written as follows (Kokusho and Kabasawa 2004).

$$-\delta E_p + E_{EQ} = E_{DP} + E_k \tag{6.4.1}$$

where $-\delta E_p$ = potential energy, $E_{EQ}$ = earthquake energy to be used for slope failure, $E_{DP}$ = dissipated energy, $E_k$ = kinetic energy of sliding soil debris. The minus sign in the

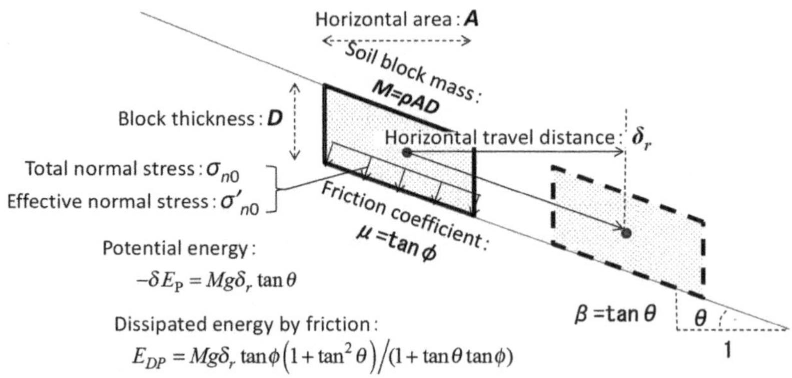


*Figure 6.4.1* A simple model of a rigid block on a straight slip plane for energy-based evaluation.

potential energy is necessary because $\delta E_p$ is normally decreasing during slope sliding. In the incremental form it is written as:

$$-\Delta\delta E_p + \Delta E_{EQ} = \Delta E_{DP} + \Delta E_k \tag{6.4.2}$$

Once failure starts, the amount of dissipated energy is critical to determine if it develops as a flow-type failure of long runout distance or not and how far it flows. In the time increment when the earthquake shaking has already ceased $\Delta E_{EQ} = 0$, the next equation holds;

$$\Delta E_k = (-\Delta\delta E_p) - \Delta E_{DP} \tag{6.4.3}$$

If $\Delta E_{DP}$ is smaller than $-\Delta\delta E_p$, then $\Delta E_k$ is positive and the sliding soil accelerates. It can also be inferred that the shift from slow slide to fast flow may occur not only due to the increase in $-\Delta\delta E_p$ but also due to the decrease of $\Delta E_{DP}$ associated with shear resistance reduction in liquefiable non-plastic soils or high-sensitivity clays. In the fast flow failures, the debris will keep flowing until the kinetic energy plus the subsequent potential energy is all gone. Namely, if $-\Delta\delta E_p$ is smaller than $\Delta E_{DP}$, then $\Delta E_k$ is negative, hence the debris decreases its speed and comes to a halt when the reserved energy is all consumed. Thus, provided that the earthquake energy and the energy dissipation in sliding soil are known, it is possible to evaluate the runout-distance in the flow-type slides by the energy approach.

At the end of sliding $E_k = 0$, Eq. (6.4.1) becomes;

$$-\delta E_p + E_{EQ} = E_{DP} \tag{6.4.4}$$

This equation is applied to a simple model of a rigid block shown in Fig. 6.4.1 with density $\rho$, horizontal area $A$, thickness $D$, and mass $M = \rho AD$ sliding on a straight slip plane of the slope angle $\theta$ and the friction coefficient $\mu = \tan\phi$ (cohesion $c = 0$) (Kokusho and Ishizawa 2007). The associated energies are calculated for the same sliding block as already addressed in the Newmark method for the ground acceleration

$\ddot{z} = kg$ and the residual displacement $\delta_r$ as shown in Fig. 6.2.2(a). It is clear that the potential energy is:

$$-\delta E_P = Mg\delta_r \tan\theta \tag{6.4.5}$$

The dissipated energy during sliding between the block and the slip plane consists of two parts; $\mu M(-kg\sin\theta + g\cos\theta) \times u_0/\cos\theta$ for the former part when the friction under the constant ground acceleration of $\ddot{z} = kg$ causes the displacement $u_0$, and $\mu Mg\cos\theta \times (u_1 - u_0)/\cos\theta$ for the latter part after the acceleration returns to zero $\ddot{z} = 0$ and the displacement increases from $u_0$ to $u_1 \equiv \delta_r$ eventually. Adding these two parts and using $u_1 \equiv \delta_r = ku_0/\tan(\phi - \theta)$ from Eq. (6.2.14), the total dissipated energy can be formulated as follows.

$$E_{DP} = Mg\delta_r \frac{\tan\phi(1 + \tan^2\theta)}{(1 + \tan\theta\tan\phi)} \tag{6.4.6}$$

Hence using Eq. (6.4.4), the earthquake energy for slope failure is expressed as:

$$E_{EQ} = -(-\delta E_P) + E_{DP} = Mg\delta_r \tan(\phi - \theta) \tag{6.4.7}$$

The above equation is also written as follows using $\delta_r = ku_0/\tan(\phi - \theta)$, again.

$$E_{EQ} = Mgk \times u_0 \tag{6.4.8}$$

This confirms the energy principle that the earthquake energy $E_{EQ}$ to be used in the slope failure is equal to the earthquake inertial force $Mgk$ times the displacement $u_0$ undergoing during the force application. If $E_{EQ}$ is normalized by $Mg$ in Eq. (6.4.7),

$$\frac{E_{EQ}}{Mg} = \delta_r \tan(\phi - \theta) \tag{6.4.9}$$

When the slip plane is saturated, Eqs. (6.4.6), (6.4.7), (6.4.9) are modified as follows based on the principle of effective stress as mentioned in Sec. 6.1.2.

$$E_{DP} = \sigma'_{n0} A\delta_r \frac{\tan\phi}{\cos^2\theta} = Mg\delta_r \frac{\sigma'_{n0}}{\sigma_{n0}} \tan\phi = Mg\delta_r \tan\phi^* \tag{6.4.10}$$

$$E_{EQ} = -(-\delta E_P) + E_{DP} = Mg\delta_r(\tan\phi^* - \tan\theta) \tag{6.4.11}$$

$$\frac{E_{EQ}}{Mg} = \delta_r(\tan\phi^* - \tan\theta) \tag{6.4.12}$$

Here, $\sigma_{n0}$ and $\sigma'_{n0}$ are the total and effective stresses, respectively, normal to the slip plane defined in Eq. (6.1.10), and $\phi^*$ is the modified friction angle defined in Eq. (6.1.13).

In the case of ground acceleration $\ddot{z}(t)$ changing with time $t$, $\ddot{z}(t)$ is divided into slices with the same time increment $\Delta t$ and approximated by a multi-step function with constant $\ddot{z}(t)$ in each time increment. Then, it is clear that the energy equations derived above are also valid to slope failures by any arbitrary ground motions, because they are expressed as the superposition of individual energy equations for the rectangular-shaped acceleration shown in Fig. 6.2.2(a).

The ratio between the potential energy $-\delta E_P$ and the earthquake energy $E_{EQ}$ is written for the unsaturated and saturated slip planes, respectively, as:

$$\frac{-\delta E_P}{E_{EQ}} = \frac{\tan\theta}{\tan(\phi - \theta)} \tag{6.4.13}$$

$$\frac{-\delta E_P}{E_{EQ}} = \frac{\tan\theta}{(\tan\phi^* - \tan\theta)} \tag{6.4.14}$$

Thus, the energy ratio derived in the simple model as a function of the friction angle $\phi$ or $\phi^*$ and slope angle $\theta$ indicates that the contribution of the potential energy $-\delta E_P$ becomes greater in comparison with the earthquake energy $E_{EQ}$ with increasing $\theta$.

### 6.4.2 Model shaking table test

In order to know the applicability of the above-energy principle in slope failures, a series of tests on a model slope made from dry sand conducted by Kokusho and Ishizawa (2007) using a spring-supported shaking table shown in Fig. 6.4.2(a) are addressed here. In the tests, the slope angle was parametrically changed as 29, 20, 15 and 10°, considering the angle of repose of the model slope (35.4°) determined from a statically inclining test. The table was initially pulled to several different horizontal displacements and then released to generate decayed free vibrations with parametrically varying frequencies of 2.7, 2.5, 2.2 and 2.0 Hz. In order to single out the energy used for the slope failure, not only the sand slope (Model-A) but a pile of rigid concrete columns of exactly the same weight (Model-B), were tested in the same way as shown in Fig. 6.4.2(b). The decayed vibrations were measured in both Model-A and B, from that the earthquake energy increment used in the model slope$\Delta E_{EQ}$ in Eq. (6.4.2) was evaluated from the loss energies per cycle in Model-A and B, $\Delta W_A$, $\Delta W_B$, respectively as $\Delta E_{EQ} = \Delta W_A - \Delta W_B$. This is because the energy in the two models can be assumed identical except that used for the slope deformation. The total energy $E_{EQ}$ calculated as the sum of $\Delta E_{EQ}$ in all cycles to the end of the slope failure represents the amount of vibration energy contributed to the residual displacement in the model slope (Kokusho and Ishizawa 2007).

In order to correlate the energy $E_{EQ}$ with the slope displacement, residual horizontal slope displacement $\delta_r$ were quantified by means of video images and laser scanning as the average over the slope surface. In Fig. 6.4.3, the residual displacements $\delta_r$ measured in individual tests are plotted versus the vibration energies $E_{EQ}$ contributed to slope failures based on a great number of tests for 4 different slope angles of 29, 20, 15 and 10° under 4 different input frequencies. For each slope angle, all the plots may be represented by a single curve even under different shaking frequencies, indicating that the earthquake energy can serve as a unique determinant for the slope displacement. Fig. 6.4.3 also indicates that the gentler the slope, the greater the energy $E_{EQ}$ is to attain the same residual displacement $\delta_r$. It is further noted that there seems to be a threshold energy, corresponding to each slope angle pointed by the dashed arrow, below which

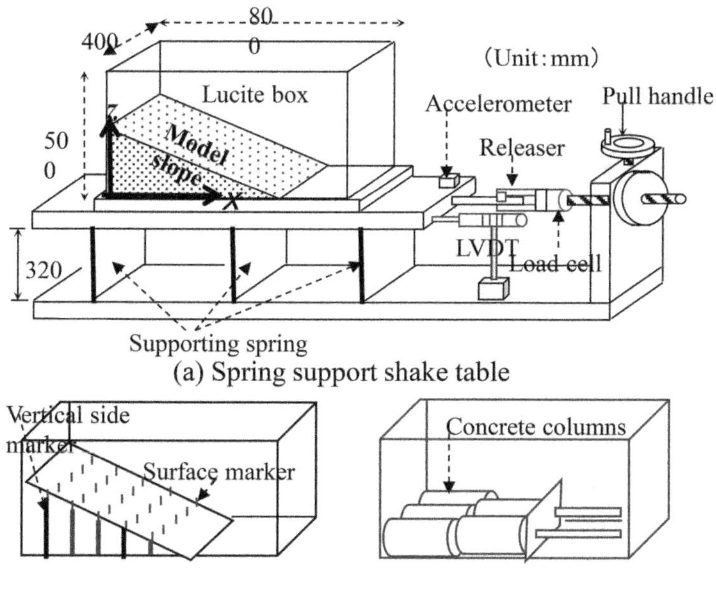


(b) 2 models compared; Model-A (left) & Model-B (right)

*Figure 6.4.2* Shaking table test studying energy balance in seismic slope failures (Kokusho & Ishizawa 2007) with permission from ASCE.

no residual displacement occurs, indicating that the energy uniquely determines not only the post-failure residual displacements but also the initiation of slope failures.

In Fig. 6.4.4, the residual displacements $\delta_r$ for the slope angle $\theta = 29°$, the same as those in Fig. 6.4.3, are plotted versus maximum accelerations in the first cycle of the free-decay vibrations measured from the same tests. Obviously, the same acceleration causes different residual displacements under different input frequencies for the same slope angle. Furthermore, threshold accelerations for the initiation of slope failure are obtained differently due to different input frequencies, indicating that the acceleration unlike the energy cannot serve as a unique determinant not only for the residual slope displacement but also for the failure initiation. This indicates that, unlike the current design practice, slope failures may be actually governed by the energy principle instead of the acceleration or inertial force (Kokusho and Ishizawa 2007). As for the energy-dependency of slope failure initiation, another basic study suggests that the strain energy up to the peak resistance of sliding mass governs the start of slope failures (Kokusho et al. 2014c). Thus, the basic model tests suggest that the energy-based slope failure evaluation may have a stronger physical basis than the acceleration-based evaluation currently employed.

The model test data may be utilized to develop a simplified energy-based evaluation method for slope deformations by comparing with the energy balance of the rigid block sliding on the straight slip plane already discussed in Sec. 6.4.1. In Fig. 6.4.5, the measured residual displacements $\delta_r$ already used in Fig. 6.4.3 are replotted in

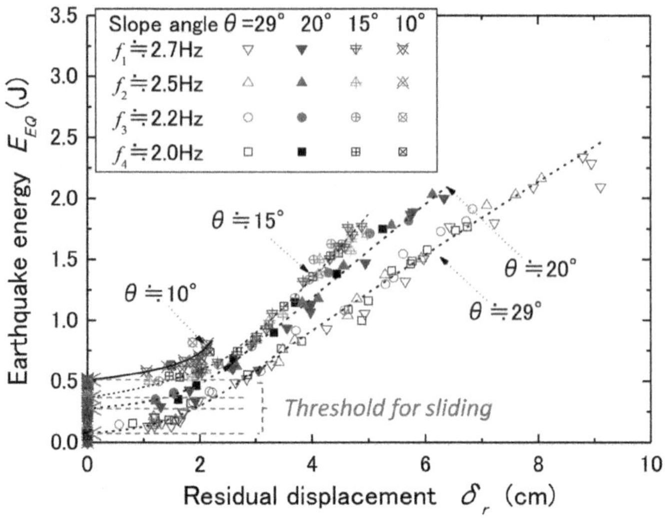


*Figure 6.4.3* Test results on slope displacement $\delta_r$ versus earthquake energy $E_{EQ}$ for 4 slope angles with 4 input frequencies (Kokusho et al. 2011a).

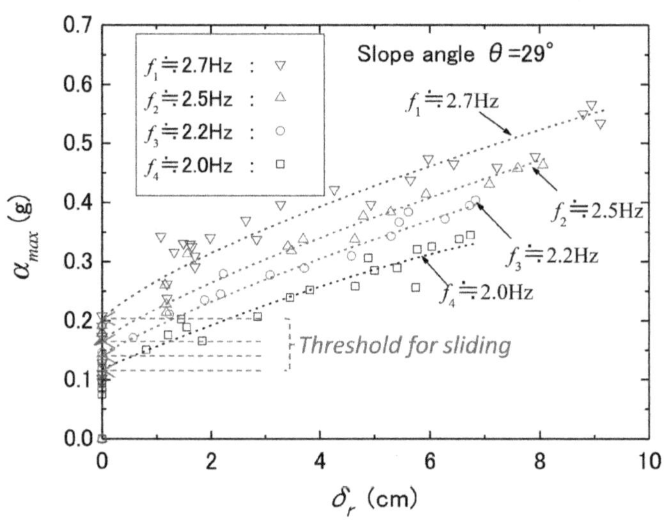


*Figure 6.4.4* Test results on slope displacement $\delta_r$ versus maximum acceleration $\alpha_{max}$ for slope angle $\theta = 29°$ with 4 input frequencies (Kokusho & Ishizawa 2007) with permission from ASCE.

the horizontal axis versus the normalized earthquake energies $E_{EQ}/Mg$, which has the dimension of length, in the vertical axis for the different slope angles and input frequencies. Here, the weight of the displaced soil mass $Mg$ was calculated from Eq. (6.4.5) using the measured potential energy $-\delta E_P$ and the measured displacement $\delta_r$ to comply with the rigid block theory in each test.

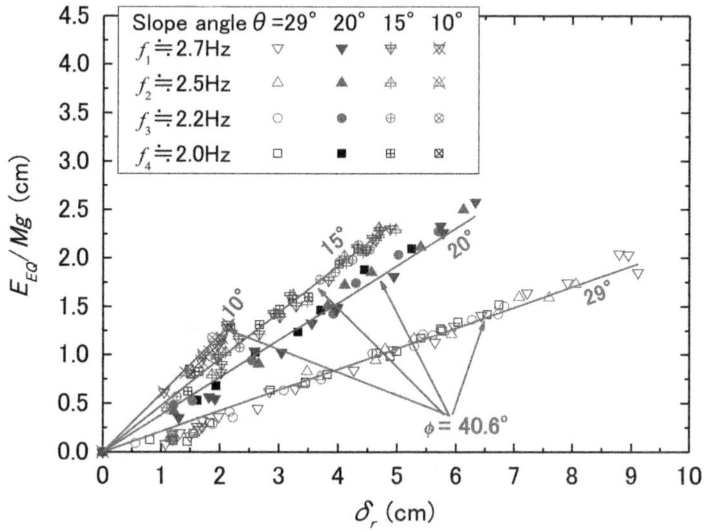


*Figure 6.4.5* Normalized earthquake energy $E_{EQ}/Mg$ versus slope displacement $\delta_r$ for 4 slope angle and 4 input frequencies compared with rigid block model (Kokusho et al. 2011a).

On the other hand, the normalized earthquake energy $E_{EQ}/Mg$ can be theoretically correlated with the block displacement in Eq. (6.4.9), if it is assumed to be identical to the measured residual displacement $\delta_r$ in the model test. If the unknown variable in Eq. (6.4.9), the friction angle $\phi$, is determined here as $\phi = 40.7°$ ($\mu = \tan\phi = 0.86$), almost perfect matching can be obtained between the shaking table test and the theory for the 4 slope angles as depicted with a set of solid straight lines in Fig. 6.4.5. This indicates that if an appropriate friction coefficient is known in advance, the rigid block model with a single slip plane can successfully predict more realistic failure modes with complicated shear mechanisms in the sand slope.

### 6.4.3 Energy-based travel distance evaluation

Based on the model test results and their interpretation by the rigid block theory, the travel distance of earthquake-induced slope failure may be readily evaluated by Eq. (6.4.9) or (6.4.12). Furthermore, the following formula can be derived by combining the potential energy $-\delta E_P$ in Eq. (6.4.5) and the earthquake energy $E_{EQ}$ in Eq. (6.4.7) or (6.4.11).

$$\frac{-\delta E_P/Mg + E_{EQ}/Mg}{\delta_r} = \tan\theta + \tan(\phi - \theta) \tag{6.4.15}$$

$$\frac{-\delta E_P/Mg + E_{EQ}/Mg}{\delta_r} = \tan\phi^* \tag{6.4.16}$$

It is clear in Fig. 6.4.6 that Eqs. (6.4.15) and (6.4.16) for unsaturated and saturated slip planes, respectively, are conveniently used to devise a graphical evaluation of the

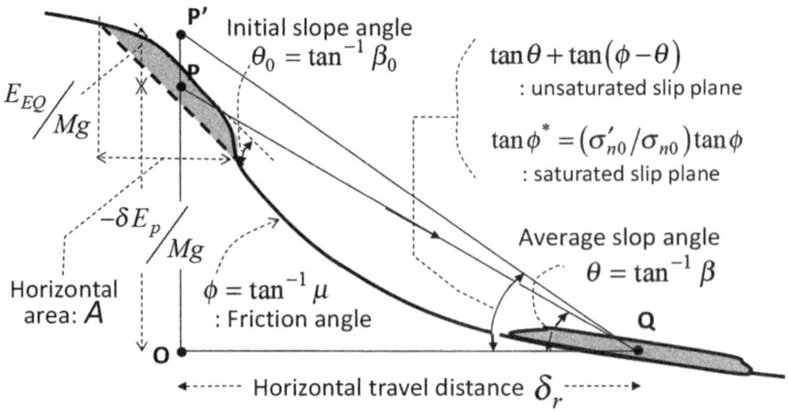


*Figure 6.4.6* Energy-based graphical evaluation on travel distance of failed slope soil mass (Kokusho et al. 2011a).

travel distance of soil mass for arbitrarily-shaped slopes. Namely, the travel distance of a soil mass in a failed slope can be evaluated by first determining the dimension and weight of a potential sliding soil mass and its centroid, Point P. Then, locate Point P′, which is higher than P by the length $E_{EQ}/Mg$, and draw a line from there having an inclination of $\tan\theta + \tan(\phi - \theta)$ or $\tan\phi^*$ for the unsaturated or saturated condition, respectively, until it crosses the downslope surface at Point Q. Thus, the horizontal residual displacement $\delta_r$ can readily be obtained from the slope geometry based on Eq. (6.4.15) or Eq. (6.4.16).

For slopes that are curved as in Fig. 6.4.6, the previous energy equations developed for a straight slip surface can still be used, if $\beta = \tan\theta$ is taken as the global inclination of the straight line PQ (directly connecting the centroids of soil mass before and after failure), different from the initial slope gradient $\beta_0$, and the mobilized friction coefficient $\mu$ is looked upon as the average over the entire travel distance $\delta_r$. The soil weight $Mg$ may be determined by a conventional slip surface analysis, where a potential slip surface having the lowest factor of safety is found. Instead, in quite a few natural slopes, the potential slip surface may be more reasonably assumed to coincide with a bedding plane or a weak seam found in field geological/soil investigations.

This very simple procedure may be conveniently used to evaluate the runout-distance for seismically induced slope failures provided that the earthquake energy $E_{EQ}$ and the mobilized friction coefficient $\mu = \tan\phi$ or $\mu^* = \tan\phi^*$ of a particular slope is known. Although the proper friction coefficient may not be easily determined by laboratory soil tests in many actual slope failures, it may be possible as a more robust way to back-calculate a number of case histories by means of the present energy-based approach and prepare a large database wherein the obtained friction coefficients are correlated with pertinent slope parameters as will be discussed later.

It is also necessary to properly evaluate the site-dependent earthquake energy $E_{EQ}$. As explained in Sec. 4.6, the input energy per unit area $E_{IP}$ at a base layer underlying a slope may be roughly determined by the empirical formulas Eqs. (4.6.19) and (4.6.20)

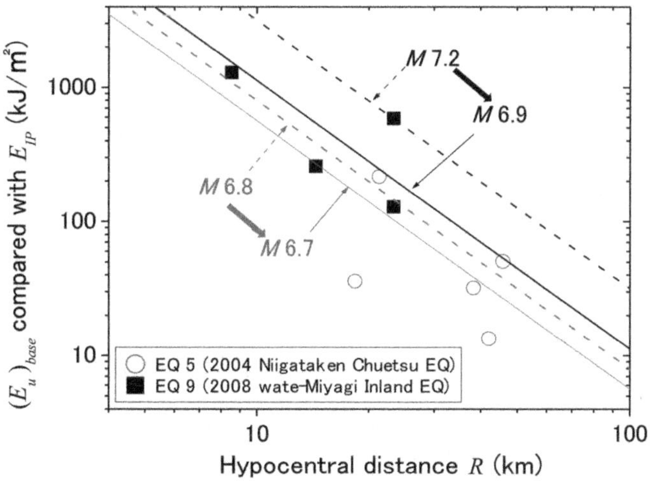


*Figure 6.4.7* Incident wave energy $E_{IP}$ plotted versus hypocentral distance $R$ for two earthquakes (EQ5 and EQ9) (Kokusho et al. 2011a).

from the earthquake magnitude $M$ and focal distance $R$. In order to have a better estimate of the incident energy by such a simple empirical formula, the earthquake magnitudes $M$ to be used may not be officially announced values but modified ones based on energies evaluated from multiple downhole strong motion records nearby (Kokusho et al. 2011a). In Fig. 6.4.7, incident energies per unit area $E_{IP}$ for the two earthquakes (EQ5 and EQ9 already addressed in Sec. 4.6) are plotted versus hypocenral distances $R$. The plots may be approximated better by the empirical formulas assuming $M=6.7$ and 6.9 (the solid lines) instead of the official values $M_J=6.8$ and 7.2 (the dashed lines), respectively.

Then the earthquake energy in the sloping layer near the ground surface may be determined by $\beta=\alpha^{0.70}$ in Eq. (4.6.17) where $\alpha=\rho_1 V_{s1}/\rho_2 V_{s2}$ is the impedance ratio between the sloping layer and the underlying base layer, and $\beta=E_u/(E_u)_{base}$ is the ratio of upward energies between the corresponding layers. $E_u/(E_u)_{base}=0.71$ is tentatively used in the following analyses assuming the impedance ratio as $\alpha=0.61$. Thus the earthquake energy arriving at the slope is calculated as $E_{EQ}=E_u A$, where $A$ is the horizontal area of slope indicated in Fig. 6.4.6 through that the upward energy is coming up.

On the other hand, another series of shaking table tests (Kokusho et al. 2014b) indicates that, out of the arriving earthquake energy $E_{EQ}$, only the energy denoted here as $E^*_{EQ}$ associated with time sections when the seismic inertia is directing downslope contributes to the slope failure. The energy ratio $\beta_E=E^*_{EQ}/E_{EQ}$ was actually calculated for many slope failures during the 2004 Niigataken Chuetsu earthquake (EQ5) and 2008 Iwate-Miyagi earthquake (EQ9) using nearby earthquake records. Fig. 6.4.8(a) and (b) show the variations of $\beta_E$ depending on the azimuth of individual slope failures from the north (anticlockwise) when the two acceleration time histories recorded nearby were used for the two earthquakes. The figure indicates that $\beta_E$ varies

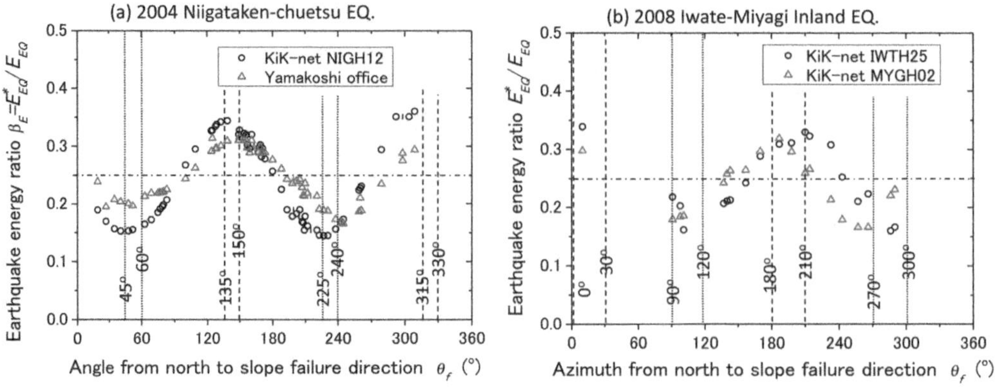


*Figure 6.4.8* Earthquake energy ratio $\beta_E$ calculated for a number of slope failures using two earthquake records (Kokusho et al. 2014b): (a) 2004 Niigataken Chuetsu EQ., (b) 2008 Iwate-Miyagi Inland EQ.

almost periodically with sloping directions from the minimum 0.15 to the maximum 0.35 approximately, and the average $\beta_E$-value takes about 0.25 in the both earthquakes (Kokusho et al. 2014b). This indicates that a quarter of the earthquake energy on average is evenly divided into four directions (up/downslope slope and right/left normal to that). Hence, $E_{EQ}$ in Eqs. (6.4.9), (6.4.11), (6.4.15), (6.4.16) and in Fig. 6.4.6 should be substituted by $E^*_{EQ} = E_{EQ}/4$, and in the following data analyses $E^*_{EQ}$ is used accordingly.

## 6.5 CASE HISTORIES AND BACK-CALCULATIONS BY ENERGY-BASED METHOD

Earthquake-induced slope failures are strongly dependent on site-specific topography, geology, soil conditions as well as earthquake motions. In order to prepare for earthquake-induced slope failures, it is essential not only to depend on the theoretical evaluation methods already mentioned but also to study actual behavior of slopes during previous earthquakes together with their site-specific pertinent parameters. For that goal, case histories on a great number of slope failures during recent earthquakes when digital mapping technologies have become available need to be addressed to statistically analyze them by utilizing reliable air-survey data before and after the earthquakes. Furthermore, those case histories are back-calculated to form the database for the energy approach and discuss on the key mechanism of actual slope failures in the energy perspective.

### 6.5.1 Slope failures during recent earthquakes

Let us take a look at slope failures during two recent large earthquakes occurred in Japan; the 2004 Niigata-ken Chuetsu earthquake and the 2008 Iwate-Miyagi Inland earthquake (EQ5 and EQ9 already addressed in Sec. 4.6), wherein reliable DEM (Digital Elevation Map) data became available before and after the earthquakes.

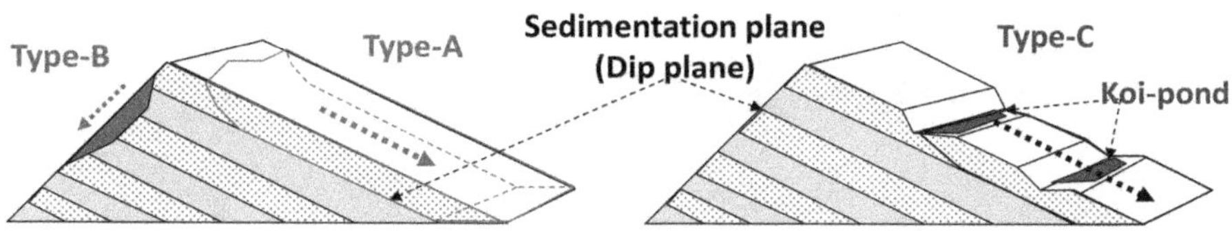


*Figure 6.5.1* Types of slope failures during 2004 Niigataken-Chuetsu earthquake.

#### 6.5.1.1 *2004 Niigataken Chuetsu earthquake*

During the 2004 Niigata-ken Chuetsu earthquake ($M_J = 6.8$, thrust fault, focal depth 13 km), more than 4000 slope failures occurred 200 km north of Tokyo in the main island of Japan. The damaged area belongs geologically to "Green Tuff region" and is known as landslide-prone with geological structures of active folding (JSCE 2007). Slopes were composed of weak sedimented rock of Neogene, consisting of interbedded layers of strongly weathered sandstones and mudstones, and the bedding planes had a considerable effect on the slope failures. The failures were classified into three types, as in Fig. 6.5.1 (Kokusho et al. 2009a).

Type-A: Deep slips parallel to bedding planes (dip planes), in gentle slopes of around 20 degrees. They glided as rigid bodies along slip planes at the bottom of the weathered sandstones. The displaced soil thickness and volume were very large and the soil blocks showed little surface disturbance after sliding.

Type-B: Shallow slips of 1–2 m deep not parallel to bedding planes in slopes of around 30 degrees or steeper. The moving mass was highly disrupted internally, and sometimes left trees with deep roots in their original locations. These failures far outnumbered the Type-A failures, but the individual soil volumes were not very large.

Type-C: Slope failures in highly weathered colluvial soils underlain by dip mudstones in places where Koi-ponds (numerous ponds were there for Koi-fish cultivations as the major local industry) and terraced paddy rice fields were located. The failures were obviously associated with the ponds in causing delayed flow-type failures due to internal erosions, involving colluvial soils of high water-content with long travel distance.

In the most of slope failures, the sandstones were largely responsible mainly because of their weakness due to strong weathering. Unconfined compression test results on intact samples taken out from failed slope scarps are shown in Fig. 6.5.2 with the open circles on the $q_u$ (unconfined compression strength) versus $F_c$ (fines content) diagram. The strengths of sandstones ($F_c \approx 0 \sim 30\%$) are $q_u = 0.1$ MPa or smaller, considerably weaker than those of interbedded mudstones ($F_c \approx 100\%$) with $q_u \approx 0.8$ MPa. Also noted is that the sandstones consisting of poorly graded fine particles have higher permeability (of the order of $10^{-3}$ cm/s) than that of mudstones (of the order of $10^{-4} \sim 10^{-6}$ cm/s) and hence may have served as aquifers (Kokusho et al. 2009b).

#### 6.5.1.2 *2008 Iwate-Miyagi Inland Earthquake*

The 2008 Iwate-Miyagi Inland earthquake ($M_J = 7.2$, thrust fault, focal depth 8 km) occurred 400 km north of Tokyo in the main island of Japan. During the earthquake,

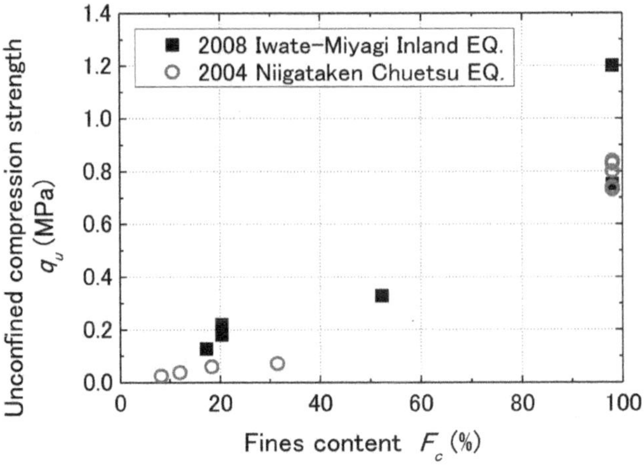


*Figure 6.5.2* Unconfined compression strength $q_u$ versus fines content $F_c$ for failed slopes during 2004 Niigataken-Chuetsu EQ. and 2008 Iwate-Miyagi Inland EQ (Kokusho et al. 2011a).

very strong ground motions were measured in the near fault zone; PGA of 2.4 g and PGV of more than 50 cm/s. About 1800 slope failures occurred mostly in the hanging wall of the fault and along several river valleys. The geology was of volcanic rocks of Miocene and Pleiocene ages; consisting predominantly of welded/non-welded tuff, sandstone, and siltstone. The unconfined compression strengths of intact samples from failed slopes are shown in Fig. 6.5.2 with the close squares. The strengths are very variable ($q_u = 0.2 \sim 1.2$ MPa) and seems to increase with increasing fines content. It may be judged that the rocks in this area are stronger than those of the 2004 Chuetsu earthquake and that sandy soils with lower $F_c$ and lower strengths served as slip planes during this earthquake, too.

Slope failures during the 2008 earthquake were also classified into 3 types as follows, although the classification may not be so clear as the 2004 earthquake, because the rocks of volcanic origin without clear bedding plane are prevalent in this area.

Type-a: Large scale slides moving almost as a rigid body along a deep slip plane.
Type-b: Medium size slides with characteristics in between Type-a and Type-c.
Type-c: Small size shallow slides with disintegrated debris.

In the largest landslide in Aratozawa classified as Type-a, a 1.2 km long by 0.8 km wide mountain body slid in a horizontal distance of almost 350 m along a deep-seated slip plane with a dip angle of around 5 degrees toward a man-made reservoir. The total volume was about 35 million m$^3$ according to DEM data (Kokusho et al. 2011a).

#### *6.5.1.3 Statistics of failed slopes in two earthquakes*

All slope failures during the two earthquakes (4321 and 1821 slopes, each) were statistically analyzed based on air-photographs taken just after the earthquakes. Figs. 6.5.3(a) and (b) show variations of the numbers of failed slopes and affected

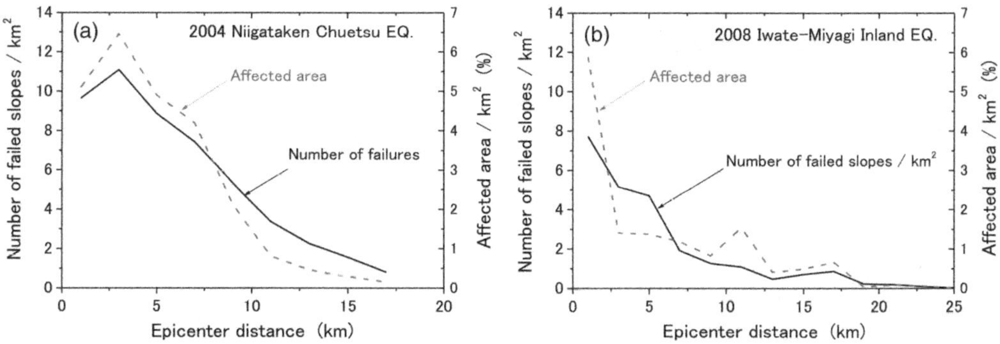


*Figure 6.5.3* Number or affected areas of failed slopes versus epicentral distance: (a) 2004 Niigataken Chuetsu EQ ($M_J = 6.8$). (b) 2008 Iwate-Miyagi EQ ($M_J = 7.2$).

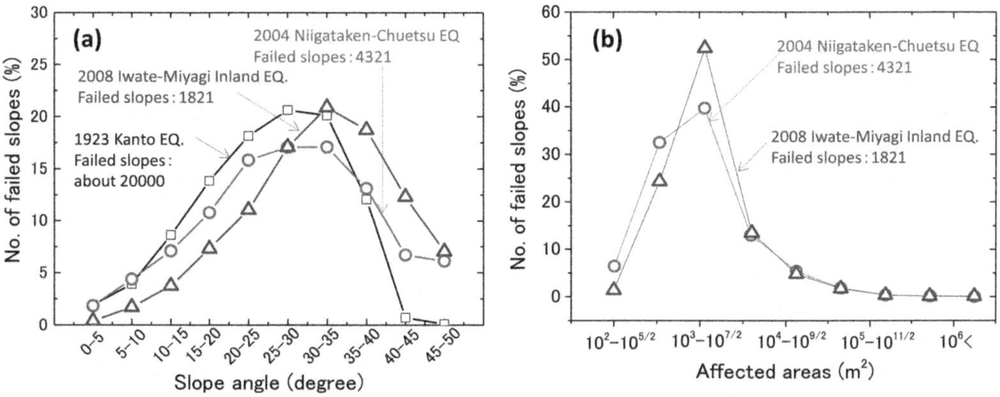


*Figure 6.5.4* Percentage in number of failed slopes: (a) Plotted versus slope angles, (b) Plotted versus affected areas (Kokusho et al. 2011a).

areas per 1 km$^2$ in the concentric circles of stepwise epicentral distances for the 2004 and 2008 earthquakes, respectively. Despite somewhat different trends near the epicenters, the density of failed slopes for the two earthquakes is highest near the epicenters, 8$\sim$11 per 1 km$^2$ in number and 7$\sim$8% in the affected area, and reduces to almost zero at about 16–18 km far.

Fig. 6.5.4(a) shows the number of failed slopes in percentage out of the total number of failed slopes versus the initial slope angles, $\theta_0 = 0$ to $50°$ stepwise, during the two earthquakes. In addition, similar data available for the slope failures occurred during the 1923 Kanto earthquake ($M_J = 7.9$) in mountainous areas of volcanic rocks 40 km west of Tokyo are added here (JSCE 2007). It is interesting to see a common trend that slopes with $\theta_0 = 30$ to $35°$ seem to have the highest number of failed slopes despite the differences in topography, geology and seismic intensity for the three earthquakes. In Fig. 6.5.4(b), the same data as in (a) are plotted versus the affected areas, $10^2$ to

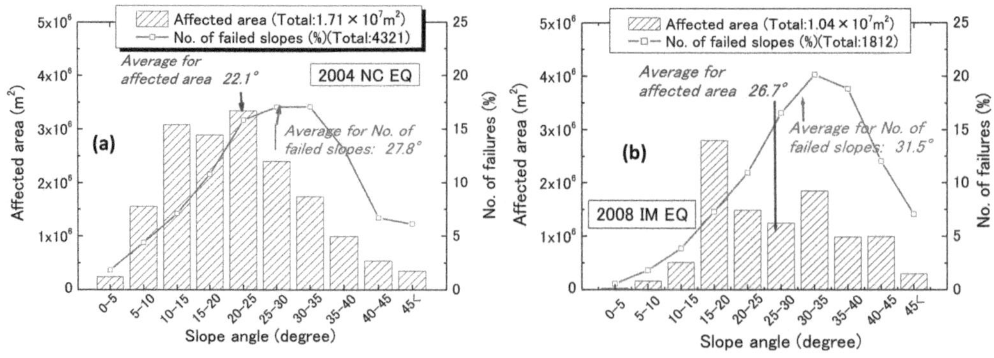


*Figure 6.5.5* Histograms of failed slopes in terms of numbers of landslides and affected areas for different slope angle groups: (a) 2004 Niigataken Chuetsu earthquake, (b) 2008 Iwate-Miyagi Inland earthquake (Kokusho et al. 2011a).

$10^6$ m$^2$ stepwise. The affected area defined here covers the entire area encompassing slope sliding; scar, path and deposit. Despite the different conditions, the two earthquakes are similar in percentage of the number of failed slopes, too. There exists a clear peak at the area of 1000 to 3200 m$^2$, below which the number of failed slopes decreases suddenly. This may somehow reflect technical limitations that slope failures smaller than a certain threshold (around 1000 m$^2$) may not be easy to identify in the air-surveys.

Figs. 6.5.5(a) and (b) depict the number of landslides in percentage by plots and connected lines, and the affected areas of failed slopes by histograms versus stepwise slope angles for the two earthquakes. The average angles are 27.8° and 31.5° in terms of the number and 22.1° and 26.7° in terms of the affected area for the 2004 and 2008 earthquake, respectively. The differences between the two earthquakes seem to reflect the different soil strength in the two areas as already shown by the unconfined compression strengths in Fig. 6.5.2. Also noted for the two earthquakes commonly is that the average slope angle for the affected areas is lower than that for the number of slopes. This indicates very interestingly that failures with larger scale tend to occur in gentler slopes.

For a number of slope failures in the damaged areas, ground surface elevations before and after the earthquakes were compared (Kokusho et al. 2009b) to quantify 3-dimensional topographical changes by using post-earthquake DEM data and prior air-photographs. Cross-sectional changes in failed slopes were developed from them. The slip surface in a failed slope, difficult to detect from DEM, was determined from the exposed scarp or slip plane in the upslope side, the original valley profile in the downslope side, and the global change of slope configuration. In slope failures chosen for case studies where reliable DEM data were available, the sliding soil mass was idealized by a pair of rectangular blocks before and after sliding. Fig. 6.5.6 exmplifies representative slides during the 2004 earthquake. Thus, the horizontal dimensions of the soil block and its thickness before and after the failure, the initial slope inclination $\beta_0 = \tan\theta_0$, the horizontal displacements of the centroid $\delta_{rn}$, and the global inclination

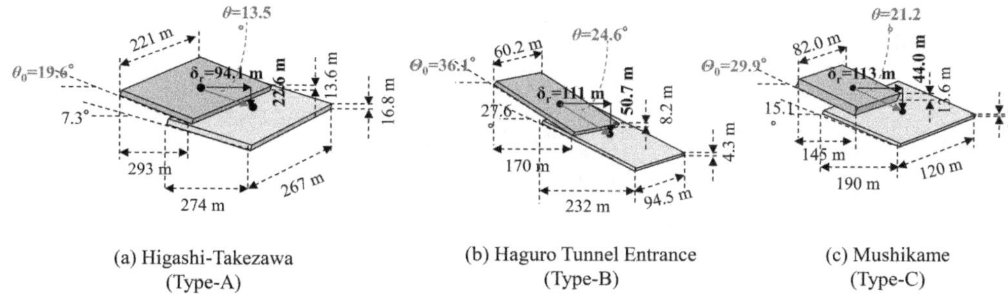

*Figure 6.5.6* Idealization of slope failures by rectangle blocks before and after sliding during 2004 Niigataken Chetsu EQ: (a) Type-A, (b) Type-B, (c) Type-C (Kokusho et al. 2011a).

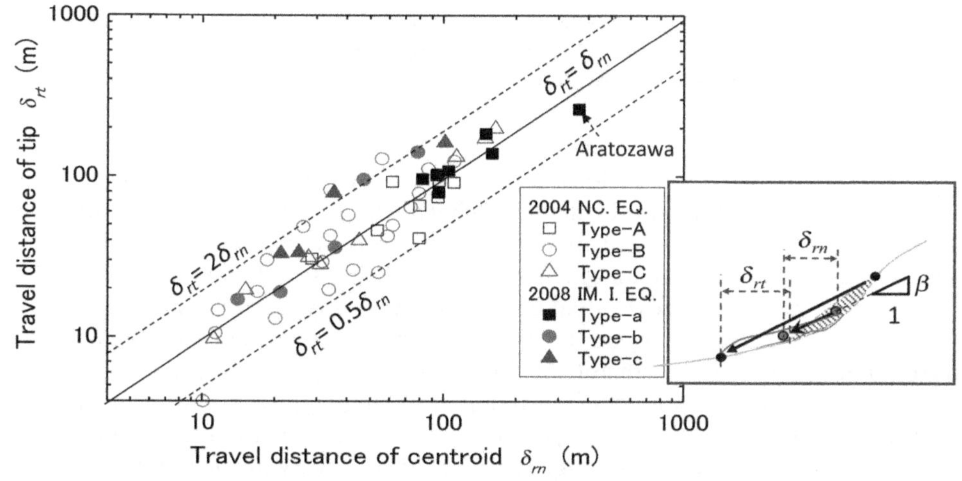


*Figure 6.5.7* Comparison of travel distance at a tip and centroid of failed soil mass for 2004 Niigataken-Chuetsu earthquake and 2008 Iwate-Miyagi Inland earthquake (Kokusho et al. 2011a).

$\beta = \tan\theta$ of the line connecting the centroids of the block before and after the failure were specified.

From the viewpoint of disaster mitigation, the runout-distance at the tip of debris $\delta_{rt}$ is more important than that at the centroid, $\delta_{rn}$. Hence, the two values were determined from the 3-dimensional changes of failed slopes and plotted in the horizontal and vertical axes respectively on the log-log chart in Fig. 6.5.7. The data points for the 2004 earthquake and the 2008 earthquake are classified into Type-A, B, C and Type-a, b, c, respectively, in accordance to the characteristics previously mentioned. No big systematic difference between them can be observed because the plots including Aratozawa spread out almost randomly around the diagonal line, $\delta_{rt} = \delta_{rn}$ (mostly within the two lines of $\delta_{rt} = 0.5\delta_{rn}$ and $\delta_{rt} = 2\delta_{rn}$), indicating that the distance of centroid $\delta_{rn}$ may be used as a representative travel distance.

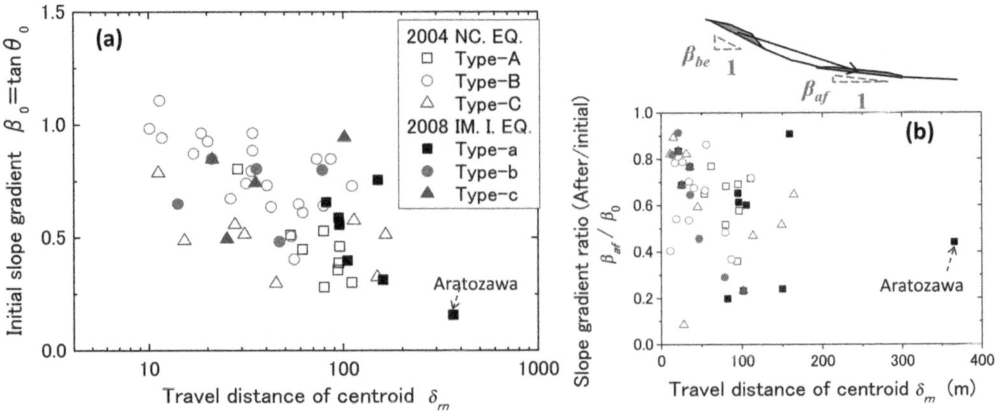


*Figure 6.5.8* Initial slope gradient $\beta_0$ versus travel distance (a), and Slope gradient ratio $\beta_{af}/\beta_0$ versus travel distance (b), for 2004 Niigataken-Chuetsu earthquake and 2008 Iwate-Miyagi Inland earthquake (Kokusho et al. 2011a).

In Fig. 6.5.8(a), the initial slope gradient $\beta_0$ is correlated with the travel distance at the centroid $\delta_{rn}$ on the semi-logarithmic chart. The data points for the 2004 earthquake, despite the significant dispersions, indicate an unexpected trend in that $\delta_{rn}$ increases with decreasing $\beta_0$ not only for Types-A, B and C individually, but also globally. The plots for the 2008 earthquake including Aratozawa seem to indicate the same unexpected trend, though they are more randomly dispersed. In order to examine that this trend may have some correlations with downslope topographies, the ratio of slope gradients for the sliding blocks before and after sliding $\beta_{af}/\beta_{be}$ is plotted versus travel distance $\delta_{rn}$ in Fig. 6.5.8(b). Here, $\beta_{af}/\beta_{be} = 1.0$ means that the local slope gradient was the same before and after sliding. The plots, though very much scattered, show no evidence of systematically increasing trends of $\delta_{rn}$ with increasing $\beta_{af}/\beta_{be}$, suggesting that this unexpected results are related not with local cross-sectional slope profiles but possibly with a more basic slope failure mechanism.

In Fig. 6.5.9, the travel distance of centroid $\delta_{rn}$ is correlated with the failed debris volume $V_f$ on the log-log diagram. Obviously, the distance $\delta_{rn}$ increases with $V_f$ as a whole as guided by the dashed lines despite large data scatters, and the trend is consistent between the two earthquakes and also for all the failure types. The Aratozawa plot, though apart from the others, does not seem to be different in the global trend. Thus, it may be summarized from the above findings that larger slope failures tend to occur in gentler slopes and travel longer distance, becoming much more hazardous than smaller ones.

Energy ratios $-\delta E_p/E^*_{EQ}$, the potential energies $-\delta E_p$ divided by the earthquake energies in the downslope direction $E^*_{EQ}$ as mentioned previously are plotted versus the volumes of failed slopes $V_f$ in Figs. 6.5.10(a) on the log-log diagram for slopes failed during the two earthquakes. The same energy ratios $-\delta E_p/E^*_{EQ}$ for the two earthquakes are plotted again versus the runout-distances $\delta_{rn}$ this time in Figs. 6.5.10(b). In addition to the natural slopes, plots for man-made fills failed during the 2004 earthquake and

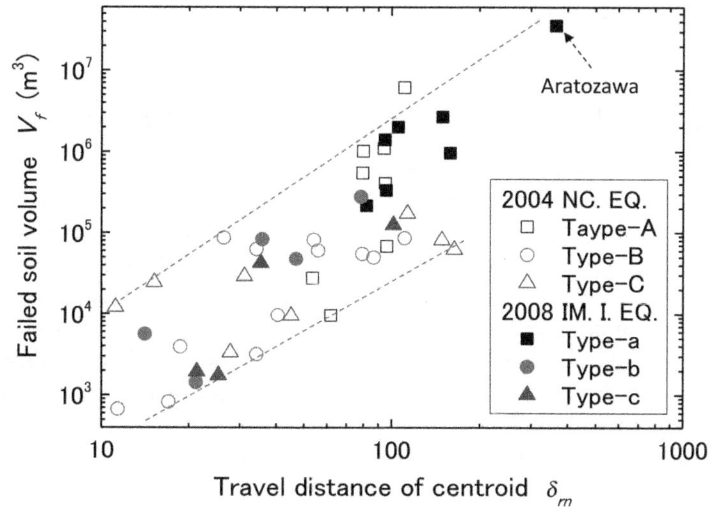


*Figure 6.5.9* Failed soil volume $V_f$ versus travel distance $\delta_m$ for 2004 Niigataken-Chuetsu earthquake and 2008 Iwate-Miyagi Inland earthquake (Kokusho et al. 2011a).

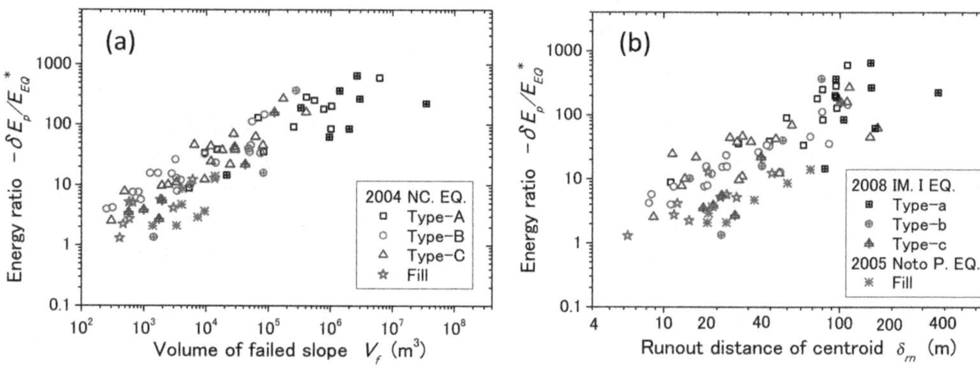


*Figure 6.5.10* Ratios of potential energy to earthquake energy $-\delta E_p/E^*_{EQ}$: (a) Plotted versus failed slope volumes $V_f$, (b) Plotted versus runout distances $\delta_m$ (Kokusho et al. 2014b).

the 2005 Noto-Peninsula earthquake ($M_J = 6.8$) in Japan are also superposed in (a) and (b). In all these plots, the earthquake energy $E^*_{EQ}$ for the 2004 and 2008 earthquakes were calculated from $E_{EQ}$ in each slope considering the site-specific downslope sliding direction, while the average energy ratio $\beta_E = E^*_{EQ}/E_{EQ} = 0.25$ was used for the 2005 earthquake as explained in Fig. 6.4.8. A clear increasing trend of $-\delta E_p/E^*_{EQ}$ is obvious with increasing failed soil volume or runout-distance irrespective of the types of slope failures despite data dispersions. It is remarkable that, even for a small slope failure with the volume of a few hundred cubic-meters or the runout-distance of several meters, the energy ratio is $-\delta E_p/E^*_{EQ} > 1.0$, while it is $-\delta E_p/E^*_{EQ} > 10$ in the medium-size slides and $-\delta E_p/E^*_{EQ} > 100$ for the large slides. This indicates that the earthquake energy serves just as a trigger and the most of the driving energy comes from the gravitational potential energy for slope failures of larger volume and longer runout-distance.

### 6.5.2 Back-calculated mobilized friction coefficients

In predicting travel distance of failed slopes in the present energy approach, it is essential that the mobilized friction coefficient $\mu$ be properly determined in advance. It may be possible in some cases to test it directly in situ or in the laboratory for man-made slopes in particular. However, due to the complexity of actual slope failures in the field, a more robust method may be to collect as many case histories as possible and back-calculate the friction coefficients to correlate them with pertinent slope parameters. Thus, the back-calculations were carried out for a number of slope failures during the two earthquakes (Kokusho et al. 2011a, Kokusho et al. 2014b) using the simplified rectangular block models before and after the earthquakes constructed from DEM data such as in Fig. 6.5.6 as well as the earthquake energies $E^*_{EQ}$. Saturated slip planes and hence $\sigma'_{n0} \approx \sigma_{n0}$ were assumed for all the slope failures based on site investigations, and the mobilized friction coefficients $\mu = \tan\phi = \tan\phi^*$ were back-calculated.

In general, the shear resistance in slip planes depends not only on $\tan\phi^*$ but also on cohesion $c$. If this effect is considered, the friction coefficient changes from Eq. (6.1.13), $\mu = \tan\phi^* = (\sigma'_{n0}/\sigma_{n0})\tan\phi$, to $\mu = (\sigma'_{n0}/\sigma_{n0})\tan\phi + (c/\sigma_{n0})$ as indicated in Eq. (6.1.12). In the following back-calculations of $\mu = \tan\phi^*$ incorporating Eq. (6.4.12), however, the effect of cohesion $c$ is neglected for simplicity in back-calculation. Actually, the key soil material of the slope failures during the two earthquakes was mostly weak sandy soils with small fines content as actually substantiated in Fig. 6.5.2. Hence, the back-calculated $\mu$-values essentially seem to represent the friction coefficients without significant contributions of cohesion. This simplicity will however may result in higher friction coefficients to a certain extent for shallow slip planes under low confining pressures in particular than the case where the cohesion is explicitly considered.

Fig. 6.5.11 shows the plots of friction coefficients $\mu = \tan\phi^*$ back-calculated from Eq. (6.4.12) versus the runout-distance of centroid $\delta_{rn}$ on the semi-log chart for a number of slopes failed during the 2004 and 2008 earthquakes, wherein the earthquake energies only in down-slope directions $E^*_{EQ}$ are used (Kokusho et al. 2014b). The $\mu$-value tends to increase with decreasing runout-distance for all the failure types of the two earthquakes. Though the Aratozawa plot is located far to the right of the others with the largest runout-distance and the lowest friction coefficient, it appears to be consistent with the overall trend. It is also noted that the $\mu$-values in man-made fill slopes are mostly smaller and located at the bottom of other plots and less dependent on the runout-distance $\delta_{rn}$ than in the natural slopes.

In Fig. 6.5.12, the same back-calculated friction coefficients $\mu = \tan\phi^*$ are plotted versus the initial slope gradients $\beta_0 = \tan\theta_0$. The plots seem to be highly dependent on the initial slope gradients. This may indicate that the friction coefficients of natural slopes strongly reflect their long-term exposures to previous natural disasters; namely, steeper slopes survived previous seismic or rainfall events because of their higher mobilized friction coefficients. In contrast, the plots for manmade fill slopes have smaller $\mu$-values and appear to be less dependent on initial slope gradient $\beta_0$. Also

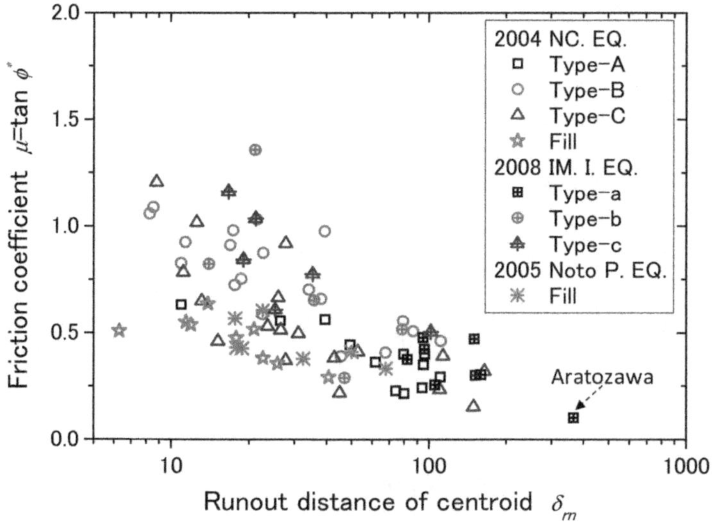


*Figure 6.5.11* Back-calculated friction coefficients $\mu = \tan\phi^*$ versus runout distances $\delta_{rm}$ in case studies during recent earthquakes (Kokusho et al. 2014b).

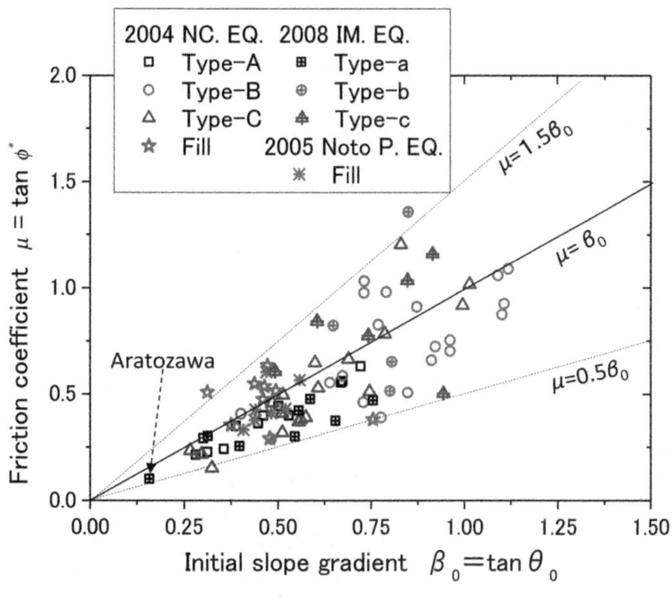


*Figure 6.5.12* Back-calculated friction coefficients $\mu = \tan\phi^*$ versus initial slope gradient $\beta_0 = \tan\theta_0$ in case studies during recent earthquakes (Kokusho et al. 2014b).

note that, for smaller values of $\beta_0$ corresponding to Type-A, C of the 2004 earthquake and Type-a of the 2008 earthquake, all of the back-calculated $\mu$-values are below or near the diagonal line of $\mu = \beta_0$. This indicates that the back-calculated friction coefficients $\mu$ originally larger than $\beta_0$ because the slope was stable before the earthquake,

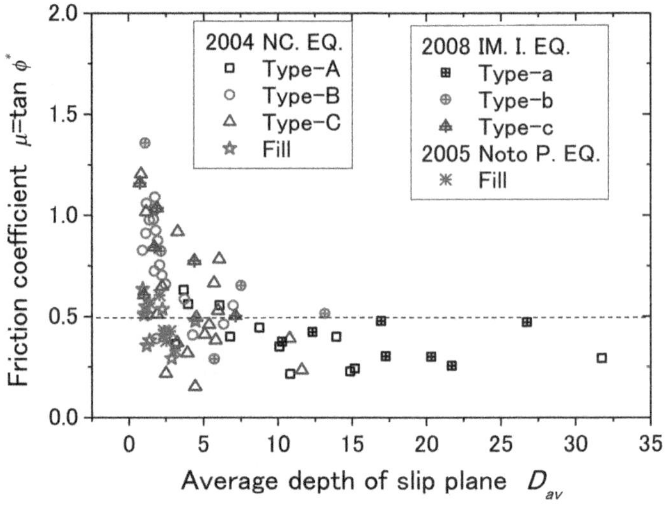


*Figure 6.5.13* Back-calculated friction coefficients $\mu = \tan\phi^*$ versus average depths of slip plane $D_{av}$ in case studies during recent earthquakes (Kokusho et al. 2014b).

decreased during earthquake shaking and subsequent sliding. As already discussed using Eq. (6.4.3), $\mu$-values smaller than $\beta_0$ imply that $\Delta E_k = (-\Delta\delta E_p) - \Delta E_{DP} > 0$ and failed debris accelerates first and then decelerates due to gentler or reverse slope angles in down-slope sections. In contrast, the data points with higher values of $\beta_0$ (typically belonging to Type-B and Type-c) are plotted on both sides of the diagonal line $\mu = \beta_0$. They tend to be above the line with increasing $\beta_0$ despite large data scatters.

The exact mechanism associated with the friction coefficients lower than the initial slope gradients is yet to be clarified. In Type-A of the 2004 earthquake in particular, seismically induced pore-pressure buildup or liquefaction in highly weathered sandstone near the slip planes seems to have occurred (Kokusho 2011a). In Aratozawa during the 2008 earthquake, however, the slip plane was very deep (several tens to more than 100 m), and it is normally considered difficult for seismically induced liquefaction to occur at such a great depth. The cause for the low back-calculated friction coefficient there may have something to do with the earthquake fault system, a portion of that is believed to have crossed the site. Presumably it may have caused the pressure increase on the slip plane. In Type-C failures, the high water content may have transitioned soil debris into high-speed mudflows due to pore-pressure buildup.

In Fig. 6.5.13, the back-calculated friction coefficients $\mu = \tan\phi^*$ are correlated with the average depths of slip planes $D_{av}$. Though there exists some differences between the two earthquakes, it clearly shows similar $D_{av}$-dependent changes in the friction coefficients. The $\mu$-values tend to be lower than 0.5 for the depth larger than $D_{av} = 5{\sim}7$ m for the rigid block type failures such as A and a-type in particular, while it can be much larger than that with decreasing $D_{av}$. For the fill slopes wherein $D_{av}$ is limited as smaller than 5 m, the back-calculated $\mu$-values do not seem to show a clear increasing trend with decreasing $D_{av}$, possibly reflecting shear strength properties different from natural slopes. In many field conditions, the depths of slope failures

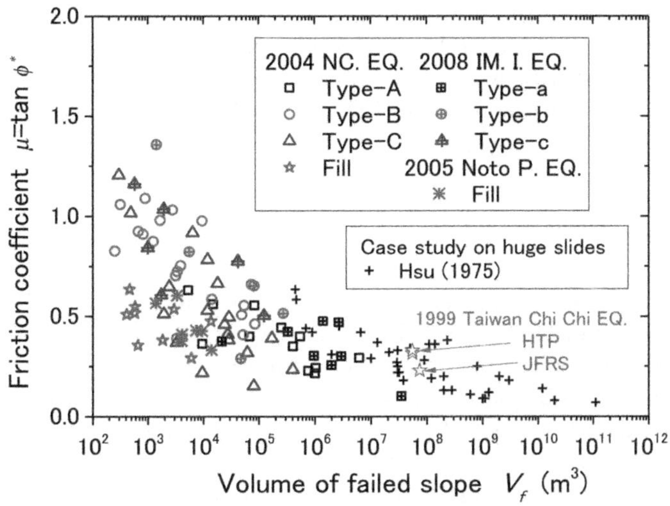


*Figure 6.5.14* Back-calculated friction coefficients $\mu = \tan\phi^*$ versus volumes of failed slope $V_f$ in case studies during recent earthquakes compared with previous studies (Kokusho et al 2014b).

seem to be easier to predict in soil investigations than other variables, hence $D_{av}$ may serve as a convenient parameter in evaluating the mobilized friction coefficient. As mentioned earlier, the shear resistance of slope materials is represented solely by the friction coefficient $\mu = \tan\phi^*$ implicitly including the effect of cohesion. Hence, the cohesion effect tends to be larger relative to the friction effect as the overburden stress or $D_{av}$ decreases with decreasing slip plane depth, even in sandy materials in which the cohesion is not large compared to the friction under normal overburden stresses. However, it is not necessary to take this effect into account in back-calculating friction coefficients if the same energy-based method is used to evaluate the travel distance.

The back-calculated $\mu$-values $\mu = \tan\phi^*$ are plotted again versus the volumes of failed slopes $V_f$ on the semi-logarithmic diagram in Fig. 6.5.14. Despite large scatters in the data, a clear decreasing trend of $\mu$ can be seen as $V_f$ increases from $10^3$ to $10^7\,\mathrm{m}^3$ irrespective of the failure types in natural slopes. The $V_f$-dependency of $\mu$-value is particularly clear for smaller-volume failures, Type-B for the 2004 earthquake and Type-c for the 2008 earthquake. For the fill slopes, however, the $\mu$-values, being smaller than other types, seem to be less dependent on $V_f$, possibly reflecting some different shear strength properties.

A similar relationship based on case histories of huge landslides not necessarily associated with earthquakes (Hsu 1975) is superposed on the same diagram with the cross symbols. The two studies show a remarkable consistency in the upper bounds of the data points in the wide range of the failed soil volume $10^3{\sim}10^{11}\,\mathrm{m}^3$. In addition, two other $\mu$-values back-calculated from huge landslides during 1999 Chi-Chi earthquake, JFES and HTP (Dong et al. 2007, Ishizawa et al. 2008), are also plotted on the chart, which seem to be compatible with Hsu's data and also with the data from the two recent earthquakes.

Thus, the back-calculated friction coefficients are found to have clear dependency on the runout-distances, the initial slope gradients, the failed soil volumes and the depths of slip planes. These correlations, though more or less affected by specific site conditions, may possibly be applicable to slope failures in general and serve as a data base on the mobilized friction coefficients in predicting runout-distances using the energy-based method.

## 6.6 SUMMARY

1. Slip surface analyses considering seismic coefficients are normally used in evaluating safety factors for slope failures, and post-failure displacement is out of the scope. While the shear resistance along the slip plane varies due to the seismic coefficient if the slip plane is unsaturated, it stays the same as the initial value (the consolidated undrained strength) if the slip plane is saturated as far as the principle of effective stress is valid.
2. The Newmark method can evaluate the residual displacement of a sliding soil mass during a given ground motion along a prescribed linear or circular slip plane. It is combined with a dynamic response analysis of a slope to consider the effect of seismic amplification on the slope displacement. Care is needed however that the dynamic response in calculating the residual displacement depends on a continuum model without considering the discontinuity along the slip plane, and may change considerably once the sliding failure occurs. Thus, one should be aware that the applicability is limited to a relative small displacement around 1 m.
3. In the self-weight slope deformation analysis, the seismic effect is taken account not directly by inertial force but indirectly by the degraded equivalent soil moduli caused by earthquake shaking. Considerable engineering judgments have to be incorporated in the hybrid method combining static and dynamic analyses of slope models with corresponding laboratory soil tests under various stress conditions to prescribe the equivalent soil moduli.
4. Statistical studies on a great number of slope failures during recent earthquakes indicate that larger slope failures tend to occur in gentler slopes and travel longer distance, becoming more devastating than smaller and steeper slopes. These observational tendencies from case histories are essential in considering seismically-induced slope hazard mitigations.
5. The slope failure mechanism can be formulated as a balance of four energies; potential energy, kinetic energy, dissipated energy and earthquake energy. In the statistical study of actual slope failures, the potential energy is becoming much larger than the earthquake energy in slope failures with larger volumes and longer runout-distances. This indicate that, for large slope failures in particular, the earthquake energy works just as a trigger and the potential energy is the major player to drive the failed soil debris in long runout-distance.
6. Shaking table tests demonstrate that the runout-distance can be uniquely determined from the energy balance of a simple rigid block model if a mobilized friction coefficient $\mu$ can be given. This energy-based evaluation of runout-distance can be applicable to arbitrary slope cross-sectional profiles in

a graphical evaluation method, if the earthquake energy $E^*_{EQ}$ and the $\mu$-value is appropriately given.

7 In order to determine appropriate $\mu$-values in the energy method, the back-calculations of previous case histories are essential to form robust database. The earthquake energy $E^*_{EQ}$ is averagely a quarter of the total earthquake energy $E_{EQ}$ coming up to the slope, and $E_{EQ}$ may be evaluated from the incident energy at the base layer using available empirical equations.

8 The back-calculations of actual slope failures during recent earthquakes using this energy-based method indicate that the mobilized friction coefficients $\mu$ in natural slopes are obviously increasing with their slope gradients. This suggests that the gradients have been determined according to their long-term exposures to natural disasters such as heavy rains and earthquakes; steeper slopes survived previous events because of their higher friction coefficients, unlike manmade fill slopes constructed with nearly constant strengths.

9 Some of the back-calculated friction coefficients $\mu$ are lower than the initial slope gradients, indicating that some mechanisms are involved to reduce them such as pore-pressure buildup during seismic loding and monotonic shearing during sliding. Despite much smaller contributions of earthquake energy than potential energy in long runout-distance, the earthquake energy greatly contributes in reducing the soil strength by pore-pressure buildup or other mechanisms so that the potential energy can make the work.

10 The back-calculated friction coefficients $\mu$ are correlated well with the average depths of failed slopes, the parameter relatively easy to determine by field investigations. The $\mu$-values are also decreasing with increasing failed soil volumes, which seems compatible with the previous research on huge slides. The back-calculated $\mu$-value may be applicable in the energy-based method to predict the runout-distance for earthquake-induced slope failures that are difficult to evaluate by other simplified methods.

# References


Acacio A.A., Kobayashi, Y., Towhata, I., Bautista, R.T. and Ishihara, K. (2001): Subsidence of building foundation resting upon liquefied subsoil: Case studies and assessment, *Soils and Foundations*, 41 (6), 111–128.

Adalier, K., Elgamal, A., Meneses, J. and Baez, J.I. (2003): Stone columns as liquefaction counter-measure in non-plastic silty soils, *Soil Dynamics & Earthquake Eng.*, Elsevier, 23 (7), 571–584.

Adalier, K., Zeghal, M. and Elgamal, A.-W. (1997): Liquefaction mechanism and countermeasures, *Seismic Behaviour of Ground & Geotech. Structures*, Balkema, 155–162.

Afifi, S.S. and Richart, F.E. (1973): Stress-history effects on shear modulus of soils, *Soils & Foundations*, 13 (1), 77–95.

AIJ (2001): *Japanese architectural standard specification for building foundations (in Japanese)*, Architectural Institute of Japan.

Akiyama, H. (1999): *Earthquake-resistant design method for buildings based on energy balance (in Japanese)*, Giho-do Publishing Co.

Alarcon-Guzman, A., Leonards, G.A. and Chameau, J.L. (1988): Undrained monotonic and cyclic strength of sands, *Journal of Geotech, Eng.* ASCE, 114 (GT10), 1089–1109.

Anderson, D.G. and Woods, R.D. (1976): Time-dependent increase in shear modulus of clays, *Journal of Geotech. Eng. Div.*, ASCE, 102 (GT8), 525–537.

Andrus, R.D. and Youd, T.L. (1989): Penetration tests in liquefiable gravels, *Proc. 12th International Conference on SMFE*, Rio de Janeiro, 1, 679–682.

Andrus, R.D. (1994): In situ characterization of gravelly soils that liquefied in the 1983 Borah Peak Earthquake, *PhD. Dissertation*, University of Texas at Austin.

Andrus, R.D. and Stokoe, K.H. (2000): Liquefaction resistance of soils from shear-wave velocity, *Journal of Geotech. & Geoenv. Eng.* ASCE, 126 (11), 1015–1025.

Annaki, M. and Lee, K.L. (1977): Equivalent uniform cycle concept for soil dynamics, *Journal of Geotech. Eng. Div.*, ASCE, 103 (GT6), 549–564.

Aoyagi, T. (2000): Inversion analysis for soil properties based on vertical array record using the extended Bayesian method, *Master's Thesis (in Japanese)*, Graduate School of Science & Engineering, Chuo University, Tokyo, Japan.

Arai, R., Kokusho, T. and Kusaka, T. (2015): Effect of initial shear stress on liquefaction failure and shear strain development by hollow cylindrical torsional shear tests, *Geotechnical Journal of JGS (in Japanese)*, 10 (2), 213–223.

Arias, A. (1970): *A measure of earthquake intensity in seismic design for nuclear power plants.* The MIT Press, Cambridge, MA, USA, 438–483.

Ashford, S.A. and Juirnarongrit, T. (2002): Response of single piles and pipelines in liquefaction-induced lateral spreads using controlled blasting, *Earthquake Eng. and Eng. Vibration*, Springer, 1 (2), 181–193.

Ashford, S.A., Boulanger, R.W. and Brandenberg, S.J. (2011): Recommended design practice for pile foundations in laterally spreading ground, *PEER Report 2011/04*, Pacific Earthquake Engineering Research Center, College of Engineering, University of California, Berkeley.

ASTM (1985): Classification of soils for engineering purposes: *Annual Book of Standards*, D 2487-83, 04.08, American Society for Testing and Materials, 1395–1408.

ASTM (2001): *ASTM C127-07 Standard test method for density, relative density (Specific gravity), and absorption of coarse aggregate,* American Society for Testing and Materials.

Bakir, B.S., Yilmaz, M.T., Yakut, A. and Gulkan, P. (2005): Re-examination of damage distribution in Adapazari: Geotechnical considerations, *Engineering Structures*, Elsevier, 27, 1002–1013.

Bath, M. (1956): Earthquake energy and magnitude, *Physics and Chemistry of the Earth*, 23 (10), 115–165.

Been, K. and Jefferies, M.G. (1985): A state parameter for sands, *Geotechnique*, 35 (1), 99–112.

Berrill, J.B. and Davis, R.O. (1985): Energy dissipation and seismic liquefaction of sands: Revised model, *Soils and Foundations*, 25 (2), 106–118.

Berrill, S.A., Christensen, S.A., Keenan, R.J. Okada, W. and Pettinga, J.R. (1997): Lateral-spreading loads on a piled bridge foundation, *Proc. of Discussion Session; Seismic behaviour of Ground and Geotechnical Structures, ICSMGE, Hamburg*, 173–183.

Biot, M.A. (1956): Theory of propagation of elastic waves in fluid saturated porous solid, *Journal of the Acoustic Society of America*, 28 (2), 168–178.

Bishop, A.W. and Blight, G.E. (1963): Some aspects of effective stress in saturated and partly saturated soils, *Geotechnique*, 13, 177–197.

Borcherdt, R., Wentworth C.M., Janssen, A., Fumal, T. and Gibbs, J. (1991): Methodology for predictive GIS mapping of special study zones for strong ground shaking in the San Francisco Bay Region, *Proc. 4th Intern. Conf. on Seismic Zonation*, 3, 545–552.

Boulanger, R.W. and Truman, S.P. (1996): Void redistribution in sand under post-earthquake loading. *Canadian Geotech. Journal*, 33, 829–833.

BRI (Building Research Institute) (1965): Niigata earthquake and damage of reinforced concrete buildings in Niigata city, *Report of Building Research Institute (in Japanese)*, No. 42, Building Research Institute, Ministry of Construction, Japan.

BSSC (Building Seismic Safety Council) (2003): *NEHRP Recommended Provisions for Seismic Regulations for new buildings and other structures*, Building Seismic Safety Council, National Institute of Building Sciences, 2003 Edition, FEMA 450.

Buckingham, E. (1914): On physically similar systems; illustrations of the use of dimensional equations, *Physical Review*, 4 (4), 345–376.

Casagrande, A. (1971): On liquefaction phenomena, *Geotechnique*, London, England, XXI (3), 197–202.

Castro, G. (1975): Liquefaction and cyclic mobility of saturated sands, *Journal of Geotech. Eng. Div.* ASCE, 101 (GT6), 551–569.

CDIT (Coastal Development Institute of Technology) (1997): *Handbook on liquefaction mitigation in reclaimed lands (Revised version) (in Japanese)*, Ministry of Infrastructure, Land, Transportation and Tourism.

Chang, Y.L. (1937): Discussion on lateral pile loading tests by Feagin, *Transaction ASCE*, Paper No. 1959, 272–278.

Chen, X.L., Ran, H.L. and Yang, W.T. (2012): Evaluation of factors controlling large earthquake-induced landslides by the Wenchuan earthquake, *Natural Hazards & Earth System Science*, Vol. 12, 3645–3657.

Close, U. and McCormick, E. (1922): "Where the mountains walked" An account of the recent earthquake in Kansu Province, China, which destroyed 100,000 lives, *The National Geographic Magazine*, XLI (5).

Clough, R.W. and Pirtz, D. (1956): Earthquake resistance of rockfill dams, *Transaction ASCE,* Vol. 123, paper No. 2939, 792–81.

Committee on the Alaskan Earthquake (1971): The Great Alaskan Earthquake of 1964, Geology, *Committee Report on the Alaskan Earthquake of the Division of Earth Sciences,* National Research Council, National Academy of Sciences.

Coulter, H.W. and Migliaccio, R.R. (1966): Effects of the Earthquake of March 27, 1964 at Valdez, Alaska, *Geological Survey Professional Paper* 542-C, U. S. Department of the Interior.

Davis, R.O. and Berrill, J.B. (1982): Energy Dissipation and Seismic Liquefaction of Sands, *Earthquake Engineering & Structural Dynamics*, Vol. 10, 59–68.

De Alba, P., Seed, H.B. and Chan C.K. (1976): Sand liquefaction in large-scale simple shear tests, *Journal of Geotech. Eng. Div.* ASCE, 102 (GT9), 909–927.

Desrues, J., Chambon, R., Mokni, M. and Mazerolle, F. (1996): Void ratio evolution inside shear bands I triaxial sand specimens studied by computed tomography, *Geotechnique*, 46 (2), 529–546.

Dobry, R., Taboada, V. and Liu, L. (1995): Centrifuge modeling of liquefaction effects during earthquakes, *Proc. 1st International conference on Earthquake Geotechnical Engineering*, Balkema, Vol. 3, 1291–1324.

Dong, J.-J., Lee, W.-R., Lin, M.-L., Huang, A.-B. and Lee, Y.-L. (2007): Effects of seismic anisotropy and geological characteristics on the kinematics of the neighboring Jiufenger-shan and Hungtsaiping landslides during Chi-Chi earthquake, *Tectonophysics*, Elsevier, No. 466, 438–457.

Editing Committee of JGS (1998): *Remedial measures against soil liquefaction: from investigation and design to implementation*, Japanese Geotechnical Society, A.A. Balkema.

Elgamal, A. W., Dobry, R. and Adalier, K. (1989): Study of effect of clay layers on liquefaction of sand deposits using small scale models, *Proc. 2nd US-Japan Workshop on Liquefaction, Large Ground Deformation and Their Effects on Lifelines*, NCEER, SUNY-Buffalo, 233–245.

Evans, M.D. and Zhou, S. (1995): Liquefaction behavior of sand-gravel composites, *Journal of Geotech. Eng.* ASCE, 121 (3), 287–298.

Ewing, W.M., Jardetzky, W.S. and Press, F. (1957): *Elastic waves in layered media*, Chap. 4 A layered half space, McGraw-Hill Series in the Geological Sciences, 144.

Field, M.E., Gardner, J.V., Jennings, A.E. and Edwards, B.D. (1982): Earthquake-induced sediment failures on a 0.25° slope, Klamath River delta, California. *Geology*, V. 10, 542–546.

Figueroa, J.L., Saada, A.D., Liang, L. and Dahisaria, N.M. (1994): Evaluation of soil liquefaction by energy principles, *Journal of Geotech. Eng.*, ASCE, 120 (9), 1554–1569.

Finn, W.D.L. (1982): Soil liquefaction studies in the People's Republic of China, *Soil Mechanics-Transient and Cyclic Loads*, John Wiley & Sons, Ltd., Ch. 22, 609–626.

Finno, R.J. and Rechenmacher, A.L. (2003): Effects of Consolidation history on critical state of sand, *Journal of Geotech. & Geoenv. Eng.* ASCE, 129 (4), 350–360.

Frankel, A. and Clayton, R.W. (1986): Finite difference simulations of seismic scattering: Implications for the propagation of short-period seismic waves in the crust and models of crustal heterogeneity, *Journal of Geophysical Research*, 91 (B6), 6465–6489.

Fujita, K. (2001): Possibility of water film generation in liquefied sand during the Niigata earthquake, *Master's Thesis (in Japanese),* Graduate School of Science & Engineering, Chuo University, Tokyo, Japan.

Fukuoka, M. (1966): Damage to civil engineering structures, *Soil and Foundation*, Vol. VI, 45–52, Japanese Society of Soil Mechanics and Foundation Engineering, 45–52.

Fukushima, Y. and Midorikawa, S. (1994): Evaluation of site amplification factors based on average characteristics of frequency dependent Q-1 of sedimentary strata, *Journal of Structural Division (in Japanese),* Japan Architectural Institute, Vol. 460, 37–46.

Gibbs, H.J. and Holtz, W.G. (1957): Research on determining the density of sand by spoon penetration test, *Proc. 4th international Conference on SMFE*, ISSMFE. Vol. 1, 35–39.

Goto, S., Tatsuoka, F., Shibuya, S., Kim, Y.S. and Sato, T. (1991): A simple gauge for small strain measurements in the laboratory, *Soils & Foundations*, 31 (1), 169–180.

Green, R.A., Mitchell, J.K. and Polito, C.P. (2000): An energy-based excess pore pressure generation model for cohesionless soils, *Proc. John Booker Memorial Symposium*, Sydney, Australia, Balkema Publishers.

Green, R.A. and Terri, G.A. (2005): Number of equivalent cycles concept for liquefaction evaluations-revisited, *Journal of Geotech. and Geoenv. Eng.*, 131 (4), 477–488.

Guo, D., He, C., Xu, C. and Hamada, M. (2015): Analysis of the relations between slope failure distribution and seismic ground motion during the 2008 Wenchuan earthquake, *Soil Dynamics and Earthquake Engineering*, Elsevier, Vol. 72, 99–107.

Gutenberg, B. (1956): The energy of earthquakes, *Quarterly Journal of the Geological Society of London*, CXII (455), 1–14.

Gutenberg, B. and Richter, C.F. (1942): Earthquake magnitude, intensity, energy and acceleration, *Bulletin of Seismological Society of America*, Vol. 32, 163–191.

Gutenberg, B. and Richter, C.F. (1956): Earthquake magnitude, intensity, energy and acceleration (Second paper), *Bulletin of Seismological Society of America*, Vol. 46, 105–145.

Hamada, M. (1992): *Large ground deformations and their effects on lifelines: 1964 Niigata earthquake. Case Studies of Liquefaction and Lifeline Performance during Past Earthquakes*, Vol. 1, Japanese Case Studies, 3.1–3.123.

Hamada, N., Yasuda, F., Nakahira, A. and Tazoh, T. (2009): Damage investigation on the foundations of the Hanshin Expressway Route 5 caused by the 1995 Hyogoken-Nambu earthquake, *Earthquake Geotechnical Case Histories for Performance-Based Design*, Taylor & Francis Group, London, 357–371.

Hara, T., Kokusho, T. and Kochi, Y. (2009): Effect of degree of saturation on cyclic undrained shear strength of sands containing non-plastic fines, *Journal of Japan Society for Civil Engineers (in Japanese)*, C, 65 (3), 587–596.

Harder, L.F. Jr. and Seed, H.B. (1986): Determination of penetration resistance for coarse-grained soils using the Becker Hammer Drill, *Report No. UCB/EERC-86/06*, University of California, Berkeley, 118 pages.

Hardin, B.O. (1965): The nature of damping in sands, *Proc. SMFD*, ASCE, 91 (SM1), 63–97.

Hardin, B.O. and Black, W.L. (1968): Vibration modulus of normally consolidated clay, *Proc. of SMFD*, ASCE, 94 (SM2), 353–369.

Hardin, B.O. and Black, W.L. (1969): Discussion; Vibration modulus of normally consolidated clay, *Proc. of SMFD*, ASCE, 94 (SM6), 1531–1537.

Hardin, B.O. and Drnevich, V.P. (1972a): Shear modulus and damping in soils: Measurement and parameter effects, *Journal of SMFD*, ASCE, 98 (SM6), 603–624.

Hardin, B.O. and Drnevich, V.P. (1972b): Shear modulus and damping in soils: Design equations and curves, *Journal of SMFD*, ASCE, 98 (SM7), 667–692.

Hardin, B.O. and Richart, F.E. (1963): Elastic wave velocities in granular soils, *Proc. SMFD*, ASCE, 89 (SM1), 33–65.

Harkrider, D.G. (1964): Surface waves in multilayered elastic media I. Rayleigh and Love waves from buried sources in a multilayered elastic half-space, *Bulletin of the Seismological Society of America*, 54 (2), 627–679.

Haskell, N.A. (1953): The dispersion of surface waves on multilayered media, *Bulletin of Seismological Society of America*, Vol. 43, 17–34.

Hiraoka, R. (2000): The effect of physical properties on liquefiability of gravelly soils, *Master's Thesis (in Japanese)*, Graduate School of Science & Engineering, Chuo University, Tokyo, Japan.

Horike, M., Zhao, B. and Kawase, H. (2001): Comparison of site response characteristics inferred from microtremors and earthquake shear waves, *Bulletin of the Seismological Society of America*, 91, 6, 1526–1536.

Hoshiya, M. and Yamazaki, T. (1979): Response analysis of structure based on earthquake energy, *Journal of Japan Society for Civil Engineers (in Japanese)*, No. 291, 1–14.

Hsu, J. (1975): Catastrophic debris streams generated by rockfalls, *Geological Society of America Bulletin*, Vol. 86, Doc. no. 50117, 129–140.

Huang, D., Yanagisawa, E. and Sugano, T. (1993): Shear characteristics of silt containing sand, *Journal of Japan Society for Civil Engineers (in Japanese)*, No. 463/III-22, 25–33.

Hyodo, M. and Uchida, K. (1998): Characterization of dynamic problems in cohesive soils, Lecture Series on Dynamic Problems in Cohesive Soils 2, *Tsuchi-to-Kiso (in Japanese)*, Japanese Geotechnical Society, 46 (6).

Hyodo, M., Adrian, F.L.H., Yamamoto, Y. and Fujii, T. (1999): Cyclic shear strength of undisturbed and remolded marine clays, *Soils and Foundations*, 39 (2), 5–58.

Iai, S. (1989): Similitude for shaking table tests on soil-structure-fluid model in 1g gravitational field, *Soils and Foundations*, 29 (1), 105–118.

Iai, S., Morita, T., Kameoka, T., Matsunaga, Y. and Abiko, K. (1995): Response of a Dense Sand Deposit During 1993 Kushiro-Oki Earthquake, *Soils & Foundations*, 35 (1), 115–131.

Idriss, I.M. (1990): Response of soft soil sites during earthquakes, *Proc. H. Bolton Seed Memorial Symposium*, 273–290.

Idriss, I.M. and Boulanger, R. (2008): Soil liquefaction during earthquakes, Earthquake Engineering Research Institute, MNO-12.

Iida, K. (1938): The velocity of elastic waves in sand, *Bulletin of Earthquake Research Institute*, University of Tokyo, No. 16, 131–144.

Imamura, A. (1925): Report on Great Kanto earthquake, *Report by Committee on Earthquake Disaster Mitigation (in Japanese)*, Report No. 100-Kou, 21–66.

Inada, M. (1960): On the use of Swedish weight sounding test results, Tsuchi-to-Kiso, *Journal of Japanese Geotechnical Society (in Japanese)*, 8 (1), 13–18.

Inagaki, H., Iai, S., Sugano, T., Yamazaki, H. and Inatomi, T. (1996): Performance of caisson type quay walls at Kobe Port, Special Issue on Geotechnical Aspects of the January 17, 1995 Hyogoken Nambu Earthquake, *Soils and Foundations*, 119–136.

Ishihara, K. (1971): On the longitudinal wave velocity and Poisson's ratio in saturated soils, *Proc. 4th Asian Regional Conference of ISSMFE*, Bangkok, Vol. 1, 197–201.

Ishihara, K. and Yasuda, S. (1975): Sand liquefaction in hollow cylinder torsion under irregular excitation, *Soils and Foundations*, 15 (1), 45–59.

Ishihara, K., Tatsuoka, F. and Yasuda, S. (1975): Undrained deformation and liquefaction of sand under cyclic stresses, *Soils and Foundations*, 15 (1), 29–44.

Ishihara, K. (1977): Simple method of analysis for liquefaction of sand deposits during earthquake, *Soils and Foundations*, 17 (3), 1–17.

Ishihara, K., Iwamoto, S., Yasuda, S. and Takatsu, H. (1977): Liquefaction of anisotropically consolidated sand, *Proc. 9th International Conference on SMGE*, Tokyo, Vol. 2, 261–264.

Ishihara, K. and Takatsu, H. (1979): Effects of overconsolidation and $K_0$ conditions on the liquefaction characteristics of sands, *Soils and Foundations*, 19 (4), 59–68.

Ishihara K. (1985): Stability of natural deposits during earthquakes, *Proc. 11th International Conference of SMFE*, San Francisco, Vol. 1, 21–376.

Ishihara, K., Okusa, S., Oyagi, N. and Ischuk, A. (1990): Liquefaction-induced flow slides in the Collapsible Loess deposit in Soviet Tajik, *Soils and Foundations*, 30 (4), 73–89.

Ishihara, K. and Yoshimine, M. (1992): Evaluation of settlements in sand deposits following liquefaction during earthquakes, *Soils and Foundations*, 32 (1), 173–188.

Ishihara, K., Kokusho, T. and Silver, M. (1992): State of the art report: Recent developments in evaluating liquefaction characteristics of local soils, *Proc. 12 International Conference on SMFE*, Rio de Janeiro, 2719–2732.

Ishihara, K. (1993): Liquefaction and flow failure during earthquakes, 33rd Rankine Lecture, *Geotechnique*, 43 (3), 351–415.

Ishihara, K., Acacio, A. and Towhata, I. (1993): Liquefaction-induced ground damage in Dagupan in the July 16, 1990 Luson earthquake, *Soils and Foundations*, 33 (1), 133–154.

Ishihara, K. (1996): Soil behavior in earthquake geotechnics, Oxford Engineering Science Series, 46, Oxford Science Publication,

Ishihara, K., Yasuda, S. and Nagase, H. (1996): Soil characteristics and ground damage, Special Issue on Geotechnical Aspects of the January 17, 1995 Hyogoken Nambu Earthquake, *Soils and Foundations*, 109–118.

Ishizawa, T., Kokusho, T. and Nshida, K. (2008): Evaluation of seismically induced slope displacement in terms of energy and a case study during Chi-Chi earthquake, *Proc. 3rd Taiwan-Japan Joint Workshop on Geotechnical hazards from large earthquakes and heavy rainfall*, Keelung, Taiwan, 273–280.

Ito, F. (2011): Liquefaction characteristics of sand containing non-plastic fines with initial shear stresses by hollow cylindrical torsional shear apparatus, *Master's Thesis (in Japanese)*, Graduate School of Science & Engineering, Chuo University, Tokyo, Japan.

Iwamoto, I., Kokusho, T. and Nakano, T. (2003): Volume change characteristics of gravelly sands by means of monotonic and cyclic shear test, *Journal of Japan Society for Civil Engineers (in Japanese)*, No. 736/III-63, 205–215.

Iwasaki, T., Tatsuoka, F. and Takagi, Y. (1978a): Shear moduli of sands under cyclic torsional shear loading, *Soils & Foundations*, 18 (1), 39–56.

Iwasaki, T., Tatsuoka, F., Tokida, K. and Yasuda, S. (1978b): A practical method for assessing soil liquefaction potential based on case studies at various sites in Japan, *Proc. 2nd International Conf. on Microzonation*, San Francisco, CA, USA, 885–896.

Jeffery, M.G. and Davies, M.P. (1993): Use of CPT to estimate equivalent SPTN60, *ASTM Geotechnical Testing Journal*, 16 (4), 458–467.

JGS (2000): Investigations of the 1999 Kocaeli earthquake, *Reports on the Investigations of the 1999 Kocaeli Earthquake in Turkey and the 1999 Chi Chi Earthquake in Taiwan (in Japanese)*, Japanese Geotechnical Society.

JGS Committee (2001): JGS Committee report on mechanical properties of gravelly soils, *Proc. Symposium on Mechanical Properties of Gravelly Soils (in Japanese)*, Japanese Geotechnical Society, 151–164.

JGS Soil Investigation Editing Committee (2004): Chap. 6; Sounding, *Soil Investigations-Methods & explanations- (in Japanese)*, Japanese Geotechnical Society, Maruzen Publishing Co. Ltd.

JGS (2008): JGS 0161-2008 *Test method of minimum and maximum densities of sands (in Japanese)*, Japanese Geotechnical Society.

JGS (2009): JGS 0162-2009 *Test method of minimum and maximum densities of gravels (in Japanese)*, Japanese Geotechnical Society.

JGS Reconnaissance Committee (2009): *Reconnaissance Report on 2007 Niigataken Chuetsu-oki earthquake (in Japanese)*, Japanese Geotechnical Society.

Joseph, P.J., Einstein, H.H. and Whitman, R.V. (1988): A literature review of Geotechnical centrifuge modeling with particular emphasis on rock mechanics, *Report of Department of Civil Engineering MIT*, Airforce Engineering & Services Center, ESL-TR-87-23.

Joyner, W.B. and Fumal, T.E. (1984): Use of measured shear-wave velocity for predicting geologic site effects on strong ground motion, *Proc. of 8th World Conference on Earthquake Engineering*, Vol. 2, 777–783.

JRA (2002): *Design specifications for highway bridges-Part V Seismic Design-*, Japan Road Association.

JSCE (Japan Society for Civil Engineers) (2007): Earthquake damage in active-folding areas – Creation of a comprehensive data archive and suggestions for its application to remedial measures for civil-infrastructure systems. *Report of JSCE by Special Coordination Funds for Promoting Science and Technology (in Japanese)*. Japan Science & Technology Agency.

JSCE committee (1966): *Reconnaissance report on earthquake damage of Niigata earthquake 1964 (in Japanese)*, Japan Society for Civil Engineers, 904 pages.

Kagawa, T. (1978): On the similitude in model vibration tests of earth-structures, *Journal of Japan Society for Civil engineers (in Japanese)*, No. 275, 69–77.

Kamikawa, T. (2004): Mechanism for water film generation by model tests in 1-dimensional soil container, *Master's Thesis (in Japanese)*, Graduate School of Science & Engineering, Chuo University, Tokyo, Japan.

Kanai, K. (1951): On the group velocity of dispersive surface waves, *Bulletin, Earthquake Research Institute*, University of Tokyo, 3, 1–18.

Kanai, K. and Tanaka, T. (1954): Measurement of microtremor, *Bull. Earthquake Research Institute*, Vol. 32, 199–209.

Kanai, K., Tanaka, T., Yoshizawa, S. (1959): Comparative studies of earthquake motions on the ground and underground, *Bulletin of the Earthquake Research Institute*, Tokyo University, Vol. 37, 53–87.

Kanai, K. (1966): A short note on the seismological features of the Niigata earthquake, *Soils and Foundations*, VI (2), 8–13.

Kanai, K., Tanaka, T., Yoshizawa, S., Morishita, T., Osada, K. and Suzuki, T. (1966): Comparative studies of earthquake motions on the ground and underground II, *Bulletin of the Earthquake Research Institute*, University of Tokyo, Vol. 44, 609–643.

Kanatani, M., Okamotom T., Kokusho, T. and Matsui, I. (1989): Experimental study on dynamic properties of stiff clay, *Research Report (in Japanese)*, Central research Institute of Electric Power Industry, Japan, Rep. U89010.

Kaneko, Y. (2015): Energy based analysis of liquefaction using hollow cylinder tests: Influence of irregular loading and confining pressure, *Master's Thesis (in Japanese)*, Graduate School of Science & Engineering, Chuo University, Tokyo, Japan.

Kato, R. (2011): Triaxial tests on the effect of initial static shear stress for liquefaction of sand, *Master's Thesis (in Japanese)*, Graduate School of Science & Engineering, Chuo University, Tokyo, Japan.

Kato, T. and Kokusho, T. (2012): Rate-dependent pullout bearing capacity of piles by similitude model tests using seepage force, *Journal of Japan Society for Civil Engineers (in Japanese)*, C, 68 (1), 117–126.

Kawakami, F. and Asada, A. (1966): Damage to the ground and earth structures by the Niigata earthquake of June 16, 1964, *Soils and Foundations*, VI (1).

Kazama, M., Suzuki, T. and Yanagisawa, E. (1999): Evaluation of dissipated energy accumulated in surface ground and its application to liquefaction prediction, *Journal of Japan Society for Civil Engineers (in Japanese)*, JSCE, No. 631/III-48, 161–177.

Keefer, D.K. (1984): Landslides caused by earthquakes, *Geological Society of America Bulletin*, Vol. 95, 406–421.

Kinoshita, S. (1983): A study for damping characteristics of surface layers, *Journal of Japan Society for Civil Engineers (in Japanese)*, Vol. 330, 15–25.

Kishida, H. (1969): Characteristics of liquefied sands during Mino-Owari, Tohnankai and Fukui earthquakes, *Soils and Foundations*, 9 (1), 75–92.

Kochi, Y. (2008): Effect of non-plastic fines on undrained cyclic shear strength of unsaturated sand, *Master's Thesis (in Japanese)*, Graduate School of Science & Engineering, Chuo University, Tokyo, Japan.

Kojima, T. (2000): Experiment and numerical analysis on mechanism for water film generation in liquefied ground, *Master's Thesis (in Japanese)*, Graduate School of Science & Engineering, Chuo University, Tokyo, Japan.

Kokusho, T., Iwatate, T. and Ooaku, S. (1979): Scaled model tests and numerical analyses on nonlinear dynamic response of soft grounds, *Proc. Japan Earthquake Engineering Symposium*, JAEE, Paper No. 96, 761–768.

Kokusho, T. (1980): Cyclic triaxial test of dynamic soil properties for wide strain range, *Soils & Foundations*, 20 (2), 45–60.

Kokusho, T., Kato, S., Shimada, M. (1981): Undrained cyclic shear behavior of dense sand under initial shear stress, *Proc. 16th National Conference on Soil Mechanics (in Japanese)*, JGS, 581–584.

Kokusho, T., Esashi, Y. and Yoshida, Y. (1982): Dynamic properties of soft clay for wide strain range, *Soils and Foundations*, 22 (4), 1–18.

Kokusho, T. (1982): Dynamic soil properties and nonlinear seismic response of ground. *PhD thesis (in Japanese)*, The University of Tokyo, *General Report of CRIEPI (in Japanese)*, Central Research Institute of Electric Power Industry, No. 301.

Kokusho, T. Yoshida, Y. Nishi, K. and Esashi, Y. (1983a): Evaluation of seismic stability of dense sand layer (Part 1) -Dynamic strength characteristics of dense sand-, *Research Report of CRIEPI (in Japanese)*, Central Research Institute of Electric Power Industry, Japan, No. 383025.

Kokusho, T., Yoshida, Y. and Esashi, Y. (1983b): Evaluation of seismic stability of dense sand layer (Part 2) – Evaluation method by Standard Penetration Test, *Research report of CRIEPI (in Japanese)*, Central Research Institute of Electric Power Industry, No. U87019.

Kokusho, T. Yoshida, Y. and Nagasaki, K (1985): Liquefaction strength evaluation of dense sand layer, *Proc. 11th Intern. Conf. on SMFE*, San Francisco, Vol. 4, 1897–1900.

Kokusho, T. (1987): In-situ dynamic soil properties and their evaluations, *Proc. 8th Asian Regional Conf. on SMFE* (Theme Lecture in Kyoto), Vol. 2, 215–235.

Kokusho, T., Nishi, N., Honsho, S., Yoshida, Y., Kataoka, T., Okamoto, T., Tanaka, Y., Kudo, K., Ikemi, M., Kanatani, M., Kusunoki, K., Nakagawa, K., and Ishida, K. (1991): Study on Quaternary ground siting of nuclear power plant – Part 1 Geological/Geotechnical investigation methods and seismic stability evaluation methods of foundation ground, *General Report (in Japanese)*, Central Research Institute of Electric Power Industry, Japan, No. U19.

Kokusho, T. and Tanaka, Y. (1994): Dynamic properties of gravel layers investigated by in-situ freezing sampling, *ASCE Geotechnical Eng. Special Publication*, ASCE Convention (Atlanta), 121–140.

Kokusho, T., Tanaka, Y., Kawai, T., Kudo, K., Suzuki, K., Tohda, S. and Abe, S. (1995): Case study of rock debris avalanche gravel liquefied during 1993 Hokkaido-Nansei-Oki Earthquake, *Soils and Foundations*, 35 (3), 83–95.

Kokusho, T. and Yoshida, Y. (1997): SPT N-value and S-wave velocity for gravelly soils with different grain size distribution, *Soils & Foundations*, 37 (4), 105–113.

Kokusho, T., Matsumoto, M., Aoyagi, T., Takahashi, Y., Honma, M. and Motoyama, R. (1998): Nonlinear site amplification in vertical array records during Hyogoken Nambu earthquake, *Proc. 10th Japan Society of Earthquake Engineering Symposium (in Japanese)*, Paper No. C5-6.

Kokusho, T. and Matsumoto, M. (1998): Nonlinearity in site amplification and soil properties during the 1995 Hyogoken-Nambu Earthquake, Special Issue, *Soils and Foundations*, 1–9.

Kokusho, T., Hiraoka, R., Yoshida, Y., Kuwabara, H. Seshimo, Y. (1999): Evaluation of liquefaction strength of Masa Decomposed Granite by SPT in soil container Part 2): Triaxial test on intact samples from soil container, *Proc. 34th Annual Conference of Japanese Geotechnical Society (in Japanese)*, 127–128.

Kokusho, T. (1999): Formation of water film in liquefied sand and its effect on lateral spread, *Journal of Geotech. and Geoenv. Eng.*, ASCE, 125 (10), 817–826.

Kokusho, T. (2000a): Correlation of pore-pressure B-value with P-wave velocity and Poisson's ratio for imperfectly saturated sand or gravel, *Soils and Foundations*, 40 (4), 95–102.

Kokusho, T. (2000b): Mechanism for water film generation and lateral flow in liquefied sand layer, *Soils and Foundations*, 40 (5), 99–111.

Kokusho, T. and Kojima, T. (2002): Mechanism for post-liquefaction water film generation in layered sand, *Journal of Geotech. and Geoenv. Eng.*, ASCE, 128 (2), 129–137.

Kokusho, T. and Tsutsumi, Y. (2002): Study on degree of Liquefaction and damage to RC buildings during the Niigata earthquake, *Proc. 11th Japan Earthquake Engineering Symposium (in Japanese)*, 891–896.

Kokusho, T. and Motoyama, R. (2002): Energy dissipation in surface layer due to vertically propagating SH wave, *Journal of Geotech. and Geoenv. Eng.*, ASCE, 128 (4), 309–318.

Kokusho, T. and Fujita, K. (2002): Site investigation for involvement of water films in lateral flow in liquefied ground, *Journal of Geotech. and Geoenv. Eng.*, ASCE, 128 (11), 917–925.

Kokusho, T. and Mantani, S. (2002): Seismic amplification evaluation in a very deep down-hole, *Proc. 12th European Conference on Earthquake Engineering*, Paper Reference 797.

Kokusho, T. (2003): Current state of research on flow failure considering void redistribution in liquefied deposits, *Soil Dynamics and Earthquake Engineering*, Elsevier, Vol. 23, 585–603.

Kokusho, T., Yoshikawa, T., Suzuki, K. and Kishimoto, T. (2003): Post-liquefaction shear mechanism in layered sand by torsional shear tests, *Proc. 12th Pan-American Conference on Soil Mechanics and Geotechnical Engineering*, Vol. 1, 1045–1050.

Kokusho, T., Hara, T. and Hiraoka, R. (2004): Undrained shear strength of granular soils with different particle gradations, *Journal of Geotech. and Geoenv. Eng.*, ASCE, 130 (6), 621–629.

Kokusho, T. and Kabasawa, K. (2004): Slope failure evaluation by energy approach in hydraulic fill dams due to liquefaction-induced water films, *Proc. 13th World Conf. on Earthquake engineering*, Vancouver, Canada, Paper No. 131.

Kokusho, T., Aoyagi, T. and Wakunami, A. (2005a): In situ soil-specific nonlinear properties back-calculated from vertical array records during 1995 Kobe Earthquake, *Journal of Geotech. and Geoenv. Eng.*, ASCE, 131 (12), 1509–1521.

Kokusho, T., Hara, T. and Murahata, K. (2005b), Liquefaction strength of fines-containing sands compared with cone-penetration resistance in triaxial specimens, *Proc. 2nd Japan-US Workshop on Geomechanics*, ASCE Geo-Institute Publication No. 156, 356–373.

Kokusho, T. (2006): Recent developments in liquefaction research learned from earthquake damage, *Journal of Disaster Research*, 1 (2), 226–243.

Kokusho, T. and Ishizawa, T. (2007): Energy approach to earthquake-induced slope failures and its implications, *Journal of Geotech. and Geoenv. Eng.*, ASCE, 133 (7), 828–840.

Kokusho, T. (2007): Liquefaction strengths of poorly-graded and well-graded granular soils investigated by lab tests, *Proc. 4th International Conference on Earthquake Geotechnical Engineering*, Thessaloniki, 159–184, Springer.

Kokusho, T., Motoyama, R. and Motoyama, H. (2007): Wave energy in surface layers for energy-based damage evaluation, *Soil Dynamics and Earthquake Engineerng*, Elsevier, Vol. 27, 354–366.

Kokusho, T. and Sato, K. (2008): Surface-to-base amplification evaluated from KiK-net vertical array strong motion records, *Soil Dynamics and Earthquake Engineering*, Vol. 28, 707–716.

Kokusho, T. (2009): PBD in earthquake geotechnical engineering and energy-based design, Special Discussion Session – Future directions of performance-based design-, Performance-Based Design in Earthquake Geotechnical Engineering – from Case History to Practice, *Proc. International Conference on Performance Based Design in Earthquake Geotechnical Engineering (IS-Tokyo 2009)*, Balkema, 359–362.

Kokusho, T., Ishizawa, T. and Nishida, K. (2009a): Travel distance of failed slopes during 2004 Chuetsu earthquake and its evaluation in terms of energy, *Soil Dynamics and Earthquake Engineering*, Elsevier, 29, 1159–1169.

Kokusho, T., Ishizawa, T. and Hara, T. (2009b): Slope failures during the 2004 Niigataken Chuetsu earthquake in Japan, *Earthquake Geotechnical Case Histories for Performance-Based Design*, Balkema, CRC Press, 47–70.

Kokusho, T., Ishizawa, T. and Koizumi, K. (2011a): Energy approach to seismically induced slope failure and its application to case histories, *Engineering Geology*, Elsevier, Vol. 122, Isuues 1–2, 115–128.

Kokusho, T., Ito, F. and Nagao, Y. (2011b): Aging effect on liquefaction strength and cone resistance of fines-containing sand investigated in triaxial apparatus, *Proc. 5th Intern. Conf. on Earthquake Geotechnical Engineering* Santiago, ISSMGE, 509–516.

Kokusho, T. and Suzuki, T. (2011): Energy flow in shallow depth based on vertical array records during recent strong earthquakes, *Soil Dynamics & Earthquake Engineering*, Elsevier, Vol. 31, 1540–1550.

Kokusho, T. and Suzuki, T. (2012): Energy flow in shallow depth based on vertical array records during recent strong earthquakes (Supplement), *Soil Dynamics and Earthquake Engineering*, Elsevier, Vol. 42, 138–142.

Kokusho, T., Ito, F., Nagao, Y. and Green, R. (2012a): Influence of non/low-plastic fines and associated aging effects on liquefaction resistance, *Journal of Geotech. and Geoenv. Eng.*, ASCE, 138 (6), 747–756.

Kokusho, T., Nagao, Y. Ito, F. and Fukuyama, T. (2012b): Sand liquefaction observed during recent earthquake and basic laboratory study on aging effect, *Proc. 2nd International Conf. on Performance-based Design in Earthquake Geotechnical Engineering*, Taormina, Italy, GGEE 28, Springer, 75–92.

Kokusho, T., Nakashima, S., Kubo, A. and Ikeda, K. (2012c): Soil investigation of fly ash deposit improved by heavy compaction method, *Journal of Geotech. and Geoenv. Eng.*, ASCE, 138 (6), 738–746.

Kokusho, T. (2013a): Site amplification formula using average Vs in Equivalent Surface Layer Based on Vertical Array Strong Motion Records, *Proc. International Conf. on Earthquake Geotechnical Engineering in honor of Prof. Kenji Ishihara*, Springer GGEE 37, 141–160.

Kokusho, T. (2013b): Liquefaction potential evaluation-energy-based method versus stress-based method-, *Canadian Geotechnical Journal*, 50, 1–12.

Kokusho, T. (2014): Seismic base-isolation mechanism in liquefied sand in terms of energy, *Soil Dynamic and Earthquake Engineering*, Elsevier, Vol. 63, 92–97.

Kokusho, T., Mukai, A. and Kojima, T. (2014a): Liquefaction behavior in Urayasu and physical properties of fines, *Proc. 14th Japan Earthquake Engineering Symposium (in Japanese)*, Japanese Association of Earthquake Engineering, G06, Thu-2.

Kokusho, T., Koyanagi, T. and Yamada, T. (2014b): Energy approach to seismically induced slope failure and its application to case histories -Supplement-, *Engineering Geology*, Elsevier, Vol. 181, 290–296.

Kokusho, T., Yamamoto, Y., Koyanagi, T., Saito, Y. and Yamada, T. (2014c): Model tests on threshold energy for slope failure and associated case studies, *Geotechnical Journal (in Japanese)*, JGS, 9 (4), 721–737.

Kokusho, T. and Mimori, Y. (2015): Liquefaction potential evaluations by energy-based method and stress-based method for various ground motions, *Soil Dynamics and Earthquake Engineering*, Elsevier, Vol. 75, 130–146.

Kokusho, T. (2016): Applicability of energy-based liquefaction potential evaluation method compared with FL-method Supplement-, *Geotechnical Journal (in Japanese)*, Japanese Geotechnical Society, 11 (3), 283–293.

Konagai, K., Orense, R.P. and Johansen, J. (2009): Las Kolinas landslide caused by the 2001 El Salvador earthquake, *Earthquake Geotechnical Case Histories for Performonce Design*, CRC Press, 227–244.

Kong, X.J., Pradhan, T.B.J., Tatsuoka, F. Tamura, C. (1986): Dynamic deformation properties of sand at extremely low pressures -Test results-, *Institute of Industrial Science Report (in Japanese)*, the University of Tokyo, Vo. 38, No. 2, 28–31.

Koseki, J. (1997): Uplift of sewer manholes during thc 1993 Kushiro-Oki earthquakes, *Soils and Foundations*, 37 (4), 109–121.

Koseki, K. lshihara, K. and Fujii, M. (1986): Cyclic triaxialtests of sand containing fines, *Proc. 21st Annual Conference of Japanese Geotechnical Society (in Japanese)*, 595–596.

Kovacs, W. D., Salomone, L.A. and Yokel, F.Y. (1983): Comparison of energy measurements in the Standard Penetration Test using cathead rope method, National Bureau of Standards, *Report of the US Nuclear Regulatory Commission.*

Kudo, K. (1976): *Mesurement of wave attenuation – Field experiment – Experiment on earthquake wave generation and propagation (in Japanese)*, Research group on seismic wave exploration.

Kudo, K. (1995): Practical estimates of site response, State-of-the-Art report, *Proc. 5th International Conference on Seismic Zonation*, Nice, France, Vol. 3, 1878–1907.

Kusaka, T. (2012): Liquefaction characteristics of sand containing non-plastic fines with initial shear stresses: Investigation by Cyclic and Monotonic Loading Torsional Shear Test, *Master's Thesis (in Japanese),* Graduate School of Science & Engineering, Chuo University, Tokyo, Japan.

Kusaka, T., Kokusho, T. and Arai, R. (2013): Liquefaction behavior of loose sand containing non-plastic fines under initial shear stresses – Investigation by cyclic and monotonic loading torsional shear tests, *Geotechnical Journal (in Japanese)*, Japanese Geotechnical Society, C, 69 (1), 80–90.

Lade, P.V. and Hernandez, S.B. (1977): Membrane penetration effects in undrained tests, *Journal of Geotech. Eng. Div.*, ASCE, 103 (GT2), 109–125.

Lee, K.L. and Albeisa, A. (1974): Earthquake induced settlements in saturated sands, *Journal of Geotechnical Eng.* ASCE, 100 (GT4), 387–406.

Lee, C.J., Abdoun, T. and Dobry, R. (2000): Movements of a quay wall during an earthquake, *Proc. International Workshop on Annual Commemoration of Chi-Chi Earthquake*, Taipei, 324–335.

Lee, J.S. and Santamarina J.C. (2005): Bender Elements: Performance and Signal Interpretation, *Journal of Geotech. and Geoenv. Eng.*, ASCE, 131 (9), 1063–1070.

Lee, K. and Roth, W. (1977): Seismic stability analysis of Hawkins hydraulic fill dam, *Journal of Geotech. Eng. Div.*, ASCE, 103 (GT6), 627–644.

Lee, K. and Seed, H.B: (1967) Cyclic stress conditions causing liquefaction of sand, *Journal of SMFE*, ASCE, 92 (SM6), 47–70.

Lee, K.L. (1974): Seismic permanent deformations in earth dams, *Report to the National Science Foundation*, Project GI 38521, University of California, California.

Lemke, R.W. (1967): Effects of the Earthquake of March 27, 1964 at Seward, Alaska, *Geological Survey Professional Paper* 542-E, US Depart of Interior.

Liao, S.C. and Whitman, R.V. (1986): Overburden correction factors for SPT in sand, *Journal of Geotech. Eng. Div.,* ASCE, 112 (3), 373–377.

Lin, M.L., Wang, K.L. and Chen, T.C. (2009a): Chiufenerhshan landslide in Taiwan during 1999 Chi-Chi earthquake, Earthquake *Geotechnical Case Histories for Performance-Based Design*, CRC Press, 259–272.

Lin, M.L., Wang, K.L. and Chen, T.C. (2009b): Taoling landslide in Taiwan during 1999 Chi-Chi earthquake, Earthquake *Geotechnical Case Histories for Performance-Based Design*, CRC Press, 273–287.

Lysmer, J. and Kuhlemeyer (1969): Finite dynamic model for infinite media, *Journal of EMD,* ASCE, 95 (EM4), 859–877.

Lysmer, J. and Waas, G. (1972): Shear waves in plane infinite structures, *Journal of EMD*, ASCE, 98 (EM1), 85–105.

Lysmer, J., Udaka, T., Tsai, C.F. and Seed, H.B. (1975): FLUSH – A computer program for approximate 3-D analysis of soil structure interaction problems, *Report EERC 75–30*, University of California, Berkeley.

Makdisi, F.I. and Seed H.B. (1978): Simplified procedure for estimating dam and embankment earthquake-induced deformations, *Journal of Geotech. Eng. Div.*, ASCE, 104 (GT7), 849–867.

Mantani, S. (2002): Interpretation of seismic ground motion in terms of energy by means of multi-reflection theory, *Master's Thesis (in Japanese)*, Graduate School of Science & Engineering, Chuo University, Tokyo, Japan.

Marcuson, W.F. and Wahls, H.E. (1972): Time effects on dynamic shear modulus of clays, *Proc. of SMFD*, ASCE, 98 (SM12), 1359–1373.

Martin, G.R., Finn, W.D.L. and Seed, H.B. (1975): Fundamentals of liquefaction under cyclic loading, *Journal of Geotech. Eng. Div.*, ASCE, 101 (GT5), 423–438.

Martin, G. R., Finn, W.D.L. and Seed, H.B. (1978): Effect of system compliance on liquefaction tests, *Journal of Geotech. Eng. Div.*, ASCE, 104 (GT4), 463–479.

Matsui, T. and Oda, K. (1996): Foundation damage of structures, Special Issue on Geotechnical Aspects of the January 17, 1995 Hyogoken Nambu Earthquake, *Soils and Foundations*, 189–200.

Matsumoto, M., Oh-ishi T. and Shimada, R. (1998): Investigation on design seismic motions considering deep soil profiles in Osaka plane, *Journal of Electric Power Civil Engineering (in Japanese)*, 277, 87–92.

Matsuo, O. (1997): Liquefaction potential evaluation and earthquake-resistant design, *Kiso-ko (in Japanese)*, 25 (3), 34–39.

Matsuo, O. and Murata, K. (1997): A proposal of simplified evaluation methods of liquefaction resistance on gravelly soils, *Proc. 32nd Annual Conference of Japanese Geotechnical Society (in Japanese)*, 775–776.

McCulloch, D.S. and Bonilla, M.G. (1970): Effects of the earthquake of March 27, 1964, on the Alaska Railroad, *USGS Professional Paper 545-D*, Government Printing Office, Washington, D. C.

Mendoza, M.J. (1987): Foundation engineering in Mexico City: Behavior of foundations, *Proc. International symposium on Geotechnical Engineering of Soft Soils*, Vol. 2, 351–367.

Meneses, J., Ishihara, K. and Towhata, I. (1998): Effects of superimposing cyclic shear stress on the undrained behavior of saturated sand under monotonic loading, *Soils and Foundations*, 38 (4), 115–127.

Meyerhof, G.G. (1957): Discussion, *Proc. 4th international Conference on SMFE*, Vol. 3, 110.

Midorikawa, S. (1987): Prediction of isoseismal map in the Kanto plain due to hypothetical earthquake, *Journal of structural engineering (in Japanese)*, Architectural Institute of Japan, Vol. 33B, 43–48.

Miyake, N., Shamoto, Y., Goto, S. (2003): Preventive measure of liquefaction by decreasing saturation ratio – Part 1: The concept and evaluation of undisturbed sandy soil, *Proc. 38th Annual Conference of Japanese Geotechnical Society (in Japanese)*, 1975–1976.

Mori, S. Numata, A., Sakaino, N. and Hasegawa, M. (1991): Characteristics of liquefied sands on reclaimed lands during earthquakes, *Tsuchi-to-kiso (in Japanese)*, Japanese Geotechnical Society, 39, No. 2, 17–22.

Mulilis, J.P., Seed, H.B., Chan, C.K., Michell, J.K. and Arulanandan, K. (1977): Effect of sample preparation on sand liquefaction, *Journal of Geotech. Eng. Div.*, ASCE 103 (GT2), 91–108.

Naesgaard, E. and Byrne, P.M. (2005): Flow liquefaction due to mixing of layered deposits, *Proc. of Geotech. Earthquake Eng.* Satellite Conference, TC4 Committee, ISSMGE, Osaka, Japan.

Naesgaard, E., Byrne, P.M. and Wijewickreme, D. (2007): Is P-Wave Velocity an Indicator of Saturation in Sand with Viscous Pore Fluid? *International Journal of Geomechanics*, ASCE. 7 (6), 437–443.

Naesgaard, E. (2011): A hybrid effective stress – total stress procedure for analyzing soil embankments subjected to potential liquefaction and flow, *PhD Thesis*, The Faculty of Graduate Studies, The University of British Columbia, Canada.

Nagase, H. and Ishihara, K. (1988): Liquefaction-induced compaction and settlement of sand during earthquakes, *Soils and Foundations*, 28 (1), 65–76.

Nakamura, Y. (1989): A method for dynamic characteristics estimations of subsurface using microtremors on the ground surface, *Quarterly Report of Railway Technology Research Institute (in Japanese)*, 2 (4), 18–27.

Nakamura, Y. (2000): Clear identification of fundamental idea of Nakamura's technique and its applications, *Proc. 12th World Conference on Earthquake Engineering*, Paper No. 2656.

Nakamura, Y., Sato, T. and Saita J. (2003): Evaluation of the amplification characteristics of subsurface using microtremor and strong motion -The studies at Mexico City-, *Proc. 27th Earthquake Engineering Symposium (in Japanese)*, Japan Society for Civil Engineers.

Nakamura, Y. (2008): Basic structure of the H/V spectral ratio, *Proc. 3rd Symposium on Earthquake Disaster Mitigation (Keynote lecture) (in Japanese)*, The Society of Exploration Geophysicists of Japan, 1–6.

Nakazawa, H., Ishihara, K., Tsukamoto, Y., Kamata, K. and Ooyama, A. (2001): Effect of degree of saturation on P-wave velocity and liquefaction behavior, *Proc. 26th Earthquake Engineering Conference of JSCE (in Japanese)*, Japan Society for Civil Engineers, 625–628.

Nemat-Nasser, S. and Shokooh, A. (1979): A unified approach to densification and liquefaction of cohesionless sand in cyclic shearing, *Canadian Geotechnical Journal*, 16, 659–678.

Newmark, N.M. (1965): Effects of earthquakes on dams and embankments, Fifth Rankine Lecture, *Geotechnique*, Vol. 15, 139–159.

Newmark, N.W. and Rosenblueth, E. (1971): *Fundamentals of Earthquake Engineering*, Prentice-Hall, Englewood Cliffs, N.J., 162–163.

Nishida, K., Takamura, T., Nakajima, M. Tanaka, H. and Tanaka, M. (1999): Case studies of seismic cone test and its amplitude characteristics on in situ data, *Journal of Japan Society for Civil Engineering (in Japanese)*, 631 (III-48), 329–338.

Nishio, S., Tamaoki, K., Shamoto, Y., Goto, S. and Baba, K. (1987): Measurement of S-wave velocity in large scale triaxial cell, *Proc. 22nd Annual Conference of Japanese Geotechnical Society (in Japanese)*, 511–514.

NIST (National Institute of Standards and Technology) (2012): *Soil-Structure Interaction for Building Structures*, NEHRP Consultants Joint Venture, U.S. Department of Commerce, NIST GCR 12-917-21.

Nogoshi, M. and Igarashi, T. (1971): On the amplitude characteristics of microtremor (Part 2), *Journal of Seismological Society of Japan (in Japanese)*, Vol. 24, 26–40.

NRC (National Research Council) (1985): *Liquefaction of soils during earthquakes*, Committee of Earthquake Engineering, Commission of Engineering and Technical Systems, National Academy Press, Washington, D.C.

Ohkawa, T. (2003): Basic experiments on mechanical properties of cement mixed flowable fill under seismic loading, *Master's Thesis (in Japanese)*, Graduate School of Science & Engineering, Chuo University, Tokyo, Japan.

Ohmachi, T., Konno, K., Endoh, T. and Toshinawa, T. (1994): Refinement and application of an estimation procedure for site natural periods using microtremor, *Journal of Japan Society for Civil Engineers (in Japanese)*, No. 489, I-27, 251–260.

Oh-oka, H. (1984): Comparison of SPT N-values by Cone & Pulley method and Tonbi method (for Pleistocene sands), *Proc. 19th Annual Conference of Japanese Geotechnical Society (in Japanese)*, 117–118.

Okamura, M. and Soga, Y. (2006): Effects of pore fluid compressibility on liquefaction resistance of partially saturated sand, *Soils and Foundations*, 46 (5), 93–104.

Okamura, M. and Noguchi, K. (2009): Liquefaction resistances of unsaturated non-plastic silt, *Soils and Foundations*, 49 (2), 221–229.

Okamura, M., Takebayashi, M., Nishida, K., Fujii, N., Jinguji, M., Imasato, T, Yasuhara, H. and Nakagawa, E. (2009): In-situ test on desaturation by air injection and its monitoring, *Journal of Japan Society for Civil Engineers (in Japanese)*, 65 (3), 756–766.

Okumura, T., Narita, K. and Ohne, Y. (1985): Dynamic deformation characteristics of sandy soils under low confining pressure, *Journal of Japan Society for Civil Engineers (in Japanese)*, Vol. 364, III-1, 67–76.

Olson, S.M. and Stark, T.M. (2002): Liquefied strength ratio from liquefaction flow failure case histories, *Canadian Geotechnical Journal*, 39, 629–647.

Papadopoulou, A. and Tika, T. (2008), The effect of fines on critical state and liquefaction resistance characteristics of non-plastic silty sands, *Soils & Foundations*, Japanese Geotechnical Society, 48, No. 5, 713–725.

Pickering D.J. (1973): Drained liquefaction testing in simple shear, *Journal of SMFE*, ASCE, 99 (SM12), 1179–1184.

Plafker, G., Erickson, G.E. and Concha, J.F. (1971): Goelogical aspects of the May 31, 1970, Peru earthquake, *Seismological Society America Bulletin*, 61 (3), 543–578.

Polito, C.P. and Martin, J.R.II (2001): Effect of nonplastic fines on the liquefaction resistance of sands, *Journal of Geotech. Eng. Div.*, ASCE, 127 (5), 408–415.

Poulos, S.T., Castro, G. and France, J.W. (1985): Liquefaction evaluation procedure, *Journal of Geotech. Eng.*, ASCE, 111 (6), 772–792.

Pyke, R. (1979): Nonlinear soil models for irregular cyclic loadings, *Journal of Geotech. Eng.*, ASCE, 105 (GT6), 715–726.

Richart, F.E., Hall, J.R. and Woods, R.D. (1970): *Vibrations of soils and foundations*, Prentice Hall Inc.

Robertson, P.K. and Campanella, R. (1985). Liquefaction Potential of Sands Using the CPT, *Journal of Geotech. Eng.*, ASCE, 111 (3), 384–403.

Robertson, P. K., Campanella, R. G., Gillespie, D. and Rice, A. (1986): Seismic CPT to measure in situ shear wave velocity, *Journal of Geotech. Eng. Div.*, ASCE, 112 (8), 791–803.

Robertson, P.K. (1990): Soil classification using the cone penetration test, *Canadian Geotechnical Journal*, 1990, 27 (1): 151–158.

Robertson, P.K. and Wride, C.E. (1998): Evaluating cyclic liquefaction potential using the cone penetration test, *Canadian Geotechnical Journal*, 35 (3), 442–459.

Robinson, K., Cubrinovski, M. and Bradley, A.B. (2014): Lateral spreading displacements from the 2010 Darfield and 2011 Christchurch earthquakes, *International Journal of Geotechnical Engineering*, 8 (4), 441–448.

Rocha, M. (1957): The possibility of solving soil mechanics problems by the use of models, *Proc. 4th Int. Conf. on SMFE*, ISSMFE, 183–188.

Roesler, S.K. (1979): Anisotropic shear modulus due to stress anisotropy, *Journal of Geotech. Eng. Div.*, ASCE, 105 (No. GT7), 871–880.

Rollins K.M. and Seed, H.B. (1990): Influence of buildings on potential liquefaction damage, *Journal of Geotech. Eng. Div.*, ASCE, 116 (GT2), 165–185.

Romo, M.P. (1995): Clay behavior, ground response and soil-structure interaction studies in Mexico City, *Proc. 3rd International Conference on Recent Advances in Geotech. Earthquake Eng. and Soil Dynamic*, St. Louis Missouri, Vol. II, 1039–1051.

Sarma, S.K. (1971): Energy Flux of Strong Earthquakes, *Techtonophysics*, Elsevier Publishing Company, 159–173.

Sarma, S.K. (1975): Seismic stability of earth dams and embankments, *Geotechnique*, 25, No. 4, 743–761.

Sasaki, T., Tatsuoka, F. and Yamada, S. (1982): Estimation of settlement during reconsolidation in sand ground, *Proc. 17th Annual Conference of Japanese Geotechnical Society (in Japanese)*, 1661–1664.

Sato, T. and Kawase, H. (1992): Finite element simulation of seismic wave propagation in near-surface random media, *Proc. International Symposium on the effect of Surface Geology on Seismic Motion*, ESG 1992 ODAWARA, 257–262.

Sato, K., Kokusho, T., Matsumoto, M. and Yamada, E. (1996): Nonlinear seismic response and soil property during strong motion, Special Issue on Geotechnical Aspects of the January 17, 1995 Hyogoken Nambu Earthquake, *Soils and Foundations*, 41–52.

Sato, M., Oda, M., Kazama, H. and Kozeki, K. (1997): Fundamental study on the effect of fines on liquefaction strength of reclaimed ground, *Journal of Japan Society for Civil Engineers (in Japanese)*, 463 (III-38), 271–282.

Schmertmann, J.S. and Palacios, A. (1979): Energy dynamics of SPT, *Journal of SMF Div.*, ASCE 105 (GT8), 909–926.

Schnabel, P.B., Lysmer, J. and Seed, H.B. (1972): SHAKE – A computer program for earthquake response analysis of horizontally layered sites, *Report EERC 72–12*, University of California, Berkeley.

Scott, R.F. and Zuckerman K.A. (1972): Sand blows and liquefaction. *Proc. The Great Alaska Earthquake of 1964-Engineering Publication 1606*; National Academy of Sciences, Washington, D.C., 179–189.

Seed H.B. and Lee, K.L. (1966): Liquefaction of saturated sands during cyclic loading, *Journal of SMFD*, ASCE, 92 (6), 105–134.

Seed, H.B. (1968): Landslides during earthquakes due to soil liquefaction, *Journal of SMFD*, ASCE, 94 (5), 1055–1122.

Seed, H.B. and Idriss, I.M. (1971): Simplified procedure for evaluating soil liquefaction potential, *Journal of SMFD,* ASCE, 97 (SM9), 1249–1273.

Seed, H.B. and Peacock, W.H. (1971): Test procedures for measuring soil liquefaction characteristics, *Journal of SMFD*, ASCE, 97 (8), 1099–1119.

Seed, H. B., Idriss, I.M., Makdisi, F. and Benerjee, N. (1975a): Representation of irregular stress time histories by equivalent uniform stress series in liquefaction analyses, Report No. EERC 75-29, University of California, Berkeley.

Seed, H.B.M., Lee, K.L., Idriss, I.M. and Makdisi, A.M. (1975b): The slides in the San Fernando Dams during the earthquake of February 9, 1971, *Journal of SMFD*, ASCE, 101 (GT7), 651–688.

Seed, H.B., Martin, P.P. and Lysmer, J. (1976): Pore-water pressure changes during soil liquefaction, *Journal of SMFD*, ASCE, 102 (GT4), 323–346.

Seed, H.B., Mori, K. and Chan C.K. (1977): Influence of Seismic history on Liquefaction of sands, *Journal of SMFD*, ASCE, 103 (GT4), 257–270.

Seed, H.B. and Booker, J.R. (1977): Stabilization of potentially liquefiable sand deposits using gravel drains, *Journal of SMFD*, ASCE, 103 (GT7), 757–768.

Seed, H.B., Pyke, R.M. and Martin, G.R. (1978): Effects of multidirectional shaking on pore pressure development in sands, *Journal of SMFD*, ASCE, 104 (GT1), 27–44.

Seed, H.B. (1979): Considerations in the earthquake-resistant design of earth and rockfill dams, *Geotechnique,* 29 (3), 215–263.

Seed, H.B. and Idriss, I.M. (1981): Evaluation of liquefaction potential of sand deposits based on observations of performance in previous earthquakes, *Preprint 81-544, In Situ Testing to Evaluate Liquefaction Susceptibility*, ASCE National Convention.

Seed, H.B. and De Alba, P. (1984): Use of SPT and CPT tests for evaluating the liquefaction resistance of sands, *Proc. In-situ '86*, ASCE Geotech. Special Publication, No. 6, 281–302.

Seed, H.B. Tokimatsu, K. Harder, L.F. Jr. and Chung, R. (1985): Influence of SPT procedures in soil liquefaction resistance evaluations, *Journal of. Geotech. Eng.*, ASCE, 111 (12), 1425–1445.

Seed, H.B., Wong, R.T., Idriss, I.M. and Tokimatsu, K. (1986): Moduli and damping factors for dynamic analyses of cohesionless soils, *Journal of Geotech. Eng.*, ASCE, 112 (11), 1016–1032.

Seed, H.B. (1987): Design problems in soil liquefaction, *Journal of Geotechnical Eng.* ASCE, 113 (8), 827–845.

Sezawa, K. (1927): Dispersion of elastic waves propagated on the surface of stratified bodies and on curved surfaces, *Bulletin, Earthquake Research Institute*, University of Tokyo, Vol. 29, 49–60.

Shima, E. (1978): Seismic Microzoning map of Tokyo, *Proc. 2nd Intern. Conf. on Seismic Zonation*, Vol. 1, 433–443.

Shirley, D.J. and Hampton, L.D. (1977): Shear wave measurements in laboratory sediments, *Journal of Acoustic Society of America*, 63 (2), 607–613.

Skempton, A.W. and Brogan, J.M. (1994): Experiments on piping in sandy gravels, *Geotechnique*, 44 (3), 449–460.

Sladen, J.A., D'Hollander, R.D. & Krahn, J. (1985): The liquefaction of sands, a collapse surface approach, *Canadian Geotechnical Journal*, 22, 564–578.

Somerville, P. (1996): Forward rupture directivity in the Kobe and Northridge earthquakes and implications for structural engineering, *Proc. International Workshop on Site Response subjected to Strong Earthquake Motions*, Yokosuka, Japan, Vol. 2, 324–342.

Stokoe, K.H. and Nazarian, S. (1984): In situ shear wave velocity from spectral analysis of surface waves, *Proc. 8th World Conference on Earthquake Engineering*, Vol. 3, 31–38.

Suda, Y., Hayashi, H., Kuroyanagi, I., Morimoto, I. and Kokusho, T. (2007a): Investigation of forces applied to piles due to liquefaction-induced lateral flow, *Journal of Japan Society for Civil Engineers (in Japanese)*, III, 63 (2), 487–501.

Suda, Y., Sato, M., Tamari, T,. and Kokusho, T. (2007b): Design method for pile foundations subjected to liquefaction-induced lateral flow, *Journal of Japan Society for Civil Engineers (in Japanese)*, III, 63 (2), 467–486.

Suetomi, I. and Yoshida, N. (1998): Nonlinear behavior of surface deposit during the 1995 Hyogoken-Nambu earthquake, Special Issue on Geotechnical Aspects of the January 17, 1995 Hyogoken-Nambu Earthquake No. 2, *Soils and Foundations*, 11–22.

Sugito, M. (1995): Frequency Dependent Equivalent Strain for Equi-Linearized Technique, *Proc. 1st International Conference on Earthquake Geotechnical Engineering (IS-Tokyo)*, Balkema, 655–660.

Suwa, M., Kokusho, T., Hiraoka, R., Suga, Y., Yoshida, Y. and Numata, J. (2000): Considerations on hammer impact efficiency in SPT laboratory tests, *Proc. 55th Annual Convention (in Japanese)*, Japan Society of Civil Engineers, III-A70, 140–141.

Suzuki, Y. and Tokimatsu, K. (2003): Correlations between CPT data and liquefaction resistance of in situ frozen samples, *Journal on Structure Construction Eng. Architectural Institute of Japan (in Japanese)*, No. 566, 81–88.

Suzuki, Y., Tokimatsu, K., Taya, Y. and Makihara, Y. (1994): Correlation between Cone penetration test results and dynamic properties of in situ frozen samples, *Proc. 9th Japan Earthquake Engineering Symposium (in Japanese)*, JAEE, Paper No. 141, 841–846.

Suzuki, Y., Tokimatsu, K., Taya, Y. and Kubota, Y. (1995): Correlation between CPT data and dynamic properties of in situ frozen samples, *Proc. 3rd Intern. Conf. on Recent Advances in Geotech. Earthquake Eng. & Soil Dynamics*, Vol. 1, 249–252.

Sy, A. and Campanella (1993): Dynamic performance of the Becker hammer drill and penetration test, *Canadian Geotechnical Journal*, 30, 607–619.

Tajimi, H. (1965): Radiation damping, *Architect vibration (in Japanese)*, Corona Publishing Company, Chap. 5.4, 125–131.
Tanaka, Y., Kudo, K., Yoshida, Y. and Ikemi, M. (1987): A study on the mechanical properties of sandy gravel -Dynamic properties of reconstituted samples-, *Research report of CRIEPI (in Japanese)*, Central Research Institute of Electric Power Industry, No. U87019.
Tanaka, Y., Kokusho, T., Yoshida, Y. and Kudo, K. (1991): A method for evaluating membrane compliance and system compliance in undrained cyclic shear tests, *Soils and Foundations*, 31 (3), 30–42.
Tanaka, Y., Kudo, K., Yoshida, Y. and Kokusho, T. (1992): Undrained cyclic strength of gravelly soil and its evaluation by penetration resistance and shear modulus, *Soils and Foundations*, 32 (4), 128–142.
Tanaka, Y. (2001): Modeling of anisotropic behavior of gravelly layer in Hualien, Taiwan, *Soils and Foundations*, 41 (3), 73–86.
Tani, K., Kaneko, S. and Sakai, K. (2007): Undisturbed sampling method using thick water-soluble polymer solution, *Proc. 13th Asian Conference on Soil Mechanics and Geotechnical Engineering*, Kolkata, Vol. 1, 93–96.
Taobada U.V., Martinez, H. and Romo, M.P. (1999): Evaluation of dynamic soil properties in Mexico City using downhole array records, *Soils and Foundations*, 39 (5), 81–92.
Tatsuoka, F., Iwasaki, T., Tokita, K., Yasuda, S., Hirose, M., Imai, T. and Imano, H. (1978): A method for estimating undrained cyclic strength of sandy soils using Standard Penetration N-values, *Soils and Foundations*, 18 (3), 43–58.
Tatsuoka, F., Iwasaki, T., Yoshida, S., Fukushima, S. and Sudo, H. (1979): Shear modulus and damping by drained tests on clean sand specimens reconstituted by various methods, *Soils and Foundations*, 19 (1), 39–54.
Tatsuoka, F. and Silver, M. (1980): New Method for the Calibration of the Inertia of Resonant Column Devices, *Geotechnical Testing Journal*, 3 (1), 30–34.
Tatsuoka, F., Iwasaki, T., Tokida, K. and Konno, M. (1981): Cyclic undrained triaxial strength of sampled sand affected by confining pressure, *Soils and Foundations*, 21 (2), 115–120.
Tatsuoka, F., Muramatsu, M. and Sasaki, T. (1982): Cyclic undrained stress-strain behavior of dense sands by torsional simple shear test, *Soils and Foundations*, 22 (2), 55–70.
Tatsuoka, F., Sasaki, T. and Yamada, S. (1984): Settlements in saturated sand induced by cyclic undrained simple shear, *Proceedings of 8th World Conference on Earthquake Engineering*, San Francisco, Vol. 3, 95–102.
Tatsuoka, F., Ochi, K., Fujii, S. and Okamoto, M. (1986a): Cyclic undrained triaxial shear strength of sands for different sample preparation methods, *Soils and Foundations*, 26 (3), 23–41.
Tatsuoka, F., Sakamoto, M., Kawamura, T. and Fukushima, S. (1986b): Strength and deformation characteristics of sand in plane strain compression at extremely low pressures, *Soils and Foundations*, 26 (1), 65–84.
Tatsuoka, F., Kato, H., Kimra, M. and Pradhan, T.B.S. (1988): Liquefaction strength of sands subjected to sustained pressure, *Soils and Foundations*, 28 (1), 119–131.
TCEGE (Technical Committee for Earthquake Geotechnical Engineering) (1999): *Manual for zonation on seismic geotechnical hazards (Revised Version)*: Japanese Geotechnical Society.
Timoshenko, S. and Goodier, J.N. (1951): Theory of Elasticity, McGraw-Hill.
Tokimatsu, K. & Midorikawa, S. (1981): Time histories of shear moduli of soil during strong earthquakes, *Proc. 16th Annual Conference of Japanese Geotechnical Society (in Japanese)*, D-7, 685–688.
Tokimatsu, K. and Yoshimi, Y. (1982): Liquefaction of sand due to multidirectional cyclic shear, *Soils and Foundations*, 22 (3), 126–130.
Tokimatsu, K. and Yoshimi, Y. (1983): Empirical correlation of soil liquefaction based on SPT N-value and fines content, *Soils and Foundations*, 23 (4), 56–74.

Tokimatsu, K. and Nakamura, K. (1986): A liquefaction test without membrane penetration effects, *Soils & Foundations*, 26 (4), 127–138.

Tokimatsu, K., Yamazaki, T. and Yoshimi, Y. (1986): Soil liquefaction evaluations by elastic shear moduli, *Soils & Foundations*, Japanese Geotechnical Society, Vol. 26, No. 1, 25–35.

Tokimatsu, K. and Nakamura, K. (1987): A simplified correction for membrane compliance in liquefaction tests, *Soils & Foundations*, 27 (4), 111–122.

Tokimatsu, K. (1988): Penetration tests for dynamic problems, *Penetration Testing*, Balkema, 117–136.

Tokimatsu, K. (1990): System compliance correction from pore pressure response in undrained cyclic triaxial tests, *Soils & Foundations*, 30 (2), 14–22.

Tokimatsu, K., Kuwayama, S., Tamura, S. and Miyadera, Y. (1991): Vs determination from steady state Rayleigh wave method, *Soils and Foundations*, 31 (2), 153–163.

Tokimatsu, K. (1995): Geotechnical site characterization using surface waves, *Proc. 1st International Conference of Earthquake Geotechnical Engineering (IS-Tokyo)*, Balkema, Vol. 3, 1333–1368.

Tokimatsu, K., Mizuno, H. and Kakurai, M. (1996): Building damage associated with geotechnical problems, Special Issue on Geotechnical Aspects of the January 17, 1995 Hyogoken Nambu Earthquake, *Soils and Foundations*, 219–234.

Tokimatsu, K. and Asada, Y. (1998): Effects of liquefaction-induced ground displacements on pile performance in the 1995 Hyogoken-Nambu earthquake, Special Issue on Geotechnical Aspects of the January 17, 1995 Hyogoken Nambu Earthquake No. 2, *Soils & Foundations*, 163–177.

Tokimatsu, K. and Suzuki, H. (2009): Seismic soil-pile-structure interaction based on large shaking table tests, *Proc. IS-Tokyo, Performance-Based Design in Earthquake Geotechnical Engineering*, Taylor & Francis Group, 77–104.

Tokimatsu, K., Tamura, S., Suzuki, H. and Katsumata, K. (2012): Building damage associated with geotechnical problems in the 2011 Tohoku Pacific Earthquake, *Soils & Foundations*, 52 (5), 956–974.

Toksoz, M.N., Dainty, A.M., Reiter, E. and Wu, R.-S. (1988): A model for attenuation and scattering in the earth's crust, *PAGEOPH*, 28 (1/2), 81–100.

Towhata, I. and Ishihara, K. (1985): Shear work and pore water pressure in undrained shear, *Soils & Foundations*, 25 (3), 73–84.

Trifunac, M.D. and Todorovska, M.I. (2004): 1971 San Fernando and 1994 Northridge, California, earthquakes: Did the zones with severely damaged buildings reoccur?, *Soil Dynamics and Earthquake Engineering*, Elsevier, 24, 225–239.

Tsukamoto, Y., Ishihara, K. and Sawada, S. (2004): Correlation between penetration resistance of Swedish weight sounding tests and SPT blow counts in sandy soils, *Soils and Foundations*, 44 (3), 13–24.

Tsukamoto, Y., Ishihara, K., Kokusho, Hara, T. and Tsutsumi, Y. (2009): Fluidisation and subsidence of gently sloped farming fields reclaimed with volcanic soils during 2003 Tokachi-oki earthquake in Japan, *Geotechnical Case History Volume*, Balkema, 109–118.

Tsurumi, T., Nakazawa, H., Mizumoto, K. and Watanabe, H. (2003): Post liquefaction process based on the sedimentation, *Journal of Japan Society for Civil Engineers (in Japanese)*, No. 743/III-64, 35–45.

Unno, T., Kazama, M., Uzuoka, R. and Sento, N. (2008): Liquefaction of unsaturated sand considering the pore air pressure and volume compressibility of the soil particle skelton, *Soils and Foundations*, 48 (1), 87–99.

Urayasu City Office (2012): http://www.city.urayasu.lg.jp/shisei/johokoukai/shingikai/shichoukoushotsu/1002796/ 1002934.html (in Japanese).

Vaid, Y.P. and Finn, W.D.L. (1979): Static shear and liquefaction potential, *Journal of SMFD*, ASCE, 195 (10), 1233–1246.

Vaid, Y.P. and Chern J.C. (1983): Effect of static shear on resistance to liquefaction, *Soils and Foundations*, 23 (1), 47–60.

Vaid, Y.P. and Chern, J.C. (1985): Cyclic and monotonic undrained response of saturated sands, Advances in the art of testing soils under cyclic conditions, *Proc.,* ASCE Convention, Detroit, Mich., 120–147.

Vaid, Y.P. and Sivathayalan, S. (1996): Static and cyclic liquefaction potential of Frazer Delta sand in simple shear and triaxial tests, *Canadian Geotechnical Journal,* 33, 281–289.

Watanabe, H. Sato, S. and Murakami, K. (1984): Evaluation of earthquake-induced sliding in rockfill dams, *Soils and Foundations*, 24 (3), 1–14.

Wu, R-S. (1982): Attenuation of short period seismic waves due to scattering, Geophysical Research Letters, 9 (1), 9–12.

Yagi, N. (1978): Volume change and excess pore pressure in sands under repeated shear stress, *Journal of Japan Society for Civil Engineers (in Japanese)*, No. 275, 79–90.

Yamamizu, F., Goto, N., Ohta, Y. and Takahashi, H. (1983): Attenuation of shear waves in deep soil deposits as revealed by down-hole measurements in the 2,300 meter borehole of the Shimohsa Observatory, Japan, *Journal of Physics of Earth*, 31 (2), 139–157.

Yamashita, S. and Toki, S. (1992): Effects of fabric anisotropy of sand on cyclic undrained triaxial and torsional strengths, *Soils and Foundations*, 33 (3), 92–104.

Yamashita, S., Kawaguchi, T. Nakata, T., Mikami, Y.T. Fujiwara and Shibuya, S. (2009): Interpretation of international parallel test on the measurement of $G_{max}$ using bender elements, *Soils & Foundations*, 49 (4), 631–650.

Yanagisawa, E. and Sugano, T. (1994): Undrained shear behaviors of sand in view of shear work, *Proc. Intern. Conf. on SMFE (Special Volume on Performance of Ground and Soil Structures during Earthquakes)*, New Delhi, India, Balkema Publishers, 155–158.

Yasuda, S. and Kiku, H. (2006): Uplift of buried manholes and pipes due to liquefaction of replaced soils, *Proc. 2nd Japan-Taiwan Joint Workshop*, Nagaoka, Japan, ATC3 Committee, ISSMGE, 146–149.

Yasuhara, K. (1994): Postcyclic undrained strength for cohesive soils. *Journal of Geotech. Eng.*, ASCE, 120 (11), 1961–1979.

Yasuhara, K. and Hyde, F.L. (1997): Method for estimating post-cyclic undrained secant modulus of clays. *Journal of Geotech. and Geoenv. Eng.*, ASCE, 123 (3), 204–211.

Yoshida, N., Tokimatsu, K., Yasuda, S., Kokusho, T. and Okimura, T. (2001): Geotechnical Aspects of Damage in Adapazari City during 1999 Kocaeli, Turkey Earthquake, *Soils & Foundations*, 41 (4), 25–45.

Yoshida, Y. and Kokusho, T. (1987): A proposal on application of penetration tests on gravelly soils, *Research Report of CRIEPI (in Japanese)*, Central Research Institute of Electric Power Industry, No. U87080.

Yoshikuni, H. and Nakanodo (1974): Consolidation of soils by vertical drain wells with finite permeability, *Soils & Foundations*, 14 (2), 35–46.

Yoshimi, Y. (1970): An outline of damage during the Tokachi-oki earthquake, *Soils & Foundations*, X (2), 1–14.

Yoshimi, Y. and Oh-oka, H. (1975): Influence of degree of shear stress reversal on the liquefaction potential of saturated sand, *Soils and Foundations*, 15 (3), 27–40.

Yoshimi, Y. and Tokimatsu, K. (1977): Settlement of buildings on saturated sand during earthquakes, *Soils and Foundations*, 17 (1), 23–38.

Yoshimi, Y. and Tokimatsu, K. (1983): SPT practice survey and comparative tests, Technical Notes, *Soils and Foundations*, 23 (3), 105–111.

Yoshimi, Y., Tanaka, K. and Tokimatsu, K. (1989): Liquefaction resistance of a partially saturated sand, *Soils and Foundations*, 29 (3), 157–162.

Yoshimi, Y. (1991): Particle grading and plasticity of hydraulically filled soils and their liquefaction resistance, *Tsuchi-to-Kiso (in Japanese)*, Japanese Geotechnical Society, 39, No. 8, 49–50.

Yoshimi, Y., Tokimatsu, K. and Ohara, J. (1994): In situ liquefaction resistance of clean sands over a wide density range, *Geotechnique,* England, 44 (3), 479–494.

Yoshimi, Y. (1994): Relationship among liquefaction resistance, SPT N-value and relative density for undisturbed samples of sands, *Tsuchi-to-Kiso (in Japanese)*, Japanese Geotechnical Society, 42, No. 4, 63–67.

Yoshimi, Y. (1998): Simplified design of structures buried in liquefiable soil, *Soils and Foundations*, Japanese Geotechnical Society, 17 (1), 23–38.

Yoshio, Y. (2002): Basic experiments on mechanical properties of cement mixed flowable fill under static and seismic loading, *Master's Thesis (in Japanese),* Graduate School of Science & Engineering, Chuo University, Tokyo, Japan.

Yu, P. and Richart, F.E. (1984): Stress ratio effects on shear modulus of dry sands, *Journal of Geotech. Div.,* ASCE, 110 (GT3), 331–345.

Zelikson, A. (1969): Geotechnical Models Using the Hydraulic Gradient Similarity Method, *Geotechnique*, 19 (4), 495–508.

Zen, K., Umehara, K. and Hamada, K. (1978): Laboratory tests and in-situ seismic survey on vibratory shear modulus of clayey soils with various plasticities, *Proc. of 5th Japan Earthquake Engineering Symposium*, 721–728.

Zienkiewicz, O.C. and Bettess, P. (1982): Soils and other saturated media under transient, dynamic conditions; General formulation and the validity of various simplifying assumptions, *Soil Mechanics: Transient and Cyclic loads*, John Wiley & Sons, 1–16.

# Index